Springer Earth System Sciences

Series editors

Philippe Blondel, Bath, UK
Eric Guilyardi, Paris, France
Jorge Rabassa, Ushuaia, Argentina
Clive Horwood, Chichester, UK

More information about this series at http://www.springer.com/series/10178

Andrés Folguera · Eduardo Contreras-Reyes
Nemesio Heredia · Alfonso Encinas
Sofía B. Iannelli · Verónica Oliveros
Federico M. Dávila · Gilda Collo
Laura Giambiagi · Andrei Maksymowicz
María Paula Iglesia Llanos
Martín Turienzo · Maximiliano Naipauer
Darío Orts · Vanesa D. Litvak
Orlando Álvarez · César Arriagada
Editors

The Evolution of the Chilean-Argentinean Andes

Springer

Editors

see next page

ISSN 2197-9596 ISSN 2197-960X (electronic)
Springer Earth System Sciences
ISBN 978-3-030-09811-7 ISBN 978-3-319-67774-3 (eBook)
https://doi.org/10.1007/978-3-319-67774-3

Softcover re-print of the Hardcover 1st edition 2018

Printed on acid-free paper

This Springer imprint is published by the registered company Springer International Publishing AG part of Springer Nature
The registered company address is: Gewerbestrasse 11, 6330 Cham, Switzerland

Editors
Andrés Folguera
Consejo Nacional de Investigaciones Científicas y Técnicas (CONICET)
Universidad de Buenos Aires
Buenos Aires, Argentina

Eduardo Contreras-Reyes
Universidad de Chile
Santiago, Chile

Nemesio Heredia
Instituto Geológico y Minero de España
Madrid, Spain

Alfonso Encinas
Universidad de Concepción
Concepción, Chile

Sofía B. Iannelli
Instituto de Estudios Andinos (IDEAN), Consejo Nacional de Investigaciones Científicas y Técnicas (CONICET)
Universidad de Buenos Aires
Buenos Aires, Argentina

Verónica Oliveros
Universidad de Concepción
Concepción, Chile

Federico M. Dávila
CICTERRA
CONICET—FCEFyN—Universidad Nacional de Córdoba
Córdoba, Argentina

Gilda Collo
CICTERRA
CONICET—FCEFyN—Universidad Nacional de Córdoba
Córdoba, Argentina

Laura Giambiagi
Consejo Nacional de Investigaciones Científicas y Técnicas (CONICET)
Ianiglia, Mendoza
Argentina

Andrei Maksymowicz
Universidad de Chile
Santiago, Chile

María Paula Iglesia Llanos
Consejo Nacional de Investigaciones Científicas y Técnicas (CONICET)
Universidad de Buenos Aires
Buenos Aires, Argentina

Martín Turienzo
Consejo Nacional de Investigaciones Científicas y Técnicas (CONICET)
Universidad de Bahía Blanca
Bahía Blanca, Argentina

Maximiliano Naipauer
Consejo Nacional de Investigaciones Científicas y Técnicas (CONICET)
Universidad de Buenos Aires
Buenos Aires, Argentina

Darío Orts
Consejo Nacional de Investigaciones Científicas y Técnicas (CONICET)
Universidad de Río Negro
Río Negro, Argentina

Vanesa D. Litvak
Instituto de Estudios Andinos (IDEAN), Consejo Nacional de Investigaciones Científicas y Técnicas (CONICET)
Universidad de Buenos Aires
Buenos Aires, Argentina

Orlando Álvarez
Instituto Geofísico Sismológico Ing. Volponi (IGSV)
Universidad de San Juan
San Juan, Argentina

and

Consejo Nacional de Investigaciones Científicas y Técnicas (CONICET)
Buenos Aires, Argentina

César Arriagada
Universidad de Chile
Santiago, Chile

For our partners of marvelous voyages through the Andes

Preface

This book analyzes the tectonic evolution of the Argentinean and Chilean Andes through four parts: the first dealing with the structure of the fore-arc from bathymetric, gravimetric, and seismic data; the second part that discusses the Paleozoic evolution of this sector, first showing the paleomagnetic behavior of the continent during this period and then discussing the different hypothesis associated with the accretion of continental slivers and consequent closure of ocean basins, producing deformation and metamorphism, reconstructing the geometry of the Early Paleozoic orogens and arcs across South American western sector; the third part that analyzes the proto-Andean arc evolution, showing the paleomagnetic path of the margin, clues about its early arc activity, its relation to LIP activity associated with Pangea supercontinent desintegration and subduction zone; the fourth part that discusses early development of the Andes that started as a non-organized collage of within-plate deformation sectors that lately established next to the subduction zone when the south Atlantic Ocean started to expand; and finally, the fifth part that discusses different aspects of the Cenozoic evolution of the Andes and its contemporaneous volcanic arc, such as development of shallow subduction settings through time, the role of dynamic subsidence, opening and closure of intra- and back-arc basins.

In detail, the first part is composed of three chapters. The first chapter by Contreras Reyes et al. analyzes from wide-angle seismic data the structure of the submerged fore-arc as a result of the balance between sediment accretion and sediment subduction and crustal erosion. Then, different segments are subdivided that are characterized by dominant frontal erosion or accretion and are defined by climate conditions, sediment dispersal pattern along the trench, and subduction of bathymetric highs. The second chapter by Maksymowicz and Tassara analyzes the geometry of the submerged frontal prism as a function of basal and poral pressure conditions. Those systems that are subjected to frontal crustal erosion become oversteepened since these are affected by normal faulting that triggers internal saturation with seawater. Additionally, this chapter highlights the role of these morphological variations on seismic segmentation along the subduction zone. The third chapter in this first part belongs to Álvarez et al. and analyzes vertical gradients in gravity anomalies and their relation to co- and post-seismic rupture zones of giant earthquakes along the subduction zone and their complex internal structure. Additionally, satellite measurements can show changes in the gravity field that could be used to predict rupture propagation along the subduction zone. Then, variable fore-arc structure revealed by gravity data shows to have an important influence in the way that ruptures propagate and therefore in seismic segmentation and their potential.

The second part starts with Rapalini et al. chapter that deals with paleomagnetic reconstructions that describe South American path as part of Gondwana during the Paleozoic. The second chapter by Heredia et al. discusses the complex collage of continental slivers that have formed the western Gondwana border that is being reconstructed on the basis of the recognition of different deformational vergences, superimposed metamorphic and deformational events, and recognition of internal and external parts of collisional orogens. The third chapter by Ramos is a reconstruction of the Early Paleozoic arc and consequently the western border of Gondwana, allowing identifying subsequent crustal accretions though differential exhumation levels preserved along the margin.

The third part starts with Oliveros et al. chapter that analyzes magmatism in northern Chile showing that arc activity had already established since the Triassic, a period that had been considered devoid of subduction processes in western Gondwana. A short lapse characterized by a decrease in magmatic activity appeared at the time when the arc established nearer the subduction zone, at the time when the crust became less assimilated by mantle derived products. Naipauer et al. discuss this same magmatic stage in Northern Patagonia discussing its relation to the proto-Pacific border and LIP activity in the Karoo-Ferrar anomaly, concluding that at these latitudes plume activity influenced considerably composition of near-trench magmatism.

The fourth part starts with Iglesia Llanos chapter that analyzes using paleo-magnetic data the path of southern Gondwana in Jurassic times. During this time the supercontinent experiments important latitudinal variations, first displacing to the north, and then returning to southern latitudes. Continental absolute displacements together with the opening of the Weddell Sea at southern Gondwana are the clue to explain within-plate deformations in Patagonia, analyzed in the second chapter by Navarrete et al. This chapter also describes early deformations through the western border of Patagonia at the time of South Atlantic opening. These proto-Andean deformations are also recorded in northern Chile, lately reactivated during the development of the Neogene Chilean-Pampean flat subduction zone, as it is described in the third chapter made by Martínez et al., Gianni et al. in the fourth chapter study synorogenic sedimentation in Northern Patagonia and southern Central Andes showing a diachronism in the uplift of the Cretaceous proto-Andes. This is interpreted as due to the subduction of active ocean ridges that are depicted in recent plate models. The next chapter by Arriagada analyzes first-order inflections of the South American subduction border using paleomagnetic data, determining their origin as oroclines formed diachronously and linked to basement heterogeneities associated with crustal amalgamation in Proterozoic to Paleozoic times. Finally in this part, the chapter of Iannelli et al. calculates ancient crustal thicknesses and degree of influence of the subduction slab on arc-related rocks of Paleogene age, determining that the Andean roots produced during the proto-Andean evolution times were practically lost during Eocene to Oligocene extensional stages along the Southern Andes. This crustal stretching could have developed in relation to the combination of an opening of a slab window during mid-ocean ridge subduction and trench rollback.

The fifth part starts with Dávila et al. chapter that quantifies subsidence and uplift influenced by the mantle dynamics, superimposed to tectonic topography. This chapter predicts within-plate subsidence next to flat subduction settings and discusses subsidence components in ancient rift systems that cannot be fully explained by thermal decay, such as the Triassic rift systems in central Argentina. Additionally, this chapter analyzes mantle upwelling component associated with Neopaleozoic regional uplift and glacial activity in southern Gondwana. The second chapter by Lossada et al. is a review centered on the Frontal Cordillera uplift constituting the major mountain system developed in central Argentina whose origin was thought to be associated with the development of Chilean-Pampean flat subduction zone. Thermochronological data show that exhumation of this system is previous to the development of the flat subduction and is most likely influenced by preexisting heterogeneities linked to the Gondwana breakup. The third chapter by Turienzo et al. analyzes mechanics of deformation of fold and thrust belt at the Southern Central Andes suggesting that shortenings have been somehow underestimated and that the structure could be considerably more complex than previously assumed. The fourth chapter by Encinas et al. discusses subsidence mechanisms associated with the Late Oligocene to Early Miocene transgressions in Patagonia as part of the rifting activity that started in Eocene times and continued in the Early Neogene, as broad sag basins that flooded most of the ancient collapsed Andes. The fifth chapter by Folguera et al. analyzes Cenozoic deformations of the North Patagonian Andes from the study of synorogenic sedimentation. This concludes that an intra-arc basin closure and a shallowing of the subducted slab could have contributed together to the Neogene uplift of western Patagonia. In the sixth chapter, Litvak et al. study arc-related composition of magmatism associated with the development of the Chilean-Pampean flat subduction zone and Payenia shallow subduction setting to the south, concluding that most likely these behaved as one single system in the 15–4 Ma period. Finally, the seventh chapter by Collo et al. constitutes a compilation of heat flow from wells that illustrate how the Chilean-Pampean flat subduction zone has refrigerated the lithosphere, expelling the mantle asthenosphere off the shallow configuration as shown in Fig. 1.

Buenos Aires, Argentina — Andrés Folguera et al.

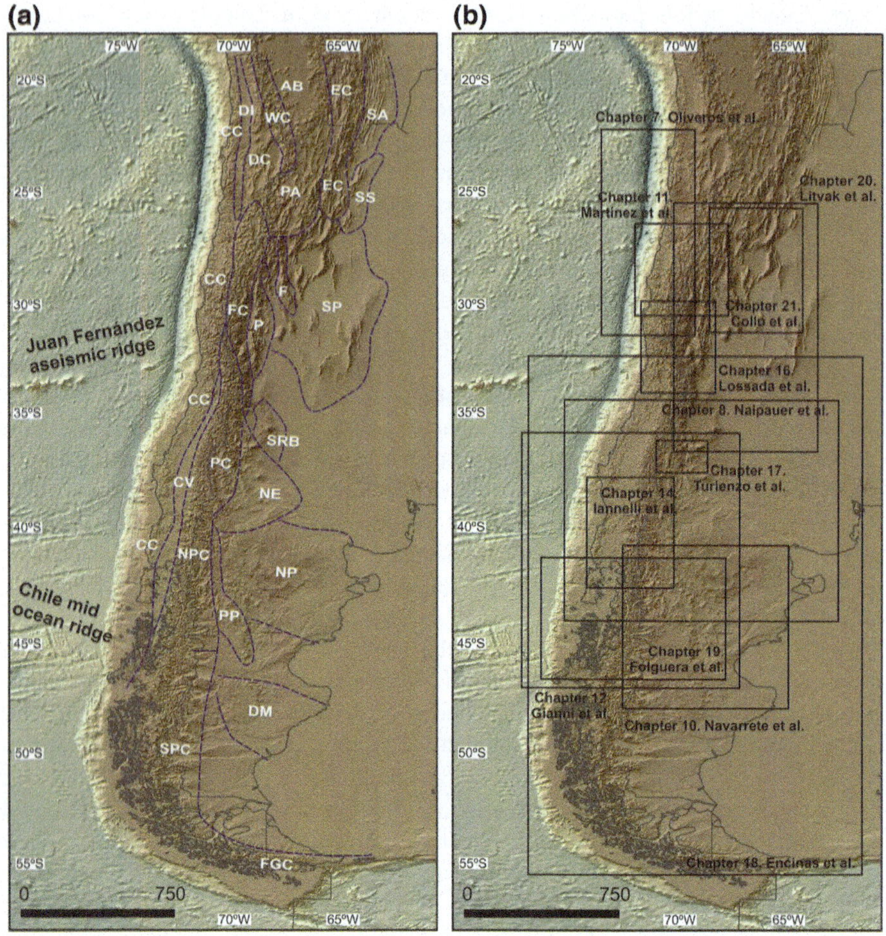

Fig. 1 To the left: Main morphostructural provinces in which the Southern Andes are divided. CC: Cordillera de la Costa, DI: Depresión Intermedia, DC: Cordillera de Domeyko, WC: Cordillera Occidental; PA: Puna; AB: Altiplano Boliviano; EC: Cordillera Oriental; SA: Sierras Subandinas; SS: Sistema de Santa Bárbara; FC: Cordillera Frontal; P: Precordillera; F: Sistema de Famatina; SP: Sierras Pampeanas; CV: Valle Central; PC: Cordillera Principal; NE: Engolfamiento Neuquino; SRB: Bloque de San Rafael; NPC: Cordillera Norpatagónica; PP: Precordillera Patagónica; NP: Macizo Norpatagónico; SPC: Cordillera Surpatagónica; DM: Macizo del Deseado; FGC: Cordillera Fueguina. To the right: book structure (empty rectangles indicate areas analyzed in each chapter)

Acknowledgements

The authors acknowledge Jorge Rabassa for the opportunity to show their research through this book.

Contents

Contributors

Juan Juis Alonso Departamento de Geología, Universidad de Oviedo, Oviedo, Spain

Orlando Álvarez Instituto Geofisico Sismológico Ing. Volponi (IGSV), Universidad de Nacional San Juan, San Juan, Argentina; Consejo Nacional de Investigaciones Científicas y Técnicas (CONICET), Buenos Aires, Argentina

César Arriagada Departamento de Geología, Facultad de Ciencias Físicas y Matemáticas, Universidad de Chile, Santiago, Chile

Pilar Ávila CICTERRA, CONICET—FCEFyN—Universidad Nacional de Córdoba, Córdoba, Argentina

Sebastián Bascuñán Departamento de Geología, Facultad de Ciencias Físicas y Matemáticas, Universidad de Chile, Santiago, Chile

Francisco Bastias Departamento de Ciencias de la Tierra, Universidad de Concepción, Concepción, Chile

Florencia Bechis Instituto de Investigaciones en Diversidad Cultural y Procesos de Cambio (IIDyPCa), CONICET—Universidad Nacional de Río Negro, San Carlos de Bariloche, Argentina

Pablo Bernabé Departamento de Ciencias de la Tierra, Universidad de Concepción, Concepción, Chile

Carla Braitenberg Dipartimento di Matematica e Geoscienze, Universita di Trieste, Trieste, Italy

Luis Buatois Department of Geological Sciences, University of Saskatchewan, Saskatoon, Canada

Pere Busquets Facultad de Geología, Universidad de Barcelona, Barcelona, Spain

Ricardo Calegari YPF, Buenos Aires, Argentina

Horacio Canelo CICTERRA, CONICET—FCEFyN—Universidad Nacional de Córdoba, Córdoba, Argentina

Raúl Cardó Universidad de San Juan y Servicio Geológico Minero Argentino, San Juan, Argentina

Reynaldo Charrier Departamento de Geología, Universidad de Chile, Santiago, Chile

Pilar Clariana Instituto Geológico y Minero de España, Madrid, Spain

Gilda Collo CICTERRA, CONICET—FCEFyN—Universidad Nacional de Córdoba, Córdoba, Argentina

Ferrán Colombo Facultad de Geología, Universidad de Barcelona, Barcelona, Spain

Eduardo Contreras-Reyes Universidad de Chile, Santiago, Chile

Christian Creixell Servicio Nacional de Geología y Minería, Providencia, Santiago, Chile

Andrés Cuesta Departamento de Geología, Universidad de Oviedo, Oviedo, Spain

José Cuitiño Instituto Patagónico de Geología y Paleontología, CCT CENPAT-CONICET, Puerto Madryn, Chubut, Argentina

Federico M. Dávila CICTERRA, CONICET—FCEFyN—Universidad Nacional de Córdoba, Córdoba, Argentina

Luis Dimieri Departamento de Geología, Instituto Geológico del Sur (INGEOSUR, UNS-CONICET), Universidad Nacional del Sur, Bahía Blanca, Argentina

Andrés Echaurren Instituto de Estudios Andinos Don Pablo Groeber, UBA—CONICET. Departamento de Ciencias Geológicas, FCEN, Universidad de Buenos Aires, Buenos Aires, Argentina

Alfonso Encinas Departamento de Ciencias de la Tierra, Universidad de Concepción, Concepción, Chile

Mauricio Espinoza Vargas Departamento de Ciencias de la Tierra, Universidad de Concepción, Concepción, Chile

Miguel Ezpeleta CICTERRA, CONICET—FCEFyN—Universidad Nacional de Córdoba, Córdoba, Argentina

Pedro Farias Departamento de Geología, Universidad de Oviedo, Oviedo, Spain

Lucas Fennell Instituto de Estudios Andinos (IDEAN), Consejo Nacional de Investigaciones Científicas y Técnicas (CONICET), Universidad de Buenos Aires, Buenos Aires, Argentina

Lucía Fernández Paz Instituto de Estudios Andinos (IDEAN), Consejo Nacional de Investigaciones Científicas y Técnicas (CONICET), Universidad de Buenos Aires, Buenos Aires, Argentina

Kenneth L. Finger University of California Museum of Paleontology, Berkeley, California, USA

Andrés Folguera Instituto de Estudios Andinos (IDEAN). Instituto de Estudios Andinos "Don Pablo Groeber," Consejo Nacional de Investigaciones Científicas y Técnicas (CONICET), Departamento de Ciencias Geológicas, FCEN, Universidad de Buenos Aires, Buenos Aires, Argentina

Pablo R. Franceschinis IGEBA-CONICET, Universidad de Buenos Aires, Buenos Aires, Argentina

Gloria Gallastegui Instituto Geológico y Minero de España, Madrid, Spain

Jorge Gallastegui Departamento de Geología, Universidad de Oviedo, Oviedo, Spain

Ezequiel García Morabito Instituto de Estudios Andinos "Don Pablo Groeber" (UBA—CONICET, Argentina), Buenos Aires, Argentina; Swiss Federal Institute of Technology (ETH Zurich), Zurich, Switzerland

Joaquín García-Sansegundo Departamento de Geología, Universidad de Oviedo, Oviedo, Spain

Silvana E. Geuna IGEBA-CONICET, Universidad de Buenos Aires, Buenos Aires, Argentina

Raúl Giacosa Universidad de Río Negro y Servicio Geológico y Minero Argentino, General Roca, Río Negro, Argentina

Laura Giambiagi Unidad de Tectónica. IANIGLA-CONICET, Mendoza, Argentina; Centro Regional de Investigaciones Científicas y Tecnológicas, IANIGLA, CCT Mendoza, Mendoza, Argentina

Guido M. Gianni Instituto Geofísico Sismológico Ing. Volponi (IGSV), Universidad de Nacional San Juan, San Juan, Argentina; Consejo Nacional de Investigaciones Científicas y Técnicas (CONICET), Buenos Aires, Argentina

Mario Giménez Instituto Geofísico Sismológico Ing. Volponi (IGSV), Universidad de Nacional San Juan, San Juan, Argentina; Consejo Nacional de Investigaciones Científicas y Técnicas (CONICET), Buenos Aires, Argentina

Lizet De Girolamo Del Mauro Departamento de Ciencias de la Tierra, Universidad de Concepción, Concepción, Chile

Javiera González Departamento de Ciencias de la Tierra, Universidad de Concepción, Concepción, Chile

Luis González-Menéndez Instituto Geológico y Minero de España, Madrid, Spain

Nemesio Heredia Instituto Geológico y Minero de España, Madrid, Spain

Gregory Hoke Department of Earth Sciences, Syracuse University, Syracuse NY, USA

Fernando Hongn IBIGEO (CONICET-UNSA), Rosario de Lerma, Salta, Argentina

Sofía B. Iannelli Instituto de Estudios Andinos (IDEAN), Consejo Nacional de Investigaciones Científicas y Técnicas (CONICET), Universidad de Buenos Aires, Buenos Aires, Argentina

María Paula Iglesia Llanos Instituto de Geociencias Básicas, Ambientales y Aplicadas, Departamento de Ciencias Geológicas, Facultad de Ciencias Exactas y Naturales, Universidad de Buenos Aires–Consejo Nacional de Investigaciones Científicas y Técnicas, Buenos Aires, Argentina

Rosemary E. Jones Department of Earth Sciences, University of Oxford, Oxford, UK

Linda A. Kirstein School of GeoSciences, University of Edinburgh, Edinburgh, United Kingdom

Federico Lince Klinger Instituto Geofísico y Sismológico Ing. Volponi, Universidad Nacional de San Juan, San Juan, Argentina; Consejo Nacional de Investigaciones Científicas y Técnicas (CONICET), Buenos Aires, Argentina

Fernando Lebinson Departamento de Geología, Instituto Geológico del Sur (INGEOSUR, UNS-CONICET), Universidad Nacional del Sur, Bahía Blanca, Argentina

Oscar Limarino Facultad de Geología, Universidad de Buenos Aires, Buenos Aires, Argentina

Carolina Lithgow-Bertelloni University College London, London, England

Vanesa D. Litvak Instituto de Estudios Andinos (IDEAN), Consejo Nacional de Investigaciones Científicas y Técnicas (CONICET), Universidad de Buenos Aires, Buenos Aires, Argentina

Ana C. Lossada Centro Regional de Investigaciones Científicas y Tecnológicas, IANIGLA, CCT Mendoza, Mendoza, Argentina

Juan Lovecchio YPF, Buenos Aires, Argentina

Andrei Maksymowicz Departamento de Geofísica, Facultad de Ciencias Físicas Y Matemáticas, Universidad de Chile, Santiago, Chile

Marcelo Manassero Centro de Investigaciones Geológicas (UNLP—CONICET, Argentina), Buenos Aires, Argentina

Federico Martina CICTERRA, CONICET—FCEFyN—Universidad Nacional de Córdoba, Córdoba, Argentina

Fernando Martínez Facultad de Ingeniería y Ciencias Geológicas, Departamento de Geología, Universidad Católica del Norte, Angamos, Antofagasta, Chile

Patricia Martínez Instituto Geofisico Sismológico Ing. Volponi (IGSV), Universidad Nacional de San Juan, San Juan, USA

Fidel Martín-González Área de Geología ESCET, Universidad Rey Juan Carlos, Mostoles, Madrid, Spain

Manuela Mazzitelli Centro Regional de Investigaciones Científicas y Tecnológicas, IANIGLA, CCT Mendoza, Mendoza, Argentina

José Mescua Centro Regional de Investigaciones Científicas y Tecnológicas, IANIGLA, CCT Mendoza, Mendoza, Argentina

Maximiliano Naipauer Instituto de Estudios Andinos "Don Pablo Groeber" (UBA—CONICET, Argentina), Buenos Aires, Argentina

César R. Navarrete Departamento de Geología, Universidad Nacional de la Patagonia San Juan Bosco, Comodoro Rivadavia, Chubut, Argentina

Sven N. Nielsen Instituto de Ciencias de la Tierra, Universidad Austral de Chile, Valdivia, Chile

Julieta Nóbile CICTERRA, CONICET—FCEFyN—Universidad Nacional de Córdoba, Córdoba, Argentina

Verónica Oliveros Departamento de Ciencias de la Tierra, Universidad de Concepción, Concepción, Chile

Darío Orts Instituto de Investigación en Paleobiología y Geología, Universidad Nacional de Río Negro—CONICET, General Roca, Argentina

David Pedreira Departamento de Geología, Universidad de Oviedo, Oviedo, Spain

Stella Poma Instituto de Geociencias Básicas y Aplicadas de Buenos Aires (IGEBA), Universidad de Buenos Aires-Conicet, Buenos Aires, Argentina

Felipe Pérez Departamento de Ciencias de la Tierra, Universidad de Concepción, Concepción, Chile

Paulo Quezada Universidad Andres Bello, Santiago, Chile

Luis Quintana Departamento de Geología, Universidad de Oviedo, Oviedo, Spain

Miguel E. Ramos Instituto de Estudios Andinos (IDEAN), Consejo Nacional de Investigaciones Científicas y Técnicas (CONICET), Universidad de Buenos Aires, Buenos Aires, Argentina

Victor A. Ramos Instituto de Estudios Andinos "Don Pablo Groeber," Departamento de Ciencias Geológicas, FCEN, Universidad de Buenos Aires–CONICET, Buenos Aires, Argentina

Augusto E. Rapalini IGEBA-CONICET, Universidad de Buenos Aires, Buenos Aires, Argentina

Ricardo Riffo Departamento de Ciencias de la Tierra, Universidad de Concepción, Concepción, Chile

Luis Roberto Rodríguez-Fernández Instituto Geológico y Minero de España, Madrid, Spain

Pablo Rossel Universidad Andres Bello, Concepción, Chile

Álvaro Rubio-Ordóñez Departamento de Geología, Universidad de Oviedo, Oviedo, Spain

Francisco Ruiz Instituto Geofisico Sismológico Ing. Volponi, Universidad Nacional de San Juan, San Juan, USA

Mario Schiuma YPF, Buenos Aires, Argentina

Raúl Seggiaro Universidad de Salta y Servicio Geológico y Minero Argentino, Salta, Argentina

Daniel Sellés Aurum Consultores, Santiago de Chile, Chile

Fernando Sepúlveda Departamento de Ciencias de la Tierra, Universidad de Concepción, Concepción, Chile

Samanta Serra-Varela Instituto de Investigación en Paleobiología y Geología, General Roca, Argentina

Santiago Soler Instituto Geofísico-Sismológico Volponi, San Juan, Argentina

Cecilia M. Spagnuolo Consejo Nacional de Investigaciones Científicas y Técnicas (CONICET), Universidad Nacional de Tucumán, San Miguel de Tucumán, Argentina

Mauro Spagnuolo Instituto de Estudios Andinos (IDEAN), Consejo Nacional de Investigaciones Científicas y Técnicas (CONICET), Universidad de Buenos Aires, Buenos Aires, Argentina

Luis Spalletti Centro de Investigaciones Geológicas, Universidad de La Plata, La Plata, Argentina

Julieta Suriano IGeBA—FCEyN, UBA-CONICET, Buenos Aires, Argentina

Francisco Sánchez CICTERRA, CONICET—FCEFyN—Universidad Nacional de Córdoba, Córdoba, Argentina

Natalia Sánchez Departamento de Geología, Instituto Geológico del Sur (INGEOSUR, UNS-CONICET), Universidad Nacional del Sur, Bahía Blanca, Argentina

Francisca Tapia Departamento de Ciencias de la Tierra, Universidad de Concepción, Concepción, Chile

Andres Tassara Departamento de Ciencias de La Tierra, Universidad de Concepción, Casilla, Concepción, Chile

Jonathan Tobal Instituto de Estudios Andinos (IDEAN), Consejo Nacional de Investigaciones Científicas y Técnicas (CONICET), Universidad de Buenos Aires, Buenos Aires, Argentina

José María Tubía Departamento de Geodinámica, Universidad del País Vasco (UPV/EHU), Lejona, Biscay, Spain

Martín Turienzo Departamento de Geología, Instituto Geológico del Sur (INGEOSUR, UNS-CONICET), Universidad Nacional del Sur, Bahía Blanca, Argentina

Victor V. Valencia School of the Environment, Washington State University, Pullman, WA, USA

Paulina Vásquez Servicio Nacional de Geología y Minería, Providencia, Santiago, Chile

Patricio Zambrano Universidad Andres Bello, Facultad de Ingeniería, Geología, Concepción, Chile

Part I
Crustal and Seismic Structure of the Chilean Fore-Arc

Structure and Tectonics of the Chilean Convergent Margin from Wide-Angle Seismic Studies: A Review

Eduardo Contreras-Reyes

Abstract Based on a compilation of published 2-D velocity-depth models along the Chilean margin (22°–48° S), I review the structure and tectonic processes that govern this convergent margin in terms of subduction erosion and sediment accretion/subduction. North of the collision point between the Juan Fernández Ridge with the overriding continental South American plate (Chile at ~32.5° S), subduction erosion has been active since Jurassic resulting in large-scale crustal thinning and long-term subsidence of the outermost forearc. Published 2-D velocity–depth models show a prominent lateral velocity contrast that propagates deep into the continental crust defining a major lateral seismic discontinuity (interpreted as the volcanic-continental basement contact of the submerged Coastal Cordillera characterized by a gravitational collapse of the outermost fore arc). Between the Juan Fernández Ridge and the Chile Triple Junction (CTJ) of the Nazca-Antarctic-South American plates (Chile at ~46.5° S), an accretionary prism 5–50 km wide has been formed due to an increase of trench sedimentation triggered by denudation processes of the Andes after the last Pleistocene Glaciation. However, the relatively small size of the accretionary prism is not compatible with an efficient history of sediment accretion since the Pleistocene, and sediment subduction is a dominant process especially south of the oceanic Mocha Fracture Zone (Chile at ~38° S) and north of the CTJ. In the overriding plate, seismic studies reveal two prominent velocity transition zones characterized by steep lateral velocity gradients, resulting in a seismic segmentation of the marine fore arc. The southern central Chilean margin is composed of three main domains: (1) a frontal prism at the toe of the continental slope, (2) a paleoaccretionary complex, and (3) the seaward edge of the Paleozoic continental framework that forms part of the Coastal Cordillera. Near the CTJ, where the Nazca-Antarctic spreading center (Chile Rise) collides with the margin, subduction erosion is active, and rapid uplift followed by subsidence of the forearc area, normal faulting and intensive sedimentary mass wasting are documented. South of the CTJ, the convergence between the oceanic Antarctic and continental South American plate is slow allowing more

E. Contreras-Reyes (✉)
Universidad de Chile, Santiago, Chile
e-mail: econtreras@dgf.uchile.cl

A. Folguera et al. (eds.), *The Evolution of the Chilean-Argentinean Andes*,
Springer Earth System Sciences, https://doi.org/10.1007/978-3-319-67774-3_1

time sediment accumulation at the trench enhancing the formation of relatively large accretionary prisms (width of 70–90 km).

Keywords Wide-angle seismic studies · Fore-arc · Crustal erosion Accretion · Subsidence · Uplift

1 Introduction

The tectonics and structure of the Chilean convergent margin varies significantly from the arid Atacama Desert in the north to the glacial climate of Patagonia in the south. Among the main factors controlling the tectonic segmentation of the outer fore-arc are the morphology of the subducting oceanic Nazca and Antarctic plates, the variability in sediment supply to the trench, and the sediment transport within the trench. The most prominent feature of the oceanic lithosphere is the Nazca-Antarctic spreading center known as the Chile Rise or Chile Ridge, which is currently colliding with the margin off Península de Taitao at ~46.5° S (Fig. 1). North of ~34° S, the oceanic Nazca plate was formed at the East Pacific Rise (Nazca-Pacific spreading center) more than 38 Myr ago, whereas between 34° and 46° S it was formed at the Chile Rise within the past 35 Ma (Herron et al. 1981; Cande et al. 1987) (Fig. 1). The relative convergence rate between the oceanic Nazca and South American plate is currently ~67 km/Myr (Angermann et al. 1999; Khazaradze and Klotz 2003), although the average convergence rate during the last 20 Myr is about 85 km/Mr (DeMets et al. 2010). The relative convergence rate between the oceanic Antarctic and continental South American plate is ~20 km/Myr in a direction perpendicular to the trench between ~46° S and 53° S, but southwards, the direction becomes quite oblique to the margin (DeMets et al. 2010).

The collision of the Chile Rise with the Chilean margin configures the Chile Triple Junction (CTJ) where the Nazca, Antarctic, and South American plates join at the trench. Fracture zones (FZs) cut the Chile Rise into several segments, resulting in abrupt plate age and morphological changes along the plate boundary (Fig. 1). One of the most striking features is the Valdivia FZ-system composed of ten individual fracture zones (Tebbens et al. 1997) with a total offset of about 600 km which orientate roughly perpendicular to the trench axis (Fig. 1). The seafloor from the CTJ up to the Valdivia FZ system is characterized by an uplifted topography associated to the young, hot and buoyant oceanic Nazca lithosphere formed during the last 20 Ma (Tebbens et al. 1997). Another striking oceanic feature is the NE (55°) trending Mocha FZ which intersects the trench at ~38° S (Fig. 1). Due to its obliqueness related to the trench axis, the Mocha FZ intersection with the trench migrated southward at about 3.1 cm/a over the past millions years relative to South America (Contreras-Reyes et al. 2008). The Mocha Block is known as the triangulation of the Chile Trench, Valdivia FZ system, and the Mocha FZ. North of the Mocha Block, the oceanic lithosphere approaching the trench is significantly older and the seafloor becomes deeper (Tebbens et al. 1997).

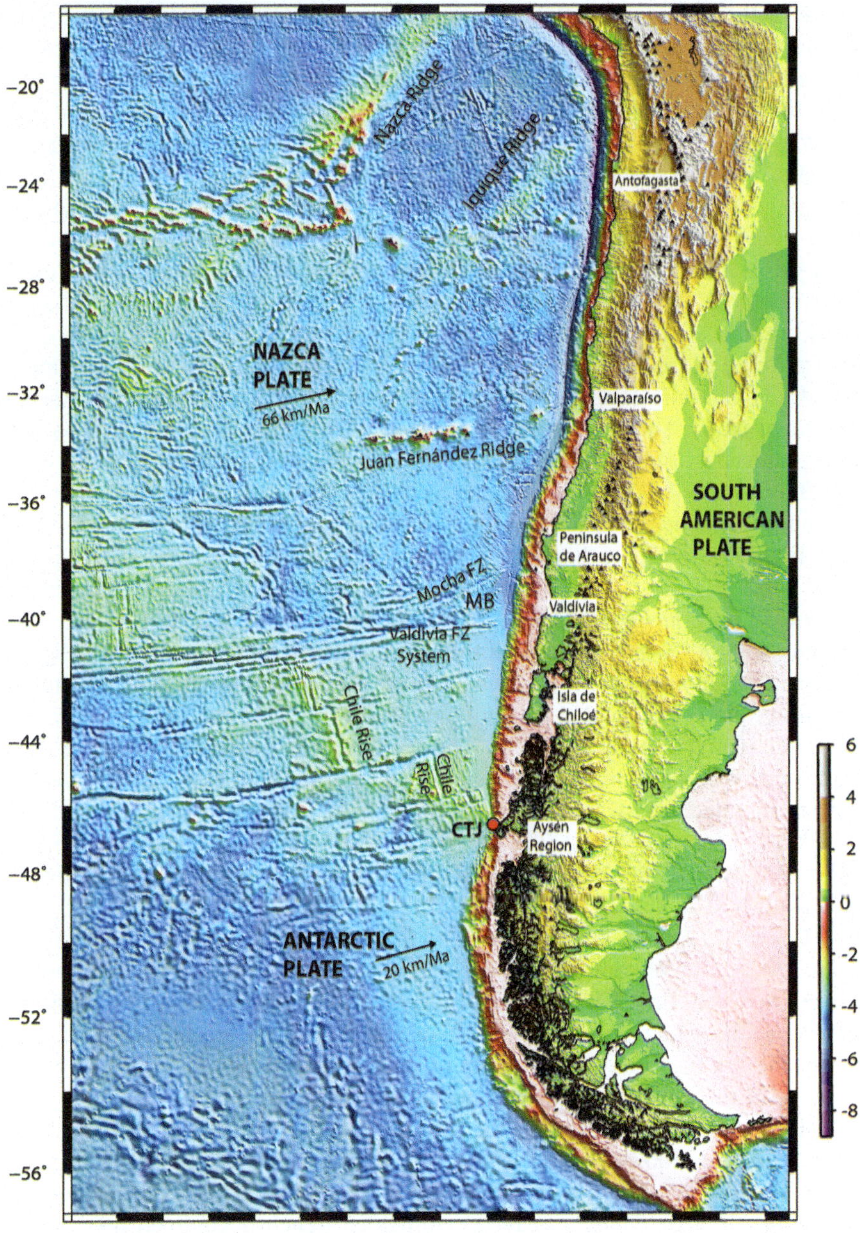

Fig. 1 Geodynamic setting of the Nazca, Antarctic and South American plates. These plates meet at the Chile Triple Junction, where the Chile Rise is currently colliding with the Chilean margin at ~46.5° S. The oceanic Nazca and Antarctic plates are segmented by several fracture zones, resulting in a strong variability in the age of the incoming oceanic plate. MB: Mocha Block

Because of the buoyancy of the hot and young oceanic Nazca lithosphere formed south of the Mocha Block, the seafloor is approximately 500–1000 m shallower than north of the Mocha Block.

The first collision of the Chile Rise with the continental South American plate occurred south of 48° S at ~14 Ma, and since then, the CTJ has been migrating northwards (Cande and Leslie 1986; Maksymowicz et al. 2012). The high relief topography in the area of the CTJ has caused trench sedimentation to be preferentially transported towards deeper topography of the trench (northward and southward of the CTJ as the oceanic Nazca and Antarctic plates become older, respectively). At the CTJ, the trench is almost devoid of sediments (Fig. 2), in contrast to the north (35°–44° S) where the trench is heavily sedimented in response to the sediment supplied by the rivers and rapid glaciation denudation of the Andes (Thornburg et al. 1990; Blumberg et al. 2008). The trenchward transport of sediment that is sourced by the denudation of the Andes is controlled by submarine landslides, submarine canyons, and turbidity currents (Thornburg et al. 1990; Blumberg et al. 2008; Voelker et al. 2013).

Another prominent feature on the oceanic Nazca lithosphere is the Juan Fernández Ridge (JFR), a hot spot track formed at the Juan Fernández Hotspot located some 900 km west of the Chile Trench (Fig. 1). The easternmost, yet unsubducted portion of the JFR is composed by two prominent seamounts: the O'Higgins Guyot (~4,000 m above the surrounding seafloor) and the O'Higgins seamount (~3,500 m above the surrounding seafloor) (Fig. 2). The O'Higgins seamount group is surrounded by a topographic swell obscured by the bending of the oceanic plate (Kopp et al. 2004). The first collision of the JFR with the Chile margin occurred approximately 22 Myr ago in north Chile at ~22° S. Then the ridge has migrated southwards to the current collision zone located off Valparaíso at 32°–33° S (Yáñez et al. 2001). The O'Higgins seamount group behaves as a barrier to the northward transport of turbidites within the trench, separating a sediment-starved trench to the north from a sediment flooded trench to the south (Fig. 2).

Thus, the southern-central Chilean margin is characterized by a filled trench confined between the JFR and the Chile Rise. Figure 2 shows the sediment thickness at the Chile trench. Sediment within the trench are redistributed from south to the north of the CTJ (Thornburg et al. 1990), which is explained by the deepening of the trench floor (e.g., Voelker et al. 2013).

North of the JFR (18°–33° S), the incoming oceanic crust carries only a thin sediment drape (<500 m), and the trench axis is poorly sedimented (Fig. 2). In contrast, south of the CTJ, the trench is filled by an up to 3 km thick thick sequence of turbidites (Fig. 2). Thus, a heavily sedimented trench is confined between the collision zones of the JFR and Chile Rise with the margin, respectively. The subduction of these two large topographic features on the oceanic plate defines the most important borders of tectonic segments along the Chilean margin in terms of the structure of the margin, continental slope morphology, trench fill thickness, and size of the accretionary prism.

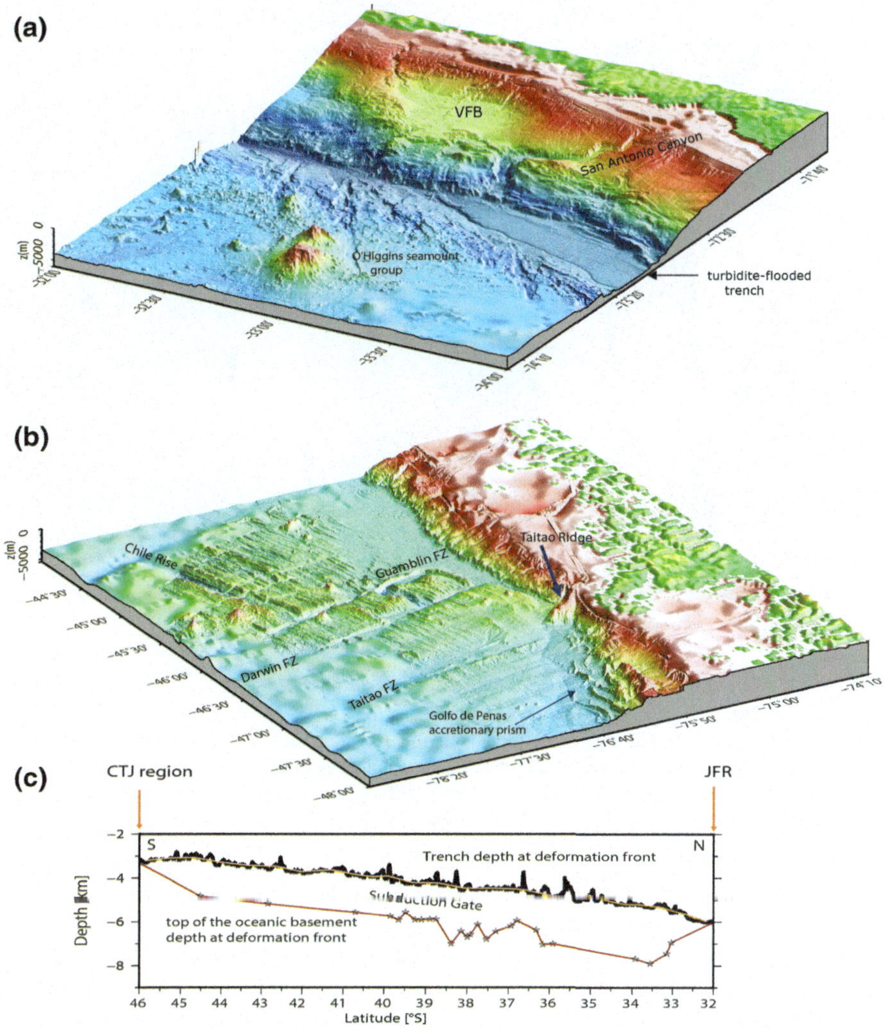

Fig. 2 Swath bathymetric image of the seafloor. **a** Collision zone between the Juan Fernández Ridge and the Chilean margin. VFB: Valparaíso Fore-arc Basin. **b** Collision zone between the Chile Rise and the Chilean margin. The Chile Triple Junction is coincident with the subduction of the Taitao Ridge. **c** Depth of the trench axis off Chile. Trench fill thickness values were computed by a compilation of published seismic reflection lines imaging the southern-central Chilean trench basin (Voelker et al. 2013)

In this chapter, I review the structure of the Chilean margin from Tocopilla (northern Chile at ~22° S) to Golfo de Penas (Chilean Patagonia at ~48° S) using a compilation of nine 2-D wide angle seismic velocity-depth models published during the last 12 years (see Fig. 3 for map location). The presented results

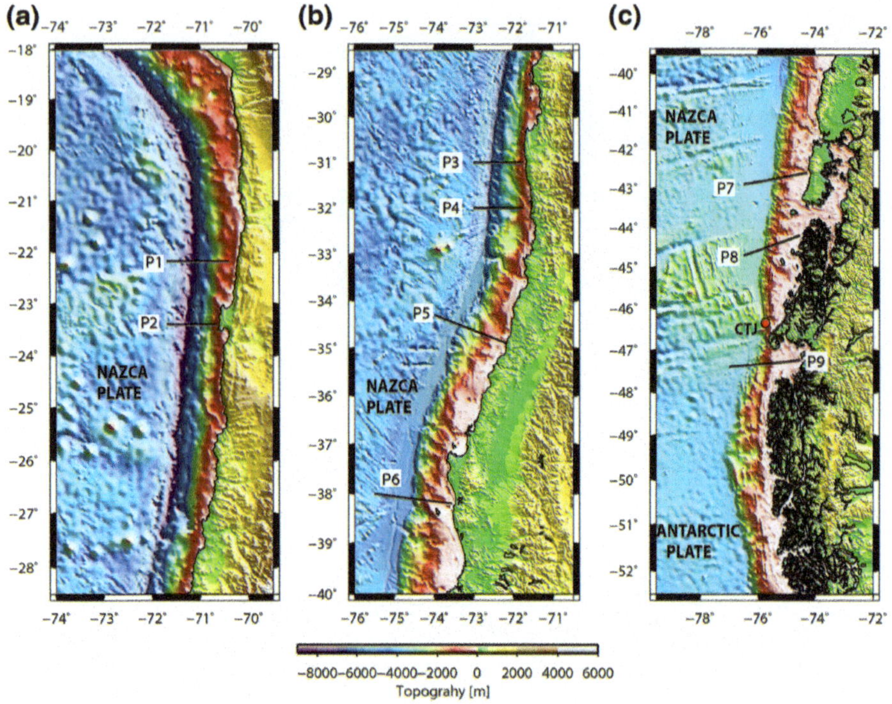

Fig 3 Black lines denote the wide-angle seismic profiles studied by Contreras-Reyes et al. (2012), **P1**; Sallares and Ranero (2005), **P2**; Contreras-Reyes et al. (2015), **P3** and **P4**; Moscoso et al. (2011), **P5**; Contreras-Reyes et al. (2008), **P6**; Scherwath et al. (2009), **P7** and **P8** and Maksymowicz et al. (2012), **P9**. CTJ: Chile Triple Junction

should be considered as the short-term structure of the margin that is the current outcome of long term geological processes.

Likewise, the tectonics segments defined in this chapter corresponds to large-scale regions of the Chilean margin (several hundred kilometers). It is worth-mentioning that at small scale (10 km-scale), subduction erosion can be coeval within segments of tectonic accretion where sediment layer are not thick enough to smooth the oceanic high features that collide with the margin (Clift and Vannucchi 2004; Stern 2011). Similarly, small frontal prisms made of debris material can be coeval in segments where subduction erosion dominates (i.e., von Huene et al. 1999; Laursen et al. 2002; Ranero et al. 2006). In this chapter, the distinction between subduction erosion, sediment accretion and subduction refers to the most dominant feature of the margin at large scale. For the next sections, I will refer to the following regions of the Chilean margin as the first-order tectonic segments: (i) the northern Chilean margin (22°–32° S) that is bounded at the south by the collision of the JFR with the Chilean trench, (ii) the south-central Chilean margin (33°–46.5° S) bounded in the north by the JFR and in the south by the CTJ, and (iii) the austral Chilean margin (south of the CTJ).

2 The Northern Chilean Margin (22°–32° S)

The northern Chilean margin is characterized by a poorly sedimented trench as a consequence of little sediment supplies from the Andes to the trench offshore the extremely arid region of the Atacama Desert. The continental slope is steep as the toe of the margin lacks a well-developed accretionary prism. In addition, the JFR behaves as a barrier to trench sediment transport from the sediment-flooded southern Chile trench to the north (von Huene et al. 1997; Voelker et al. 2013). Subduction erosion of the northern Chilean margin has been inferred from the eastward migration of the volcanic arc since the Mesozoic (e.g. Rutland 1971; Stern 2011) as well as the long-term arcward retreat of the trench, crustal thinning and subsidence of the outer forearc (e.g., Kukowski and Oncken 2006; Contreras-Reyes et al. 2015). Long term subduction rates for northern Chile have been estimated by several authors with relatively similar values: 50 km^3/km/my (Stern 2011), 45–50 km^3/km/my (von Huene et al. 1999); 40–45 km^3/km/my (Kukowski and Oncken 2006); and 28–40 km^3/km/my (Scholl and von Huene 2009).

Figures 4 and 5 show the 2-D velocity-depth models for the northern Chilean margin off the Antofagasta District at ~22.1° S (off Tocopilla) and at ~23.5° S (off southern Peninsula de Mejillones). This region is characterized by the Coastal Cordillera, a trench-parallel morphostructural system that was formed by the exhumation of a Mesozoic paleomagmatic arc that overlies the coupled zone of the Nazca-South American subduction interface (e.g. Charrier et al. 2007; Folguera et al. 2015). The Coastal Cordillera in north Chile is composed of Palaeozoic intrusive rocks and metamorphosed remnants of a late Palaeozoic accretionary prism, as well as Jurassic-Triassic intrusive, sedimentary and volcanic rocks constituting the primary geologic basement unit of the northern Chilean margin (SERNAGEOMIN 2003). Some authors have suggested that underplating at the base of the continental crust caused by subduction erosion may be responsible for the uplift of the Coastal Cordillera (e.g., Hartley et al. 2000). Alternatively, Melnick (2016) proposed that the rise of the Coastal Cordillera correlates with slip during earthquakes straddling the continental Moho.

The northern Chilean coast presents a spectacular coastal scarp, a major trench-parallel morphological feature which extends for more than 1,000 km along the coast with an average height of 1 km (e.g., Armijo and Thiele 1990, Marquardt et al. 2004). The continental slope off north Chile is interrupted by another prominent scarp with a vertical offset of more than 1,000 m located ~50 km seawards of the coast (Fig. 4). This abrupt slope change in the continental slope overlies a velocity discontinuity seen in the 2-D velocity-depth model of profile P1 (Contreras-Reyes et al. 2012). The frontal part of the overriding plate is characterized by a wedge-shaped body 30–40 km-wide, with velocities in the range of 4.0–6.0 km/s (Fig. 4). This body is interpreted as the front of the margin, fluid saturated, metamorphosed and disaggregated by fracturing as a consequence of subduction erosion (Sallares and Ranero 2005; Contreras-Reyes et al. 2012; Fig. 4).

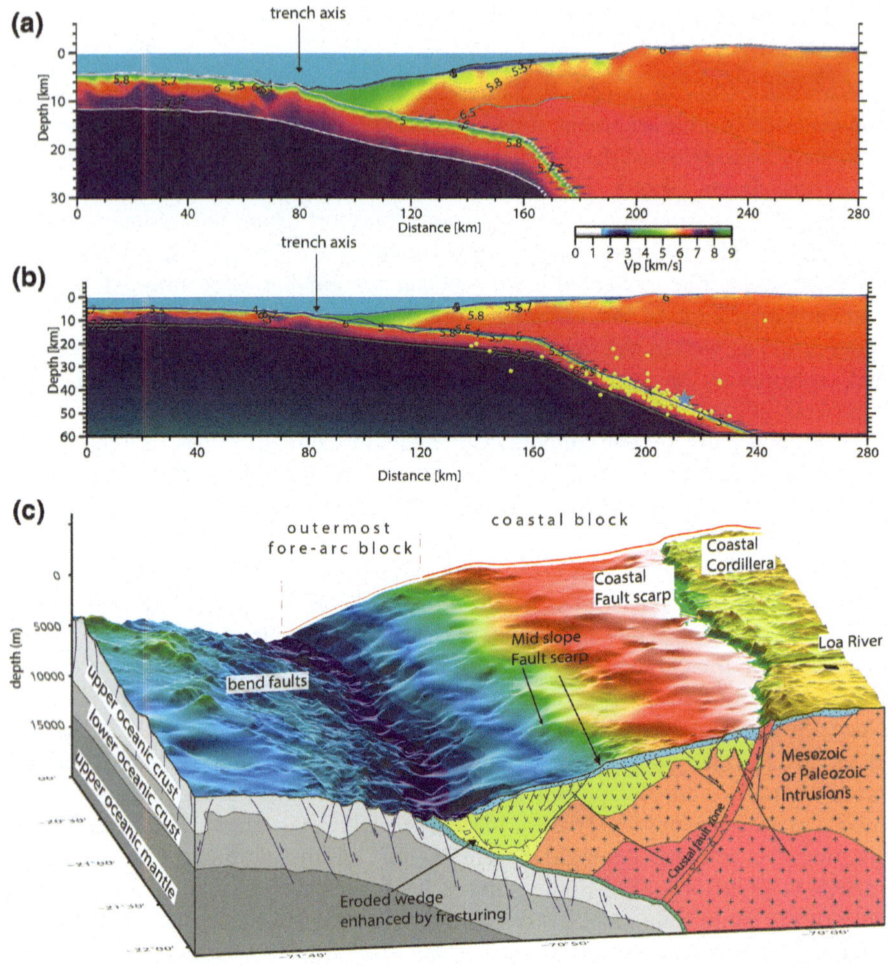

Fig. 4 Seismic velocity structure and interpretation of the upper subduction zone of Tocopilla, north Chile (profile P1). **a** 2-D velocity-depth model obtained from tomographic inversion of travel times (Contreras-Reyes et al. 2012). The vertical exageration is ~3:1. **b** The 2-D velocity-depth model shown in (a) but at true scale. The light blue star denotes the Tocopilla 2007 hypocentre. Yellow circles indicate the seismicity distribution of the Tocopilla earthquake between 21.5° and 22.5° S from Fuenzalida et al. (2013). **c** Summarized interpretation of the 2-D velocity–depth model. The front of the margin consists of a metamorphosed and disaggregated wedge as a consequence of subduction erosion. This region spatially correlates with the continental slope scarp implying gravitational collapse of the trenchward part of the margin. The coastal zone of the upper plate is dominated by crustal faults, which are probably related to the change of plate-coupling degree around the abrupt subducting plate-dip change

Reduced seismic velocities in the frontal part of the margin were also reported along profile P2 (offshore southern Peninsula de Mejillones), and it was interpreted in terms of hydrofracturing of the upper plate (Fig. 5; Sallares and Ranero 2005).

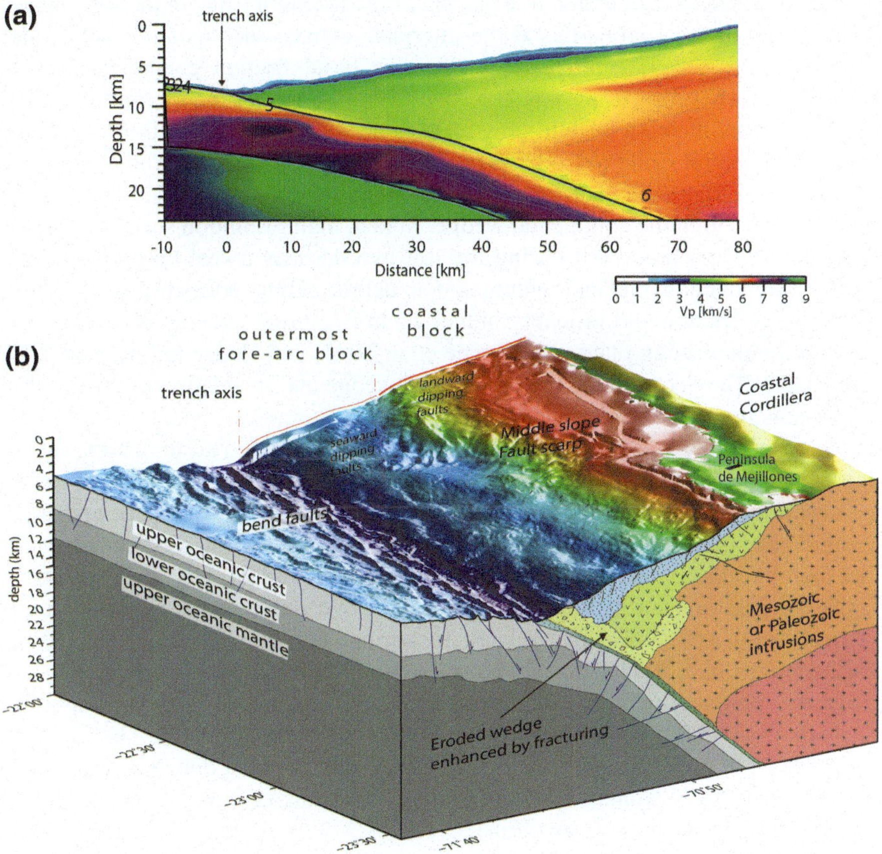

Fig. 5 Seismic velocity structure and interpretation of the upper subduction zone off Peninsula de Mejillones, north Chile (profile P2). **a** 2-D velocity-depth model obtained from tomographic inversion of travel times (Sallares and Ranero 2005). **b** The outermost fore-arc block is composed of volcanic rocks that are largely hydrofractured as a consequence of subduction erosion. The landward limit of the outermost fore-arc block is coincident with the location of the continental slope scarp suggesting subsidence of the outermost fore-arc block

Along seismic line P01, the 2-D velocity-depth model presents an abrupt lateral velocity gradient at 30–50 km from the trench axis, from which velocities increase landward suggesting that vigorous subduction erosion has ceased and the igneous framework is not very fractured or hydrated. The 2-D velocity-depth model across line P02 does not present an abrupt lateral velocity gradient. Nevertheless, Sallares and Ranero (2005) derived porosity and density values from their 2-D velocity-depth model, finding low porosity values at depths greater than 20 km. The low porosity at theses depths was interpreted in terms of fluid depletion in the subduction channel and low fracturing of the upper plate.

The lithological segmentation along dip of the northern Chilean margin suggests a change in physical properties at the interplate contact such as the basal friction. Maksymowicz (2015) estimated the effective basal friction coefficients for the Chilean forearc wedge using non cohesive Coloumb wedge theory, and his results show a general increase in fluid pressure and decrease in basal friction coefficients from south-central Chile (38°–33° S) to north Chile (32°–20° S). This finding was interpreted in terms of enhanced pervasive fracturing due to subduction erosion at the base and within the continental wedge. In fact, high resolution seismic reflection data acquired and processed during the last two decades reveal that fluids carried into the subduction channel within slope debris filling underthrusting grabens reduce basal friction and probably induce hydrofracturing leading to basal erosion beneath the overriding plate (von Huene et al. 2004; von Huene and Ranero 2003) (see Chap. "The geometry of the continental wedge and its relation to the rheology and seismicity of the Chilean interplate boundary").

Northernmost Chile (18.5°–23° S) has also been identified as an historic seismic gap where the last major megathrust earthquake took place on May 9, 1877 with an estimated magnitude of 8.9 (Comte and Pardo 1991). At the southern end of the seismic gap, the Mejillones Peninsula represents a morphological anomaly along the shoreline that is under E-W extension. Trench-parallel, east dipping normal faults are predominant as well as large-scale recent deformation characterized by vertical uplift and subsidence related to normal faulting (Delouis et al. 1998). Further north, the 2007 Mw 7.7 Tocopilla earthquake ruptured the deeper seismogenic interface, whereas the coupled upper interface remained unbroken (Delouis et al. 2009; Peyrat et al. 2010; Béjar-Pizarro et al. 2013). The Tocopilla event represents only a small fraction of the seismic moment deficit acumulated in the thrust interface. Nonetheless, on April 1, 2014, the central part of the historic seismic gap broke in a large thrust event of Mw 8.1 (e.g., Ruiz et al. 2014; León-Ríos et al. 2016). This event known as the Iquique or Pisagua earthquake presented abundant microseismicity in the upper plate suggesting fore-arc fault-related seismicity (Comte et al. 2016), and most of the interplate thrust aftershock seismicity concentrates in the shallow part of the subduction interface (Alvarez et al. 2015; León-Ríos et al. 2016).

Figure 4 shows an abrupt change in the dip of the subducting plate ($\sim$12°) at $\sim$60 km landwards of the trench offshore Tocopilla. This remarkable feature is constrained by wide angle seismic refraction and reflection data from land and sea, complemented by hypocentre data recorded during the 2007 Tocopilla aftershocks (Contreras-Reyes et al. 2012; Fuenzalida et al. 2013). Likely, the abrupt slab-dip change produces a strong plate-coupling gradient affecting the stress field both within the adjacent plates and along the interplate contact. It has been proposed that the coastal scarp is the surface expression of a crustal fault zone caused by the deviatoric stresses transferred to the overriding crust from the strong plate-coupling gradient around the abrupt slab dip change (Armijo and Thiele 1990; Contreras-Reyes et al. 2012; Béjar-Pizarro et al. 2013).

It is also expected that the strong plate-coupling gradient caused by the abrupt slab-dip change affect the frictional conditions of the subduction interface.

Contreras-Reyes et al. (2012) proposed that the area around the dip change could be either a high stress region, with an abrupt increase of plate coupling, or low stress owing to high fracturing. Interestingly, seismological data show that the abrupt slab-dip change behaved as a barrier for trenchward propagation of earthquake of moderate magnitude nucleated below the Coastal Cordillera such as the 2007 Mw 7.7 and 1967 Mw 7.4 events (Béjar-Pizarro et al. 2013; Fuenzalida et al. 2013). Geodetic models constrained by satellite interferometric synthetic aperture radar (InSAR) and Global Positioning System (GPS) data are consistent with the incorporation of an abrupt slab dip change (Béjar-Pizarro et al. 2013). Furthermore, the slab dip change marks a transition from a shallow subduction interface rather coupled (locked) to a deeper subduction interface less coupled or partially locked (Béjar-Pizarro et al. 2013; Métois et al. 2013, 2016).

Because of the lack of amphibious active seismic experiments complemented with amphibious passive seismological data, evidence for abrupt dip changes are difficult to detect. Exceptions have been reported in the Japanese subduction zone (Ito et al. 2005; Fujie et al. 2006), and off central Chile (Ruiz et al. 2013; Muñoz 2015). More amphibious (active and passive) seismic data are needed to constrain the along strike prolongation and variations in the degree of the abrupt dip slab change along the northern Chilean margin. Furthermore, between Peninsula de Mejillones and Punitaqui (Chile at ~31° S), no wide angle seismic profile has ever been acquired, and the velocity structure in this segment of the Chilean margin is unconstrained.

The southern part of the northern Chilean margin is characterized by the current collision of the JFR with the margin at 32°–33° S (Fig. 2). The ridge-trench collision zone marks the transition from the poorly sedimented trench in the north to the sediment-flooded trench in the south. The collision zone is also characterized by the presence of the prominent Valparaíso Forearc Basin (VFB); a shelf basin with 3.0–3.5 km thick sediment fill of late Cenozoic age (Laursen et al. 2002). The formation of the VFB has been attributed to the large subsidence rate of the margin caused by thinning of the continental crust due to the basal erosion processes enhanced by the JFR subduction (Laursen et al. 2002).

The first collision of the JFR with the Chilean margin occurred off Iquique at 20° S at ~22 Ma, and the collision migrated southwards to its present position offshore Valparaíso ever since (Yáñez et al. 2001). Due to the small obliquity between the trend of the JFR and the convergence direction of the Nazca plate (~78°) at present, the current ridge-trench collision zone has remained quasi-stationary over the last 10 Ma (Somoza 1998). Coincidentally, subsidence has occurred in the Valparaiso forearc basin after the JFR subduction, suggesting a rapid rate of subduction erosion of 96–128 km^3/km/My for the last 10 Ma (Laursen et al. 2002; Kukowski and Oncken 2006).

Seismic lines P3 and P4 sample the Chilean margin in the region of ridge-trench post-collision zone at 31° S and 32° S, respectively (Fig. 3). The 2-D velocity-depth models along these lines display typical features of an erosive convergent margin (Figs. 6–7). That is a poorly sedimented trench, a steep continental slope, the absence of a well-developed forearc and shelf basins, and a small frontal prism. The seismic models present a 5–10 km wide wedge-shaped body of

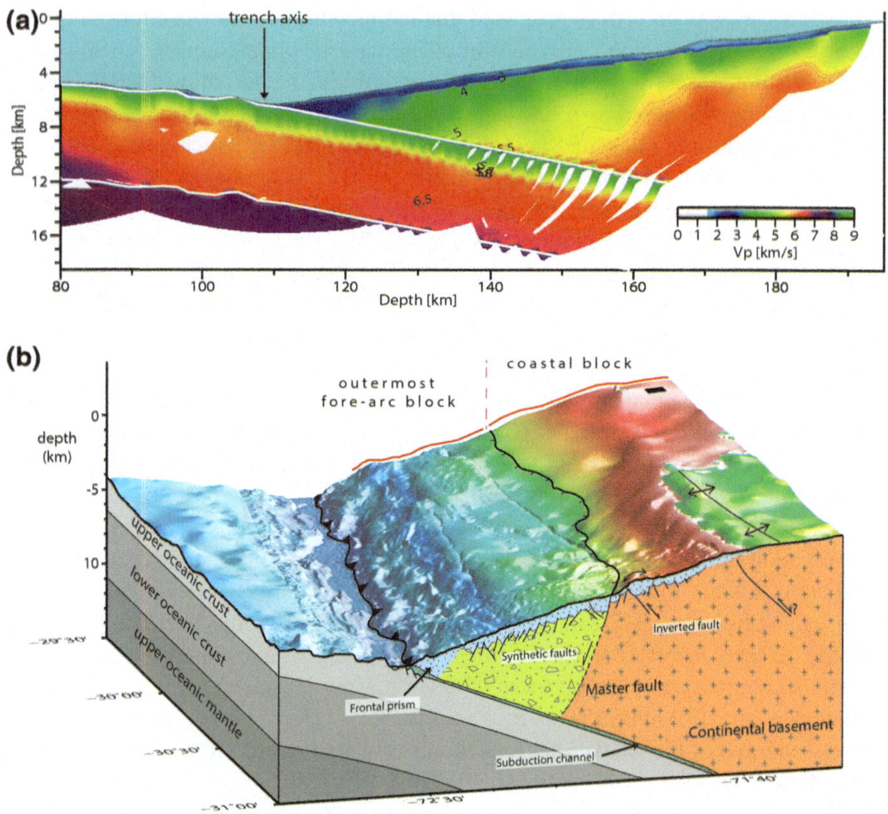

Fig. 6 Seismic velocity structure and interpretation of the upper subduction zone off central Chile at ~31° S (profile P3). **a** 2-D velocity model obtained from tomographic inversion of travel times (Contreras-Reyes et al. 2015). **b** The outermost fore-arc block is composed of volcanic rocks that are largely hydrofractured as a consequence of subduction erosion. The landward limit of the outermost fore-arc block is coincident with the location of the continental slope scarp suggesting subsidence of the outermost fore-arc block. The Coastal Block is composed of igneous and metamorphic rocks that form part of the uplifted Coastal Cordillera

reduced velocities (2.0–3.5 km/s) interpreted as the frontal prism at the base of the continental slope (Figs. 6–7). Seismic reflection data show that the landward edge of the frontal prism of the offscraped material is highlighted by a prominent reflection that separates the frontal prism from the continental basement (Contreras-Reyes et al. 2015).

Seismic velocities increase landwards from the frontal prism to the middle-upper slope transition zone (from 3.5 to ~5.0 km/s). These velocities were interpreted to display a volcanic unit of the Mesozoic magmatic arc that forms part of the outermost subsided fore-arc block (Figs. 6–7). Likely, this presumed volcanic unit is characterized by a significant degree of rock hydration, fracturation and porosity, since the outermost fore-arc block is sited at shallow lithostatic pressures where

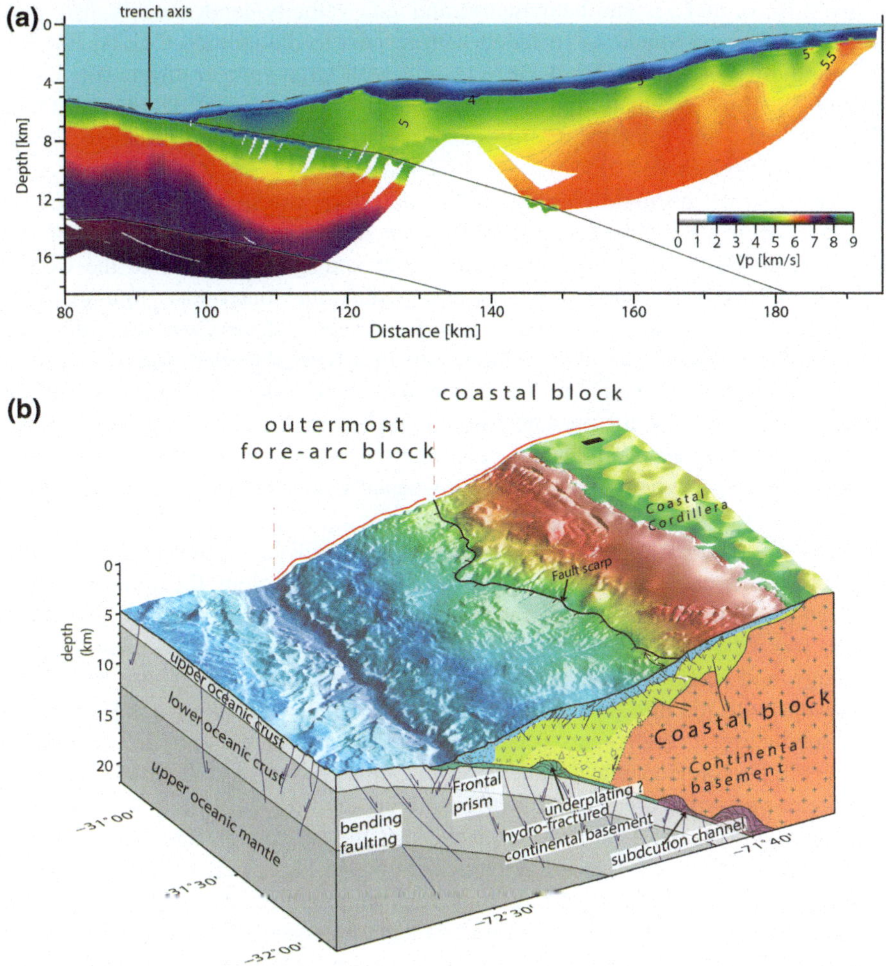

Fig. 7 Seismic velocity structure and interpretation of the upper subduction zone off central Chile at ~32° S (profile P4). **a** 2-D velocity model obtained from tomographic inversion of travel times (Contreras-Reyes et al. 2015). **b** A schematic model for the structure and tectonics off central Chile based on the seismic velocity model shown in (a). The toe of the margin presents a small frontal prism made of debris material that is undergoing underthrusting. The outermost fore-arc block is composed of eroded and fractured volcanic rocks as a consequence of subduction erosion caused by the subduction of high-relief horst and graben features poorly lubricated by sediment and hosted by the oceanic Nazca plate. Subduction erosion results in the collapse of the outermost fore-arc block and long-term subsidence. Landward of the middle-upper continental slope, the Coastal Block is composed of igneous and metamorphic rocks that form part of the uplifted Coastal Cordillera

failure requires low deviatoric stresses and therefore subduction erosion is more intense. Reduced continental crustal velocities in the frontal part of the margin are a common feature observed along the erosive northern central Chilean margin

(Figs. 4, 5, 6, and 7). The bathymetric and 2-D velocity-depth models also show that the outermost forearc is subsided with respect to the uplifted Coastal Block at the middle-upper continental slope transition with large vertical offsets (Figs. 4, 5, 6, and 7). Thus, subduction erosion has led to long-term subsidence and crustal thinning of the outermost fore-arc crust (Sallares and Ranero 2005; Contreras-Reyes et al. 2015).

Landward of the middle-upper continental slope transition, seismic velocities increase abruptly to values larger than 6.0 km/s that were interpreted as to represent the Coastal Block composed by metamorphic and igneous rocks of the Coastal Cordillera that are much less fractured than the outermost fore-arc crust (Contreras-Reyes et al. 2015). The seaward extension of the Coastal Cordillera is located in the upper continental slope domain where abundant normal landward dipping faults cut the continental basement and slope sediment sequences (Becerra et al. 2016). However, the presence of tectonic inverted features implies that thrust faults of low dip angles are also present, and there is a co-existence of normal and reverse faults in the seaward part of the Coastal Cordillera (e.g., Contreras-Reyes et al. 2015; Becerra et al. 2016).

Maksymowicz 2015) estimated basal friction coefficient values at the interplate contact for the seawardmost part of the interplate contact along the northern-central Chilean margin, and they are characterized by remarkable low basal friction values (<0.2) (see Chap. "The geometry of the continental wedge and its relation to the rheology and seismicity of the Chilean interplate boundary"). These results are consistent with the presence of slope debris accumulated in the frontal prism that fills subducting grabens and is subsequently incorporated into the subduction channel (e.g., von Huene and Ranero 2003; Contreras-Reyes et al. 2015). The high fluid pore pressure stored in the debris material should reduce the basal friction at the interplate contact. Thus, it is expected that the subduction interface below the wedge-shaped body of reduced velocities shown in Figs. 4, 5, 6, and 7 is moderately to weakly coupled due to the high fracturing degree and fluid pore pressure caused by subduction erosion. In fact, recent interseismic coupling coefficient estimates obtained by geodetic modeling of onshore GPS show that the shallow part of the seismic interface is characterized by values in the range of 0 and 50% (Métois et al. 2016).

3 The South-Central Chilean Margin (33°–46.5° S)

A fundamental change in the structure of the continental margin occurs in the current collision zone between the JFR and the central Chilean margin. In the trench axis, the JFR forms a topographic barrier behind which sediments ~2 km thick have been pounded. The lower continental slope in the collision zone with the JFR appears eroded, whereas the margin not yet affected by JFR subduction is fronted by an accretionary prism about 25–30 km wide (von Huene et al. 1997; Flueh et al. 1998; Zelt 1999). Sediment accretion is not a steady-state process south of the

Valparaíso Fore-arc Basin due to temporal variations of sediment supply to the trench and convergence rate. Actually, Kukowski and Oncken (2006) suggested that subduction erosion ceased ~5 Myr ago and it occurred since at least the late Oligocene when the trench was poorly sedimented and the main convergence rate was about 85 km/Myr. The seaward trench migration south of the Valparaíso Fore-arc Basin and the relatively thick trench fill suggest that frontal accretion processes have facilitated the formation of a well developed accretionary prism observed at present (e.g., Bangs and Cande 1997; Moscoso et al. 2011).

Off Constitución at 35° S, a high resolution 2-D velocity-depth model reveals the presence of an accretionary prism 40–50 km wide (Fig. 8; Moscoso et al. 2011). The landward edge of the accretionary prism, interpreted as the contact between the accretionary prism and the continental basement is well defined by an abrupt lateral velocity gradient (Fig. 8). In this case, the interpreted lithological contact underlies the location of the shelf break (upper slope-shelf basin).

Moscoso et al. (2011) also show that the landward part of the accretionary prism is coincident with the seismic-aseismic transition of the megathrust 2010 Maule earthquake (Mw 8.8). The accretionary prism is formed by unconsolidated sediment characterized by high porosity and high fluid content that is expected to behave aseismically in response to thrust earthquakes (Byrne et al. 1988). If so, the accretionary prism would lack sufficient competence to lock the fault and sustain large shear stresses; furthermore, it would exhibit velocity strengthening behavior that would tend to terminate rupture (Wang and Hu 2006). Therefore, it might be expected that slip would terminate on the landward extension of the accretionary prism as shown by Byrne et al. (1988) and Moscoso et al. (2011).

By comparing the size of the accretionary prism imaged by seismic refraction data from the JFR up to the CTJ, Contreras-Reyes et al. (2013) proposed that the southern-central margin can be characterized by two main sub-segments. The northern sub-segment (called the Maule segment) is located between the JFR-trench collision and the Mocha Block, and it is characterized by a relatively wide accretionary prism 20–50 km wide. The southern sub-segment (called the Chiloé segment) is situated between the Mocha Block and the CTJ, and it is characterized by a remarkably small accretionary prism as will be shown and discussed in the next section of this Chapter. Interestingly, these two segments spatially correlate with the location of two megathrust earthquakes: the Maule segment with the 2010 event (Mw 8.8), and the Chiloé segment with the 1960 event (Mw 9.5). Since the size of the accretionary prism controls the location of the up-dip limit of the seismogenic zone (e.g., Byrne et al. 1988), the differences in magnitude between these two earthquakes can be partially explained in terms of the accretionary prism size along the Chiloé and Maule segments. More wide-angle seismic data are required to constrain the up-dip seismic/aseismic transition zone along the extremely active Chilean convergent plate margin.

Seismic line P6 is located just north of the Mocha Island at the northern edge of the Mocha Block. The seismic line also intersects the oblique Mocha FZ at the trench (Fig. 3). Seismic results show that the overriding plate presents two prominent velocity transition zones characterized by abrupt lateral velocity

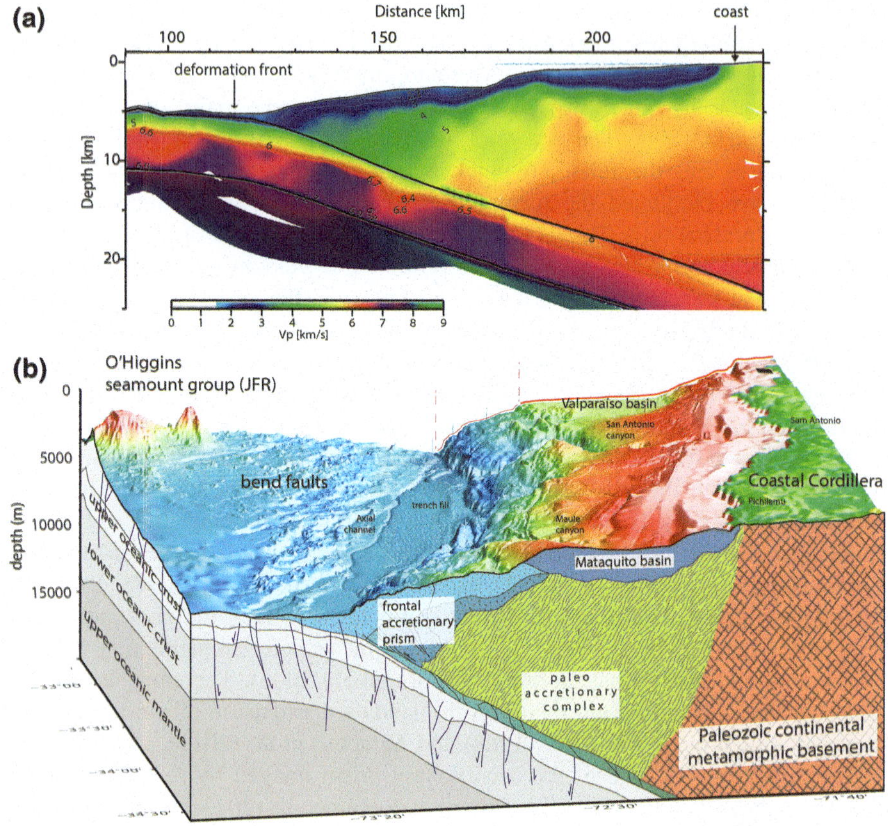

Fig. 8 Seismic velocity structure and interpretation of the upper subduction zone off Maule (profile P5). **a** 2-D velocity model obtained from tomographic inversion of travel times (Moscoso et al. 2011). **b** Interpretation based on the seismic model shown in (a). The frontal prism is 40–50 km wide and its landward edge is highlighted by the backstop that is coincident with an abrupt lateral velocity contrast. The backstop also appears to coincide with the shelf break

gradients defining three segments of the marine fore-arc: (1) in the frontal part of the margin, an accretionary prism ($\sim$20 km wide) below the continental slope with seismic velocities in the range of 2.0–3.5 km/s, (2) a $\sim$50 km area with seismic velocities of 4.5–5.5 km/s, interpreted as a paleo-accretionary complex located just landward of the accretionary prism, and (3) the seaward edge of the Paleozoic continental framework with seismic velocities higher than 6.0 km/s (Fig. 9; Contreras-Reyes et al. 2008).

The 2-D velocity-depth model shown in Fig. 9 samples the margin structure near the nucleation area of the1960 megathrust earthquake, the largest ever recorded event of Mw 9.5 (Cifuentes 1989; Barrientos and Ward 1990; Scherwath et al. 2006). The rupture area of this event is located between the Mocha Block and CTJ, and hosts relatively young oceanic lithosphere 0–30 Ma (Tebbens et al. 1997;

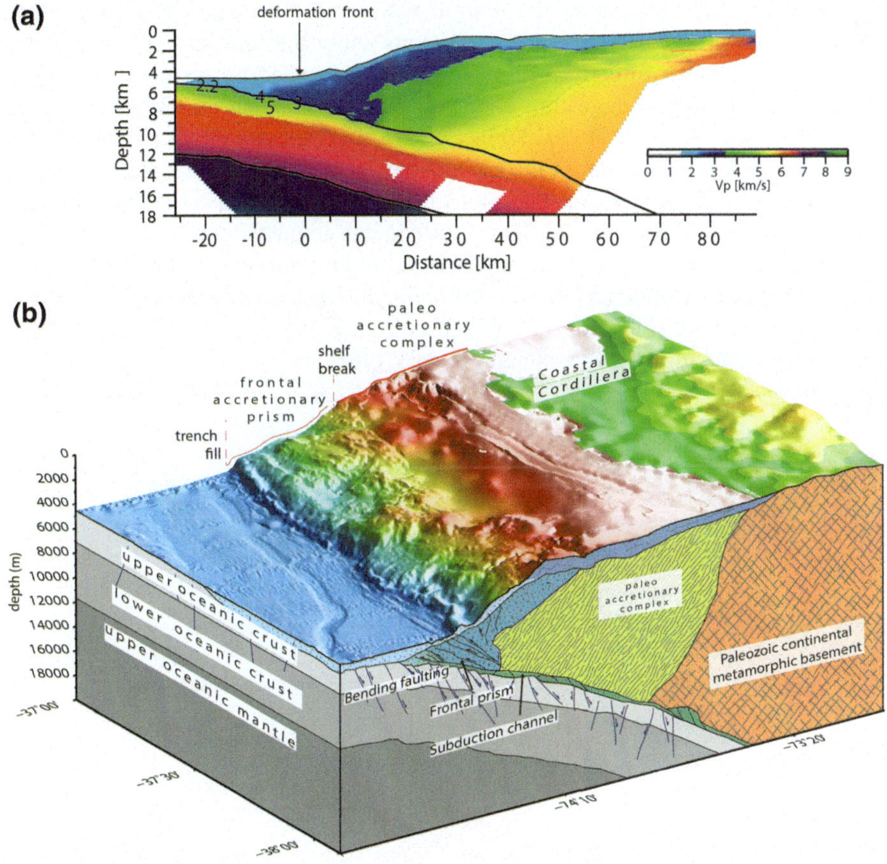

Fig. 9 Seismic velocity structure and interpretation of the upper subduction zone off Peninsula de Arauco (profile P6). **a** 2-D velocity model obtained from tomographic inversion of travel times (Contreras-Reyes et al. 2008). **b** Interpretation based on the seismic model shown in (a). The frontal prism is ~25 km wide and its landward edge is highlighted by the backstop that is coincident with an abrupt lateral velocity contrast. The backstop also appears to coincide with the shelf break

Voelker et al. 2011). Fault slip models show that the northern limit of the 1960 earthquake rupture area is coincident with the subducting Mocha FZ, while in the south fault slip ceased a few kilometres north of the CTJ (e.g., Moreno et al. 2009). Particularly, the maximum coseismic slip (30–40 m) is located between the Mocha and Valdivia FZ system (Barrientos and Ward 1990; Moreno et al.2009). In general, the main slip patches are bounded by fractures zones suggesting that these high oceanic features present a reduction of fault slip or increase of coupling (e.g., Contreras-Reyes and Carrizo 2011). Unfortunately, no wide-angle seismic profile constrain the structure of the margin in the maximum 1960 slip area, that is located south of Mocha FZ (~38.5° S) and north of Isla de Chiloé (42° S).

Seismic profile P7 samples the margin between the oceanic Chiloé and Guafo fracture zones at ~43° S. Seismic constraints show that the seawardmost part of the southern Chilean forearc (43°–46° S) is characterized by a small acrretionary prism (<10 km wide) suggesting efficient sediment subduction (Scherwath et al. 2009; Contreras-Reyes et al. 2013). In fact, the 2-D velocity-depth models of lines P7 and P8 show the location of the contact between the continental framework and the frontal accretionary prism (backstop) that is well resolved by an abrupt lateral velocity gradient (Figs. 10–11). This lithological contact defines the size of the frontal accretionary prism. Scherwath et al. (2009) presented the 2-D velocity models off Isla de Chiloé and Aysén displaying a rather narrow frontal accretionary prism 5–7 km wide (Figs. 10–11), which is limited landward by an abrupt lateral velocity gradient suggesting a change in rock type.

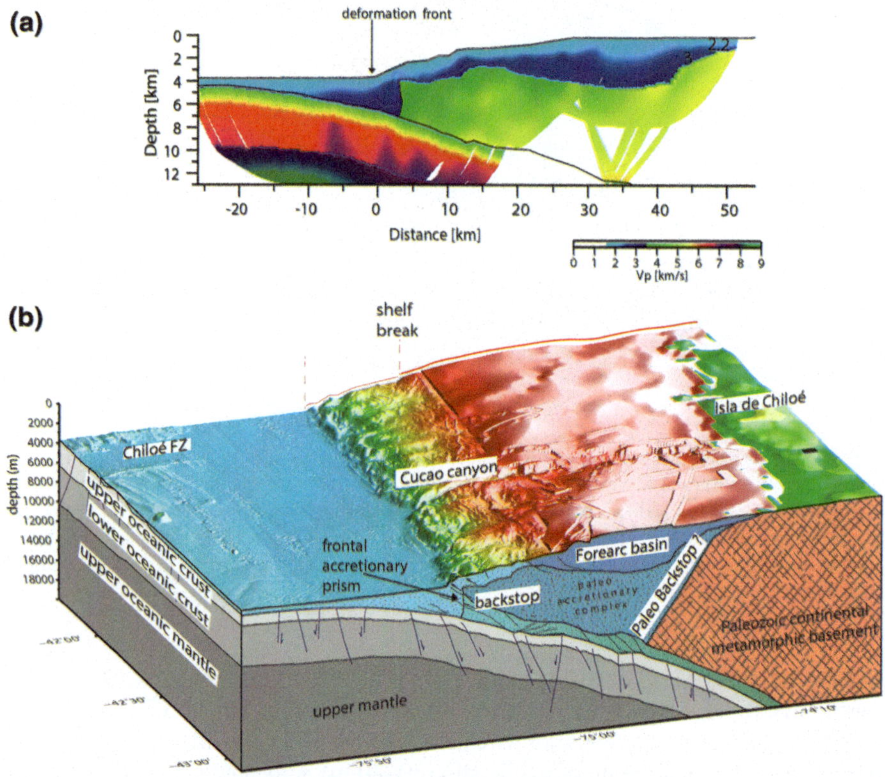

Fig. 10 Seismic velocity structure and interpretation of the upper subduction zone off Isla de Chiloé (profile P7). **a** 2-D velocity model obtained from tomographic inversion of travel times (Scherwath et al. 2009). **b** Interpretation based on the seismic model shown in (a). The trench is heavily sedimented, and turbidites span up to 250 km seaward of the trench and smooth the seafloor for a broad region. The frontal prism is ~10 km wide

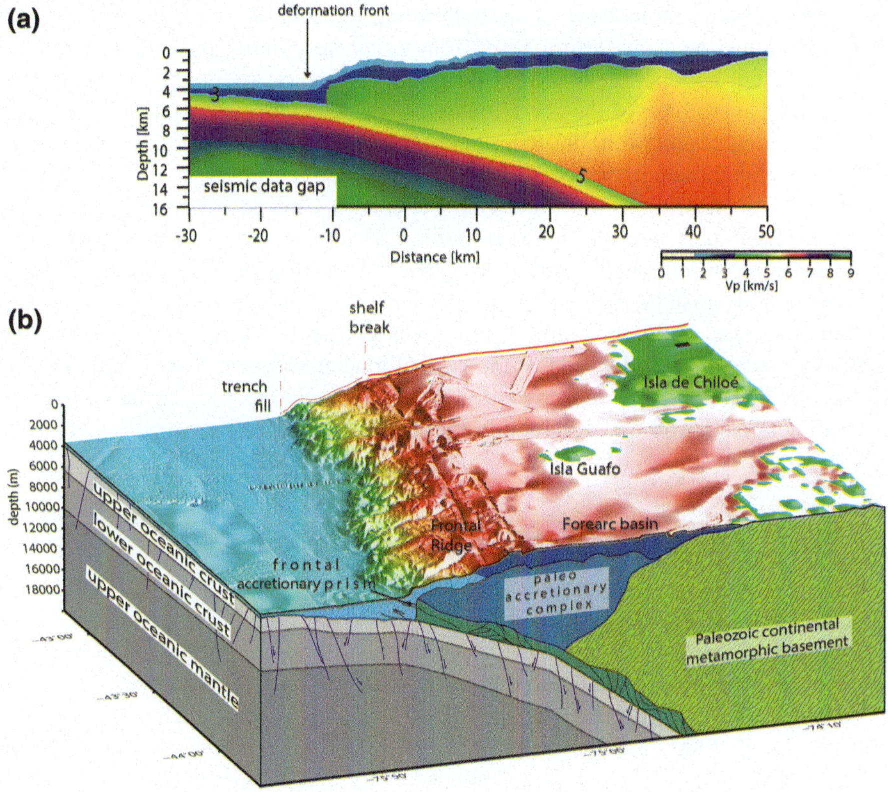

Fig. 11 Seismic velocity structure and interpretation of the upper subduction zone off Aysen at ~44.5° S (profile P8). **a** 2-D velocity model obtained from tomographic inversion of travel times (Scherwath et al. 2009). **b** Interpretation based on the seismic model shown in (a). The trench is heavily sedimented, mass wasting events are also evident at the base of the continental slope. The frontal prism is approximately 5 km wide

It is worth noting that the size of the accretionary prism is not defined using a specific velocity range, but instead the location of abrupt jumps of the lateral velocity gradient that should define the landward edge of the accretionary prism (Contreras-Reyes et al. 2013). Generally, large accretionary prisms comprise faster velocities at their landward edges than small ones due to sediment compaction.

The small size of the accretionary prism is an indirect evidence for the prevalence of sediment subduction over sediment accretion. Consequently, several seismic reflection data lines show thick subduction channels near the deformation front off Chile between ~43° S and the CTJ (e.g., Bangs and Cande 1997; Díaz-Naveas 1999; Geersen et al. 2011; Scherwath et al. 2009). However, the seismic reflection data presented by the latter authors are usually characterized by shallow penetration and they image the subduction channel only in the first 5–10 km of the shallow subduction interface. At these depths, a large variation in subduction channel thickness exists both along dip and strike. Thus, a thick subduction channel near the deformation front is not necessarily an evidence for efficient sediment subduction in a long term scale. In fact, sediment can be further underplated at the base of the accretionary prism (e.g., Contardo et al. 2008). Thus, seismic velocities and pronounced velocity gradients provide an indirect but a better clue for sediment subduction on a long term scale.

It is worth mentioning that the southern-central Chilean margin over the long term (i.e., the past several hundred million years) should be considered as an erosional margin instead of a site of net growth. This explains the close proximity of continental crust to the current trench and the small size of the accretionary prism. However, if this convergent margin is considered since the late Cenozoic glaciation characterized by the fast increase of glacial sediment flux (Kukowski and Oncken 2006), then the southern-central Chilean margin has been mainly governed by sediment accretion and subduction rather than subduction erosion.

4 The Austral Chilean Margin (South of the CTJ)

The 2-D seismic structure off Golfo de Penas (at $\sim 48°$ S) differs considerably from the margin structure north of the CTJ. Particularly, the accretionary prism off Golfo of Penas is 70–90 km wide (one of the largest acrecionary prisms off Chile), and it displays a wide accretionary ridge complex parallel to the trench axis (Fig. 12; Bourgois et al. 2000; Maksymowicz et al. 2012). The size of the accretionary prism is inferred from the imaged wedge-shaped body (70–90 km wide) with seismic velocities in the range of 2.5–6.0 km/s observed from the deformation front up to the continental shelf that is interpreted as the accretionary prism (Fig. 12). The landward limit of this feature coincides with a pronounced lateral velocity-gradient near the shelf break (Fig. 12) which marks the contact between the accretionary prism and framework rock (backstop). The framework rock off Golfo de Penas might be composed of Paleozoic paleo-accretionary complex, Cretaceous plutonic rocks of the northern Patagonian Batholith and Pliocene intrusions (Lagabrielle et al. 2000; SERNAGEOMIN 2003). The seismic model also shows a seismic discontinuity at 30–40 km landward of the deformation front interpreted as the contact between the frontal (poorly consolidated sedimentary unit) and middle (more compacted sedimentary unit) accretionary prism (Maksymowicz et al. 2012). Both the frontal and middle prisms form part of the accretionary prism or acretionary complex (also called outer wedge; von Huene et al. 2009).

The current CTJ region (at 46°–47° S) is characterized by a small frontal prism made of removed crustal debris (<5 km wide; Bourgois et al. 2000; Behrmann et al. 1994) and it has an estimated subduction erosion rate of 450 km^3/km/Myr (Guivel et al. 2003). The collision of the Chile Rise with the Chilean margin in the region of Golfo de Penas ocurred about 4 Myr before present and it is expected that the intensity of subduction erosion was similar to the conditions facing in the current CTJ region. If this is correct, then the relatively large accretionary prism (70–90 km wide)

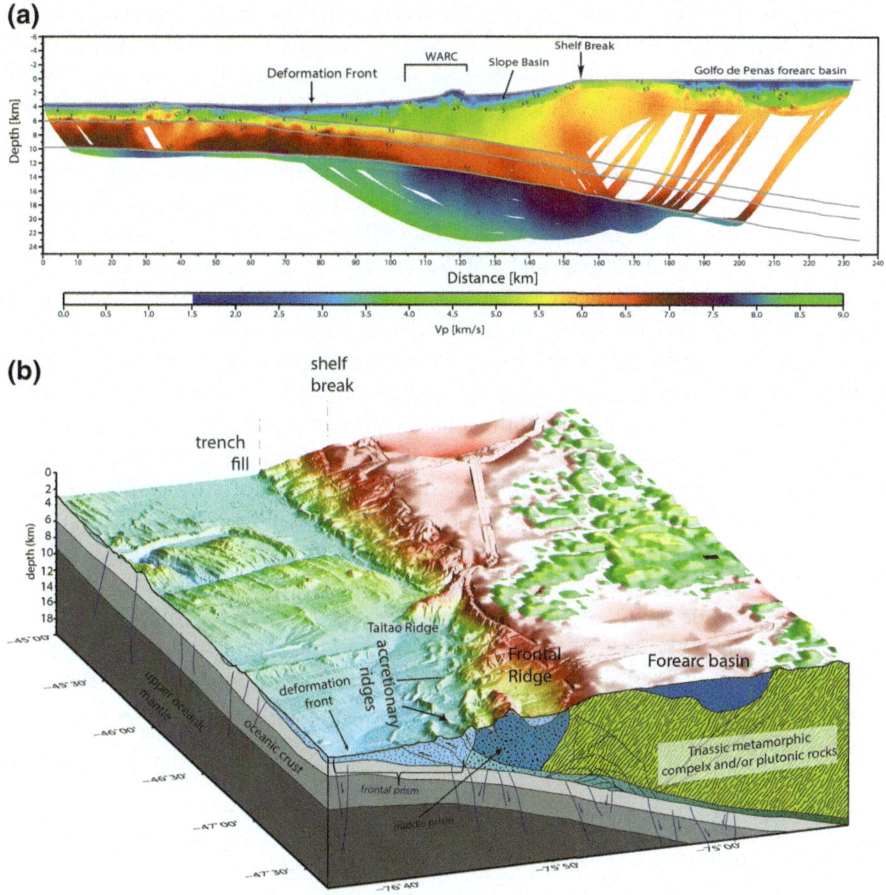

Fig. 12 Seismic velocity structure and interpretation of the upper subduction zone off Golfo de Penas at ~48° S (profile P9). **a** 2-D velocity model obtained from tomographic inversion of travel times (Maksymowicz et al. 2012). **b** Interpretation based on the seismic model shown in (a). The trench is heavily sedimented, and the frontal part of the margin is characterized by well developed accretionary ridges. The accretionary prism is 70–90 km wide and its landward edge is highlighted by the backstop that is coincident with an abrupt lateral velocity contrast. The fore-arc basin is composed of two sedimentary units with a total thickness of 3.7 km (Maksymowicz et al. 2012)

off Golfo de Penas was formed in the last 4 Ma indicating efficient sediment accretion. Efficient sediment accretion is consistent with the slow convergence between the oceanic Antarctic plate and the continental South American plate (~1.9 cm/a; DeMets et al. 1994) that facilates sediment acumulation at the trench (von Huene et al. 2009). The heavily sedimented trench enhances sediment accretion as shown by the growth of accretionary ridges as the Chile Ridge migrates to the north.

North of the CTJ, in contrast, the relatively small acrretionary prism reported between the Mocha Block and the CTJ can be partially explained in terms of the fast convergence rate between the oceanic Nazca plate and the continental South American plate (7.0–6.6 cm/a during the last 5 Myr; Kukowski and Oncken 2006). The rapid convergence facilates sediment subduction (e.g., von Huene et al. 2009), such as the case of the accretionary prism off Isla de Chiloé that is only ~10 km wide (Fig. 10). However, north of the Mocha Block, seismic studies reported an accretionary prism ~25 km wide off Peninsula de Arauco (Fig. 9; Contreras-Reyes et al. 2008) and >50 km wide off the Maule District (Fig. 8; Moscoso et al. 2011) indicating that sediment accretion is more efficient in this segment of the southern-central Chilean margin. This segment also has a convergence rate of about ~6.6 cm/a (e.g., Contreras-Reyes et al. 2013), so efficient sediment accretion strongly depends on other factors such as the sediment input to the trench, flexure of the oceanic lithosphere, and variations of frictional conditions (Contardo et al. 2008; Contreras-Reyes et al. 2013; Maksymowicz 2015).

The 2-D velocity model off Golfo de Penas also shows a wide relatively low velocity anomaly in the upper mantle under the continental slope and shelf (Fig. 12a).This low velocity zone might be related to high temperatures (>650°C) around the astenospheric window of the subducted Antarctic–Nazca spreading center (Lagabrielle et al. 2000; Maksymowicz et al. 2012). This shallow high temperature anomaly has been considered as the primary cause for melting at relatively shallow depths (e.g., Lagabrielle et al. 2000). Additionally, the uplift caused by the buoyancy of the hot subducted spreading center enhances frontal and basal subduction erosion beneath the overriding fore-arc wedge.

5 Summary

Subduction erosion, sediment accretion and subduction are active processes along the Chilean margin, and are controlled by climatic conditions, sediment supply to the trench, subduction of high oceanic features, and rheology of the subducting oceanic plate among other factors. The results from 9 wide-angle seismic profiles acquired along the Chilean margin conclude that processes of subduction accretion and erosion are mainly segmented by climatic conditions and high oceanic features such as the Juan Fernández Ridge and the Nazca-Antarctic spreading center (Chile Rise). The main large-scale segments can be summarized as follows:

- North of the current collision of the Juan Fernández Ridge with the margin at ~33° S, subduction erosion has been active since Jurassic times resulting in the eastward migration of the trench and magmatic arc. Few hundred meters of slope debris reach the sediment-starved trench and is further subducted. Typically, the middle-upper continental slope transition is characterized by a gravitational collapse highlighted by a prominent trenchward dipping normal scarp. Subduction erosion has, most likely, caused large-scale crustal thinning and long-term subsidence of the outermost fore-arc.

- Between the collision of the Juan Fernández Ridge with the margin and the Chile Triple Junction of the Nazca-Antarctic-South American plates, an accretionary prism 5–50 km wide has been formed due to an increase of trench sedimentation triggered by denudation processes of the Andes after the last Pleistocene Glaciation. However, the relative small size of the accretionary prism is not compatible with an efficient history of accretion, and sediment subduction is also an important process in this segment of the margin, especially between the Mocha Block and the Chile Triple Junction. Additionally, if the southern-central Chilean margin is analyzed since the Pleistocene glaciation, it should be classified as accretionary, while if it is viewed over the long term (i.e., the past several hundred million years) should be considered as an erosional margin.
- In the vicinity of the Chile Triple Junction, where the Chile Ridge collides with the margin, subduction erosion is active, and rapid uplift followed by subsidence of the forearc area and normal faulting and intensive sedimentary mass wasting are oberved.
- South of the CTJ, the convergence rate between the oceanic Antarctic and continental South American plate is slow ($\sim$1.9 cm/a), and the trench basin is thick and wide. These conditions have favored sediment accretion and subduction resulting in the formation of a large accretionary prism 70–90 km wide off Golfo de Penas.

Acknowledgements This work was supported by the Chilean National Science Foundation (FONDECYT) projects 1130004 and 1170009. I gratefully acknowledge Juan Becerra for the illustrative graphics, and greatly appreciate the constructive reviews of David Voelker and Orlando Álvarez.

References

Alvarez O, Nacif S, Spagnotto S, Folguera A, Gimenez M, Chlieh M, Braitenberg C (2015) Gradients from GOCE reveal gravity changes before Pisagua Mw = 8.2 and Iquique Mw = 7.7 large megathrust earthquakes. J South Am Earth Sci 64(2):273–287

Angermann D, Klotz J, Reigber C (1999) Space-geodetic estimation of the Nazca-South America Euler vector. Earth Planet Sci Lett 171(3):329–334

Armijo R, Thiele R (1990) Active faulting in northern Chile: ramp stacking and lateral decoupling along a subduction plate boundary? Earth Planet Sci Lett 98(1):40–61

Bangs NL, Cande SC (1997) Episodic development of a convergent margin inferred from structures and processes along the southern Chile margin. Tectonics 16:489–503

Barrientos SE, Ward SN (1990) The 1960 Chile earthquake: inversion for slip distribution from surface deformation. Geophys J Int 103:589–598. https://doi.org/10.1111/j.1365-246X.1990.tb05673.x

Becerra J, Arriagada C, Contreras-Reyes E, Bascuñan S, De Pascale G, Reichert C, Díaz-Naveas J, Cornejo N (2016) Gravitational deformation and inherited structural control on slope

morphology in the subduction zone of north-central Chile (~29°–33° S). Basin Res. https://doi.org/10.1111/bre.12205
Behrmann JH, Lewis SD, Cande SC (1994) Tectonics and geology of spreading ridge subduction at the Chile triple junction: a synthesis of results from leg 141 of the ocean drilling program. Geol Rundsch 83(4):832–852
Béjar-Pizarro M, Socquet A, Armijo R, Carrizo D, Genrich J, Simons M (2013) Andean structural control on interseismic coupling in the North Chile subduction zone. Nat Geosci 6(6):462–467
Blumberg S, Lamy F, Arz HW, Echtler HP, Wiedicke M, Haug GH, Oncken O (2008) Turbiditic trench deposits at the south-Chilean active margin: a Pleistocene-Holocene record of climatic and tectonics. Earth Planet Sci Lett 268:526–539
Bourgois J, Guivel C, Lagabrielle Y, Calmus T, Boulegue J, Daux V (2000) Glacial–interglacial trench supply variation, spreading-ridge subduction, and feedback controls on the Andean margin development at the Chile triple junction area (45–48° S). J Geophys Res 105 (B4):8355–8386
Byrne DE, Davis DM, Sykes LR (1988) Loci and maximum size of thrust earthquakes and the mechanics of the shallow region of subduction zones. Tectonics 7(4):833–857
Cande SC, Leslie RB (1986) Late cenozoic tectonics of the southern Chile trench. J Geophys Res: Solid Earth 91(B1):471–496
Cande SC, Leslie RB, Parra JC, Hobart M (1987) Interaction between the Chile ridge and Chile trench: geophysical and geotermal evidence. J Geophys Res 92(B1):495–520
Charrier R, Pinto L, Rodríguez MP (2007) Tectonostratigraphic evolution of the Andean Orogen in Chile. In: The Geology of Chile
Cifuentes IL (1989) The 1960 Chilean earthquakes. J Geophys Res 94:665–680. https://doi.org/10.1029/JB094iB01p00665
Clift P, Vannucchi P (2004) Controls on tectonic accretion versus erosion in subduction zones: implications for the origin and recycling of the continental crust. Rev Geophys 42(RG2001). https://doi.org/10.1029/2003RG000127
Comte D, Carrizo D, Roecker S, Ortega-Culaciati F, Peyrat S (2016) Three-dimensional elastic wave speeds in the northern Chile subduction zone: variations in hydration in the supraslab mantle. Geophys J Int 207(2):1080–1105. https://doi.org/10.1093/gji/ggw318
Comte D, Pardo M (1991) Reappraisal of great historical earthquakes in the northern Chile and southern Peru seismic gaps. Nat Hazards 4(1):23–44
Contardo X, Cembrano J, Jensen A, Díaz-Naveas J (2008) Tectono-sedimentary evolution of marine slope basins in the Chilean forearc (33 30′–36 50′ S): insights into their link with the subduction process. Tectonophysics 459(1):206–218
Contreras-Reyes E, Carrizo D (2011) Control of high oceanic features and subduction channel on earthquake ruptures along the Chile–Peru subduction zone. Phys Earth Planet Inter 186(1):49–58
Contreras-Reyes E, Grevemeyer I, Flueh ER, Reichert C (2008) Upper lithospheric structure of the subduction zone offshore of southern Arauco peninsula, Chile, at 38° S. J Geophys Res 113 (B07303). https://doi.org/10.1029/2007JB005569
Contreras-Reyes E, Jara J, Maksymowicz A, Weinrebe W (2013) Sediment loading at the southern Chile trench and its tectonic implications. J Geodyn 66:134–145. https://doi.org/10.1016/j.jog.2013.02.009
Contreras-Reyes E, Jara J, Grevemeyer I, Ruiz S, Carrizo D (2012) Abrupt change in the dip of the subducting plate beneath north Chile. Nat Geosci 5:342–345. https://doi.org/10.1038/ngeo1447
Contreras-Reyes E, Ruiz J, Becerra J, Kopp H, Reichert C, Maksymowicz A, Arriagada C (2015) Structure and tectonics of the central Chilean margin (31°–33° S): implications for subduction erosion and shallow crustal seismicity. Geophys J Int 653(2):776–791. https://doi.org/10.1093/gji/ggv309
DeMets C, Gordon RG, Argus DF (2010) Geologically current plate motions. Geophys J Int 181 (1):1–80
DeMets C, Gordon RG, Argus DF, Stein S (1994) Effect of recent revisions to the geomagnetic reversal time scale on estimates of current plate motions. Geophys Res Lett 21(20):2191–2194

Delouis B, Philip H, Dorbath L, Cisternas A (1998) Recent crustal deformation in the Antofagasta region (northern Chile) and the subduction process. Geophys J Int 132(2):302–338

Delouis B, Pardo M, Legrand D, Monfret T (2009) The Mw 7.7 Tocopilla earthquake of 14 November 2007 at the southern edge of the northern Chile seismic gap: rupture in the deep part of the coupled plate interface. Bull Seismol Soc Am 99(1):87–94

Díaz-Naveas JL (1999) Sediment subduction and accretion at the Chilean convergent margin, between 35°and 40S.° Doctoral dissertation, Christian-Albrechts-Universität

Flueh ER, Vidal N, Ranero CR, Hojka A, Von Huene R, Bialas J, Hinz K, Cordoba D, Danobeitia JJ, Zelt C (1998) Seismic investigation of the continental margin off-and onshore Valparaiso, Chile. Tectonophysics 288(1):251–263

Folguera A, Gianni G, Sagripanti L, Rojas Vera E, Novara I, Colavitto B, Alvarez O, Orts D, Tobal J, Giménez M, Introcaso A, Ruiz F, Martínez P, Ramos VA (2015) A review about the mechanisms associated with active deformation, regional uplift and subsidence in southern South America. J South Am Earth Sci 64(2):511–529

Fuenzalida A, Schurr B, Lancieri M, Sobiesiak M, Madariaga R (2013) High-resolution relocation and mechanism of aftershocks of the 2007 Tocopilla (Chile) earthquake. Geophys J Int 194 (2):1216–1228

Fujie G, Ito A, Kodaira S, Takahashi N, Kaneda Y (2006) Confirming sharp bending of the Pacific plate in the northern Japan trench subduction zone by applying a traveltime mapping method. Phys Earth Planet Int 157:72–85

Geersen J, Behrmann JH, Völker D, Krastel S, Ranero CR, Diaz-Naveas J, Weinrebe W (2011) Active tectonics of the South Chilean marine fore arc (35 S–40 S). Tectonics 30(3)

Guivel C, Lagabrielle Y, Bourgois J, Martin H, Arnaud N, Fourcade S, Cotten J, Maury RC (2003) Very shallow melting of oceanic crust during spreading ridge subduction: origin of near-trench quaternary volcanism at the Chile Triple Junction. J Geophys Res 108:2345–2463

Hartley AJ, May G, Chong G, Turner P, Kape SJ, Jolley EJ (2000) Development of a continental forearc: a Cenozoic example from the central Andes, northern Chile. Geology 28:331–334

Herron EM, Cande SC, Hall BR (1981) An active spreading center collides with a subduction zone: a geophysical survey of the Chile margin triple junction. Geol Soc Am Mem 154:683–702

Ito A, Fujie G, Miura S, Kodaira S, Kaneda Y, Hino R (2005) Bending of the subducting oceanic plate and its implication for rupture propagation of large interplate earthquakes off Miyagi, Japan, in the Japan trench subduction zone. Geophys Res Lett 32(5). https://doi.org/10.1029/2004GL022307

Khazaradze G, Klotz J (2003) Short- and long-term effects of GPS measured crustal deformation rates along the south central Andes. J Geophys Res 108(B6). https://doi.org/10.1029/2002JB001879

Kopp H, Flueh E, Papenberg C, Klaeschen D (2004) Seismic investigations of the O'Higgins Seamount Group and Juan Fernández Ridge: aseismic ridge emplacement and lithosphere hydration. Tectonics 23(2). https://doi.org/10.1029/2003TC001590

Kukowski N, Oncken O (2006) Subduction erosion: the normal mode of forearc material transfer along the Chilean margin? In: Oncken O et al (eds) The Andes: Active Subduction Orogeny. Frontiers in Earth Sci 3, pp 217–236

Lagabrielle Y, Guivel C, Maury RC, Bourgois J, Fourcade S, Martin H (2000) Magmatic–tectonic effects of high thermal regime at the site of active ridge subduction: the Chile triple junction model. Tectonophysics 326(3):255–268

Laursen J, Scholl DW, von Huene R (2002) Neotectonic deformation of the central Chile margin: deepwater forearc basin formation in response to hot spot ridge and seamount subduction. Tectonics 21(5):1038. https://doi.org/10.1029/2001TC901023

León-Ríos S, Ruiz S, Maksymowicz A, Leyton F, Fuenzalida A, Madariaga R (2016) Diversity of the 2014 Iquique's foreshocks and aftershocks: clues about the complex rupture process of a Mw 8.1 earthquake. J Seismology, 1–15. https://doi.org/10.1007/s10950-016-9568-6

Maksymowicz, A (2015) The geometry of the Chilean continental wedge: Tectonic segmentation of subduction processes off Chile. Tectonophysics 659:183–196

Maksymowicz A, Contreras-Reyes E, Grevemeyer I, Flueh ER (2012) Structure and geodynamics of the post-collision zone between the Nazca-Antartic spreading center and South America. Earth Planet Sci Lett 345–318:27–37. https://doi.org/10.1016/j.epsl.2012.06.023

Marquardt C, Lavenu A, Ortlieb L, Godoy E, Comte D (2004) Coastal neotectonics in Southern Central Andes: uplift and deformation of marine terraces in Northern Chile (27 S). Tectonophysics 394(3):193–219

Melnick D (2016) Rise of the central Andean coast by earthquakes straddling the Moho. Nat Geosc 9:401–407

Métois M, Socquet A, Vigny C, Carrizo D, Peyrat S, Delorme A, Maureira E, Valderas-Bermejo M-C, Ortega I (2013) Revisiting the north Chile seismic gap segmentation using gps-derived interseismic coupling. Geophys J Int 194(3):1283–1294

Métois M, Vigny C, Socquet A (2016) Interseismic coupling, megathrust earthquakes and seismic swarms along the Chilean subduction zone (38°–18° S). Pure Appl Geophys 173(5):1431–1449. https://doi.org/10.1007/s00024-016-1280-5

Moreno MS, Bolte J, Klotz J, Melnick D (2009) Impact of megathrust geometry on inversion of coseismic slip from geodetic data: Application to the 1960 Chile earthquake. Geophys Res Lett 36(16)

Moscoso E, Grevemeyer I, Contreras-Reyes E, Flueh ER, Dzierma Y, Rabbel W, Thorwart M (2011) Revealing the deep structure and rupture plane of the 2010 Maule, Chile Earthquake (Mw = 8.8) using wide angle seismic data. Earth Planet Sci Lett 307(1–2):147–155. https://doi.org/10.1016/j.epsl.2011.04.025

Muñoz P (2015) Caracterización sísmica del antearco marino en la zona epicentral del mega-terremoto del Maule 2010. M. S. C. thesis. Departamento de Geofísica. Chile University

Peyrat S, Madariaga R, Buforn E, Campos J, Asch G, Vilotte JP (2010) Kinematic rupture process of the 2007 Tocopilla earthquake and its main aftershocks from teleseismic and strong-motion data. Geophys J Int 182(3):1411–1430

Ranero CR, von Huene R, Weinrebe W, Reichert C (2006) Tectonic processes along the Chile Convergent Margin. In: Oncken et al (eds) The andes: active subduction Orogeny Frontiers in Earth Sci. 3, pp 91–121

Ruiz S, Grandin R, Dionicio V, Satriano C, Fuenzalida A, Vigny C, Madariaga R (2013) The Constitución earthquake of 25 March 2012: a large aftershock of the Maule earthquake near the bottom of the seismogenic zone. Earth Planet Sci Lett 377:347–357

Ruiz S, Metois M, Fuenzalida A, Ruiz J, Leyton F, Grandin R, Vigny C, Madariaga R, Campos J (2014). Intense foreshocks and a slow slip event preceded the 2014 Iquique Mw 8.1 earthquake. Science 345(6201):1165–1169

Rutland RWR (1971) Andean orogeny and ocean floor spreading. Nature 233:252–255

Sallares V, Ranero CR (2005) Structure and tectonics of the erosional convergent margin off Antofagasta, north Chile (23° 30'S). J Geophys Res 110(B6). https://doi.org/10.1029/2004JB003418

Scherwath M, Contreras-Reyes E, Flueh ER, Grevemeyer I, Krabbenhoeft A, Papenberg C, Petersen CJ, Weinrebe RW (2009) Deep lithospheric structures along the southern central Chile margin from wide-angle P-wave modelling. Geophys J Int 179(1):579–600. https://doi.org/10.1111/j.1365-246X.2009.04298.x

Scherwath M, Flueh E, Grevemeyer I, Tilmann F, Contreras-Reyes E, Weinrebe W (2006) Investigating subduction zone processes in Chile. Eos Trans AGU 87(27). https://doi.org/10.1029/2006EO270001

Scholl DW, von Huene R (2009) Implications of estimated magmatic additions and recycling losses at the subduction zones of accretionary (non-collisional) and collisional (suturing) orogens. Geol Soc 318(1):105–125 Special Publications, London

SERNAGEOMIN: Chilean Geological and Mining Service (2003) Geologic map of Chile: Digital version, scale 1:1.000.000, Santiago, Chile

Somoza R (1998) Updated Nazca-South America relative motions during the last 40 My: implications for mountain building in the central Andean region. J South Am Earth Sci 11:211–215. https://doi.org/10.1016/S0895-9811(98)00012-1

Stern CR (2011) Subduction erosion: rates, mechanisms, and its role in arc magmatism and the evolution of the continental crust and mantle. Gondwana Res 20:284–308. https://doi.org/10.1016/j.gr.2011.03.006

Tebbens SF, Cande SC, Kovacs L, Parra JC, LaBrecque JL, Vergara H (1997) The Chile ridge: a tectonic framework. J Geophy Res 102(B6):12035–12060. https://doi.org/10.1029/96JB02581

Thornburg TM, Kulm DM, Hussong DM (1990) Submarine-fan development in the southern Chile trench: a dynamic interplay of tectonics and sedimentation. Geol Soc Am Bulletin 102:1658–1680

von Huene R, Corvalán J, Flueh ER, Hinz K, Korstgard J, Ranero CR, Weinrebe W, CONDOR Scientists (1997) Tectonic control of the subducting Juan Fernández Ridge on the Andean margin near Valparaíso, Chile. Tectonics 16(3):474–488

von Huene R, Ranero CR. (2003) Subduction erosion and basal friction along the sediment-starved convergent margin off Antofagasta, Chile. J Geophys Res: Solid Earth 108(B2)

von Huene R, Ranero CR, Scholl DW (2009) Convergent margin structure in high-quality geophysical images and current kinematic and dynamic models. In Subduction Zone Geodynamics. Springer, Berlin, Heidelberg, pp. 137–157

von Huene R, Ranero CR, Vannucchi P (2004) Generic model of subduction erosion. Geology 32 (10):913–916

von Huene R, Weinrebe W, Heeren F (1999) Subduction erosion along the north Chile margin. J Geodyn 27:345–358

Voelker D, Geersen J, Contreras-Reyes E, Reichert C (2013) Sedimentary fill of the Chile Trench (32°–46° S): volumetric distribution and causal factors. J Geol Soc London 170(5):723–736. https://doi.org/10.1144/jgs2012-119

Wang K, Hu Y (2006) Accretionary prisms in subduction earthquake cycles: the theory of dynamic Coulomb wedge. J Geophys Res 111(B6). https://doi.org/10.1029/2005JB004094

Yáñez G, Ranero CR Díaz J (2001) Magnetic Anomaly interpretation across the southern central Andes (32°–34° S): the role of the Juan Fernández Ridge in the late Tertiary evolution of the margin. J Geophys Res 106:6325–6345

Zelt CA (1999) Modeling strategies and model assessment for wide-angle seismic travel time data. Geophys J Int 139:183–204

The Geometry of the Continental Wedge and Its Relation to the Rheology and Seismicity of the Chilean Interplate Boundary

Andrei Maksymowicz and Andres Tassara

Abstract A latitudinal tectonic segmentation along the Chilean subduction margin is defined by the modeling of the continental wedge geometry. The segments are characterized by different effective basal friction coefficients or Hubbert–Rubey fluid pressure ratio and are limited by the subduction of oceanic features and seaward continental prolongations. The analysis of the modeled parameters indicates that the process of tectonic erosion probably is associated with high levels of overpressure in the *decóllement* and inside the continental wedge. The observed segmentation shows a spatial correlation with the distribution of large earthquake ruptures, which suggest a link between the long-term and short-term deformation process. Joint interpretation of the results with the b-value analysis and the density-depth models in the 2010 Maule Mw8.8 earthquake zone shows the importance of these studies to understand the geodynamics of the subduction zones.

Keywords Accretionary wedge · Basal friction · Fluid pressure
Tectonic erosion · Coulomb wedge model · Large earthquake ruptures

1 Introduction

In 1994, Serge E. Lallemand and coauthors published a key observation (Lallemand et al. 1994) comparing the geometry of the continental wedge in several margins worldwide. The continental wedge was analyzed by considering a slope angle (α), which represents the continental surface from the trench to shelf break, and a basal angle (β) of the wedge, which corresponds to the interplate boundary angle in the

A. Maksymowicz (✉)
Departamento de Geofísica, Facultad de Ciencias Físicas Y Matemáticas,
Universidad de Chile, Santiago, Chile
e-mail: andrei.maksymowicz.jeria@gmail.com

A. Tassara
Departamento de Ciencias de La Tierra, Universidad de Concepción, 160-C Casilla,
Concepción, Chile

A. Folguera et al. (eds.), *The Evolution of the Chilean-Argentinean Andes*,
Springer Earth System Sciences, https://doi.org/10.1007/978-3-319-67774-3_2

case of subduction zones. Plotting these angles in a graphic α versus β (Fig. 1a), they observed a clear pattern where the margins can be separated into three clusters according to their tectonic style: (1) "typical accretionary wedge" with low slope and basal angles, characterized by the presence of a large unit of accreted sediments (accretionary prism) in the frontal portion of the continental wedge, (2) "intermediate accretionary wedge" with higher α and β angles (but lower than 10°), characterized by a relatively small accretionary prism, and (3) "non-accretionary wedge" that presented β values higher than 10° and some α values larger than the observed for the accretionary groups. The margins in the non-accretionary wedge group are located in subduction zones with small sedimentary trench fill which favored tectonic erosion. According to this classification, the geometry of the continental wedge is directly associated with the tectonic style of the margin, agreeing with predictions of the non-cohesive Coulomb wedge model (NCCW), developed at the beginnings of the 80's, and published in three seminal papers (Davis et al. 1983; Dahlen et al. 1984; Dahlen 1984).

The NCCW is a simple theoretical model that allows the prediction of the wedge shape (α and β angles) and the principal critical rupture direction inside de wedge (Fig. 1b) as a function of four constant parameters: internal density (ρ_s), internal friction coefficient (μ), internal Hubbert–Rubey fluid pressure ratio (λ), and the effective basal friction coefficient (μ_b^*). For a set of these parameters, there is an associated rupture envelope (red line in Fig. 1c) that represents all possible wedges (α and β angles) reaching the Mohr–Coulomb failure criterion (without cohesion) inside and at the base. It is important to note that sectors in the lower branch and in the upper branch of rupture envelope correspond to wedges under contractional deformation and normal faulting, respectively (black lines in Fig. 2c). Then, the margins under tectonic erosion (non-accretionary wedge group) are near a critical normal faulting state, as was suggested by Lallemand et al. (1994), (see red lines in Fig. 1a). It is important to note that according to numerical experiments the results of the NCCW model are highly dependent on two parameters, μ_b^* and λ, and that there is a tradeoff between them (Maksymowicz 2015).

2 Tectonic and Geology of the Chilean Continental Wedge

The Chilean margin is characterized by the interaction of the Nazca, Antarctic, South America, and Scotia plates (Fig. 2). Between ~18° S and 46.5° S, the rapid subduction of the Nazca plate under South America at a convergence rate >66 km/Myr has been slightly oblique to the trench during the last ~25 Myr (Cande and Leslie 1986; Angermann et al. 1999), and it determines a general compressive tectonics in the zone. In the southern portion of the margin (~52° S–46.5° S), the Antarctica plate subducts beneath South America with a low convergence rate of ~20 km/Myr in a direction that has been perpendicular to the trench from at least the early Eocene (Cande and Leslie 1986), while South of ~52° S, the transform boundary between South America and Scotia plates accommodates part of the

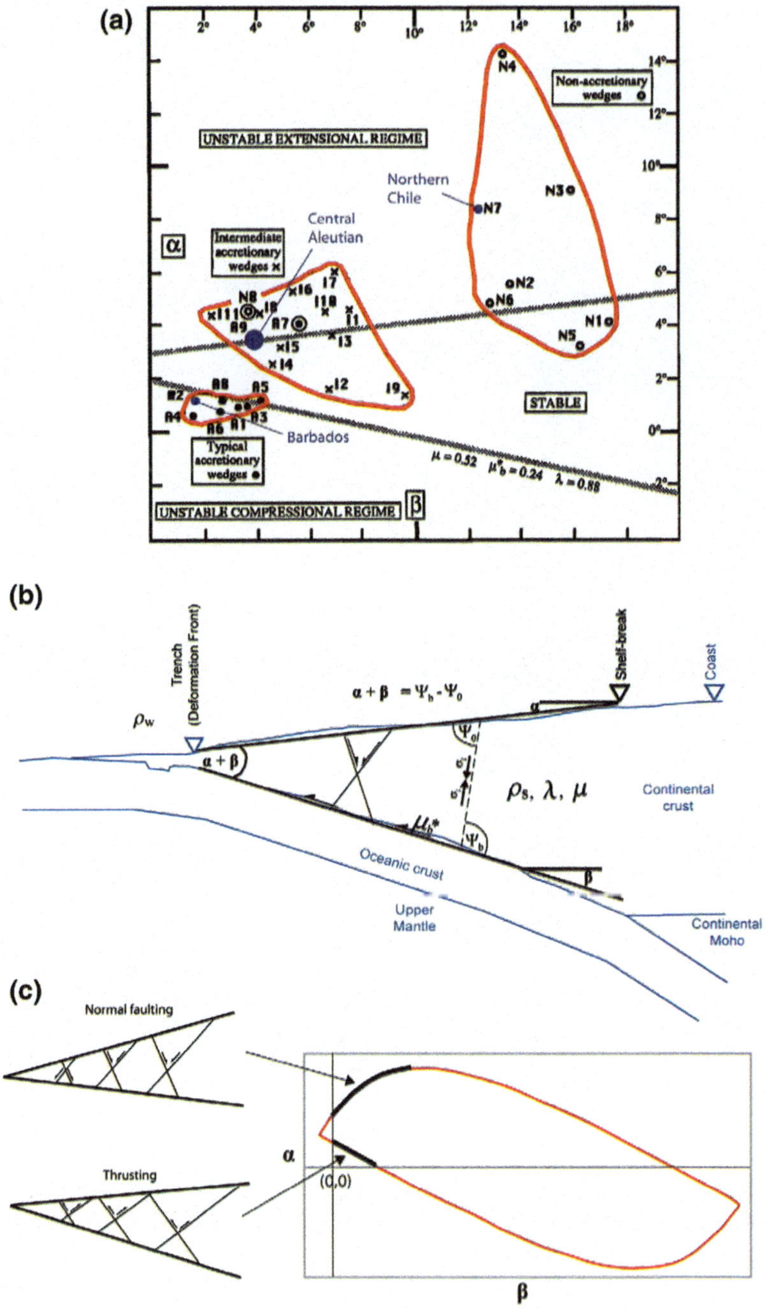

Fig. 1 Elements of the NCCW theory applied to the continental wedge analysis. **a** α versus β graphic for the margins worldwide. Blue dots highlight margins in different tectonic settings. Figure extracted from Lallemand et al. (1994). **b** Main aspects of NCCW theory. The figure shows the relation between the wedge geometry (α and β angles) and the direction of principal internal stress (σ_1). ρ_w is the water density, ρ_s is the density of the wedge, λ is the Hubbert–Rubey fluid pressure ratio, μ is the internal friction coefficient of the wedge, and μ_b^* is the effective basal friction of the wedge. **c** Rupture envelope of the NCCW Theory. The black lines are the sectors of the rupture envelope where thrusting faults and normal faulting are observed

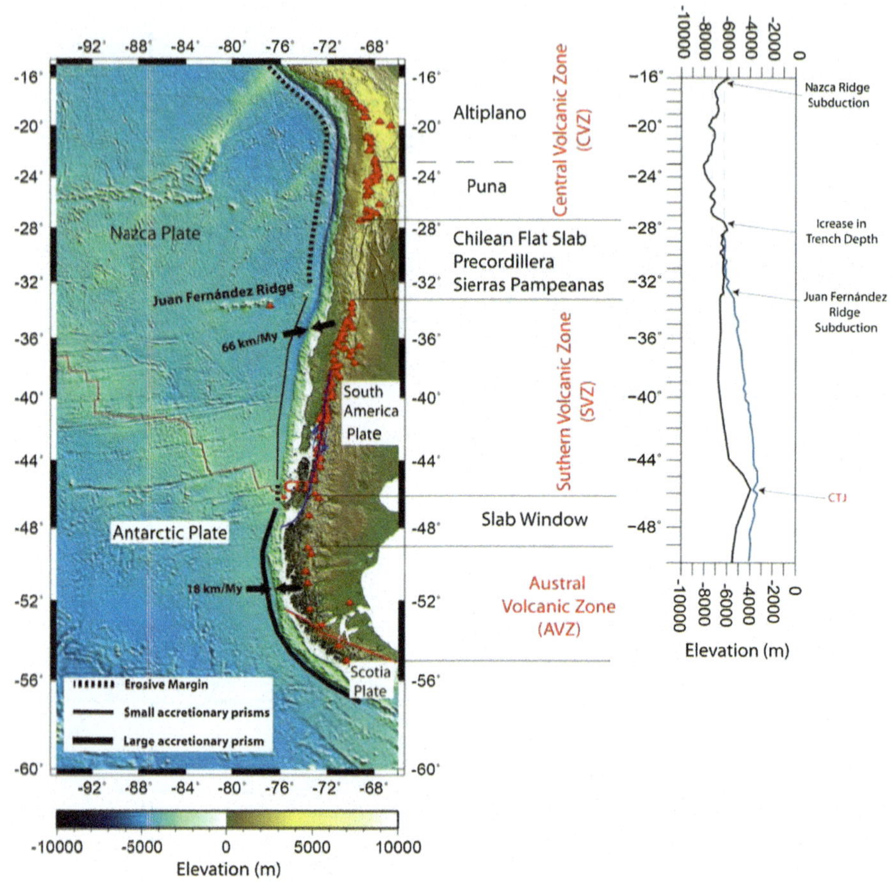

Fig. 2 First-order tectonic segmentation of the Chilean continental wedge. Red triangles are the location of Quaternary volcanoes, and the blue line corresponds to the intra-crustal Liquiñe–Ofqui fault system. CTJ indicates the location of the Chile triple junction. The limits of the major tectonic segments of the Andes and Quaternary volcanic zones are also indicated. Black line in the right panel shows the elevation of the slab at the Peru–Chile trench in function of the latitude (modified from Contreras-Reyes and Osses 2010), and the blue line corresponds to the seafloor elevation at the same latitudes. The difference between these curves is the sedimentary thickness of the Peru–Chile trench

convergence resulting in a low convergence rate for the Antarctica–Scotia subduction of about 13 km/Myr (Thomas et al. 2003). The Nazca–South America convergence has generated large megathrust earthquakes such as: Valdivia 1960 Mw9.6, Maule 2010 Mw8.8, Iquique 2014 Mw8.1, Illapel Mw8.3; Valparaiso 1730 ~M8.7, Iquique 1877 ~M8.5, Vallenar 1922 ~M8.4 (Comte and Pardo 1991; Beck et al. 1998; Cisternas et al. 2005; Ruegg et al. 2009; Bilek 2010; Moreno et al. 2012; Ruiz et al. 2014; Tilmann et al. 2016).

As was pointed out by Lallemand et al. (1994) and Clift and Vannucchi (2004), the amount of trench sedimentary infill and the convergence rate are key factors to generate accretion or tectonic erosion. Then, the variability of these parameters along the Chilean margin determines a tectonic segmentation in the continental wedge which has been tested by numerous geophysical studies (mainly, seismic reflection profiles and wide-angle seismic profiles). The Chilean continental wedge can be divided into three main segments that present different tectonic styles. The northern segment extended to the north of the Juan Fernández Ridge subduction point (JFR) is a typical erosive margin characterized by a poorly sedimented trench, normal faulting in the continental wedge, and the presence of a small frontal accretionary wedge (Sallares and Ranero 2005; Ranero et al. 2006; Contreras-Reyes et al. 2012, 2014). The starved trench observed in the northern segment results from the arid climate and from the subduction of JFR that blocked the northward redistribution of sediments through the trench axial channel (Thornburg and Kulm 1987; Völker et al. 2013). The second segment is extended between the JFR and the Nazca–Antarctica–South America triple junction (CTJ in Fig. 2) and is characterized by a largely sedimented trench and an active accretionary prism of small to medium size in the continental wedge (<50 km width) that presents low-angle west-vergent reverse faults (Bangs and Cande 1997; Zelt 1999; Diaz-Naveas 1999; Contreras-Reyes et al. 2008, 2010; Scherwath et al. 2009; Moscoso et al. 2011). In this segment, the relatively small size of the accretionary prism is explained by rapid Nazca–South America convergence rate. Finally, the third segment located southward of the CTJ presents a wide accretionary prism (>50 km) in response to the low Antarctica–South America convergence rate (~20 km/My) and large trench infill. In this southern segment, the deformation style is characterized by a general west-vergent reverse high-angle faulting associated to quasi-symmetric folding (Ranero et al. 2006; Polonia et al. 2007). Additionally to the main tectonic segments, the subduction of the rough bathymetry of the Nazca–Antarctica spreading center erodes the continental margin generating a local segment of very intense tectonic erosion (Bangs and Cande 1997; Bourgois et al. 2000; Maksymowicz et al. 2012).

The regional structure of the continental wedge has been recognized by marine geophysical studies. The geology offshore is interpreted based on the units observed in the coast. In the accretionary segment of the Chilean margin, the structure of the continental wedge (below the continental slope) comprises the current accretionary prism with variable width, generally larger than 40 km depending on the latitude (Polonia et al. 2007; Maksymowicz et al. 2012; Scherwath et al. 2009; Contreras-Reyes et al. 2008, 2010; Moscoso et al. 2011; Maksymowicz et al. 2015) and excluding the short erosive segment around the Chile triple junction (CTJ, Fig. 2). From ~53° to ~49° S, the geology at the coast is characterized by the presence of Paleozoic to early Cretaceous metamorphic complexes (Hervé et al. 2008) immediately to the west of the Mesozoic Patagonian Batholith (Bell and Suárez 2000). Similarly, to the north of ~45° S, extensive outcrops of late Paleozoic/Mesozoic metamorphic rocks are observed in the Chilean coast immediately to the west of the Patagonian Batholith and late Paleozoic granitoids in the northern area. These units correspond to paired metamorphic belts observed along the Chonos Archipelago, Chiloé Island, and the coast of central Chile where these are referred as Western and Eastern series (Hervé and Fanning 2000; Glodny et al. 2005; Willner et al. 2005). To the north of ~33° S, in the erosive segment of the margin, sparse outcrops of the Paleozoic metamorphic complexes are observed in the coast (Bell 1984; Willner et al. 2012), where the volcano-sedimentary units and granitoids associated with the Mesozoic magmatic arc are the most common units along the northern Chile margin (Geol, 1:1000000, SERNAGEOMIN 2002). According to the geological characteristic and the small frontal accretionary prism observed in the erosive northern Chilean margin (<20 km width, Zelt 1999; Contreras-Reyes et al. 2015; Sallares and Ranero 2005 and Contreras-Reyes et al. 2012), it is possible to interpret a continental wedge mainly formed by fractured Mesozoic volcano-sedimentary rocks that cover a core of a Paleozoic metamorphic complex, which probably has a more rigid response in comparison to the continental wedge of the accretionary segment (to the south of ~30° S), where the sediments of the relatively large accretionary prism are deformed against the metamorphic complexes (Fig. 3).

3 The Geometry of the Chilean Continental Wedge

Maksymowicz (2015) calculated the slope angle α every two kilometers along the Chilean margin (Fig. 3), observing the positions and elevations of the deformation front (DF) and shelf break (SB) (blue lines in Fig. 3). The bathymetric data base used came from of different sources: ETOPO1 bathymetry data (Amante and Eakins 2009), multi-beam bathymetry acquired from ~20° S to ~48° S (Bourgois et al. 2000; GEOMAR TIPTEQ Project: The Incoming Plate to mega-Thrust EarthQuake, Flueh and Grevemeyer 2005). To avoid artifacts caused by the margin curvatures, a maximum α is calculated by selecting, for each DF point, the SB point that maximizes the α value. This maximum α is preferred because it is representative of the slope angle in zones where the DF and SB are parallel, and it is less sensitive to local decreases caused by the presence of submarine canyons and local strike rotations of the DF and the SB.

Figure 3 shows a high-frequency variation of the slope angle due to the presence of submarine canyons, landslides, slope basins, and local mass transfer processes, but at regional scale a clear pattern emerges. To the south of ~33° S, the Chilean margin presents values of $\alpha < 5°$ and higher values to the north. The limit between these two segments can be associated with the principal tectonic change in the Chilean wedge, where the accretionary segment (south of JFR) has lower slope angles than the erosive margin to the North, as it is predicted by the NCCW theory. At smaller scale, there is a clear correlation between the points where oceanic features (fault zones and aseismic ridges) are subducted and local changes in the slope angle (see location of FZ, JFR, Copiapo Ridge, and Taltal Ridge in Fig. 3). In general, these structures correlate with strong decreases or abrupt changes of α, with

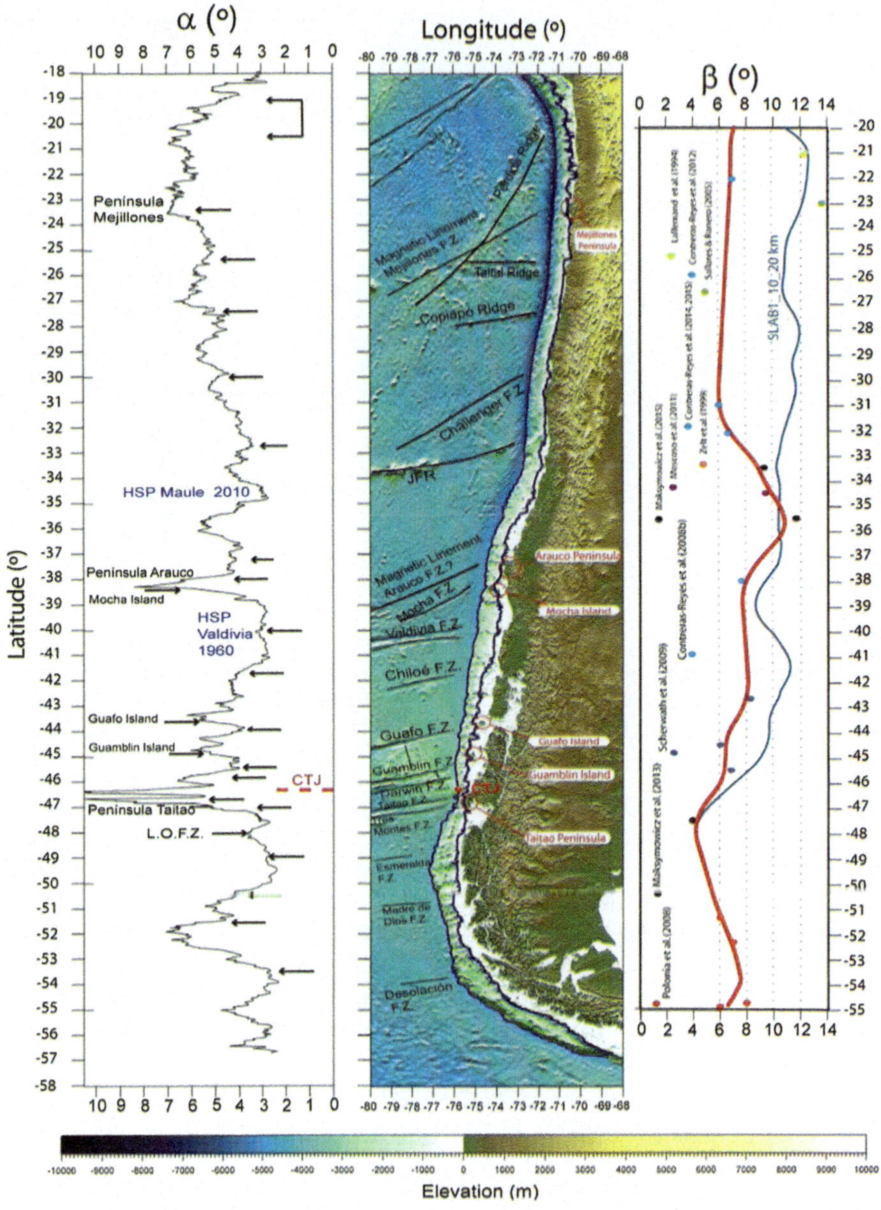

Fig. 3 α and β values along the Chilean Margin. The left panel shows the values of the slope angle as a function of the latitude. Arrows indicate the subduction of the oceanic features and the location of the Guamblin, Guafo, and Mocha islands. The dotted red line shows the location of the Chile triple junction (CTJ). The high slip patches (HSP) of the Maule 2010 and Valdivia 1960 earthquakes are indicated in blue. The central panel shows elevation map of the Chilean margin. Blue lines correspond to the deformation front (DF) and shelf break (SB) along the margin; black lines are the oceanic bathymetric features. The red dot indicates the position of the CTJ, and the red circles show the location of the Guamblin, Guafo, and Mocha islands. The right panel corresponds to the values of the subduction angle β of the continental wedge as a function of the latitude. Blue line is the version derived from the SLAB1 model from 10 to 20 km depth, and thick red line is the preferred model derived from several seismic and gravimetric studies (modified from Maksymowicz 2015)

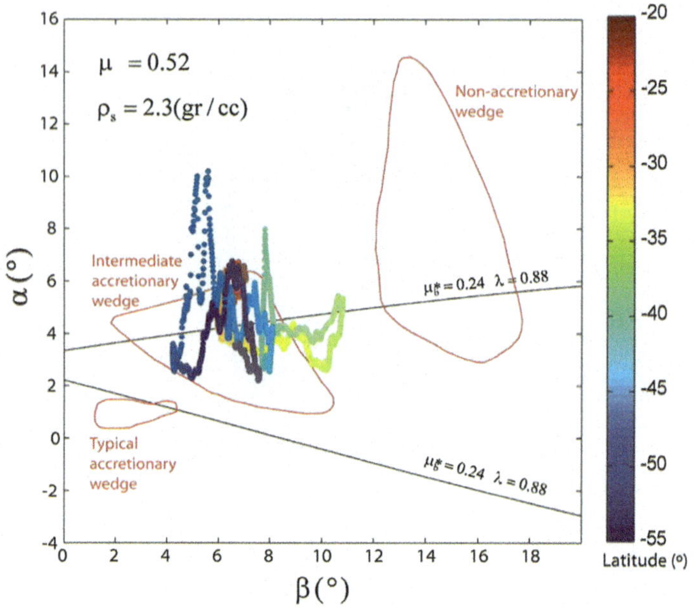

Fig. 4 Slope angle α verses basal angle β for the entire Chilean margin (latitudes are color coded). Solid black lines correspond to the NCCW rupture envelope for the parameters analyzed by Lallemand et al. (1994), and the red lines are the classification of the convergent margins of the same authors. Figure extracted from Maksymowicz (2015)

the exception of Taitao FZ, Darwin FZ, Mocha FZ, and Mejillones FZ that seem to be hidden by the presence of the Taitao, Arauco, and Mejillones peninsulas. In fact, bathymetric highs in the continental wedge are systematically associated with local increases of slope angle (Mejillones Peninsula, Arauco Peninsula, Taitao Peninsula, Guafo Island, Guamblin Island, and Mocha island).

In order to generate a characterization of the geometry of the Chilean continental wedge, the α angle must be associated with the subduction angle (β) below the slope. Maksymowicz (2015) shows that subduction angles derived from seismological data (Hayes et al. 2012; Lallemand et al. 1994, among others) overestimate the β angle below the slope due to the general onshore distribution of seismological networks. Then, it is necessary to generate a β model with direct marine geophysical observations. Figure. 3 shows an upgrade of the β value along the Chilean margin presented by Maksymowicz et al. (2015). The curve was obtained by a smooth interpolation of β values observed in the seismic and gravimetric models presented by Polonia et al. (2007), Maksymowicz et al. (2012, 2015), Scherwath et al. (2009), Contreras-Reyes et al. (2008, 2014, 2015), Moscoso et al. (2011), Zelt (1999), Sallares and Ranero (2005). As a reference, Fig. 3 also shows β values derived from SLAB1 model (Hayes et al. 2012) between depths of 10 km and 20 km (blue line). It is worth noting that the high β derived from Zelt (1999), Sallares and Ranero (2005), and Lallemand et al. (1994) were not considered because they are ~15°, ~7°, and ~5° higher than the ones obtained by Contreras-Reyes et al. (2012, 2014, 2015) with a well constrained wide-angle seismic tomography (see Chap. "Structure and Tectonics of the Chilean Convergent Margin from Wide-Angle Seismic Studies: A Review"). However, the comparisons with the SLAB1 model suggest that the anomalous large β angles are representative of deeper sectors of the interplate boundary, which agree with the model presented by Contreras-Reyes et al. (2012), that shows an abrupt increase of β (similar to kink geometry) at ~20 km depth. Then, this increase of subduction angle with depth seems to be characteristic for the northern Chile margin.

Based on the slope and subduction angles, Fig. 4 shows the α versus β graphic for the Chilean margin, i.e., the same visualization presented by Lallemand et al. (1994) to characterize the tectonic nature of the margins worldwide. As it is observed, the entire Chilean margin can be classified inside or near the "intermediate accretionary wedge" group proposed initially by Lallemand et al. (1994), which means that even the erosive northern Chilean margin is classified in this group as a consequence of a correct definition of the subduction angle below the slope. This analysis shows the importance of a precise determination of the subduction angle β below the wedge and raises the need of a more detailed tectonic interpretation of the wedge geometry variability for the Chilean and other subduction margins. Analyzing the latitudinal variation of the wedge geometry in Fig. 4, local strong variations and regional trends are observed, which suggest that different segments of the margin are characterized by different rheological parameters, and that the study of this complex behavior becomes important to understand the tectonic evolution of the margin. In this view, an specific accretionary or erosive margin falls in a broad spectra of tectonic behaviors according to the spatial and temporal variations of numerous parameters like subduction rate, sedimentary trench fill, roughness of the incoming plate, fluid transport, lithology of the continental wedge, porosity and compaction of the subduction channel (Lallemand et al. 1994; Kukowski and Oncken 2006; Ranero et al. 2008; Vannucchi et al. 2012).

4 Model of the Effective Basal Friction Coefficient and the Hubbert–Rubey Fluid Pressure Ratio (μ_b^* and λ) Along the Chilean Continental Wedge

The definition of the Chilean continental wedge geometry allows modeling the effective basal friction coefficient and the Hubbert–Rubey fluid pressure ratio (λ) as a function of the latitude using the NCCW theory. Maksymowicz (2015) calculated μ_b^* along the entire margin assuming in general a constant value of $\lambda = 0.67$ (Dahlen 1984) for the accretionary segment and $\lambda = 0.8$ for the erosive segment. Due to the tradeoff between these parameters, this is one of the many possible options to model the continental wedge geometry, but the objective here is to analyze the variability of the wedge rheology, and not the determinations of exact values. It is worth noting that the variation of these parameters can have a similar tectonic interpretation since μ_b^* decreases when the fluid pressure increases at the base of the wedge, and λ increases when the fluid pressure increases within the wedge. Hence, if the tectonics of the margin determines a general change in the fracturing or in the porosity, both at the base and within the wedge, these changes must be reflected in the estimated parameters.

Figures 5 and 6 correspond to examples of the NCCW modeling of the Chilean continental wedge. The segment of the margin extending between 46° and 39° S (Fig. 5) is representative of the Chilean accretionary margin. As it is observed, the β angles in this region vary from ~6° to ~8° and the geometry of the continental wedge south of the Arauco Peninsula is explained by μ_b^* = ~0.35, while southward

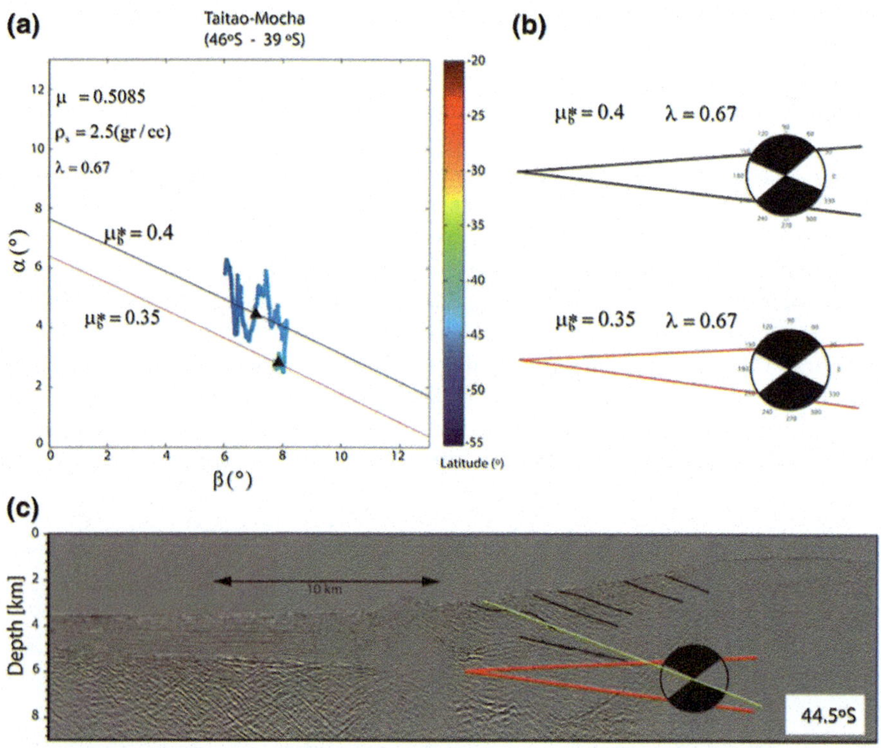

Fig. 5 NCCW analyses for the central Chile margin. **a** The dots are the $\alpha\beta$ data from 46° S to 39° S (the latitudes are shown in the color bar). Solid lines correspond to the NCCW rupture envelopes for different values of μ_b^*, considering constant values of λ = 0.67, ρ_s = 2.5 g/cc, and μ = 0.5085. **b** Schemes of the predicted continental wedge geometry and the principal rupture direction for each black triangle in (**a**). **c** Comparison between two predicted geometries for the continental wedge and the rupture directions with the reflection seismic profile at the latitude 44.5° S (extracted from Contreras-Reyes et al. 2010). Yellow lines indicate the predicted faulting inclination (modified from Maksymowicz 2015)

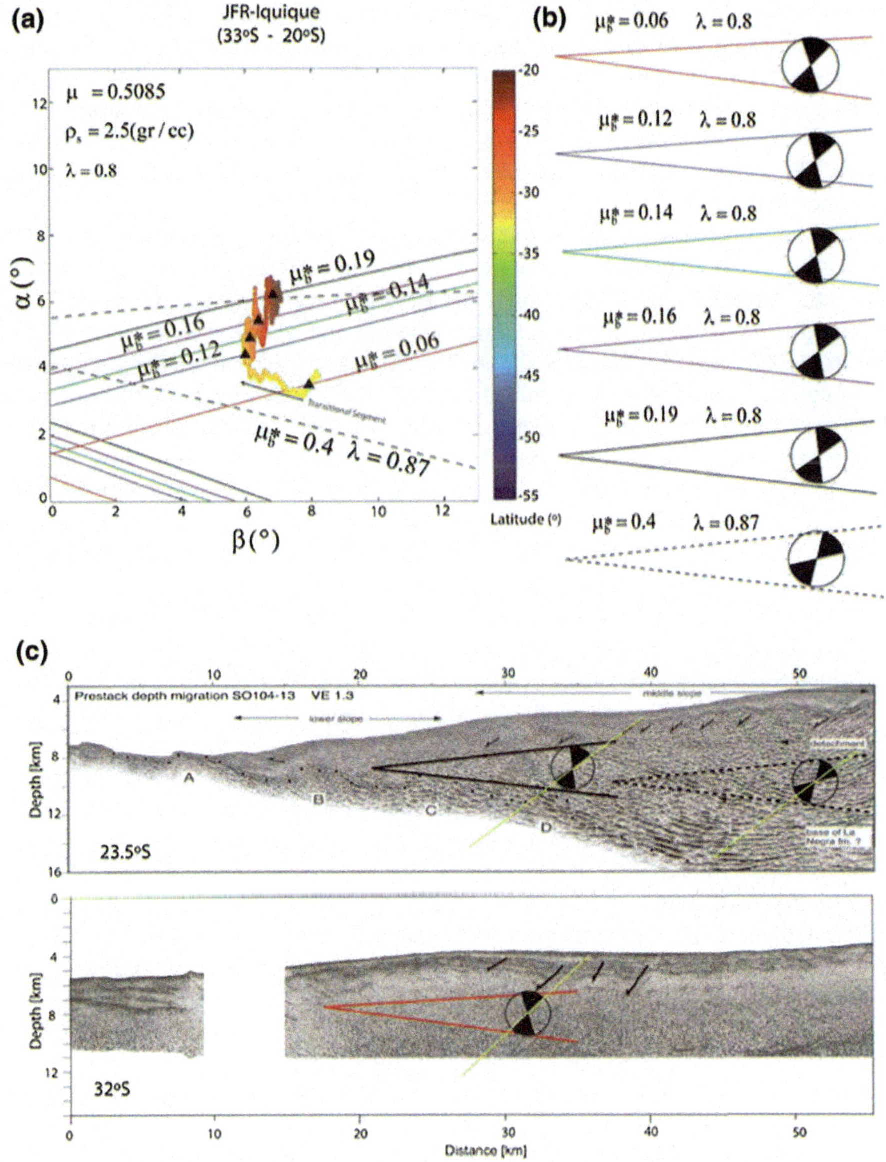

Fig. 6 NCCW analysis for the northern Chile margin. **a** The dots are the $\alpha\beta$ data from 33° S to 20° S (the latitudes are shown in the color bar). Solid lines correspond to the NCCW rupture envelopes for different values of μ_b^* considering constant values of $\lambda = 0.8$, $\rho_s = 2.5$ g/cc, and $\mu = 0.5085$. Dashed line corresponds to the NCCW rupture envelope for $\mu_b^* = 0.4$ considering constant values of $\lambda = 0.87$, $\rho_s = 2.5$ g/cc, and $\mu = 0.5085$. **b** Schemes of the predicted continental wedge geometry and the principal rupture direction for each black triangle in (**a**). **c** Comparison between two predicted geometries for the continental wedge and the rupture directions with the reflection seismic profiles at 23.5° S (extracted from Ranero et al. 2006) and 32° S (extracted from Contreras-Reyes et al. 2015). Yellow lines indicate the predicted faulting inclination (modified from Maksymowicz 2015)

from Chiloé FZ ($\sim$41.5° S) a complex variations in the α versus β graphic correlate with the presence of numerous fracture zones and isolated islands in the shelf, where the regional trend is explained by values of $\mu_b^* \sim 0.4$ (Fig. 5a). The predicted rupture directions for the modeled parameters indicate a deformation style described by west-dipping (37° to 40°) reverse faults and east-dipping (20° to 25°) reverse faults (Fig. 5b), which is consistent with the general deformation observed in seismic lines for this zone (see Fig. 5c).

Fig. 6a shows that for the erosive segment of the margin (northward from $\sim$33° S) it is necessary to consider μ_b^* values from 0.06 to 0.19 and to increase the λ value to 0.8. Regionally, a systematic northward increase of μ_b^* suggests a sub-segmentation of the northern Chilean margin characterized by an increment of μ_b^* from 0.06 to 0.12 between the JFR and $\sim$29.5° S, and to the north, three sub-segments with values of about 0.14, 0.16 and 0.19. However, the definition and interpretation of these sub-segments are only approximate due to the rapid slope angle variations. The predicted deformation style is characterized by west-dipping (35° to 45°) normal faults and east-dipping (70° to 90°) normal faults (Fig. 6b), and these directions agree with the normal faulting observed in reflection seismic profiles at different latitudes of the northern Chile margin (Fig. 6c, see also Ranero et al. 2006; Contreras-Reyes et al. 2014) (see Chap. Structure and Tectonics of the Chilean Convergent Margin from Wide-Angle Seismic Studies: A Review).

It is worth noting that the first sub-segment (from JFR to $\sim$29.5° S) presents a different trend in the α verses β graphic and latitudinal variation of the wedge geometry. Recently, Becerra et al. (2016) showed that in this segment ($\sim$31° S) the frontal accretionary prism is $\sim$20 km width, which also is showed by Contreras-Reyes et al. (2015) at $\sim$32° S (see Fig. 6c) and by Ranero et al. (2006) few kilometers to the south of JFR. Then, the frontal accretionary prism in this segment is clearly larger than the observed far to the north (Fig. 6c, Sallares and Ranero 2005). In particular, Ranero et al. (2006) interpreted that the normal faulting overprinted to the currently inactive accretionary prism is the erosive effect of the JFR subduction. Therefore, as the JFR migrated from the north to reach the current position at $\sim$10 Ma. (Yáñez et al. 2001), the recent erosive effect of the JFR and the subsequent deactivation of the accretionary process give a transitional character (between accretion and tectonic erosion) to the segment immediately to the north of JFR, which is reflected in the continental wedge geometry.

The black segmented lines in Fig. 6 show an alternative solution for the northernmost portion of the margin with $\mu_b^* = 0.4$ similar to the observed in the accretionary zone. However, to fit the α and β data, a higher value of $\lambda = 0.8$ must be considered, and the resulting deformation style does not reproduce the general deformation observed in the seismic profiles (Fig. 6c), which shows that the decrease of the μ_b^* and the increase of the λ in the erosive segment of the margin, in comparison to the accretionary segment, are supported by the geometry of the continental wedge and also by the deformation observed inside the wedge.

The result of the NCCW modeling for the entire Chilean margin is presented in Fig. 7 where segments of different wedge geometry, μ_b^* and λ parameters have been individualized. According to this, the tectonic latitudinal variation of the Chilean continental wedge can be described as a sequence of segments with different values of μ_b^* and/or λ, limited by the presence of fracture zones, bathymetric highs, and seaward continental prolongations, which locally modify the geometry of the continental wedge.

5 Tectonic Segmentation Along the Chilean Continental Wedge

The model for the Chilean margin shows an important change in the μ_b^* and λ parameters between the accretionary segment of the margin, southward of ~33° S, and the erosive segment to the north (Fig. 7). The effective basal friction coefficient decreases to the north from values around 0.3–0.4 to values of about 0.05–0.2, and the Hubbert–Rubey fluid pressure ratio changes from 0.67 to 0.8 in the erosive portion of the margin. This result seems to be inconsistent with the tectonic settings because μ_b^* is lower in the erosive segment which is supposed to be under high basal friction (Dahlen 1984; Lallemand et al. 1994; Adam and Reuther 2000). However, an analysis of the published geophysical models in different margins, and considerations on the NCCW simplifications, supports the obtained values. In the first place, in order to associate the NCCW models with nature, it is necessary to assume that the obtained values are representative for the long-term deformation processes, and that modeled homogeneous parameters (μ_b^* and λ) correspond to the bulk properties of the wedge. According to this, the properties at the base are representative for the transition between the wedge and the subduction channel (*decóllement*), which is a long-term feature of the margin. In the case of the accretionary systems, the subduction channel is formed mainly by pelagic sediments and terrigenous turbiditic material previously accumulated at the trench, with a thicknesses around ~0.5 to 1 km below Barbados accretionary wedge (Bangs et al. 1990; DiLeonardo et al. 2002) and Cascadia margin (MacKay 1995; Booth-Rea et al. 2008), and thicker than 1 or 1.5 km for the Chilean accretionary segment southward of ~33° S (Scherwath et al. 2009; Contreras-Reyes et al. 2010; Polonia et al. 2007). On the other hand, in the case of tectonic erosion, the subduction channel is formed by a merge of pelagic sediments accumulated on the oceanic crust, fractured rocks, and/or sediments previously housed in the small frontal accretionary prism, and fractured material removed from the base of continental wedge (Clift and Vannucchi 2004). The thickness of the subduction channel is about ~0.2–0.5 km in the erosive NE Japanese margin (Boston et al. 2014; Koge et al. 2014), ~0.2–0.7 in Central Ecuador (Sage et al. 2006), and thinner than ~1 km in the northern Chile segment (Sallares and Ranero 2005; Geersen et al. 2015; Contreras-Reyes et al. 2015). It is important to note that according to the penetration of the seismic reflection techniques, the observed thicknesses of the subduction channel are in general representative for the shallower portion of the subduction zone near the trench.

The differences in thickness and rheological properties of the different subduction channels, between accretionary and erosive margins, are key factors to interpret the values of the modeled friction at the base of the continental wedge. Likewise, changes in the homogeneous parameters inside the wedge can be associated with the rheological and lithological variability of the continental units along the margin.

Another important element to discuss in the models is that, by definition, both parameters (μ_b^* and λ) are highly dependent of the fluid pressures at the base and inside the wedge, respectively. According to Dahlen (1984):

$$\mu_b^* = \mu_b \frac{(1 - \lambda_b)}{(1 - \lambda)} \tag{1}$$

and

$$\lambda = \frac{P_f - \rho_w g D}{|\sigma_z| - \rho_w g D} \tag{2}$$

$$\lambda_b = \frac{P_f^b - \rho_w g D}{|\sigma_z| - \rho_w g D} \tag{3}$$

where μ_b is the friction coefficient in the base of the wedge, P_f and P_f^b are the pore fluid pressure inside the wedge and below the wedge, respectively, ρ_w is the water density, g is the gravity acceleration, and D is the thickness of the sea water column. According to the definition of Davis et al. (1983) for $|\sigma_z|$ and introducing the concept of pore fluid overpressure (u) we can write:

$$\mu_b^* = \mu_b \frac{[(\rho_s - \rho_w) g \cos(\alpha) z - u_b]}{[(\rho_s - \rho_w) g \cos(\alpha) z - u]} \tag{4}$$

$$\lambda = \frac{\rho_w}{\rho_s} + \frac{u}{\rho_s g \cos(\alpha) z} \tag{5}$$

$$\lambda_b = \frac{\rho_w}{\rho_s} + \frac{u_b}{\rho_s g \cos(\alpha) z} \tag{6}$$

The term "$\cos(\alpha)z$" is the depth below the top of the wedge where z is a coordinate perpendicular to the slope, ρ_s and ρ_w are the wedge and water densities, g is the gravity acceleration, and the parameters u and u_b are the pore fluid overpressure inside and below the wedge, respectively. From these expressions it is clear that the dependence of μ_b^* and λ parameters are functions of the hydrostatic/lithostatic pore fluid pressure ratio, and also respond to the pore fluid overpressure inside and below the wedge (Suppe 2014). Note that if u and u_b are linear functions of depth (z), μ_b^*, λ and λ_b are constants.

Considering the geophysical and geological characteristics of the margins, the observed low μ_b^* in the erosive segment can be explained by high overpressure in the subduction channel in comparison to the accretionary margin. The fluids in the subduction channel are provided by the subducted sediments previously cumulated in the deep sedimentary layers of the trench. As the compaction increases with depth in the trench sedimentary fill, the porosity of the subducted sediments in accretionary margins can be higher in comparison to the erosive margins. Then, even considering the thinner subduction channel in erosive segments, the amount of fluids subducted and released by compaction in the subduction channel could be similar or higher than those subducted below accretionary wedges (Le Pichon et al. 1993). Additionally, the subduction of the large oceanic crustal faults, that transport fluids from the surface to the oceanic upper mantle (Ranero et al. 2003; Naif et al. 2015; Moscoso and Grevemeyer 2015), also can provide fluids to the subduction channel (von Huene and Ranero 2003).

Considering the case of a large release of fluid to the *decóllement* in the erosive margins due to the subduction of poorly compacted sediments, the model of Le Pichon et al. (1993) suggests that the fluid pressure inside the *decóllement* can be higher than in the accretionary margins and similar to the lithostatic pressure, which implies large overpressure and low basal friction coefficient below the erosive wedge, as it is observed in the Chilean margin. However, thickness, porosity, and permeability of the subduction channels can be extremely variable, making difficult to extract direct conclusions from simple models (Sage et al. 2006). In fact, in one of the few seismic velocity-depth models for subduction channels, Calahorrano et al. (2008) show profiles located in Ecuador where the Vp velocities are consistent with strong variations of porosity along the channel. The model shows a shallow zone, below the frontal accretionary prism, with relatively large porosity (~ 0.3) related to low overpressure and low values of $\lambda_b \sim 0.5$. Landward, an intermediate zone below the wedge basement of high density presents a small decrease of porosity (~ 0.25) related to an increase of the overpressure and high values of $\lambda_b \sim 1$. Finally, the model shows a deeper zone with low porosity (~ 0.1) related to decreases of the overpressure and λ_b.

The generalization of the model of Calahorrano et al. (2008) could be the best explanation for the values obtained in the Chilean margin. The transition from the frontal accretionary prism to a dense wedge, accompanied by an increase of the overpressure, is analog to the transition from accretionary segments, with large accretionary prism and paleo-accretionary complexes (metamorphic complex), to a dense continental wedge in the northern Chile erosive margin (Tassara 2010). Additionally, the transition between the second zone to the deeper zone in the model of Calahorrano et al. (2008) shows the effect of subduction of poor compacted sediments in the erosive segments which favored the increases of the overpressure, and therefore, the reduction of the effective basal friction coefficient.

An important factor involved in the effective basal friction coefficient variation is indeed the basal friction coefficient (μ_b in Eq. 4) that depends, among other parameters, on the lithologies and roughness of the surfaces in contact. However, the basal friction coefficient in the interplate boundary is almost unknown due to the impossibility to obtain in situ observations of the convergence process at depth and during the long time of the seismic cycle. In this situation, analog models can provide details of the friction properties showed by different material and physical conditions that can be related to the behavior at real scale. In particular, Corbi et al. (2011) present an experiment consisting of a gelatin block driven at constant velocity over sand paper. After several experiments using sand papers of different roughnesses, the authors show that in a range of low roughnesses, the static friction coefficient decreases with roughness increments. This suggests, for instance, that a general increase of the slab roughness in the northern Chile segment, due to its thin sedimentary coverage, can generate a general decrease of the basal friction coefficient, in comparison to the accretionary segment of the margin.

The low values of μ_b^* obtained for the northern Chile margin agree with the conceptual model proposed for erosive margins, where a weak interplate boundary has been invoked to explain the basal removal of wedge material by an intense hydrofracturing above the subduction channel (von Huene and Ranero 2003; Sage et al. 2006; Ranero et al. 2008). According to this model, the basal hydrofracturing process transfers overpressurized fluids from the subduction channel to the overriding plate creating a thin layer of fractured rock at the base of the wedge (Sage et al. 2006; Ranero et al. 2008). The hydrofracturing at the base and the subsequent upward migration of fluids inside the continental wedge can be invoked to explain anomalous low velocities observed inside the Chilean continental wedge at ~23° 30′S (Sallares and Ranero 2005). In the same way, the pervasive long-term fracturing of the rigid wedge, and the upward rise of overpressurized fluid from the interplate boundary, explain the general low velocities ($Vp < 6$ km/s) observed systematically in the frontal 40–50 km of the continental wedge in the northern Chile margin (Contreras-Reyes et al. 2012; Sallares and Ranero 2005; Contreras-Reyes et al. 2015). The pervasive long-term fracturing of the wedge also explains the high values of the Hubbert–Rubey fluid pressure ratio in the erosive portion of the margin (Fig. 7). On the other hand, the high porosity in the sedimentary units inside the accretionary wedge (that probably decrease with depth), and the long-term compaction process related to the contractional deformation (Maksymowicz et al. 2012), determines pore fluids' pressures closer to the hydrostatic pressure in the accretionary segment of the Chilean margin, which can explain the relatively low values of λ obtained in the model. In this view, the deformation process in the accretionary margins can be seen as a process of the long-term fluid release of the continental wedge material, and in the case of tectonic erosion as process of long-term capture of overpressure fluids that migrates upward from the interplate boundary.

Within the two major segments of the Chilean margin, characterized by accretion and tectonic erosion, respectively, a second-order tectonic segmentation can be defined as a sequence of segments with different values of μ_b^* and/or λ, limited by the presence of fracture zones, bathymetric highs, and seaward continental prolongations (Fig. 7). A physic interpretation of this variability is extremely complex, and a complete understanding of the processes involved on this needs further tectonic, geological, and geophysical studies. However, according to Maksymowicz (2015), some factors that could explain the observed segmentation can be summarized as:

(1) Local increases of fluid content in the subducted sediment and/or in fractures of the subducted oceanic lithosphere can rise the pore fluid pressure of the interplate boundary, decreasing the effective basal friction coefficient in some segments of the margins, which also can explain the local decreases of α that correlate to the subduction of oceanic fracture zones (Figs. 4 and 7). An example of this interpretation is the segment corresponding to the high slip patch of the 2010 Maule earthquake (HSP in Fig. 2), which is associated to low μ_b^* and to a high fluid content (see the next section).
(2) The subduction of a seamount or an aseismic ridge (JFR, Copiapo Ridge, and Taltal Ridge) has a small-scale local impact on the continental wedge tectonics, fracturing the continental basement and modifying the slope angle during the subduction process. In this case, the frontal slope angle increases during the seamount subduction, but afterward the continental wedge experiences subsidence and the frontal slope angle decreases (Park et al. 1999; Kodaira et al. 2000; Dominguez et al. 2000). The pervasive fracturing due to seamount subduction also can increase the pore fluid inside the continental wedge, which can lead to a local increase of the λ parameter.
(3) The increase of the slope angle near continental peninsulas and islands (Guafo, Guamblin, Mocha islands, and Arauco Peninsula) can be explained by a seaward extension of the rigid continental basement which generates high slope angle compared to the adjacent zones (Ruh et al. 2013). A complementary explanation is the local increase of the basal friction coefficient which can generate high values of α and low-angle inverse faulting that favors the formation of duplexes at the base of the wedge, and then, the uplift of the basement below the forearc basins (Gutscher et al. 1996).

Finally, an important factor that can be explored in the future to explain the observed second-order tectonic segmentation is the roughness of the subducted oceanic plate. The slab roughness in a specific segment of the margin depends on the particularities of the magmatic and tectonic process in the spreading center where the slab is generated (Eason et al. 2016), and the changes on the fracturing in the outer-rise zone related to the elastic thickness variability (Contreras-Reyes and Osses 2010). Then, each segments of the oceanic lithosphere limited by fracture zones could be associated to an specific roughness, which can explain in part the correlation between fracture zones and the abrupt changes of the μ_b^* and/or λ parameters.

6 Seismotectonic Implications of the Chilean Continental Wedge Geometry

The left panel in Fig. 7 shows that rupture extensions of the historic and instrumentally registered earthquakes (Comte and Pardo 1991; Beck et al. 1998; Cisternas et al. 2005; Ruegg et al. 2009; Bilek 2010; and the catalog of Chilean

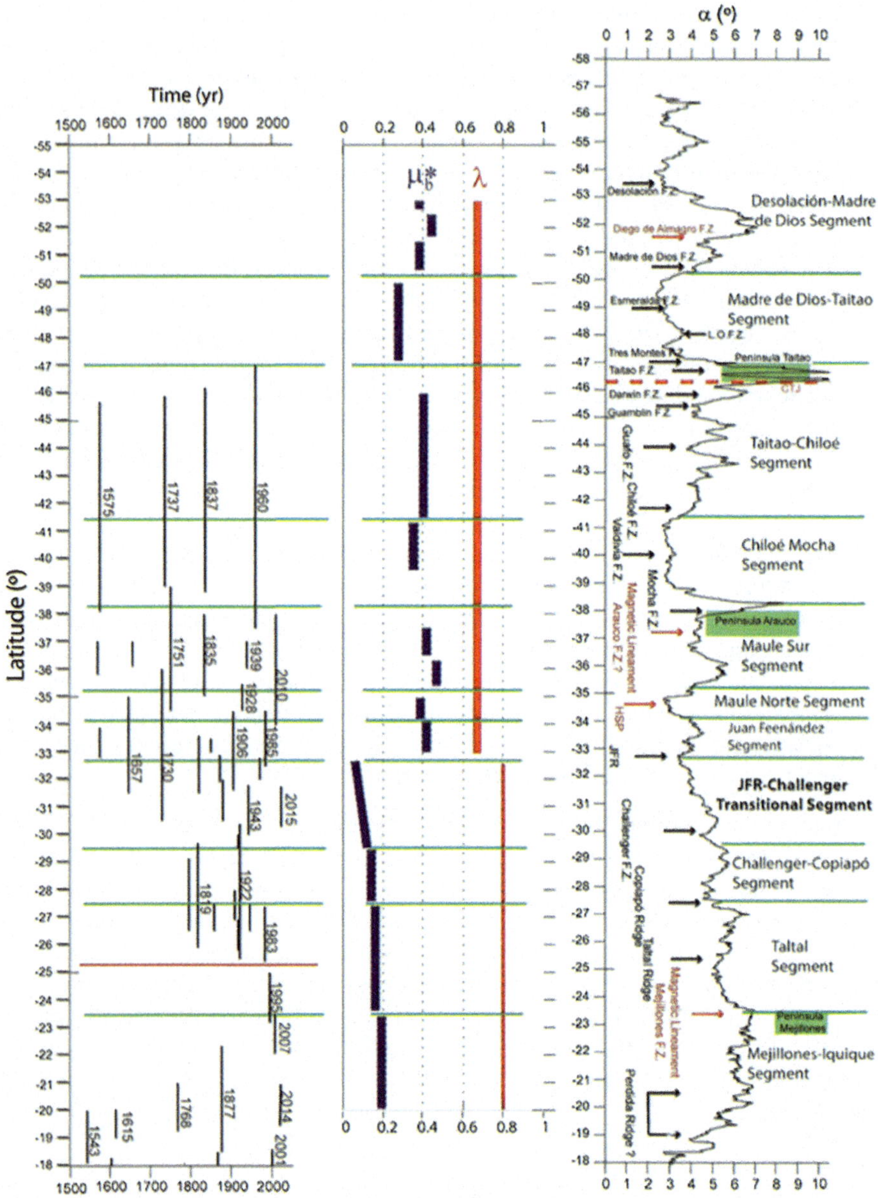

Fig. 7 Definition of tectonic segmentation of the Chilean continental wedge. The central panel shows the parameters derived from the NCCW model. Blue lines are the values of effective basal friction coefficient μ_b^*, and red lines are the λ values, and green lines correspond to the limits between tectonic segments of the continental wedge. The right panel shows the values of the slope angle as a function of the latitude. The suggested names for the tectonic segments of the continental wedge are shown between the green lines. The left panel shows the correlation between tectonic segmentation of the Chilean continental wedge and the rupture zones of large Chilean earthquakes. The blue lines correspond to the historical and instrumentally registered large earthquakes in Chile; green lines are the limits between tectonic segments of the continental wedge, and red line indicates a seismogenic limit correlated to the position of Taltal Ridge (modified from Maksymowicz 2015)

National Seismological Center, CSN) can be roughly correlated with latitude with the limits of the proposed long-term tectonic segmentation of the margin. Interesting cases are in particular: (1) the Chiloé-Mocha segment that coincides with the northern portion of the Valdivia 1960 earthquake where the high slip patch was developed (see Fig. 2), (2) the Maule–Norte segment that also correlates with the high slip patch of the 2010 Maule earthquake and seems to be an approximated limit between the ruptures south of the Maule region (1835–1751) and north of the Maule region (1730, 1906, 1985), and (3) the northern limit of the Transitional JFR-Challenger segment that correlates with the southern end of the large 1819 and 1922 earthquakes. On the other hand, the limit between the 1983 and 1995 earthquakes is well defined around the Taltal Ridge (red line in the left panel of Fig. 7), but there is not a clear variation of slope angle associated with that feature. The observed correlation between the proposed segmentation and the earthquake distribution suggests that segments with different long-term tectonics parameters, μ_b^* and/or λ, can have different short-term parameters, μ_b^* and/or λ, that can affect the stress state during the seismic cycle and the complex sequence of coseismic activation, but also the variation of these parameters can be related to the slip inhomogeneities during large coseismic ruptures.

The explanation of this short-term/long-term tectonic spatial correlation is not clear yet. Contreras-Reyes and Carrizo (2011) proposed that the high oceanic features locally increase the interplate coupling, according to the buoyancy of these structures, generating strong barriers for the earthquakes. In the case of subducted aseismic ridges (JFR, Copiapó Ridge, and Taltal Ridge), the relation between the increase in coupling and the collapse of the continental wedge seems to be direct. However, some fracture zones without an important bathymetric expression produce local decreases in the slope angle (e.g., Challenger FZ and Diego de Almagro FZ), which suggests a key role of the fluids provided by the fracture zones in the generation of weak seismic barriers. In general, a coherent physical explanation of the observed spatial correlation should consider the long-term rheological conditions inside the segments (static friction coefficient, fluid overpressure, slab roughness, among others), and the mechanical properties of the borders (week and strong barriers).

An interesting example that could illustrate the possible relationship between the long-term and short-term tectonic process is observed in the zone of the high slip patch (HSP) of the Maule 2010 Mw8.8 earthquake (see Fig. 8). Here, a decrease of the slope angle to the north of the ~35.5° S correlates with a segment of low effective basal friction coefficient (Maule–Norte Segment), with a low free-air gravity signal in the marine forearc (Song and Simons 2003; Wells et al. 2003; Tassara 2010; Álvarez et al. 2014; Maksymowicz et al. 2015), and a northward increase of the *b*-value derived from the analysis of aftershock seismicity using the Gutenberg–Richter law as calculated by Tassara et al. (2016).

During the last decades, strong efforts were dedicated to relate spatial and temporal variations of *b*-value in seismogenic zones. Aki (1981), Hirata (1989), Main et al. (1994), Legrand (2002), among others, studied the link between the *b*-value and fractal dimension of the earthquake distributions, characterizing the

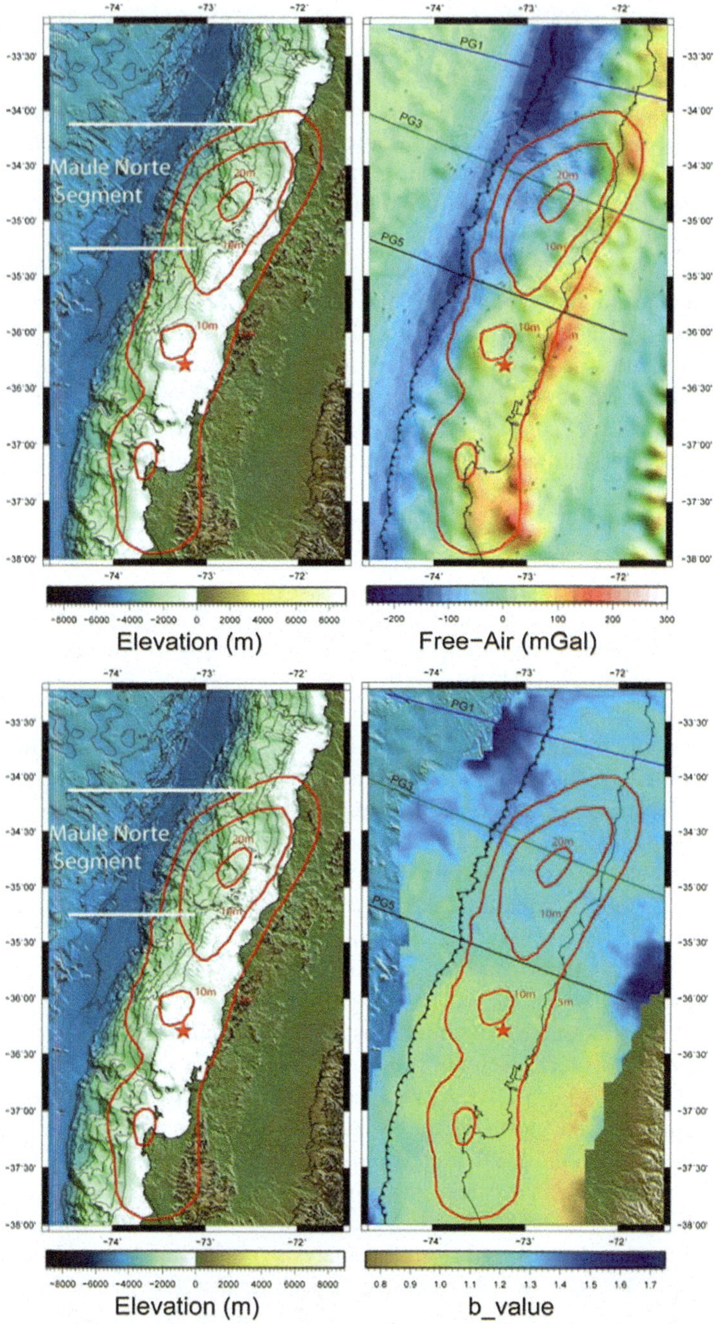
Maule Norte Segment
20m
10m
5m
PG1
PG3
PG5
Elevation (m)
Free−Air (mGal)
Maule Norte Segment
Elevation (m)
b_value

◂**Fig. 8** Comparison of the coseismic slip of the Maule Mw8.8 earthquake with free-air gravity anomaly and b-value. The upper panels show the regional bathymetry/topography and free-air gravity anomaly (grid extracted from Maksymowicz et al. 2015). The lower panels show the regional bathymetry/topography and the *b*-value of the zone (grid extracted from Tassara et al. 2016). The red star shows the epicenter of the Maule earthquake determined by the Centro Sismológico Nacional, University of Chile (CSN). The red curves correspond to iso-contours of slip during the Maule earthquake according to Moreno et al. (2012). Profiles PG1, PG3, and PG5 correspond to density-depth models analyzed in Fig. 9

seismic process as a self-organized dynamical system. Mogi (1962) concluded that the *b*-value increases with the degree of material heterogeneities and the spatial variations in the stress distribution. Studying subduction margins, crustal deformation, and volcanic systems, Wyss (1973), Schorlemmer et al. (2005), Ghosh et al. (2008), Farrell et al. 2009, Legrand et al. (2012), Chen et al. (2016), among others, highlighted the relation between zones of low *b*-values and high tectonic stress, and also the correlation between low tectonic stresses, pore fluid pressure increases and high *b*-values. Therefore, according to the observed *b*-values, the HSP zone of the Maule earthquake correlates with a region characterized by high fluid pressure, and probably low tectonic stress, in the seismogenic contact (Tassara et al. 2016), which supports in turns, a connection between high pore fluid pressure and the low effective basal friction coefficient (obtained here to model the decrease of the slope angle).

It is important to note that the *b*-value presented in Fig. 8 corresponds to the aftershock seismicity of the NEIC catalog during the first year after the earthquake (Tassara et al. 2016). However, the decrease of *b*-value in the northern portion of the Maule earthquake rupture is observed in the longer catalog analyzed by Legrand et al. (2012), which covered seismicity between the years 1973 and 2010, suggesting that the observed correlation is preserved during the different stages of the seismic cycle.

The HSP zone of the Maule earthquake is characterized also by a general decrease of vertical load over the seismogenic contact northward of ~35.5° S (Tassara 2010; Maksymowicz et al. 2015), which explains the free-air gravity low (Fig. 8). The vertical load over the seismogenic contact was derived in base of 2-D density-depth profiles generated by gravity modeling (three of them are presented in Fig. 9a). The procedure involved the vertical integration of the weights of all mass elements inside the continental crust and the oceanic water column above, i.e., considering the weight of the mass column over the seismogenic contact for each horizontal position. In Fig. 9c, a profile located south of ~35.5° S (PG5 in Figs. 8 and 9) has a vertical load up to 200 MPa that is higher than in the zone of the HSP zone (profiles PG1 and PG3), which represents an important change in normal stresses supported by the seismogenic contact. Again, different methodologies (gravity modeling and *b*-values analysis) agree with the results of a low tectonic stress in the zone of the high slip patch of the earthquake, which in turn is characterized by low effective friction coefficient (see Fig. 9b, c). In conclusion, this

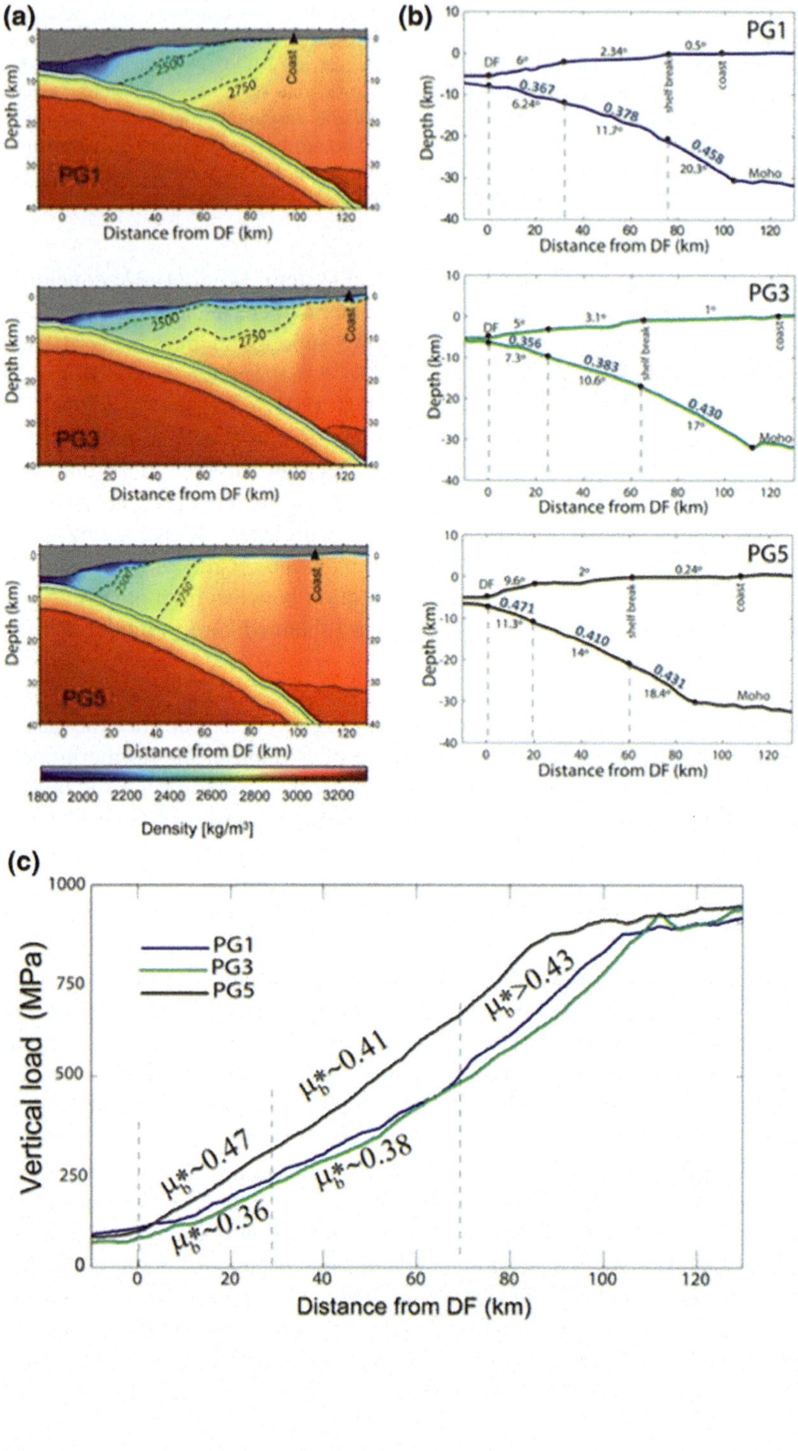
(a)
(b)
(c)
PG1
PG3
PG5
Coast
2500
2750
Depth (km)
Distance from DF (km)
Density [kg/m³]
DF
shelf break
coast
Moho
0.367
0.378
0.458
0.356
0.383
0.430
0.471
0.410
0.431
Vertical load (MPa)
μ*b~0.47
μ*b~0.36
μ*b~0.41
μ*b~0.38
μ*b>0.43

◂**Fig. 9** Density-depth models, effective basal friction coefficients, and vertical load profiles PG1, PG3, and PG5. **a** Density-depth model of the profiles PG1, PG3, and PG5 (see Fig. 8 for locations). Dashed curves correspond to the contours of 2500 kg/m^3 and 2750 kg/m^3 at the frontal part of continental wedge. The zero position of the horizontal scale corresponds to the location of the deformation front. **b** Estimation of the effective basal friction coefficient according to the NCCW theory. The analysis was performed for profiles PG1, PG3, and PG5. The wedge was divided into three sectors (1, 2, and 3), which were simplified and modeled as independent wedges with constant basal and taper angles. Blue numbers correspond to the estimated effective basal friction coefficients in the case of a wedge characterized by hydrostatic internal pressure. **c** Vertical loading over the slab for each of three modeled profiles (see the explanation in the text). Figures modified from Maksymowicz et al. (2015)

particular zone corresponds to a weak segment of the interplate contact during the last seismic cycle, and probably, during the long-term tectonic process.

The case study presented above signals the importance of the wedge geometry analysis and modeling to understand the geodynamic of the subduction margin. By integrating these techniques with high-resolution seismic profiles and new marine geological information, future work in the Chilean and other subduction margins will improve the knowledge of the seismic cycle characteristics, the long-term stress field of the continental wedge, and the complex process of deformation observed in the forearc zone.

Acknowledgements Andrei Maksymowicz was supported by project FONDECYT 3150160 of the Chilean National Science Cooperation (CONICYT). Andres Tassara thanks project FONDECYT 1151175.

References

Adam J, Reuther CD (2000) Crustal dynamics and active fault mechanics during subduction erosion: Application of frictional wedge analysis on to the North Chilean Forearc. Tectonophysics 321:297–325

Aki K (1981) A probabilistic synthesis of precursory phenomena. In: Simpson DW, Richards PG (eds) Earthquake prediction: An international review, Maurice Ewing Set., vol 4. AGU, Washington, D.C, pp 566–574

Álvarez O, Nacif S, Gimenez M, Folguera A, Braitenberg C (2014) GOCE de-rived vertical gravity gradient delineates great earthquake rupture zones along the Chilean margin. Tectonophysics (ISSN0040-1951). http://dx.doi.org/10.1016/j.tecto.2014.03.011

Amante C, Eakins BW (2009) NOAA, Technical Memorandum. NESDIS; 2009. ETOPO1 1 arc-minute global relief model: procedures, data sources and analysis. p 19. NGDC-24

Angermann D, Klotz J, Reigber C (1999) Space-geodetic estimation of the Nazca–South America Euler vector. Earth Planet Sci Lett 171(3):329–334

Bangs N, Cande S (1997) Episodic development of a convergent margin inferred from structures and processes along the Southern Chile margin. Tectonics 16(3):489–503

Bangs N, Westbrook GK, Ladd JW, Buhl P (1990) Seismic velocities from the Barbados Ridge complex: Indicators of high pore fluid pressures in an accretionary complex. J Geophys Res 95:8767–8782

Becerra J, Arriagada C, Contreras-Reyes E, Bascuñan S, De Pascale GP, Reichert C, Díaz-Naveas J, Cornejo N (2016) Gravitational deformation and inherited structural control on slope morphology in the subduction zone of north-central Chile (ca. 29–33° S). Basin Res. https://doi.org/10.1111/bre.12205

Beck S, Barrientos S, Kausel E, Reyes M (1998) Source characteristics of historic earthquakes along the central Chile subduction zone. J South Am Earth Sci 11:115–129

Bell CM (1984) Deformation produced by the subduction of a paleozoic turbidite sequence in northern Chile. J Geol Soc London 141:339–347

Bell CM, Suárez M (2000) The Río Lácteo Formation of Southern Chile. Late Paleozoic orogeny in the Andes of southernmost South America. J S Am Earth Sci 13(1–2):133–145

Bilek SL (2010) Seismicity along the South American subduction zone: Review of large earthquakes, tsunamis, and subduction zone complexity. Tectonophysics 495(1–2). https://doi.org/10.1016/j.tecto.2009.02.037

Booth-Rea G, Klaeschen D, Grevemeyer I, and Reston T (2008) Heterogeneous deformation in the Cascadia convergent margin and its relation to thermal gradient (Washington, NW USA). Tectonics 27(TC4005). https://doi.org/10.1029/2007TC002209

Boston B, Moore GF, Nakamura Y, Kodaira S (2014) Outer-rise normal fault development and influence on near-trench décollement propagation along the Japan Trench, off Tohoku. Earth Planets Space 66(1):135

Bourgois J, Guivel C, Lagabrielle Y, Calmus T, Boulague J, Daux V (2000) Glacial-interglacial trench supply variation, spreading-ridge subduction, and feedback controls on the Andean margin development at the Chile triple junction area (45-48° S). J Geophys Res 105(B4):8355–8386

Calahorrano A, Sallares V, Sage F, Collot JY, Ranero CR (2008) Nonlinear variations of the physical properties along the Southern Ecuador subduction channel: Results from depth-migrated seismic data. Earth Planet Sci Lett 267:453–467. https://doi.org/10.1016/j.epsl.2007.11.061

Cande SC, Leslie RB (1986) Late Cenozoic tectonics of the southern Chile Trench. J Geophys Res 91:471–496

Cisternas M, Atwater BF, Torrejón F, Sawai Y, Machuca G, Lagos M, Eipert A, Youlton C, Salgado I, Kamataki T, Shishikura M, Rajendran CP, Malik JK, Rizal Y, Husni M (2005) Predecessors of the giant 1960 Chile earthquake. Nature 437:404–407. https://doi.org/10.1038/nature03943

Clift P, Vannucchi P (2004) Controls on tectonic accretion versus erosion in subduction zones: Implications for the origin and recycling of the continental crust. Rev Geophys 42(RG2001). https://doi.org/10.1029/2003RG000127

Comte D, Pardo M (1991) Reappraisal of great historical earthquakes in the northern Chile and southern Peru seismic gap. Nat Hazards 4(23–44). https://doi.org/10.1007/BF00126557

Contreras-Reyes E and Carrizo D (2011) Control of high oceanic features and subduction channel on earthquake ruptures along the Chile–Peru subduction zone. Phys Earth Planet Int 186(49–58). doi:https://doi.org/10.1016/j.pepi.2011.03.002

Contreras-Reyes E, Osses A (2010) Lithospheric flexure modeling seaward of the Chile trench: Implications for oceanic plate weakening in the Trench Outer Rise region. Geophys J Int 182 (1):97–112. https://doi.org/10.1111/j.1365-246X.2010.04629.x

Contreras-Reyes E, Grevemeyer I, Flueh ER, Reichert C (2008) Upper lithospheric structure of the subduction zone offshore of southern Arauco peninsula, Chile, at 38° S. J Geophys Res 113 (B07303). https://doi.org/10.1029/2007JB005569

Contreras-Reyes E, Flueh ER, and Grevemeyer I (2010) Tectonic control on sediment accretion and subduction off south central Chile: Implications for coseismic rupture processes of the 1960 and 2010 megathrust earthquakes. Tectonics 29(TC6018). https://doi.org/10.1029/2010TC002734

Contreras-Reyes E, Jara J, Grevemeyer I, Ruiz S, Carrizo D (2012) Abrupt change in the dip of the subducting plate beneath north Chile. Nat Geosci 5(342–345). https://doi.org/10.1038/ngeo1447

Contreras-Reyes E, Becerra J, Kopp H, Reichert C, Díaz-Naveas J (2014) Seismic structure of the north-central Chile convergent margin: Subduction erosion of a paleomagmatic arc. Geophys Res Lett 41(5):1523–1529. https://doi.org/10.1002/2013GL058729

Contreras-Reyes E, Ruiz J, Becerra J, Kopp H, Reichert C, Maksymowicz A, Arriagada C (2015) Structure and tectonics of the central Chilean margin (31°–33° S): Implications for subduction erosion and shallow crustal seismicity. Geophys J Int 653(2):776–791. https://doi.org/10.1093/gji/ggv309

Corbi F, Funiciello F, Faccenna C, Ranalli G, Heuret A (2011) Seismic variability of subduction thrust faults: Insights from laboratory models. J Geophys Res 116(B06304). https://doi.org/10.1029/2010JB007993

Dahlen FA (1984) Noncohesive critical Coulomb wedges: An exact solution. J Geophys Res 89 (10):125–133

Dahlen FA, Suppe J, Davis DM (1984) Mechanics of fold-and thrust belts and accretionary wedges: Cohesive Coulomb theor. J Geophys Res 8(9):10087–10101

Davis DM, Suppe J, Dahlen FA (1983) Mechanics of fold-and thrust belts and accretionary wedges. J Geophys Res 88:1153–1172

Diaz-Naveas J (1999) Sediment subduction and accretion at the Chilean Convergent Margin between 35′ and 40′S, Dissertation zur Erlangung des Doktorgrades. Christian-Albrechts-Universitat zu Kiel, 1999. 130 p

DiLeonardo CG, Moore JC, Nissen S, Bangs N (2002) Control of internal structure and fluid-migration pathways within the Barbados Ridge decollement zone by strike-slip faulting: Evidence from coherence and three-dimensional seismic amplitude imaging. Bull Geol Soc Am 114(1):51–63

Dominguez S, Malavieille J, Lallemand SE (2000) Deformation of accretionary wedges in response to seamount subduction: Insights from sandbox experiments. Tectonics 19. https://doi.org/10.1029/1999TC900055. issn: 0278-7407

Eason DE, Dunn RA, Canales JP, Sohn R (2016) Segment-scale variations in seafloor volcanic and tectonic processes from multibeam sonar imaging, mid-Atlantic Ridge Rainbow region (35°45′–36°35′N). Geochem Geophys Geosyst. In Press. https://doi.org/10.1002/2016GC006433

Farrell J, Husen S, Smith RB (2009) Earthquake swarm and *b*-value characterization of the Yellowstone volcano-tectonic system. J Volcanol Geotherm Res 188:260–276. https://doi.org/10.1016/j.jvolgeores.2009.08.008

Flueh ER, Grevemeyer I (eds) (2005) TIPTEQ SONNE Cruise SO-181, from the Incoming Plate to mega Thrust EarthQuakes. Geomar Rep 102, Geomar, Kiel, Germany

Geersen J, Ranero C, Barkhausen U, Reichert C (2015) Subducting seamounts control interplate coupling and seismic rupture in the 2014 Iquique earthquake area. Nat Comm 6(8267). https://doi.org/10.1038/ncomms9267

Ghosh A, Newman AV, Thomas AM, Farmer GT (2008) Interface locking along the subduction megathrust from *b*-value mapping near Nicoya Peninsula, Costa Rica. Geophys Res Lett 35: L01301. https://doi.org/10.1029/2007GL031617

Glodny J, Lohrmann J, Echtler H, Grafe K, Seifert W, Collao S, Figueroa O (2005) Internal dynamics of apaleoaccretionary wedge: Insights from combined isotope tectonochronology and sandbox modelling of the South-Central Chilean for earc. Earth Planet Sci Lett 231:23–39

Gutscher MA, Kukowski K, Malavieille J, Lallemand S (1996) Cyclical behavior of thrust wedges: Insights from high basal friction sandbox experiments. Geology 24(2):135–138

Hayes GP, Wald DJ, Johnson RL (2012) Slab1.0: A three-dimensional model of global subduction zone geometries. J Geophys Res 117(B01302). https://doi.org/10.1029/2011JB008524

Hervé F, Fanning CM (2000) Late Triassic zircons in metaturbidites of the Chonos Metamorphic Complex, southern Chile. Revista Geológica de Chile 28(1):91–104

Hervé F, Calderón M, Faúndez V (2008) The metamorphic complexes of the Patagonian and Fuegian Andes. Geologica Acta 6(1):43–53

Hirata T (1989) A correlation between the *b* value and the fractal dimension of earthquakes. J Geophys Res 94:7505–7514

Kodaira S, Takahashi N, Nakanishi A, Miura S, Kaneda Y (2000) Subducted seamount imaged in the rupture zone of the 1946 Nankaido earthquake. Science 289:104–106

Koge H et al (2014) Friction properties of the plate boundary megathrust beneath the frontal wedge near the Japan Trench: An inference from topographic variation. Earth Planets Space 66 (1):153. https://doi.org/10.1186/s40623-014-0153-3

Kukowski N, Oncken O (2006) Subduction erosion—is this the "normal" mode of fore-arc material transfer along the Chilean margin? In: Oncken O, Chong G, Franz G, Giese P, Götze HJ, Ramos V, Strecker M, Wigger P (eds) Andean Geodynamics, Frontiers in Geosciences, vol 1. Springer, pp 217–236

Lallemand SE, Schnurle P, Malavieille J (1994) Coulomb theory applied to accretionary and nonaccretionary wedges—Possible causes for tectonic erosion and/or frontal accretion. J Geophys Res 99:12033–12055

Le Pichon X, Henry P, Lallemand S (1993) Accretion and erosion in subduction zones: The role of fluids. Annu Rev Earth Planet Sci 21:307–331

Legrand D (2002) Fractal dimensions of small, intermediate and large earthquakes. Bull Seismol Soc Am 92:3318–3320

Legrand D, Tassara A, Morales D (2012) Megathrust asperities and clusters of slab dehydration identified by spatiotemporal characterization of seismicity below the Andean margin. Geophys J Int 201 191(3):923–931

MacKay ME (1995) Structural variation and landward vergence at the toe of the Oregon accretionary prism. Tectonics 14:1309–1320

Mogi K (1962) Magnitude-frequency relationship for elastic shocks accompanying fractures of various materials and some related problems in earthquakes, vol 40. Bulletin of the Earthquake Research Institute, University of Tokyo, pp 831–853

Main I, Henderson JR, Meredith PG, Sammonds PR (1994) Self-organised criticality and fluid-rock interactions in the brittle field. Pure Appl Geophys 142:529–543

Maksymowicz A (2015) The geometry of the Chilean continental wedge: Tectonic segmentation of subduction processes off Chile. Tectonophysics 659:183–196. https://doi.org/10.1016/j.tecto.2015.08.007

Maksymowicz A, Contreras-Reyes E, Grevemeyer I, Flueh ER (2012) Structure and geodynamics of the post-collision zone between the Nazca–Antarctic spreading center and South America. Earth Planet Sci Lett 345–348:27–37. https://doi.org/10.1016/j.epsl.2012.06.023

Maksymowicz A, Tréhu A, Conterras-Reyes E, Ruiz S (2015) Density-depth model of the continental wedge at the maximum slip segment of the Maule Mw8.8 megathrust earthquake. Earth Planet Sci Lett 409:265–277. https://doi.org/10.1016/j.epsl.2014.11.005

Moreno M, Melnick D, Rosenau M, Báez JC, Klotz J, Oncken O, Tassara A, Bataille K, Chen J, Socquet A, Bevis M, Bolte J, Vigny C, Brooks B, Ryder I, Grund V, Smalley R, Carrizo D, Bartsch M, Hase H (2012) Toward understanding tectonic control on the Mw8.8 2010 Maule Chile earthquake. Earth Planet Sci Lett 321:152–165

Moscoso E, Grevemeyer I (2015) Bending-related faulting of the incoming oceanic plate and its effect on lithospheric hydration and seismicity: A passive and active seismological study offshore Maule, Chile. J Geodyn 90:58–70, ISSN 0264-3707. http://dx.doi.org/10.1016/j.jog.2015.06.007

Moscoso E, Grevemeyer I, Contreras-Reyes E, Flueh ER, Dzierma Y, Rabbel W, Thorwart M (2011) Revealing the deep structure and rupture plane of the 2010 Maule, Chile earthquake (Mw = 8.8) using wide angle seismic data. Earth Planet Sci Lett 307:147–155. https://doi.org/10.1016/j.epsl.2011.04.025

Naif S, Key K, Constable S, Evans RL (2015) Water-rich bending faults at the Middle America Trench. Geochem Geophys Geosyst 16:2582–2597. https://doi.org/10.1002/2015GC005927

Park J, Tsuru T, Kaneda Y, Kono Y, Kodaira S, Takahashi N, Kinoshita H (1999) A subducting seamount beneath the Nankai Accretionary Prism off Shikoku, southwestern Japan. Geophys Res Lett 26. https://doi.org/10.1029/1999GL900134. issn: 0094-8276

Polonia A, Torelli L, Brancolini G, Loreto MF (2007) Tectonic accretion versus erosion along the southern Chile trench: Oblique subduction and margin segmentation. Tectonics 26(TC3005). https://doi.org/10.1029/2006TC001983

Ranero CR, Morgan JP, McIntosh KD, Reichert C (2003) Bending, faulting, and mantle serpentinization at the Middle America trench. Nature 425:367–373

Ranero CR, von Huene R, Weinrebe W, Reichert C (2006) Tectonic Processes along the Chile Convergent Margin. In: Onken O et al. (eds) The Andes: Active Subduction Orogeny. Berlin, Springer-Verlag, pp 91–122

Ranero CR, Grevemeyer I, Sahling H, Barckhausen U, Hensen C (2008) Hydrogeological system of erosional convergent margins and its influence on tectonics and interplate seismogenesis. Geochem Geophys Geosyst 9(Q03S04)

Ruegg JC, Rudloff A, Vigny C, Madariaga R, de Chabalier JB, Campos J, Kausel E, Barrientos S, Dimitrov D (2009) Interseismic strain accumulation measured by GPS in the seismic gap between Constitución and Concepción in Chile. Phys Earth Planet In 175:78–85. https://doi.org/10.1016/j.pepi.2008.02.015

Ruh J, Gerya T, Burg JP (2013) High resolution 3D numerical modeling of thrust wedges: Influence of décollement strength on transfer zones. Geochem Geophys Geosyst 14(4):1131–1155

Ruiz S, Metois M, Fuenzalida A, Ruiz J, Leyton F, Grandin R, Vigny C, Madariaga R, Campos J (2014) Intense foreshocks and a slow slip event preceded the 2014 Iquique Mw8.1 earthquake. Science 345:1165–1169. https://doi.org/10.1126/science.1256074

Sage F, Collot JY, Ranero CR (2006) Interplate patchiness and subduction-erosion mechanisms: Evidence from depth-migrated seismic images at the central Ecuador convergent margin. Geology 34:997–1000. https://doi.org/10.1130/G22790A.1

Sallares V, Ranero CR (2005) Structure and tectonics of the erosional convergent margin off Antofagasta, north Chile (23°30′S). J Geophys Res 110(B06101)

Scherwath M, Contreras-Reyes, Flueh ER, Grevemeyer I, Krabbenhoeft A, Papenberg C, Petersen CJ, Weinrebe RW (2009) Deep lithospheric structures along the southern central Chile margin from wide-angle P-wave modeling. Geophys J Int 179(1):579–600. https://doi.org/10.1111/j.1365-246X.2009.04298.x

Schorlemmer D, Wiemer S, Wyss M (2005) Variations in earthquake-size distribution across different stress regimes. Nature 437:539–542

SERNAGEOMIN (2002) Mapa Geológico de Chile. Escala 1:1000000 Map M61, Servicio Nacional de Geología y Minería

Song TA, Simons M (2003) Large trench-parallel gravity variations predict seismo-genic behavior in subduction zones. Science 301:630–633

Suppe J (2014) Fluid overpressures and strength of the sedimentary upper crust. J Struct Geol 69 (2014):481–492. https://doi.org/10.1016/j.jsg.2014.07.009

Tassara A (2010) Control on forearc density structure on megathrust shear strength along the Chilean subduction zone. Tectonophysics 495:34–47. https://doi.org/10.1016/j.tecto.2010.06.004

Tassara A, Soto H, Bedford J, Moreno M, Baez JC (2016) Contrasting amount of fluids along the megathrust ruptured by the 2010 Maule earthquake as revealed by a combined analysis of aftershocks and afterslip. Tectonophysics 671:95–109, ISSN 0040-1951. http://dx.doi.org/10.1016/j.tecto.2016.01.009

Thomas C, Livermore R, Pollitz F (2003) Motion of the Scotia Sea plates. Geophys J Int 155:789–804

Thornburg TM, Kulm LD (1987) Sedimentation in the Chile Trench: Depositional morphologies, lithofacies, and stratigraphy. Geol Soc Am Bull 98:33–52

Tilmann F et al. (2016) The 2015 Illapel earthquake, central Chile: A type case for a characteristic earthquake? Geophys Res Lett 43:574–583

Vannucchi P, Sage F, Morgan JP, Remitti F, Collot JY (2012) Toward a dynamic concept of the subduction channel at erosive convergent margins with implications for interplate material transfer. Geochem Geophys Geosyst 13(Q02003). https://doi.org/10.1029/2011GC003846

Völker D, Geersen J, Contreras-Reyes E, Reichert C (2013) Sedimentary fill of the Chile Trench (32°–46° S): Volumetric distribution and causal factors. J Geol Soc London. https://doi.org/10.1144/jgs2012-119

von Huene R, Ranero CR (2003) Subduction erosion and basal friction along the sediment starved convergent margin off Antofagasta, Chile. J Geophys Res 108(B2):2079. https://doi.org/10.1029/2001JB001569

Wells RE, Blakely RJ, Sugiyama Y, Scholl DW, Dinterman PA (2003) Basin-centered asperities in great subduction zone earthquakes: A link between slip, subsidence, and subduction erosion? J Geophys Res 108(B10):2507. http://dx.doi.org/10.1029/2002JB002072

Willner AP, Thomson SN, Króner A, Wartho JA, Wrjbrans J, Hervé F (2005) Time markers for the evolution and exhumation history of a late Palaeozoic paired metamorphic beltin central Chile (34°–35°30′S). J Petrol 46:1835–1858

Willner AP, Massonne HJ, Ring U, Sudo M, Thomson SN (2012) P-T evolution and timing of a late Palaeozoic fore-arc system and its heterogeneous Mesozoic overprint in north-central Chile (latitudes 31–32 degrees S). Geol Mag 149:177–207. https://doi.org/10.1017/S0016756811000641

Wyss M (1973) Towards a physical understanding of the earthquake frequency distribution. Geophys J Roy Astron Soc 31:341–359

Yáñez GA, Ranero CR, von Huene R, Díaz J (2001) Magnetic anomaly interpretation across the southern central Andes (32–34° S): The role of the Juan Fern_andez ridge in the Late Tertiary evolution of the margin. J Geophys Res Solid Earth 106(B4):6325–6345

Chen YL, Hung SH, Jiang JS, Chiao LY (2016) Systematic correlations of the earthquake frequency-magnitude distribution with the deformation and mechanical regimes in the Taiwan orogen. Geophys Res Lett 43(10):5017–5025

Zelt CA (1999) Modelling strategies and model assessment for wide-angle seismic travel time data. Geophys J Int 139:183–204

The Peru-Chile Margin from Global Gravity Field Derivatives

Orlando Álvarez, Mario Giménez, Federico Lince Klinger, Andrés Folguera and Carla Braitenberg

Abstract Deformation along the 3,500 km subduction Pacific margin of the Peru-Chile trench is partially controlled by ocean bathymetric heterogeneities and sediments. Oceanic highs (e.g. ridges, fracture zones, plateaus) influence deformation in the fore-arc zone where collision occurs, and control turbiditic flow dispersal and consequently the amount of sediments accreted at the frontal accretionary prism and subduction channel, compartmentalizing the trench into segments linked to seismic segmentation. Recent satellite missions (CHAMP, GRACE and GOCE) have introduced an extraordinary improvement in the reconstruction of the global gravity field. Earth gravity field models, mainly derived from satellite measurements, reflect mass inhomogeneities of the earth. This chapter focuses on the determination of mass heterogeneities over the oceanic plate and their relation to general distribution of sediments over the Peru-Chilean margin, seismic segmentation along the margin, and the relationship between trench sediment thickness and the variable Andean orogenic volume, by means of a gravimetric analysis. Using the gravity potential model EGM2008 and satellite GOCE data we calculated two functionals of the geopotential: the Bouguer anomaly and the vertical gravity

O. Álvarez (✉) · M. Giménez · F. L. Klinger
Instituto Geofísico Sismológico Ing. Volponi (IGSV),
Universidad de Nacional San Juan, San Juan, Argentina
e-mail: orlando_a_p@yahoo.com.ar

O. Álvarez · M. Giménez · F. L. Klinger · A. Folguera
Consejo Nacional de Investigaciones Científicas y Técnicas (CONICET),
Buenos Aires, Argentina

A. Folguera
Instituto de Estudios Andinos (IDEAN), University of Buenos Aires,
Buenos Aires, Argentina

C. Braitenberg
Dipartimento di Matematica e Geoscienze, Universita di Trieste,
Via Weiss 1, 34100 Trieste, Italy

A. Folguera et al. (eds.), *The Evolution of the Chilean-Argentinean Andes*,
Springer Earth System Sciences, https://doi.org/10.1007/978-3-319-67774-3_3

gradient, both corrected for the topographic effect. The vertical gravity gradient field is of special interest as it highlights main geological features, and allows unraveling unknown structures that are concealed by sediments. From these, different features can be clearly depicted such as the contact between the Pacific oceanic crust and the South American plate, the Nazca Ridge, the Juan Fernandez Ridge and the Chile Rise, among others. The segmentation between a filled trench south of Juan Fernandez Ridge, a partially filled trench to the north up to the Copiapo Ridge, and a completely starved trench north of this latitude is depicted. Finally, the relationship between gravity derived fields, high oceanic features and seismic segmentation is discussed for the last megathrust earthquakes that affected this subductive plate boundary.

Keywords GOCE · EGM2008 · Megathrust earthquakes · Vertical gravity gradient

1 Introduction

The interface between Nazca and South American plates coincides with the Peru-Chile trench (Fig. 1), a major tectonic feature located at about 150 km off the coast of Peru and Chile, approximately 5,900 km long with a mean width of 64 km. The trench deepens from 4,100 m at southern Chile (44° S) to 6,400 m off the coast of central Chile, reaching 8,200 m off northern Chile (Lindquist et al. 2004; Völker et al. 2006).

The Peru-Chile trench is affected by the subduction of several bathymetric highs (Fig. 1). The most outstanding determine a morphological segmentation in the trench and are the following: in the north, at about 15° S the Nazca Ridge (NR), at 33° S the Juan Fernández Ridge (JFR), both associated with hot spot activity, and to the south the Chile Rise, an active spreading center which is subducting at 46.5° S and separates the Antarctica and Nazca plates.

The age of the oceanic subducting plate increases northwards along the Peru-Chile trench (Fig. 2) from 0 Ma at the Chile triple junction (46.5° S), 10 Ma at 44° S, 20 Ma at 41° S, 33 Ma at 35° S and to a maximum of ~48 Ma around 20° S (Müller et al. 1997). Along-strike, several jumps in the age of the Nazca plate are associated with fracture zones recognizable as pronounced bathymetric lows with a general azimuth of ~70° (Tassara et al. 2006).

The bathymetry of the segmented oceanic lithosphere influences the distribution of sediments along the trench (Ranero et al. 2006). Since the oceanic floor ages to the north, its isostatic bathymetry deepens in this direction along the trench, and consequently the sediments are dispersed northwards by a parallel-to-the-trench turbiditic axial channel. This channel is dammed by the JFR which forms a topographic barrier within the trench separating two distinct domains. South of the JFR to the Chile Triple Junction, the trench is partly filled with sediments forming a 2.0–2.5 km turbiditic succession that produces a flat bottom bathymetry (Völker et al. 2006).

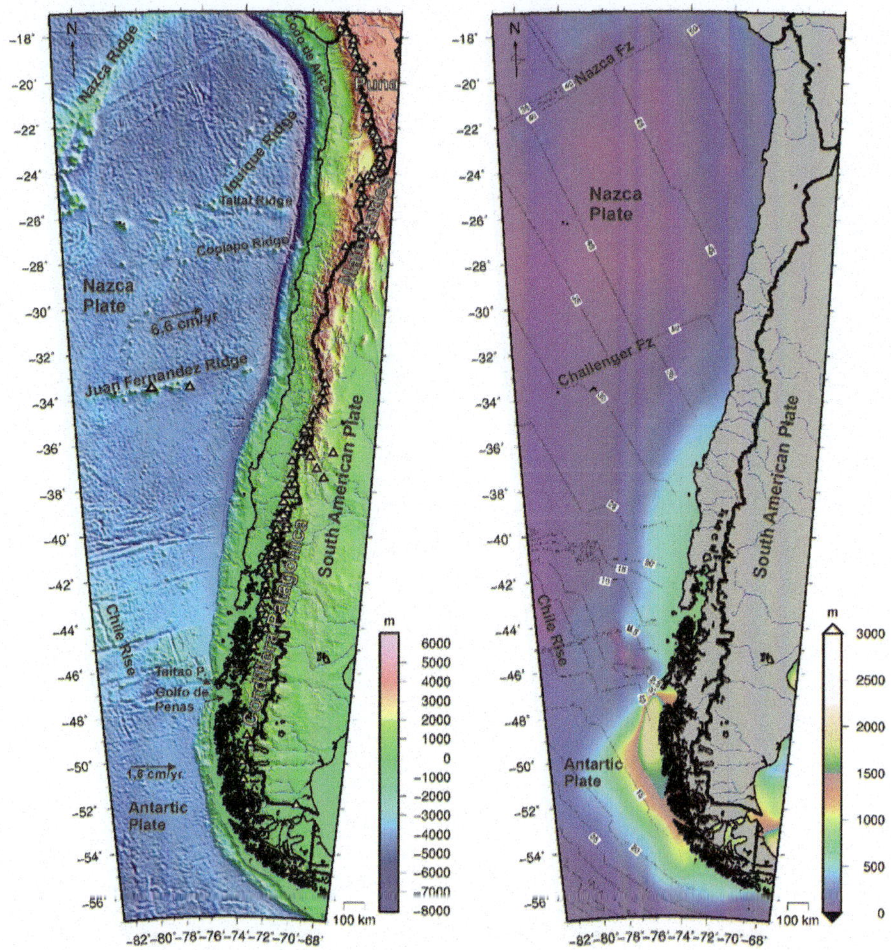

Fig. 1 *Left* Morphology of the oceanic Nazca plate with main bathymetric features (Amante and Eakins 2009). Nazca/South America convergence rate is approx. 6.8 cm/yr with a convergence angle of 78° N. The Antarctic plate moves at considerably lower rates of 1.8 cm/yr between ~46 and 53° S rather perpendicular to the trench (DeMets et al. 1990). *Right* Sediment thickness distribution offshore (Whittaker et al. 2013) and oceanic crust ages (Müller et al. 2008)

Contrastingly, north of the JFR, the trench becomes a narrower depression with steep walls, ranging in depth from 6,100 m to more than 7,000 m and starving of sediments (Schweller et al. 1981; von Huene et al. 1997; Laursen et al. 2002) (Fig. 2).

Sediment infill has been considered one of the parameters that control the coupling at the plate interface, affecting the strength balance between the upper plate lithosphere and the Nazca/South American Plate interface (Oncken et al. 2006). Under this hypothesis, a lower volume of trench sediments increases the effective frictional strength of the plate boundary through time and results in an

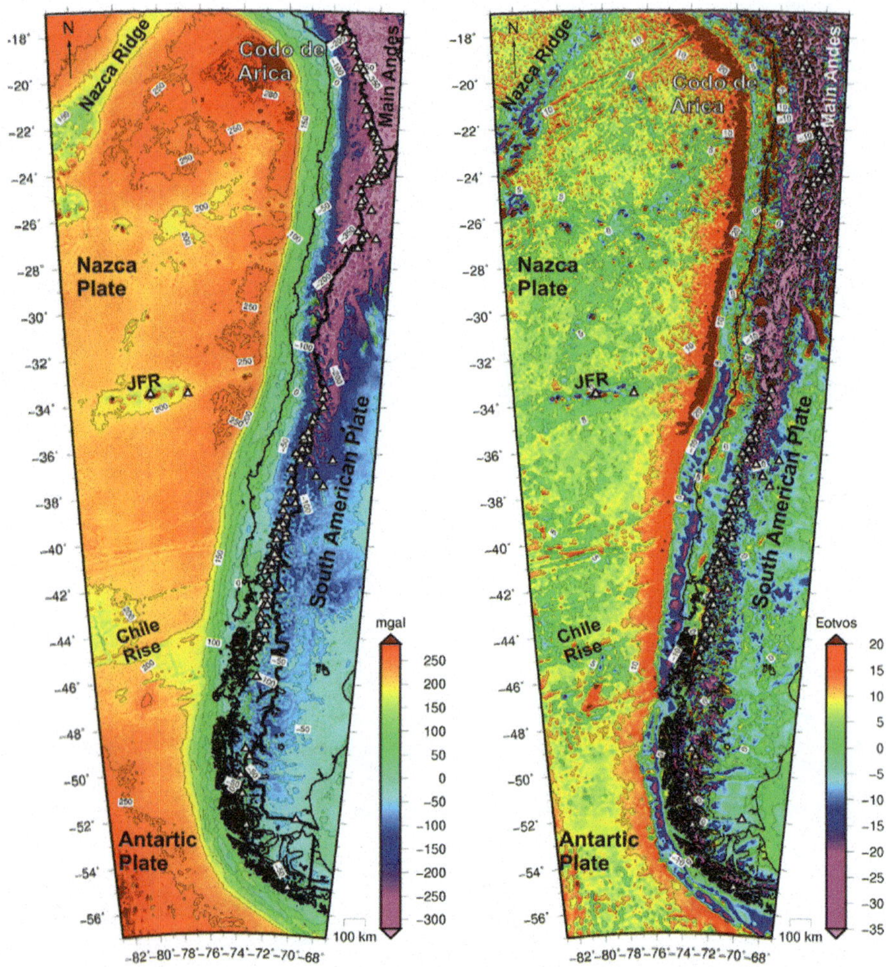

Fig. 2 *Left* Map of the Bouguer anomaly corrected by topography for EGM2008 model up to degree/order $N = 2159$. The most prominent features of the oceanic crust as the Nazca Ridge, the Juan Fernandez Ridge and the Chile Rise are depicted, and are expressed by lower gravity values. Note the relation between the maximum values of the off-shore gravity anomalies and the age of the oceanic crust (Fig. 2). In the continental area the Bouguer anomaly reflects the southern reduction of the Andean topography (Fig. 1) and consequently the thinning of mountain roots with a shallower mantle. Coastal borders: Thin black line. National borders: Thick black line. *Right* Map of the vertical gravity gradient corrected by topography for the model EGM2008 up to degree and order 2159. The main features of the oceanic crust are depicted, being notorious the low gravity gradient values at the Chile trench south of Juan Fernandez Ridge, expression of low density sediments

increased coupling and accelerated shortening in the upper plate. In this line, trench fill thickness variations through the convergent margin and consequent variations in the friction coefficient or lubrication of the plate interface have been associated with sediment accretion or tectonic fore-arc erosion (von Huene and Scholl 1991; Adam and Reuther 2000; Vietor and Echtler 2006; Völker et al. 2006). Lamb and Davis (2003) suggested that the high topography associated with the Central Andes is partly explained by a high plate coupling that increased shear stresses resulting from the lack of sediment input to the Peru-Chile trench. Across the trench, the orogenic volumes of the Central Andes compared with the volumes of the Patagonian Andes have been attributed to variable sediment trench infill that reduces coupling between the upper plate and the subducted slab (Ramos and Ghiglione 2008).

This work by means of the Earth gravity field model EGM2008 (Pavlis et al. 2008, 2012) and the satellite only model obtained from GOCE (Pail et al. 2011) characterizes main bathymetric anomalies of the Nazca plate and discusses their influence over the general distribution of the trench sediment infill and the interplate seismic segmentation.

2 Tectonic Setting

The oceanic Nazca Plate presents a rough morphology defined by seamounts, aseismic ridges and oceanic plateaus (Fig. 1). These oceanic highs are progressively buried by trench sediments as they get closer to the frontal accretionary prism (Völker et al. 2006).

The trench sedimentary infill is sourced from the high Andes and fore-arc uplifts (Völker et al. 2006), through a variable volume supply that depends on the latitudinal development of the topography, the rock erodibility, and climate conditions (precipitation) (Ranero et al. 2006; Zachos et al. 2001). In northern Chile and southern Perú, moisture-bearing winds that originate in the Amazon region are blocked by the Central Andes. This process joined to atmospheric and ocean circulation effects, leads to pronounced contrasts between humid sectors along the eastern flanks of the orogen and extreme arid conditions (<5 cm/yr) within the Puna-Altiplano Plateau and western fore-arc of Chile and Perú (Lamb and Davis 2003; Alonso et al. 2006).

Contrastingly, on the Patagonian Andes, the windward western side of the orogen exhibits high rainfall rates (200–800 cm/yr) (New et al. 1999, 2002; Völker et al. 2006), glacier development and vigorous stream flow rates that contrast with the arid climate to the east (Lenters and Cook 1997). These dominant southwestern wet winds produce a rain shadow that erodes the Patagonian cordillera increasing the sediment supply into the trench (Ramos and Ghiglione 2008).

These effects determine strong differences on erosion and sediment supply in both regions (e.g., Bangs and Cande 1997; Haselton et al. 2002; Hartley 2003; Alonso et al. 2006). The upper plate denudes in relation to climate conditions that control the sediment supplied to the trench (Vietor and Echtler 2006). Within the trench, general sediment transport occurs by turbiditic currents in a south to north direction, with an axial channel integrating lateral sediment sources (Völker et al. 2006).

3 Global Earth Gravity Field Models

Through new high resolution gravity field models, regional gravity modeling is feasible with higher accuracy and spatial resolution. Different satellite only GOCE +GRACE+LAGEOS models (Pail et al. 2011) were developed up to degree/order $N = 250/280/300$.

EGM2008 model (Pavlis et al. 2008, 2012) combines GRACE-derived satellite solutions plus a worldwide free-air Bouguer anomaly database of $5' \times 5'$ resolution. This model is presented as sets of coefficients of a spherical harmonic approximation up to degree/order $N = 2160$ with some additional terms up to degree/order 2190. The spatial resolution of the model $\lambda_{min} \approx 2\pi R/N_{max} \approx 19$ km depends on the maximum degree and order of the harmonic expansion N_{max}, with R meaning earth radius (Hofmann-Wellenhof and Moritz 2005; Li 2001; Barthelmes 2009).

Pavlis et al. (2012) exposed that the grid of surface free-air gravity anomalies was formed by merging terrestrial, altimetry-derived, and airborne gravity data. The spectral content was supplemented with the gravitational information obtained from the global set of RTM-implied gravity anomalies over areas where only lower resolution gravity data were available and over areas void of any gravity observations. This explains discrepancies over the south Central Andes with satellite only models (Alvarez et al. 2012). These differences are mainly due to the sparseness of the terrestrial data in large regions, especially those of difficult access, and to a non-unified height system used in different campaigns.

To estimate an ocean-wide set of free-air gravity anomalies, Pavlis et al. (2012) used satellite altimeter data, along with the estimated Dynamic Ocean Topography model. Since the grid entering the model is based on the altimeter data for oceanic areas, which has higher resolution of $2' \times 2'$, the EGM2008 model presents better resolution over the ocean than inland.

This chapter focuses in the interplay between prominent ocean features of the ocanic Nazca plate and the general distribution of trench sediments, whose areal distribution and volume are delineated by gravity data. This requires a gravity data set with a regional coverage plus the high frequency data. The model EGM2008 fulfills both requirements over the ocean but presents greater differences over the south Central Andes. Here, GOCE remains the best field although with less spatial resolution.

4 Methodology

In the present work we calculate the functionals of the geopotential in terms of the spherical harmonic coefficients up to degree/order $N = 2159$ for EGM2008 (Pavlis et al. 2008, 2012) and up to degree/order $N = 300$ for GOCE (Bruinsma et al. 2013), on a regular grid of 0.05° grid cell size. From the Earth gravity field model we obtained the observed potential, and then the disturbing potential (T) by subtracting the potential field of the reference ellipsoid (Janak and Sprlak 2006). The Bouguer anomaly is obtained as the first spatial derivative of T plus a correction term. The gravity gradient tensor (Marussi tensor) is obtained as the second derivatives of the disturbing potential (e.g. Hofmann-Wellenhof and Moritz 2005).

With the aim of reducing the gravity anomalies for the gravity topographic effect, we used spherical prisms (Anderson 1976; Heck and Seitz 2007; Wild-Pfeiffer 2008) in order to discretize the digital terrain model and contemplate the Earth's curvature (Uieda et al. 2010, 2016) as was applied in previous works (see Braitenberg et al. 2011 and Alvarez et al. 2012, 2013 for more detail). The potential generated by the topography is calculated from the Digital Elevation Model (ETOPO1, Amante and Eakins 2009), using the software *Tesseroids* (Uieda et al. 2010, 2016), where the calculation height is 7,000 m to ensure that all values are above the topography. The topographic correction amounts up to tens of Eötvös (10^{-4} mGal/m) for the vertical gradient and up to a few hundreds of mGal for gravity. It is greatest over the highest topographic elevations and over the lower bathymetric depressions as the Chile trench.

5 Results

The topography corrected Bouguer anomaly (or Bouguer anomaly) obtained from the model EGM2008 is mapped in Fig. 2 (left) and the topography corrected vertical gravity gradient is shown in Fig. 2 (right). A comparison between both fields reveals an optimal correlation in the location of the anomalies, with the Tzz resolving more accurately.

The Bouguer anomaly reflects the mass inhomogeneities of the Nazca and Antarctica oceanic plates, being particularly notorious the path of the Nazca Ridge, the Juan Fernandez Ridge (JFR), and the Chile Rise expressed by gravity values lower than the surrounding oceanic floor (Fig. 2). Seaward off the trench, the outer rise in the downgoing Nazca plate is marked by a positive Bouguer anomaly greater than +280 mGal in the northern analyzed section, of about +250 mGal in the central region, decreasing to the south up to the Chile Rise where orogenic volumes are lower and consequently flexure of the Nazca plate is less pronounced. However, in the southernmost region, south of the Chile Rise, the Bouguer anomaly increases in the southwest direction as oceanic lithosphere age increases as well (Fig. 1 right). This relation between the age of the oceanic crust (Müller et al. 2008) and gravity, also observed in the vicinity of the Codo de Arica, was reported previously by Wienecke (2006) who found a clear correlation between old crustal ages of the oceanic crust and the positive offshore Bouguer Anomaly. However, this trend is modified by anomalies related to ocean bathymetric highs and the ocean floor flexure (Götze et al. 2003; Wienecke 2006).

The vertical gravity gradient (Fig. 2 right) presents a maximum positive anomaly, greater than +20 Eötvös in the northern region, at the fore-bulge, which diminishes to the south of the Juan Fernandez Ridge. At these latitudes, towards the coastline, the Chile trench is marked by a negative Tzz anomaly of less than −10 Eötvös resulting from the presence of low density sediments. The gravimetric effect of the Andes diminishes in the south direction reflecting the lower elevation of the mountain system and the consequent reduction of the crustal thickness as explained by Hackney et al. (2006).

6 Gravity Derivatives from GOCE Satellite Data

The topography corrected Bouguer anomaly and Tzz for the satellite only model GOCE (*GO_CONS_GCF_2_DIR_R5*) up to degree/order $N = 300$ are presented in Fig. 3. The resolution of the geological structures is lower than that expected for EGM2008, about of $\lambda min \approx 2\pi R/Nmax \approx 130$ km. The Bouguer anomaly exhibits a similar pattern to that obtained with EGM2008, with anomalies smoothed out due to the lower spatial resolution of the data. The Nazca, Iquique, and Juan Fernandez ridges and the Chile Rise are detected over the oceanic lithosphere. The outer rise whose magnitude decreases in the southern direction, and the maximum anomaly (more than 300 mGal) in the Codo de Arica and in the south-westernmost region, coincides with the previous interpretation. The negative gravimetric response of the Andean root is also depicted.

The vertical gravity gradient obtained from GOCE (Fig. 3 right) also depicts the Nazca, Iquique, and Juan Fernandez ridges. The effect of the outer rise decreases to the south as indicated previously, and the Tzz anomalies eastward to the trench present the same pattern. These become more positive north to the JFR, presenting the same gaps at 38° S, at 54° S, and between 46° S and 48° S where the Chile Rise subducts beneath the continental shelf.

7 2D Profiles Across the Margin

Four profiles were traced over the topography corrected Bouguer anomaly obtained with EGM2008 and with GOCE (for location of profiles see Fig. 3 left).

The first profile (profile 1) was traced in the northernmost region with an EW-direction approximately. In this profile (Fig. 4a), the outer rise of the oceanic lithosphere does not present a significative topographic expression, and is revealed

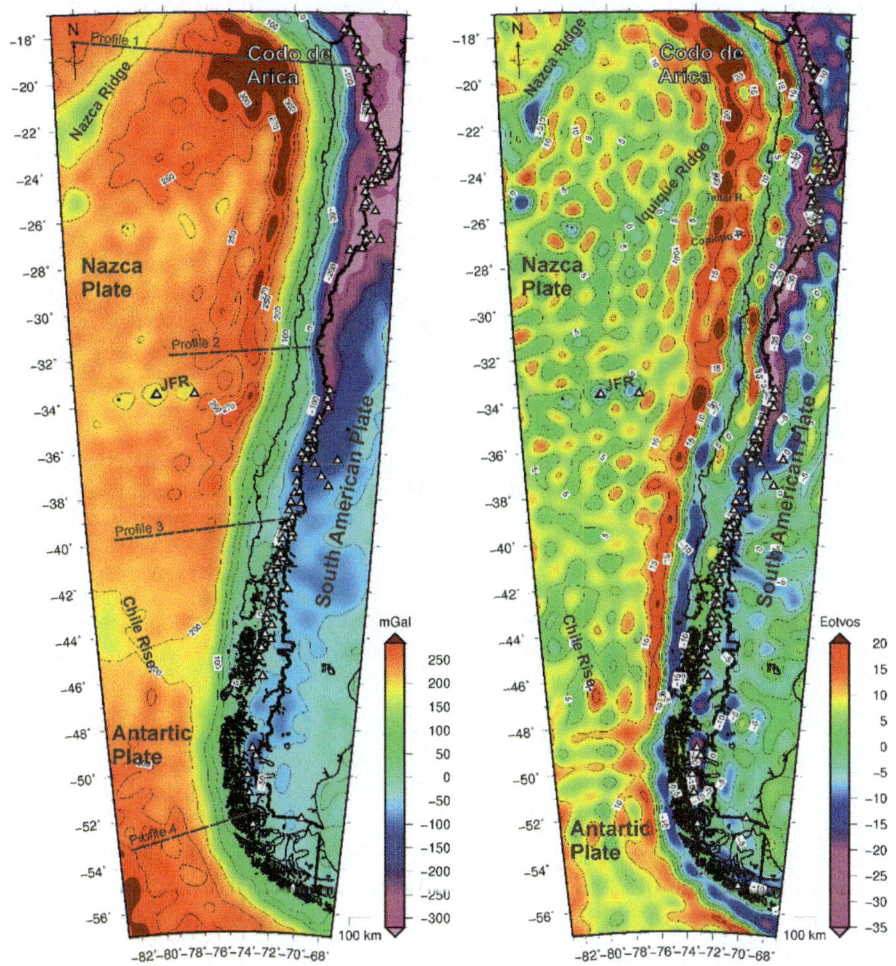

Fig. 3 *Left* Bouguer anomaly map corrected by topography for GOCE model up to degree/order 300. Coastal borders: Thin black line. National borders: Thick black line. Dashed lines are profiles traced at different latitudes. *Right* Vertical Gravity gradient map corrected by topography for GOCE model up to degree/order 300. Coastal borders: Thin black line. National borders: Thick black line

by a maximum in the GA signal of approximately +300 mGal. The Nazca Ridge exhibits a lower gravimetric effect than the surrounding oceanic floor, expressed in both signals. Inland, the Andes (with > 4,000 m) are expressed by a negative gravimetric response representative of the Andean roots deeper than 65 km. The profile 2 (Fig. 4b) traced north of the JFR presents a similar pattern, and also exhibits a good correspondence between both gravity signals.

The most outstanding characteristic of profiles 3 and 4 is the lower elevation of the Andes (<2,000 m) and the corresponding lower gravimetric response of the

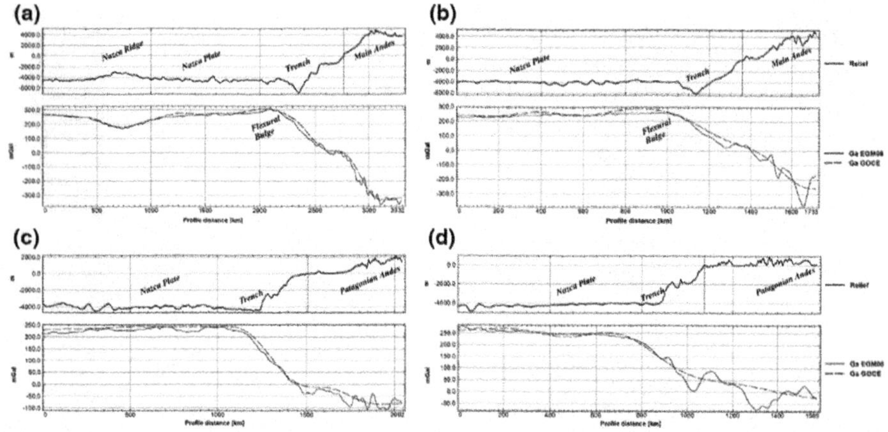

Fig. 4 Profiles over the topography corrected Bouguer anomaly obtained from EGM2008 and GOCE (for location of profiles see Fig. 3 left). Profiles across the northern region of the JFR (**a** and **b**) show a sediment starved trench, the high relief of the Main Andes with more than 4,000 m, a notorious positive gravimetric response over the outer rise (>+300 mGal) and the negative gravimetric response of the Andean root (>−300 mGal). Profiles across the southern region of the JFR (**c** and **d**) show a trench full of sediments, the lower relief of the Patagonian Andes with less than 2,000 m, a minor positive gravimetric response over the outer rise (<+250 mGal) and a less pronounced negative gravimetric response of the Andean roots

Andean root. In these profiles there is no significant bathymetric expression of the trench which is filled with sediments. Despite this, the trench is detected as an inflexion of the EGM2008 gravity signal. The gravimetric signal of the forebulge reaches up to +250mGal, 50 mGal less than the profiles 1 and 2.

The gravimetric response of both gravity potential models shows a good agreement, especially at long wavelengths and over the ocean, with GOCE in general having the more positive signal. Greater differences are present in the continental area at the higher frequencies of the gravimetric signal where the EGM2008 presents more variability. Results obtained from the EGM2008 model for the Andes region are influenced by the lack of a homogeneous terrestrial Bouguer anomaly database, as explained previously.

8 Seismotectonics and Gravity

The EGM2008 model has high resolution of the field and is well correlated with the satellite only GOCE model over the ocean. This allows correlation with the morphology of the oceanic plate in an optimal way as made in Braitenberg et al (2011) and Alvarez et al. (2012).

The continental margin is divided for the analysis in three sections according to morphology, amounts of sediments and seismogenic behaviour: (**a**) a northern region comprised between the Nazca and the Copiapo ridges that presents a starved trench, **(b)** a central zone between the Copiapo Ridge and the JFR that presents an intermediate to low amount of sediments, **(c)** south of the Juan Fernandez Ridge, a southern region characterized by a trench full of sediments.

8.1 Northern Chile

The northern Chilean margin has been characterized as an erosive margin (von Huene and Scholl 1991; Laursen et al. 2002; Adam and Reuther 2000; Völker et al. 2006) with a trench that is virtually starved of sediments (up to 500 m) (Lamb and Davis 2003; Oncken et al. 2006). Subduction erosion processes in this region of the Andes are related to the morphological roughness of the oceanic plate imprinted in the outer rise and to the climate-related lack of sediments within the trench (Ranero et al. 2006). In addition the JFR blocked sediment transport along the trench axis from the south (Yañez et al. 2001) (see Chap. "The Geometry of the Continental Wedge and Its Relation to the Rheology and Seismicity of the Chilean Interplate Boundary").

The gravimetric effect of the oceanic lithosphere rises in the northeast direction with plate age, to a maximum value at the Codo de Arica. The location of the Nazca Ridge is defined by a lower gravimetric signal with respect to the surrounding Nazca Plate, from −20 to +200 mGal approximately (Fig. 5a). The Tzz values range from −10 to +0 Eötvös and become more positive away from the ridge. The Nazca fracture zone is better defined by the Tzz, since it presents a straight gradient signal that extends obliquely in the NE direction for more than 800 km, from the Nazca Ridge to the trench (Fig. 5b).

The Iquique Ridge is also detected by the gravimetric signal, presenting a higher value than the Nazca Ridge. Seamounts over the Iquique Ridge present more than +270 mGal, and more than +25 Eötvos for Tzz. The Taltal and Copiapo ridges are detected in the GA and in the Tzz; seamount chains over these ridges present more than +250 mGal and +25 Eötvos.

North of the Copiapo Ridge up to 20° S, no significant negative anomalies appear in Tzz, which would indicate the presence of low density materials between the trench and the coastline. Instead, several positive anomalies of more than +20 Eötvös appear extending beyond the continental margin. Previous works have shown from seismic lines that between 21° and 24° S the trench is almost empty of sediments (Sick et al. 2006).

North of approximately 20° S up to the Nazca Ridge, in the Codo de Arica region, some negative anomalies in the GA signal and in the Tzz are depicted most likely indicating the existence of low density materials (black dashed ellipses in Fig. 5a, b). These fore-arc basins reflect relatively reduced surficial erosion during the past 20 Ma (Scholl et al. 1970, Hartley and Jolley 1995), as explained Ranero et al. (2006).

Alvarez et al. (2015) compared the topography and sediment corrected vertical gravity gradient (Tzz) computed from GOCE (TIM_R5) model with slip models of

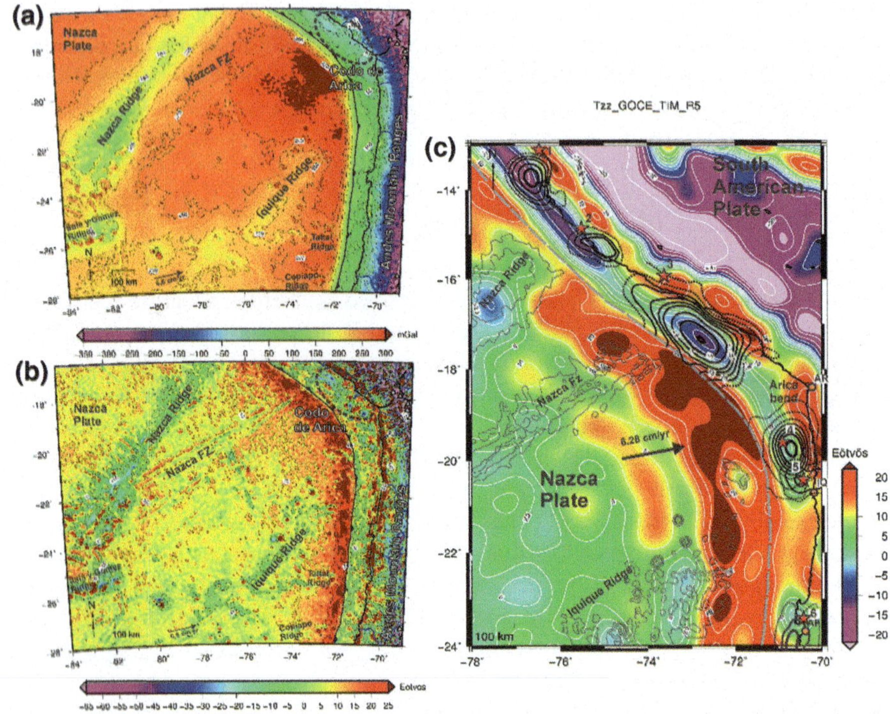

Fig. 5 **a** Topography corrected Bouguer anomaly map for northern Chile trench from EGM2008 model. **b** Topography corrected Vertical gravity gradient map for northern Chile trench from EGM2008 model. **c** Topography corrected vertical gravity gradient (white contours) from GOCE satellite-only model GO_CONS_GCF_2_TIM_R5 from Brockmann et al. (2014). Convergence between Nazca-South American plates (black arrow) is from Kendrick et al. (2003) (Gray dashed line: Perú-Chilean trench, thin black lines: bathymetric highs). Superimposed slip distributions for the 2007 Mw = 8.0 Pisco (1); 1996 Mw = 7.7 Nazca (2), 2001 Mw = 8.4 Arequipa (3), 1995 Mw = 8.1 Antofagasta (6) from Chlieh et al. (2004, 2011) and for the 1st April 2014 Mw = 8.2 Pisagua (4) and 3rd April 2014 Mw = 7.7 Iquique (5) from Schurr et al. (2014). References: Red stars are the epicenters for the different earthquakes shown in this figure. AR: Arica, IQ: Iquique, AF: Antofagasta (modified from Alvarez et al. 2015)

the main great earthquakes in the region (Fig. 5c), finding a good spatial correlation between areas of minimum relative Tzz and high seismic slip, especially for the Mw > 8.0 events. Main rupture zones for the events of Pisagua and Iquique tend to be concentrated over positive Tzz values although this region presents relatively lower values than to the south. Then the authors analyzed the spatial-temporal variations of the gravity field in this region prior to the main rupture as the last two GOCE models (GO_CONS_GCF_2_TIM_R4 and GO_CONS_GCF_2_TIM_R5), which include data measurements from the last four years before these earthquakes occurred. From the comparison between the Tzz derived from both models they proposed that temporal variations in the gravity field are probably either related to mass redistribution changes or variations in fluid pressure before the 2014

Mw = 8.2 Pisagua earthquake. They found a good agreement between the main slip patch for this earthquake and its main shock with a gravity fall and an increase at its northern and southern terminations, presenting the gravity signal a relative higher value close to the area of the foreshocks, probably indicating the presence of an asperity. These High Oceanic Features (HOF's) have been related to an increase in the seismic activity with low magnitude (swarms) and upper plate basal erosion. These mechanisms could explain the foreshock sequence and the gravity fall at least in the last two years prior to the main shock of April 1st 2014, being probably related to upper plate extensional collapse due to the basal erosion (see Alvarez et al. 2015).

8.2 *Central Chile*

The Central Chile sector ranges between the Copiapo Ridge and Juan Fernandez Ridge. The last is a hot spot chain formed with a source ~900 km west off the trench, that first collided with the Chile margin in the north (~20°) at about 22 Ma, and then moved progressively southwards to its present position at about 32–33° S (Yañez et al. 2001), producing tectonic erosion, extensional deformation (von Huene et al. 1997) and local uplift on the continental slope (Ranero et al. 2006). Additionally, as mentioned, the ridge restrains the transport of sediment along the trench axis separating a heavily sedimented trench to the south that extends almost to the triple junction from a trench north of 32.5° S that contains less than 1 km of turbidites confined to a narrow axial zone (Schweller et al. 1981; Bangs and Cande 1997). Sediment distribution south of 33° S ranges from 2.2 to 3.5 km in thickness, at the time the trench flat floor widens from 25 km at 33° S to 80 km at 41° S (Völker et al. 2006).

The JFR is delimited by a well defined gravimetric signal (Fig. 6a) that is lower than the surrounding oceanic floor. This is an indication of a thickened oceanic crust beneath JFR, as reported by von Huene et al. (1997) based on wide-angle seismic data. East to the present hot spot, west of Alexander Selkirk Island, lower values on the gravity and negative values on Tzz appear. Sandwell and Smith (1997) related negative satellite derived gravity anomalies to a crustal root indicative of crustal flexure from loading associated with seamounts. The gravity reaches its maximum value over seamount chains with more than +250 mGal. The Tzz signal (Fig. 6b) also exhibits different positive aligned anomalies of more than +25 Eötvös, expression of the numerous hot-spot derived seamounts over the JFR.

Seismic profiles (von Huene et al. 1997) indicate a 1.5–2 km thicker oceanic crust in the north than to the south of the ridge axis, and shallower next to the trench axis than elsewhere around the ridge (Sandwell and Smith 1997), as is expressed in the gravity and Tzz maps. Next to the trench, over the oceanic plate, the O`Higgins Seamounts are expressed by more than +270 mGal and more than +20 Eötvös. To the south, the O`Higgins fracture is depicted in the gravity by a narrowing in the +250 mGal contour. The O`Higgins ridge is located to the north of the seamounts

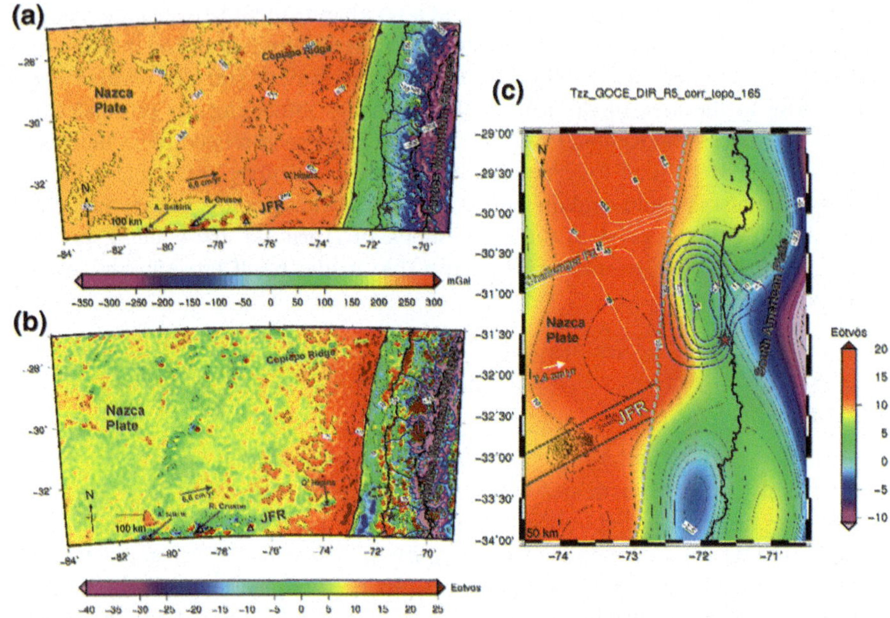

Fig. 6 **a** Topography corrected Bouguer anomaly for Central Chile from EGM2008 model. **b** Topography corrected Vertical gravity gradient for Central Chile trench from EGM2008 model. **c** Topography and sediment corrected vertical gravity gradient obtained from GOCE satellite-only model GO_CONS_GCF_2_DIR_R5 (Bruisma et al. 2013) up to $N = 165$ (modified from Álvarez et al. 2016). Solid black contours: Juan Fernandez Ridge, Ocean floor age: solid white line. Red star indicates the 2015 Mw = 8.2 Illapel earthquake epicenter, slip distribution from Tilmann et al. (2016): blue dashed contours

and expressed by the negative values of Tzz (less than −10 Eötvös). There is a significative Tzz anomaly eastward of the trench, of about +10 Eötvös, presented as a continuation of this seamount chain (black dashed ellipse in Fig. 6b).

Between the trench axis and the continental slope, the Tzz signal exhibits a well defined contour of +0 Eötvös, with lower values inside, indicating the abrupt decease in the sediment infill. The lower values of the Tzz anomaly are located westward towards the trench axis, except in the vicinity of the JFR where the contours are shifted to the east. Different authors (Bangs and Cande 1997; von Huene et al. 1997; Ranero et al. 2006; Lohrmann et al. 2006; among others) coincide that in this area sediment infill thickness is <1 km, and is confined to a narrow zone at the trench axis. Ranero et al. (2006) reported that the continental slope is formed by a 5–10 km wide sediment prism.

South of the prolongation of the JFR, eastward of the trench, the Valparaiso Basin is depicted by negative values in both the gravity and in the Tzz maps (Fig. 6a and b, see also Fig. 3) (see Chap. "Structure and Tectonics of the Chilean Convergent Margin from Wide-Angle Seismic Studies: A Review").

The 16 September 2015 Mw = 8.3 Illapel earthquake occurred in a region with intermediate to low trench sediment infill (Ranero et al. 2006; Völker et al. 2006) (estimated thickness below 500 m). Preliminary slip models along the Illapel rupture zone depict a gross latitudinal slip distribution ranging from 30° to 32.5° S. The refined model from Tilmann et al. (2016) presents a homogeneous and smaller rupture, centered between the extrapolated paths of the Challenger Fz to the north and the Juan Fernandez Ridge to the south. In this region the Tzz signal is mainly positive hiding the relationship between low Tzz and high slips such as it was done for other events along the subduction zone. However, we expanded the limit of observation exploring at depth and analyzing deeper sources by limiting the harmonic expansion to lower degrees (Álvarez et al. 2016). Therefore to analyze the combined effect of both bathymetric highs we calculated Tzz by cutting the harmonic expansion finding that for $N = 165$ a high Tzz is observed at the extrapolated path of the subducted Challenger Fz. Additionally, an eastward deflection of the positive Tzz contours (mainly the +5 Eötvös) coincides with the extrapolation of the Juan Fernandez Ridge. Both Tzz relative maximums roughly correlate with rupture endings. The maximum slip distribution from Tilmann et al. (2016) is located between these two relative maximums of Tzz, being the main peak located in coincidence with a relative minimum inland (see Álvarez et al. 2016 for a detailed analysis).

8.3 South Central Chile

In the southern region of the study area the downgoing oceanic lithosphere presents a prominent feature, the Chile Rise. This active spreading center marks the boundary between the Nazca Plate to the north and the Antarctic Plate to the south. The spreading center trends roughly NNW and approaches the trench at a low angle, colliding with the continent at ~46° S. This feature produces a broad topographic swell of young oceanic lithosphere extending to the north to about 41° S, where the Valdivia fracture zone offsets the spreading center (Ranero et al. 2006).

The region comprehended between the JFR and the Chile Rise, is characterized by an abrupt change in the Tzz signal (lower than −10 Eötvös) between the coast and the trench axis (Fig. 7b). The gravimetric response south of JFR becomes more negative with values lower than +50 mGal, indicating the existence of low density materials (Fig. 7a). Here, the trench axis widens up to 40 km, with a thick turbidity infill (up to 2.5 km thick), and a 50–60 km wide accretionary prism (Ranero et al. 2006; Bangs and Cande 1997). South of the Valparaiso basin, shallow forearc basins (1 km thick; von Huene et al. 1997) extend on the shelf (Bangs and Cande 1997). South of the zone where the JFR collides, frontal sediment accretion dominates extending to the region where the Chile Rise subducts (Bangs and Cande 1997; Sandwell and Smith 1997; Laursen et al. 2002).

The gravimetric response of the flexural bulge in the downgoing Nazca plate diminishes south to JFR to lower Tzz values, being this effect more notorious in the

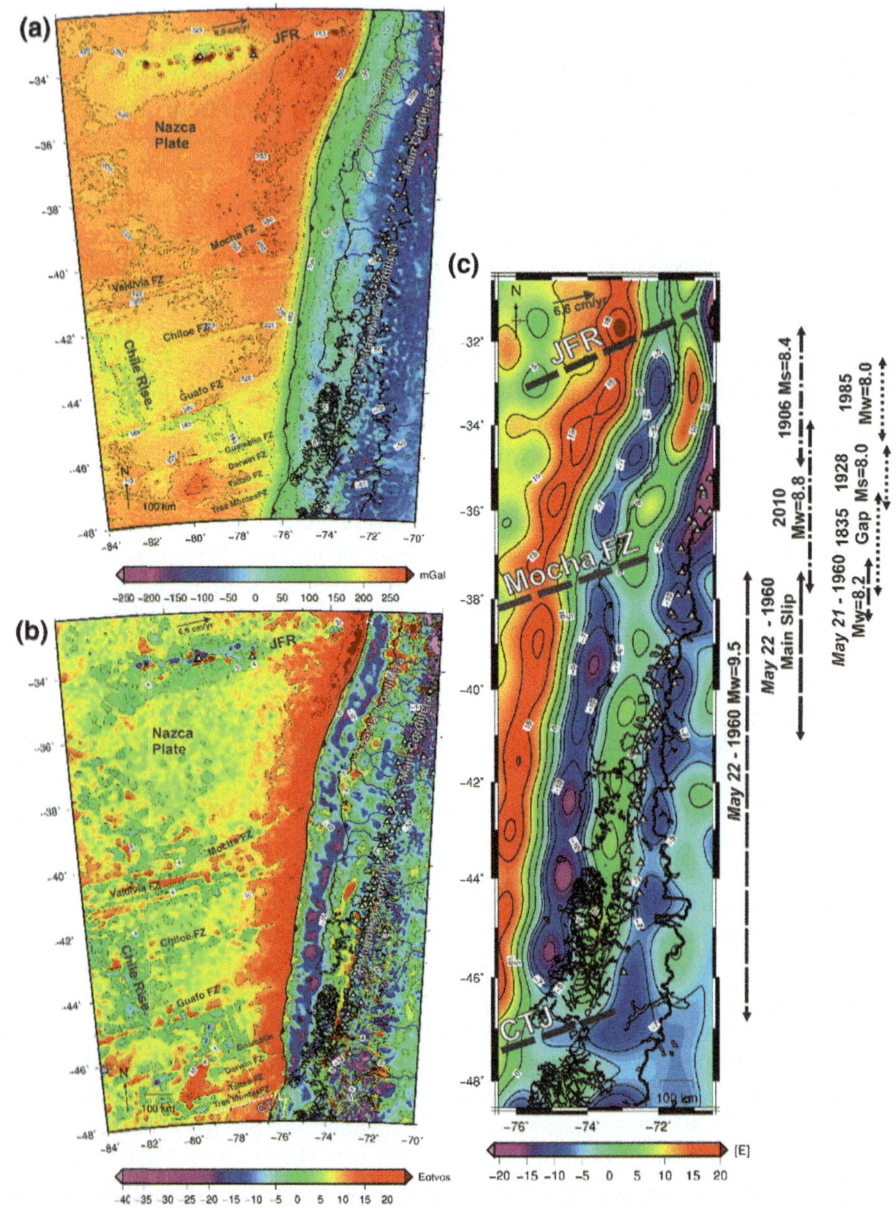

Fig. 7 **a** Bouguer anomaly map for south-central Chile trench. **b** Vertical gravity gradient map for south-central Chile trench. **c** Topography corrected Vertical Gravity Gradient (Tzz) obtained from GOCE model GO_CONS_GCF_2_TIM_R4 (Pail et al. 2011) and Nazca plate Hof's (dashed lines) extrapolation (modified from Alvarez et al. 2014). To the right the dimensions of the approximate rupture areas of the largest earthquakes (Mw $\geq$ 8)

gravity signal. Fracture zones over the Nazca plate are well defined by both the gravity and the Tzz.

These fracture zones affect the sediment distribution at the trench being notorious in the Tzz signal. South of the Mocha Island, the trend of negative Tzz anomalies shifts to the shelf giving place to two positive anomalies of more than +10 Eötvös. In the vicinity of the triple junction, the Tzz signal shows a narrowing of the sediment infill, reaching +0 Eötvös in the CTJ collision zone.

To the South-East of the Taitao Peninsula, the Golfo de Penas exhibits low gravity values (<0 mGal) and less than −10 Eötvös indicating a low density body in this area. South of this, the Tzz shows a truncated sedimentary prism consequence of the collision of the Chile Rise with the margin at around 3–6 Ma (Ranero et al. 2006).

Segmentation of the Tzz signal along the trench from satellite GOCE data is related to irregularities in the oceanic floor. Particularly, the JFR, the Mocha Fz and the Chile Rise are related to high Tzz signals (Fig. 7c).

In a recent work Métois et al. (2016) explored the link between coupling and seismicity over the Chilean margin from 38° to 18° S obtaining the first nearly continuous map of interseismic coupling variations on the subduction interface based on GPS networks. They found at least six relative low coupling zones, five of which correlate with the subduction of ridges or fracture zones of the Nazca plate that enter the trench. They found that for the recent Mw > 8 events, co-seismic asperities correlate well with highly coupled segments, while low coupling zones (LCZ) behaved as barriers and stopped the ruptures.

When comparing the Tzz with the main slip distribution for different megathrust earthquakes in the region, a loose spatial correlation between high Tzz values and rupture limits for some events is found by Alvarez et al. (2014). Low Tzz values are roughly correlated with the location of the main slip patches, showing for the Maule 2010 event a good correlation between high slip (Moreno et al. 2012) and low Tzz.

9 Concluding Remarks

Direct modelling of Bouguer anomaly and vertical gravity gradient using the EGM2008 model and satellite GOCE data allowed delineating different anomalies over the Nazca Plate and over the Chile margin. The gravimetric effect related to high oceanic features and to the general distribution of sediments over the trench can be clearly depicted. The Tzz signal is an optimal tool to delineate areas with significant density contrasts as mass heterogeneities detected over the margin, complemented by means of the Bouguer anomaly.

Of particular interest is the relationship between the high Bouguer anomaly and vertical gravity gradient value at the outer rise north of the Juan Fernandez Ridge, the lack of sediments in the trench expressed by high Tzz values and the higher elevation of the Andes. Contrastingly, south of 32° S the effect of the outer rise diminishes and Tzz values reflect the presence of low density materials filling the trench. Inland, the effect of the Andean roots diminishes to the south where the Andes loose height and amplitude.

Our data imply that sediment distribution in the Chile trench is strongly related to the incoming oceanic features, such as aseismic ridges, fracture zones and ridges. The delineation of different anomalies such as seamounts over the oceanic plate is of special interest since these are related to earthquake generation producing asperities and could also act as barriers to seismic energy propagation. The identification of minor density areas over the margin, corresponding to trench sediment infill or basins developed over the accretionary prism shows a striking correlation with rupture zones, becoming a key issue to infer seismogenic segmentation along the subduction zone.

Acknowledgements Authors acknowledge the use of the GMT-mapping software of Wessel & Smith (1998). The authors would like to thank to CONICET.

References

Adam J, Reuther CD (2000) Crustal dynamics and active fault mechanics during subduction erosion. Application of frictional wedge analysis on to the North Chilean Forearc. Tectonophysics 321:297–325

Alonso R, Bookhagen B, Carrapa B, Coutand I, Haschke M, Hilley G, Schoenbohm L, Sobel E, Strecker M, Trauth M, Villanueva A (2006) Tectonics, climate, and landscape evolution of the Southern Central Andes: the Argentine Puna Plateau and adjacent regions between 22 and 30° S. In: Oncken O, Chong G, Franz G, Giese P, Götze H-J, Ramos VA, Strecker MR, Wigger P (eds) The Andes active subduction orogeny. Frontiers in earth science series. Springer, Berlin Heidelberg New York, pp 265–284

Alvarez O, Gimenez ME, Braitenberg C, Folguera A (2012) GOCE satellite de rived gravity and gravity gradient corrected for topo graphic Effect in the South Central Andes region. Geophys J Int 190(2):941–959. https://doi.org/10.1111/j.1365-246X.2012.05556.x

Alvarez O, Gimenez ME, Braitenberg C (2013) Nueva metodología para el cálculo del efecto topográfico para la corrección de datos satelitales. Rev Asoc Geol Arg 70(4):422–429

Alvarez O, Nacif S, Gimenez M, Folguera A, Braitenberg A (2014) GOCE drived vertical gravity gradient delineates great earthquake rupture zones along the Chilean margin. Tectonophysic 622:198–215. https://doi.org/10.1016/j.tecto.2014.03.011

Alvarez O, Nacif S, Spagnotto S, Folguera A, Gimenez M, Chlieh M, Braitenberg C (2015) Gradients from GOCE reveal gravity changes before Pisagua Mw = 8.2 and Iquique Mw = 7.7 large megathrust earthquakes. J South Am Earth Sci 64(2):15–29. https://doi.org/10.1016/j.jsames.2015.09.014

Álvarez O, Pesce A, Gimenez M, Folguera A, Soler S, Wenjin C (2016) Analysis of the Illapel Mw = 8.3 thrust earthquake rupture zone using GOCE derived gradients. Pure Appl Geophys, online first August, 2016. https://doi.org/10.1007/s00024-016-1376-y

Amante C, Eakins BW (2009) ETOPO1 1 arc-minute global relief model: procedures, data sources and analysis. NOAA Technical Memorandum NESDIS NGDC-24, pp 19

Anderson EG (1976) The effect of topography on solutions of Stokes' problem. Unisurv S-14, Rep, School of Surveying, University of New South Wales, Kensington

Bangs NL, Cande SC (1997) Episodic development of a convergent margin inferred from structures and processes along the southern Chile margin. Tectonics 16(3):489–505

Barthelmes F (2009) Definition of functionals of the geopotential and their calculation from spherical harmonic models theory and formulas used by the calculation service of the International Centre for Global Earth Models (ICGEM), Scientific Technical Report STR09/02, GFZ German Research Centre for Geosciences, Postdam, Germany. http://icgem.gfz-postdam.de

Braitenberg C, Mariani P, Ebbing J, Sprlak M (2011) The enigmatic Chad lineament revisited with global gravity and gravity gradient fields, Spec. Pub. Geol. Soc. London, 'The formation and evolution of Africa from the Archaean to Present'

Bruinsma SL, Förste C, Abrikosov O, Marty JC, Rio MH, Mulet S, Bonvalot S (2013) The new ESA satellite-only gravity field model via the direct approach. Geophys Res Lett 40:3607–3612

Cande SC, Leslie RB, Parra JC, Hobart M (1987) Interaction between the Chile ridge and the Chile trench: geophysical and geothermal evidence. J Geophys Res 92:495–520

Contreras-Reyes ER (2008) Evolution of the seismic structure of the incoming/subducting oceanic Nazca plate off south-central Chile (Ph.D. thesis): Kiel, Germany, Christian-Albrechts-Universität zu Kiel, p 143

Contreras-Reyes ER, Flueth E, Grevemeyer I (2010) Tectonic control on sediment accretion and subduction off south-central Chile: Implications for coseismic rupture processes of the 1960 and 2010 megathrust earthquakes. Tectonics 29 (TC6018). https://doi.org/10.1029/2010TC002734

DeMets C, Gordon RG, Argus DF, Stein S (1990) Current plate motions. Geophys J Int 101:425–478

Götze HJ, Schmidt S, Wienecke S, Braitenberg C, Schreckenberger B (2003) Regional gravity offshore Chile—new insight into crustal structures. Poster presentation SFB 267 Workshop, Pucón, Chile, October 2003

Hackney R, Echtler HP, Franz G, Götze HJ, Lucassen F, Marchenko D, Melnick D, Meyer U, Schmidt S, Tašárová Z, Tassara A, Wienecke S (2006) The segmented overriding plate and coupling at the South-Central Chilean Margin (36°–42° S). In: Oncken O, Chong G, Franz G, Giese P, Götze H- J, Ramos VA, Strecker MR, Wigger P (eds) The Andes—active subduction orogeny. Frontiers in earth science series. Springer, Berlin Heidelberg New York, pp 355–375

Hartley AJ (2003) Andean uplift and climate change. J Geol Soc Lond 160:7–10

Hartley AJ, Jolley EJ (1995) Tectonic implications of Late Cenozoic sedimentation from the Coastal Cordillera of northern Chile (22°–24° S). J Geol Soc London 152:51–63

Haselton K, Hilley G, Strecker MR (2002) Average Pleistocene climatic patterns in the southern Central Andes: controls on mountain glaciations and palaeoclimate implications. J Geol 110:211–226

Heck B, Seitz K (2007) A comparison of the tesseroid, prism and point mass approaches for mass reductions in gravity field modeling. J Geod 81(2):121–136. https://doi.org/10.1007/s00190-006-0094-0

Hofmann-Wellenhof B, Moritz H (2005) Physical Geodesy. ISBN-10 3-211-33544-7, SpringerWien, NewYork

Janak J, Sprlak M (2006) New software for gravity field modelling using spherical armonic. Geodetic and Cartographic Horizon 52:1–8

Kendrick E, Bevis M, Smalley R Jr, Brooks B, Vargas RB, Lauría E, Fortes LPS (2003) The Nazca—South America Euler vector and its rate of change. J South Am Earth Sci 16:125–131

Lamb S, Davis P (2003) Cenozoic climate change as a possible cause for the rise of the Andes. Nature 425:792–797

Laursen J, Scholl D, von Huene R (2002) Neotectonic deformation of the central Chile margin: deepwater forearc basin formation in response to hot spot ridge and seamount subduction. Tectonics 21. doi:https://doi.org/10.1029/2001TC901023

Lenters JD, Cook KH (1997) On the origin of the Bolivian High and related circulation features of the South American climate. J Atm Sci 54:656–677

Li X (2001) Vertical resolution: gravity versus vertical gravity gradient. Lead Edge 20(8):901–904

Lindquist K, Engle K, Stahlke D, Price E (2004) Global topography and bathymetry grid improves research efforts. EOS 85(19). https://doi.org/10.1029/2004EO190003

Lohrmann J, Kukowski N, Krawczyk CM, Oncken O, Sick C, Sobiesiak M, Rietbrock A (2006) Subduction channel evolution in brittle fore-arc wedges a combined study with scaled sandbox experiments, seismological and reflection seismic data and geological field evidence. In: Oncken O, Chong G, Franz G, Giese P, Götze H-J, Ramos VA, Strecker MR, Wigger P (eds) The Andes—Active subduction orogeny. Frontiers in earth science series. Springer, Berlin Heidelberg New York, pp 237–262

Métois M, Vigny C, Socquet A (2016) Interseismic coupling, megathrust earthquakes and seismic swarms along the Chilean subduction zone (38°–18° S). Pure Appl Geophys 173(5):1431–1449. https://doi.org/10.1007/s00024-016-1280-5

Moreno MS, Melnick D, Rosenau M, Baez J, Klotz J, Oncken O, Tassara A, Chen J, Bataille K, Bevis M, Socquet A, Bolte J, Vigny C, Brooks B, Ryder I, Grund V, Smalley B, Carrizo D, Bartsch M, Hase H (2012) Toward understanding tectonic control on the Mw 8.8 2010 Maule Chile earthquake. Earth Planet Sci Lett 321–322:152–165. https://doi.org/10.1016/j.epsl.2012.01.006

Müller RD, Roest WR, Royer J, Gahagan LM, Sclater JG (1997) Digital isochrons of the world's ocean floor. J Geophys Res 102:3211–3214

Müller RD, Sdrolias M, Gaina C, Roest WR (2008) Age, spreading rates and spreading symmetry of the world's ocean crust. Geochem Geophys Geosyst 9:Q04006. https://doi.org/10.1029/2007GC001743

New MG, Hulme M, Jones PD (1999) Representing 20th century spacetime climate variability. I: Development of a 1961–1990 mean monthly terrestrial climatology. J Climate 12:829–856

New M, Lister D, Hulme M, Makin I (2002) A high-resolution data set of surface climate over global land areas. Climate Res 21:1–25

Oncken O, Hindle D, Kley J, Elger K, Victor P, Schemmann K (2006) Deformation of the Central Andean upper plate system—facts, fiction, and constraints for plateau models. In: Oncken O, Chong G, Franz G, Giese P, Götze H-J, Ramos VA, Strecker MR, Wigger P (eds) The Andes—active subduction orogeny. Frontiers in earth science series. Springer, Berlin Heidelberg New York, pp 3–28

Pail R, Bruisma S, Migliaccio F, Förste C, Goiginger H, Schuh WD, Höck E, Reguzzoni M, Brockmann JM, Abrikosov O, Veicherts M, Fecher T, Mayrhofer R, Krasbutter I, Sansò F, Tscherning CC (2011) First GOCE gravity field models derived by three different approaches. J Geod 85:819–843

Pavlis NK, Holmes SA, Kenyon SC, Factor JK (2008) An earth gravitational model to degree 2160: EGM2008, paper presented at the 2008 General Assembly of the European Geosciences Union, Vienna, Austria

Pavlis NK, Holmes SA, Kenyon SC, Factor JK (2012) The development and evaluation of the Earth Gravitational Model 2008 (EGM2008). J Geophys Res 117:B04406. https://doi.org/10.1029/2011JB008916

Ramos VA, Ghiglione M (2008) Tectonic evolution of the Patagonian Andes. Developments in quaternary sciences vol. 11: ISSN 1571-0866

Ranero C, von Huene R, Weinrebe W, Reichert C (2006) Tectonic processes along the Chile convergent margin. In: Oncken O, Chong G, Franz G, Giese P, Götze H-J, Ramos VA, Strecker MR, Wigger P (eds) The Andes—active subduction orogeny. Frontiers in earth science series. Springer, Berlin Heidelberg New York, pp 91–121

Sandwell DT, Smith WHF (1997) Marine Bouguer anomaly from Geosat and ERS-1 satellite altimetry. J Geophys Res 102:10039–10050

Scholl DW, Christensen MN, von Huene R, Marlow MS (1970) Peru-Chile trench sediments and sea-floor spreading. Geol Soc Am Bull 81:1339–1360

Schweller WJ, Kulm LD, Prince RA (1981) Tectonics structure, and sedimentary framework of the Perú-Chile Trench. In: Kulm LD, et al. (eds) Nazca Plate: Crustal formation and Andean convergence. Mem Geol Soc Am 154: 323–349

Sick C, Yoon MK, Rauch K, Buske S, Lüth S, Araneda M, Bataille K, Chong G, Giese P, Krawczyk C, Mechie J, Meyer H, Oncken O, Reichert C, Schmitz M, Shapiro S, Stiller M, Wigger P (2006) Seismic images of accretive and erosive subduction zones from the Chilean margin. In: Oncken O, Chong G, Franz G, Giese P, Götze H-J, Ramos VA, Strecker MR, Wigger P (eds) The Andes—active subduction orogeny. Frontiers in earth science series. Springer, Berlin Heidelberg New York, pp 147–169

Tassara A, Götze H, Schmidt S, Hackney R (2006) Three dimensional density model of the Nazca plate and the Andean continental margin. J Geophys Res 111:B09404. https://doi.org/10.1029/2005JB003976

Tebbens SF, Cande SC (1997) Southeast Pacific tectonic evolution from early Oligocene to present. J Geophys Res 102(B6):12061–12084

Tilmann F, Zhang Y, Moreno M, Saul J, Eckelmann F, Palo M, Deng Z, Babeyko A, Chen K, Baez JC, Schurr B, Wang R, Dahm T (2016) The 2015 Illapel earthquake, central Chile: a type case for a characteristic earthquake? Geophys Res Lett 43:574–583. https://doi.org/10.1002/2015GL066963

Uieda L, Ussami N, Braitenberg CF (2010) Computation of the gravity gradient tensor due to topographic masses using tesseroids, Eos Trans. AGU, 91(26), Meet. Am. Suppl., Abstract G22A-04. http://code.google.com/p/tesseroids/

Uieda L, Barbosa V, Braitenberg C (2016) Tesseroids: Forward-modeling gravitational fields in spherical coordinates. Geophysics: F41–F48. https://doi.org/10.1190/geo2015-0204.1

Vietor T, Echtler H (2006) Episodic Neogene southward growth of the Andean subduction orogen between 30° S and 40° S—plate motions, mantle flow, climate, and upper-plate structure. In: Oncken O, Chong G, Franz G, Giese P, Götze H-J, Ramos VA, Strecker MR, Wigger P (eds) The Andes—Active Subduction Orogeny. Frontiers in earth science series, Vol 1. Springer, Berlin Heidelberg New York, pp 375–400

Völker D, Wiedicke M, Ladage S, Gaedicke C, Reichert C, Rauch K, Kramer W, Heubeck C (2006) Latitudinal variation in sedimentary processes in the Peru-Chile trench off central Chile. In: Oncken O, Chong G, Franz G, Giese P, Götze H-J, Ramos VA, Strecker MR, Wigger P (eds), The Andes—active subduction orogeny. Frontiers in Earth Science Series, Part II. Springer, Berlin Heidelberg New York, pp 193–216. https://doi.org/10.1007/978-3-540-48684-8_9

von Huene R, Corvalán J, Flueh ER, Hinz K, Korstgard J, Ranero CR, Weinrebe W, CONDOR scientists (1997) Tectonic control of the subducting Juan Fernández Ridge on the Andean margin near Valparaiso, Chile. Tectonics 16(3):474–488

von Huene R, Scholl DW (1991) Observations at convergent margins concerning sediment subduction, subduction erosion, and the growth of continental crust. Rev Geophys 29:279–316

von Huene R, Weinrebe W, Heeren F (1999) Subduction erosion along the north Chile margin. J Geodyn 27:345–358

Wild-Pfeiffer F (2008) A comparison of different mass element for use in gravity gradiometry. J Geod 82:637–653. https://doi.org/10.1007/s00190-008-0219-8

Wienecke S (2006) A new analytical solution for the calculation of flexural rigidity: significance and applications. PhD Thesis, Free University Berlin, Berlin, p 126. World Wide Web Address: http://www.diss.fuberlin.de/

Whittaker J, Goncharov A, Williams S, Müller RD, Leitchenkov G (2013) Global sediment thickness dataset updated for the Australian-Antarctic Southern Ocean. Geochem Geophys Geosystems 14:3297–3305. https://doi.org/10.1002/ggge.20181

Yañez GA, Ranero CR, von Huene R, Díaz J (2001) A tectonic interpretation of magnetic anomalies across a segment of the convergent margin of the Southern Central Andes (32°–34° S). J Geophys Res 106:6325–6345

Zachos J, Pagani N, Sloan L, Thomas E, Billups K (2001) Trends, rhythms, and aberrations in the global climate 65 Ma to present. Science 292:686–693

Part II
The Paleozoic Evolution of the Chilean-Argentinean Margin

Paleogeographic and Kinematic Constraints in the Tectonic Evolution of the Pre-Andean Basement Blocks

Augusto E. Rapalini, Silvana E. Geuna, Pablo R. Franceschinis and Cecilia M. Spagnuolo

Abstract Formation of the south Central Andes basement cannot be unlinked from the final stages of Western Gondwana amalgamation. This process is still matter of much debate and continuous research. In the last two decades, a picture has emerged that suggests that the accretion and/or displacement of discrete crustal blocks in the Ediacaran and Early Paleozoic was a major process that finally configured the basement upon which three hundred million years later the Andean orogen developed. A main crustal fragment of the basement of central Argentina is the Pampia terrane. Recent, but still scarce, paleomagnetic results on Cambrian rocks from this terrane show anomalous pole positions whose interpretation is still ambiguous. Whether Pampia is a cortical fragment linked to the Amazonian craton that collided against the Rio de la Plata craton coetaneously with the closure of the hypothetical Clymene Ocean, whether it was attached previously and subsequently collided with a western crustal sliver (Western Pampia) associated with the Arequipa-Antofalla terrane, or whether it was a fragment detached from southern Kalahari that was displaced along the margin of the Río de la Plata craton in the Cambrian, cannot be unambiguously solved due to the scarcity of paleomagnetic data. It is generally accepted that the very long magmatic belt along the western boundary of Pampia in the Ordovician (the Famatinian arc) developed on stretched continental crust. Systematic large clockwise rotation of Ordovician magmatic and sedimentary rocks along this belt has been confirmed in recent years, but the originally para-autochthonous rotated terrane model has been replaced by one in which a systematic pattern of small crustal block rotations accompanied deformation due to collision of the allochthnous Cuyania terrane. The very large counter-clockwise rotations in the Western Puna (Antofalla terrane) of Chile and Argentina are still interpreted by some authors as evidence of closure of a V-shaped back-arc basin between Antofalla (and Arequipa?) and the Western Gondwana margin in the Late Ordovician. However, the interpretation of such rotations as due to tectonic

A. E. Rapalini (✉) · S. E. Geuna · P. R. Franceschinis
IGEBA-CONICET, Universidad de Buenos Aires, Buenos Aires, Argentina
e-mail: rapalini@gl.fcen.uba.ar

C. M. Spagnuolo
CONICET, Universidad Nacional de Tucumán, San Miguel de Tucumán, Argentina

A. Folguera et al. (eds.), *The Evolution of the Chilean-Argentinean Andes*,
Springer Earth System Sciences, https://doi.org/10.1007/978-3-319-67774-3_4

escape of the Antofalla block during Cuyania collision may be more compatible with geochemical signatures that suggest crustal links between the Sierras Pampeanas and the Puna basements. Paleomagnetic support for the Laurentian origin of the Cuyania terrane has endured the significant improvement of the apparent polar wander path for Gondwana in the Ediacaran–Cambrian. Alternative models suggesting a para-autochthonous origin of Cuyania are difficult to reconcile with different lines of evidence (biogeographic, isotopic) and are not supported by available paleomagnetic data for this terrane. Whether Chilenia is a truly allochtonous terrane or a fragment of Cuyania that rifted apart to collide back in the Devonian is still controversial. Unfortunately, paleomagnetic data are not available to test these models yet.

Keywords Paleomagnetic data · Paleozoic · Amalgamation · Allochtonous Para-authochthonous

1 Introduction

The development of the Andean orogen must have had a major control associated with the tectonic and geologic evolution of its basement before the Andes became into existence. Unraveling such history is a daunting task considering the extension of the Andean chain and the long and complex history that the main basement blocks apparently had. Therefore, it is not surprising that after several decades of research in many different disciplines, the pre-Andean evolution of its basement is still matter of many controversies and opposing views. Starting points are never easy to be established in science (or in many other human activities) as almost always precursor ideas or models are to be found, and the same may apply to the "plate tectonics" view on the pre-Andean evolution of South America. However, the publication by Ramos (1988) may be taken as a landmark in the slow but steady progress in our knowledge of this complex and fascinating history. In that work, the main building pieces of the South American basement were depicted in a way that has significantly influenced most of the research that was carried out in the following three decades. The idea that continents are generally built by a complex mosaic of several crustal blocks, either cratons or terranes (sensu Monger 1984), although not novel at those times, was presented in a simple fashion. However, since then, not all the proposed terranes could stand the scientific scrutiny and also new ones were proposed, while others saw their boundaries changed or disputed. Nevertheless, the idea of a composite South American basement built by the collision and tectonic juxtaposition of crustal blocks with independent geologic evolutions and kinematic histories has remained. In Fig. 1, we present a sketch of the distribution of the major terranes that constitute the basement of southern South America and that are the subject of this chapter. No sketch is able to satisfy all researchers, since definition of the terranes varies according to different authors, as well as their boundaries, which remain mostly covered, resulting in different

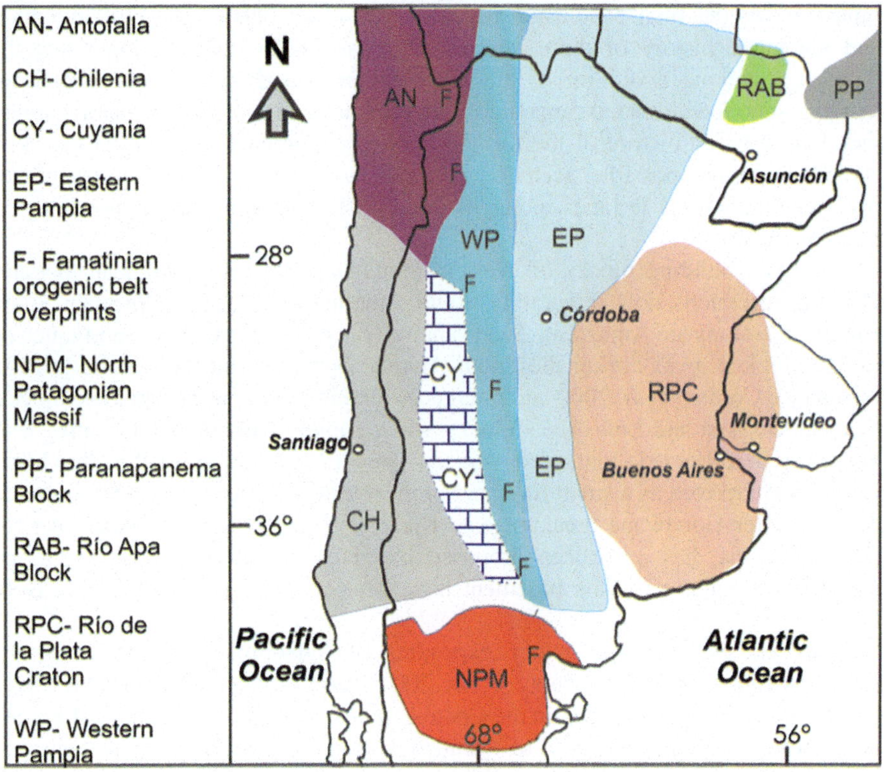

Fig. 1 Main crustal blocks that constitute the basement of central southern South America. Modified and simplified from Rapalini (2005), Ramos et al. (2010), Rapela et al. (2015)

interpretations of their respective origin and evolution. Recent use of geophysical techniques to identify and trace the terrane boundaries (Chernicoff and Zappettini 2004; Favetto et al. 2008, 2015; Ramé and Miró 2011; Álvarez et al. 2012; Peri et al. 2013) contributed substantially to locate some of the major crustal discontinuities that constitute these boundaries.

To unravel the kinematic and paleogeographic history of the Pre-Andean basement blocks, different disciplines must be involved, such as geochronology, petrology, paleontology, stratigraphy, geophysics, and paleomagnetism, among others. The sometimes contradictory interpretations that arise from different disciplines mainly due to the complex histories of these Andean basement blocks, their variable exposure quality and extension and the pervasive Andean deformation that affects most of them. Further and better quality data generally tend to resolve such controversies, although achieving such goal is not always straightforward. Rapalini (2005) presented a concise review on these matters with the main focus placed on the constraints posed by the paleomagnetic data known at those times. In a way, this contribution may be considered an update from that publication. As stated at that time, besides the final conclusion regarding the exact paleogeographic evolution and kinematic history of these terranes, there seems to be little doubt that the Paleozoic tectonic evolution of the southwestern Gondwana (South American) margin cannot be ultimately separated from the final stages of Gondwana assembly; therefore, a broad picture of the formation of this supercontinent should take into consideration the possible accretion and displacement of allochthonous and/or para-authochthonous terranes along its southwestern margin during the Early to Late Paleozoic.

The main building blocks of Western Gondwana are a few Archean to Early Proterozoic cratons now integrating South America and Africa. The main South American cratons are Amazonia, Sao Francisco (as part of the Congo-Sao Francisco craton), broken apart during the South Atlantic opening, and the Río de la Plata craton (see Cordani et al. 2000 and references therein). Other minor cratonic blocks, like the Río Apa and Luis Alves blocks, show complex and disputable links with Amazonia and Río de la Plata, respectively. The Sao Luis craton, on the other hand, has been interpreted as a small fragment of the Western African craton left in South America after Gondwana breakup (e.g., Klein et al. 2005). Other blocks, like the Paraná-Panema, has only been proposed based on geophysical evidence as no recognizable outcrops of its basement have been reported (Mantovani and Brito Neves 2005).

With the exception of minor exposures of Paleoproterozoic rocks in the Arequipa block in Perú (Loewy et al. 2004), no Archean or Paleoproterozoic basement has been recognized in the Central Andes. Most of its basement, if not all, is therefore probably composed of "younger" accreted terranes. Their tectonic and kinematic history undoubtedly conditioned the way the Andean orogen developed in Meso-Cenozoic times.

2 Pampia

To the west of the Río de la Plata craton, there is the Pampia terrane (Ramos et al. 2010 and references therein) extending about 1500 km long and circa 250 km wide (without palinpastic restoration, Fig. 1). This terrane represents the basement of most of the Sierras Pampeanas of Argentina. Its structure and lithologic content is better exposed along the Sierras de Córdoba around 32° S (Martino et al. 1995; Baldo et al. 1996), but it is inferred that it extends over a thousand kilometers to the north, up to the Tucavaca Aulacogen in Bolivia (around 18° S, Ramos et al. 2010). On the other hand, it continues several hundred kilometers south into La Pampa province (38° S, Chernicoff and Zappettini 2003). Its eastern boundary has been determined with some certainty thanks to magnetotelluric (Favetto et al. 2008, 2015; Orozco et al. 2013; Peri et al. 2013, 2015) and gravimetric surveys (Ramé and Miró 2011; Sigismondi and Fantin 2014) as well as to the findings of Paleoproterozoic rocks in samples from boreholes to the East of the Sierras de Córdoba by Rapela et al. (2007). Such findings permitted to extend the Río de la

Plata craton to the western margin of the Chaco-Pampean plains. The magnetotelluric data allowed tracing a nearly rectilinear N–S contact with the Río de la Plata craton south of Córdoba (Fig. 1). To the north, the contact must be unexposed basement, which should constitute either a northern continuation of the Río de la Plata craton or the unexposed and hypothetical block of Paraná-Panema (Mantovani and Brito Neves 2005). Peri et al. (2013) and Favetto et al. (2015) have found evidence that the eastern boundary takes a NE trend in northern Argentina. To the south, the boundary between Pampia and Río de la Plata has been proposed on the basis of aeromagnetic anomalies (Chernicoff and Zappettini 2004).

The basement of Pampia is mainly represented by thousands of meters of metasedimentary rocks with some intercalations of metavolcanics from the Puncoviscana Fm (e.g., Aceñolaza and Toselli 1981; Escayola et al. 2011) spanning the late Ediacaran and Early Cambrian. These rocks are better exposed in NW Argentina, while those representing deeper structural levels of the same basin are exposed toward the south (Rapela et al. 1998a). Early Cambrian magmatism recognized in the Eastern Cordillera and Eastern Sierras Pampeanas is known as the Pampean magmatic arc, with an age peak in the Early Cambrian (ca. 530 Ma; Rapela et al. 1998a, b; Omarini et al. 2008; Hauser et al. 2011; Iannizzotto et al. 2013; von Gosen et al. 2014; Dahlquist et al. 2016).

The tectonic framework for the development of the Puncoviscana Fm (and equivalent and related units) has been interpreted in several different ways. Originally, it was understood as a passive margin depositional sequence (Jezek et al. 1985) on the western margin of the Río de la Plata craton, with sediment sources located mostly to the east. This interpretation can be ruled out considering that the Puncoviscana (s.l.) sediments are restrained within the Pampia terrane. Conversely, Keppie and Bahlburg (1999) considered that this unit was deposited in a foreland basin associated with a Pampean orogen developed to the east. Omarini et al. (1999) and later Ramos (2008) proposed that the Puncoviscana sediments were part of the infill of a rift-related basin that resulted from the rifting between Pampia and the Antofalla terrane. Collo et al. (2009) proposed a similar tectonic setting but between the so-called Cordoba arc in the east and the Pampia terrane on the west. In Rapela et al. (2007), the Puncoviscana Fm was apparently deposited as platform sediments on a passive continental margin in southern Kalahari. Finally, Escayola et al. (2011) proposed a more complex evolution in which this sequence developed in an arc-trench gap between the Arequipa-Antofalla block and West Gondwana (Río de la Plata + Amazonia) followed by foreland deposits after their collision at around 530 Ma. In this model, Pampia itself loses its identity as a single terrane.

The Puncoviscana Fm was tightly folded in the Early Cambrian (Tilcaric event, see Astini 2003) showing a first-class angular unconformity with the Middle to Late Cambrian clastic sediments of the Mesón Group in NW Argentina. This deformation event caused crustal thickening and high-grade metamorphism to the south (ca. 525 Ma). Immediately following that, isothermal uplift and widespread low-P anatexis took place in the Sierras de Córdoba at around 520 Ma (Rapela et al. 1998b). The most straightforward explanation for the data presented so far would be a collision (Rapela et al. 1998a, b; von Gosen and Prozzi 2010) or transpressive "docking" (Iannizzotto et al. 2013) of Pampia against the Río de la Plata craton, dated as Early Cambrian. Ramos (1988), Kraemer et al. (1995), and Escayola et al. (2007) have argued in favor of a collision of Pampia with the Río de la Plata craton. A belt of ultramafic and mafic rocks dated as Late Neoproterozoic in the Sierras de Córdoba (Ramos et al. 2000, 2010) extends discontinuously for 150 km and has been interpreted as an ophiolite (Villar 1975) that may constitute relicts of a suture of Pampia-Río de la Plata collision (Kraemer et al. 1995; Escayola et al. 1996). A second belt of ophiolites, located some 50 km to the east, shows back-arc geochemical signature (Escayola et al. 1996) and has been interpreted as remnants of a back-arc basin closed on the western margin of the Río de la Plata craton (Ramos et al. 2000; Escayola et al. 2007).

However, the Pampean orogen suffered the effects of later deformation events along large-scale shear zones in the Eastern Sierras Pampeanas (e.g., Martino 2003; Whitmeyer and Simpson 2003), which might evidence a complicated history for Pampia accretion to Gondwana margin. In addition, provenance studies by Schwartz and Gromet (2004), Escayola et al. (2007, 2011), Rapela et al. (2007, 2016), and Adams et al. (2011) also show a complex pattern for U–Pb ages on detrital zircons in different Neoproterozoic to Cambrian sedimentary and metasedimentary rocks from the Sierras Pampeanas (Pampia). Rapela et al. (2016) remarked a striking difference in the provenance pattern of pre-Pampean (pre-525 Ma) rocks from the Eastern Sierras Pampeanas with respect to those from the Western Sierras Pampeanas. The first show a clear bimodal pattern with peaks at ca. 600 and 1000 Ma, which according to these authors, strongly suggest provenance from the Natal-Namaqua Belt in southern Africa or the Dom Feliciano Belt in Uruguay and southern Brazil. On the other hand, those in the Western Sierras Pampeanas show a major peak between 1030 and 1330 Ma that correlates with the Arequipa-Antofalla block and the Grenville and the southern Granite-Rhyolite provinces of southeastern Laurentia. According to these authors, such differences between Eastern and Western Pampia (Fig. 1) indicate that the latter was accreted in Cambrian times by a collision that caused the Pampean Orogeny, while Eastern Pampia was probably still attached to Kalahari (Rapela et al. 2007). Furthermore, similarities between the Western Pampia and the Arequipa-Antofalla terrane led these authors to propose the occurrence of a single block called MARA which includes both land masses. Escayola et al. (2011) presented a broadly similar scenario in which Western Pampia and the Arequipa-Antofalla terrane would constitute a single block detached from Laurentia in the middle to late Ediacaran.

Rapalini (2005) mentioned that at those times no paleomagnetic data were available to constrain the paleogeographic and kinematic evolution of Pampia. In the last decade, however, some paleomagnetic studies were carried out on late Ediacaran to Cambrian rocks of this terrane and also in the adjacent Río de la Plata craton. Accordingly, the final closure of an ocean in Cambrian times between Pampia and the Río de la Plata craton, recorded in magmatism along the Eastern Sierras de Cordoba (Escayola et al. 2007; von Gosen and Prozzi 2010;

D'Eramo et al. 2013), is consistent with the paleogeographic model of a large Ediacaran–Cambrian Ocean across South America ("Clymene Ocean") as proposed by Trindade et al. (2006) and Tohver et al. (2010, 2012). This hypothetical ocean separated Amazonia, Rio Apa, and Pampia blocks on one side from the Río de la Plata and Congo-Sao Francisco cratons on the other. This model has been strongly disputed by Cordani et al. (2013) who suggested that no such ocean existed since at least 600 Ma. Ediacaran to Cambrian paleomagnetic data from the Río de la Plata and Congo-Sao Francisco cratons (Rapalini et al. 2015) are consistent with the existence of the Clymene Ocean at ca. 570 Ma (Fig. 2a), although alternative interpretations are still possible due to the lack of coetaneous data from Amazonia. Figure 2c illustrates the location of the proposed suture line indicating the closure of the late Ediacaran Clymene Ocean separating Amazonia, Rio Apa, and Pampia from the cratonic blocks of proto-Western Gondwana. Final closure of this ocean should have occurred in Early Cambrian times (Tohver and Trindade 2014).

Spagnuolo et al. (2008a) were the first to publish a study on the early Late Cambrian Campanario Fm, which integrates the Mesón Group that unconformably overlies the Puncoviscana Fm. This was followed by new studies from the same unit at a different locality as well as preliminary results from the earliest Ordovician

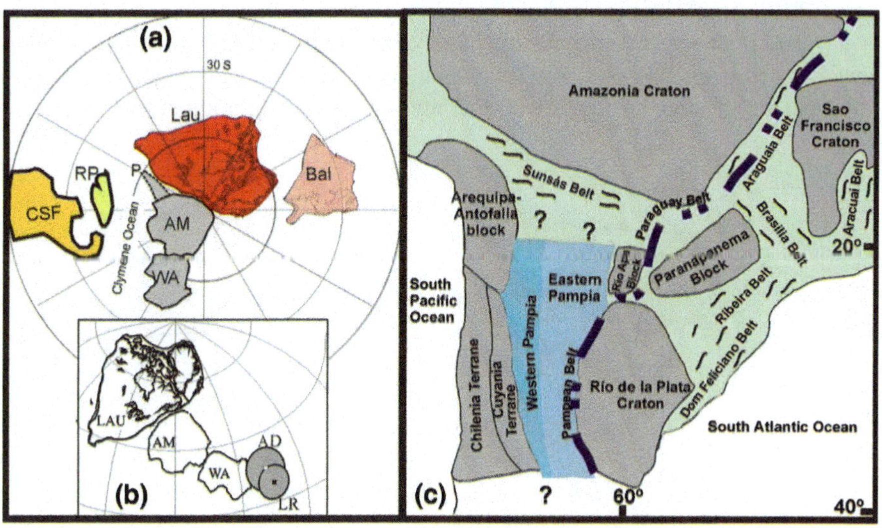

Fig. 2 **a** Mid-Ediacaran paleomagnetically controlled paleogeographic reconstruction of the main Western Gondwana crustal blocks (CSF: Congo-Sao Francisco, RP: Río de la Plata, WA: Western Africa; AM: Amazonia, Lau: Laurentia, Bal: Baltica, P: Pampia). The position of Pampia is shown although no independent paleomagnetic data is available for that time. **b** Paleomagnetically controlled paleogeographic reconstruction between Amazonia, West Africa, and Laurentia for the Early Ediacaran (AD: Adma Diorite pole for WA, LR: Long Range Dykes, for Lau). Both figures modified from Rapalini et al. (2015). **c** Main tectonic blocks of South America. Dark blue dashed line indicates location of the proposed suture of the Late Ediacaran to Early Cambrian Clymene Ocean (modified from Tohver and Trindade 2014)

Santa Rosita Fm (Spagnuolo et al. 2012). Very recently, Franceschinis et al. (2016) published a third study on the Campanario Fm and the first results on volcanic units intercalated within the Puncoviscana Fm in its type locality. The three studies report apparent anomalous pole positions for the Campanario Fm. A non-unique solution for those anomalies has been interpreted (Franceschinis et al. 2016), since Andean tectonic rotations affecting the three sampling localities (located several tens of kilometers apart from each other) cannot be definitely ruled out. However, the data allow the alternative interpretation of a fast displacement of Pampia along the Río de la Plata craton margin, from close to the Kalahari craton to its present-day relative position during the Late Cambrian (Spagnuolo et al. 2012). This model is similar to that proposed by Rapela et al. (2007), although the displacement of Pampia is somewhat younger (Fig. 3). Paleomagnetic results from the volcanic and pyroclastic rocks intercalated in the Puncoviscana Formation and dated as 537 Ma (Escayola et al. 2011) only yielded two virtual geomagnetic poles which fall relatively close to the most likely late Ediacaran–Cambrian position on the apparent polar wander path for the Río de la Plata craton (Rapalini et al. 2015), although we are aware that they are not enough to test any specific paleogeographic reconstruction for Pampia. Paleomagnetic studies were reported on plutonic rocks intruding the metamorphic basement to the south, in Sierras de Córdoba; they are the Calmayo tonalite (Geuna et al. 2010), dated as Early Cambrian by D'Eramo et al. (2013) and Devonian Achala batholith (Geuna et al. 2008, updated by Geuna et al. 2011). The resultant paleomagnetic poles are not fully consistent with the coeval paleomagnetic poles for Gondwana. This inconsistency might be related to improper knowledge of the paleohorizontal surface at the moment of remanence

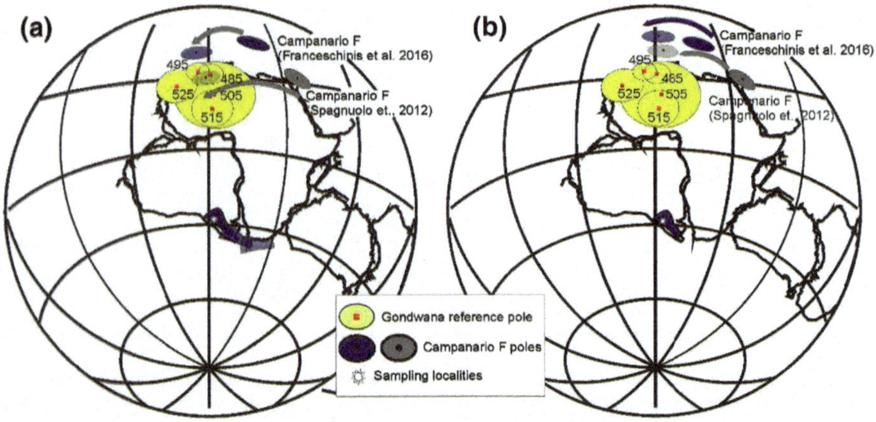

Fig. 3 **a** Anomalous positions of the paleomagnetic poles from the early late Cambrian Campanario Fm obtained by Spagnuolo et al. (2012) and Franceschinis et al. (2016) and their rotation into a better concordance with the mean Cambrian to Ordovician reference poles for Gondwana following different displacements of Pampia along the Western Gondwana margin. **b** Idem (a) but considering in situ rotations due to Andean tectonics. Numbers indicate mean age for reference poles in Ma. Taken from Franceschinis et al. (2016)

acquisition, which at this time precludes giving a definite tectonic interpretation to these results.

The models that involve a strike-slip displacement of Pampia along the Río de la Plata margin, as proposed by Rapela et al. (2007) and Spagnuolo et al. (2012), are contradicted by the model of Ramos et al. (2010) in which Pampia was attached to Amazonia in a similar position as today, since collision in the Meso-Proterozoic. Spagnuolo et al. (2012) further speculated on the possibility that the displacement of Pampia may have involved the Arequipa-Antofalla block, for which the only Cambrian paleomagnetic pole (the Choschas pluton, Forsythe et al. 1993) shows a significant paleolatitude anomaly.

As described above, the tectonic setting of the basement of the Sierras Pampeanas, the proper constitution of Pampia, its paleogeographic, tectonic, and kinematic evolution are still matter of hot debate. Yet, some consensus has been reached in relation to the occurrence of a major boundary separating two very different basements running close to the eastern margin of the Eastern Sierras Pampeanas. Those two different blocks should have attained their respective present-day relative position probably in the Cambrian.

To the west, Pampia confines with the Laurentian-derived Cuyania terrane through a suture located along the Valle Fértil megashear (Introcaso et al. 2004) which is characterized by a belt of mylonitic rocks with conspicuous Early Paleozoic deformation (e.g., Ramos et al. 1998), related to docking of the Cuyania terrane.

3 Famatinian Magmatic Arc

An outstanding geologic feature in northwest Argentina is a N–S trending belt of Lower Ordovician magmatic rocks (F, Fig. 1) that comprises the Famatina System, continues along the Sierras Pampeanas to the south (Chernicoff et al. 2010) and may have reached the North Patagonian Massif (Pankhurst et al. 2006) (see Chap. "The Famatinian Orogen Along the Protomargin of Western Gondwana: Evidence for a Nearly Continuous Ordovician Magmatic Arc Between Venezuela and Argentina"). To the north, Lower Ordovician magmatism is present along two belts: the Western and Eastern Puna Eruptive Belts ("Faja eruptiva"), lying probably on top of the Pampia terrane and the Antofalla block, respectively. This volcanism has been assigned to a magmatic arc on the basis of its geochemical signature and geotectonic setting (Coira et al. 1982; Ramos 1986; Mannheim 1993; Toselli et al. 1996; Saavedra et al. 1998; Rapela et al. 1998b; Miller and Söllner 2005; Otamendi et al. 2009; Grosse et al. 2011, and many others). Since the work of Astini et al. (1995), the Famatina magmatic arc has been generally considered as the product of east-dipping (present-day coordinates) oceanic crust subduction under Western Gondwana margin associated with the approach of the Cuyania terrane prior to its collision.

Conti et al. (1996) proposed that the Famatina Belt constituted a para-authochthonous terrane that rotated some 50° clockwise before accretion to the

Western Gondwana margin. According to these authors, this event preceded the Cuyania collision. This model arose from consistent paleomagnetic directions from four localities of Early Ordovician age for the Eastern Puna, Western Pampean Ranges, and the Famatina System. The same magnetization direction was found in different lithologies carried by different magnetic minerals. The Early Ordovician paleomagnetic poles from all these localities show no paleolatitudinal anomaly with respect to Gondwana, but a large declination anomaly suggesting such clockwise rotation. This model implies a very large V-shaped back-arc basin that should have developed oceanic crust. This hypothetical basin should be wider toward the north (present-day coordinates). The model portrayed by Conti et al. (1996) for the Famatina arc is inconsistent with geochemical evidence that strongly support the development of such magmatic arc on continental crust (Pankhurst et al. 1998; Saavedra et al. 1998; Rapela et al. 1998b; Astini 2003; Miller and Söllner 2005; Viramonte et al. 2007, and many others). Furthermore, several evidences suggest that the Famatina Complex extends northward into the Western Puna and not along the Eastern Puna Eruptive Belt. The latter has been interpreted as having been formed in an extensional or transtensional setting (Coira et al. 1999, 2009), probably in a back-arc environment, instead of a magmatic arc. This signature, however, seems to be well established along the belt in the Western Puna.

Rapalini (2005) acknowledged the inconsistency between the paleomagnetically based tectonic model for the Famatina Belt and the geochemical evidence. This author claims that new and more systematic paleomagnetic data were needed for a definite answer to such controversy.

Spagnuolo et al. (2008b) published a detailed paleomagnetic study on late Early to Middle Ordovician volcanics of the Molles Formation and Cerro Morado Group in the Famatina System, about 100 km to the south of the southernmost locality studied by Conti et al. (1996). A high-quality pre-tectonic remanence found in these rocks confirmed a large (39°) cw rotation with no significant paleolatitude anomaly, reproducing the results previously obtained in other localities by Conti et al. (1996). Furthermore, Spagnuolo et al. (2008b) were able to determine that the rotation found was pre-Permian. Figure 4 illustrates the anomalous pole positions found at different localities and rocks of Ordovician age along the Famatina Belt and the Eastern Puna. More recent paleomagnetic data by Rapalini et al. (2010) on the Ordovician Sierra del Valle Fértil Magmatic Complex, exposed about 200 km south from the former study location, support a clockwise rotation of the area of similar magnitude than those reported by Conti et al. (1996) and Spagnuolo et al. (2008b). These new results confirm the robustness of the Famatina (and Eastern Puna) Ordovician rotated poles. Although the consistent directions found at different localities located several hundred kilometers apart seem compelling in favor of a single coherent crustal block (Fig. 4), Spagnuolo et al. (2011) analyzed the viability of the para-autochthonous rotated terrane model of Conti et al. (1996), suggesting that it is unlikely on the basis of the geologic evidence. This includes (see appropriate references in Spagnuolo et al. 2011): (i) no evidence of a wide ocean-floored back-arc basin to the east of the Eastern Puna magmatic belt; (ii) different evidence pointing to the fact that the Famatina magmatic Belt continues northward into the Western Puna and not the

Eastern Puna; (iii) Eastern Puna magmatism not consistent with a magmatic arc environment; (iv) lack of high P–low T metamorphism in Eastern Puna that could be compatible with a subduction-accretion regime. In order to avert these problems, Spagnuolo et al. (2011) proposed a model of systematic rotations, in a domino-like style, that affected the Famatina Belt and the Eastern Puna produced by the collision of the Cuyania terrane in the Middle to Late Ordovician (Fig. 4c). The presence of a back-arc basin to the east (present-day coordinates) of the Famatina magmatic arc (Miller et al. 2003) probably provided the crustal discontinuity that bounded the rotational domains. It is well known that in situ rotations around vertical axes are a very effective method to achieve large amounts of crustal shortening (e.g., Garfunkel and Ron 1985). Abrupt displacement of the Famatina arc toward the east with respect to the Western Puna arc due to the Cuyania collision is evident (Fig. 4) and has been noted by Astini (2003), among others. This displacement may account for closure of a Famatinian back-arc basin as well as crustal shortening obtained by systematic rotations.

Spagnuolo et al. (2011) also provided an explanation for the Eastern Puna clockwise rotations and the large Western Puna ccw rotations (see below) that had been previously reported. Both deformational patterns have been associated with the Cuyania collision and interpreted as due to "escape tectonics" in a similar fashion to present-day relations between the Anatolian plate and the Arabian Peninsula (Adiyaman et al. 2001).

4 Western Puna (Antofalla Terrane)

The existence of Early to Middle Proterozoic crystalline rocks along the southern coast of Peru (Dalmayrac et al. 1977; Shackleton et al. 1979) that were generally referred to as the Arequipa Massif, produced speculations on its possible exotic origin (Ramos 1988). In several models, this block has been extended into northern Chile to include outcrops of Middle to Late Proterozoic metamorphic rocks (see Ramos 2008, for a recent update on this). Ramos (1988) labeled this enlarged Arequipa block as the Arequipa-Antofalla Massif. Outcrops in northern Chile are small and disconnected and younger than those in Peru, leading Bahlburg and Hervé (1997) to suggest that there may be two separate tectonic blocks: an Early to Middle Proterozoic Arequipa block in the North and a Late Proterozoic Antofalla block in the South (Figs. 1 and 2). This has been lately supported by Loewy et al. (2004) and accepted by Ramos (2008).

It is now accepted that the Ordovician magmatism along the Western Puna (Western Puna Eruptive Belt, Palma et al. 1986) is the northern continuation of the Famatinian arc (Astini 2003; Escayola et al. 2011) (see Chap. "The Famatinian Orogen Along the Protomargin of Western Gondwana: Evidence for a Nearly Continuous Ordovician Magmatic Arc Between Venezuela and Argentina"). This belt is located close to the eastern border of the Antofalla terrane (Bahlburg and Hervé 1997) and is separated from the Eastern Puna of Argentina by basic to

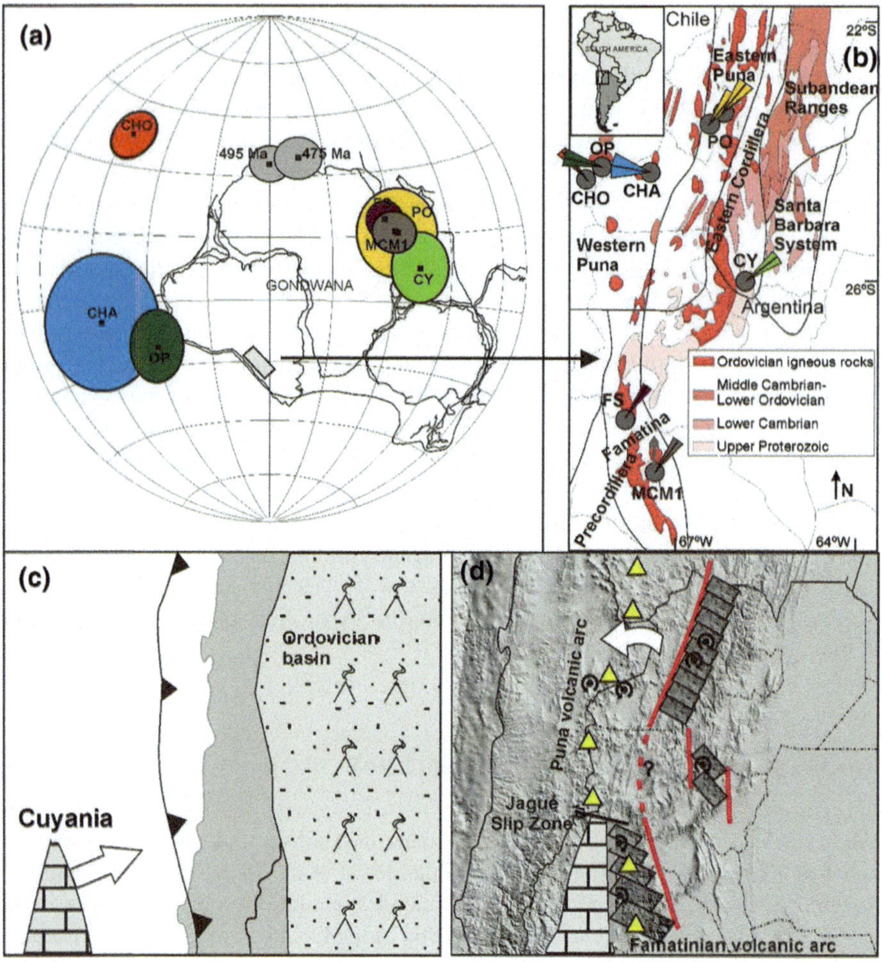

Fig. 4 **a** Paleomagnetic pole positions for different units and localities of Early Ordovician rocks from the Suri Formation (FS), Molles and Cerro Morado Formations (MCM1), Cuchiyaco Granodiorite (CY), and Eastern Puna volcanics (PO). Late Cambrian pole from the Choschas pluton (CHO), Early to Middle Ordovician Chachas pluton (CHA) and Middle to Late Ordovician Tucucaro, Alto del Inca and Tilopozo plutons (OP) from the Western Puna are also shown. Mean reference poles for Gondwana for 495 and 475 Ma are shown in gray. **b** Paleomagnetically determined rotations of localities with Late Cambrian to Late Ordovician rocks from the Famatina System, Eastern Cordillera, Eastern and Western Puna. Rotations are represented respect to the north–south directions with their respective confidence angles. **c** Tectonic sketch of the Early Ordovician Famatinian magmatic arc and approach of the Cuyania allochtonous terrane. **d** sketch of the complex pattern of rotations produced by the Cuyania collision against the SW Gondwana margin in the Middle Ordovician, according to Spagnuolo et al. (2011). All figures are taken from Spagnuolo et al. (2011)

ultrabasic rocks that were originally interpreted by some authors as ophiolites (Allmendinger et al. 1983; Blasco et al. 1996; Ramos 2000). However, geochemical signature of these rocks from the southern and western Puna (Zimmermann and Bahlburg 1999; Bock et al. 2000; Zimmermann et al. 2014) is interpreted as more consistent with an arc setting and recycling of older crust rather than accretion of juvenile exotic lithospheric material.

Forsythe et al. (1993), based on the first paleomagnetic data from Early Paleozoic rocks of the Antofalla terrane, in Sierra de Almeida, Chile, proposed the well-known "para-autochthonous" model for the Arequipa-Antofalla terrane, in which this block partially rifted apart from the proto-Western Gondwana margin in the late Proterozoic, opening a V-shaped small oceanic basin that broadened to the south (present-day coordinates). This basin was closed in the Ordovician through a ccw rotation of the whole block that re-accreted to the margin of Gondwana, producing the Late Ordovician deformational event known as the "Ocloyic" orogenic phase, well represented in the Puna. This rotation is supported on the anomalous pole positions of the Late Cambrian Choschas pluton (CHO, Fig. 4) and the Tucucaro, Alto del Inca and Tilopozo plutons (OP, Fig. 4) of Early to Middle Ordovician age in Chile. This Early Paleozoic paleomagnetic anomaly of the Antofalla terrane was confirmed by the studies of Rapalini et al. (2002) on the Early to Middle Ordovician volcanics of Vega Pinato, the Chachas magmatic Complex, and the Taca-Taca Batholith (469 ± 4 Ma, CHA pole, Fig. 4) in the Western Puna of Argentina. In a more recent review, Ramos (2008) supported the "para-autochthonous" model, but only for the Antofalla terrane. As already mentioned, Spagnuolo et al. (2011) proposed a radically different explanation for the anomalous paleomagnetic poles of Antofalla (Western Puna), which were attributed to Late Ordovician rotations associated with tectonic escape of the Antofalla block during collision of the Cuyania terrane (Fig. 4d). The paleomagnetic database of this region has not improved or enlarged in the last decade, and available dataset is not as robust as it should be, among other things due to the lack of precise paleohorizontal control on most sites in plutonic rocks. Spagnuolo et al. (2012) also considered the fact that the Late Cambrian CHO pole shows a significant paleolatitude anomaly, already noted by Rapalini et al. (1999), that could not be explained by the para-autochthonous model and claimed that it would disappear if the Antofalla (and Arequipa?) block would have been already attached to Pampia close to the Kalahari craton in the Cambrian (Fig. 3).

The tectonic interpretation of the available Early Paleozoic paleomagnetic data is therefore ambiguous. However, the striking systematic difference between the Late Cambrian to Ordovician poles from the Western and Eastern Puna (see Fig. 4a) strongly suggests a major crustal discontinuity between both domains that must have played a significant role in the Early Paleozoic construction of this part of the Western Gondwana continental margin. Such major tectonic discontinuity between Western and Eastern Puna magmatic belts had already been suggested by many authors, either as consumed oceanic crust (Coira et al. 1982; Allmendinger et al. 1983; Forsythe et al. 1993; Blasco et al. 1996) or as a major strike-slip fault (Coira et al. 1999). The western Puna magmatic belt is approximately coincident as well with a conspicuous positive gravity anomaly (Goetze and Kirchner 1997,

Omarini et al. 1999), interpreted as an Early Paleozoic or Late Proterozoic terrane boundary (Gangui and Götze 1996; Omarini et al. 1999). However, Lucassen et al. (2000, 2001), on the basis of isotopic and geochemical data, interpreted that all Early Paleozoic magmatic rocks in the region were produced from reworked continental crust of similar characteristics with minor addition of juvenile material. Zimmermann and Bahlburg (1999) and Astini (2003) also interpreted an autochthonous and ensialic Paleozoic tectonic evolution for this region, suggesting crustal continuity between the Arequipa-Antofalla terrane and the Sierras Pampeanas (Pampia). Escayola et al. (2011) and Rapela et al. (2016) have proposed that the Arequipa-Antofalla and Western Sierras Pampeanas (here considered as Western Pampia) constitute a single "ribbon-like" terrane of Laurentian affinities that was accreted against the proto-Gondwana margin during the Pampean (Early Cambrian) orogeny.

The conflicting evidence and models just mentioned indicate that a precise reconstruction of the paleogeographic and tectonic evolution of the Western Puna in the Early Paleozoic is still elusive. Higher quality and more systematic paleomagnetic results on Early Paleozoic rocks from this large region are clearly necessary.

5 Cuyania

The Cuyania terrane (see Ramos 2004 and references therein, Fig. 1) is the most studied and best example of well-defined allochthonous terrane accreted to SW Gondwana in Early Paleozoic times. Although there had been precursor ideas about the allochtonous nature of the Cambrian-Ordovician carbonate platform of the Argentine Precordillera (e.g., Dalla Salda et al. 1992, Benedetto 1993), it was Astini et al. (1995) who presented the first relatively complete evolutionary model that portrayed the Argentine Precordillera as a Laurentian-derived terrane that had rifted apart from Laurentia in Cambrian times, drifted across the Iapetus Ocean in the Late Cambrian to Middle Ordovician and collided with SW Gondwana margin in the Middle to Late Ordovician (Fig. 5). Thomas and Astini (1996) completed the model proposing that the origin of the terrane was in the Ouachita Embayment in SE Laurentia (Fig. 5a). This model soon received support from several different lines of evidence, from biogeographic considerations (Benedetto 1993, 1998, 2004; Benedetto et al. 1999, 2009; Cañas 1999; Vaccari and Waisfeld 2008, etc.) and stratigraphic interpretations (e.g., Astini 1998) to the very similar age and isotopic signature of the basement (Kay et al. 1996; Sato et al. 2000; Cingolani et al. 2005) with respect to the SE Laurentian basement provinces (Mosher 1998; Slagstad et al. 2009; among others). A full review of the whole evidence in favor of this model is out of the scope of this article, and the reader is referred to Ramos (2004) for the most relevant literature on this topic. Aceñolaza et al. (2002), Finney et al. (2002), and Finney (2007) challenged this model interpreting most of the evidence used to support the Laurentian origin of the Cuyania in favor of a Gondwanan origin.

In this alternative model, the terrane was portrayed as para-autochthonous and displaced along the margin for about two thousand kilometers during the Early Paleozoic (Fig. 5f), instead of drifting across the southern Iapetus Ocean. According to this alternative model, Cuyania originated in a kind of embayment along the Western Gondwana margin between Kalahari and East Antarctic cratons, as part of the hypothetical continent of SAFRAN (Fig. 5f).

Rapalini and Astini (1998) produced the first robust paleomagnetic data on Early Paleozoic rocks from the Argentine Precordillera. These were obtained on the late Early Cambrian Cerro Totora Fm. and strongly supported the Laurentian-derived model for Cuyania. A second paleomagnetic pole, obtained from early Late Ordovician clastic rocks of the Pavón Fm. exposed in the San Rafael Block (Rapalini and Cingolani 2004) provided a more ambiguous result as it could be interpreted either as supporting an already accreted Cuyania for the Late Ordovician or Cuyania as part of the "Texas plateau" (Dalziel 1997) and therefore still linked to Laurentia. No further Early Paleozoic paleomagnetic data from Cuyania have been published since then. However, significant changes in the Cambrian apparent polar wander path of Gondwana were reported in the last decade (e.g., Trindade et al. 2006) that may affect the original interpretation of the Cerro Totora paleomagnetic pole. This has been assessed and discussed in detail by Rapalini (2012) and is illustrated in Fig. 6.

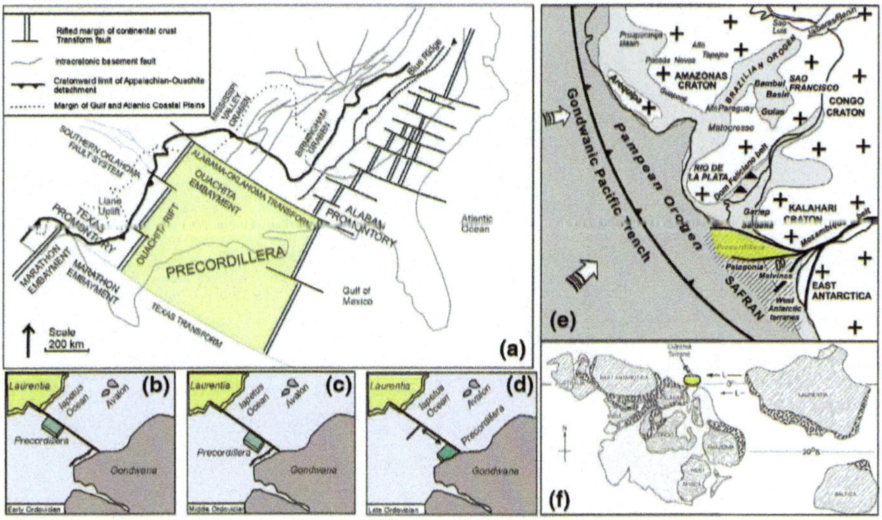

Fig. 5 **a** Tectonic sketch of southeastern North America with the modeled location of the Argentine Precordillera (or Cuyania) from Thomas and Astini (1996). **b–d** Kinematic sketch of the hypothetical transfer of this terrane from Laurentia to Gondwana during Ordovician times (redrawn and simplified from Astini et al. 1995). **e** Proposed original position of the Argentine Precordillera (or Cuyania) in the SAFRAN hypothetical microcontinent (modified from Aceñolaza et al. 2002). **f** Paleogeographic reconstruction for Early Cambrian times with Cuyania located along the Gondwana margin, according to the SAFRAN hypothesis (modified from Finney et al. 2005). Taken from Rapalini (2012)

Figure 6b illustrates the Late Ediacaran to Cambrian (570–500 Ma) apparent polar wander path for Gondwana (see Rapalini 2012, for a detailed discussion and list of poles) and the position of the late Early Cambrian pole for Cuyania according to its present position in South America (light blue pole) or as part of the SAFRAN

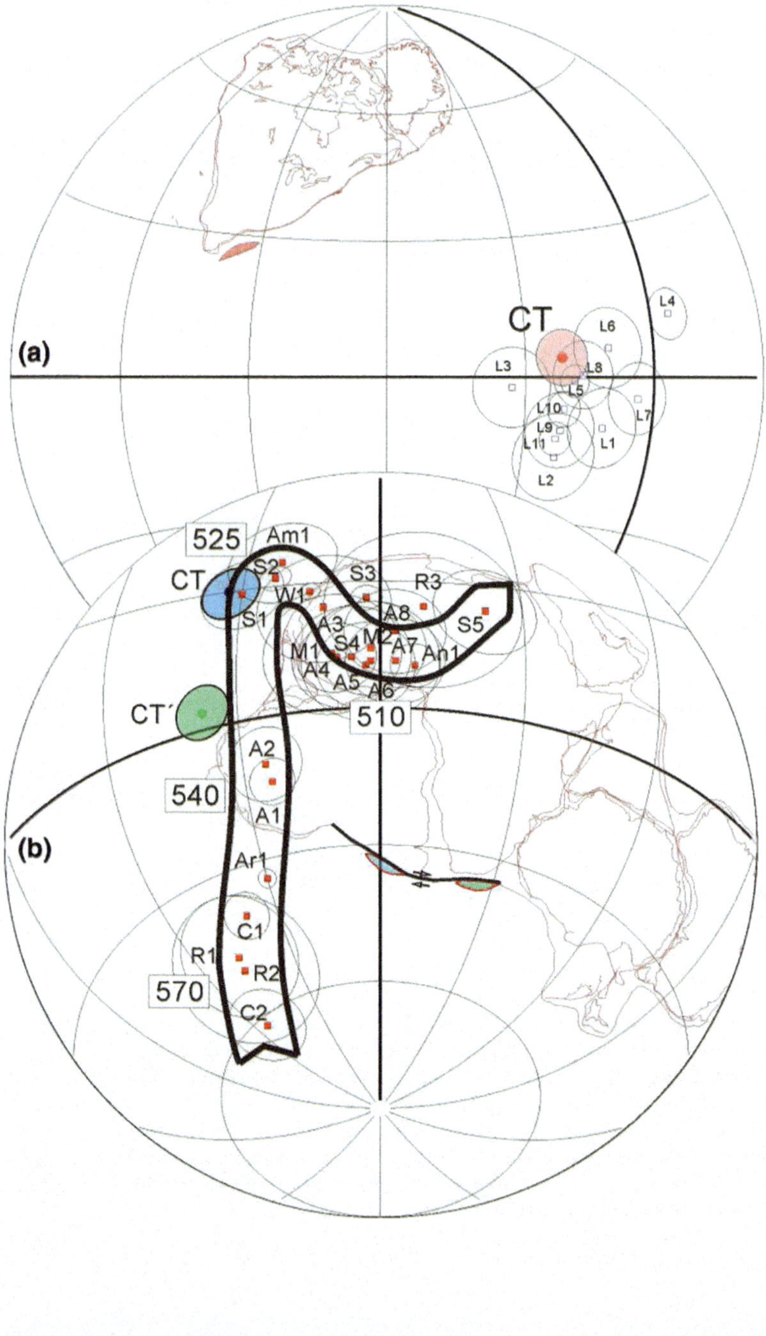

◂**Fig. 6** **a** Comparison of the Cerro Totora (CT) paleomagnetic pole (510–515 Ma) with the approximately coeval poles for Laurentia placing Cuyania as the conjugate margin of the Ouachita Embayment. **b** Comparison of the Cerro Totora paleomagnetic pole with an updated apparent polar wander path for the "assembling Gondwana" (indicated with wide black solid line). Cuyania is shown in two different positions: present-day location in South America (light blue) and the hypothetical position in the South American–African–East Antarctic (SAFRAN) embayment (light green). CT (blue) and CT′ (green) ellipses are the respective positions of the pole in each reconstruction. All poles are represented with their A95. Numbers along the Gondwana path indicate approximate ages in Ma. Figure modified from Rapalini (2012). The reader is referred to that publication for identification of each paleomagnetic pole

continent (light green pole). Since the most likely age of the Cerro Totora pole is around 510 Ma (see discussion in Rapalini 2012), the displacement of Cuyania from a position close to Kalahari is not supported by the paleomagnetic data, as the pole becomes discordant in age and position with the Gondwana APWP.

More curious is the position of CT keeping Cuyania in its present-day position in South America, as it is consistent with the 525 Ma elbow in the Gondwana path. However, after the proposal of González Bonorino and González Bonorino (1991) 25 years ago, there has been no other claim for a full autochthony of the Precordillera Early Paleozoic carbonate platform, which becomes incompatible in its present position with an Early Ordovician magmatic arc (Famatinian arc) located to the East. Furthermore, Rapalini (2012) indicated that the *Olenellus* fauna found on top levels of the Cerro Totora Fm. and isotopic composition of evaporites restrict its age to the 515–510 Ma interval making the CT pole inconsistent in age with the Gondwana APWP as it falls in a position significantly older than the age of the rocks from which it was computed.

On the other hand, the position of the late Early Cambrian Cerro Totora paleomagnetic pole (CT, Fig. 6a) continues to be consistent with the Cambrian poles of Laurentia when Cuyania is placed in the Ouachita Embayment as proposed by Thomas and Astini (1996). As indicated by Rapalini (2012), the paleomagnetic data available for the Cuyania terrane continues to be evidence in favor of the Laurentian origin of this terrane. Kinematics of its transfer process from Laurentia to Gondwana continues to be unrestricted by paleomagnetic data due to the lack of Middle Cambrian to Early Ordovician poles from this terrane.

6 Chilenia

Nearly three decades ago, Ramos et al. (1986) proposed that the basement of the Central Andes of Western Argentina and Chile, to the west of the Argentine Precordillera, was an allochthonous terrane (Fig. 1) that was accreted to the western Gondwana margin in the Middle Paleozoic. This hypothesis was based on the oceanic (or near-oceanic) character (E-Morb signature) of Late Ordovician basaltic pillow lavas, dikes, and gabbros exposed along the western border of the Precordillera (Kay et al. 1984, Boedo et al. 2013) that has been interpreted as an ophiolitic succession (see Ramos et al. 2000). The paleogeographic interpretation was that an open ocean existed in the Late Ordovician to the west of Cuyania (and Gondwana) thus rendering Chilenia a truly allochthonous terrane. Exposures of pre-Late Paleozoic rocks of Chilenia are very scarce, and there are no available biostratigraphical data that permit any comparison with the Early Paleozoic successions of Cuyania and Antofalla. However, U–Pb ages on zircons of 1069 ± 36 Ma (Ramos and Basei 1997) and Nd model ages of 1.4 to 1.7 Ga suggest a Grenvillian basement and point to a possible Laurentian link for Chilenia (Keppie and Ramos 1999). More recently, Álvarez et al. (2012) analyzed three Paleozoic accretionary complexes in northern Chile by means of U–Pb dating of detrital zircons and suggested that the basement of Chilenia could be characterized by a large amount of Neoproterozoic to Early Cambrian igneous and/or metamorphic rocks. Hyppolito et al. (2014) have claimed to have found evidence of fragments of accreted oceanic crust along the present-day continental margin of South America in Chile that may represent vestiges of the oceanic basin opened between Chilenia and its parent continent (Laurentia?) in Early Paleozoic times. A main tectonic phase in the Late Devonian ("Chañic Orogeny"), very well represented in several areas of western Argentina, has been interpreted as due to the collision of this terrane. Willner et al. (2011) dated the collision of Chilenia as the peak of metamorphism in the Guarguaraz complex in the Frontal Cordillera of Mendoza as 390 ± 2 Ma and described a relatively slow cooling and exhumation reaching 350–400 °C by 353 ± 1 Ma.

The exact extension of Chilenia also has been matter of debate. Martínez et al. (2012) stated that the Chilenia terrane would extend southwards to at least 41° S into the North Patagonian Massif. Meanwhile, Tomezzoli (2012) suggested a more radical model in which Chilenia and the North Patagonian Massif might have behaved as a single plate in the Devonian before its joint collision with southwestern Gondwana.

However, since the work of Dalla Salda et al. (1992), many authors have questioned the "truly" allochthonous model, by interpreting that the Late Ordovician ophiolites are the product of extensional processes that led to the development of a small oceanic basin (e.g., Keller et al. 1998, Davis et al. 1999). Such assumption has recently been supported by González Menéndez et al. (2013), based on the geochemical signature of the Late Ordovician lavas and gabbros in the Western Precordillera, interpreting that Chilenia rifted apart from Cuyania in the Late Ordovician, opening a V-shaped basin (open to the south), that was closed again in the Late Devonian. Similar tectonic setting was suggested by Boedo et al. (2013) from a more regional study in which a similar geochemical signature with E-Morb characteristics was found for the so-called Early Paleozoic mafic-ultramafic belt of the Cuyania terrane along around 500 km on the western margin of Cuyania.

Based on the first geochemical data, Rapela et al. (1998b) claimed that Chilenia may be part of Cuyania and not an independent terrane. Similar ideas were published by Lopez and Gregori (2004). In this model, Chilenia and Cuyania originally integrated a single terrane that collided with Gondwana in the Middle to Late Ordovician.

The brief and incomplete description of recent studies on Chilenia confirms that there is a consensus that it collided with the Gondwana margin in the Late Devonian, as originally proposed by Ramos et al. (1986). Whether Chilenia is a fully allochthonous terrane independent from Cuyania or a fragment that rifted apart in the Ordovician, to collide again in the Devonian is yet to be established. Even more disputable is its extension into the North Patagonian Massif. Since no paleomagnetic data are available for Chilenia, independent kinematic controls are not viable. The lack of exposed undeformed pre-Devonian rocks in this terrane turns the possibility of getting those constraints into a very hard task.

7 Conclusions

In the last two decades, the classical view of South America as a unique plate in pre-Gondwana times was slowly replaced by models in which numerous cratonic blocks and terranes interacted in a complex manner before Gondwana was completely assembled. In particular, this "terrane collage" must have played a significant role in the development of the Andean Orogen three hundred million years later.

Definition of the boundaries of the Pampia terrane has improved significantly in the last decade, mainly due to geophysical surveys. However, models regarding its relations to the Río de la Plata and Amazonia cratons are still controversial and widely different. Reliable paleomagnetic data from Cambrian rocks of this terrane came out only in the last decade but their interpretation is still ambiguous.

A systematic pattern of clockwise rotations was confirmed along the Famatinian magmatic belt, including the Eastern Puna, for Ordovician rocks. Considering the lack of geologic support for the para-autochthonous model of the Famatinian arc, these systematic anomalies could be explained by "in domino" crustal block rotations caused by deformation produced by the collision of the Cuyania terrane in the Middle to Late Ordovician.

The very large counterclockwise rotations that are widespread in Western Puna (Antofalla terrane) can be used to support the classical para-autochthonous model, although an escape tectonics scenario also associated with the Cuyania collision may skip the problems posed by lack of ophiolitic signature of Early Paleozoic basic rocks of the southern Puna.

Despite major changes in the apparent polar wander path of Gondwana for the latest Proterozoic-Cambrian in the last decade, paleomagnetism still supports the Laurentian origin of Cuyania.

The origin and evolution of Chilenia is controversial. The lack of paleomagnetic data from this terrane precludes any independent evaluation of its kinematic history.

Acknowledgements The Universidad de Buenos Aires (UBACyT 20020130100465BA), the Consejo Nacional de Investigaciones Científicas y Técnicas (CONICET, Argentina), and the Agencia Nacional de Investigaciones Científicas y Técnicas (ANPCyT) gave institutional support to these investigations. We thank the contribution of different colleagues who helped us understand better this complex and still enigmatic geological history. Among others, we would like to thank R. Astini, C. Cingolani, D. Poiré, M. Escayola, M. Lopez de Luchi, V. Ramos, C. Rapela, R. Trindade, and C. Vasquez. Constructive review by M.P. Iglesia Llanos is gratefully acknowledged.

References

Aceñolaza FG, Toselli AJ (1981) Geología del noroeste argentino. Facultad de Ciencias Naturales, Universidad Nacional de Tucumán, Argentina, Publicación Especial, pp 212

Aceñolaza FG, Miller H, Toselli AJ (2002) Proterozoic–early Paleozoic evolution in western South America: A discussion. Tectonophysics 354:121–137

Adams CJ, Miller H, Aceñolaza FG, Toselli AJ, Griffin WL (2011) The Pacific Gondwana margin in the late Neoproterozoic–early Paleozoic: Detrital zircon U–Pb ages from metasediments in northwest Argentina reveal their maximum age, provenance and tectonic setting. Gondwana Res 19:71–83

Adiyaman Ö, Chorowicz J, Arnaud ON, Gündogdu MN, Gourgaud A (2001) Late Cenozoic tectonics and volcanism along the North Anatolian Fault: New structural and geochemical data. Tectonophysics 338:135–165

Allmendinger RW, Ramos VA, Jordan TE, Palma M, Isacks BL (1983) Paleogeography and Andean structural geometry, NW Argentina. Tectonics 2:1–16

Álvarez J, Mpodozis C, Arriagada C, Astini R, Morata D, Salazar E, Valencia VA, Vervoort JD (2011) Detrital zircons from late Paleozoic accretionary complexes in north-central Chile (28°–32° S): Possible fingerprints of the Chilenia terrane. J South Am Earth Sci 32(4):460–476

Álvarez O, Gimenez M, Braitenberg C, Folguera A (2012) GOCE satellite derived gravity and gravity gradient corrected for topographic effect in the South Central Andes region. Geophys J Int 190(2):941–959

Astini RA (1998) Stratigraphic evidence supporting the rifting, drifting and collision of the Laurentian Precordillera terrane of western Argentina. In: Pankhurst RJ, Rapela CW (eds) The Proto-Andean margin of Gondwana. Geol Soc London, Special Publication 142:11–33

Astini RA (2003) The Ordovician Proto-Andean basins. In: Benedetto JL (ed) Ordovician fossils of Argentina. Secretaría de Ciencias y Tecnología, Universidad Nacional de Córdoba, pp 1–74

Astini RA, Benedetto JL, Vaccari NE (1995) The early Paleozoic evolution of the Argentine Precordillera as a Laurentian rifted, drifted and collided terrane: A geodynamic model. Geol Soc Am Bulletin 107:253–273

Bahlburg H, Hervé F (1997) Geodynamic evolution and tectonostratigraphic terranes of northwestern Argentina and northern Chile. Geol Soc Am Bulletin 109:869–881

Baldo E, Casquet C, Galindo C (1996) El metamorfismo de la Sierra Chica de Córdoba (Sierras Pampeanas), Argentina. Geogaceta 19:48–51

Benedetto JL (1993) La hipótesis de la aloctonía de la Precordillera Argentina: un test estratigráfico y biogeográfico. In: XII° Congreso Geológico Argentino, Actas 3, pp 375–384

Benedetto JL (1998) Early Palaeozoic brachiopods and associated shelly faunas from western Gondwana: Their bearing on the geodynamic history of the pre-Andean margin. In: Pankhurst RJ, Rapela CW (eds) The Proto-Andean margin of Gondwana. Geol Soc London, Special Publication 142:57–83

Benedetto JL (2004) The allochthony of the Argentine Precordillera ten years later (1993–2003): A new paleobiogeographic test of the microcontinental model. Gondwana Res 7(4):1027–1039

Benedetto JL, Sanchez TM, Carrera MG, Brussa ED, Salas MJ (1999) Paleontological constraints on succesive paleogeographic positions of Precordillera terrane during the early Paleozoic. In: Ramos VA, Keppie JD (eds) Laurentia–Gondwana connections before Pangea. Geol Soc Am, Special Paper 336:21–42

Benedetto JL, Vaccari NE, Waisfeld BG, Sánchez TM, Foglia RD (2009) Cambrian and Ordovician biogeography of the South American margin of Gondwana and accreted terranes. Geol Soc London, Special Publications 325:201–232

Blasco G, Villar L, Zappettini EO (1996) El complejo ofiolítico desmembrado de la Puna argentina. Provincias de Jujuy, Salta y Catamarca. In: XIII° Congreso Geológico Argentino and III° Congreso de Exploración de Hidrocarburos, Actas 3, pp 653–667

Bock B, Bahlburg H, Wörner G, Zimmermann U (2000) Tracing crustal evolution in the southern Central Andes from late Precambrian to Permian with geochemical and Nd and Pb isotope data. J Geology 108:515–535

Boedo FL, Vujovich GI, Kay SM, Ariza JP, Pérez Luján SB (2013) The E-MORB like geochemical features of the early Paleozoic mafic-ultramafic belt of the Cuyania terrane, western Argentina. J South Am Earth Sci 48:73–84

Cañas FL (1999) Facies and sequences of the late Cambrian–early Ordovician carbonates of the Argentine Precordillera: A stratigraphic comparison with Laurentian platforms. In: Ramos VA, Keppie JD (eds) Laurentia–Gondwana connections before Pangea. Geol Soc Am, Special Paper 336:43–62

Chernicoff CJ, Zappettini EO (2003) Evidencias geofísicas para la delimitación de terrenos preGondwánicos en la región centro-austral argentina. In: X° Congreso Geológico Chileno, Concepción, Actas CD

Chernicoff CJ, Zappettini EO (2004) Geophysical evidence for terrane boundaries in south-central Argentina. Gondwana Res 7:1105–1116

Chernicoff CJ, Zappettini EO, Santos JO, Allchurch S, McNaughton NJ (2010) The southern segment of the Famatinian magmatic arc, La Pampa Province, Argentina. Gondwana Res 17:662–675

Cingolani CA, Santos JOS, McNaughton NJ, Hartmann LA (2005) Geocronología U–Pb SHRIMP sobre circones del Granitoide Montecristo. Tandil, Provincia de Buenos Aires, Argentina. In: XVI° Congreso Geológico Argentino, Actas 1, La Plata, pp 299–302

Coira B, Davidson J, Mpodozis C, Ramos V (1982) Tectonic and magmatic evolution of the Andes of northern Argentina and Chile. Earth Sci Rev 18:303–332

Coira B, Perez B, Flores P, Mahlburgh Kay S, Woll B, Hanning M (1999) Magmatic sources and tectonic setting of Gondwana margin Ordovician magmas, northern Puna of Argentina and Chile. In: Keppie D, Ramos V (eds) Laurentia–Gondwana connections before Pangea. Geol Soc Am, Special Paper 336:143–170

Coira B, Koukharsky M, Guevara SR, Cisterna CE (2009) Puna (Argentina) and northern Chile Ordovician basic magmatism: A contribution to the tectonic setting. J South Am Earth Sci 27:24–35

Collo G, Astini R, Cawood PA, Buchan C, Pimentel M (2009) U–Pb detrital zircón ages and Sm–Nd isotopic features in low grade metasedimentary rocks of the Famatina belt: Implications for the late Neoproterozoic–early evolution of the proto-Andean margin of Gondwana. J Geol Soc 166:303–319

Conti CM, Rapalini AE, Coira B, Koukharsky M (1996) Paleomagnetic evidence of an early Paleozoic rotated terrane in NW Argentina. A clue for Gondwana–Laurentia interaction? Geology 24:953–956

Cordani UG, Sato K, Teixeira W, Tassinari CCG, Basei MAS (2000) Crustal evolution of the South American platform. In: Cordani UG, Milani EJ, Thomaz Filho A, Campos DA (eds) Tectonic evolution of South America, 31st International Geological Congress, Rio de Janeiro, Brazil, pp 19–40

Cordani UG, Pimentel MM, Ganade de Araujo CE, Basei MAS, Fuck RA, Girardi VAV (2013) Was there an Ediacaran Clymene Ocean in central SouthAmerica? Am J Sci 313:517–539

Dahlquist JA, Verdecchia SO, Baldo EG, Basei MAS, Alasino PH, Urán GA, Rapela CW, da Costa Campos Neto M, Zandomeni PS (2016) Early Cambrian U–Pb zircon age and Hf-isotope data from the Guasayán pluton, Sierras Pampeanas, Argentina: implications for the northwestern boundary of the Pampean arc. Andean Geol 43(1):137–150

Dalla Salda L, Cingolani C, Varela R (1992) Early Paleozoic orogenic belt of the Andes in southwestern South America: Result of Laurentia–Gondwana collision? Geology 20:617–620
Dalmayrac B, Lancelot JR, Leyreloup A (1977) Two billion year granulites in the late Precambrian metamorphic basement along the southern Peruvian coast. Science 198:49–51
Dalziel IWD (1997) Neoproterozoic–Paleozoic geography and tectonics: Review, hypothesis, environmental speculation. GSA Bull 109:16–42
Davis JS, Roeske SM, McClelland WC, Snee LW (1999) Closing the ocean between the Precordillera terrane and Chilenia: Early Devonian ophiolite emplacement and deformation in the southwest Precordillera. In: Ramos V, Keppie D (eds) Laurentia–Gondwana connections before Pangea. Geol Soc Am, Special Paper 336:115–138
D'Eramo F, Tubía JM, Pinotti L, Vegas N, Coniglio J, Demartis M, Aranguren A, Basei M (2013) Granite emplacement by crustal boudinage: example of the Calmayo and El Hongo plutons (Córdoba, Argentina). Terra Nova 25:423–430
Escayola MP, Pimentel MM, Armstrong R (2007) A Neoproterozoic Back-Arc Basin: SHRIMP U–Pb and Sm–Nd isotopic evidence from the Eastern Pampean Ranges, Argentina. Geology 35:495–498
Escayola MP, Ramé GA, Kraemer PE (1996) Caracterización y significado geotectónico de las fajas ultramáficas de Córdoba. In: XIII° Congreso Geológico Argentino, Actas 3, Buenos Aires, pp 421–438
Escayola MP, Van Staal CR, Davis WJ (2011) The age and tectonic setting of the Puncoviscana Formation in northwestern Argentina: An accretionary complex related to early Cambrian closure of the Puncoviscana Ocean and accretion of the Arequipa-Antofalla block. J South Am Earth Sci 32:438–459
Favetto A, Pomposiello C, de Luchi MGL, Booker J (2008) 2D Magnetotelluric interpretation of the crust electrical resistivity across the Pampean terrane–Río de la Plata suture, in central Argentina. Tectonophysics 459:54–65
Favetto A, Rocha V, Pomposiello C, García R, Barcelona H (2015) A new limit for the NW Río de la Plata Craton Border at about 24° S (Argentina) detected by Magnetotellurics. Geol Acta 13 (3):243–254
Finney SC (2007) The parautochthonous Gondwanan origin of the Cuyania (greater Precordillera) terrane of Argentina: A re-evaluation of evidence used to support an allochthonous Laurentian origin. Geol Acta 5(2):127–158
Finney S, Gleason J, Gehrels G, Peralta S, Aceñolaza G (2002) Early Gondwanan connection for the Argentine Precordillera terrane. Earth Planet Sci Lett 76:1–11
Finney S, Peralta S, Gehrels G, Marsaglia K (2005) The early Paleozoic history of the Cuyania (greater Precordillera) terrane of western Argentina: Evidence from geochronology of detrital zircons from Middle Cambrian sandstones. Geol Acta 3(4):339–354
Forsythe R, Davidson J, Mpodozis C, Jesinkey C (1993) Lower Paleozoic relative motion of the Arequipa block and Gondwana: Paleomagnetic evidence from Sierra de Almeida of northern Chile. Tectonics 12:219–236
Franceschinis P, Rapalini A, Escayola M, Luppo T (2016) Paleomagnetic studies on the late Ediacaran–early Cambrian Puncoviscana and the late Cambrian Campanario formations, NW Argentina: New paleogeographic constraints for the Pampia terrane. J South Am Earth Sci 70:145–161
Gangui A, Götze HJ (1996) The deep structures of the Northern Argentina: constraints from 2D seismic data 3D gravity modelling. In: XIII° Congreso Geológico Argentino, Actas 5, pp 545–565
Garfunkel Z, Ron H (1985) Block rotation and deformation by strike-slip faults: 2. The properties of a type of macroscopic discontinuous deformation. J Geophysical Res Solid Earth 90:8589–8602
Geuna SE, Escosteguy LD, Miró R (2008) Paleomagnetism of the late Devonian–early Carboniferous Achala Batholith, Córdoba, central Argentina: Implications for the apparent polar wander path of Gondwana. Gondwana Res, Special Issue "The Western Gondwana Margin: Proterozoic to Mesozoic" 13:227–237

Geuna SE, D'Eramo F, Pinotti L, Di Marco A, Mutti D, Escosteguy L (2010) Preliminary results of a paleomagnetic study on the Ordovician Calmayo granitoid, Sierras de Córdoba, Argentina. In: Tassone et al. (eds) Scientific contributions of the Geosur, International Geological Congress on the Southern Hemisphere, Bollettino di Geofisica Teorica ed Applicata 51:137–140

Geuna SE, López de Luchi M, Escosteguy LD (2011) Caracterización paleomagnética y de magnetofábrica de las dos secuencias que componen el batolito de Achala, Devónico, Sierras Pampeanas (Argentina). In: II° Biennial Meeting of Asociación Latinoamericana de Paleomagnetismo y Geomagnetismo, Tandil, Latinmag Letters 1, Special Issue B25:1–6

Goetze HJ, Kirchner A (1997) Interpretation of gravity and geoid in the central Andes between 20° and 29° S. J South Am Earth Sci 10:179–188

González Bonorino F, González Bonorino G (1991) Precordillera de Cuyo y Cordillera Frontal en el Paleozoico temprano: terrenos bajo sospecha de ser autóctonos. Revista Geológica de Chile 18:97–107

González Menéndez L, Gallastegui G, Cuesta A, Heredia N, Rubio-Ordóñez A (2013) Petrogenesis of Early Paleozoic basalts and gabbros in the western Cuyania terrane: Constraints on the tectonic setting of the southwestern Gondwana margin (Sierra del Tigre, Andean Argentine Precordillera). Gondwana Res 24(1):359–376

Grosse P, Bellos LI, de los Hoyos CR, Larrovere MA, Rossi J, Toselli AJ (2011) Across-arc variation of the Famatinian magmatic arc (NW Argentina) exemplified by I-, S- and transitional I/S-type early Ordovician granitoids of the Sierra de Velasco. J South Am Earth Sci 32(1):110–126

Hauser N, Matteini M, Omarini R, Pimentel M (2011) Combined U–Pb and Lu–Hf isotope data on turbidites of the Paleozoic basement of NW Argentina and petrology of associated igneous rocks: Implications for the tectonic evolution of western Gondwana between 560 and 460 Ma. Gondwana Res 19:100–127

Hyppolito T, Juliani C, García-Casco A, Tieppo Meira V, Bustamante Hervé F (2014) The nature of the Palaeozoic oceanic basin at the southwestern margin of Gondwana and implications for the origin of the Chilenia terrane (Pichilemu region, central Chile). Int Geol Rev 56(9):1097–1121

Iannizzotto NF, Rapela CW, Baldo EG, Galindo C, Fanning CM, Pankhurst RJ (2013) The Sierra Norte-Ambargasta batholith: Late Ediacaran–early Cambrian magmatism associated with Pampean transpressional tectonics. J South Am Earth Sci 42:127–143

Introcaso A, Martínez MP, Giménez ME, Ruiz F (2004) Geophysical study of the Valle Fértil lineament between 28°45′S and 31°30′S: Boundary between the Cuyania and Pampia terranes. Gondwana Res 7:1117–1132

Jezek P, Wilner AP, Aceñolaza F, Miller H (1985) The Puncoviscana through—a large basin of Late Precambrian to Early Cambrian age on the Pacific edge of the the Brasilian shield. Geol Rundsch 74:573–584

Kay SM, Orrell S, Abruzzi JM (1996) Zircon and whole rock Nd–Pb isotopic evidence for a Grenville age and Laurentian origin for the basement of the Precordilleran terrane in Argentina. J Geol 104:637–648

Kay SM, Ramos VA, Kay R (1984) Elementos mayoritarios y trazas de las vulcanitas ordovícicas de la Precordillera Occidental: basaltos de rift oceánico temprano (?) próximos al márgen continental. In: IX° Congreso Geológico Argentino, Actas 2, S.C. Bariloche, pp 48–65

Keller M, Buggisch W, Lehnert O (1998) The stratigraphical record of the Argentine Precordillera and its plate-tectonic background. In: Pankhurst RJ, Rapela CW (eds) The Proto-Andean margin of Gondwana. Geol Soc London, Special Publication 142:35–56

Keppie JD, Bahlburg H (1999) Puncoviscana formation of northwestern and central Argentina: Passive margin or foreland basin deposit? In: Ramos V, Keppie D (eds) Laurentia–Gondwana connections before Pangea. Geol Soc Am, Special Paper 336:139–143

Keppie JD, Ramos VA (1999) Odyssey of terranes in the Iapetus and Rheic oceans during the Paleozoic. In: Ramos V, Keppie D (eds) Laurentia–Gondwana connections before Pangea,

Geol Soc Am, Special Paper 336:267–276

Klein EV, Moura CAV, Pinheiro BLS (2005) Paleoproterozoic crustal evolution of the São Luís Craton, Brazil: Evidence from zircon Geochronology and Sm–Nd isotopes. Gondwana Res 8:177–186

Kraemer PE, Escayola MP, Martino RD (1995) Hipótesis sobre la evolución tectónica neoproterozoica de las Sierras Pampeanas de Córdoba (30°40′–32°40′). Rev Asoc Geo Arg 50:47–59

Loewy SL, Connelly JN, Dalziel IWD (2004) An orphaned basement block: The Arequipa-Antofalla Basement of the central Andean margin of South America. Geol Soc Am Bull 116:171–187

Lopez VL, Gregori DA (2004) Provenance and evolution of the Guarguaraz Complex, Cordillera Frontal, Argentina. Gondwana Res 7:1197–1208

Lucassen F, Becchio R, Wilke H, Franz G, Thirlwall M, Viramonte J, Wemmer K (2000) Proterozoic–Paleozoic development of the Central Andes (18–26° S)—a mobile belt of the South American craton. J South Am Earth Sci 13:697–715

Lucassen F, Becchio R, Harmon R, Kasemann S, Franz G, Trumbull R, Wilke HG, Romer RL, Dulski P (2001) Composition and density model of the continental crust in an active continental margin—the Central Andes between 18° and 27° S. Tectonophysics 341:195–223

Mannheim R (1993) Génesis de las volcanitas eopaleozoicas del sistema de Famatina. In XII° Congreso Geológico Argentino, Actas 4, Mendoza, pp 147–155

Mantovani MSM, Brito Neves BB (2005) The Paranapanema Lithospheric Block: Its importance for Proterozoic (Rodinia, Gondwana) supercontinent theories. Gondwana Res 8(3):303–315

Martínez JC, Dristas JA, Massonne HJ (2012) Palaeozoic accretion of the microcontinent Chilenia, North Patagonian Andes: High-pressure metamorphism and subsequent thermal relaxation. Int Geol Rev 54(4):472–490

Martino R (2003) Las fajas de deformación dúctil de las Sierras Pampeanas de Córdoba: Una reseña general. Rev Asoc Geol Arg 58:549–571

Martino R, Kraemer P, Escayola M, Giambastianai M, Arnosio M (1995) Transecta de las Sierras Pampeanas de Córodoba a los 32° S. Rev Asoc Geol Arg 50:60–77

Miller H, Hockenreiner M, Sollner F (2003) Opening and closure of an island arc—back arc system in the early Paleozoic at the Gondwana margin: Famatinian system and Las Termas belt, NW Argentina. In: X° Congreso Geológico Chileno, Actas CD, Concepción

Miller H, Söllner F (2005) The Famatina complex (NW Argentina): Back-docking of an island arc or terrane accretion? Early Palaeozoic geodynamics at the western Gondwana margin. Geol Soc London, Special Publications 246:241–256

Monger JW (1984) Cordilleran tectonics; a Canadian perspective. Bull Soc Geol Fr 7:255–278

Mosher S (1998) Tectonic evolution of the southern Laurentian Grenville orogenic belt. Geol Soc Am Bull 110:1357–1375

Omarini R, Götze HJ, Sureda RJ, Seilacher A, Pflüger F (1999) Puncoviscana folded belt in NW Argentina: Testimony of late Proterozoic Rodinia fragmentation and pre-Gondwana collisonal episodes. Geol Rundschau 88:76–97

Omarini RH, Sureda RJ, López de Azarevich VL, Hauser N (2008) El basamento Neoproterozoico- Cámbrico inferior en la provincia de Jujuy. In: Coira B, Zappettini E (eds) Relatorio del Congreso Geológico Argentino. Geología y Recursos Naturales de la provincia de Jujuy, No. 17, San Salvador de Jujuy, pp 17–28

Orozco LA, Favetto A, Pomposiello C, Rossello E, Booker J (2013) Crustal deformation of the Andean foreland at 31°30′S (Argentina) constrained by magnetotelluric survey. Tectonophysics 582:126–139

Otamendi JE, Ducea MN, Tibaldi AM, Bergantz GW, Jesús D, Vujovich GI (2009) Generation of tonalitic and dioritic magmas by coupled partial melting of gabbroic and metasedimentary rocks within the deep crust of the Famatinian magmatic arc, Argentina. J Petrology 0:1–33

Palma MA, Párica PD, Ramos VA (1986) El granito Archibarca: su edad y significado tectónico, provincia de Catamarca. Rev Asoc Geol Arg 41:414–419

Pankhurst RJ, Rapela CW, Saavedra J, Baldo E, Dahlquist J, Pascua I, Fanning CM (1998) The Famatinian magmatic arc in the central Sierras Pampeanas: an early to mid-Ordovician continental arc on the Gondwana margin. In: Pankhurst RJ, Rapela CW (eds) The Proto-Andean margin of Gondwana. Geol Soc London, Special Publication 142:343–367

Pankhurst RJ, Rapela CW, Fanning CM, Márquez M (2006) Gondwanide continental collision and the origin of Patagonia. Earth Sci Rev 76:235–257

Peri VG, Pomposiello MC, Favetto A, Barcelona H, Rossello EA (2013) Magnetotelluric evidence of the tectonic boundary between the Río de La Plata Craton and the Pampean terrane (Chaco-Pampean Plain, Argentina): The extension of the Transbrasiliano Lineament. Tectonophysics 608:685–699

Peri VG, Barcelona H, Pomposiello MC, Favetto A (2015) Magnetotelluric characterization through the Ambargasta-Sumampa Range: The connection between the northern and southern trace of the Río de La Plata Craton–Pampean Terrane tectonic boundary. J South Am Earth Sci 59:1–12

Ramé GA, Miró RC (2011) Modelo geofísico de contacto entre el orógeno Pampeano y el cratón del Río de la Plata en las provincias de Córdoba y Santiago del Estero. Serie Correlación Geológica 27(2):111–123

Ramos V (1986) El diastrofismo Oclóyico: un ejemplo de tectónica de colisión durante el Eopaleozoico en el Noroeste Argentino. Revista del Instituto Geológico y Minero, Jujuy, 6: 13–28

Ramos VA (1988) Late Proterozoic–early Paleozoic of South America—a collisional history. Episodes 11(3):168–174

Ramos VA (2000) The Southern Central Andes. In: Cordani UG, Milani EJ, Thomaz Filho A, Campos DA (eds) Tectonic evolution of South America, 31st International Geological Congress, Rio de Janeiro, Brazil, pp 561–604

Ramos VA (2004) Cuyania, an exotic block to Gondwana: Review of a historical success and the present problems. Gondwana Res 7:1009–1026

Ramos VA (2008) The basement of the central Andes: the Arequipa and related terranes. Annu Rev Earth Planet Sci 36:289–324

Ramos VA, Basei M (1997) The basement of Chilenia: An exotic continental terrane to Gondwana during the early Paleozoic. In: Symposium on Terrane Dynamics, New Zeland, pp 140–143

Ramos VA, Dallmeyer RD, Vujovich GI (1998) Time constraints on the early Paleozoic docking of the Precordillera, central Argentina. In: Pankhurst RJ, Rapela CW (eds) The Proto-Andean margin of Gondwana. Geol Soc London, Special Publication 142:143–158

Ramos VA, Escayola M, Mutti DI, Vujovich GI (2000) Proterozoic–early Paleozoic ophiolites of the Andean basement of southern South America. In: Dilek Y, Moores EM, Elthon D, Nicolas A (eds) Ophiolites and Oceanic Crust: New insights from field studies and Ocean Drilling program. Geol Soc Am, Special Paper 349:331–349

Ramos VA, Jordan TE, Allmendinger RW, Mpodozis C, Kay SM, Cortés JM, Palma M (1986) Paleozoic terranes of the Central Argentine–Chilean Andes. Tectonics 5(6):855–880

Ramos VA, Vujovich G, Martino R, Otamendi J (2010) Pampia: A large cratonic block missing in the Rodinia supercontinent. J Geodyn 50:243–255

Rapalini AE (2005) The accretionary history of southern South America from the latest Proterozoic to the late Paleozoic: Some paleomagnetic constraints. In: Vaughan APM, Leat PT, Pankhurst RJ (eds) Terrane processes at the margins of Gondwana. Geol Soc London, Special Publication 246:305–328

Rapalini AE (2012) Paleomagnetic evidence for the origin of the Argentine Precordillera, fifteen years later: What is new, what has changed, what is still valid? Latinmag Letters 2(1):1–20. http://www.geofisica.unam.mx/LatinmagLetters/files/LM12-0102Rv.pdf

Rapalini AE, Astini RA (1998) Paleomagnetic confirmation of the Laurentian origin of the Argentine Precordillera. Earth Planet Sci Lett 155:1–14

Rapalini AE, Cingolani CA (2004) First late Ordovician paleomagnetic pole for the Cuyania (Precordillera) terrane of western Argentina: A microcontinent or a Laurentian plateau? Gondwana Res 7(4):1089–1104

Rapalini AE, Astini RA, Conti CM (1999) Paleomagnetic constraints on the tectonic evolution of Paleozoic suspect terranes from southern South America. In: Ramos V, Keppie D (eds) Laurentia–Gondwana connections before Pangea. Geol Soc Am, Special Paper 336:171–182

Rapalini A, Pinotti L, D'Eramo F, Otamendi J, Vegas N, Tubía J, Singer S, Vujovich G (2010) A case of paleohorizontal restoration of plutonic bodies using paleomagnetic data: The Sierra de Valle Fértil, Magmatic Complex, Western Argentina. Bollettino di Geofísica Teorica ed Applicata, Italia, Sp Vol Geo Sur 51:156–159

Rapalini AE, Tohver E, Sanchez Bettucci L, Lossada AC, Barcelona H, Perez C (2015) The late Neoproterozoic Sierra de las Animas Magmatic Complex and Playa Hermosa Formation, southern Uruguay, revisited: paleogeographic implications of new paleomagnetic and precise geochronologic data. Precambrian Res 259:143–155

Rapalini AE, Velasco MS, Koukharsky M (2002) New paleomagnetic data from the Western Puna of Argentina: Some tectonic speculations on its Early Paleozoic evolution. In: 5th International Symposium on Andean Geodynamics, Extended Abstract, Toulouse, France, pp 505–508

Rapela CW, Pankhurst RJ, Casquet C, Baldo E, Saavedra J, Galindo C, Fanning CM (1998a) The Pampean orogeny of the southern proto-Andes: Cambrian continental collision in the Sierras de Córdoba. In: Pankhurst RJ, Rapela CW (eds) The Proto-Andean margin of Gondwana. Geol Soc London, Special Publication 142:181–217

Rapela CW, Pankhurst RJ, Casquet C, Baldo E, Saavedra J, Galindo C (1998b) Early evolution of the Proto-andean margin of South America. Geology 26:707–710

Rapela CW, Pankhurst R, Casquet C, Fanning C, Baldo E, González Casado J, Galindo C, Dahlquist J (2007) The Rio de la Plata Craton and the assembly of SW Gondwana. Earth Sci Rev 83:49–82

Rapela CW, Verdecchia SO, Casquet C, Pankhurst RJ, Baldo EG, Galindo C, Murra JA, Dahlquist JA, Fanning CM (2016) Identifying Laurentian and SW Gondwana sources in the neoproterozoic to early Paleozoic metasedimentary rocks of the Sierras Pampeanas: paleogeographic and tectonic implications. Gondwana Res 32:193–212

Saavedra J, Toselli A, Rossi J, Pellitero E, Durand F (1998) The early Paleozoic magmatic record of the Famatina System: A review. In: Pankhurst RJ, Rapela CW (eds) The Proto-Andean Margin of Gondwana. Geol Soc London, Special Publication 142:283–295

Sato AM, Tickyj H, Llambías ES, Sato K (2000) The Las Matras tonalitic-trondjhemitic pluton, central Argentina: Grenvillian-age constraints, geochemical characteristics and regional implications. J South Am Earth Sci 13:587–610

Schwartz JJ, Gromet LP (2004) Provenance of a late Proterozoic–early Cambrian basin, Sierras de Córdoba, Argentina. Precambrian Res 219:1–21

Shackleton RM, Ries AC, Coward MP, Cobbold PR (1979) Structure, metamorphism and geochronology of the Arequipa Massif of coastal Perú. Geological Society, Memoir 12, London, pp 1–21

Sigismondi ME, Fantín FE (2014) Estructura cortical y características geodinámicas. In: Martino RD, Guereschi AB (eds) Geología y recursos naturales de la provincia de Córdoba. Relatorio del XIX° Congreso Geológico Argentino, pp 939–961

Slagstad T, Culshaw NG, Daly JS, Jamieson RA (2009) Western Grenville Province holds key to midcontinental Granite Rhyolite Province enigma. Terra Nova 21(3):181–187

Spagnuolo CM, Rapalini AE, Astini RA (2008a) Paleogeographic and tectonic implications of the first paleomagnetic results from the middle–late Cambrian Meson Group: NW Argentina. J South Am Earth Sci 25:86–99

Spagnuolo C, Rapalini AE, Astini RA (2008b) Palaeomagnetic confirmation of Palaeozoic clockwise rotation of the Famatina Ranges (NW Argentina): Implications for the evolution of the SW margin of Gondwana. Geophys J Int 173(1):63–78

Spagnuolo C, Rapalini AE, Astini RA (2011) Reinterpretation of the Ordovician rotations found in NW Argentina and Northern Chile: A consequence of the Precordillera collision? Inter J Earth Sci 100:603–618

Spagnuolo CM, Rapalini AE, Astini RA (2012) Assembly of Pampia to the SW Gondwana margin: A case of strike-slip docking? Gondwana Res 21:406–421
Thomas W, Astini RA (1996) The Argentine Precordillera: A traveler from the Ouachita embayment of North American Laurentia. Science 273:752–757
Tohver E, Cawood PA, Rossello EA, Jourdan F (2012) Closure of the Clymene Ocean and formation of West Gondwana in the Cambrian: Evidence from the Sierras Australes of the southernmost Rio de la Plata craton, Argentina. Gondwana Res 21:394–405
Tohver E, Trindade RIF, Solum JG, Hall CM, Riccomini C, Nogueira AC (2010) Closing the Clymene ocean and bending a Brasiliano belt: Evidence for the Cambrian formation of Gondwana, southeast Amazon craton. Geology 38:267–270
Tohver E, Trindade RI (2014) Comment on "Was there an Ediacaran Clymene Ocean in central South America?" By UG Cordani and others. Am J Sci 314:805–813
Tomezzoli RN (2012) Chilenia y Patagonia: ¿un mismo continente a la deriva? Rev Asoc Geol Arg 69:222–239
Toselli A, Durand F, Rossi de Toselli, Saavedra J (1996) Esquema de la evolución geotectónica y magmática del sistema eopaleozoico de Famatina y sectores de Sierras Pampeanas. In: XIII° Congreso Geológico Argentino, Actas 5, Buenos Aires, pp 443–462
Trindade RIF, D'Agrella-Filho MS, Epof I, Brito Neves BB (2006) Paleomagnetism of early Cambrian Itabaiana mafic dikes (NE Brazil) and the final assembly of Gondwana. Earth Planet Sci Lett 244:361–377
Vaccari NE, Waisfeld BG (2008) The Proto-Andean margin of Gondwana and accreted terranes: Contrasting biogeographics signatures based on late Cambrian–early Ordovician trilobites. Advances in trilobite research. Cuadernos del Museo Geominero 9:403–409
Villar L (1975) Las fajas y otras manifestaciones ultrabásicas en la República Argentina y su significado metalogénetico. In: II° Congreso IberoAmericano de Geología Económica, Actas 3, Buenos Aires, pp 135–155
Viramonte JM, Becchio RA, Viramonte JG, Pimentel MM, Martino RD (2007) Ordovician igneous and metamorphic units in southeastern Puna: New U–Pb and Sm–Nd data and implications for the evolution of northwestern Argentina. J South Am Earth Sci 24:167–183
Von Gosen W, Prozzi C (2010) Pampean deformation in the Sierra Norte de Córdoba, Argentina: Implications for the collisional history at the western pre-Andean Gondwana margin. Tectonics 29
Von Gosen W, McClelland WC, Loske W, Martínez JC, Prozzi C (2014) Geochronology of igneous rocks in the Sierra Norte de Córdoba (Argentina): Implications for the Pampean evolution at the western Gondwana margin. Lithosphere 6:277–300
Whitmeyer SJ, Simpson C (2003) High strain-rate deformation fabrics characterize a kilometers-thick Paleozoic fault zone in the Eastern Sierras Pampeanas, central Argentina. J Struct Geol 25:909–922
Willner AP, Gerdes A, Massonne HJ, Schmidt A, Sudo M, Thomson SN, Vujovich G (2011) The geodynamics of collision of a microplate (Chilenia) in Devonian times deduced by the pressure–temperature–time evolution within part of a collisional belt (Guarguaraz Complex, W-Argentina). Contrib Miner Petrol 162(2):303–327
Zimmermann U, Bahlburg H (1999) The evolution of the Ordovician southern Puna retro-arc basin (NW Argentina): Provenance analysis and paleotectonic setting. In: 4th International Sypsium of Andean Geodynamics, Abstracts, Göettingen, Germany, pp 830–831
Zimmermann U, Bahlburg H, Mezger K, Berndt J, Kay SM (2014) Origin and age of ultramafic rocks and gabbros in the southern Puna of Argentina: an alleged Ordovician suture revisited. Inter J Earth Sci 103(4):1023–1036

The Pre-Andean Phases of Construction of the Southern Andes Basement in Neoproterozoic–Paleozoic Times

Nemesio Heredia, Joaquín García-Sansegundo, Gloria Gallastegui, Pedro Farias, Raúl Giacosa, Fernando Hongn, José María Tubía, Juan Juis Alonso, Pere Busquets, Reynaldo Charrier, Pilar Clariana, Ferrán Colombo, Andrés Cuesta, Jorge Gallastegui, Laura Giambiagi, Luis González-Menéndez, Oscar Limarino, Fidel Martín-González, David Pedreira, Luis Quintana, Luis Roberto Rodríguez-Fernández, Álvaro Rubio-Ordóñez, Raúl Seggiaro, Samanta Serra-Varela, Luis Spalletti, Raúl Cardó and Victor A. Ramos

Abstract During the late Neoproterozoic and Paleozoic times, the southern Andes of Argentina and Chile (21°–55° S) formed part of the southwestern margin of Gondwana. During this period of time, a set of continental fragments of variable

N. Heredia (✉) · G. Gallastegui · P. Clariana
L. González-Menéndez · L. R. Rodríguez-Fernández
Instituto Geológico y Minero de España, C/Ríos Rosas 23, E28003 Madrid, Spain
e-mail: n.heredia@igme.es

J. García-Sansegundo · P. Farias · J. J. Alonso · A. Cuesta · J. Gallastegui
D. Pedreira · L. Quintana · Á. Rubio-Ordóñez
Departamento de Geología, Universidad de Oviedo, C/Jesús Árias de Velasco s/n, E33005 Oviedo, Spain

R. Giacosa
Universidad de Río Negro y Servicio Geológico y Minero Argentino, Parque Industrial, 8332 General Roca, Río Negro, Argentina

F. Hongn
IBIGEO (CONICET-UNSA), 9 de julio 14, 4405 Rosario de Lerma, Salta, Argentina

J. M. Tubía
Departamento de Geodinámica, Universidad del País Vasco (UPV/EHU), Lejona, Biscay, Spain

P. Busquets · F. Colombo
Facultad de Geología, Universidad de Barcelona, C/Martí i Franqués s/n, E08028 Barcelona, Spain

R. Charrier
Departamento de Geología, Universidad de Chile, Plaza Ercilla 803, Santiago, Chile

A. Folguera et al. (eds.), *The Evolution of the Chilean-Argentinean Andes*, Springer Earth System Sciences, https://doi.org/10.1007/978-3-319-67774-3_5

extent and allochtony was successively accreted to that margin, resulting in six Paleozoic orogenies of different temporal and spatial extension: Pampean (Ediacaran–early Cambrian), Famatinian (Middle Ordovician–Silurian), Ocloyic (Middle Ordovician–Devonian), Chanic (Middle Devonian–early Carboniferous), Gondwanan (Middle Devonian–middle Permian), and Tabarin (late Permian–Triassic). All these orogenies culminate with collisional events, with the exception of the Tabarin and a part of the Gondwanan orogenies that are subduction-related.

Keywords Paleozoic · Southern Andes · Pampean orogen · Ocloyic orogen Famatinian orogen · Chanic orogen · Gondwanan orogen · Tabarin orogen

1 Introduction

In the southern part of the Andean Cordillera (21°–55° S, Fig. 1a) and nearby areas, there are Neoproterozoic (Ediacaran)–Paleozoic basement relicts of variable extension. This basement has been involved in orogenic events prior to the Andean orogeny, which is related to the current configuration of the Andean chain (active since the Cretaceous).

L. Giambiagi
Unidad de Tectónica. IANIGLA-CONICET, Avda. Ruiz Leal s/n,
5500 Mendoza, Argentina

O. Limarino
Facultad de Geología, Universidad de Buenos Aires, C/Intendente Güiraldes 2160,
Ciudad Universitaria, Pabellón II, C1428EGA Buenos Aires, Argentina

F. Martín-González
Área de Geología ESCET, Universidad Rey Juan Carlos, C/Tulipán s/n,
28933 Mostoles, Madrid, Spain

R. Seggiaro
Universidad de Salta y Servicio Geológico y Minero Argentino,
Avda. Bernardo Houssay 1099, Barrio Castañeras, 4400 Salta, Argentina

S. Serra-Varela
Instituto de Investigación en Paleobiología y Geología, General Roca, Argentina

L. Spalletti
Centro de Investigaciones Geológicas, Universidad de La Plata,
Diagonal 113 n° 275, B1904DPK La Plata, Argentina

R. Cardó
Universidad de San Juan y Servicio Geológico Minero Argentino,
Sargento Cabral, 685 Oeste, 5400 San Juan, Argentina

V. A. Ramos
Instituto de Estudios Andinos "Don Pablo Groeber", Departamento de Ciencias
Geológicas, FCEN, Universidad de Buenos Aires–CONICET, Buenos Aires, Argentina

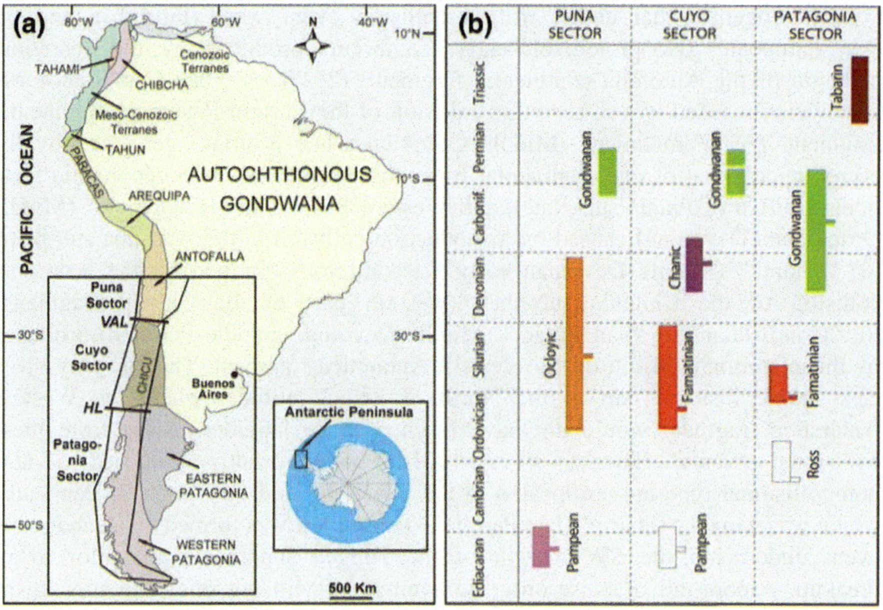

Fig. 1 **a** Paleozoic terranes and continental fragments of South America and location of the Puna, Cuyo, and Patagonia sectors. Modified from Ramos (2009). The Western Antarctica fragment is mainly preserved in the Antarctic Peninsula. Lineaments: VAL-Valle Ancho, HL-Huincul. **b** Sketch with the time distribution of the Paleozoic Andean orogenies. Uncolored bars, orogenies outside the Andes: Pampean in the Cuyo Sector and Ross in the Patagonia Sector. The small rectangles mark the beginning of the collisions

The Paleozoic orogens were developed in the SW Gondwana paleomargin by the docking of different continental fragments, named, from N to S: Antofalla, Chi-Cu, Eastern Patagonia, Western Patagonia, and Western Antarctica (Fig. 1a). These continental fragments have different extension, lithology, and origin (allochthonous and parautochthonous). Some of them could be exotic and therefore considered as terranes: Chi-Cu and Western Antarctica. All these different fragments were formed by the Neoproterozoic rifting of the Rodinia supercontinent. After separation, these fragments traveled forming part of lithospheric plates and sub-plates such as Chilenia and Cuyania, included in Chi-Cu, and Atacama belonging to the Antofalla continental fragment (Heredia et al. 2016). The Paleozoic orogens were formed, in some cases, by a subduction event prior to collision (non-collisional type) and in other cases by direct continental fragment/terrane collision with the SW Gondwana paleomargin (collisional type). All these orogenic events were sequential in time, non-collisional followed by collisional ones, or overlapped in space or time (Ramos et al. 1986; Ramos 1988a, b; Ramos 2009; Rapela et al. 2001; Ramos and Naipauer 2014; Heredia et al. 2016; among others).

The orogenies that ended with a collision event were (Figs. 1 a and b): (1) "Pampean" (Neoproterozoic–early Cambrian) produced by the accretion/collision of the Antofalla continental fragment. (2) "Ross" (late Cambrian–Early Ordovician) caused by the accretion/collision of the Eastern Patagonia continental fragment. (3) "Famatinian" (Middle Ordovician–late Silurian) generated by the accretion/collision of two continental fragments at different paleogeographic locations: Chi-Cu (Cuyania sub-plate), and Western Patagonia. (4) "Ocloyic" (Middle Ordovician–Devonian) caused by a subduction/collision in the Atacama sub-plate. (5) "Chanic" (Middle Devonian–early Carboniferous) formed by the accretion/collision of the Chilenia sub-plate (western part of the Chi-Cu fragment). (6) "Gondwanan" or "San Rafael" (Middle Devonian–middle Permian) produced by the accretion/collision of the Western Antarctica fragment. This orogeny has a clear non-collisional part to the North of 38° S latitude, where the Western Antarctica fragment would not have reached those latitudes. Apart from these collisional events, Heredia et al. (2016) have recently proposed a new non-collisional type orogen located in the Patagonia and the Antarctic Peninsula, which was named "Tabarin" (late Permian–Triassic). It was formed by a subduction event underneath the SW margin of the Pangea supercontinent, prior to its breakup. Among all of these orogenic events, we will not refer hereafter to the "Ross" orogeny, because it is not recognized in the cordilleran territory, but only in the northern part of the Atlantic Patagonian coast (González et al. 2011; Ramos and Naipauer 2014).

2 Main Features of the Paleozoic Orogenies in the Southern Andes

The imprints of the different Paleozoic orogenies are not ubiquitous along the Argentinean–Chilean Andes; only the Gondwanan orogeny had a widespread effect, although with significant variations from North to South, both in age and type (i.e., collisional or non-collisional) (Fig. 1b). These observations led Heredia et al. (2016) to separate three distinct sectors in the Paleozoic basement of the southern Andes, whose boundaries are approximately coincident with important tectonic lineaments: the Valle Ancho lineament (28° S, VAL in Fig. 1a) (Ramos 1999) and the Huincul lineament (38° S, HL in Fig. 1a) (Ploszkiewicz et al. 1984). Therefore, to the north of 28° S, the Puna Sector was affected by the Pampean, Ocloyic, and Gondwanan (non-collisional) orogenies. Then, between 28° and 38° S, the Cuyo Sector records the Famatinian, Chanic, and Gondwanan (non-collisional) orogenies. Finally, south of 38° S the Famatinian, Gondwanan (collisional), and Tabarin orogenies are identified in the Patagonia Sector.

3 The Pampean Orogeny

The Pampean orogeny has been a source of deep debate about its age, spatial distribution and characteristics since its initial proposal (Aceñolaza and Toselli 1976). The key geological units identifying the first-order discordance that separate the Pampean orogeny from the following tectonic or orogenic cycle (i.e., Ediacaran–lower Cambrian Puncoviscana Formation/Complex and middle to upper Cambrian Mesón Group, characterized by Turner 1960; Turner and Mon 1979; Aceñolaza and Aceñolaza 2005) crop out extensively in the Eastern Cordillera and not in the nearby Pampean Ranges (see discussion in Sureda and Omarini 1999). The main records of the Pampean orogeny are broadly represented from the Argentina–Bolivia border (Puna and Eastern Cordillera) up to the Pampean Ranges showing a progressive deepening of the exposed crustal levels southward (Rapela et al. 1998; Ramos 2008). The Puncoviscana basin was originally interpreted as a passive margin basin (Jezek et al. 1985), but lately it is interpreted as a basin related to an active margin linked to variable geotectonic scenarios (Keppie and Bahlburg 1999; Zimmermann 2005; Escayola et al. 2011).

The subduction along this active margin, followed by the collision of the Antofalla continental fragment (Fig. 1a) against the western border of the Gondwana continent (Pampia block, previously amalgamated to the Río de la Plata craton), generated the complex structure of the Puncoviscana Complex, showing superposed deformations with fold interferences and related cleavages. The available data are abundant but not enough for getting an unequivocal characterization of the structural evolution of the Puncoviscana Complex. Regional variations of structural data, like vergence and fold plunge, gave place to proposals of internal unconformities (e.g., Mon and Hongn 1991). Although recent studies on the age and provenance of detrital zircons are also consistent with the existence of internal discontinuities in the Puncoviscana Complex (Adams et al. 2011; Aparicio González et al. 2014), there is still no consensus about such internal unconformities. Lower to middle Cambrian granitoids, such as Cañani and Tastil batholiths, are part of the Pampean arc, although at least one of the facies of the Tastil batholith postdated the deposition of the shelf clastic sedimentary sequences filling the early Cambrian–Lower Ordovician basins (Hongn et al. 2010b).

Outcrops of units preserving the Pampean metamorphic and magmatic events at low crustal levels are very sparse in the Puna and the Eastern Cordillera. In contrast, they are largely represented in the Pampean Ranges (Grissom et al. 1998; Rapela et al. 1998).

4 The Ocloyic Orogeny

The Ocloyic deformational phase was defined by Turner and Méndez (1975) after identifying the deformational events that generate the discontinuity between Ordovician and Silurian successions in the Eastern Cordillera. New researches strongly improved the knowledge of the Paleozoic rocks in NW Argentina by recognizing and characterizing deformational, magmatic, metamorphic, and stratigraphic features related to Paleozoic orogenic events. However, the original definition of the Ocloyic phase has lost its identity and it was used in several senses in order to define orogenies or deformational phases at different areas and times between the Ordovician and the Devonian (Ramos 1986, 1988b, 1999; Mon and Hongn 1987; Bahlburg and Hervé 1997; Astini and Dávila 2004; Collo et al. 2008; Heredia et al. 2012; Moya 2015).

Here we describe the main events related to the Ocloyic orogeny in the sense of Heredia et al. (2016). Based on the age, geotectonic evolution, paleogeographic situation, and deformational features, these authors propose to restrict the use of Ocloyic orogeny to the Puna Sector. The Ocloyic orogeny started in the Middle Ordovician and ended in the Late Devonian (Fig. 1b); however, in this chapter, we also describe ductile pre-orogenic extensional deformations, developed in Early Ordovician times, which can be confused with some contractional deformations related to subsequent orogenic processes.

After the accretion of the Antofalla terrane to the Gondwana continent during the early Cambrian (Pampean orogeny), a subduction zone was installed along the new western Gondwana margin. The western border of the Puna preserves the related magmatic arc (WP in Fig. 2a), where the oldest components probably are of late Cambrian age (Moya 2015), although the arc achieved its climax during the Early Ordovician (Tremadocian–Floian) (Coira et al. 1999; Zimmermann et al. 2010; Niemeyer et al. 2014) reaching the middle Silurian (ca. 431 Ma) in the Chilean Andes (Vergara 1978) (see Chap. "The Famatinian Orogen Along the Protomargin of Western Gondwana: Evidence for a Nearly Continuous Ordovician Magmatic Arc Between Venezuela and Argentina").

Changes in plate dynamics have led to major tectonomagmatic events in the Puna and surrounding areas during the upper Tremadocian and lower Floian times. The thermal anomaly related to the back-arc extension gave place to metamorphic, magmatic, and deformation processes well defined along the eastern border of the Puna (Hongn et al. 2014). Imprints of these processes are well preserved mainly in the Ordovician units northward of 24° S and also in the Pampean basement southwards of this latitude. The Ordovician magmatic belt along the eastern border of the Puna (EP in Fig. 2a) (Méndez et al. 1973) is currently thought to be the result of the back-arc thermal anomaly (Coira et al. 2009; Sola et al. 2010) in contrast to some proposals interpreting these rocks as a magmatic arc related to a second subduction zone (Ramos 2008). A high-T and low-P metamorphism transformed the Puncoviscana Complex in schists, gneisses, and migmatites, mainly in the eastern border of the southern Puna, the SW area of the Eastern Cordillera (Sola et al. 2013; Hongn et al. 2014) and their southern prolongations in the northern Pampean Ranges (Finch et al. 2015) (Fig. 2a). The Early Ordovician metamorphic and magmatic complexes display well-defined structures, such as folds of several orders with related minor structures (foliations and lineations) and mylonitic belts, at different scales and with different kinematics (Hongn and Riller 2007; Hongn et al. 2014; Finch et al. 2015).

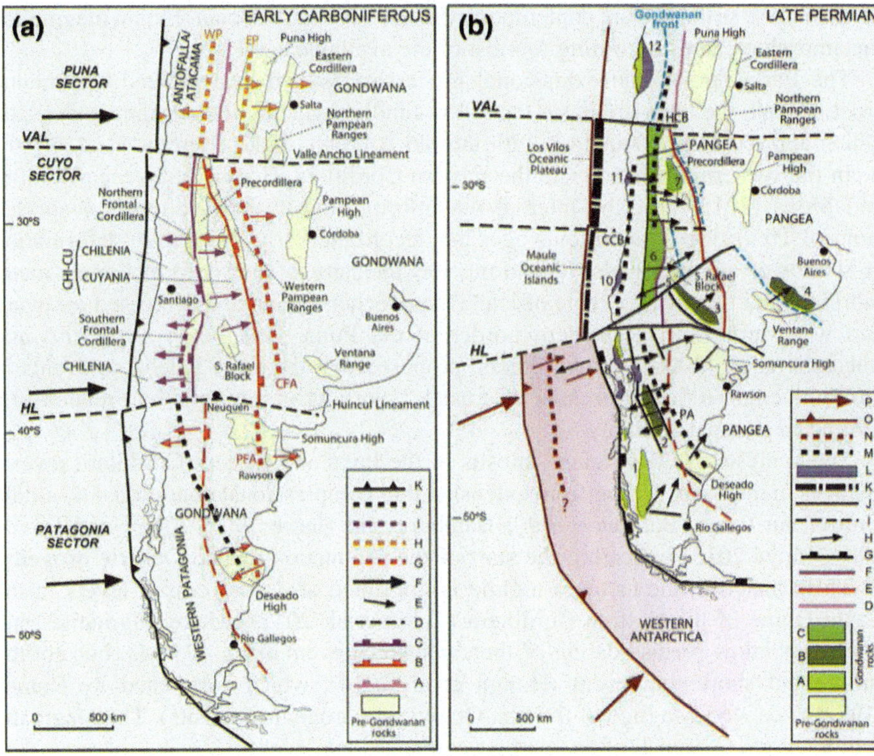

Fig. 2 Main geotectonic features in the Paleozoic basement of the Argentinean–Chilean Andes and their foreland in: **a** Early Carboniferous times. Sutures (little rectangles mark the upper plate): A-Pampean, B-Famatinian, C-Chanic. D-Limits between the Paleozoic Andean sectors that coincide with Andean lineaments: VAL-Valle Ancho, HL-Huincul. E-Vergences of Paleozoic structures (Pink: Pampean, Orange: Ocloyic, Red: Famatinian, Purple: Chanic). F-Displacement sense of the proto-Pacific lithosphere relative to the Gondwana margin. Magmatic Arcs: G-Famatinian magmatic arcs: CFA-Cuyo Famatinian arc (late Cambrian–Middle Ordovician), PFA-Patagonian Famatinian arc (Ordovician), H-Ocloyic magmatic arcs: EP-Eastern Puna arc (Ordovician), WP-Western Puna arc (late Cambrian–middle Silurian), I-Chanic magmatic arc (Devonian), J-Gondwanan magmatic arc of the Patagonia Sector (late Silurian–late Carboniferous). K-Gondwanan subduction zone (Orange triangle: former Ocloyic subduction in the Puna Sector). **b** Late Permian times. Main Gondwanan rocks outcrops: A-Fore-arc pre- and synorogenic rocks: 1-Eastern Andes metamorphic complex, 8 to 12-Eastern Series of the Coastal Cordillera (same names as the Western Series described in L). B-Back-arc pre-orogenic rocks and peripheral foreland basin synorogenic rocks: 2-Tecka–Tepuel, 3-Southern San Rafael, 4-Claromecó, 5-Northern San Rafael. C-Retrowedge pre-orogenic rocks and retroarc foreland basin synorogenic rocks: 6-Río Blanco, 7-Paganzo. Pre-Gondwanan sutures: D-Pampean, E-Famatinian, F-Chanic. G-Gondwanan suture (little black rectangles mark the upper plate). H-Vergences of Paleozoic structures (Black-Gondwanan, Maroon-Tabarin). I-Limits between the Paleozoic Andean sectors that coincide with Andean lineaments: VAL-Valle Ancho, HL-Huincul. J-Trace of Gondwanan structures. K-Oceanic reliefs accreted to Gondwana. L-HP metamorphic rocks related to the Gondwanan basal accretionary prism (Western Series) emplaced over the fore-arc basin (Eastern Series): 8-Puerto Montt–Chiloé, 9-Bariloche, 10-Pichilemu–Constitución, 11-Los Vilos, 12-Antofagasta. M-Gondwanan magmatic arcs: PA-Gondwanan magmatic arc of the Patagonia Sector (late Silurian–late Carboniferous), CCB-Coastal Cordillera batholith (late Carboniferous), HCB-High Cordillera batholiths (early Permian). N-Tabarin magmatic arc (Triassic). O-Southwestern margin of Pangea: triangles mark the position of the Tabarin subduction. P-Displacement sense of the last proto-Pacific lithosphere relative to the Pangea margin

Most of these structures are contemporary to the late Tremadocian–Floian magmatic–metamorphic events according to most of the available ages.

The Early Ordovician extensional processes resulted in a thinned continental crust beneath the back-arc basin that allows individualizing the Atacama sub-plate, which approximately coincides with the old Antofalla plate (Heredia et al. 2016).

In the westernmost Puna and the Eastern Cordillera, Hongn and Vaccari (2008) and Moya (2015) describe internal unconformities between sedimentary successions of Tremadocian and Floian ages that record the Early Ordovician deformation at shallower crustal levels. Therefore, the intensity of the deformational, metamorphic, and magmatic events of late Tremadocian–Floian age decreased eastward and westward from the eastern border of the Puna. Becchio et al. (1999) and Lucassen et al. (2000), among others, proposed that the early Paleozoic events in the Puna could reflect processes of crustal reworking in a mobile belt related to a protracted thermal anomaly.

The analysis of Ordovician deposits of the Puna and Eastern Cordillera reveals changes in the basin setting from extensional to compressional conditions in Middle Ordovician times (Bahlburg 1990; Bahlburg and Hervé 1997; Astini and Dávila 2004; Moya 2015) indicating the start of the subduction-related Ocloyic orogeny.

This change in the tectonic regime is imprinted at deeper crustal levels in the Cachi Dome of the Eastern Cordillera (Tubía et al. 2012), where migmatites and granitoids allow precise dating of the extensional event at ca. 475 Ma, but not the subsequent shortening event (Hongn et al. 2014), which was dated by Ramos (1972) at ca. 465 Ma (age of the first Ocloyic synorogenic deposits). The magmatic activity starts to vanish after this first compressional event.

The thick turbidite successions exposed in the central Puna were deposited in a compressional setting, related to subduction, during the Middle Ordovician and early Silurian times (Ramos 1986; Bahlburg and Hervé 1997; Heredia et al. 2016). These levels are strongly folded and metamorphosed at very low-grade metamorphic conditions (Mon and Hongn 1987; Astini 2008) during latter phases of the Ocloyic deformation.

The Ocloyic subduction probably ends in the middle Silurian (ca. 431 Ma), when the magmatism was drastically reduced or disappeared and the inversion of the back-arc basin starts, as a result of the approach and later collision of the Atacama sub-plate with Gondwana (Heredia et al. 2016). This last orogenic episode finally forms a N–S and double-vergent orogenic belt (Fig. 2a) with scarce internal zones and clear east vergences in the Eastern Cordillera (Alonso et al. 2012), to the east of the older Pampean suture (Fig. 2a). In the Puna and the Chilean Andes (western branch of the Ocloyic orogen), the structures related to the Ocloyic orogeny do not show a clear vergence in the late Ordovician levels. This allows identifying two well-defined segments, the northern one without a clear generalized vergence and a southern portion (Salta and Catamarca provinces) with a marked west-vergence (Fig. 2a). These variations argue for models including superposed deformations (subduction- and collision-related), structural systems with main and secondary vergences (retro-vergent structures), transpression and, probably, a lack of systematic research. The Ocloyic orogeny finished with the accretion of the Atacama sub-plate to Gondwana in the Late Devonian (Fig. 1b), although some authors, like Díaz-Martínez (1996), consider that reached the early Carboniferous.

The dominant N–S structural trend of the Ocloyic structures enhanced their reactivation during the following orogenies (Hongn et al. 2010a; Alonso et al. 2012). Thus, the Ocloyic eastern front seems to coincide with the Eastern Cordillera Andean front and the western orogenic front probably coincides with the eastern border of the Coastal Cordillera (Heredia et al. 2016) (Fig. 2a).

5 The Famatinian Orogeny

The Famatinian orogeny (Aceñolaza and Toselli 1976) can be recognized in the Cuyo and Patagonia sectors (Fig. 1b) and in both areas it ends with a collisional event. This orogeny can be related to the accretion/collision of two fragments with a different paleogeographic and allochthonous position: Chi-Cu and Western Patagonia (Fig. 1a). According to some authors, the Chi-Cu continental fragment (Cuyo Sector) is an allochthonous fragment, a terrain with a peri-Laurentic origin (Ramos et al. 1986; Dalla Salda et al. 1992; Astini et al. 1995; Keller 1999; Thomas and Astini 2003; Ramos 2004, among others). For some other authors, this same fragment could not be considered a terrain but could be attributed to a peri-Gondwanic origin (Baldis et al. 1989; Aceñolaza et al. 2002; Finney et al. 2003; López and Gregori 2004; Finney 2007; González-Menéndez et al. 2013). As regards the Western Patagonia (Patagonia Sector), it seems to be a peri-Gondwanic fragment (Heredia et al. 2016).

Although this orogeny presents N–S variations in age, it could be considered as Middle Ordovician–Silurian (Fig. 1b). In the Cuyo Sector, east-vergent structures suggest a relationship to a non-collisional orogen (Andean Type), generated in the Middle Ordovician (Fig. 1b) close to the active margin of Gondwana (ca. 460 Ma) (Astini and Dávila 2004), where the subduction was already active since the late Cambrian (ca 500 Ma). The collision between Chi-Cu (Cuyania sub-plate) and Gondwana took place from the Late Ordovician (ca. 460 Ma) (Astini et al. 1995; Pankhurst et al. 1998) and extended up to the Silurian–Devonian limit (ca. 420 Ma) (Mulcahy et al. 2011) (Fig. 1b). In the Patagonia Sector, there is no evidence of pre-collisional contractional deformation related to the subduction that was originated underneath the Gondwana margin. The Famatinian deformation in Patagonia affects pre-orogenic granitoids related to the Famatinian magmatic arc (PFA in Fig. 2a), which reach the Upper Ordovician (ca 452 Ma), according Loske et al. (1999). Thus, the Famatinian orogeny was generated by the subsequent collision of the Western Patagonia continental fragment in the Late Ordovician (Heredia et al. 2016) (Fig. 1b). This orogeny ends in this sector in the Wenlock (middle Silurian, ca 430 Ma), the age of the fossils from the post-orogenic sequence (Müller 1965; Manceñido and Damborenea 1984).

The Famatinian orogenic belt has an N–S trend that rotates toward NNW–SSE in its northern part (Fig. 2a). This belt presents a double vergence, although only its western branch, developed over Cuyania (Fig. 2a), is located in the surroundings of the Andean Cordillera (Astini et al. 1995; Astini and Dávila 2004). In this sense, Famatinian structures can be recognized in the Cuyo Sector in a broad part of the northern border of the Precordillera range, while toward the south, these structures are restricted to the Eastern Precordillera (Heredia et al. 2016). Furthermore, no Famatinian structures can be seen in the San Rafael Block (Heredia et al. 2016) (Fig. 2a), which constitutes the extension of the Precordillera toward the south. In the Patagonia Sector, these structures can be recognized in most part of the Argentinean side of the Patagonia.

This orogen presents some quite well-developed internal zones in which high-grade metamorphic conditions and pervasive tectonic foliations can be seen, especially in the Patagonia Sector (Loske et al. 1999; Giacosa et al. 2012, 2010; Serra-Varela et al. 2016; Heredia et al. 2016), where up to three cleavages are developed (Serra-Varela et al. 2016; Heredia et al. 2016). These cleavages seem to be related to both folds and shear zones. In the internal zones of the Cuyo Sector, close to the Famatinian suture (northern Precordillera, Fig. 2a), the Grenvillian Mesoproterozoic basement (Varela et al. 2011) and mid-crustal fragments of the Famatinian magmatic arc (CFA in Fig. 2a) are involved in the Famatinian deformation (Otamendi et al. 2009). However, in both areas, scarce synorogenic granites were identified. As regards the external zones and foreland basins, they are only well preserved in the western branch of the Cuyo Sector (Precordillera).

6 The Chanic Orogeny

The Chanic orogeny (Ramos et al. 1986) has a Middle Devonian to early Carboniferous age (Davis et al. 1999, 2000; Willner et al. 2011; Heredia et al. 2012, 2016; Colombo et al. 2014). It is only recognized in the Cuyo Sector (Fig. 1b), where it has been related to the accretion/collision of the westernmost Chi-Cu sub-plate (Chilenia) with Gondwana (Fig. 2a).

This orogeny shows different characteristics from N to S. In the north, due to the lack of oceanic crust separating the Cuyania and Chilenia sub-plates, subduction was not required to produce their collision (González-Menéndez et al. 2013; Heredia et al. 2012, 2016). In the south, related to the Devonian Chanic subduction, an incipient magmatic arc was developed (Fig. 2a) and also pre-collisional contractional deformations took place under high-pressure metamorphic conditions (Massonne and Calderón 2008; Willner et al. 2011; García-Sansegundo et al. 2016). These processes only affected the eastern Chilenia margin in Devonian (Tickyj et al. 2011) and Middle Devonian times (ca. 390 Ma, according to Willner et al. 2011). During the subsequent collision of Chilenia with Gondwana (western margin of the previously accreted Cuyania sub-plate) in Late Devonian times (ca. 374 Ma, according to Heredia et al. 2016), the high-pressure rocks were exhumed and emplaced over Chilenia. Moreover, remains of the ancient oceanic crust that separated the two sub-plates (ophiolitic klippes described by Davis et al. 2000) and some fragments of Chilenia were emplaced over the Gondwana margin (Giambiagi et al. 2014; Farias et al. 2016), marking the Chanic suture (Fig. 2a). In this southern part of the Chanic orogen, the internal zones are well developed, mainly on the eastern Frontal Cordillera and the western part of the Precordillera. In these zones, the deformation is developed in medium-grade metamorphic conditions, recognizing shear zones, up to two pervasive cleavages related to folds and some synorogenic granites (Heredia et al. 2012; Giambiagi et al. 2014; Giacosa et al. 2014; García-Sansegundo et al. 2014b, 2016; Farias et al. 2016). By contrast, the northern part of the Chanic orogen is produced by inversion of the rift that separated Cuyania and Chilenia (Fig. 2a), resulting in a marine basin developed over a thin continental crust. This rift inversion produced a less intense deformation, which resulted in a narrower orogen with barely internal zones, where a locally pervasive cleavage, associated with very low metamorphic conditions (von Gosen 1997), was developed. In this context, thrust and related folds are the main structures, which form a well-preserved fold and thrust belt in the eastern branch of the Chanic orogen, located in the Precordillera (Álvarez-Marrón et al. 2006). The foreland areas of the Chanic orogen are only well preserved in this eastern branch (over Cuyania/Gondwana), and their main synorogenic basins have been dated as Tournaisian–Visean (Gallastegui et al. 2014; Colombo et al. 2014).

The Chanic orogenic belt trends N–S and shows a double vergence (Fig. 2a): to the east in the branch developed over Cuyania, and to the west in the branch developed over Chilenia. The western branch is recognized throughout the High Andes (Alta Cordillera de los Andes) between 28°–38° S, while the structures associated with its eastern branch are recognized in the Central Precordillera and in the San Rafael Block (Heredia et al. 2016). Both the Chilean Coastal Cordillera and the Eastern Precordillera are not affected by this orogeny, which suggests that the current western boundary of the High Cordillera (Alta Cordillera de los Andes) at this latitude is a Chanic heritage and can be correlated with the western limit of this Paleozoic orogen.

7 The Gondwanan Orogeny

The Gondwanan orogeny (Ramos 1988b) can be recognized in all the pre-Andean basement of the Argentinean–Chilean Andes but shows significant variations from N to S. In the Patagonia Sector, this orogeny is related to the accretion–collision of Western Antarctica with Gondwana (Heredia et al. 2016) (Figs. 1b and 2b). In the Cuyo and Puna sectors, it is related to subduction developed under the Gondwanan margin (Figs. 1b and 2b), which started in the early Carboniferous (Hervé 1988; Rebolledo and Charrier 1994), specifically in the Visean (ca. 337 Ma) (Heredia et al. 2016). In the Puna Sector, the Gondwanan subduction is a reactivation of the

Ocloyic subduction. Therefore, in those sectors, the Gondwanan orogeny took place from the late Carboniferous to the middle Permian (Fig. 1b).

The Gondwanan orogen is very wide and shows an arcuate shape (Fig. 2b), so that in the northern part it is oriented N–S and in the southern part almost E–W (Giacosa et al. 2012). In the Patagonia Sector, the subduction process that allowed the accretion of Western Antarctica to Gondwana began in the late Silurian and probably led to the development of an Andean-type orogenic belt from the Middle Devonian (ca. 391 Ma) to the early Carboniferous (335 Ma) (Fig. 1b) as proposed by Heredia et al. (2016) based on a reinterpretation of data provided by Pankhurst et al. (2006) and Martínez et al. (2012). During the subsequent collision (late Carboniferous-early Permian) this orogenic belt acquires its characteristic double vergence. So, in the branch developed over the Gondwanan margin, the vergence of the structures is to the internal (concave) part of the Gondwanan arc (Fig. 2b), while the structures of the branch developed in the Western Antarctica (less preserved) show a vergence pointing to the external (convex) arc. However, the entire Patagonian region is located in the Gondwanan branch, as the branch developed on Western Antarctica has moved to the south since the Mesozoic and it can only be recognized in the Antarctic Peninsula (Fig. 1a).

The internal zones of this orogen show evidence of up to three deformational episodes; two of them developed up to high-grade metamorphic conditions accompanied by shear zones and anatectic synorogenic granites. In these zones, the remains of a basal accretionary prism, previously deformed under HP–LT metamorphic conditions (Willner et al. 2004), are also preserved (Fig. 2b). These rocks were thrusted and placed over the margin of Gondwana, hundreds of kilometers away from the original subduction zone (García-Sansegundo et al. 2009) (Fig. 2b). In the Antarctic Peninsula (Western Antarctica plate), Loske et al. (1990) dated the Gondwanan metamorphism as late Carboniferous–early Permian (ca. 315-291 Ma).

The external zones and foreland basins of the Gondwanan collisional orogen in the Gondwanan branch are well preserved away from the Andean Cordillera, for all the Patagonian territory, and the orogenic front is located close to the Ventana Range, to the north of the Huincul lineament (38° S) (Fig. 2b), where the deformation would have a late early Permian age (López-Gamundi et al. 1995).

In the non-collisional part of the Gondwanan orogen (north of 38° S), three different segments can be distinguished (Fig. 2b) according to (i) the variation of the subduction angle through time and (ii) the characteristics of the oceanic reliefs that reached the trench at the end of the orogenic process. Thus, in the Puna Sector and in the southern part of the Cuyo Sector, the subduction angle remained fairly inclined during all the orogenic process, while in the northern part of the Cuyo Sector the subduction was of flat-slab type (Ramos and Folguera 2009) during the late Carboniferous–early Permian (ca. 300-280 Ma) (Alonso et al. 2005, 2014; García-Sansegundo et al. 2014a). The development of the flat-slab subduction allowed the migration of deformation, arc magmatism (Parada 1990; Hervé et al.

2014; Sato et al. 2015) and synorogenic retroarc depocenters (Busquets et al. 2005, 2013) to far more eastward positions than in the two other segments, so that the deformation reached the western Pampean Ranges, nearby the Famatinian suture (Fig. 2b). In the flat-slab segment, the Gondwanan orogeny ended in the middle Permian (ca. 265 Ma, Fig. 1b) when an "oceanic plateau" reached the Gondwana continental margin (Fig. 2b) and forced the migration of the deformation to the trench (García-Sansegundo et al. 2014a). The approach of this submarine relief to the continent led to the steepening of the subduction slab, and its final docking led to the exhumation of the Gondwanan basal accretionary prism. This accretionary prism was thrusted for tens of kilometers over the fore-arc basin, reaching the vicinity of the volcanic arc (García-Sansegundo et al. 2014a).

In the southern part of the Cuyo Sector, the Gondwanan orogeny ends earlier, in the early Permian (ca. 280 Ma, Fig. 1b) (Ramos et al. 2011; Giacosa et al. 2014), with the arrival of oceanic islands to the Gondwanan margin (Hyppolito et al. 2014) (Fig. 2b). This process also led to the exhumation of the basal accretionary prism, although it is not placed so close to the arc as in the central segment (Heredia et al. 2016) (Fig. 2b). The Gondwanan deformation in this segment does not seem to have surpassed eastward of the San Rafael Block.

In the Puna Sector, the Gondwanan deformation also ended in the middle Permian (Fig. 1b). Here the basal accretionary prism was also exhumed and placed far away from the old margin (after Lucassen et al. 1999), suggesting that some sort of oceanic relief could have docked to this region of the Gondwana margin. However, the Gondwanan deformation is practically limited to the Chilean Andes (Fig. 1b), indicating that (i) this subduction was the steepest of all of them and/or (ii) the relief that approached the Puna margin did not exist or was smaller than those accreted in the adjacent Cuyo Sector. Furthermore, the late Carboniferous and Permian sediments of the eastern Argentinean Andes at this latitude (Eastern Cordillera) have been related to extensional basins.

The internal zones of the Gondwanan subduction orogen (named Western Series by the Chilean authors) are restricted to the exhumed remains of the basal accretionary prism, where three deformation events can be recognized, mainly represented by tectonic foliations. The first of these foliations developed in the basal accretionary prism under HP–LT metamorphic conditions and is preserved as relictic into albite and/or garnet porphyroblasts (Willner et al. 2004). The other two deformation events were developed in relation to the exhumation of the basal accretionary prism (García-Sansegundo et al. 2014a) and, excluding the last one, they are pervasive. As in all the Andean-type orogens, this orogen shows a single vergence (Fig. 2b), in this case to the east, although the presence of structures with an opposite vergence is common near the accretionary prism zone.

8 The Tabarin Orogeny

The Tabarin orogeny is only recognized in the Patagonia Sector and has been related to an early Permian–Triassic subduction episode (Heredia et al. 2016). The subduction zone was located along the western margin of the Western Antarctica continental fragment (Heredia et al. 2016), previously accreted to Gondwana during the Gondwanan cycle, which is now part of the Pangea supercontinent (Fig. 2b). The Tabarin orogen is late Permian–Triassic in age (ca. 265-208 Ma) (Hervé 1992; Heredia et al. 2016) and is best preserved in the Antarctic Peninsula. In this area, several authors reported compressive deformations of this age (Thomson 1975; Smellie 1991; Hervé 1992; Muñoz et al. 1992; del Valle et al. 2007) as well as the presence of retroarc synorogenic deposits, typical of subduction-related orogens (del Valle et al. 2007). In the surroundings of the Antarctic Peninsula, the main structures related to the Tabarin orogeny are folds and thrusts with NW-oriented trend and NE vergences (Fig. 2b), developed under very low-grade to non-metamorphic conditions (Muñoz et al. 1992; del Valle et al. 2007). However, these structures and deposits were initially attributed to the Gondwanan orogeny.

Since the late Permian (ca. 265 Ma, Tickyj et al. 1997) and especially during the Triassic, the entire Patagonian territory, including the Andes, changed from a post-Gondwanan extensional regime to a new setting in which compressional stresses are transmitted from a new subduction zone developed to the W and SW, underneath the margin of the new Pangea supercontinent (Heredia et al. 2016) (Fig. 2b). In this context, probably transpressive, the Tabarin orogeny resulted in several NW–SE sinistral shear zones (Coira et al. 1975; Proserpio 1978; Nullo 1978; Volkheimer and Lage 1981; Llambías et al. 1984; von Gosen 2003; von Gosen and Loske 2004; Pankhurst et al. 2006, among others), which overlapped the sub-parallel Gondwanan structures, affecting also some post-Gondwanan plutons with ages up to ca. 208 Ma (Rapela et al. 1992).

Acknowledgements This chapter shows the main results of the InverAndes (BTE2002-04316-C03), PaleoAndes I (CGL2006-12415-CO3/BTE), PaleoAndes II (CGL2009-13706-CO3), TORANDES (CGL2012-38396-C03) projects, supported by the I+D+i Spanish Plan and FEDER Funds from the European Union. We are especially grateful to Romina Sulla for reviewing the English text. Thanks also due to P. González and M. Naipauer for their suggestions and valuable comments.

References

Aceñolaza FG, Aceñolaza G (2005) La Formación Puncoviscana y unidades estratigráficas vinculadas en el Neoproterozoico-Cámbrico Temprano del Noroeste Argentino. Latin American J Sedimentology and Basin Analysis 12:65–87

Aceñolaza FG, Toselli AJ (1976) Consideraciones estratigráficas y tectónicas sobre el Paleozoico inferior del Noroeste Argentino. Actas 2° Congreso Latinoamericano de Geología (Caracas) 2, pp 755–763

Aceñolaza FG, Miller H, Toselli AJ (2002) Proterozoic-Early Paleozoic evolution in western South America—a discussion. Tectonophysics 354:121–137

Adams CJ, Miller H, Aceñolaza FG, Toselli A, Griffin WL (2011) The Pacific Gondwana margin in the late Neoproterozoic-early Paleozoic: Detrital zircon U-Pb ages from metasediments in northwest Argentina reveal their maximum age, provenance and tectonic setting. Gondwana Res 19:71–83

Alonso JL, Gallastegui J, Rodríguez Fernández LR, García-Sansegundo J (2014) Stratigraphy and structure of the Punta Negra Anticline. Implications on the structural evolution of the Argentine Precordillera. Journal of Iberian Geology 40(2):283–292

Alonso JL, Rodríguez Fernández LR, García-Sansegundo J, Heredia N, Farias P and Gallastegui J (2005) Gondwanic and Andean structure in the Argentine Central Precordillera: The Río San Juan section revisited. 6th Internacional Symposium on Andean Geodynamics. IRD Editions, París, pp 36–39

Alonso JL, Seggiaro R, Quintana L, Gallastegui J, Bulnes M, Poblet J, Heredia N, Rodríguez-Fernández LR (2012) Paleozoic deformations in the Eastern Cordillera of the Andes at 23° S (NW Argentina). Geo-Temas 13:1844–1847

Álvarez-Marrón J, Rodríguez-Fernández LR, Heredia N, Busquets P, Colombo F, Brown D (2006) Neogene structures overprinting Palaeozoic thrust systems in the Andean Precordillera at 30° S latitude. J Geol Soc London 163:949–964

Aparicio González PA, Pimentel MM, Hauser N, Moya MC (2014) U-Pb LA-ICP-MS geochronology of detrital zircón grains from low-grade metasedimentary rocks (Neoproterozoic-Cambrian) of the Mojotoro Range, northwest Argentina. J South American Earth Sci 49:39–59

Astini RA (2008) Sedimentación, facies, discordancias y evolución paleoambiental durante el Cámbrico-Ordovícico. In: Coira B, Zappettini E (eds.) Geología y Recursos Naturales de la Provincia de Jujuy. Relatorio 17th Congreso Geológico Argentino, Asociación Geológica Argentina, Jujuy, pp 50–73

Astini RA, Benedetto JL, Vaccari NE (1995) The early Paleozoic evolution of the Argentine Precordillera as a Laurentia rifted, drifted and collided terrane: a geodynamic model. Geological Society of America Bulletin 107:253–273

Astini RA, Dávila FM (2004) Ordovician back arc foreland and Ocloyic thrust belt development on the western Gondwana margin as a response to Precordillera terrane accretion. Tectonics 23:1–19

Bahlburg H (1990) The Ordovician basin in the Puna of NW Argentina and N Chile: geodynamic evolution from back-arc basin to foreland basin. Geotekton Forsch 75:1–107

Bahlburg H, Hervé F (1997) Geodynamic evolution and tectonostratigraphic terranes of Northwestern Argentina and Northern Chile. Geological Society of America Bulletin 109 (7):869–884

Baldis BA, Peralta S, Villegas R (1989) Esquematizaciones de una posible transcurrencia del terrane de Precordillera como como fragmento continental procedente de áreas pampeano-bonaerenses. Instituto Superior de Correlación Geológica 5:81–100

Becchio R, Lucassen F, Franz G, Viramonte J, Wemmer K (1999) El basamento paleozoico inferior del noroeste de Argentina (23°–27° S)-metamorfismo y geocronologia. In: González Bonorino G, Omarini R, Viramonte J (eds) Geología del Noroeste Argentino, Relatorio 19th Congreso Geológico Argentino 1, pp 58–72

Busquets P, Colombo F, Heredia N, De Porta NS, Rodríguez-Fernández LR, Álvarez Marrón J (2005) Age and tectonostratigraphic significance of the Upper Carboniferous series in the basement of the Andean Frontal Cordillera: Geodynamic implications. Tectonophysics 399(1–4):181–194

Busquets P, Limarino CO, Cardo R, Méndez-Bedia I, Gallastegui G, Colombo F, Heredia N, Césari SN (2013) The neopaleozoic of the Sierra de Castaño (Andean Cordillera Frontal, San Juan, Argentina): Tectonic and paleoenviromental reconstruction. Andean Geology 40(1):172–195

Coira B, Kay SM, Pérez B, Woll B, Hanning M, Flores P (1999) Magmatic sources and tectonic setting of Gondwana margin Ordovician magmas, northern Puna of Argentina and Chile. In: Ramos VA, Keppie JD (eds) Laurentia-Gondwana Connections before Pangea. Geological Society of America, Special Paper 336, pp 145–170

Coira B, Kirschbaum A, Hongn F, Pérez B, Menegatti N (2009) Basic magmatism in northeastern Puna, Argentina: Chemical composition and tectonic setting in the Ordovician back-arc. J South American Earth Sci 28:374–382

Coira B, Nullo F, Proserpio C, Ramos VA (1975) Tectónica de basamento de la región occidental del Macizo Nordpatagónico (Prov. de Río Negro y Chubut) República Argentina. Revista de la Asociacón Geológica Argentina 30(3):361–383

Collo G, Astini RA, Cardona A, do Campo MD, Cordani U (2008) Edades de metamorfismo en las unidades con bajo grado de la región central del Famatina: la impronta del ciclo orogénico oclóyico (Ordovícico). Revista Geológica de Chile 35(2):191–213

Colombo F, Limarino CO, Spalletti LA, Busquets P, Cardó R, Méndez-Bedia I, Heredia N (2014) Late Palaeozoic lithostratigraphy of the Andean Precordillera revisited (San Juan Province, Argentina). J Iberian Geology 40(2):241–259

Dalla Salda LH, Cingolani C, Varela R (1992) Early Palaeozoic Orogenic belt of the Andes in southwestern South-America—result of Laurentia-Gondwana collision. Geology 20(7):617–620

Davis JS, Roeske SM, McClelland WM, Kay SM (2000) Mafic and ultramafic crustal fragments of the SW Precordillera terrane and their bearing on tectonic models of the early Paleozoic in Western Argentina. Geology 28:171–174

Davis JS, Roeske SM, McClelland WC, Snee LW (1999) Closing the ocean between the Precordillera terrane and Chilenia: Early Devonian ophiolite emplacement and deformation in the SW Precordillera. Ramos VA, Keppie JD (eds), Laurentia-Gondwana Connections before Pangea. Geological Society of America, Special Paper 336:115–138

del Valle R, Heredia N, Montes M, Nozal F, Martín-Serrano A (2007) El Grupo Trinity Peninsula en la Península Tabarín, extrtemo norte de la Península Antártica. Revista de la Asociación Geológica Argentina 62(4):498–505

Díaz-Martínez E (1996) Síntesis estratigráfica y geodinámica del Carbonífero de Bolivia. Memorias del XII Congreso Geológico de Bolivia (Tarija, Bolivia), pp 355–367

Escayola MP, van Staal CR, Davis WJ (2011) The age and tectonic setting of the Puncoviscana Formation in northwestern Argentina: an accretionary complex related to Early Cambrian closure of the Puncoviscana Ocean and accretion of the Arequipa-Antofalla block. J South American Earth Sci 32:437–458

Farias P, García-Sansegundo J, Rubio-Ordóñez Á, Clariana P, Cingolani C, Heredia N (2016) La deformación Chánica en el Bloque San Rafael (Provincia de Mendoza, Argentina): implicaciones tectónicas. Geo-Temas 16(2):411–414

Finch M, Weinberg R, Fuentes G, Hasalová P, Becchio R (2015) One kilometre-thick ultramylonite, Sierra de Quilmes, Sierras Pampeanas, NW Argentina. J Structural Geology 72:33–54

Finney SC (2007) The parautochthonous Gondwanan origin of Cuyania (greater Precordillera) terrane of Argentina: A reevaluation of evidence used to support an allochthonous Laurentian origin. Geologica Acta 5:127–158

Finney SC, Gleason JD, Gehrels GG, Peralta SH, Aceñolaza G (2003) Early Gondwanan Connection for the Argentine Precordillera Terrane. Earth and Planetary Sciences Letters 205:349–359

Gallastegui G, González-Menéndez L, Rubio-Ordóñez A, Cuesta A, Gerdes A (2014) Origin and provenance of igneous clasts from late Palaeozoic conglomerate formations (Del Ratón and El Planchón) in the Andean Precordillera of San Juan, Argentina. J Iberian Geology 40(2):261–282

García-Sansegundo J, Farias P, Gallastegui G, Giacosa R, Heredia N (2009) Structure of the North-Patagonian Gondwanic basement in the Bariloche area. Int J Earth Sci 98:1599–1608
García-Sansegundo J, Farias P, Heredia N, Gallastegui G, Charrier R, Rubio-Ordóñez A, Cuesta A (2014a) Structure of the Andean Palaeozoic basement in the Chilean coast at 31°30'S: Geodynamic evolution of a subduction margin. J Iberian Geology 40(2):293–308
García-Sansegundo J, Farias P, Rubio-Ordóñez Á, Heredia N (2014b) The Palaeozoic basement of the Andean Frontal Cordillera at 34° S (Cordón del Carrizalito, Mendoza Province, Argentina): Geotectonic implications. J Iberian Geology 40(2):321–330
García-Sansegundo J, Gallastegui G, Farias P, Rubio-Ordóñez A, Cuesta A, Heredia N, Giambiagi L, Clariana P (2016) Evolución tectono-metamórfica Chánica del Complejo Guarguaraz, Cordillera Frontal de los Andes (Mendoza, Argentina). Geo-Temas 16(2):427–430
Giacosa RE, Zubia M, Sanchez M, Allard J (2010) Meso-Cenozoic tectonics of the southern Patagonian foreland: structural evolution and implications for Au-Ag veins in the eastern Deseado Region (Santa Cruz, Argentina). J S Am Earth Sci 30(3–4):134–150. https//:doi.org. 10.1016/j.jsames.2010.09.002
Giacosa R, Allard J, Foix N, Heredia N (2014) Stratigraphy, structure and geodynamic evolution of the Paleozoic rocks in the Cordillera del Viento (37° S latitude, Andes of Neuquén, Argentina). J Iberian Geology 40(2):331–348
Giacosa R, Fracchia D, Heredia N (2012) Structure of the Southern Patagonian Andes at 49° S, Argentina. Geologica Acta 10(3):265–282
Giambiagi LB, Mescua JF, Heredia N, Farias P, García-Sansegundo J, Fernández C, Stier S, Pérez DJ, Bechis F, Moreiras SM, Lossada A (2014) Reactivation of Paleozoic structures during Cenozoic deformation in the Cordon del Plata and Southern Precordillera ranges (Mendoza, Argentina). J Iberian Geology 40(2):309–320
González PD, Tortello MF, Damborenea SE (2011) Early Cambrian archaeocyathan limestone blocks in low-grade meta-conglomerate from El Jagüelito Formation (Sierra Grande, Río Negro, Argentina). Geologica Acta 9(2):159–173
González-Menéndez L, Gallastegui G, Cuesta A, Heredia N, Rubio-Ordóñez A (2013) Petrogenesis of Early Paleozoic basalts and gabbros in the western Cuyania terrane: Constraints on the tectonic setting of the southwestern Gondwana margin (Sierra del Tigre, Andean Argentine Precordillera). Gondwana Res 24(1):359–376
Grissom G, Debari S, Snee L (1998) Geology of the Sierra de Fiambalá, nothwestern Argentina: Implications for Early Paleozoic Andean tectonics. In: Pankhurst B, Rapela C (eds) The Proto Andean margin of Gondwana. Geological Society, Special Publication 142:297–323
Heredia N, Farias P, García-Sansegundo J, Giambiagi L (2012) The Basement of the Andean Frontal Cordillera in the Cordón del Plata (Mendoza, Argentina): Geodynamic Evolution. Andean Geology 39(2):242–257
Heredia N, García-Sansegundo J, Gallastegui G, Farias P, Giacosa R, Alonso JL, Busquets P, Charrier R, Clariana P, Colombo F, Cuesta A, Gallastegui J, Giambiagi L, González-Menéndez L, Limarino CO, Martín-González F, Pedreira D, Quintana L, Rodríguez-Fernández LR, Rubio-Ordóñez A, Seggiaro R, Serra-Varela S, Spalletti L, Cardó R, Ramos VA (2016) Evolución Geodinámica de los Andes de Argentina, Chile y la Península Antártica durante el Neoproterozoico superior y el Paleozoico. Trabajos de Geología 35:000–000
Hervé F (1988) Late Palaeozoic subduction and accretion in Southern Chile. Episodes 11:183–188
Hervé F (1992) Estado actual del conocimiento del metamorfismo y plutonismo en la Península Antártica al norte de los 65° S y el archipiélago de las Shetlands del Sur: Revisión y problemas. In: López-Martínez J (ed) Geología de la Antártida Occidental. Simposios del III Congreso Geológico de España y VIII Congreso Latinoamericano de Geología 3, pp 19–31

Hervé F, Fanning M, Calderón MR, Mpodozis C (2014) Early Permian to Late Triassic batholiths of the Chilean Frontal Cordillera (28°–31° S): SHRIMP U-Pb zircon ages and Lu–Hf and O isotope systematics. Lithos 184–187:436–446

Hongn F, Mon R, Petrinovic I, del Papa C, Powell J (2010a) Inversión y reactivación tectónicas cretácico-cenozoicas en el noroeste argentino: influencia de las heterogeneidades del basamento neoproterozoico-paleozoico inferior. Revista de la Asociación Geológica Argentina 66:38–53

Hongn F, Riller U (2007) Tectono-magmatic evolution of the western margin of Gondwana inferred from syntectonic emplacement of Paleozoic granitoid plutons in NW-Argentina. J Geology 115:163–180

Hongn F, Tubía JM, Aranguren A, Vegas N, Mon R, Dunning G (2010b) Magmatism coeval with lower paleozoic shelf basins in NW-Argentina (Tastil batholith): constraints on current stratigraphic and tectonic interpretations. J South American Earth Sci 29:289–305

Hongn F, Tubía JM, Esteban JJ, Aranguren A, Vegas N, Sergeev S, Larionov A, Basei M (2014) The Sierra de Cachi (Salta, NW Argentina): geological evidence about a Famatinian back-arc at mid crustal levels. J Iberian Geology 40(2):225–240

Hongn F, Vaccari NE (2008) La discordancia tremadociano superior-arenigiana inferior en vega Pinato (Salta): Evidencia de deformación intraordovícica en el borde occidental de la Puna. Actas 17 Congreso Geológico Argentino III, pp 1299–1300

Hyppolito T, Juliani C, García-Casco A, Meira VT, Bustamante A, Hervé F (2014) The nature of the Palaeozoic oceanic basin at the southwestern margin of Gondwana and implications for the origin of the Chilenia terrane (Pichilemu region, central Chile). Int Geology Rev 56(9):1097–1121

Jezek P, Wilner AP, Aceñolaza FG, Miller H (1985) The Puncoviscana trough-a large basin of Late Precambrian to Early Cambrian age on the Pacific edge of the Brazilian shield. Geologische Rundsch 4:573–584

Keller M (1999) Argentine Precordillera: sedimentary and plate tectonic history of a Laurentia crustal fragment in South America. Geological Society of America, Special Paper 341:1–131

Keppie JD, Bahlburg H (1999) Puncoviscana formation of northwestern and central Argentina: Passive margin or foreland basin deposit? Geological Society of America, Special Paper 336:139–143

Llambías EJ, Llano JA, Rossa N, Castro CE, Puigdomenech HH (1984) Petrografía de la Formación Mamil Choique en la Sierra del Medio-Departamento Cushamen-Provincia del Chubut. Actas IX Congreso Geológico Argentino (San Carlos de Bariloche/Río Negro) 2, pp 554–567

López VL, Gregori DA (2004) Provenance and evolution of the Guarguaraz Complex, Cordillera Frontal, Argentina. Gondwana Res 7:1197–1208

López-Gamundi OR, Conaghan PJ, Rosello EA, Cobbold PR (1995) The Tunas Formation (Permian) in the Sierras Australes Fold belt, East-Central Argentina: evidence of syntectonic sedimentation in a Varisican foreland basin. J South American Earth Sci 8(2):129–142

Loske W, Márquez M, Giacosa R, Pezzuchi H, Fernández MI (1999) U/Pb geochronology of pre-Permian Basement Rocks in the Macizo del Deseado, Santa Cruz province, Argentine Patagonia. Actas XV Congreso Geológico Argentino (Salta-Argentina), pp 1–102

Loske W, Miller H, Milne A, Hervé F (1990) U-Pb zircon ages of xenoliths from Cape Dubouzet, Antarctic Peninsula. Zentralblatt fúr Geologie und Paláontologie 1(1–2):87–95

Lucassen F, Franz G, Laber A (1999) Permian high pressure Rocks. The basement of the Sierra de Limón Verde in northern Chile. J South American Earth Sciences 12(2):183–199

Lucassen F, Becchio R, Wikle HG, Franz G, Thirlwall MF, Viramonte J, Wemmer K (2000) Proterozoic-Paleozoic development of the basement of the Central Andes (18°–26° S)—a mobile belt of the South American craton. J South American Earth Sci 13:697–715

Manceñido M, Damborenea S (1984) Megafaunas de invertebrados paleozoicos y mesozoicos. In: Ramos, VE (ed) Relatorio del IX Congreso Geológico Argentino: Geología y recursos naturales de la provincia de Río Negro (San Carlos de Bariloche-Argentina), pp 413–466

Martínez JC, Dristas JA, Massonne HJ (2012) Palaeozoic accretion of the microcontinent Chilenia, North Patagonian Andes: high-pressure metamorphism and subsequent thermal relaxation. Int Geology Rev 54(4):472–490

Massonne HJ, Calderón M (2008) P-T evolution of metapelites from the Guarguaraz Complex, Argentina: evidence for Devonian crustal thickening close to the western Gondwana margin. Revista Geológica de Chile 35(2):215–231

Méndez V, Navarini A, Plaza D, Viera O (1973) Faja eruptiva de la Puna oriental. Actas 5° Congreso Geológico Argentino (Buenos Aires) 4, pp 89–100

Mon R, Hongn F (1987) Estructura del Ordovícico de la Puna. Revista de la Asociación Geológica Argentina 42:31–38

Mon R, Hongn F (1991) The structure of the Precambrian and Lower Paleozoic Basement of the Central Andes between 22° and 32° S Lat. Geol Rundsch 83:745–758

Moya C (2015) La "Fase Oclóyica" (Ordovícico Superior) en el noroeste argentino. Interpretación histórica y evidencias en contrario. Serie Correlación Geológica, Contribuciones a la Geología Argentina 31:73–110

Mulcahy SR, Roeske S, McClelland WC, Jourdan F, Iriondo A, Renne PR, Vervoot JD, Vujovich G (2011) Structural evolution of a composite middle to lower crustal section: The Sierra de Pie de Palo, northwest Argentina. Tectonics 30(1):TC1005

Müller H (1965) Zur Alterfrage der eisenerzlagerstätte Sierra Grande/Río Negro in Nordpatagonien Aufgrund neuer Fossilfunde. Geologische Rundschau 54:715–732

Muñoz JA, Sabat F, Pallás R (1992) Estructura pre-cretácica de la Península Hurd, Isla Livingston, Islas Shetland del Sur. In: López-Martínez J (ed) Geología de la Antártida Occidental. Simposios del III Congreso Geológico de España y VIII Congreso Latinoamericano de Geología 3, pp 127–140

Niemeyer H, Meffre S, Guerrero R (2014) Zircon U-Pb geochronology of granitic rocks of the Cordón de Lila and Sierra de Almeida ranges, northern Chile: 30 m.y. of Ordovician plutonism on the western border of Gondwana. J South American Earth Sci 56:228–241

Nullo FE (1978) Descripción Geológica de la Hoja 41d, Lipetrén, Provincia de Río Negro. Boletın del SEGEMAR 158:1–88

Otamendi JE, Vujovich GI, de la Rosa JD, Tibaldi AM, Castro A, Martino RD, Pinotti LP (2009) Geology and petrology of a deep crustal zone from the Famatinian paleo-arc, Sierras de Valle Fértil and La Huerta, San Juan, Argentina. J South American Earth Sci 27(4):258–279

Pankhurst RJ, Rapela CW, Fanning CM, Márquez M (2006) Gondwanide continental collision and the origin of Patagonia. Earth Sci Rev 7:235–257

Pankhurst RJ, Rapela CW, Saavedra J, Baldo E, Dahlquist J, Pascua I, Fanning CM (1998) The Famatinian magmatic arc in the Central Sierras Pampeanas: An early to mid-Ordovician continental arc on the Gondwana margin. In: Pankhurst RJ, Rapela CW (eds) The Proto-Andean Margin of Gondwana. Geological Society of London, Special Publication 142, pp 343–367

Parada MA (1990) Granitoid plutonism in Central Chile and its geodynamic implications: a review. Geological Society of America, Special Paper 241, pp 51–65

Ploszkiewicz JV, Orchuela IA, Vaillard JC, Viñes R (1984) Compresión y desplazamiento lateral en la zona de la Falla Huincul, estructuras asociadas, Provincia del Neuquén. Actas 9° Congreso Geológico Argentino (Bariloche) 2, pp 163–169

Proserpio CA (1978) Descripción Geológica de la Hoja 42d, Gastre, Provincia del Chubut. Boletín del SEGEMAR 159:1–75

Ramos VA (1972) El Ordovícico fosilífero de la Sierra de Lina. Provincia de Jujuy. Revista de la Asociación Geológica Argentina 27(1):84–94

Ramos VA (1986) El diastrofismo oclóyico: un ejemplo de tectónica de colisión en el Eopaleozoico del noroeste Argentino. Revista del Instituto de Ciencias Geológicas 6:13–28

Ramos VA (1988a) Tectonics of the Late Proterozoic-early Paleozoic: A collisional history of southern South America. Episodes 11(3):168–174

Ramos VA (1988b) The tectonics of the Central Andes; 30° to 33° S latitude. In: Clark S, Burchfiel D (eds) Processes in continental lithospheric deformation. Geological Society of America, Special Paper 218:31–54

Ramos VA (1999) Rasgos estructurales del territorio argentino. In: Caminos R (ed) Geología de Argentina. Servicio Geológico y Minero Argentino, Anales 29:715–784

Ramos VA (2004) Cuyania, an exotic block to Gondwana: review of a historical success and the present problems. Gondwana Res 7:1009–1026

Ramos VA (2008) The Basement of the Central Andes: The Arequipa and Related Terranes. Annual Review of Earth and Planetary Sci 36:289–324

Ramos VA (2009) Anatomy and global context of the Andes: Main geologic features and the Andean orogenic cycle. In: Kay SM, Ramos VA, Dickinson W (eds) Backbone of the Americas: Shallow Subduction, Plateau Uplift, and Ridge and Terrane Collision. Geological Society of America Memoirs 204, pp 31–65

Ramos VA, Folguera A (2009) Andean flat-slab subduction through time. In: Murphy B (ed) Ancient Orogens and Modern Analogues. Geological Society London, Special Publication 327, pp 31–54

Ramos VA, Jordan TE, Allmendinger RW, Mpodozis C, Kay SM, Cortés JM, Palma MA (1986) Paleozoic Terranes of the Central Argentine-Chilean Andes. Tectonics 5(6):855–880

Ramos VA, Mosquera A, Folguera A, García Morabito E (2011) Evolución tectónica de los Andes y del engolfamiento neuquino adyacente. In: Leanza H et al. (eds) Geología y Recursos Naturales de la Provincia del Neuquén. Relatorio 18° Congreso Geológico Argentino (Neuquén), pp 335–344

Ramos VA, Naipauer M (2014) Patagonia: where does it come from? J Iberian Geology 40 (2):367–379

Rapela CW, Casquet C, Baldo EGA, Dalquist J, Pankhurst RJ, Galindo C, Saavedra J (2001) Las Orogénesis del Paleozoico Inferior en el margen proto-andino de América del Sur, Sierras Pampeanas, Argentina. J Iberian Geology 27:23–41

Rapela CW, Pankhurst RJ, Casquet C, Baldo EGA, Saavedra J, Galindo C (1998) Early evolution of the Proto-Andean margin of South America. Geology 26(8):707–710

Rapela CW, Pankhurst RJ, Harrison R (1992) Triassic Gondwanan granites of the Gastre district, North Patagonian Massif. Transaction of the Royal Society of Edimburg, Earth Sciences 83:291–304

Rebolledo S, Charrier R (1994) Evolución del basamento paleozoico en el área de Punta Claditas, Región de Coquimbo, Chile (31°–32° S). Revista Geológica de Chile 21(1):55–69

Sato AM, Llambías EJ, Basei MAS, Castro CE (2015) Three stages in the Late Paleozoic to Triassic magmatism of southwestern Gondwana, and the relationships with the volcanogenic events in coeval basins. J South American Earth Sciences 63:48–69

Serra-Varela S, González PD, Giacosa R, Heredia N, Pedreira D, Martín-González F (2016) Geología y geocronología del basamento paleozoico de los Andes Norpatagónicos en el área de San Martín de los Andes. Geo-Temas 16(2):431–434

Smellie J (1991) Stratigraphy, provenance and tectoning setting of Late-Paleozoic-Triassic sedimentary sequences in northern Graham Land and South Scotia Ridge. In: Thomson, MRA, Crame, JA, Thomson, JW (eds) Geological Evolution of Antarctica. Cambridge University Press, pp 411–417

Sola AM, Becchio RA, Pimentel MM (2010) Leucogranito Pumayaco: anatexis cortical durante el ciclo orogénico Famatiniano en el extremo norte de la Sierra de Molinos, provincia de Salta. Revista de la Asociación Geológica Argentina 66:206–224

Sola AM, Becchio RA, Pimentel MM (2013) Petrogenesis of migmatites and leucogranites from Sierra de Molinos, Salta, Northwest Argentina: A petrologic and geochemical study. Lithos 177:470–491

Sureda RJ, Omarini R (1999) Evolución geológica y nomenclatura pre-Gondwánica en el Noroeste de Argentina (1800-160 Ma). Acta Geológica Hispánica 34(2–3):197–225

Thomas WA, Astini RA (2003) Ordovician accretion of the Argentine Precordillera terrane to Gondwana: a review. J South American Earth Sci 16:67–79

Thomson MRA (1975) New palaeontological, lithological observations on the Legoupil Formation, northwest Antarctic Peninsula. British Antarctic Survey Bulletin 41–42:169–185

Tickyj H (2011) Granitoides calcoalcalinos tardío-Famatinianos en el Cordón del Carrizalito, Cordillera Frontal, Mendoza. Actas 18 Congreso Geológico Argentino (Neuquén, Argentina): Petrología Ígnea y Metamórfica, pp 1–2

Tickyj H, Dimieri LV, Llambías EJ, Sato AM (1997) Cerro de Los Viejos (38°28′S–64°26′O): Cizallamiento ductil en el sudeste de La Pampa. Revista de la Asociación Geológica Argentina 52(3):311–321

Tubía JM, Hongn FD, Aranguren A, Vegas N (2012) Inversión tectónica en la corteza dúctil: Basamento neoproterozoico-paleozoico inferior de la Sierra de Cachi (Salta, Argentina). Geo-Temas 13:1895–1898

Turner JCM (1960) Estratigrafía de la Sierra de Santa Victoria y adyacencias. Boletín de la Academia Nacional de Ciencias de Córdoba 41(2):163–196

Turner JCM, Méndez V (1975) Geología del sector oriental de los Departamentos de Santa Victoria e Iruya, Provincia de Salta, República Argentina. Boletín de la Academia Nacional de Ciencias de Córdoba 51(1–2):11–24

Turner JCM, Mon R (1979) Cordillera Oriental. In: Turner JCM (ed) Geología Regional Argentina, Academia Nacional de Ciencias de Córdoba, pp 57–94

Varela R, Basei MAS, González PD, Sato AM, Naipauer M, Campos Neto M, Cingolani CA, Meira VT (2011) Accretion of Grenvillian terranes to the southwestern border of the Río de la Plata craton, western Argentina. Int J Earth Sci 100:243–272

Vergara H (1978) Carta Geológica de Chile a E. 1:50.000, Cuadrángulo Ujina, Región de Tarapacá. Servicio Nacional de Geología y Minería, pp 1–33

Volkheimer W, Lage J (1981) Descripción Geológica de la Hoja 42c, Cerro Mirador, Provincia del Chubut. Boletıín del SEGEMAR 181:1–71

von Gosen W (1997) Early Paleozoic and Andean structural evolution in the Río Jáchal section of the Argentine Precordillera. J South American Earth Sci 10:361–388

von Gosen W (2003) Thrust tectonics in the North Patagonian massif (Argentina): implications for a Patagonian plate. Tectonics 22(1):5-1–5-33

von Gosen W, Loske W (2004) Tectonic history of the Calcatapul Formation, Chubut province, Argentina, and the "Gastre fault system". J South American Earth Sci 18:73–88

Willner AP, Gerdes A, Massonne HJ, Schimidt A, Sudo M, Thomson SN, Vujovich G (2011) The geodynamics of collision of a microplate (Chilenia) in Devonian times deduced by the pressure-temperature time evolution within part of a collisional belt (Guarguaraz Complex, W-Argentina). Contribution to Mineralogy and Petrology 162:303–327

Willner AP, Glodny J, Gerya TV, Godoy E, Massonne HJ (2004) A counterclockwise PTt path of high-pressure/low-temperature rocks from the Coastal Cordillera accretionary complex of south-central Chile: constraints for the earliest stage of subduction mass flow. Lithos 75:283–310

Zimmermann U (2005) Provenance studies of very low-to low-grade metasedimentary rocks of the Puncoviscana complex, northwest Argentina. In: Vaughan AP, Leat, PT, y Pankhurst, RJ (eds) Terrane Processes at the Margins of Gondwana. Geological Society of London, Special Publications 246:381–416

Zimmermann U, Niemeyer H, Meffre S (2010) Revealing the continental margin of Gondwana: the Ordovician arc of the Cordón de Lila (northern Chile). Int J Earth Sci 99:39–56

The Famatinian Orogen Along the Protomargin of Western Gondwana: Evidence for a Nearly Continuous Ordovician Magmatic Arc Between Venezuela and Argentina

Victor A. Ramos

Abstract The continental protomargin of Western Gondwana in South America records an important Early-Middle Ordovician magmatic activity associated with the development of the Famatinian orogen. Almost the entire margin has evidence of a magmatic arc preserved as orthogneisses in the high-grade metamorphic domains up to volcanic rocks of the same age interfingered with sedimentary facies. These subduction-related calcalkaline rocks have new U-Pb zircon dates that show striking similar ages bracketed between 490 and 460 Ma. The different domains along the continental margin are compared taking the western Sierras Pampeanas as the type locality, showing an alternation among high-grade metamorphic—greenschist facies—sedimentary facies. There are three deeply exhumed segments preserved as orthogneisses in high-grade amphibolite facies, the Sierras Pampeanas, the Marañón, and the Santander-Mérida domains. These domains are flanked by greenschist facies such as the Quetame in Colombia, the Vilcabamba in Perú, and the Puna Eruptive Belt in northern Argentina. Some segments are characterized by sedimentary facies as the Altiplano domain of Bolivia and the Olmos-Loja domain between Perú and Ecuador. The location and metamorphic grade are controlled by the amount of shortening and uplift, responsible for the different crustal levels exposed, as a consequence of the characteristics of the distinct terranes that collided against the continental margin. As a final remark, the time span of the Famatinian episode when globally compared has a widespread development in Laurentia, Baltica, and Australia, as a consequence of a period of high mobility of the plates during Early-Middle Ordovician times.

Keywords Magmatic arc · Terrane collision · Subduction related Metamorphic grade · Protomargin · Paleozoic orogen

---

V. A. Ramos (✉)
Instituto de Estudios Andinos "Don Pablo Groeber", Departamento de Ciencias Geológicas, FCEN, Universidad de Buenos Aires–CONICET, Buenos Aires, Argentina
e-mail: andes@gl.fcen.uba.ar

A. Folguera et al. (eds.), *The Evolution of the Chilean-Argentinean Andes*, Springer Earth System Sciences, https://doi.org/10.1007/978-3-319-67774-3_6

1 Introduction

Along the protomargin of Western Gondwana, there is evidence of subduction-related arc rocks during the Early Paleozoic (Fig. 1). These rocks are emplaced in the present metamorphic basement exposed along the western continental margin of South America. Although numerous authors described the geology, the geochemistry, and the geochronological data in different segments of Venezuela, Colombia, Perú, Chile, and Argentina, there is not a comprehensive view of their importance, time constraints, and paleogeographic significance. The understanding of these magmatic rocks of mainly Ordovician age is important to constrain the tectonic setting of the protomargin of Western Gondwana and several processes that have affected almost the entire margin at that time.

The reconstruction of the Paleozoic protomargin requires the recognition of the different episodes of terrane accretion that affected the margin during Mesozoic and Cenozoic times. The accretion of oceanic terranes is very conspicuous in the Northern Andes, and the ancient continental margin has to be reconstructed by removing the terranes accreted after the Ordovician (Restrepo and Toussaint 1988; Aleman and Ramos 2000; Ordóñez-Carmona et al. 2006; Ramos 2009; Gómez Tapias et al. 2015). The Early Paleozoic restoration of Western Gondwana also needs the removal of Patagonia, which is considered an allochthonous terrane accreted during the Late Paleozoic (Ramos 1984, 2008; Pankhurst et al. 2006). Although these last two proposals have different paleogeographic boundaries for northern Patagonia, after the finding of Early Cambrian archeocyatids in northern Patagonia, there is more consensus in the boundary indicated in Fig. 1 based on the correlation proposed by González et al. (2011) and Ramos and Naipauer (2014).

The objective of the present study is to review and correlate the different segments where Famatinian rocks have been recognized along the continental margin of South America in order to understand the tectonic evolution of the Early Paleozoic protomargin, based on previous work of the author, and recent data and new geochronological studies performed along the Andes by different research groups.

2 The Famatinian Orogen in Its Type Locality

The Famatinian orogenic cycle was defined in northwestern Argentina by Aceñolaza and Toselli (1976). The original definition encompasses a larger time span, which included all the sequences above the Tilcárica angular unconformity at the base of the Cambrian produced by their Pampean orogeny, up to the Late Devonian, leaving out the Late Paleozoic lesser deformed successions. However, the evidence provided by later tectonic studies demonstrated that a major orogeny

Fig. 1 Reconstruction of the Paleozoic protomargin of Western Gondwana (modified from Ramos 2009). Note the occurrence of oceanic terranes obducted during Mesozoic and Cenozoic times in the Northern and Southern Andes. The southern limit of the South American platform sensu Almeida et al. (1976) coincides with the Patagonia terrane as defined by Ramos and Naipauer (2014)

ended at Middle Ordovician times, and as a consequence, subsequent studies restricted the Famatinian orogeny from the base of the Cambrian up to the Middle Ordovician (Haller and Ramos 1984; Astini and Benedetto 1993; Astini et al. 1995; Pankhurst et al. 1996, 1998). This time span is generally accepted nowadays (Chew et al. 2007; Romero et al. 2013; among others).

The type locality for the Famatinian orogeny and the Famatinian magmatic arc is well exposed in the Sierra de Famatina, located in the western side of Sierras Pampeanas (Fig. 2) in central western Argentina (Ramos 1988; Astini et al. 1995). The tectonic evolution of this area comprises the docking and amalgamation against the protomargin of Western Gondwana of the allochthonous Cuyania terrane, derived from Laurentia (Ramos et al. 1986; Astini and Benedetto 1993; Astini et al. 1995, 1996; Ramos 2004).

As a consequence of the approach and accretion of Cuyania, the Famatinian orogen recorded a subduction-related magmatic arc, where mainly plutonic and metamorphic rocks were described along the western Sierras Pampeanas (Aceñolaza et al. 1996; Pankhurst et al. 1998; Quenardelle and Ramos 1999; Otamendi et al. 2008, 2009a, b, 2010). The magmatic arc was active from 490 to 460 Ma with a magmatic peak at about 470 Ma; main deformation took place close to 460 Ma, and postectonic granites were emplaced in Late Ordovician and Early Silurian times (see Chernicoff and Ramos 2004, and cites therein).

There are many studies that characterize the geochemical and petrological behavior of these magmatic arc rocks. In the Famatinian magmatic belt of the Sierras Pampeanas, Pankhurst et al. (2000) identified three distinct granite types: dominant I-type, S-type, and small-scale tonalite–trondhjemite–granodiorite (TTG) type which were confined further to the east in the Sierras de Córdoba, as pointed out by Dahlquist et al. (2013). These three granite types were emplaced within the 484–463 Ma interval, which characterizes the Famatinian rocks (Dahlquist et al. 2008). Detailed petrological and geochemical studies have been presented by Rapela et al. (1990), Aceñolaza et al. (1996), Saavedra et al. (1998), Pankhurst et al. (1998, 2000), Dahlquist et al. (2005, 2007, 2008), Ducea et al. (2010), and Otamendi et al. (2012), among others.

These igneous rocks were emplaced in a metamorphic basement characterized by typical low-P and high-T conditions reported all along the Famatinian belt of the Sierras Pampeanas (e.g., Toselli et al. 1987; Pankhurst et al. 1998; Dahlquist et al. 2005, 2007; Otamendi et al. 2008; Collo et al. 2009; Verdecchia et al. 2011).

These rocks are limited to the west by a basic and ultrabasic belt (Fig. 3) described as the Famatinian ophiolites (Ramos et al. 2000) associated with paired metamorphic belts, with high pressure—low temperature belts ocean-wards (see Boedo et al. 2016).

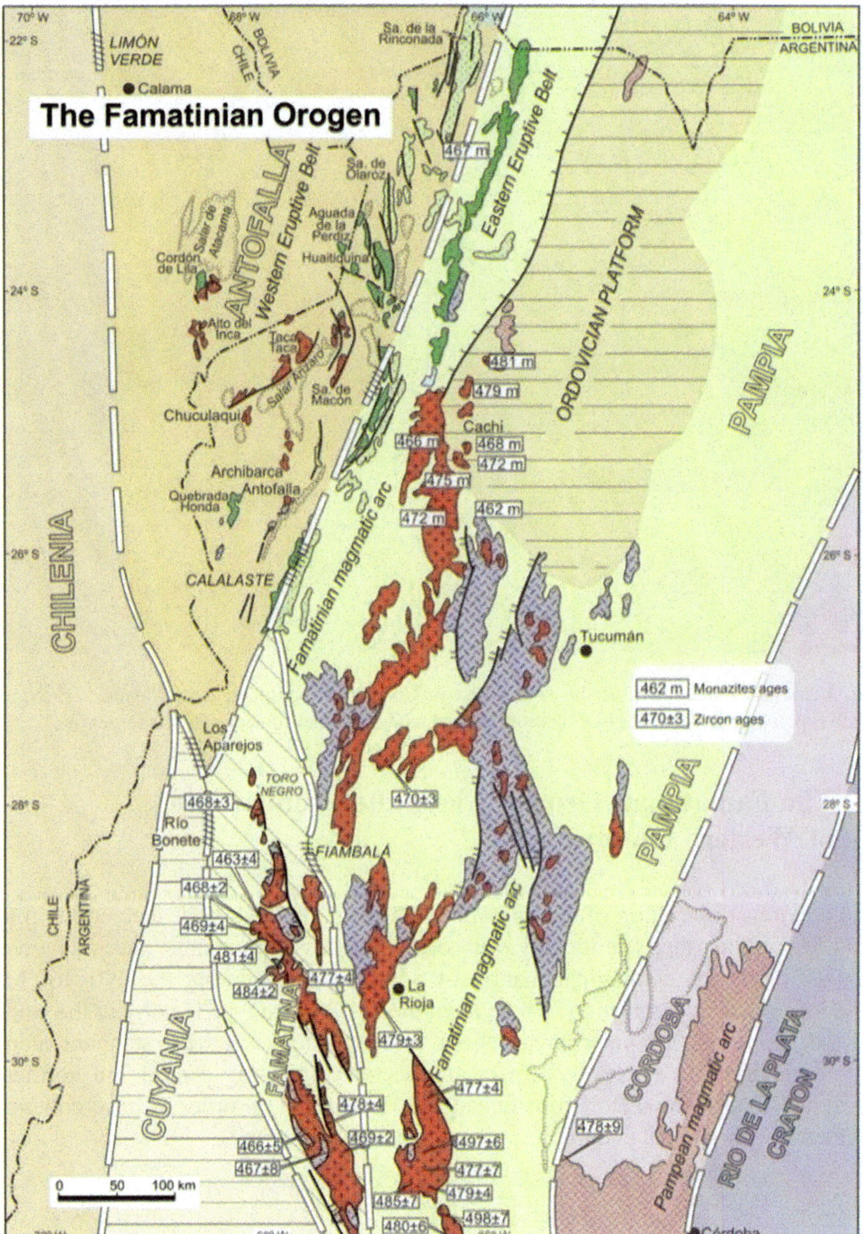

Fig. 2 Type locality of the Famatinian orogen in Sierras Pampeanas. Some ages are indicated in millions of years (Ma) as a guide based on different data sets (see Pankhurst et al. 1998; Quenardelle and Ramos 1999; Otamendi et al. 2008, 2009a, b, 2010). The terrane boundaries are based on Ramos (2009) and Ramos et al. (2010)

Fig. 3 Metamorphic rocks of the sole thrust in Cortaderas ophiolitic belt, Mendoza Province (after Haller and Ramos 1984). Foliation is enhanced by spinel minerals in the serpentine

3 The Famatinian Orogen Along the Protomargin of Western Gondwana

The previous description of the Famatinian orogen, although very succinct, will be used to correlate the type locality with other igneous and metamorphic rocks recognized along the continental margin of South America during the 490–460 Ma time interval. The description will be from the best known type locality in the south toward the north. The margin has been divided into a southern segment along northern Argentina and Bolivia, a central sector comprising central and northern Perú, and a northern sector from Ecuador to Venezuela, in order to have coherent segments with similar evolution.

3.1 Southern Sector

Since the early work of Coira et al. (1982), it is known that the Famatinian Ordovician arc granitoids of Sierras Pampeanas transitionally pass to the *Faja Eruptiva de la Puna* (Eruptive belt of the Puna). These igneous rocks defined by

Méndez et al. (1973) have been divided by Niemeyer (1989) in two volcanic belts: the eastern and western eruptive belts (see Fig. 2). A complete description of the geological, geochemical, and isotopic characteristics of these rocks is presented by Coira (2008), who interpreted these rocks as having a weak arc signature. The eastern belt of volcanic rocks is interbedded with clastic sediments bearing graptolites of Floian (Arenig) age. The biostratigraphic control, based on the graptofauna, indicates that these volcanic rocks span between 470 and 477 Ma (ICS-IUGS Geological Time Table, 2016). This belt has also some minor intrusives ranging in age from 476 ± 1 and 478 ± 3.5 Ma (U-Pb in monazite, Lork and Bahlburg 1993; Coira 2008).

The new data presented by Bahlburg et al. (2016) show two discrete phases of intrusion in the *Faja Eruptiva de la Puna Oriental* (Eastern Eruptive Belt). The first phase coincides with the main Famatinian activity, which according to these authors is around 465 Ma old. This phase is related to mafic and intermediate volcanic rocks that are interfingered with graptolite bearing sediments which have been recently revised by Brussa et al. (2008). Several studies confirm the old Arenigian age of the fauna, but with more detailed work, it is now known that the time span of these graptolites can be assigned to different biozones ranging from Late Floian to mainly Dapingian–Early Darriwilian age (Brussa et al. 2008; Toro et al. 2006; Martínez et al. 1999). These biozones encompass a time interval between 477 and 465 Ma (Ogg et al. 2016), which coincides with the main activity of the Famatinian arc. The second main phase of Bahlburg et al. (2016) corresponds to 444.9 ± 2.3 Ma, almost at the Ordovician-Silurian boundary (443.8 Ma, according to Ogg et al. 2016). Based on the new precise data, the second phase is attributed by Bahlburg et al. (2016) to the Ocloyic phase, which according to Turner and Méndez (1975) and Ramos (1986) occurred at the end of the Ordovician. It is worth mentioning that this phase was disregarded by Moya (2015) that after a complete analysis of the present evidence, assumed that the main orogenic phase occurred in Northern Argentina in Dapingian times, and not along the Ordovician-Silurian boundary as previously proposed. Based on this regional evidence, the second phase of Bahlburg et al. (2016) is assumed to be a reactivation, probably controlled by extension or transtension along a crustal weakness zone. The occurrence of large amounts of ultrahigh temperature melts in the eastern eruptive belt was interpreted by Fernández et al. (2008) as evidence of extension in the back-arc of the western eruptive belt. The control of this extension is linked to the crustal weakness zone, which was interpreted as an Ordovician suture by Coira et al. (1982). However, this Ordovician suture has been put to rest by Bahlburg et al. (2016) based on the demonstration of Zimmermann et al. (2014) that gabbros on this suture were emplaced at 543.4 ± 7.2 Ma. These authors interpreted the Calalaste mafic and ultramafic rocks as Precambrian, failing to recognize the Late Cambrian age of a rhyolite dated by U-Pb in zircons and associated with the mafic and ultramafic complexes (Pinheiro et al. 2008). This important weakness zone, which continues further north in Pocitos (see Zappettini et al. 1994), is interpreted here as the Grenville suture between Antofalla and Pampia, which has been partially reactivated by extension in Famatinian (and possibly in Pampean) times.

The geochemical characteristics of the rocks of *Faja Eruptiva de la Puna* indicate a typical arc-related plutonism and volcanism (Niemeyer 1989; Coira 2008), with strong arc affinities in the western belt, which weaken eastward in the eastern eruptive belt. This activity ends by the end of the Floian (Ramos and Coira 2008), and the final products at the end of the Ordovician are probably related to extension (Bahlburg et al. 2016). The country rocks of these volcanic and subvolcanic products are clastic sediments or metasedimentary rocks with very low metamorphic grade. The anchimetamorphism field of these rocks indicates temperatures of about 275–300 °C and pressure near 1.5 kbars (Toselli 1982).

As illustrated in Fig. 4, there is a conspicuous change from south to north in the metamorphic conditions between the Sierras Pampeanas and the Eruptive Belt of the Puna, indicating shallower crustal levels, exposed to the north. The arrow at the latitude of the city of Jujuy (Fig. 4) points out the occurrence of volcanic rocks to the north, in contrast with the dominant intrusive rocks to the south of the Famatinian orogen. A second arrow along the boundary between Argentina and Bolivia (Fig. 4) coincides with the last evidence of volcanic rocks to the north. The entire Ordovician rocks of the Altiplano domain of Bolivia and southern Perú have neither arc volcanism nor metamorphism at all (Ramos 2008). However, some minor occurrences have been reported by Bahlburg et al. (2006) in the Ollantaytambo Formation and in the Umachiri beds of Early to Middle Ordovician age and interpreted as back-arc rocks.

At these latitudes, the volcanic arc is developed closer to the continental margin of northern Chile and southern Peru during Ordovician times (Fig. 4). This proximity could be a consequence of crustal erosion by subduction that affected this margin during the Andean cycle, when more than a hundred kilometers were eroded away (Kay and Coira 2009). Relics of the Famatinian arc have been identified along the present continental margin (Loewy et al. 2004; Boekhout et al. 2013; among others).

The southern segment has two regions with different tectonic evolution. The Sierras Pampeanas segment exposes the lower crust as a consequence of the collision of the Cuyania terrane against the Western Gondwana margin (Ramos 2004). This collision deformed and uplifted the continental margin uncovering the lowermost levels of the crust as described by Otamendi et al. (2012). The Altiplano domain preserves clastic deposits of the Ordovician retroarc platform without igneous rocks and no metamorphism. This fact is explained by a different Early Paleozoic tectonic setting. The Altiplano domain was the retroarc basin of an arc developed along the continental margin of northern Chile and southern Perú (see Fig. 4) and did not record any collision at Middle Ordovician times (Bahlburg 1990; Bahlburg and Hervé 1997; Egenhoff 2007; Ramos 2008; Ramos and Coira 2008).

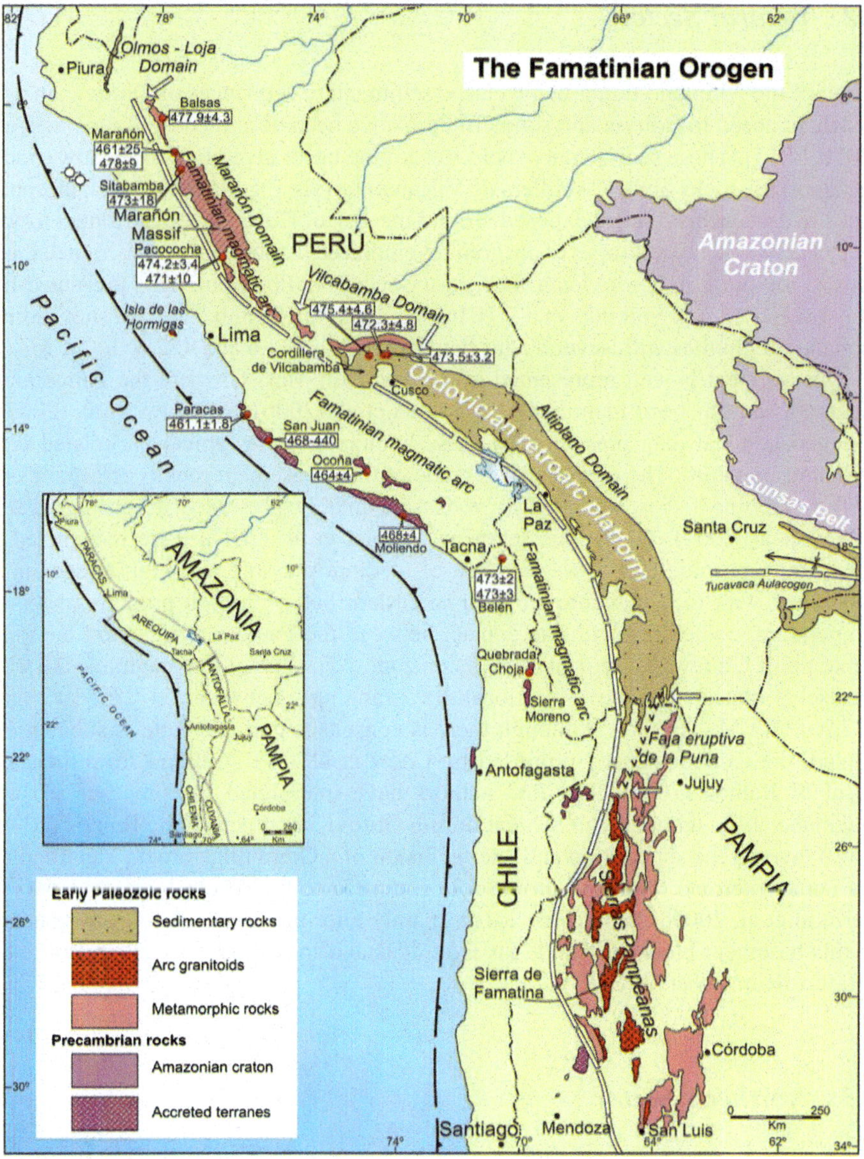

Fig. 4 Famatinian orogen along the central western part of South America shows different segments with distinct behavior. It can be recognized a southern segment where the Sierras Pampeanas, the type locality of the orogen, has been described, the Altiplano domain where an Early Paleozoic retroarc platform is well developed without magmatic arc rocks, and a central segment, where metamorphic and igneous arc rocks are again exposed in northern Perú. The inset shows the different terranes identified along the continental margin after Ramos (2010) and Romero et al. (2013). Some representative ages in millions of years (Ma) are based on Loewy et al. (2004), Cardona et al. (2006), Chew et al. (2007), Reitsma (2012), Boekhout et al. (2013). The broad dashed line between the Ordovician retroarc platform and the Famatinian magmatic arc represents the approximate location of the Grenvillian suture between the Arequipa and Antofalla terranes after Ramos (2010), which has been only partially reactivated by extension during the Famatinian orogeny

3.2 Central Sector

North of the Altiplano domain, the clastic sedimentary sequences of Ordovician age widely exposed in Bolivia and southern Perú have been described by Reimann et al. (2010, 2015). These sedimentary rocks toward the north give place to the low-grade metamorphic rocks of the Cordillera de Vilcabamba (see Fig. 4). These metamorphic rocks in greenschist facies, exposed around the city of Cusco, were attributed to the Early Paleozoic, but new ages of the granites emplaced in these rocks constrained their ages to the pre-Late Ordovician. In fact, the recent U-Pb zircon ages presented by Reitsma (2012) and Spikings et al. (2016) from the Vilcabamba domain show some Famatinian granites with several ages between 475.4 ± 4.6 and 472.3 ± 4.8 Ma.

This segment is well represented in the Marañón Massif, where the Famatinian magmatic arc has been identified by Cardona et al. (2006) and Chew et al. (2007). The geochemical data presented by these authors show a typical arc-related calcalkaline signature. The metamorphic grade of these basement rocks corresponds to the amphibolite facies, and the arc rocks are mainly preserved as orthogneisses (Chew et al. 2007). Along this segment, Willner et al. (2014) have recognized a paired metamorphic belt, a western belt of medium pressure and high temperature coinciding with the magmatic arc, and an eastern belt with high pressure and low temperature, coinciding with the contact between the Paracas terrane and the protomargin of Gondwana, typical of collision zones. The age of the magmatic arc was precisely dated in several localities ranging between 471 ± 10 and 477.9 ± 4.5 Ma (Fig. 4). Although there is consensus that this suite is subduction related, the tectonic setting proposed by Chew et al. (2007) differs from the proposal of Ramos (2008). The first authors have interpreted the presence of this magmatic arc as an open embayment during Ordovician times, while Ramos (2008, 2009) favored the subduction and later collision of a Grenvillian block, identified as the Paracas terrane. This last proposal is getting support in later local studies (see Carlotto et al. 2009; Chew et al. 2016), mainly after the identification of remnants of this basement block in Isla de las Hormigas and in several oil wells offshore of Trujillo (Romero et al. 2013).

3.3 Northern Sector

This sector encompasses the Famatinian rocks from Ecuador to Venezuela. In recent years, many studies provided new ages that allowed identifying the extension of the Famatinian magmatic arc up to the Cordillera de Mérida in Venezuela. In this sector, several domains with different metamorphic facies have been recognized.

A segment characterized by Ordovician sedimentary and low-grade metamorphic rocks is exposed north of the Cordillera de Marañón, which continues into Ecuador at the latitude of Loja. This segment has been identified by Carlotto et al. (2009) as the Olmos-Loja domain. The metamorphic grade of the Marañón domain drops north of

Balsas (Fig. 4) to the Ordovician phyllites and schists of the Olmos Complex, and the metasedimentary rocks of the Salas Group (Reyes and Caldas 1987).

North of the Olmos-Loja domain in the Cordillera Real of Ecuador, there are some inliers where Paleozoic rocks were identified based on the work of Herbert (1977), who studied the geochemistry of the exposed black phyllites and quartzites. The rocks of these greenschist facies were grouped in the Chiguinda and Agoyán units. These rocks are exposed north of Loja in a belt along the western margin of the Cordillera Real for more than a hundred kilometers (Litherland et al. 1994). Similar rocks crop out again north and south of Papallacta (see location in Litherland et al. 1994), but in this northern part they are associated with orthoamphibolites derived from basaltic precursors. These rocks have important ductile deformation near Cuyuga, where Trouw (in Litherland et al. 1994) has identified a conspicuous nappe tectonics. We had the opportunity to examine the structure of these exposures along the Papallacta Valley, confirming a pre-Jurassic intense deformation (Figs. 5 and 6). No precise ages are available, but there are some old K-Ar ages from Agoyán low-grade rocks, which yielded 417 Ma, probably indicating a cooling age of an older metamorphic rock (Litherland et al. 1994).

The Paleozoic rocks of Papallacta, southeast of Quito (Fig. 7), have K-Ar ages in biotite and hornblende that vary from 881 $\pm$ 20 to 306 $\pm$ 10 Ma (Litherland et al. 1994), but U-Pb ages have yet to be measured. The highly deformed Paleozoic rocks of the Cordillera Real have been tentatively grouped as the Cuyuga domain (?) in Fig. 7, but it is still uncertain if they are some remains of the Famatinian orogen or they are correlative with the younger Tahamí terrane of the Central Cordillera of Colombia as Martens et al. (2014) have proposed.

The basement of the Eastern Cordillera of Colombia has been interpreted as an allochthonous terrane accreted during Late Paleozoic times (Restrepo and Toussaint 1988). These authors proposed that the Chibcha terrane included the continental basement of the easternmost part of Central Cordillera, the Eastern Cordillera, the Santander Massif, and part of the Santa Marta Massif (Toussaint and Restrepo 1989). Aleman and Ramos (2000) proposed that this continental terrane was accreted by the end of the Ordovician, founded on some preliminary dates available from the Santander Massif and the age of the main deformation of the Quetame low-grade metamorphic rocks. The southern segment of the Chibcha terrane exposed the Quetame Massif (Fig. 7), composed of low-grade metamorphic rocks attributed to the Early Paleozoic.

Based on the stratigraphic relations between the greenschists and the overlying quartzites and conglomerates with intercalated shales, and the palynological assemblage described in these deposits, Grösser and Prössl (1991) separated this unconformable unit as Silurian, restricting the age of the Quetame Schists to the Ordovician. Ordóñez-Carmona et al. (2006) interpreted the collision of the Chibcha terrane to the autochthonous margin of Gondwana at around that time and defined the Quetame event as the main deformation of these metamorphic rocks. Horton et al. (2010) reported a U-Pb age in zircons of 483 $\pm$ 10 Ma from a granitic boulder

Fig. 5 Cuyuga nappes in the Papallacta Valley of the Cordillera Real of Ecuador described by Litherland et al. (1994). Red dashed line indicates the base of the overthrust

derived from La Mina Granodiorite, intruded in the phyllitic rocks of the Quetame Group. This group further south passes to Ordovician sedimentary rocks.

There is no reliable data on the age of the metamorphism in Quetame, but based on the preliminary ages obtained in the Santander Massif, Restrepo-Pace and Cediel (2010) recognized three important events in the Chibcha terrane. A Grenvillian age metamorphism ($\sim$1.0 Ga) in the basement is intruded by foliated granitoids of $\sim$477 Ma and non-foliated granites of $\sim$471 Ma. These authors identified the deformation of the Chibcha basement as produced by the Caparoensis orogenic episode, but derived from the evidence obtained in the Santander Massif. Recently, the main peak of metamorphism of equivalent rocks in the Santander Massif has been assigned to the Middle and Late Ordovician by Mantilla Figueroa et al. (2016) and correlated to the Famatinian deformation.

The low-grade metasedimentary rocks exposed in the Floresta Massif, located between the Santander and the Quetame massifs in the Eastern Cordillera (Figs. 7 and 8), have been interpreted by Forero Suárez (1990) as part of the Grenvillian basement of his allochthonous terrane. Recent studies by Horton et al. (2010) described some monzogranitic rocks intruded in this basement with U-Pb zircon ages of 482 $\pm$ 15 Ma and 464.2 $\pm$ 8.2 Ma. These zircons have inherited cluster ages at 1200–1140 Ma and 1050–1000 Ma, consistent with the assumed ages of Forero Suárez (1990).

Fig. 6 Detailed view of pyrophyllite schists of the Cayuga nappes with top-to- the-east deformation in the Papallacta Valley, Cordillera Real of Ecuador

The new studies of Van der Lelij (2013) in the Santander Massif of Colombia and in the Mérida Andes of Venezuela, further north, dated the crystalline basement exposed on these regions. The new ages in the main Mérida basement as well as in the Caparo block show that both units belong to the same terrane, modifying the original proposal of Bellizzia and Pimentel (1994) of the allochthonous Mérida terrane and the autochthonous Caparo block, followed by several authors (see Aleman and Ramos 2000). The new data indicate that the metamorphic basement protolith of the Mérida terrane, including the Caparo block of Bellizia and Pimentel (1994), has mean Grenvillian ages of 1008.6 ± 6.7 Ma and 1018.3 ± 8.9 Ma (Van der Lelij 2013). This study together with the new work of Mantilla Figueroa et al. (2016) and Van der Lelij et al. (2016) shows that the Northern Andes basement of Colombia and Venezuela was affected by an important Ordovician metamorphism. Van der Lelij et al. (2016) recognized a Barrovian metamorphism in upper amphibolite facies with peak conditions between ~489 and ~472 Ma, and a subsequent anatexis that partially melt the orthogneisses at ~454 ± 10 Ma.

The new ages bracketed an important deformation in the Mérida Andes and in the Santander Massif that fit in general terms with the Famatinian orogeny. The deformation is also constrained by the age of the Caparo Formation, a clastic sequence of Late Ordovician age that unconformably covers the metamorphic rocks (Aleman and Ramos et al. 2000; Ordóñez-Carmona et al. 2006).

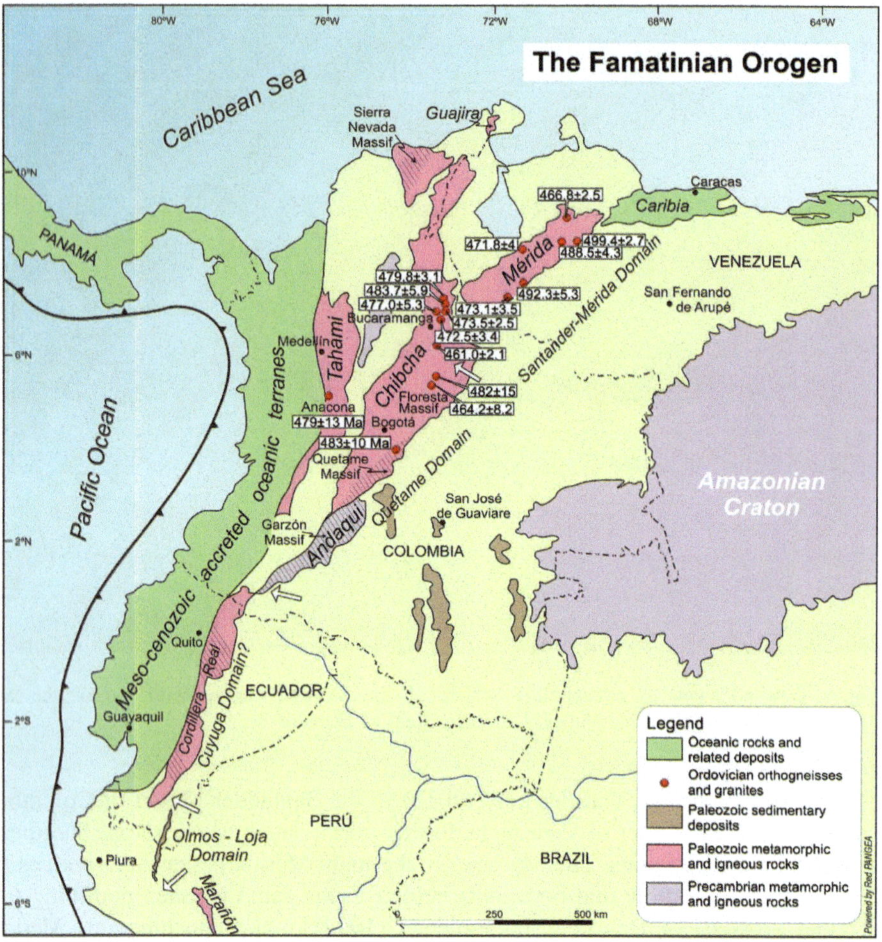

Fig. 7 Famatinian orogen in the Northern Andes. Some representative ages in millions of years (Ma) of the magmatic rocks are indicated (based on Van der Lelij 2013; Martens et al. 2014; Mantilla Figueroa et al. 2016). Note that the reconstruction of the protomargin of Gondwana requires eliminating the oceanic terranes accreted after the Cretaceous

4 The Grenvillian Suture Between Arequipa and Amazonia

There is indirect evidence of an important weakness zone between Arequipa and Amazonia. The data presented by Loewy et al. (2004) have demonstrated that on the Arequipa Massif a magmatic arc during Mesoproterozoic times was developed. Ramos (2008, 2009) has interpreted this juvenile arc as a Grenville arc coeval with the Sunsas arc, which at about ~1,000 Ma has been amalgamated during the formation of the Rodinia supercontinent. The collision between Arequipa

Fig. 8 Deformed Ordovician quartzites in the Quetame Massif, Arroyo Susumuco, Eastern Cordillera

(Antofalla) and Amazonia (Pampia) cratonic blocks produced a prominent suture between these terranes. That suture is interpreted based on the three-dimensional inversion model of the seismic profile of Dorbath et al. (1993) and the structural interpretation of Martínez et al. (1994) depicted in Figs. 9 and 10.

The boundary between the Arequipa terrane and the Amazonian craton has been interpreted as the Grenvillian suture between this continental blocks (Ramos and Jiménez 2014). This weakness zone has been several times reactivated, mainly as a extensional zone controlling the emplacement of igneous rocks during the Pampean, Famatinian, Late Paleozoic, and Andean times (Fig. 10).

This suture continues further south between Antofalla and Pampia cratonic blocks, and is responsible of the emplacement of volcanic and subvolcanic bodies of the eastern eruptive belt in the Puna (see location along the Altiplano as indicated in Fig. 3). This suture probably controlled the Calalaste, Pocitos, and other mafic and ultramafic rocks of different ages (see Pinheiro et al. 2008; Zimmermann et al. 2014). It has also controlled the emplacement of the Frailes Ignimbrite in Quaternary times (Kay and Coira 2009) and the Miocene volcanic rocks of Potosí in Bolivia. This important suture was actively controlling the inception of the Puncoviscana basin in Neoproterozoic times, the *Faja Eruptiva Oriental* of the Puna, and even is still actively controlling the present delamination in the Altiplano (see Beck and Zandt 2002).

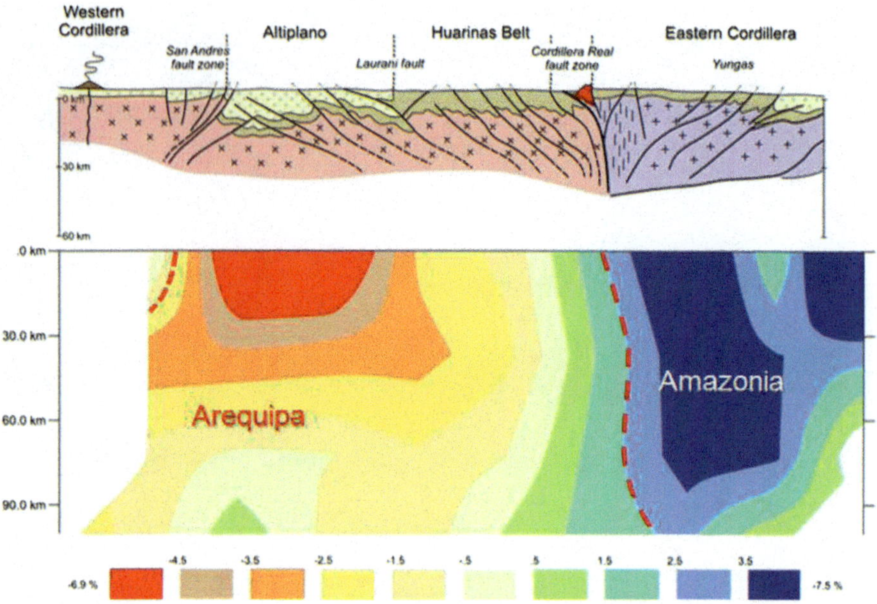

Fig. 9 Geological and structural cross section from the Altiplano to the Eastern Cordillera at the latitude of La Paz, Bolivia. Vertical cross section down to 90 km through the three-dimensional inversion model beneath a seismic profile. Notice that the high-velocity zone (> +2.5%) is limited to the southwest, at least in the crust by the Cordillera Real fault system (after Dorbath et al. 1993). This crustal contrast is interpreted as the boundary between Arequipa and Amazonia

5 Tectonic Setting of the Famatinian Orogen

There is evidence of several segments with calcalkaline magmatic activity along the protomargin of Gondwana during latest Cambrian–Middle Ordovician times. As depicted in Fig. 11, the magmatism is not continuous along the margin and the different sectors have distinctive tectonic settings. The type locality, the Famatinian magmatic arc, is the best known on petrological, geochemical, and isotopic grounds (Ramos 1988; Pankhurst et al. 1998; Fernández et al. 2008; Otamendi et al. 2012). It has an extensive development between Bolivia in the north and the northern boundary of Patagonia to the south. Arc magmatism has been described during Ordovician times in Patagonia, but it is older than Famatinian and it has been correlated by Chernicoff et al. (2013) with the Ross Orogeny developed in the Transantarctic Mountains during Cambrian to Early Ordovician times.

Lower crust levels are exposed in the central part of the Famatinian arc, as described by Otamendi et al. (2012), and these deepest levels are associated with the central part of the collision with the Cuyania terrane around 465–460 Ma. Northward, the magmatic rocks have more superficial volcanic and subvolcanic levels, up to its end in southern Bolivia.

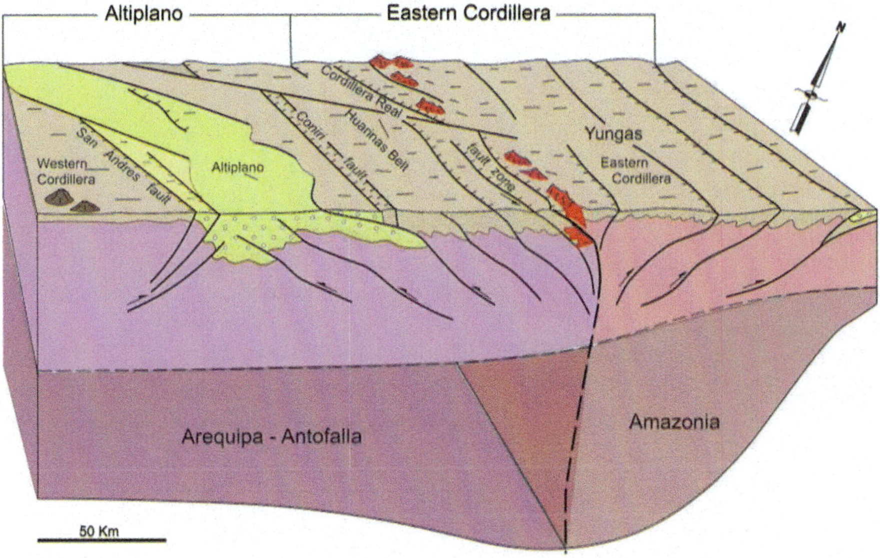

Fig. 10 Structural interpretation showing the Huarinas structural belt developed with a top-to-the-west vergence, between Amazonia and Arequipa (Antofalla) (after Martínez et al. 1994). Late Paleozoic and Oligocene granites that constitute the Cordillera Real are emplaced along the suture (see details in Ramos and Jiménez 2014)

The intense ductile deformation observed in both margins of the Cuyania exotic terrane has kinematic indicators showing a top-to-west vergence (Ramos et al. 1986; Martino et al. 1993), which together with the ophiolitic belt along the western boundary clearly points out to an east-dipping paleosubduction zone (present coordinates). This fact indicates that the magmatic belt developed on the autochthonous protomargin of Gondwana (Ramos 2004).

A branch of the Famatinian arc toward the north developed on the Arequipa-Antofalla terrane along the present margin. A large part of this margin has been eliminated by crustal erosion by subduction (Stern 1991). Along this segment, isolated patches of the magmatic arc were identified by several authors in northern Chile and along the boundary with Bolivia (Wörner et al. 2000; Coira et al. 1999, 2009; Ramos 2009). Geochemical and isotopic characteristics indicate the continuation of the Famatinian arc (Chew et al. 2007; Coira 2008). The old Grenville suture has been extensionally reactivated during Ordovician times and controlled the western slope of the clastic platform (Fig. 11) and the location of submarine tholeiitic rocks and turbiditic deposits along the Grenvillian suture in the Altiplano of Bolivia (Ávila Salinas 1992; Ramos 2008).

The location of the Grenville suture was based on geophysical grounds. The lithospheric model of Dorbath et al. (1993) based on teleseismic data from southern Peru and Bolivia, together with the tomography of Dorbath and Granet (1996) led to the structural interpretation of Martínez et al. (1994) of Figs. 9 and 10. Based on

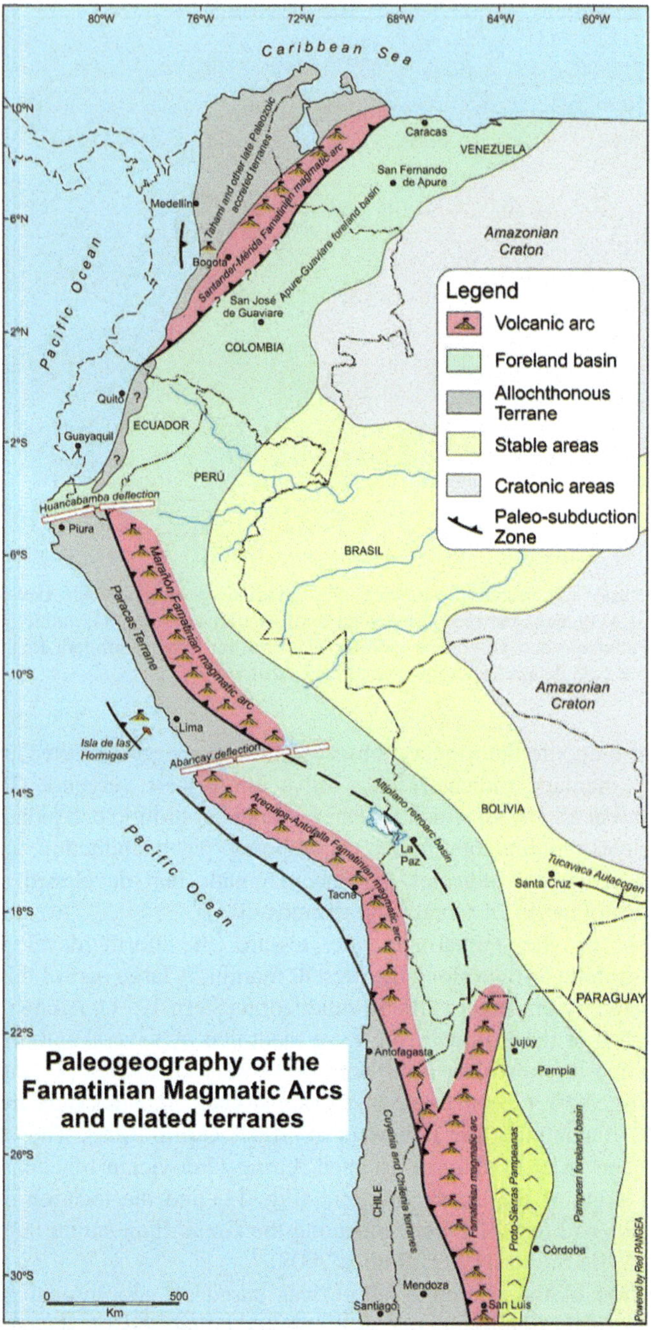

Fig. 11 Paleogeography of the Early Paleozoic along the reconstructed protomargin of Western Gondwana (based on Toussaint and Restrepo 1989; Ramos 2004, 2009; Chew et al. 2007; Coira 2008; Restrepo-Pace and Cediel 2010). Note the relationship between the Huancabamba and Abancay deflections with the paleogeography of the magmatic arc

these data, Ramos and Jiménez (2014) interpreted the location of the suture as traced in Fig. 3. It is worth noting that this weakness zone is controlling the inception of present delamination along the proposed suture east of the Titicaca Lake (Beck and Zandt 2002).

North of the Abancay deflection, Chew et al. (2007) and Cardona et al. (2009) described again the magmatic arc and correlated these rocks with the Famatinian orogen. The magmatic activity prior to the collision has similar ages as the orthogneisses of the type Famatinian locality. Again in this area, the collision with the Paracas terrane is constrained by the metamorphic pair described by Willner et al. (2014). These authors identified high-pressure conditions of 11–13 kbar/ 500º540 ºC during maximum burial derived from a garnet amphibolite in the Tapo Ultramafic Massif, west of the magmatic arc. The Famatinian arc in the western Marañón Complex has low-PT conditions at 2.4–2.6 kbar and 300–330 °C estimated for a phyllite-greenschist assemblage, representing for Willner et al. (2014) contrasting metamorphic conditions characteristic for a magmatic arc environment. The age of the peak metamorphism is 465 ± 24 Ma dated by Sm–Nd mineral-whole rock isochrone. The associated Tapo Ultramafic Complex was interpreted as a relic of oceanic crust which was subducted and exhumed in a collision zone along a suture (Castroviejo et al. 2009, 2010). Structural studies of these ophiolitic relics in Huánuco and Tarma, along the western Marañón Massif, have demonstrated the development of phyllonites and mylonites along the contacts that rule out a previous interpretation of these rocks as peridotite igneous intrusives (Rodrigues et al. 2010a). The ductile deformation D3 that affected the peridotites and serpentinites in Huánuco has a southwest vergence (Rodrigues et al. 2010b).

Based on these data, it is likely that the Paracas microcontinent, a para-utochthonous terrane, collided against the protomargin of Gondwana with a subduction zone dipping to the east (present coordinates) as depicted in Fig. 11 (Ramos 2008, 2009; Carlotto et al. 2009; Willner et al. 2014; Chew et al. 2016, among others).

North of the Huancabamba deflection, there are many uncertainties in the Cordillera Real of Ecuador derived from poor exposures due to the dense vegetation, the conspicuous strike-slip deformation that truncates the continuity of the different units, and the lesser amount of data available in comparison with the Central Andes. However, there is incomplete evidence in two different sectors of the Cuyuga domain (see Fig. 7) where deformed Paleozoic rocks are exposed (Litherland et al. 1994).

Geochronological data only indicate possible Early Paleozoic rocks, but these are not precise enough to correlate these rocks with other known areas either to the north or to the south. Martens et al. (2004) correlated these rocks with the Tahamí terrane, because they share common characteristics with some of the rocks exposed in the Central Cordillera of Colombia. One of the few unambiguous facts is that the deformation of the Cuyuga nappes has a clear vergence to the east (Trouw in Litherland et al. 1994), which may offer some clues in reconstructing the subduction polarity.

The Chibcha and Mérida terranes are important in reconstructing the Famatinian orogen in the Northern Andes (Figs. 7 and 11). The Chibcha terrane has been divided into two different domains. The southern Quetame domain has a lower metamorphic grade and absence of orthogneisses of Famatinian age, when compared with the Santander-Mérida domain (Fig. 8). The relationship between these two domains is similar to the transition among Sierras Pampeanas-Puna Eruptive Belt-Altiplano domains, the Marañón- Vilcabamba-Altiplano domains, or the Marañón-Olmos-Loja domains (Figs. 3 and 7). These transitions are characterized by different crustal levels exposed in each domain, varying from lower crust high-grade metamorphic rocks to upper crustal levels with low-grade metamorphism or just sedimentary and volcanic rocks. These different structural levels have been explained by the amount of shortening and uplift, and subsequent exhumation related to the collision that has affected some segments.

The tectonic setting of the Quetame domain has not consensus. Toussaint and Restrepo (1989) interpreted it as the result of collision of Chibcha terrane with the Gondwana protomargin, proposal that was followed with some little differences by Restrepo-Pace (1992), Restrepo-Pace et al. (1997), Aleman and Ramos (2000), Ordóñez-Carmona et al. (2006), Ramos (2009), and Restrepo-Pace and Cediel (2010). Others interpreted the Chibcha terrane as an autochthonous part of the Gondwana margin (Van der Lelij et al. 2016).

In order to evaluate the tectonic setting of the Quetame domain, it is important to understand the evolution of the Early Paleozoic protomargin. There is good evidence that during Early to Middle Cambrian times the Los Llanos Basin in Colombia and the Apure Basin in Venezuela were part of a continental passive margin, possibly a conjugate margin of the West Avalonia terrane (Murphy et al. 2006). The occurrence of the typical Acado-Baltic trilobite *Paradoxides* in the Cambrian carbonates of Serranía de Macarena, known since the early work of Harrington and Kay (1951), was confirmed by later studies of Rushton (1962), who accepted his provincialism as part of the Atlantic Realm. *Paradoxides* can also be found within the Carolina Slate belt, in eastern New England, eastern Newfoundland, New Brunswick, and in the Avalon Peninsula (Restrepo-Pace and Cediel 2010). Murphy et al. (1999, 2006) assumed that West Avalonia was detached from northern South America, based on the basement characteristics (see also Benedetto and Ramírez 1985; Benedetto et al. 1999; Benedetto 2004). As a result of that, the foreland sector of the Northern Andes was a passive margin during Cambrian to Early-Middle Ordovician times as described by Aleman and Ramos (2000), Restrepo-Pace and Cediel (2010), among others. The change to a contractional tectonic regime occurred in Middle-Late Ordovician times, when the Guaviare–Apure foreland basin was formed (Fig. 11), and for that reason the basal conglomerates of the Darriwilian Caparo Formation are unconformably overlying the Early Ordovician metamorphic rocks of the Mérida Andes. The seismic data on the westernmost Llanos Basin show a fold and thrust belt developed between Late Ordovician to Devonian times, beneath the truncation of the Cretaceous flat lying unconformity.

The Santander-Mérida domain shares in common a high amphibolite grade of its basement, and the occurrence of Famatinian orthogneisses (Restrepo-Pace et al. 1997; Van der Lelij 2013). Main deformation and important uplift took place during the Caparoensis-Famatinian orogeny in Early to Middle Ordovician times. However, two different processes have been invoked to explain the tectonic evolution. Recently, Van der Lelij et al. (2016) challenged the exotic nature of the Mérida terrane, postulating that Chibcha and Merida terranes were formed through a period of alternating compression and extension in an Andean-type setting. The Laurentian nature of the basement proposed by Forero Suárez (1990) was questioned based on a Pb isotope correlation diagram, which shows that the isotopic composition of the basement of Merida terrane was distinct from the eastern Laurentian Grenville Province (Van der Lelij 2013). A similar conclusion was obtained for the Central Andean terrane of Colombia by Ruiz et al. (1999). Cuyania seems to be the only truely exotic terrane for Gondwana, when the Pb isotope compositions of the different Grenvillian terranes associated with the continental margin are compared (Ramos 2004, 2010). Most of the terranes are para-utochthonous to Gondwana and have collided and detached from the margin several times (Ramos 2009).

Another important point is the high-grade Ordovician metamorphism of the Santander Massif and the Mérida Andes, and the subsequent uplift, which are hard to explain in Andean-type settings. The subduction along an Andean margin never exposes the lower crust, unless some extraordinary process occurs, such as the collision of a terrane. Based on these facts, we agree with Restrepo and Toussaint (1988), Toussaint and Restrepo (1989), Restrepo-Pace (1992), Aleman and Ramos (2000), and Ordóñez-Carmona et al. (2006) among many others that considered the Chibcha and Mérida terranes as continental blocks that collided with the continental margin in the Paleozoic.

The polarity of the paleosubduction zone is still a matter of debate. When all the Famatinian arcs along the protomargin are compared, the only one that has a reverse polarity is the Santander-Mérida domain. This is based on the vergence of the structure on the eastern side of the Chibcha and Mérida terranes, the location of the magmatic arc on the hanging wall of the paleosubduction, and the potential sutures described along this limit. However, more data are needed to confirm this assertion.

There are two intriguing facts which have not yet a reasonable justification. One of them is the Anacona terrane in the Central Andes of Colombia south of Medellín (Martens et al. 2004). A typical Famatinian orthogneiss (Figs. 7 and 11), known as La Miel orthogneiss, has xenocryst zircons with igneous overgrowths with ages varying from 479 to 470 Ma, very distinct from other gneisses from the Tahamí terrane. The occurrence of lower Paleozoic metamorphosed granitoids in the Central Cordillera of Colombia, although scarce, has been interpreted as an Ordovician arc formed along the margins of the Rheic Ocean, previous to the amalgamation of Pangea (Villagómez et al. 2011). Another hypothesis for these igneous rocks is that the zircons are fingerprints of the lower crust formed during the rift-related opening of the Iapetus Ocean and could be related to the bimodal igneous domain now in south Mexico and Central America (Martens et al. 2004).

The second intriguing fact is from Isla de las Hormigas along the western margin of Paracas terrane near Lima, where igneous zircons of 467.9 ± 4.5 Ma age in high-grade orthogneisses of Grenvillian age record a Famatinian igneous episode (Romero et al. 2013). The interpretation of this isolated example has the same alternatives as the Anacona terrane previously discussed.

6 Concluding Remarks

The analyses of the occurrence of Famatinian magmatic arcs along the protomargin of Western Gondwana indicate complex and variable tectonic settings, far from a unique simple Andean margin. The status of the present knowledge, although uneven, allows some interesting remarks.

1. There is an almost semicontinuous belt of arc-related rocks with ages varying from 490 to 460 Ma, which correlates with the time span of the Famatinian orogen in its type locality in western Sierras Pampeanas.
2. The metamorphic grade of these rocks has some noticeable variations controlled by the crustal levels exposed. The high-grade domains are flanked by transitional segments where the metamorphic grade changes from greenschist facies to sedimentary facies, as between the Vilcabamba and Altiplano domains, Puna Eruptive Belt and the Altiplano domains, or Marañón and Olmos-Loja domains.
3. The deep crustal level exhumed in some segments seems to be controlled by shortening and uplift produced by collision of different accreted terranes. The large, old, and exotic Cuyania terrane, which has a thick Precambrian cold crust, produced one of the most prominent uplifts and deformations.
4. The variable location of the Famatinian arcs, from far to near the continental margin, is controlled by the presence or absence of terranes colliding against the margin. The Altiplano domain is a good example of a suture of the old Grenville margin, which was not reactivated during Early Cambrian extension.
5. The quasi-uniform ages of all these episodes are not unique, as episodes of "Famatinian" age are registered in the Caledonides in Europe, the Appalachian in North American, and in the Lachlan belt of Australia. This fact obviously indicates a period of fast drift of the Ordovician plates that produced subduction-related igneous rocks all around the globe.

Acknowledgements This paper is the result of many field trips, meetings, and symposia held along the Andes from Venezuela to Patagonia. The author thanks numerous colleagues for livid discussions, hard debates, and generous friendship that fed during the years my interest to learn how the Andes work. The critical reviews of the manuscript by Juan Otamendi, *Universidad de Río Cuarto*, Roberto Martino from *Universidad Nacional de Córdoba* and Heinrich Bahlburg from *Universität Münster* are deeply acknowledged. This is the contribution R-208 of the Instituto de Estudios Andinos Don Pablo Groeber (UBA-CONICET).

References

Aceñolaza FG, Toselli A (1976) Consideraciones estratigráficas y tectónicas sobre el Paleozoico inferior del Noroeste Argentino. 2° Congreso Latinoamericano de Geología (1973), Actas 2:755–763, Caracas

Aceñolaza FG, Miller H, Toselli AJ (1996) Geología del Sistema de Famatina. Münchner Geologische Heste, A 19:1–410, München

Aleman A, Ramos VA (2000) The Northern Andes. In UJ Cordani, EJ Milani, A Thomaz Filho, DA Campos (eds.) Tectonic evolution of South America, 31° International Geological Congress, 453–480, Río de Janeiro

Astini RA, Benedetto JL (1993) A collisional model for the stratigraphic evolution of the Argentine Precordillera during the early Paleozoic. 2° International Symposium on Andean Geodynamics (Oxford), 501–504, Paris

Astini RA, Benedetto JL, Vaccari NE (1995) The early Paleozoic evolution of the Argentina Precordillera as a Laurentian rifted, drifted, and collided terrane: a geodynamic model. Geol Soc Am Bull 107(3):253–273

Astini R, Ramos VA, Benedetto JL, Vaccari NE (1996) La Precordillera: un terreno exótico a Gondwana. 13° Congreso Geológico Argentino y 3° Congreso Exploración de Hidrocarburos (Buenos Aires). Actas 5:293–324

Ávila Salinas W (1992) El magmatismo Cámbrico-Ordovícico en Bolivia. In Gutiérrez-Marco JC, Saavedra J, Rábano I (eds.) Paleozoico Inferior de Iberoamérica, Universidad de Extremadura, 241–253

Bahlburg H (1990) The Ordovician basin in the northern Puna of Argentina and Chile: geodynamic evolution from back-arc to foreland basin. Geotekton Forsch 75:1–107

Bahlburg H, Hervé F (1997) Geodynamic evolution and tectonostratigraphic terranes of NW-Argentina and N-Chile. Geol Soc Am Bull 109:869–884

Bahlburg H, Berndt J, Gerdes A (2016) The ages and tectonic setting of the Faja Eruptiva de la Puna Oriental, Ordovician, NWArgentina. Lithos 256–257:41–54

Bahlburg H, Carlotto V, Cárdenas J (2006) Evidence of Early to Middle Ordovician arc volcanism in the Cordillera Oriental and Altiplano of southern Peru, Ollantaytambo formation and Umachiri beds. J South Am Earth Sci 22:52–65

Beck SL, Zandt G (2002) The nature of orogenic crust in the central Andes. J Geophys Res 107. https://doi.org/10.1029/2000JB000124

Bellizzia A, Pimentel N (1994) Terreno Mérida: un cinturón alóctono herciniano en la Cordillera de Los Andes de Venezuela. 5º Simposio Bolivariano Exploración Petrolera en las Cuencas Subandinas, Memoria, 271–299

Benedetto JL (2004) The allochthony of the Argentine Precordillera ten years later (1993–2003): a new paleogeographic test of the microcontinental model. Gondwana Res 7:1027–1039

Benedetto JL, Ramírez PE (1985) La secuencia sedimentaria Precámbrico-Paleozoico Inferior pericratónica del extremo norte de Sudamérica y sus relaciones con las cuencas del norte de África. Quinto Congreso Latinoamericano de Geología, Actas 2:411–425

Benedetto JL, Sánchez TM, Carrera MG, Brussa ED, Salas JM (1999) Paleontological constraints on successive paleogeographic positions of Precordillera terrane during the early Paleozoic. In: Ramos VA, Keppie D (eds) Laurentia Gondwana Connections before Pangea. Geological Society of America, Special Paper 336, pp 21–42

Boedo FL, Willner AP, Vujovich GI, Massonne H -J (2016) High-pressure/low-temperature metamorphism in the collision zone between the Chilenia and Cuyania microcontinents (western Precordillera, Argentina). J South Am Earth Sci 72:227–240

Boekhout F, Sempere T, Spikings R, Schaltegger U (2013) Late Paleozoic to Jurassic chronostratigraphy of coastal southern Peru: temporal evolution of sedimentation along an active margin. J S Am Earth Sci 47:179–200

Brussa ED, Toro BA, Vaccari NE (2008) Bioestratigrafía del Paleozoico inferior en el ámbito de la Puna. In: Coira B, Zappettini EO (eds) Geología y recursos naturales de la provincia de jujuy, 17° Congreso Geológico Argentino, Relatorio, pp 93–97

Cardona A, Cordani UG, MacDonald WD (2006) Tectonic correlations of pre-Mesozoic crust from the northern termination of the Colombian Andes, Caribbean region. J S Am Earth Sci 21:337–354

Cardona A, Cordani U, Ruiz J, Valencia VA, Armstrong R, Nutman A, Sanchez A (2009) U/Pb zircon and Nd isotopic signatures of the pre-Mesozoic metamorphic basement of the Eastern Peruvian Andes: Growth and provenance of a late Neoproterozoic to Carboniferous accretionary orogen on the Northwest margin of Gondwana. J Geol 117:285–305

Carlotto V, Quispe J, Acosta H, Rodríguez R, Romero D, Cerpa L, Mamani M, Díaz Martínez E, Navarro P, Jaimes F, Velarde T, Lu S, Cueva E (2009) Dominios geotectónicos y Metalogénesis del Perú. Sociedad Geológica del Perú. Boletín 103:1–90

Castroviejo R, Pereira E, Rodrigues JF, Acosta J, Espi JA (2009) Pre-Andean serpentinite chromite orebodies in the Eastern Cordillera of Central Perú, Tarma province. In: 10th biannual society for geology applied for mineral deposits, vol 2, Townsville, pp 927–929

Castroviejo R, Rodrigues JF, Tassinari C, Pereira E, Acosta J (2010) Ophiolites in the Eastern Cordillera of the central Peruvian Andes. IMA2010: 20th. general meeting of the international mineralogical association, Budapest

Chernicoff J, Ramos VA (2004) El basamento de la Sierra de San Luis: nuevas evidencias magnéticas y sus implicancias tectónicas. Revista de la Asociación Geológica Argentina 58 (4):511–524

Chernicoff CJ, Zappettini EO, Santos JOS, McNaughton NJ, Belousova E (2013) Combined U-Pb SHRIMP and Hf isotope study of the Late Paleozoic Yaminué Complex, Rio Negro Province, Argentina: Implications for the origin and evolution of the Patagonia composite terrane. Geosci Front 4(1):37–56

Chew DM, Schaltegger U, Košler J, Whitehouse MJ, Gutjahr M, Spikings RA, Miškovíc A (2007) U-Pb geochronologic evidence for the evolution of the Gondwanan margin of the north-central Andes. Geol Soc Am Bull 119:697–711

Chew DM, Pedemonte G, Corbett E (2016) Proto-Andean evolution of the Eastern Cordillera of Peru. Gondwana Res 35:59–78

Coira BL (2008) El volcanismo del Paleozoico inferior de la Puna Jujeña. In: Coira B, Zappettini EO (eds) Geología y Recursos Naturales de la provincia de Jujuy, 17° Congreso Geológico Argentino, Relatorio, Jujuy, pp 140–154

Coira BL, Davidson JD, Mpodozis C, Ramos VA (1982) Tectonic and magmatic evolution of the Andes of Northern Argentina and Chile. Earth Sci Rev 18(3–4):303–332

Coira BL, Mahlburg Kay S, Pérez B, Woll B, Hanning M, Flores P (1999) Magmatic sources and tectonic setting of Gondwana margin Ordovician magmas, northern Puna of Argentina and Chile. In: Ramos VA, Keppie D (eds) Laurentia Gondwana Connections before Pangea. Geological Society of America, Special Paper 336, pp 145–170

Coira B, Koukharsky M, Ribeiro Guevara S, Cisterna CE (2009) Puna (Argentina) and Northern Chile Ordovician basic magmatism: A contribution to the tectonic setting. J S Am Earth Sci 27:24–35

Collo G, Astini RA, Cawood P, Buchan C, Prmentel M (2009) U-Pb detrital zircon ages and Sm-Nd isotq:ic features in low-grade metasedimentary rocks of the Famatina belt: implications for late Neoproterozoic – early Palaeozoic evolution of the proto-Andean margin of Gondwana. J Geol Soc 116:1–17

Dahlquist JA, Rapela CW, Panhurst RJ, Baldo EG, Saavedra J, Alasino PH (2005) Los granitoides de la sierra de Chepes y su comparación con granitoides paleozoicos de las Sierras Pampeanas: implicancias para el orógeno famatiniano. Asociación Geológica Argentina, Serie D: Publicación Especial 8:87–108

Dahlquist JA, Galindo C, Pankhurst RJ, Rapela CW, Alasino PH, Saavedra J, Fanning CM (2007) Magmatic evolution of the Peñón Rosado granite: petrogenesis of garnet-bearing granitoids. Lithos 95:177–207

Dahlquist JA, Panhurst RJ, Rapela CW, Galindo C, Alasino PH, Fanning CM, Saavedra J, Baldo EG (2008) New SHRIMP U-Pb data from the Famatina Complex: Constraining Early-Mid Ordovician Famatinian magmatism in the Sierras Pampeanas, Argentina. Geol Acta 6(4):319–333

Dahlquist JA, Pankhurst RJ, Gaschnig RM, Rapela CW, Casquet C, Alasino PH, Galindo C, Baldo EG (2013) Hf and Nd isotopes in Early Ordovician to Early Carboniferous granites as monitors of crustal growth in the Proto-Andean margin of Gondwana. Gondwana Res 23:1617–1630

de Almeida FFM, Hasui Y, Brito Neves BB (1976) The Upper Precambrian of South America. Universidade de Sao Paulo, Instituto de Geociencias, Boletim 7:45–80

Dorbath C, Granet M, Poupinet G, Martinez C (1993) A teleseismic study of the Altiplano and the Eastern Cordillera in northern Bolivia: new constraints on a lithospheric model. J Geophys Res 98(B6):9825–9844

Dorbath C, Granet M (1996) Local earthquake tomography of the Altiplano and the Eastern Cordillera of northern Bolivia. Tectonophysics 259:117–136

Ducea MN, Otamendi JE, Bergantz G, Stair KM, Valencia VA, Gehrels GE (2010) Timing constraints on building an intermediate plutonic arc crustal section: U-Pb zircon geochronology of the Sierra Valle Fértil–La Huerta, Famatinian arc, Argentina. Tectonics 29(TC4002):21–22

Egenhoff SO (2007) Life and death of a Cambrian-Ordovician basin: an Andean three act play featuring Gondwana and the Arequipa-Antofalla terrane. Geol Soc Am Spec Pap 423:511–524

Fernández C, Becchio R, Castro A, Viramonte JM, Moreno-Ventas I, Corretgé LG (2008) Massive generation of atypical ferrosilicic magmas along the Gondwana active margin: Implications for cold plumes and back-arc magma generation. Gondwana Res 14:451–473

Forero Suárez A (1990) The basement of the Eastern Cordillera, Colombia: an allochthonous terrane in northwestern South America. J S Am Earth Sci 3(2):141–151

Gómez Tapias J, Montes Ramírez NE, Alcárcel Gutiérrez FA, Ceballos Hernández JA (2015) Catálogo de dataciones radiométricas de Colombia en ArcGIS 63 y Google Earth. In: Gómez Tapias J, Almanza Meléndez MF (eds) Compilando la geología de Colombia: Una visión a 2015, Servicio Geológico Colombiano, Bogotá, pp 63–420

González PD, Tortello F, Damborenea S (2011) Early Cambrian archaeocyathan limestone blocks in low-grade metaconglomerate from El Jagüelito Formation (Sierra Grande, Río Negro, Argentina). Geologica Acta 9:159–173

Grösser JR, Prössl KF (1991) First evidence of the Silurian in Colombia: Palynostratigraphic data from the Quetame Massif, Cordillera Oriental. J S Am Earth Sci 4(3):231–238

Haller MA, Ramos VA (1984) Las ofiolitas famatinianas (Eopaleozoico) de las provincias de San Juan y Mendoza. 9° Congreso Geológico Argentino (S.C. Bariloche). Actas 2:66–83

Harrington JH, Kay M (1951) Cambrian and Ordovician Fauna of eastern Colombia. J Paleontol 25:655–668

Herbert H (1977) Petrochemie und Ausgangmaterial von Grünschefren aus der E-cordillere Ecuadors. Fortschr Mineral 55(1):45–46

Horton BK, Saylor JE, Nie J, Mora A, Parra M, Reyes-Harker A (2010) Linking sedimentation in the northern Andes to basement configuration, Mesozoic extension, and Cenozoic shortening: Evidence from detrital zircon U-Pb ages, Eastern Cordillera, Colombia. Geol Soc Am Bull 122 (9–10):1423–1442

ICS-IUGS Geological Time Table (2016) In: https://www.stratigraphy.org/index.php/ics-chart-timescale

Kay SM, Coira B (2009) Shallowing and Steepening Subduction Zones, Continental Lithosphere Loss, Magmatism and Crustal Flow under the Central Andean Altiplano-Puna Plateau. In: Kay S, Ramos VA, Dickinson W (eds) Backbone of the Americas, Geological Society of América, vol 204, Memoir, pp 229–259

Litherland M, Aspden JA, Jemielita RA (1994) The metamorphic belts of Ecuador. Brit Geol Surv Overseas Mem 11:1–146

Loewy SL, Connelly JN, Dalziel IWD (2004) An orphaned basement block: the Arequipa-Antofalla basement of the central Andean margin of South America. Geol Soc Am Bull 116:171–187

Lork A, Bahlburg E (1993) Precise U-Pb ages of monazites from the Faja Eruptiva de la Puna Oriental, NW Argentina. 12° Congreso Geológico Argentino y 2° Congreso Exploración de Hidrocarburos, Actas, vol 4, Buenos Aires, pp 1–6

Mantilla Figueroa LC, Ordóñez-Carmona JJ, García-Ramírez CA, Valencia VA (2016) Nuevas evidencias que soportan la escisión de la formación Silgará y propuesta de un nuevo marco estratigráfico para el basamento metamórfico del Macizo de Santander (Cordillera Oriental de Colombia). Revista Academia Colombiana de Ciencias Exactas Físicas y Naturales 40 (155):320–336

Martens U, Restrepo O, Correa-Martínez AM (2014) The Tahamí and Anacona Terranes of the Colombian Andes: Missing Links between the South American and Mexican Gondwana Margins. J Geol 122:507–530

Martínez C, Dorbath C, Lavenu A (1994) La cuenca subsidente cenozoica noraltiplánica y sus relaciones con una subducción transcurrente continental. 12° Congreso Geológico de Bolivia, Actas, Tarija, pp 3–28

Martínez M, Brussa E, Pérez B, Coira B (1999) El Ordovícico de la sierra de Quichagua (Puna nororiental argentina): litofacies volcanosedimentarias y graptofaunas. 14° Congreso Geológico Argentino (Salta). Actas 1:347–350

Martino RD, Simpson C, Law RD (1993) Taconic (Ocloyic) aged west-directed ductile thrusts in basement rocks of the Sierras Pampeanas, Argentina. Geological Society of America, Abstracts with Program A-233

Méndez V, Navarini A, Plaza D, Viera O (1973) Faja eruptiva de la puna oriental. 5° Congreso Geológico Argentino (Córdoba), Actas 4:89–100

Moya MC (2015) La "Fase Oclóyica" (Ordovícico Superior) en el noroeste argentino. Interpretación histórica y evidencias en contrario. Serie Correlación Geológica 31(1):73–110

Murphy JB, Keppie JD, Dostal J, Nance RD (1999) Neoproterozoic-Early Paleozoic evolution of Avalonia. In: Ramos VA, Keppie D (eds) Laurentia Gondwana Connections before Pangea. Geological Society of America, Special Paper 336, pp 253–266

Murphy JB, Gutiérrez-Alonso G, Nance RD, Fernández-Suárez J, Keppie JD, Quesada C, Strachan RA, Dostal J (2006) Origin of the Rheic Ocean: Rifting along a Neoproterozoic suture? Geology 34:325–328

Niemeyer RH (1989) El Complejo ígneo-sedimentario del Cordón de La Lila, Región de Antofagasta: Estratigrafía y significado tectónico. Revista Geológica de Chile 16(2):163–182

Ogg JG, Ogg GM, Gradstein FM (2016) A concise geologic time scale. Elsevier, 235 pp

Ordóñez-Carmona O, Restrepo JJ, Álvarez A, Pimentel MM (2006) Geochronological and isotopical review of pre-Devonian crustal basement of the Colombian Andes. J S Am Earth Sci 21:372–382

Otamendi JE, Tibaldi AM, Vujovich GI, Viñao GA (2008) Metamorphic evolution of migmatites from the deep Famatinian arc crust exposed in Sierras Valle Fértil e La Huerta, San Juan, Argentina. J S Am Earth Sci 25:313–335

Otamendi JE, Vujovich GI, de la Rosa JD, Tibaldi AM, Castro A, Martino RD (2009a) Geology and petrology of a deep crustal zone from the Famatinian paleo-arc, Sierras Valle Fértil-la Huerta, San Juan, Argentina. J S Am Earth Sci 27:258–279

Otamendi JE, Ducea MN, Tibaldi AM, Bergantz G, de la Rosa JD, Vujovich GI (2009b) Generation of tonalitic and dioritic magmas by coupled partial melting of gabbroic and metasedimentary rocks within the deep crust of the Famatinian magmatic arc, Argentina. J Petrol 50:841–873

Otamendi JE, Cristofolini E, Tibaldi AM, Quevedo F, Baliani I (2010) Petrology of mafic and ultramafic layered rocks from the Jaboncillo Valley, Sierra de Valle Fértil, Argentina: implications for the evolution of magmas in the lower crust of the Famatinian arc. J S Am Earth Sci 29:685–704

Otamendi JE, Ducea MN, Bergantz GW (2012) Geological, petrological and geochemical evidence for progressive construction of an arc crustal section, Sierra de Valle Fertil, Famatinian Arc, Argentina. J Petrol 53:761–800

Pankhurst R, Rapela C, Saavedra J, Baldo E, Dahlquist J, Pascua I (1996) Sierra de Los Llanos, Malanzán, and Chepes: Ordovician I and S-type granitic magmatism in the Famatinuan orogen. 13° Congreso Geológico Argentino and 3° Congreso de Exploración de Hidrocarburos, Actas 5: 415, Buenos Aires

Pankhurst R, Rapela C, Saavedra J, Baldo E, Dahlquist J, Pascua I, Fanning CM (1998) The Famatinian magmatic arc in the Central Sierras Pampeanas: an Early to mid-Ordovician continental arc on the Gondwana margin. In: Pankhurst RJ, Rapela CW (eds) The Proto-Andean margin of Gondwana, vol 142, Geological Society of London, Special Publication, pp 343–367

Pankhurst R, Rapela C, Fanning CM (2000) Age and origin of coeval TTG, I- and S-type granites in the Famatinian belt of NW Argentina. Trans R Soc Edinb Earth Sci 91:151–168

Pankhurst RJ, Rapela CW, Fanning CM, Márquez M (2006) Gondwanide continental collision and the origin of Patagonia. Earth-Sci Rev 76:235–257

Pinheiro GM, Pimentel MM, Schalamuk IB (2008) Sm–Nd and LAM-ICPMS U–Pb data for Cambrian/Ordovician rocks of the Calalaste range, NW Argentina. In: 4° South American Symposium on Isotope Geology, Actas Digitales, Bariloche, 4 p

Quenardelle S, Ramos VA (1999) The Ordovician western Sierras Pampeanas magmatic belt: record of Argentine Precordillera accretion. In: Ramos VA, Keppie D (eds) Laurentia Gondwana Connections before Pangea. Geological Society of America, Special Paper 336, pp 63–86

Ramos VA (1984) Patagonia: ¿Un continente paleozoico a la deriva? 9° Congreso Geológico Argentino (S.C. Bariloche). Actas 2:311–325

Ramos VA (1986) El Diastrofismo Oclóyico: Un ejemplo de tectónica de colisión durante el Eopaleozoico en el No-roeste Argentino. Revista del Instituto de Geología y Minería (San Salvador de Jujuy) 6:13–28

Ramos VA (1988) The Tectonics of the Central Andes: 30° 33° S latitude. In: Clark S, Burchfiel D, Suppe J (eds) Processes in Continental Lithospheric Deformation, Geological Society of America, Special paper 218, pp 31–54

Ramos VA (2004) Cuyania, an exotic block to Gondwana: review of a historical success and the present problems. Gondwana Res 7(4):1009–1026

Ramos VA (2008) The basement of the Central Andes: the Arequipa and related terranes. Annual Review on Earth and Planetary Sciences 36:289–324

Ramos VA (2009) Anatomy and global context of the Andes: Main geologic features and the Andean orogenic cycle. In: Kay SM, Ramos VA, Dickinson W (eds) Backbone of the Americas: Shallow Subduction, Plateau Uplift, and Ridge and Terrane Collision, Geological Society of America, Memoir 204:31–65

Ramos VA (2010) The Grenville-age basement of the Andes. J S Am Earth Sci 29(1):77–91

Ramos VA, Coira B (2008) Las provincias geológicas de Jujuy. In Coira B, Zappettini EO (eds.) Geología y Recursos Naturales de la Provincia de Jujuy, 17° Congreso Geológico Argentino, Relatorio, pp 11–15

Ramos VA, Jiménez N (2014) Extensión oriental del macizo de Arequipa en los Andes bolivianos: Sus implicancias tectónicas. In Simposio Tectónica preandina, 19° Congreso Geológico Argentino, Actas S21–50, Córdoba, 2 p

Ramos VA, Naipauer M (2014) Patagonia: Where does it come from? J Iberian Geology 40 (2):367–379

Ramos VA, Escayola M, Mutti D, Vujovich GI (2000) Proterozoic-early Paleozoic ophiolites in the Andean basement of southern South America. In: Dilek Y, Moores EM, Elthon D, Nicolas A (eds) Ophiolites and oceanic crust: new insights from field studies and ocean drilling program. Geol Soc Am, special paper 349, pp 331–349

Ramos VA, Jordan TE, Allmendinger RW, Mpodozis C, Kay SM, Cortés JM, Palma MA (1986) Paleozoic terranes of the central Argentine Chilean Andes. Tectonics 5(6):855–880

Ramos VA, Vujovich G, Martino R, Otamendi J (2010) Pampia: a large cratonic block missing in the Rodinia supercontinent. J Geodyn 50:243–255
Rapela CW, Toselli AJ, Heaman L, Saavedra J (1990) Granite plutonism of Sierras Pampeanas; an inner Cordilleran Paleozoic arc in the southern Andes. In: Kay SM, Rapela CW (eds) Plutonism from Antarctica to Alaska. Geological Society of America, Special paper 241, pp 77–90
Reimann CR, Bahlburg H, Kooijman E, Berndt J, Gerdes A, Carlotto V, Lopez S (2010) Geodynamic evolution of the early Paleozoic Western Gondwana margin 14°–17° S reflected by the detritus of the Devonian and Ordovician basins of southern Peru and northern Bolivia. Gondwana Res 18:370–384
Reimann CR, Bahlburg H, Carlotto V, Boekhout F, Berndt J, Lopez S (2015) Multi-method provenance model for early Paleozoic sedimentary basins of southern Peru and northern Bolivia (13°–18° S). J S Am Earth Sci 64:94–115
Reitsma MJ (2012) Reconstructing the Late Paleozoic: Early Mesozoic plutonic and sedimentary record of south-east Peru: orphaned back-arcs along the western margin of Gondwana. PhD thesis, University of Geneva, Geneve
Restrepo JJ, Toussaint J-F (1988) Terranes and continental accretion in the Colombian Andes. Episodes 11:189–193
Restrepo-Pace PA (1992) Petrotectonic characterization of the Central Andean Terrane, Colombia. J S Am Earth Sci 5(1):97–116
Restrepo-Pace PA, Cediel F (2010) Northern South America basement tectonics and implications for paleocontinental reconstructions of the Americas. J S Am Earth Sci 29:764–771
Restrepo-Pace PA, Ruiz J, Gehrels G, Cosca M (1997) Geochronology and Nd isotopic data of Grenville-age rocks in the Colombian Andes: new constraints for Late Proterozoic-Early Paleozoic paleocontinental reconstructions of the Americas. Earth Planet Sci Lett 150:427–441
Reyes L, Caldas J (1987) Geología de los cuadrángulos de Las Playas, La Tina, Las Lomas, Ayabaca, San Antonio, Chulucanas, Morropón, Huancabamba, Olmos y Pomahuaca. Boletín Ingemmet 39(A):1–83, Lima
Rodrigues J, Acosta J, Castroviejo R, Quispe J, Romero D, Uribe R, Campián M (2010a) Geología y estructura de las ultramafitas de Tapo y Acobamba (Tarma, Perú), removilización tectónica andina de un segmento ofiolítico pre-andino. Sociedad Geológica del Perú, Publicación Especial 9:79–82, Cusco
Rodrigues J, Acosta J, Macharé J, Pereira E, Castroviejo R (2010b) Evidencias estructurales de aloctonía de los cuerpos ultramáficos y máficos de la Cordillera Oriental del Perú en la región de Huánuco. Sociedad Geológica del Perú, Publicación Especial 9: 75–78, Cusco
Romero D, Valencia K, Alarcón P, Peña D, Ramos VA (2013) The offshore basement of Perú: evidence for different igneous and metamorphic domains in the forearc. J S Am Earth Sci 42:47–60
Ruiz J, Tosdal RM, Restrepo PA, Murillo-Muñetón G (1999) Pb isotope evidence for Colombia–southern México connections in the Proterozoic. Geological Society of America, Special Papers 336:183–197
Rushton AWA (1962) Paradoxides from Colombia. Geological Magazine 100:255–257
Saavedra J, Toselli A, Rossi J, Pellitero E, Durand F (1998) The Early Palaeozoic magmatic record of the Famatina System: a review. In: Pankhurst RJ, Rapela CW (eds) The Proto-Andean Margin of Gondwana, vol 142, Geological Society of London, Special Publication, pp 283–295
Spikings R, Cochrane R, Villagomez D, Van der Lelij R, Vallejo C, Winkler W, Beate B (2016) The geological history of northwestern South America: from Pangaea to the early collision of the Caribbean Large Igneous Province (290–75 Ma). Gondwana Res 27:95–139
Stern CR (1991) Role of subduction erosion in the generation of Andean Magmas. Geology 19 (1):78–81
Toro BA, Brussa ED (2003) Graptolites. In: Benedetto JL (ed) Ordovician fossils of Argentina. Secretaría de Ciencia y Tecnología, Universidad Nacional de Córdoba Chapter 11, pp 441–505

Toro BA, Brussa ED, Maletz J (2006) Implicancias bioestratigráficas y paleobiogeográficas de los graptolites de la localidad de Santa Rosa, Puna Oriental, Argentina. 9° Congreso Argentino de Paleontología y Bioestratigrafía, Actas 116, Córdoba

Toselli AJ (1982) Criterios de definición del metamorfismo de muy bajo grado. Con especial énfasis en el perfil de Falda Ciénaga, Puna de Catamarca. Asociación Geológica Argentina 37 (2):205–213

Toselli A, Rossi de Toselli JN, Saavedra J (1987) Petrological and geochemical considerations about the Lower Paleozoic Granitoids of the Pampean Ranges, Argentina. Revista Brasileira de Geociencias 17(4):619–622

Toussaint J-F, Restrepo JJ (1989) Acreciones sucesivas en Colombia: un Nuevo modelo de evolución geológica. 5° Congreso Colombiano de Geología, Memorias 1:127–146

Turner JCM, Méndez V (1975) Geología del sector oriental de los Departamentos de Santa Victoria e Iruya, Provincia de Salta, República Argentina. Boletín de la Academia Nacional de Ciencias de Córdoba 51(1–2):11–24

Van der Lelij R (2013) Reconstructing north–western Gondwana with implications for the evolution of the Iapetus and Rheic Oceans: A geochronological, thermochronological and geochemical study. Ph.D. Thesis, University of Geneve, Sc. 4581, Geneve, 247 p

Van der Lelij R, Spikings R, Ulianov A, Chiaradia M, Mora A (2016) Palaeozoic to Early Jurassic history of the northwestern corner of Gondwana, and implications for the evolution of the Iapetus, Rheic and Pacific Oceans. Gondwana Res 31:271–294

Verdecchia SO, Casquet C, Baldo EG, Pankhurst RJ, Rapela CW, Fanning M, Galindo C (2011) Docking of the Río de la Plata craton to southwestern Gondwana: age constraints from U-Pb SHRIMP detrital zircon ages from Sierras de Ambato and Velasco (Sierras Pampeanas, Argentina). J Geol Soc 168:1061–1071

Villagómez D, Spikings R, Magna T, Kammer A, Winkler W, Beltrán A (2011) Geochronology, geochemistry and tectonic evolution of the Western and Central cordilleras of Colombia. Lithos 125(3):875–896

Willner AP, Tassinari CCG, Rodrigues JF, Acosta J, Castroviejo R, Rivera M (2014) Contrasting Ordovician high- and low-pressure metamorphism related to a microcontinent-arc collision in the Eastern Cordillera of Perú (Tarma province). J S Am Earth Sci 54:71–81

Wörner G, Lezaun J, Beck A, Heber V, Lucassen F, Zinngrebe E, Rössling R, Wilcke HG (2000) Geochronology, petrology, and geochemistry of basement rocks from Belen (N. Chile) and C. Uyarani (W. Bolivian Altiplano): implication for the evolution of the basement. J S Am Earth Sci 13:717–737

Zappettini E, Blasco G, Villar L (1994) Geología del extremo sur del Salar de Pocitos, Provincia de Salta, República Argentina. 7° Congreso Geológico Chileno, Actas 1:220–224, Concepción.

Zimmermann U, Bahlburg H, Mezger K, Kay SM (2014) Origin and age of ultramafic rocks and gabbros in the southern Puna of Argentina: an alleged Ordovician suture revisited. Int J Earth Sci 103:1023–1036

Part III
The Early Andean Arc in the Chilean-Argentinean Margin

The Early Stages of the Magmatic Arc in the Southern Central Andes

Verónica Oliveros, Javiera González, Mauricio Espinoza Vargas, Paulina Vásquez, Pablo Rossel, Christian Creixell, Fernando Sepúlveda and Francisco Bastias

Abstract A temporal and tectonic framework for the initiation of the Andean magmatism at its earliest stages is presented in this work, based on the current geological, geochemical, and geochronological information available in northern Chile between 20° and 30° S. This information has been collected from Triassic and Jurassic plutonic, volcanic, and volcanosedimentary units cropping out at the Coastal, Domeyko, Frontal Cordilleras, and the Precordillera, as two roughly parallel belts that extend for more than 1,000 km in the Chilean territory. Petrological and geochemical data suggest that the magmatism during the Triassic and Jurassic was dominantly intermediate in composition and had arc or subduction affinities. Even though magma sources should have had a clear mantle component, a crustal influence was more important during the Triassic. Geochronological information shows a significant magmatic lull near the Triassic–Jurassic boundary, associated with the retraction of the arc to the present-day Coastal Cordillera. Based on these data, the early stages of the Andean magmatism can be tracked backup to the Triassic igneous units in northern Chile.

Keywords Andes · Jurassic · Triassic · Arc

V. Oliveros (✉) · J. González · M. Espinoza Vargas · F. Sepúlveda · F. Bastias
Departamento de Ciencias de la Tierra, Universidad de Concepción, Casilla 160-C, Concepción, Chile
e-mail: voliveros@udec.cl

P. Vásquez · C. Creixell
Servicio Nacional de Geología y Minería, Av. Santa María 0140, Providencia, Santiago, Chile

P. Rossel
Universidad Andres Bello, Autopista Concepción-Talcahuano 7100, Concepción, Chile

A. Folguera et al. (eds.), *The Evolution of the Chilean-Argentinean Andes*,
Springer Earth System Sciences, https://doi.org/10.1007/978-3-319-67774-3_7

1 Introduction

For the last decades, the Lower Jurassic volcanic deposits of the Coastal Cordillera between 18° and 30° S were considered the earliest volcanic products of the Andean arc (García 1967; Rogers and Hawkesworth 1989), produced by subduction of an oceanic plate under the continent after a period of within-plate extension and rifting of the western margin of the South American during the Late Permian to Early Jurassic (Charrier et al. 2007; Mpodozis and Ramos 1989, 2008). Consequently, the onset of Andean subduction would have taken place roughly at 200 Ma. After this, the voluminous magmatism developed during the Jurassic to Early Cretaceous in the Coastal Cordillera of southern Peru and northern central Chile was mainly produced during a protracted transtensional tectonic regime (Charrier et al. 2007; Scheuber and González 1999). The oblique subduction of the Phoenix plate under the continent resulted in a rollback of the oceanic plate, retreat of the trench, and progressive thinning of the continental crust (Charrier et al. 2007). This paleogeographic configuration was characterized by intensive volcanism and plutonism in the arc region (the present-day Coastal Cordillera from 17° to 35° S) and along strike back-arc extension (Mpodozis and Ramos 1989).

Contrastingly, the Permian to Triassic period for the southwestern Gondwana margin was associated with a significant magmatic lull resulting from the lack of subduction and related to the orogenic collapse that followed the Permian "San Rafael" contractional event (Mpodozis and Ramos 1989; Nasi et al. 1990; Mpodozis and Kay 1992; Kleiman and Japas 2009). In this context, the magmatic sources would have changed significantly, decreasing the crustal component, between the Late Permian and the Early Jurassic (Mpodozis and Kay 1992; Hervé et al. 2014). However, recent works dealing with igneous units in the Frontal and Coastal Cordilleras of Chile and westernmost Argentina have failed to sustain the model in which subduction ceased for most of the Triassic and instead have proposed a setting characterized by an extensional overriding plate in an active subduction setting (Vásquez et al. 2011; del Rey et al. 2016; Coloma et al. 2017).

In this chapter, we present a compilation of geological, geochemical, and geochronological data for Triassic and Jurassic (ca. 250–150 Ma) volcanic, volcanosedimentary, and plutonic rocks cropping out in the Coastal, Domeyko, and Frontal Cordilleras of northern Chile between 20° S and 30° S. We critically review these existing data with the aim of establishing the initiation of the Andean volcanic arc and the overall tectonic setting in the continental margin for the aforementioned period.

2 Geological Framework for the Early Andes

Thanks to the semiarid climate of the present-day forearc region, the Triassic and Jurassic magmatic rocks are well exposed in northern Chile, where Cretaceous to Cenozoic deformation exhumed upper crustal plutonic roots (Mpodozis and

Ramos 1989) (see Chap. "Mechanisms and Episodes of Deformation Along the Chilean–Pampean Flat-Slab Subduction Segment of the Central Andes in Northern Chile"). Different morphostructural systems in this part of the margin are N-oriented and vary in number and surficial expression along strike in response to the Chilean–Pampean flat subduction segment between 26°–30° S (Fig. 1). The westernmost system is the Coastal Cordillera that limits to the east with the central depression north of the flat subduction segment at 26° S. South of this latitude, this low-relief zone is absent and a series of E-incised valleys cut several minor mountain systems. To the east, the Domeyko Cordillera (22°–26° S) and Precordillera (26°–30° S) constitute a series of fault-bounded ranges exposing mostly Mesozoic stratified rocks and a Paleozoic basement. South of 26° S, the Frontal Cordillera considered a product of the flat subduction segment is a thick-skinned system that exposes deep Paleozoic plutonic and metamorphic rocks, covered by Mesozoic and Cenozoic volcanic and sedimentary sequences.

The studied Triassic rocks crop out mainly in the Precordillera/Domeyko and Frontal Cordilleras. Plutonic rocks correspond to middle and upper crustal intrusions that form the Montosa–Chollay Batholith between 27° and 30° S. Volcanic and volcanosedimentary rocks crop out as tilted and folded sequences that overlie the Paleozoic plutonic rocks and underlie Jurassic marine or continental sequences between 24° and 30° S (SERNAGEOMIN 2002; Salazar et al. 2013). These rocks are commonly crosscut by numerous Mesozoic dykes and less often intruded by Cenozoic plutons (Parada et al. 1991; Creixell et al. 2009; Salazar et al. 2013). Triassic rocks in the Coastal Cordillera are volumetrically restricted, although recent works indicate that these units are more widely exposed than previously thought, particularly at 20°–22° S and 25°–26° S (Sepúlveda et al. 2014; Álvarez et al. 2016). Volcanic and volcanosedimentary rocks crop out as relatively restricted homoclinal sequences, spatially associated with, although not controlled by, the N-S trending Atacama Fault System (AFS), a 1,000 km trench–parallel structure segmented by minor oblique NW-structures that accommodated strike–slip displacements during the Mesozoic (Espinoza et al. 2015), whereas the Triassic intrusions consist mostly of minor plutons and plutonic complexes at 26°30′ and 28° S (Fig. 1). These rocks overlie and intrude the Late Paleozoic Chañaral Epimetamorphic Complex and other Permian igneous units (Marinovic et al. 1995; Godoy and Lara 1998; Creixell et al. 2016). During this period, the dominant crustal stress was extensional, which led to the deposition of volcanosedimentary sequences in NW-oriented basins (Charrier et al. 2007). Fault-bounded Triassic plutonic complexes also denote extensional conditions during their emplacement (Grocott et al. 2009), while a roughly N-trending Middle Triassic mylonitic belt recognized along the Frontal Cordillera at 29° S indicates deep-seated deformation under sinistral transtensional conditions (Murillo et al. 2013). Additionally, an exhumation event would have taken place during the Middle to Late Triassic (Salazar et al. 2013; Kato and Godoy 2015).

Contrastingly, the Jurassic volcanic and plutonic rocks are widely exposed in the Coastal Cordillera between 20° and 28° S, in close relation to the AFS. The main geological units that represent the magmatic arc are the La Negra Formation, a thick

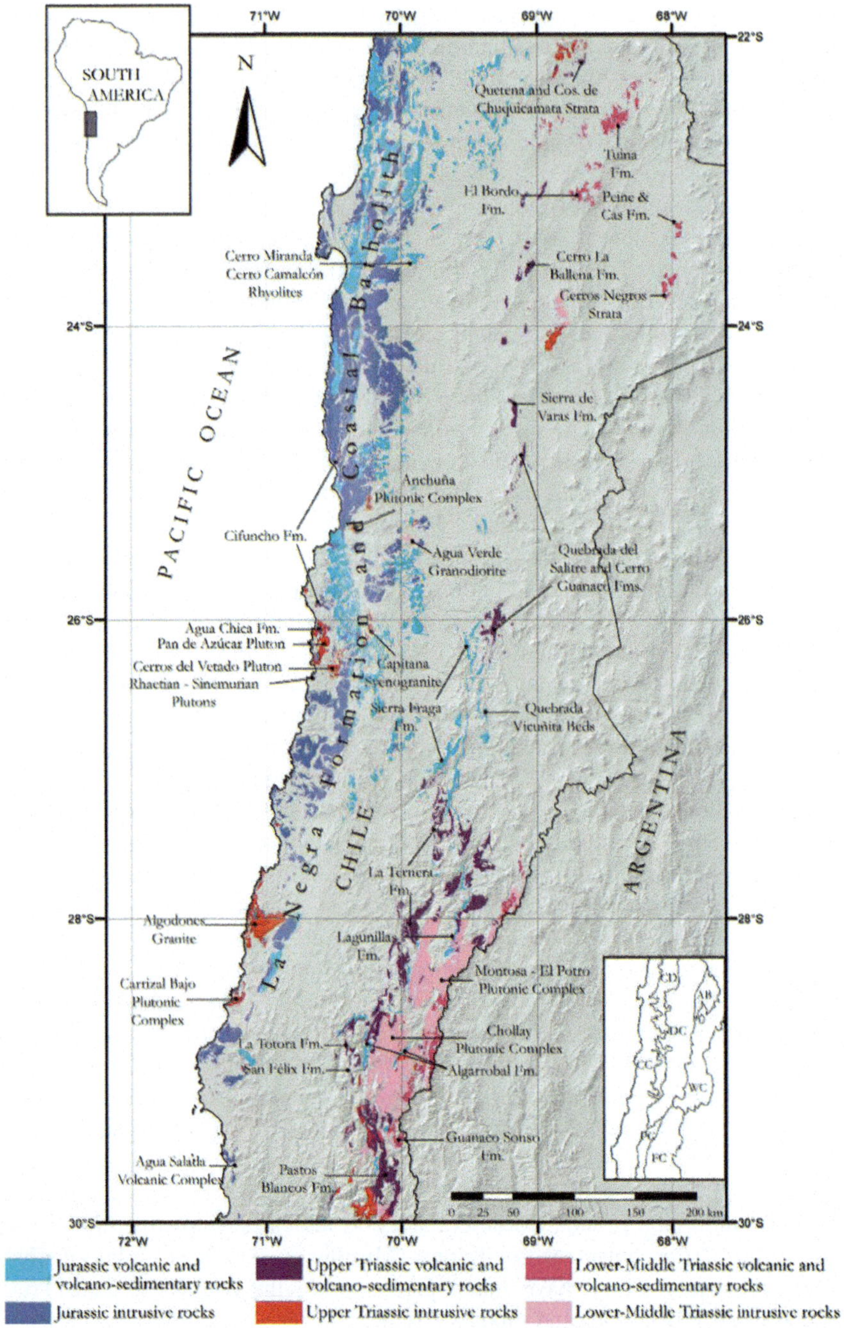

Fig. 1 Location of the outcrops of Triassic and Jurassic igneous rocks in the Southern Central Andes between 22° and 30° S. The inset shows the present-day morphotectonic units at these latitudes (CC: Coastal Cordillera; CD: Central Depression; DC: Domeyko Cordillera; PC: Precordillera; AB: Salar de Atacama Basin; FC: Frontal Cordillera and WC: Western Cordillera). The simplified geological map is based on the recently updated cartography done by the Chilean Geological and Mining Survey. For specific references see the text

homoclinal volcanic sequence, and the Coastal Batholith (Fig. 2). Marine sedimentary beds representing the back-arc domain are exposed in the eastern flank of the Coastal Cordillera, in the central depression and in the Domeyko Cordillera at the same latitudes (SERNAGEOMIN 2002; Vicente 2006). A more restricted belt of Jurassic volcanosedimentary units is exposed in the Precordillera and Domeyko Cordillera between 25° and 30° S. Structural studies carried out between 23° and 28° S suggest that throughout the Jurassic the subduction system developed under an extensional regime that later shifted to a transtensional setting (Scheuber and González 1999). The arc zone accommodated most of the extensional deformation, as evidenced by the structures observed in syntectonic plutonic bodies that were intruded at relatively deep crustal levels, whereas the back-arc basins were submitted to a slow thermal subsidence (Scheuber and González 1999; Grocott and Taylor 2002). A short episode of transpressive deformation has been detected along the Jurassic arc south of 31° S (Creixell et al. 2011; Ring et al. 2012), but this has not been documented north of that latitude. The shift from extensional to transtensional conditions would have occurred likely at the Jurassic–Cretaceous boundary (Scheuber and González 1999; Grocott and Taylor 2002) followed by the main phase of transtensive and transpressive deformation in the AFS at ~135–120 Ma (Scheuber and González 1999; Arévalo and Creixell 2009). The magmatic arc and associated rift system were aborted after ca. 118–106 Ma, when the magma foci shifted to the east and a compressive tectonic regime started during the Late Cretaceous (Grocott and Taylor 2002; Escribano et al. 2013;

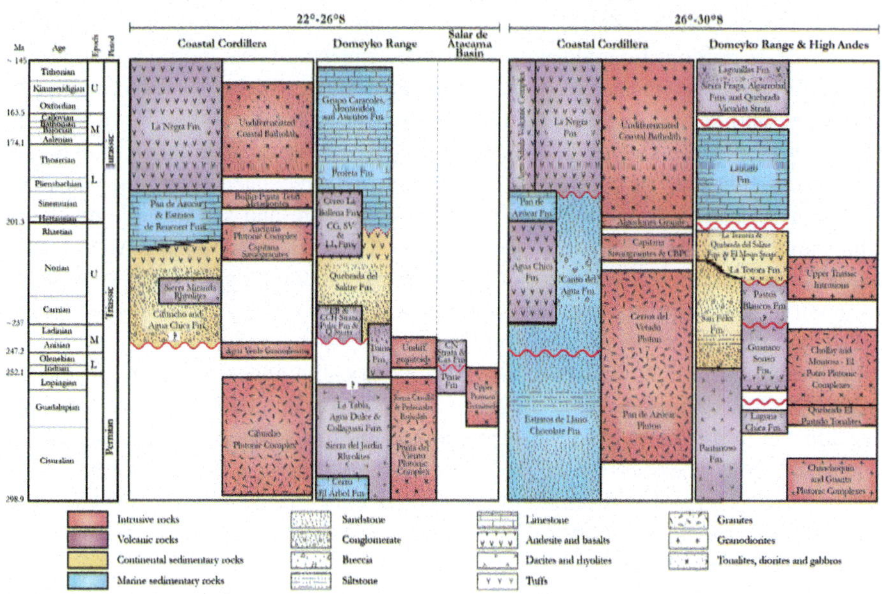

Fig. 2 Chronostratigraphic chart with Triassic and Jurassic rocks in the studied segments and morphotectonic units. Based on the recently updated cartography done by the Chilean Geological and Mining Survey. For specific references see the text

Sepúlveda et al. 2014) (see Chap. "Mechanisms and Episodes of Deformation Along the Chilean–Pampean Flat-Slab Subduction Segment of the Central Andes in Northern Chile").

3 Petrography and Geology of the Triassic–Jurassic Igneous Units in Northern Chile

3.1 *Triassic–Jurassic Igneous Units Between 20° and 22° S*

The studied volcanic units in this segment are as follows: Sierra de Lagunas Beds (Upper Triassic–Lower Jurassic), Oficina Viz Formation (Toarcian–Bajocian), and the volcanic facies from the Huantajaya Group corresponding to the Caleta Lígate Formation (Bajocian) and the El Godo Formation (Bajocian–Oxfordian). Sierra de Lagunas Beds are exposed in both Coastal Cordillera and Precordillera, whereas in the Precordillera, volcanic formations younger than Lower Jurassic are absent.

Sierra de Lagunas Beds and Oficina Viz Formation consist in monotonous homoclinal sequences of subaereous andesitic to basaltic andesitic lava flows and minor pyroclastic rocks in the Sierra de Lagunas Beds. The contact relation between these units is unclear because of late tectonic deformation. Additionally, a plutonic event took place at 180–175 Ma, represented by the Vetarrón Monzonite, which intruded the Sierra de Lagunas Beds (Sepúlveda et al. 2014). The Sierra de Lagunas Beds are located at the central and eastern portion of the Coastal Cordillera and are composed of porphyritic andesitic and basaltic andesitic lavas with a total thickness of 2,250 m. These lavas are dark gray and contain scarce lenses of laminated sandstone and crystal and lithic tuffs. Locally, they vary to amygdule-bearing and red-colored glassy andesitic lavas. Along the Precordillera, Tomlinson et al. (2015) recognized this unit as composed of massive and brecciated pyroxene and hornblende andesites with porphyrytic, glomeroporphyritic and seriate textures, minor lithic tuffs, volcanic and volcanosedimentary breccias, conglomerates, and sandstones. The Oficina Viz Formation consists mostly in a dark green to gray sequence of aphiric, porphyrytic, and amygdaloidal basaltic-andesite and andesite lavas, andesitic volcanic breccias, sandstone lenses, and scarce reddish brown tuff layers, with a minimum thickness of 1,000–1,500 m. Locally, andesitic dykes cut the lava piles (Thomas 1970; Kramer et al. 2005; Sepúlveda et al. 2014). The volcanic rocks are affected by very low-grade metamorphism, but fresh clinopyroxene, plagioclase, and amphibole phenocrysts do occur in most of the lava flows (Oliveros et al. 2006). The Oficina Viz Formation is covered concordantly by the Caleta Lígate Formation with a Bajocian age (Thomas 1970; Kossler 1998; Vásquez and Sepúlveda 2013).

The volcanic lithofacies of Caleta Lígate Formation has been recognized south of Iquique town and consist of pyroclastic breccias, basalts, and tuffs intercalated with fossiliferous sandstones (Kossler 1998; Vásquez and Sepúlveda 2013).

The pyroclastic breccias contain long rhyolithic blocks (juvenile) of decimetric size, and minor centimetric andesitic and sedimentary clasts, both included in a glassy groundmass. The pyroclastic rocks vary from breccias to glassy lapilli tuffs, with "fiamme" structures and lithics fragments mostly of andesitic composition. There are also basaltic lavas with a greenish gray color with pillow tubes. The volcanic rocks of El Godo Formation correspond mainly to basalt and basaltic andesites with pillowed or brecciated structure. Locally, olistoliths, humid intrusive dikes, and peperites associated with subaqueous lapilli tuffs, denote that volcanism interacted with water-rich sediments. Andesites and basalts have subophitic and porphyritic texture with plagioclase, pyroxene, and minor amphibole phenocrysts, in a crystal-bearing glassy groundmass. The volcanic rocks associated with the Huantajaya Group were deposited in a submarine environment, intercalated in marine sedimentary rocks, while their ages have been determined on the base of the fossiliferous content and radiometric dating (Thomas 1970; Kossler 1998; Marquardt et al. 2008; Vásquez and Sepúlveda 2013). Both formations are unconformably covered by the Lower Cretaceous Punta Barranco Formation.

3.2 *Triassic–Jurassic Igneous Units Between 22° and 26° S*

In addition to the two morphostructural units where the Triassic rocks usually crop out in northern Chile, the Triassic rocks are also found in the vicinity of the Salar de Atacama basin in this segment (Figs. 1 and 2). Near the Peine locality, volcanic sequences of reddish and greenish andesitic lava flows and breccias, vitric and crystalline tuffs, interbedded with conglomeratic sedimentary rocks are grouped in the Peine Formation (Upper Permian?—Middle Triassic). This unit is unconformably overlain by porphyritic andesitic lavas and breccias and dacitic agglomerates of the Cas Formation (Niemeyer 2013). Partially coetaneous andesitic volcanism and continental sedimentation in fluvial systems are represented by the volcanosedimentary sequences of the Cerros Negros Beds (Lower Triassic) (Niemeyer 2013) (Fig. 2).

In the Domeyko Cordillera, the Triassic units also represent volcanic activity and continental sedimentation, where they unconformably overlie Carboniferous to Lower Permian strata. Near the city of Calama, the Quetena Beds composed of conglomerates and sandstones (Middle Triassic) (Tomlinson et al. 2010) are overlain by the andesitic lavas, with minor dacitic and rhyolitic rocks, of the Chuquicamata Beds (Carnian) (Tomlinson and Blanco 2008) (Fig. 2). The latter sequence is intruded by small granodiorites plutons of Carnian age (Proffett and Dilles 2007; Tomlinson et al. 2010). In the eastern flank of the Domeyko Cordillera, the Tuina Formation crops out, which is composed of andesitic lava flows and minor dacites and rhyolites (Lopingian—Middle Triassic) (Henríquez et al. 2014). Also in this flank, andesitic and dacitic lava flows are interbedded with lacustrine deposits that correspond to the El Bordo Beds (Middle Triassic) (Basso and Mpodozis 2012) and their equivalents further south in Sierra Mariposas and Imilac

(Cortés 2012; Solari et al. 2015). In the southern part of the Domeyko Cordillera, the Exploradora, Vaquillas, and Castillo ranges, Triassic sequences of the volcanosedimentary Quebrada del Salitre Formation crop out extensivelly (Middle?—Upper Triassic) (Fig. 2). Its lower member is composed of basaltic lava flows, dacites, and pyroclastic and volcanoclastic rocks interbedded with sedimentary rocks that represent an alluvial–fluvial environment (Mpodozis and Cornejo 1997; Venegas et al. 2013), and it is intruded by rhyolitic domes dated at ca. 232 Ma (Cornejo et al. 1998, 2009). Its upper member (Norian-Rhaetian) is composed of thick piles of andesitic and dacitic lava flows, breccias, and tuffs, interbedded with fluvial and alluvial sedimentary strata that grade into shallow marine beds that mark an important transgression in the area (Venegas et al. 2013; Espinoza et al. 2015), which lasted until the Middle Jurassic and is represented by units such as the Rencoret Beds and Profeta Formation (Chong 1973; Tobar 1966; Marinovic 2007; Cornejo et al. 1998, 2009). Other units, with similar composition to the Quebrada del Salitre Formation, that represent significant volcanic activity contemporaneous with lacustrine and alluvial continental sedimentation in the area are the Las Lomas Beds, the Cerro La Ballena, Cerro Guanaco, and Sierra de Varas formations (Norian—Rahetian) (Marinovic and García 1999; Marinovic 2007; Basso and Mpodozis 2012; González et al. 2015) (Fig. 2).

In the Coastal Cordillera at $\sim$23° S, Triassic volcanic activity is recorded by tuffs and domes of rhyolitic composition and restricted volume (Basso 2004). Further south, a volcanosedimentary sequence that constitutes so far the oldest known Triassic unit (Middle to Upper Triassic) in the coast of northern Chile crops out near the Paposo town (Álvarez et al. 2016). It has been assigned to the Cifuncho Formation (Álvarez et al. 2016), although the latter is mainly Upper Triassic in age in its type locality, where it is composed mainly by braided fluvial sedimentary and pyroclastic rocks (Contreras et al. 2013). The Agua Chica Formation, the southernmost Triassic unit in the Coastal Cordillera, is composed by andesitic lava flows and subvolcanic bodies, interbedded with sandstones and conglomerates representing an alluvial and fluvial environment (Godoy and Lara 1998). Triassic magmatism in the Coastal Cordillera is also represented by small plutons and plutonic complexes, such as the Middle Triassic granodioritic intrusions (Espinoza et al. 2011), the Upper Triassic (Norian) two mica monzo and syenogranites near Taltal (Escribano et al. 2013), and minor tonalitic intrusions in Mejillones (Casquet et al. 2014) (Fig. 2). The Upper Triassic marine transgression identified in the Domeyko Cordillera is also recorded in the Coastal Cordillera by the Rencoret Beds and the Pan de Azúcar Formation (González and Niemeyer 2005; Contreras et al. 2013) (Fig. 2).

3.3 *Triassic Igneous Units Between 26° and 30° S*

The Triassic igneous units between 26° and 30° S crop out as two roughly subparallel belts in the Coastal Cordillera and in the Precordillera/Frontal Cordillera

(Fig. 1). The outcrops in the western belt, in the Coastal Cordillera, are volumetrically more restricted than their eastern counterparts (Fig. 1). Igneous rocks in this belt are mainly plutonic, and they intrude or overlie schists, metasandstones, and metapsamites of the Chañaral Epimetamorphic and Punta de Choros metamorphic complexes (Devonian—Permian) (Godoy and Lara 1998; Creixell et al. 2016). Among these units are the Cerros del Vetado Pluton composed by hornblende and two mica syenogranites and monzogranites (Triassic); the Capitana Syenogranite, a two mica suite of monzo and syenogranites (Norian); the Peralillo, Bufadero, Barquito, and Cerro Castillo plutons, which include biotite and hornblende gneissic granodiorites, diorites and tonalites (with paragneiss inclusions), and monzogranites (Rhaetian—Sinemurian) (Godoy and Lara 1998; Cruden et al. 2004; Espinoza et al. 2014) (Fig. 2). Further south in the Coastal Cordillera, outcrops of the Carrizal Bajo Plutonic Complex (Rhaetian) are composed by a bimodal suite of pyroxene–hornblende–biotite diorites, quartz-diorites and gabbronorites, and biotite–hornblende granites and granodiorites. This complex is intruded by the biotite–hornblende monzogranites and granodiorites of the Algodones Granite (Triassic—Jurassic boundary) (Cruden et al. 2004; Arévalo and Welkner 2008) (Fig. 2).

A small portion of the volcanosedimentary Agua Chica Formation (Upper Triassic) crops out in this segment, and it is composed of andesitic lava flows, crystalline and lithic tuffs, shales and basal conglomerates, and breccias. The beds are interdigitated with the Cifuncho Formation and overlaid by the marine Pan de Azúcar Formation (Godoy and Lara 1998) (Fig. 2). Further south, the Canto del Agua Formation (Middle—Upper Triassic) is exposed in two main depocenters and composed by sandstones, shales, conglomerates, minor calcareous sedimentary rocks and volcanoclastic breccias, and tuffaceous sandstones (Arevalo and Welkner 2008). Most of the sedimentary and volcanic rocks that were exposed in the southern depocenter have been recently assigned to the Carboniferous—Permian Llanos de Chocolate beds based on radiometric ages and fossil content (Welkner et al. 2006; Creixell et al. 2016).

In the Precordillera and Frontal Cordillera, a thick sequence of intermediate to acidic volcanic rocks, the Guanaco Sonso Formation (Lopingian—Middle Triassic), covers a large area near the Chile—Argentina border, overlying the Late Carboniferous basement (Las Placetas Formation). It is composed of andesitic to rhyolitic lavas flows and domes, interbedded with glass and lithic tuffs and a minor proportion of siliciclastic rocks (Ortiz and Merino 2015). This unit is temporal and spatially related to the Chollay Plutonic Complex (Lower—Middle Triassic), a NNE-elongated series of plutonic bodies that along with the Montosa—El Potro Plutonic Complex (Lopingian—Middle Triassic) represent the largest Mesozoic batholith in the Frontal Cordillera, cropping out to the west of the Guanaco Sonso Formation (Fig. 2). This plutonic unit is composed mainly by hornblende–biotite monzogranites, granodiorites and tonalites, with minor amounts of gabbros and syenogranites. It intrudes the Paleozoic metasedimentary sequences and abundant xenoliths of mid-crustal metamorphosed plutonic rocks, known as La Pampa Gneisses, are incorporated in its western flank (Álvarez et al. 2013; Salazar et al. 2013; Coloma et al. 2017). The Chollay Plutonic Complex is overlaid by the Pastos

Blancos Formation (Upper Triassic) (Fig. 2), formerly assigned to the Pastos Blancos Group along with the Guanaco Sonso Formation by Martin et al. (1999), a volcanic bimodal succession composed by a basal section of basaltic lava flows and pyroclastic rocks, and an upper section of dacitic to rhyolitic domes, lava flows, and tuffs. The Colorado Syenogranites (Upper Triassic) intrudes both the Chollay complex and the Pastos Blancos Formation (Fig. 2); it is a spatially restricted suite of biotite syeno- and monzogranites of characteristic pink color (Martin et al. 1999; Salazar et al. 2013; Coloma et al. 2017). Numerous dykes of basaltic-andesite and rhyolite (pink colored) composition cross cut the aforementioned units; their contacts with the wall rock are irregular, suggesting that they are partly contemporaneous with the Pastos Blancos Formation and the Colorado Syenogranite (Martin et al. 1999; Creixell et al. 2009).

To the west of the Chollay Plutonic Complex crops out the La Totora Formation (Upper Triassic), a volcanic sequence that discordantly overlies the Guanaco Sonso Formation, the Carboniferous—Permian intrusions, and the Permian Laguna Chica Formation (Fig. 2). It is in gradational contact with the Triassic sedimentary San Felix Formation; in fact, the upper member of the latter has been assigned to the La Totora Formation based on their lithological similarity (Coloma et al. 2017). It is composed by basaltic and basaltic-andesite lava flows, ash and lapilli tuffs, rhyolitic domes, and sedimentary beds of breccias and lithic arkose (Salazar et al. 2013; Maksaev et al. 2014). The shallow marine deposits of the Lower—Middle Jurassic Lautaro Formation overlie the Triassic volcanics in slight unconformity (Salazar et al. 2013). North of 28°30′S, the La Ternera Formation (Upper Triassic—Hettangian), discordantly overlies Late Paleozoic intrusions and older stratified units and it is in turn overlaid by the Lautaro Formation (Iriarte et al. 1996; Iriarte et al. 1999; Arévalo 2005; Matthews et al. 2006; Mpodozis et al. 2012; Peña et al. 2013). This volcanosedimentary sequence is composed by andesitic and basaltic andesitic lava flows, minor rhyolitic domes, glass and lithic tuffs, conglomerates, quartz sandstones, and shales (Iriarte et al. 1996). Further north of 27°30′, the outcrops of the El Mono Beds (Lopingian) and the Quebrada del Salitre Formation are located in a similar position as the La Ternera and La Totora within the eastern belt of Triassic rocks at the Frontal Cordillera (Fig. 2).

3.4 *Jurassic Igneous Units Between 22° and 26° S*

In contrast to the Triassic units, the present-day Coastal Cordillera of northern Chile was the main locus of Jurassic magmatism as it is attested by the vast outcrops of volcanic and plutonic rocks of the period aligned along more than 1,000 km (Fig. 1). The Jurassic volcanic and sedimentary rocks are assigned to the La Negra Formation and crop out as a homoclinal sequence that can reach thicknesses of 7,000–10,000 m (Buchelt and Tellez 1988). They mainly consist of porphyritic lavas with phenocryst contents up to 25%, with basaltic andesites and andesites as the main compositional types, although basalts, dacites, and rhyolites (ignimbrites) have been reported from several areas (Boric et al. 1990; Cortés 2000; Kramer et al. 2005). Volcanic breccias, tuffs, and sedimentary rocks as well as epiclastic sandstone lenses are less abundant. The age of volcanism has traditionally been established on the basis of contact relationship with sedimentary units or the fossiliferous content of interbedded sediments, from Early to lowermost Upper Jurassic (Rogers and Hawkesworth 1989; González and Niemeyer 2005) (Fig. 2), although some authors argue that the main volume of volcanism and sedimentation is restricted to the Middle and Upper Jurassic (Oliveros et al. 2006, 2008). Large and widely distributed plutonic bodies, the Coastal Batholith (Figs. 1 and 2), of gabbroic to granitic in composition, intrude the volcanic sequence together with dykes and stocks of mainly basic to intermediate composition (Jurassic—Early Cretaceous) (Pichowiak 1994; Dallmeyer et al. 1996; Lucassen and Thirlwall 1998; González and Niemeyer 2005; Oliveros et al. 2006, 2008).

3.5 *Jurassic Igneous Units Between 26° and 30° S*

The units that represent the Early Andean magmatism in this segment can be divided into two main areas that form two parallel belts of outcrops (Fig. 1). As in the northern segment, the first group of rocks, representing the Jurassic arc, is located in the present-day Coastal Cordillera and comprises thick sequences of volcanic and volcanoclastic material and large batholitic bodies. The second group of rocks has been interpreted as the Jurassic back-arc domain, and it is located in the Precordillera and Frontal Cordillera (Charrier et al. 2007; Rossel et al. 2013). The back-arc domain can be further subdivided into western and eastern belts of outcrops, both of them composed almost exclusively by volcanic and volcaniclastic rocks (Rossel et al. 2013, 2015).

The volcanism in the Coastal Cordillera is represented in this segment by the thick deposits of the La Negra Formation (26°–29° S) and the volcanic and subvolcanic rocks of the Agua Salada Volcanic Complex (29°30′–31° S). This last unit can be correlated with the lower member of the Punta del Cobre Formation that is intruded by 143 Ma tonalites (Creixell et al. 2012). The sequence crops out mainly as large roof pendants as a consequence of the intrusion of Upper Jurassic and Cretaceous plutonic bodies; therefore, its base is in general not well exposed. Locally, it can be observed that the La Negra Formation overlies in concordance the Pan de Azúcar (Rhaetian—Sinemurian) and Canto del Agua (Upper Triassic) formations (Naranjo and Puig 1984; Welkner et al. 2006; Arévalo and Welkner 2008, Creixell et al. 2016) or unconformably the Chañaral Epimetamorfic Complex of Permian age (Naranjo and Puig 1984; Espinoza et al. 2015) (Fig. 2). On the basis of its fossil content, stratigraphic relationships and scarce radiometric data, an age ranging from Pliensbachian to Upper Jurassic, are assigned to this unit (Charrier et al. 2007 and references therein) (Fig. 2).

Plutonic rocks of the Coastal Batholith cropping out in this segment range compositionally from gabbro to granite, but correspond mostly to diorites, monzonites, tonalites, and granodiorites with typical metaluminous mafic mineralogy, that intruded in several magmatic pulses during the Jurassic (Fig. 2). Structural analyses suggest that the intrusion of these bodies was facilitated mostly by tensional fractures and faults (Grocott et al. 1994; Creixell et al. 2011), but it is possible to observe forced intrusions induced by sequential emplacement of magmatic pulses in early stages (Rodríguez et al. 2016).

In the Precordillera and Frontal Cordillera, a series of volcanosedimentary units (Fig. 2) would represent a back-arc volcanism during the Late Jurassic (Rossel et al. 2013). The rocks are distributed in two parallel chains and, due to their position relative to the arc domain, would resemble the configuration of some of the modern island arc magmatic systems of western Pacific (Rossel et al. 2013). The westernmost chain of volcanic rocks is represented by the Sierra Fraga (26–27° S) and Algarrobal-Picudo (28–31° S) formations. These units are composed by thick piles of more than 1,500 m of volcanic and pyroclastic material, intercalated with sedimentary continental and marine beds in the case of Sierra Fraga Formation and with continental sedimentary beds in the Algarrobal Formation (Iriarte et al. 1996; Rossel et al. 2013; Salazar et al. 2013). The eastern chain is represented by the Quebrada Vicuñita Beds (26°–27° S) and the Lagunillas Formation (27°30′–30° S) (Fig. 1). These units are characterized by homogeneous sequences of basaltic and basaltic andesitic lavas with a maximum thickness of 200 m, interbedded with minor limestones and calcareous sandstones in the Quebrada Vicuñita Beds, (Cornejo et al. 1998; Iriarte et al. 1999; Salazar et al. 2013). Only two minor Jurassic intrusive bodies crop out in the Frontal Cordillera at 29° S, the Quebrada Lopez (150 Ma), and the Río Potrerillos (144 Ma.) tonalitic stocks (Merino and Ortiz 2015).

4 Age Distribution of the Triassic and Jurassic Magmatism

The Mesozoic volcanic and plutonic rocks cropping out between 22° and 30° S have been dated mainly through the K-Ar (Ar–Ar) and U-Pb methods in primary mineral phases such as zircon, amphibole, biotite, and plagioclase (Oliveros et al. 2006; Parada et al. 2007; Hervé et al. 2014; Maksaev et al. 2014, and references therein). In terms of the total volume of units and rocks sampled for geochronology, there is a slight bias toward plutonic rocks due to the higher abundance of suitable mineral phases in these rocks and the lower alteration degree that they exhibit. Volcanic rocks in general lack of amphibole and biotite, and their phenocrystals and groundmass are strongly altered and replaced by minerals typical of very low to low-grade metamorphism, which is particularly problematic when dating with the K-Ar and Ar–Ar methods (Oliveros et al. 2006, 2008). Consequently, the Coastal Batholith and the Triassic plutons in the High Andes have extensively been more dated than the Triassic and Jurassic volcanic units. Taking this into account, the volume of magmatism seems invariant throughout most of the Triassic, with a significant peak at ca. 245–250 Ma and an important decrease toward the Triassic–Jurassic boundary (Fig. 3). The Jurassic igneous rocks are volumetrically more abundant than their Triassic counterparts (Fig. 1), and the same pattern is observed in terms of age distribution (Fig. 3). The peaks in the Middle and Late Jurassic have higher magnitudes than the Triassic, with the c. 150 Ma peak being the largest of the studied time frame (Fig. 3). The opposite is observed in the Latest Triassic and Early Jurassic, where igneous ages are scarce. A relatively similar pattern of age distribution is observed in the back-arc marine and continental sedimentary units representative of the Tarapacá and Neuquén basin (Naipauer et al. 2015) (see Chap. "A Provenance Analysis from the Lower Jurassic Units of the Neuquén Basin. Volcanic Arc or Intraplate Magmatic Input?"). The detrital zircons of different localities of these Jurassic basins yield a frequency distribution with two prominent peaks at ca. 243 Ma and 144–152 Ma and a magmatic lull at ca. 197 Ma, which those authors interpreted as the onset of Andean subduction. Overall, the amount to Triassic detrital zircons is smaller than the Jurassic material (Fig. 3). Given the current geochronological database, it is therefore possible to infer that the volume of magmatism would have remained relatively constant during most of the Triassic, decreasing notoriously at ca, 200 Ma, and later increasing to its maximum for the studied period at the Late Jurassic.

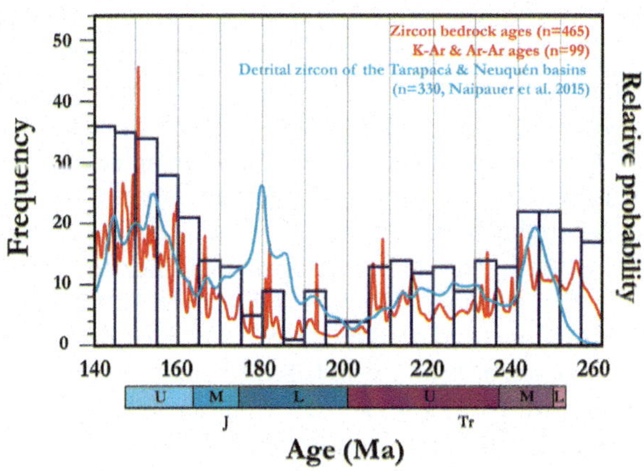

Fig. 3 Frequency distribution histogram for mineral U-Pb and K-Ar (Ar–Ar) radiometric ages from rocks of northern Chile (20°–30° S) and model distribution of detrital zircon U-Pb ages for Tr-Jr rocks of the Neuquén and Tarapacá back-arc basins. Data from Breitkreuz and Van Schmus (1996), Oliveros et al. (2006), Parada et al. (2007), Maksaev et al. (2014), Hervé et al. (2014), Naipauer et al. (2015) and references therein, and geological maps of the Chilean Geological and Mining survey between 20° and 30° S

5 Geochemistry of the Triassic and Jurassic Magmatism

The significant compositional contrasts between the Jurassic and Triassic magmatism in northern Chile are one of the main reasons for which these two periods have been associated with specific and distinctive evolutionary stages of the South American margin. From a purely petrographic perspective (Fig. 4), most of the exposed Triassic igneous rocks are intermediate to acid in composition (Fig. 4a), whereas the Jurassic magmatism was predominantly basic to intermediate (Fig. 4b). Previously, the bimodal character of the Triassic volcanism has been invoked as evidence of continental rifting during that stage (Ramos and Kay 1991; Parada et al. 1999; Morata et al. 2000). However, in spite of the more evolved character of the Triassic rocks, no bimodality in their composition is observed when integrating either the petrography or the major elements content of all the reported igneous units in northern Chile (Figs. 4 and 5a); in fact, the volcanism is largely andesitic in composition for the studied 100 my interval. The Triassic rocks are mainly subalkaline (Fig. 5a), with few samples escaping to this trend, and also follow a calk-alkaline trend in the AFM diagram (Fig. 5b). The plutonic Jurassic rocks are mostly metaluminous, whereas the majority of the Triassic intrusions are peraluminous (Fig. 5c), indicating an important crustal component in the magma source. However, the metaluminous intrusions correspond to the less evolved lithologies, which suggest that a mantle component must have also been involved in the generation of these magmas (Coloma et al. 2017). Due to the significant alteration exhibited by the volcanic rocks, the use of mobile major elements, such as K, Na, Ca, Fe, and Mg, is not recommended (Oliveros et al. 2007; Rossel et al. 2015; Coloma et al. 2017) and most of the tectonic classifications are made based on the more immobile trace elements (Fig. 5d, e). The two selected diagrams show that the vast majority of the Triassic and Jurassic rocks plot either in the field of "volcanic arc basalts" or "volcanic arc granites", with few samples in other fields. These Triassic igneous rocks with transitional to alkaline affinities have been interpreted as precursors to the establishment of the arc at the present-day Coastal Cordillera, which took place at some point after ca. 215 Ma (Vásquez and Franz 2008; Coloma et al. 2017). The "arc-signature" for the Triassic and Jurassic rocks is further reinforced by the primitive mantle-normalized trace element abundances (Fig. 5f) in which there is a marked enrichment in LIL elements over the HFS elements, with characteristics troughs in Nb, Ta, P, and Ti. High abundances of Pb and Zr would suggest an important crustal component, which is more evident for the Triassic rocks.

In terms of the REE abundances, when considering the whole dataset for each stage, it is difficult to find distinctive patterns as both groups have rather flat slopes or LREE to HREE ratios (Fig. 5g). However, a systematic decrease in the La/Yb and La/Sm parameters from the Triassic to the Jurassic has been reported (Oliveros et al. 2016). For a group of Triassic rocks, the Eu anomaly is markedly more negative than for the whole population of samples, suggesting more reduced conditions for the magmatism. There is also a particular group of Jurassic samples for

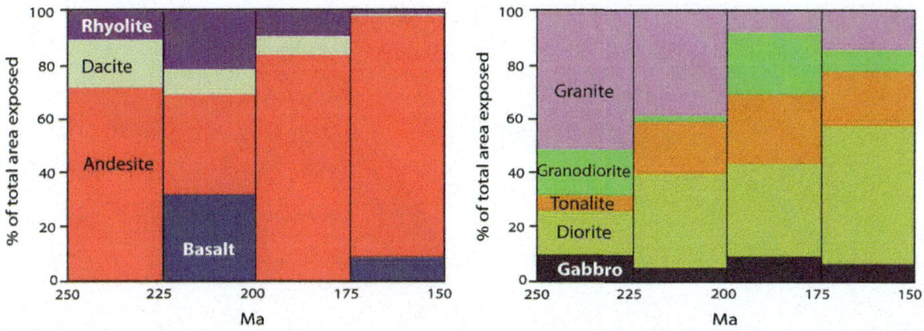

Fig. 4 Compositional distribution of the Triassic and Jurassic igneous rocks cropping out in the Chilean territory between 20° S and 30° S. The percentages of each lithology are calculated from the petrographic, stratigraphic and geochronological information available for ca. 20,000 km^2 of surficial exposure of relevant units according to the geological maps published by the Chilean Geological and Mining Survey

which the slope and abundances of REE are lower, indicating the less evolved magmas would have been generated under a thinned crust.

The isotope composition of the Mesozoic rocks suggests that the magma source must have had a mantle component as they plot along the mantle array, and most of the Jurassic and a significant portion of the Triassic rocks plot above and to the left of the "bulk silicate earth" values, in the Nd-Sr diagram (Fig. 5h). Some of the Triassic rocks have undoubtedly a crustal component in their source due to the elevated $^{87}Sr/^{86}Sr$ and low $^{143}Nd/^{144}Nd$ that they exhibit. The Jurassic rocks with the more radiogenic signature are those representing the back-arc domain (Late Jurassic volcanism), and they are thought to have a more enriched mantle source because of their inferred position relative to the mantle wedge in the subduction zone at that time (Rossel et al. 2013, 2015). The Pb isotopic composition of the Triassic rocks is very similar to the Jurassic ones and both groups plot in a field that suggest a rather restricted crustal or sediment influence in the magma sources (Fig. 5i). The Late Jurassic rocks mark a distinctive, more radiogenic group, for which their Pb signature is interpreted as a more enriched mantle source behind the main arc at that time (Rossel et al. 2013, 2015).

6 Tectonic Setting and Initiation of the Andean Magmatism

The geochemical signature of the Triassic magmatism, along with its intermediate to acid composition, would require a depleted mantle source that melted in presence of fluids enriched in mobile incompatible elements. The most likely tectonic setting to produce such magmatic products is the subduction of the oceanic plate beneath the continent (Fig. 6a, b). Although some authors have recently speculated with a

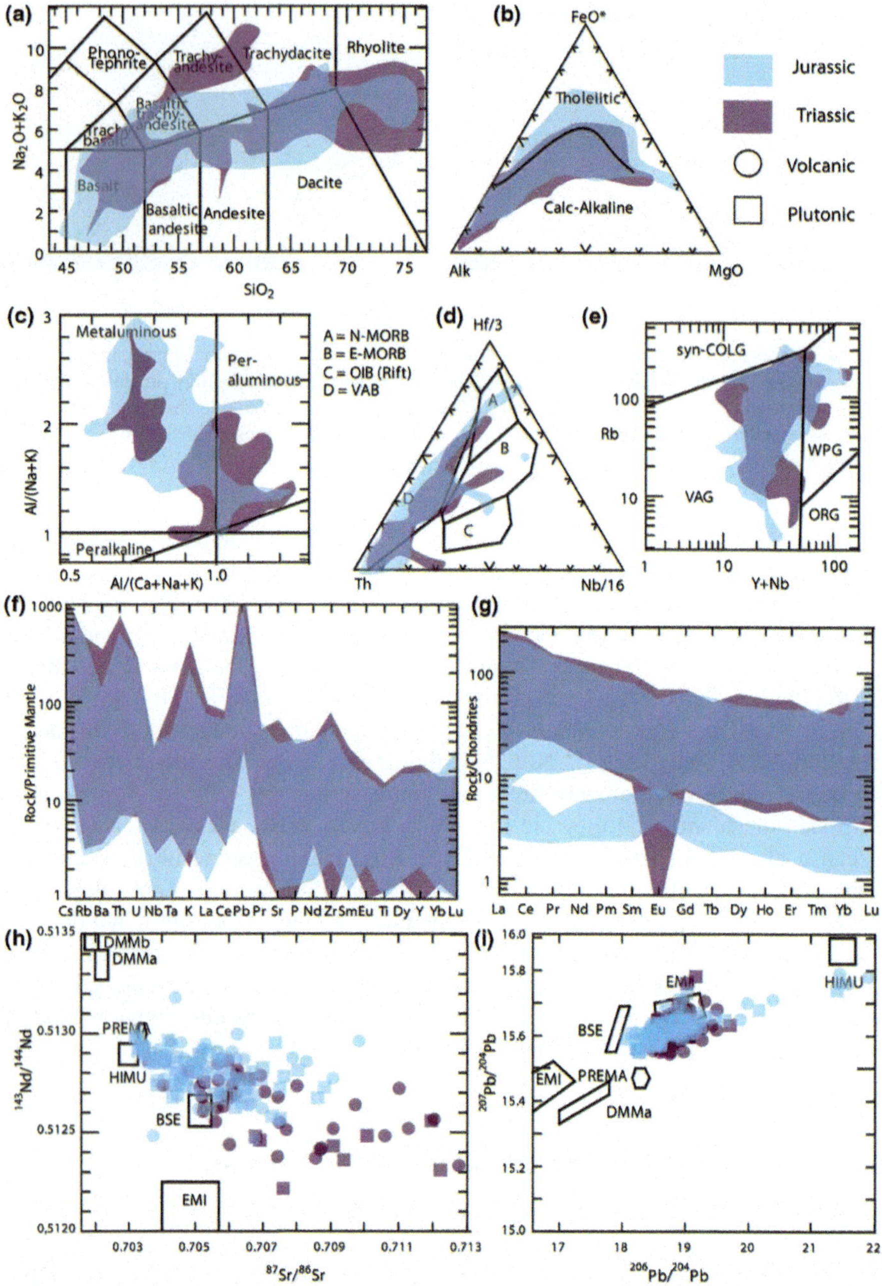
(a)
Phono-Tephrite
Trachy-andesite
Trachydacite
Rhyolite
Basaltic trachy-andesite
Trachy basalt
Basalt
Basaltic andesite
Andesite
Dacite
Na2O+K2O
SiO2
(b)
FeO*
Tholeiitic
Calc-Alkaline
Alk
MgO
Jurassic
Triassic
Volcanic
Plutonic
(c)
Metaluminous
Per-aluminous
Peralkaline
Al/(Na+K)
Al/(Ca+Na+K)
A = N-MORB
B = E-MORB
C = OIB (Rift)
D = VAB
(d)
Hf/3
Th
Nb/16
(e)
syn-COLG
WPG
VAG
ORG
Rb
Y+Nb
(f)
Rock/Primitive Mantle
Cs Rb Ba Th U Nb Ta K La Ce Pb Pr Sr P Nd Zr Sm Eu Ti Dy Y Yb Lu
(g)
Rock/Chondrites
La Ce Pr Nd Pm Sm Eu Gd Tb Dy Ho Er Tm Yb Lu
(h)
DMMb
DMMa
PREMA
HIMU
BSE
EMI
143Nd/144Nd
87Sr/86Sr
(i)
HIMU
EMII
BSE
EMI
PREMA
DMMa
207Pb/204Pb
206Pb/204Pb

◂**Fig. 5** **a** TAS (Le Bas et al. 1986), **b** AFM (Irvine and Baragar 1971) **c** alumina saturation index diagrams for plutonic rocks (Shand 1943), **d** Th-Nb-Hf for basic and intermediated volcanic rocks (Wood 1980) **e** Rb versus Y+Nb diagram for plutonic and acid volcanic rocks (Pearce et al. 1984). Spider diagrams of **f** chondrite-normalized rare earth and **g** primitive mantle-normalized trace elements abundances in plutonic and volcanic rocks. Normalizing values are from Sun and McDonough (1989). **h** $^{87}Sr/^{86}Sr$ versus $^{143}Nd/^{144}Nd$ and **i** $^{207}Pb/^{204}Pb$ versus $^{206}Pb/^{204}Pb$ diagrams for the studied rocks, showing the mantle components after Zindler and Hart (1986). Data are from Rogers and Hawkesworth (1989), Brown (1991), Vergara et al. (1991), Mpodozis and Kay (1992), Pichowiak (1994), Parada et al. (1999), Morata et al. 2000, Llambías et al. (2003), Kramer et al. (2005) Lucassen et al. (2006), Llambías et al. (2007), Oliveros et al. (2007), Vásquez and Franz (2008), D'Elia et al. (2012), Vásquez et al. (2011), Creixell et al. (2012), Parada (2013), Rossel et al. (2013), Salazar et al (2013), Coloma et al. (2017) and Oliveros et al. (2016)

scenario like this for the Triassic magmatism (Vásquez et al. 2011; del Rey et al. 2016; Coloma et al. 2017), the more generalized hypothesis is that during this period, subduction ceased and magmatism was generated merely by continental rifting (Mpodozis and Kay 1992; Charrier et al. 2007 and references therein).

At a continental scale, the tectonic forces during the Triassic were indeed dominated by the Pangea–Gondwana breakup that imposed tensional stresses along the proto-south american part of the continental plate (Frizon de Lamotte et al. 2015). This resulted firstly in extensive crustal anathexis, generating the acidic Choiyoi province, and then in the formation of NW- oriented rift basins (Llambías et al. 1993, 2003; Charrier et al. 2007; Bechis et al. 2014), with a minor component of N-S extensional stresses, such as the Ichigualasto and Cuyo basins in the present-day Argentinian territory, which accumulated large volumes of siliciclastic sediments and OIB-like or anorogenic bimodal magmatism (Ramos and Kay 1991; Llambías et al. 2007). All this features are well-documented in Argentina and allow constraining the tectonic model of orogenic collapse, continental extension, and rifting without subduction. However, Chilean units show that closer to the continental margin, volcanic and plutonic rocks do not exhibit a bimodal compositional distribution, with dominant intermediate lithologies that show systematic enrichment in LILE over HFSE. Given the evidence of a crustal component in the Triassic magmas, i.e., the peraluminous character of several plutonic units, part of the LILE enrichment could have been generated from melting or assimilation of the preexisting continental crust, during the ascent of the magmas to the surface or middle-upper crustal levels. Nevertheless, the Sr–Nd–Pb isotope signature of volcanic and plutonic rocks requires a magmatic source with a depleted mantle component (Fig. 5h, i); thus, the Triassic magmatism cannot be solely attributed to dehydration melting of a preexisting crust bearing an arc-signature nor to decompression melting of an asthenospheric mantle beneath the continent. The features of the Triassic volcanic and plutonic rocks are better explained if fluid-induced melting of a depleted mantle source generated basaltic magmas that latter evolved due to different degrees of continental crust assimilation, in the context of active subduction under tensional stresses along the margin.

The progressive stretching of the continental lithosphere due to the Gondwana breakup would have resulted, during the Jurassic, in a paleogeographic

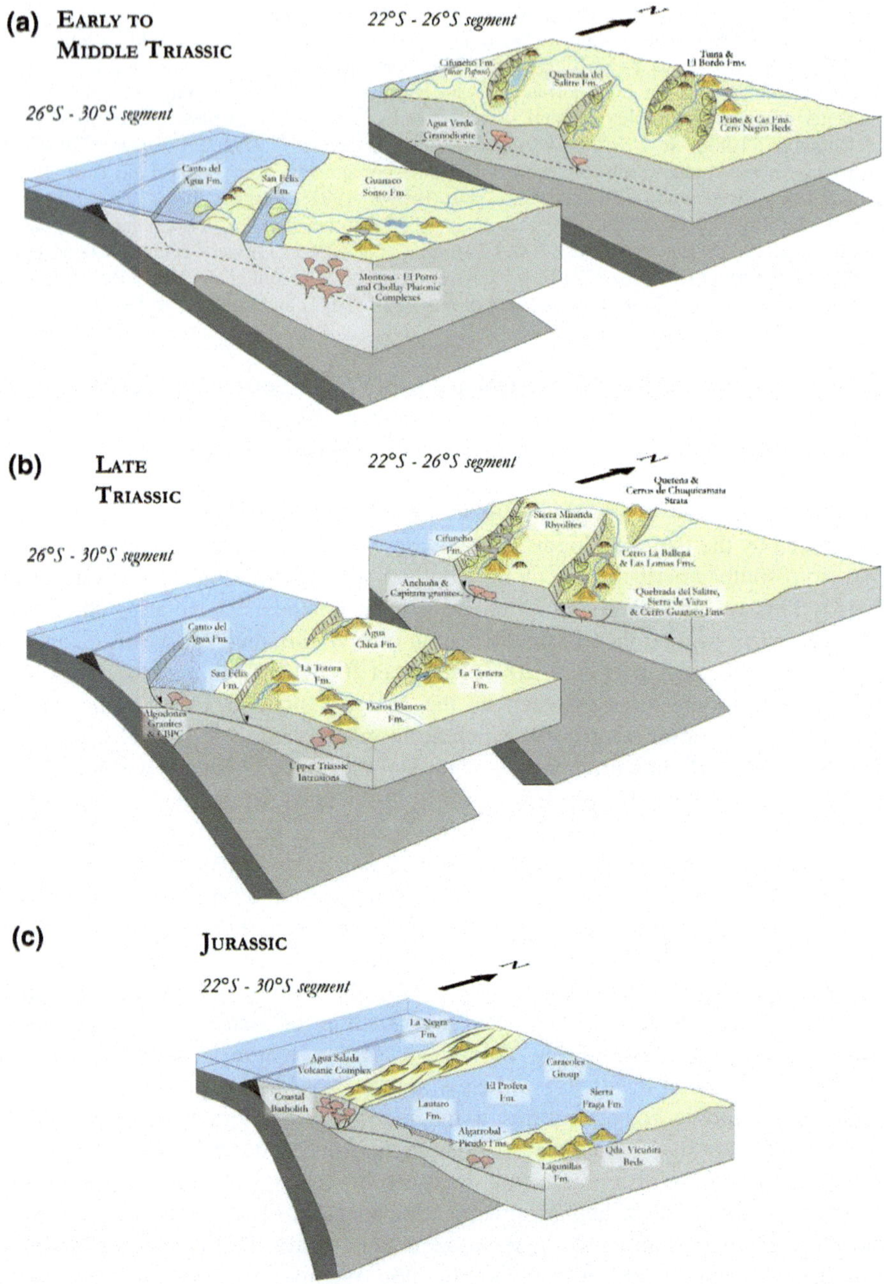

Fig. 6 Paleography and tectonic setting for the magmatism at the northern Chile margin during the Triassic and Jurassic. Location of the segments corresponds to the present-day latitudes

configuration that resembled an ocean–ocean convergent margin, with a low-relief, partly submarine, arc chain, and a back-arc marine basin (Rossel et al. 2013) (Fig. 6c). The Jurassic arc magmatism is remarkably homogeneous both lithologically and geochemically, with basaltic-andesite, andesite, and diorite being by far the most abundant type of products, while major and trace elements composition groups the samples tightly in the different trends of calc-alkaline (Fig. 5b) or arc affinities (Fig. 5d, e). The small crustal contribution to the magma sources and the protracted extensional regime during the Jurassic resulted in a large addition of juvenile magmas to the continent in this period (Lucassen et al. 2006; Oliveros et al 2007). Back-arc magmatism was much more restricted in terms of volume and duration, but also more heterogeneous in its composition, including calc-alkaline and transition to alkaline volcanism. It took place during a marine regression that left the back-arc basin partially emerged and that would have resulted from transpressional conditions in the arc region (Rossel et al 2013, 2015).

Therefore, subduction associated with extensional stresses, indeed persisted during the Triassic and the Jurassic, indicating that the beginning of the "Andean cycle" or "Andean subduction" was arbitrarily placed at the Early Jurassic. In spite of their similar origin, Triassic igneous rocks are somewhat different to the Jurassic ones, as they are more acidic (Fig. 4) and their deposits in general lack of thick piles of effusive lava flows, which is the norm in the "La Negra arc" representative units. Pyroclastic and volcanoclastic deposits are the most common occurrence of the Triassic volcanics, which reflects in part its more evolved character. The crustal component in the Jurassic magmas is significantly lower than the older magmatism, with the Jurassic volcanic rocks being partly tholeiitic having MORB-like isotopic signatures (Figs. 5b, h and 6). Thus, even though the mechanism for magma generation would have remained invariant through the ca. 100 Ma, the lithospheric contributions to the magmatic processes significantly diminished in the same period (Oliveros et al. 2016). In addition to the geochemical and petrological differences, a potential "break" in the magmatism could have been occurred near the Triassic–Jurassic boundary because both igneous and detrital ages exhibit a marked lull around this time frame (Fig. 3). The settlement of the volcanic arc in the present-day Coastal Cordillera is thought to have taken place at the Hettangian or Sinemurian age (Mpodozis and Ramos 1989; Charrier et al. 2007 and references therein), but the recent finding of the Estratos de Sierra Lagunas (Sepúlveda et al. 2014) suggests that the typical volcanism of the "La Negra arc" could have started during the Rhaetian age. Still, this event roughly coincides in time with the magmatic lull and the shift from a rifting stage to a thermal sag stage in the Neuquén basin (see Chap. "A Provenance Analysis from the Lower Jurassic Units of the Neuquén Basin. Volcanic Arc or Intraplate Magmatic Input?"), suggesting that probably a common tectonic phenomenon existed associated with the decrease in the magma production rates prior to the focus of the arc in a narrow position along the margin. Further studies are necessary to determine the nature of such phenomenon and more precisely constrain the exact time when it started. This work argues for the need to redefine the beginning of the "Andean Cycle" to be placed either at sometime during the Triassic or after the end of the Choiyoi event during the Late Permian–Early Triassic (Sato et al. 2015).

7 Pre-Andean Segmentation of the Convergent Margin

Even though Mesozoic subduction may have taken place along the entire section of the Andean margin between 22° and 30° S, this could have had different characteristics evidenced by different deformation styles affecting the uppermost crust evidenced in the geological record that could reflect a tectonic segmentation of the margin. During the Early to Middle Triassic, the magmatism in the northern segment was less abundant and closely related to continental basins, whereas in the southern segment, large volumes of both volcanic and plutonic rocks were emplaced in the crust, relatively far from the marine forearc basins (Fig. 6). During the Late Triassic, the differences between both segments are less notorious but still the magma production was likely higher in the northern region, where volcanism was closely related to rifting in continental-marine basins (Fig. 6).

The segmentation of the margin becomes especially evident during Late Jurassic in the back-arc, with exclusively marine sedimentation north of 27° S and continental deposition between 27° and 36° S. However, it is possible to observe a second-order segmentation related to magmatic activity (Rossel et al. 2013). Thus, Rossel et al. (2015) suggest the following scenario: (a) 20°–25° S, marine sedimentation without volcanism, (b) 25–27° S, mainly marine sedimentation and submarine calc-alkaline-tholeitic volcanism, (c) 27°–31° S, continental sedimentation and heterogeneous volcanism.

Acknowledgements This work has been funded by the Fondecyt grants 1108004 and 1120715 (VO), the Conicyt doctoral grant 21140774 (ME), and the Chilean Geological and Mining Survey.

References

Álvarez J, Mpodozis C, Blanco-Quintero I, García-Casco A, Arriagada A, Morata D (2013) U-Pb ages and metamorphic evolution of the La Pampa Gneisses: Implications for the evolution of the Chilenia Terrane and Permo-Triassic tectonics of north Central Chile. J S Am Earth Sci 47:100–115

Álvarez J, Jorquera R, Miralles C, Padel M, Martinez P (2016) Cartas Punta Posallaves y Sierra Vicuña Mackenna, Región de Antofagasta. Servicio Nacional de Geología y Minería. Carta Geológica de Chile, Geología Básica (1:100.000)

Arévalo C (2005) Carta Los Loros, Región de Atacama. Servicio Nacional de Geología y Minería. Serie Geología Básica 92 (1:100.000)

Arévalo C, Creixell C (2009) The Atacama Fault System and its role on the migration and deposition of Iron Oxide Copper Gold and Magnetite-Apatite ores: an evaluation from the Los Choros and Huasco valleys. In: Wörner G, Möller-McNett S (eds) International Lateinamerika – Kolloquium, Alemania, pp 23–25

Arévalo C, Welkner D (2008) Geología del área Carrizal Bajo-Chacritas, Región de Atacama. Servicio Nacional de Geología y Minería. Serie Geología Básica 111:67 (1:100.000)
Basso M (2004) Carta Baquedano, Región de Antofagasta. Servicio Nacional de Geología y Minería. Carta Geológica de Chile, Serie Geología Básica 82 (1:100.000)
Basso M, Mpodozis C (2012) Carta Cerro Quimal, Región de Antofagasta. Servicio Nacional de Geología y Minería. Carta Geológica de Chile, Serie Geología Básica 143 (1:100.000)
Bechis F, Cristallini EO, Giambiagi LB, Yagupsky DL, Guzmán CG, García VH (2014) Transtensional tectonics induced by oblique reactivation of previous lithospheric anisotropies during the Late Triassic to Early Jurassic rifting in the Neuquén basin: insights from analog models. J Geodyn 79:1–17
Boric R, Díaz F, Maksaev V (1990) Geología y yacimientos metalíferos de la Región de Antofagasta. Servicio Nacional de Geología y Minería, Santiago, Boletín 40:246
Breitkreuz C, Van Schmus WR (1996) U-Pb geochronology and significance of Late Permian ignimbrites in Northern Chile. J S Am Earth Sci 9(5):281–293
Brown M (1991) Comparative geochemical interpretation of Permian-Triassic plutonic complexes of the Coastal Range and Altiplano (25 30′ to 26 30′ S), northern Chile. Geol Soc Am Spec Pap 265:157–178
Buchelt M, Tellez C (1988) The Jurassic La Negra formation in the area of Antofagasta, northern Chile (lithology, petrography, geochemistry). In: Bahlburg H, Breitkreuz C, Giese P (eds) The Southern Central Andes, vol. 17. Springer, Heidelberg. Lecture Notes in Earth Sciences, pp 171–182
Casquet C, Hervé F, Pankhurst RJ, Baldo E, Calderón M, Fanning CM, Dahlquist J (2014) The Mejillonia suspect terrane (Northern Chile): Late Triassic fast burial and metamorphism of sediments in a magmatic arc environment extending into the Early Jurassic. Gondwana Res 25 (3):1272–1286
Charrier R, Pinto L, Rodríguez M (2007) Tectonostratigraphic evolution of the Andean Orogen in Chile. In: Gibbonds W, Moreno T (eds) The Geology of Chile. Geo Soc London, Special Publications:21–116
Chong G (1973) Reconocimiento geológico del área Catalina-Sierra de Varas y estratigrafía del Jurásico del Profeta, Provincia de Antofagasta. Memoria de Título (Unpubl.), Departamento de Geología, Universidad de Chile, p 284
Coloma F, Valin X, Oliveros V, Vásquez P, Creixell C, Salazar E, Ducea MN (2017) Geochemistry of Permian to Triassic igneous rocks from northern Chile (28°–29°30′S): implications on the dynamics of the proto-Andean margin. Andean Geology 44(2): 147–178
Contreras JP, Espinoza M, De la Cruz R, Jorquera R, Kraus S, Ramírez C, Naranjo JA, Escribano J, Martínez P (2013) Carta Cifuncho, Regiones de Antofagasta y Atacama. Servicio Nacional de Geología y Minería. Carta Geológica de Chile, Serie Geología Básica 161 (1:100.000)
Cornejo P, Mpodozis C, Tomilnson A (1998) Hoja Salar de Maricunga. Región de Atacama. Servicio Nacional de Geología y Minería. Mapas Geológicos, Serie Geología Básica 7 (1:100.000)
Cornejo P, Mpodozis C, Rivera O, Matthews S (2009) Carta Exploradora, Regiones de Antofagasta y Atacama. Servicio Nacional de Geología y Minería. Carta Geológica de Chile, Serie Geología Básica 119, p 99 (1:100.000)
Cortés J (2000) Carta Palestina, Región de Antofagasta. Servicio Nacional de Geología y Minería. Carta Geológica de Chile, Serie Geología Básica 19 (1:100.000)
Cortés J (2012) Carta Sierra Mariposa, Región de Antofagasta. Servicio Nacional de Geología y Minería. Carta Geológica de Chile, Serie Geología Básica 144, p 30 (1:100.000)
Creixell C, Parada MA, Morata D, Roperch P, Arriagada C (2009) The genetic relationship between mafic dike swarms and plutonic reservoirs in the Mesozoic of central Chile (30°–33° 45′S): insights from AMS and geochemistry. Int J Earth Sci 98(1):177–201
Creixell C, Parada MÁ, Morata D, Vásquez P, Pérez de Arce C, Arriagada C (2011) Middle-Late Jurassic to Early Cretaceous transtension and transpression during arc building in Central Chile: evidence from mafic dike swarms. Andean Geol 38(1):37–63

Creixell C, Ortiz M, Arévalo C (2012) Geología del área Carrizalillo-El Tofo, Regiones de Atacama y Coquimbo. Servicio Nacional de Geología y Minería. Carta Geológica de Chile, Serie Geología Básica 133–134, p 82 (1:100.000)

Creixell C, Oliveros V, Vásquez P, Navarro J, Vallejos D, Valin X, Godoy E, Ducea MN (2016) Geodynamics of Late Carboniferous Early Permian forearc in north Chile (28°30–29°30′S). J Geol Soc. https://doi.org/10.1144/jgs2016-010

Cruden A, Arévalo C, Davis D, Grocott J (2004) Magmatic migration and pluton construction rates in the Mesozoic chilean coastal cordillera batholith (27° to 29° S). International Geological Congress N°32, Florence in CD-ROM

D'Elia L, Muravchik M, Franzese JR, Bilmes A (2012) Syn-rift volcanism of the Neuquen Basin, Argentina: relationships with the Late Triassic-Early Jurassic evolution of the Andean margin. Andean Geol 39:106–132

Dallmeyer D, Brown M, Grocott J, Taylor G, Treloar PJ (1996) Mesozoic magmatic and tectonic events within the Andean Plate boundary zone, 26º–27º30′S, North Chile: constraints from 40Ar/39Ar mineral ages. J Geol 104:19–40

del Rey A, Deckart K, Arriagada C, Martínez F (2016) Resolving the paradigm of the late Paleozoic-Triassic Chilean magmatism: isotopic approach. Gondwana Res. https://doi.org/10.1016/j.gr.2016.06.008

Escribano J, Martínez P, Domagala J, Padel M, Espinoza M, Jorquera R, Contreras J, De la Cruz R, Calderón M (2013) Cartas Bahía Isla Blanca y Taltal, Región de Antofagasta. Servicio Nacional de Geología y Minería. Carta Geológica de Chile, Serie Geología Básica 164–165 (1:100.000)

Espinoza F, Matthews S, Cornejo P, Venegas C (2011) Carta Catalina, Región de Antofagasta. Servicio Nacional de Geología y Minería. Carta Geológica de Chile, Serie Geología Básica 129 (1:100.000)

Espinoza M, Contreras JP, Jorquera R, De La Cruz R, Kraus S, Ramírez C (2014) Carta Cerro del Pingo, Regiones de Antofagasta y Atacama. Servicio Nacional de Geología y Minería. Carta Geológica de Chile, Serie Geología Básica N° 169, p 109 (1:100.000)

Espinoza M, Oliveros V, Vásquez P, Bechis F (2015) U-Pb geochronology and kinematic preliminary analyses of Late Triassic-Early Jurassic basins in northern Chile (24.5°–26° S). XIV Congreso Geológico Chileno, La Serena. Actas, vol I, pp 840–843

Frizon de Lamotte D, Fourdan B, Leleu S, Leparmentier F, Clarens P (2015) Style of rifting and the stages of Pangea breakup. Tectonics 34(5):1009–1029

García F (1967) Geología del Norte Grande de Chile. Sociedad Geológica de Chile, Santiago, Chile, p 138

Godoy E, Lara L (1998) Hojas Chañaral y Diego de Almagro, Región de Atacama. Servicio Nacional de Geología y Minería. Mapas Geológicos 5–6 (1:100.000)

González G, Niemeyer H (2005) Cartas Antofagasta y Punta Tetas, Región Antofagasta. Servicio Nacional de Geología y Minería. Carta Geológica de Chile, Serie Geología Básica 89 (1:100.000)

González R, Wilke G-H, Menzies AH, Riquelme R, Herrera C, Matthews S, Espinoza F, Cornejo P (2015) Carta Sierra de Varas, Región de Antofagasta. Servicio Nacional de Geología y Minería. Carta Geológica de Chile, Serie Geología Básica 178 (1:100.000)

Grocott J, Taylor GK (2002) Magmatic arc fault systems, deformation partitioning and emplacement of granitic complexes in the Coastal Cordillera, north Chilean Andes (25º30′S to 27º30′S). J Geol Soc, London 159:425–442

Grocott J, Brown M, Dallmeyer RD, Taylor GK, Treloar PJ (1994) Mechanisms of continental growth in extensional arcs: an example from the Andean plate-boundary zone. Geology 22 (5):391–394

Grocott J, Arévalo C, Welkner D, Cruden A (2009) Fault-assisted vertical pluton growth: Coastal Cordillera, north Chilean Andes. J Geol Soc 166(2):295–301

Henríquez SM, Becerra J, Arriagada C (2014) Geología del área San Pedro de Atacama, Región de Antofagasta. Servicio Nacional de Geología y Minería. Carta Geológica de Chile, Serie Geología Básica 171 (1:100.000)

Hervé F, Fanning CM, Calderón M, Mpodozis M (2014) Early Permian to Late Triassic batholiths of the Chilean Frontal Cordillera (28°–31° S): SHRIMP U-Pb zicon ages and Lu–Hf and O isotope systematics. Lithos 184–187:436–446

Iriarte S, Arévalo C, Mpodozis C, Rivera O (1996) Mapa Geológico de la Hoja Carrera Pinto, Región de Atacama. Servicio Nacional de Geología y Minería. Mapas Geológicos 3 (1:100.000)

Iriarte S, Arévalo C, Mpodozis C (1999) Hoja La Guardia, Región de Atacama. Servicio Nacional de Geología y Minería. Mapas Geológicos 13 (1:100.000)

Irvine TN, Baragar WRA (1971) A guide to the chemical classification of the common volcanic rocks. Can J Earth Sci 8(5):523–548

Kato TT, Godoy E (2015) Middle to late Triassic mélange exhumation along a pre-Andean transpressional fault system: coastal Chile (26°–42° S). Int Geol Rev 57(5–8):606–628

Kleiman LE, Japas MS (2009) The Choiyoi volcanic province at 34° S–36° S (San Rafael, Mendoza, Argentina): Implications for the Late Palaeozoic evolution of the southwestern margin of Gondwana. Tectonophysics 473(3):283–299

Kossler A (1998) Der Jura in der Küstenkordillere von Iquique (Nordchile): Paläontologie, Lithologie, Stratigraphie, Paläogeographie. Berliner Geowissenschaftliche Abhandlungen (A, 197)

Kramer W, Siebel W, Romer RL, Haase G, Zimmer M, Ehrlichmann R (2005) Geochemical and isotopic characteristics and evolution of the Jurassic volcanic arc between Arica (18°30′S) and Tocopilla (22° S), North Chilean Coastal Cordillera. Chem Erde 65(1):47–78

Le Bas MJ, Maitre RWL, Streckeisen A, Zanettin B (1986) A chemical classification of volcanic rocks based on the total alkali-silica diagram. J Petrol 27(3):745–750

Llambías EJ, Kleiman LE, Salvarredi JA (1993) El Magmatismo Gondwánico. In: Ramos VA (ed) Geología y Recursos Naturales de Mendoza. Relatorio, XII Congreso Geológico Argentino y II Congreso de Exploración de Hidrocarburos, Mendoza, pp 53–64

Llambías EJ, Quenardelle S, Montenegro T (2003) The Choiyoi Group from central Argentina: a sub-alkaline transitional to alkaline association in the craton adjacent to the active margin of the Gondwana continent. J S Am Earth Sci 16:243–257

Llambías EJ, Leanza HA, Carbone O (2007) Evolución tectono-magmática durante el Pérmico al Jurásico temprano en la Cordillera del Viento (37°05′S – 37°15′S): Nuevas evidencias geológicas y geoquímicas del inicio de la Cuenca Neuquina. Revista de la Asociación Geológica Argentina 62(2):217–235

Lucassen F, Thirlwall M (1998) Sm-Nd ages of mafic rocks from the Coastal Cordillera at 24° S, northern Chile. Geol Rundsch 86:767–774

Lucassen F, Kramer W, Bartsch V, Wilke HG, Franz G, Romer RL, Dulski P (2006) Nd, Pb and Sr isotope composition of juvenile magmatism in the Mesozoic large magmatic province of northern Chile (18°–27° S): indications for a uniform subarc mantle. Contrib Miner Petrol 152:571–589

Maksaev V, Munizaga F, Tassinari C (2014) Timing of the magmatism of the paleo-Pacific border of Gondwana: U-Pb geochronology of Late Paleozoic to Early Mesozoic igneous rocks of the north Chilean Andes between 20° and 31° S. Andean Geol 41(3):447–506

Marinovic N (2007) Carta Oficina Domeyko, Región de Antofagasta. Servicio Nacional de Geología y Minería. Carta Geológica de Chile, Serie Geología Básica 105, p 41 (1: 100.000)

Marinovic N, García M (1999) Hoja Pampa Unión, Región de Antofagasta. Servicio Nacional de Geología y Minería. Mapas Geológicos 9 (1:100.000)

Marinovic N, Smoje I, Maksaev V, Hervé M, Mpodozis C (1995) Hoja Aguas Blancas, Región de Antofagasta. Servicio Nacional de Geología y Minería. Carta Geológica de Chile 70, p 150 (1:250.000)

Marquardt C, Marinovic N, Muñoz V (2008) Geología de las Ciudades de Iquique y Alto Hospicio, Región de Tarapacá. Servicio Nacional de Geología y Minería. Carta Geológica de Chile, Serie Geología Básica 133, p 33 (1:25.000)

Martin MW, Clavero RJ, Mpodozis MC (1999) Late Paleozoic to Early Jurassic tectonic development of the high Andean Principal Cordillera, El Indio Region, Chile (29-30° S). J S Am Earth Sci 12:33–49

Matthews S, Cornejo P, Riquelme R (2006) Carta Inca de Oro, Región de Atacama. Servicio de Geología y Minería. Carta Geológica de Chile, Serie Geología Básica 102, p 79 (1:100.000)

Merino R, Ortiz M (2015) Nuevos antecedentes de rocas volcánicas e intrusivas del Jurásico Superior – Cretácico Inferior de la Alta Cordillera de Vallenar. Área Río Chollay – Matancilla (29°–29°30′S). XIV Congreso Geológico Chileno, La Serena. Actas, vol I, pp 747–750

Morata D, Aguirre L, Oyarzún M, Vergara M (2000) Crustal contribution in the genesis of the bimodal Triassic volcanism from the Coastal Range, central Chile. Rev Geol Chile 27:83–98

Mpodozis C, Cornejo P (1997) El rift triásico-sinemuriano de Sierra Exploradora, Cordillera de Domeyko (25º–26º S): asociaciones de facies y reconstrucción tectónica. In VIII Congreso Geológico Chileno, Antofagasta. Actas, vol I, pp 550–554

Mpodozis C, Kay S (1992) Late Paleozoic to Triassic evolution of the Gondwana margin: Evidence from Chilean Frontal Cordilleran batholiths (28° S to 31° S). Geol Soc Am Bull 104:999–1014

Mpodozis C, Ramos VA (1989) The Andes of Chile and Argentina. In: Ericksen GE, Cañas Pinochet MT, Reinemund JA (eds) Geology of the Andes and its relation to hydrocarbon and mineral resources. Houston, Texas, Circum-Pacific Council for Energy and Mineral Resources, Earth Sciences Series 11:59–90

Mpodozis C, Ramos VA (2008) Tectónica Jurásica en Argentina y Chile: extensión, subducción oblicua, rifting, deriva y colisiones? Rev Asoc Geol Argentina 63(4):481–497

Mpodozis C, Iriarte S, Gardeweg M, Valenzuela M (2012) Carta Laguna del Negro Francisco, Región de Atacama. Servicio Nacional de Geología y Minería. Carta Geológica de Chile, Serie Geología Básica 145, p 30 (1:100.000)

Murillo I, Álvarez J, Montecinos P, Creixell C, Salazar E, Arriagada C (2013) Geochronology and kinematics of the El Portillo Mylonites: relation with San Rafael Orogeny and Middle Triassic extensión in north-Central Chile. In International Geological Congress on the Southern Hemisphere GEOSUR. Bollettino di Geofisica teorica ed applicata (Supplement 2):54

Naipauer M, Tunik M, Marques JC, Rojas Vera E, Vujovich GI, Pimentel MM, Ramos VA (2015) U-Pb detrital zircon ages of Upper Jurassic continental successions: implications for the provenance and absolute age of the Jurassic-Cretaceous boundary in the Neuquén Basin. Geol Soc, London, Special Publications 399(1):131–154

Naranjo JA, Puig A (1984). Hojas Taltal y Chañaral, Regiones de Antofagasta y Atacama. Servicio Nacional de Geología y Minería. Carta Geológica de Chile 62–63, p 140 (1:250.000)

Nasi C, Moscoso R, Maksaev V (1990) Hoja Guanta. Servicio Nacional de Geología y Minería. Carta Geológica de Chile, Serie Geología Básica 67, p 141 (1:250.000)

Niemeyer H (2013) Geología del Área Cerro Lila-Peine, Región de Antofagasta. Servicio Nacional de Geología y Minería. Carta Geológica de Chile, Serie Geología Básica 147 (1:100.000)

Oliveros V, Féraud G, Aguirre L, Fornari M, Morata D (2006) The Early Andean Magmatic Province (EAMP): 40Ar/39Ar dating on Mesozoic volcanic and plutonic rocks from the Coastal Cordillera, northern Chile. J Volcanol Geoth Res 157(4):311–330

Oliveros V, Morata D, Aguirre L, Féraud G, Fornari M (2007) Jurassic to Early Cretaceous subduction–related magmatism in the Coastal Cordillera of northern Chile (18°30′–24° S): geochemistry and petrogenesis. Rev Geol Chile 34:209–232

Oliveros V, Féraud G, Aguirre L, Ramírez LE, Fornari M. Palacios C, Parada MA (2008) Detailed 40Ar/39Ar dating of geological events associated with the Mantos Blancos copper deposit, northern Chile. Miner Deposita 43(3):281–293

Oliveros V, Vásquez P, Creixell C, Lucassen F, Ducea M, González J, Ciocca I (2016) Lithospheric evolution at the Early Andean convergent margin, Chile. Goldschmidt conference, Yokohama, Japan

Ortiz M, Merino N (2015) Geología de las áreas Río Chollay – Matancilla y Cajón del Encierro, Regiones de Atacama y Coquimbo. Servicio Nacional de Geología y Minería. Carta Geológica de Chile 175–176 (1:100.000)

Parada F (2013) Geoquímica de las rocas ígneas del Carbonífero-Triásico de la Alta Cordillera, Región de Atacama, Chile. Memoria de Título (Unpubl), Departamento de Geología, Universidad de Chile, p 93

Parada MA, Levi B, Nystrom JO (1991) Geochemistry of the Triassic to Jurassic plutonism of Central Chile (30 to 33 S); petrogenetic implications and a tectonic discussion. Geol Soc Am Spec Pap 265:99–112

Parada MA, Levi B, Nyström JO (1999) Multiple sources for the Coastal Batholith of central Chile (31–34° S): geochemical and Sr–Nd isotopic evidence and tectonic implications. Lithos 46:505–521

Parada MA, López-Escobar L, Oliveros V, Fuentes F, Morata D, Calderón M, Aguirre L, Féraud F, Espinoza F, Moreno H, Figueroa O (2007) Andean magmatism. In: Moreno T, Gibbons W (eds) The Geology of Chile. The Geological Society, London:115–146

Pearce JA, Harris NBW, Tindle AG (1984) Trace element discrimination diagrams for the tectonic interpretation of granitic rocks. J Petrol 25(4):956–983

Peña M, Becerra J, Martinez F, Arriagada C (2013) Geología del área Yerbas Buenas – Tres Morros, Región de Atacama. Servicio Nacional de Geología y Minería. Carta Geológica de Chile 155, p 78 (1:100.000)

Pichowiak S (1994) Early Jurassic to Early Cretaceous magmatism in the Coastal Cordillera and the Central Depression of North Chile. In: Reutter KJ, Scheuber E, Wigger PJ (eds) Tectonics of the Southern Central Andes. Structure and evolution of a continental margin. Springer Verlag, Stuttgart, pp 203–217

Proffett JM, Dilles JH (2007) SHRIMP-RG ion microprobe U-Pb age determinations of intrusive rock units northeast of the Chuquicamata mine, Chile. CODELCO, Informe (Unpubl.), p 15

Ramos VA, Kay SM (1991) Triassic rifting and associated basalts in the Cuyo basin, central Argentina. In Harmon RS, Rapela CW (eds) Andean magmatism and its tectonic setting. Geol Soc Am, Special Papers 265:79–91

Ring U, Willner A, Layer P, Richter P (2012) Jurassic to Early Cretaceous postaccretional sinistral transpression in north-central Chile (latitudes 31–32° S). Geol Mag 149(2):202–220

Rodríguez N, Díaz-Alvarado J, Rodríguez C, Riveros K, Fuentes P (2016) Petrology, geochemistry and thermobarometry of the northern area of the Flamenco pluton, Coastal Range batholith, northern Chile. A thermal approach to the emplacement processes in the Jurassic andean batholiths. J S Am Earth Sci 67:122–139

Rogers G, Hawkesworth CJ (1989) A geochemical traverse across the North Chilean Andes: Evidence for crust generation melt from the mantle wedge. Earth and Planet Sci Lett 91:271 285

Rossel P, Oliveros V, Ducea M, Charrier R, Scaillet S, Retamal L, Figueroa O (2013) The Early Andean subduction system as an analogue to island arcs: evidence from across-arc geochemical variations in northern Chile. Lithos 179:211–230

Rossel P, Oliveros V, Ducea MN, Hernández L (2015) Across and along arc geochemical variations in altered volcanic rocks: Evidence from mineral chemistry of Jurassic lavas in northern Chile, and tectonic implications. Lithos 239:97–113

Salazar E, Coloma F, Creixell C (2013) Geología del área El Tránsito - Lagunillas, Región de Atacama. Servicio Nacional de Geología y Minería. Carta Geológica de Chile, Serie Geología Básica 149, p 121 (1:100.000)

Sato AM, Llambías EJ, Basei MA, Castro CE (2015) Three stages in the Late Paleozoic to Triassic magmatism of southwestern Gondwana, and the relationships with the volcanogenic events in coeval basins. J S Am Earth Sci 63:48–69

Scheuber E, González G (1999) Tectonics of the Jurassic – Early Cretaceous magmatic arc of the north Chilean Coastal Cordillera (22°–26° S): A story of crustal deformation along a convergent plate boundary. Tectonics 18:895–910

Sepúlveda F, Vásquez P, Quezada A (2014) Cartas Patillos y Oficina Victoria, Región de Tarapacá. Servicio Nacional de Geología y Minería. Carta Geológica de Chile, Serie Geología Básica 167–168 (1:100.000)

SERNAGEOMIN (2002) Mapa Geológico de Chile. Servicio Nacional de Geología y Minería, Chile. Carta Geológica de Chile, Serie Geología Básica 75 (1:1.000.000)
Shand S (1943) Eruptive rocks: their genesis, composition, and classification, with a chapter on meteorites. John Wiley Sons, Inc. p 444
Solari M, Montecinos D, Venegas C, Espinoza F (2015). Hallazgo de unidades volcánicas del Triásico Medio en la Sierra de Imilac, Segunda Región de Antofagasta. XIV Congreso Geológico Chileno, La Serena. Actas, vol III, pp 398–400
Sun S-s, McDonough WF (1989) Chemical and isotopic systematics of oceanic basalts: implications for mantle composition and processes. Geol Soc London, Special Publications 42 (1):313–345
Thomas A (1970) Cuadrángulos Iquique y Caleta Molle. Instituto de Investigaciones Geológicas, Chile. Carta Geológica de Chile 21–22. 2 maps (1:50.000)
Tobar A (1966) Estratigrafía del área Baquedano-Rencoret, Provincia de Antofagasta. Memoria de Título (Unplub.), Departamento de Geología, Universidad de Chile, p 69
Tomlinson AJ, Blanco N (2008) Geología de la franja El Abra-Chuquicamata, II Región (21°45′–22°30′S). Servicio Nacional de Geología y Minería, Informe Registrado IR-08–35, p 196
Tomlinson AJ, Blanco N, Dilles JH (2010) Carta Calama, Región de Antofagasta. Servicio Nacional de Geología y Minería. Carta Geológica de Chile, Serie Preliminar 8: 3 anexos (1:50.000)
Tomlinson A, Blanco N, Ladino M (2015) Carta Mamina, Región de Tarapacá. Servicio Nacional de Geología y Minería. Carta Geológica de Chile, Serie Geología Básica (1:100.000)
Vásquez P, Franz G (2008) The Triassic Cobquecura Pluton (Central Chile): An example of a fayalite-bearing A-type intrusive massif at a continental margin. Tectonophysics 459(1):66–84
Vásquez P, Sepúlveda F (2013) Cartas Iquique y Pozo Almonte, Región de Tarapacá. Servicio Nacional de Geología y Minería. Carta Geológica de Chile, Serie Geología Básica 162–163 (1:100.000)
Vásquez P, Glodny J, Franz G, Frei D, Romer RL (2011) Early Mesozoic Plutonism of the Cordillera de la Costa (34–37° S), Chile: constraints on the onset of the Andean Orogeny. J Geol 119(2):159–184
Venegas C, Cervetto M, Astudillo N, Espinoza F (2013) Carta Sierra Vaquillas Altas, Regiones de Antofagasta y Atacama. Servicio Nacional de Geología y Minería. Carta Geológica de Chile, Serie Geología Básica 159 (1:100.000)
Vergara M, López-Escobar L, Cancino A, Levi B (1991) The Pichidangui Formation; Some geochemical characteristics and tectonic implications of the Triassic marine volcanism in central Chile (31°55′ to 32°20′S) In: Harmon R S, Rapela C W (eds) Andean magmatism and its tectonic setting. Geol Soc Am Spec Pap 265:93–98
Vicente JC (2006) Dynamic Paleogeography of the Jurassic Andean Basin: pattern of regression and general considerations on main features. Revista de la Asociación Geológica Argentina 61 (3):408–437
Welkner D, Arévalo C, Godoy E (2006) Geología del área Freirina-El Morado, Región de Atacama. Servicio Nacional de Geología y Minería. Carta Geológica de Chile, Serie Geología Básica 100 (1:100.000)
Wood DA (1980) The application of a Th-Hf-Ta diagram to problems of tectonomagmatic classification and to establishing the nature of crustal contamination of basaltic lavas of the British Tertiary Volcanic Province. Earth Planet Sci Lett 50:11–30
Zindler A, Hart S (1986) Chemical geodynamics. Ann Rev of Earth Planet Sci 14(1):493–571

A Provenance Analysis from the Lower Jurassic Units of the Neuquén Basin. Volcanic Arc or Intraplate Magmatic Input?

Maximiliano Naipauer, Ezequiel García Morabito, Marcelo Manassero, Victor V. Valencia and Victor A. Ramos

Abstract A possible signal of the Jurassic Chon Aike Igneous Province within the early infill of the Neuquén Basin is recognized in our provenance analyses. We use a combination of detrital zircon geochronology and Lu–Hf isotope analysis, along with sandstone petrography descriptions to characterize the sediments source region in the Cuyo Group. The sandstone petrographic analysis confirms important contributions from volcanic sources of different compositions. U-Pb ages and Hf isotopes obtained in the analyzed zircons indicate a more complex configuration given by multiple igneous components. The variations in the provenance patterns allow us to make some observations about the paleogeographic evolution of the basin. At the base of the sequence, the sediments were derived from local sources composed of Permian basement and Upper Triassic volcanic rocks, whereas in the top, the zircons were supplied from Lower Jurassic volcanic rocks and Lower Paleozoic and Precambrian basements suggesting more distal and ancient sources. Low values of εHf_t (−15.5 to −0.7) analyzed in Permian and Triassic detrital zircons indicated an evolved source with strong crustal contribution. These data are in agreement with

M. Naipauer (✉) · E. García Morabito
Instituto de Estudios Andinos "Don Pablo Groeber" (UBA—CONICET, Argentina), Buenos Aires, Argentina
e-mail: maxinaipauer@hotmail.com

E. García Morabito
Swiss Federal Institute of Technology (ETH Zurich), Zurich, Switzerland

M. Manassero
Centro de Investigaciones Geológicas (UNLP—CONICET, Argentina), Buenos Aires, Argentina

V. V. Valencia
School of the Environment, Washington State University, Pullman, WA, USA

V. A. Ramos
Instituto de Estudios Andinos "Don Pablo Groeber", Departamento de Ciencias Geológicas, FCEN, Universidad de Buenos Aires–CONICET, Buenos Aires, Argentina

A. Folguera et al. (eds.), *The Evolution of the Chilean-Argentinean Andes*, Springer Earth System Sciences, https://doi.org/10.1007/978-3-319-67774-3_8

the negative values of εHf_t and εNdt calculated in Permian and Triassic igneous rocks from the North Patagonian Massif. The εHf_t about −4 of the Jurassic detrital zircons indicated a crustal origin for the source rocks and are clearly compatible with the isotopic compositions of the Chon Aike Igneous Province. A volcanic source region compatible with the Andean arc is dismissed because the Jurassic arc has isotope characteristics of a mantle source.

Keywords Hf isotopes · U-Pb detrital ages · Provenance · Cuyo Group Neuquén Basin

1 Introduction

From the latest Triassic to the Early Jurassic, the proto-Pacific margin of Gondwana was affected by a generalized pre-Andean continental extension related to the breakup of the Pangea (Mpodozis and Ramos 1989; Franzese and Spalletti 2001). The associated magmatic, tectonic, and sedimentary processes strongly influenced the subsequent Andean evolution of the whole region. Widespread extension created isolated depocenters that progressively evolved into bigger basins during the thermal subsidence. The inception of the southern Neuquén Basin as a single depocenter dates back to the Pliensbachian, when subsidence led to a first general marine transgression that deposited the thick siliciclastic successions of the Cuyo Group (Gulisano 1981; Leanza et al. 2013). The preservation of several ash layers within these marine sequences was interpreted as an evidence for an Early Andean magmatic arc (Kamo and Riccardi 2009; Mazzini et al. 2010; Leanza et al. 2013; Armella et al. 2016). This exhumed arc, widely preserved along the axis of the cordillera north of 36° S (Fig. 1), was characterized by lavas of tholeiitic to calc-alkaline composition, product of a magmatic arc emplaced in a thin crust (Lucassen et al. 2004; Oliveros et al. 2006) (see Chap. "The Early Stages of the Magmatic Arc in the Southern Central Andes").

At the same time, widespread magmatism occurred in the extra-Andean Patagonian region (Fig. 1) (see Chap. "Lower Jurassic to Early Paleogene Intraplate Contraction in Central Patagonia"). This event, dominated by a bimodal association of rhyolitic ignimbrites and mafic and intermediate lavas dated between 188 and 144 Ma, originated one of the largest silicic igneous provinces known, the Chon Aike Igneous Province (Kay et al. 1989; Pankhurst et al. 1998; 2000; Féraud et al. 1999). Despite its regional importance, there are no investigations that link the evolution of the Neuquén Basin to this widespread contemporary magmatic event, and no evidence of this event within the early infill of the basin has been identified so far. The identification of the different volcanic sources, which contributed to the infill of the Neuquén Basin in the early stages, such as the arc in the west or the extra-Andean igneous province in the east, may have important tectonic and geodynamic implications.

This study aims to investigate a possible signal of the Chon Aike Igneous Province within the early infill of the basin. For this purpose, we use a combination of detrital zircon geochronology and Lu–Hf isotopic analysis, along with

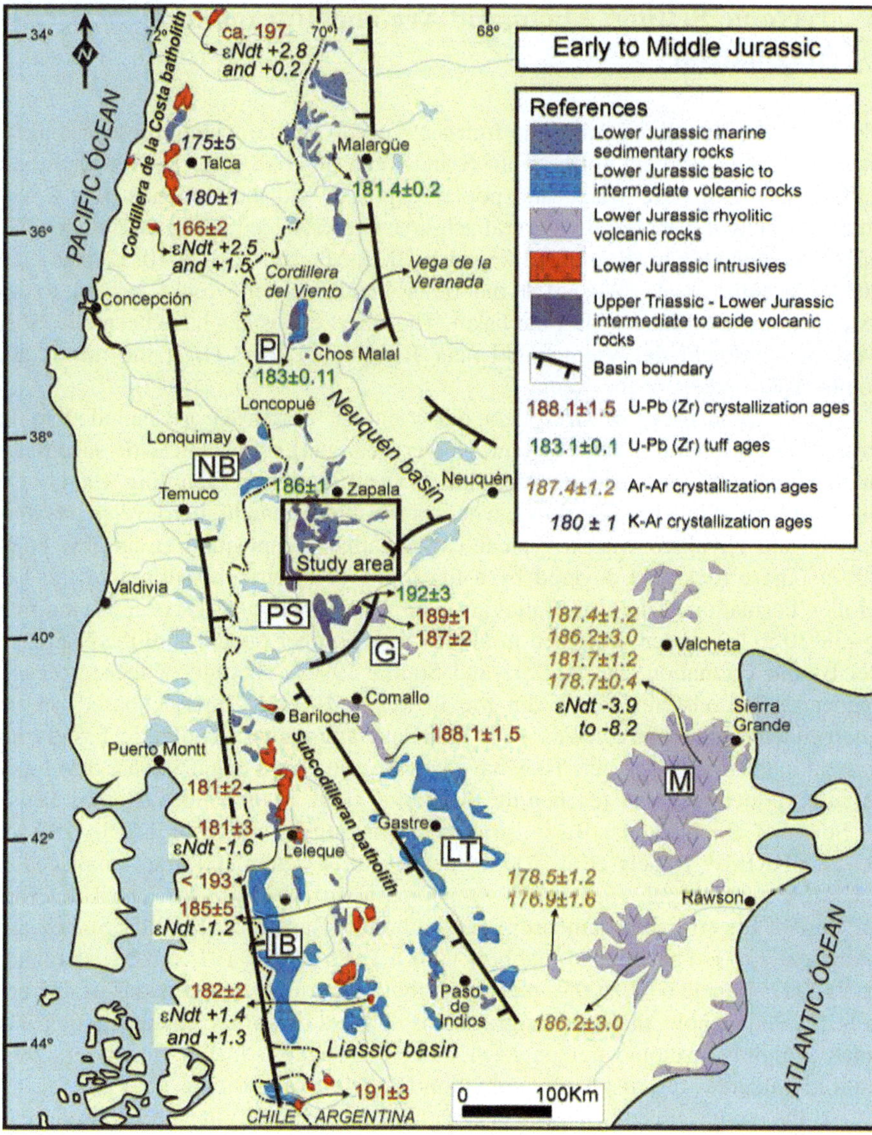

Fig. 1 Regional geological map with the Early to Middle Jurassic igneous and sedimentary units from Patagonia. Ages and isotopic data based on several authors cited in the text. M: Marifil Formation; LT: Lonco Trapial Formation; IB: Ibañez Formation; G: Garamilla Formation; PS: Piedra del Águila and Sañicó formations; NB: Nacientes del Río Biobío Formation; P: La Primavera Formation. Location of the Neuquén and Liassic basins, as well as the Subcordilleran batholith and Cordillera de la Costa batholiths are indicated

descriptions of the sandstone petrography. We focus on a marine succession spanning ~25 m.y. from the Pliensbachian to Callovian, exposed in the south-eastern margin of the Neuquén Basin (Fig. 1).

2 Tectonic Setting: Magmatic Arc and Intraplate Magmatism

Volcanic and plutonic rocks of the Jurassic Andean arc are widely preserved along the axis of the cordillera north of 36° S and south of 41° S (Fig. 1). Between these latitudes, outcrops are scarce and poorly exposed, and the evidence of a Lower Jurassic magmatic arc are the several ash layers interbedded in the marine deposits of the Neuquén Basin (Kamo and Riccardi 2009; Mazzini et al. 2010; Leanza et al. 2013; Armella et al. 2016). On the other hand, volcanic rocks related to the exhumed arc are scarce within the basin. Their best exposures have been described west of Cordillera del Viento, and near Lonquimay (Fig. 1), at the north- and southwestern edges of the basin.

The volcanic rocks of Cordillera del Viento (~37°00′S) are included in La Primavera Formation (Pliensbachian—lower Toarcian), a volcaniclastic succession interbedded with marine deposits (Gulisano and Gutiérrez Pleimling 1994). The base of this unit has basaltic flows, whereas the middle and higher levels consist of dacites and rhyolites with a typical calc-alkaline composition (Llambías et al. 2007). These rocks are covered by a thick pile of marine mudstones of the Los Molles Formation (Toarcian–Callovian). Jurassic volcanic rocks near Lonquimay (~38°30′S) have been described in the lower and upper members of the Nacientes del Biobío Formation (De la Cruz and Suárez 1997). The age of these volcanic sequences is constrained between the Lower and Upper Jurassic based on the interbedded sedimentary facies with abundant ammonites. Tholeiitic basalts are present in the lower member. They have been assigned to a magmatic arc developed in an intra-arc basin that reached the Argentine slope of the Andes (Ramos 1999).

South of the Neuquén Basin, granitoids ranging in age from 178 to 187 Ma (U-Pb SHRIMP: Rapela et al. 2005; Suarez and Márquez 2007) are emplaced in Paleozoic metamorphic rocks and Lower Jurassic marine sediments. These outcrops are part of the so called Subcordilleran batholith (Rapela et al. 2005; Suarez and Márquez 2007). The Muzzio Gabbro, with a U-Pb age of 191 ± 3 Ma (Rolando et al. 2002; Rapela et al. 2005), represents the southernmost expressions of this belt (Fig. 1). As a whole, these I-type granitoids define a belt of arc characteristics. The relative high Nd isotopic signature (+1.4 to −1.6) suggests mantle-derived magmas with low degrees of crustal contamination (Rapela et al. 2005).

At the same time, widespread magmatism occurred in the extra-Andean Patagonian region (Fig. 1). This event, dominated by a bimodal association of rhyolitic ignimbrites and mafic and intermediate lavas, has been related to an extensional tectonic setting (Pankhurst and Rapela 1995; Féraud et al. 1999). This large silicic igneous province integrates the North Patagonian Massif, which defines the southeastern margin of the Neuquén Basin. Its age has been constrained between 177 and 189 Ma by different dating methods (Rb-Sr, K-Ar, Ar–Ar, and U-Pb) (Pankhurst and Rapela 1995; Féraud et al. 1999; Pankhurst et al. 2000; Franzese et al. 2002; Cúneo et al. 2013; Benedini et al. 2014) (Fig. 1). According to Pankhurst and Rapela (1995), most of the Jurassic volcanic rocks are isotopically uniform, with εNdt = −4 (± 2) and depleted-mantle model ages of 1150–1600 Ma.

These magmatic rocks are associated with a diachronous extensional event, with a southwestward migration path, where three stages of magmatic activity are identified: V1 (189–177 Ma) in the North Patagonian Massif, V2 (172–162 Ma) in the Deseado Massif (~46º–48º LS), and V3 (153–144 Ma) along the Southern Patagonian Andes (~48º–53º LS) (Féraud et al. 1999; Pankhurst et al. 2000) (Fig. 3) (see Chap. "Lower Jurassic to Early Paleogene Intraplate Contraction in Central Patagonia").

3 Geological and Stratigraphic Framework of the Studied Region

The Neuquén Basin has a long and complex history characterized by three main stages of evolution: initial Late Triassic—Early Jurassic rifting followed by a retroarc stage dominated by thermal subsidence, which ended in the Late Cretaceous in a foreland stage (Vergani et al. 1995; Howell et al. 2005; Ramos and Folguera 2005). It contains a near-continuous Mesozoic succession including several depositional settings now widely exposed due to the Andean tectonics.

In the study area, located at the southwestern edge of the basin, basement rocks are exposed, early synrift, marine, and continental deposits of the retroarc stage (Fig. 3). Devonian to Carboniferous metamorphic rocks (e.g.: Piedra Santa Formation, Colohuincul Complex, Mamil Choique Formation; Franzese 1995; Lucassen et al. 2004; Varela et al. 2005), and Permian igneous rocks (Choiyoi magmatism; Sato et al. 2015) compose the basement. These units are exposed along the North Patagonian Andes in the west, in the southeastern North Patagonian Massif, and within the northern Huincul High (Fig. 2). Each of these areas is interpreted as source regions in the past for sediment supply to the Neuquén Basin (Naipauer et al. 2012). Upper Triassic–Lower Jurassic volcanic and volcaniclastic rocks, and coarse-grained conglomerates grouped in the synrift Precuyano cycle are very well exposed in the southwestern Neuquén Basin. They have several local denominations within the region (Fig. 3). The Precuyano strata underlie thousands of meters of a retroarc succession deposited during three principal transgressive–regressive cycles. The Cuyo Group represents the first of these cycles and has three major sequences deposited during Pliensbachian–Bathonian times: deep marine shales (Los Molles Formation), shallow-marine to deltaic deposits (Lajas Formation), and fluvial deposits (Challacó Formation). Its base, represented by the Chachil Limestone, has been dated between 186–182 Ma (U-Pb ID-TIMS: Leanza et al. 2013) (Fig. 3).

In the studied section (Arroyo Lapa—Estancia Charahuilla) the Cuyo Group starts with a succession of marine sandstones and tuffaceous levels included in the Sierra de Chacaicó Formation (Volkheimer 1973). These are overlaid by marine black shales and fine-grained sandstones levels of the Los Molles Formation (Weaver 1931). The Sierra de Chacaicó and Los Molles formations are Pliensbachian-Toarcian in age according to their fossiliferous content (mainly ammonites) (Gulisano and Gutiérrez Pleimling 1994; Burgess et al. 2000; Al-Suwaidi et al. 2010). Above Los Molles Formation, yellow calcareous,

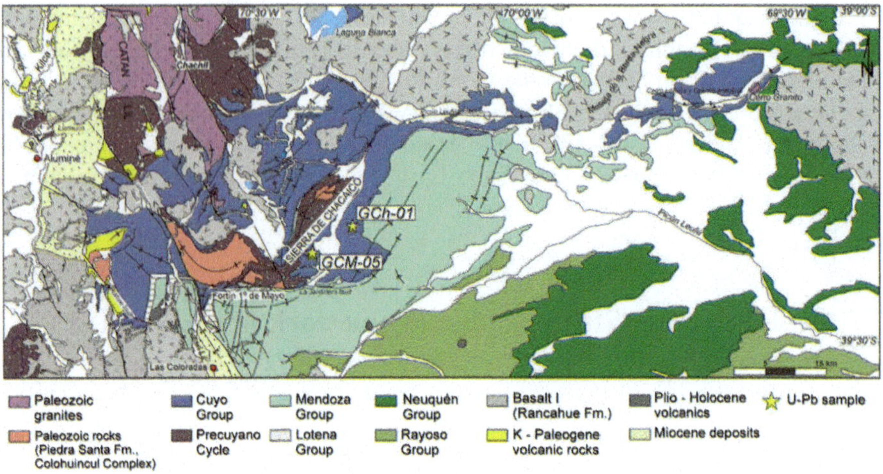

Fig. 2 Geological map of the studied region and sample location (modified from Leanza and Hugo 1997; García Morabito and Ramos 2012)

tuffaceous sandstones, and fine-grained conglomerates integrate the Lajas Formation (Weaver 1931). These facies were interpreted as deposited in coastal marine to deltaic environments. The record of the first red-beds of the Challacó Formation, composed of pelites, fine-grained sandstones, and conglomerates, indicates a regression and continentalization of the basin that culminated with the evaporitic deposits of the Tábanos Formation in the central part of the basin (Gulisano et al. 1984). Sandstone and conglomerate facies of the Lajas and Challacó formations are assigned to the Late Toarcian to Callovian and interpreted as part of the post-rift stage by Pángaro et al. (2009).

Finally, the Cuyo Group is covered in erosive discordance (intra-Malm unconformity) by red sandstones from the Tordillo Formation (Mendoza Group) in the study area, which represents another sedimentary cycle ranging from Late Jurassic to the Early Cretaceous (Figs. 2 and 3) (Gulisano and Gutiérrez Pleimling 1994).

4 Methods and Stratigraphic Section Studied

Twelve samples of sandstones were collected from the Cuyo Group for petrographic analysis. We took samples from the classic Arroyo Lapa, Estancia Charahuilla, La Jardinera and Picún Leufú sections (Fig. 2). Two additional samples were also collected from the Vega de la Veranada section, located north of the study area (Figs. 1, 2, and 3). The thin sections were studied under polarized microscope, and 300 points were counted per sample following Dickinson et al. (1983). Table A.1 (Appendix) shows the main detected minerals.

Samples for U-Pb chronology were taken from the Arroyo Lapa and Charahuilla profiles. Sample GCM-05 (GPS: 39°23′29.8″S–70°24′41.4″W) was taken about 400 m above the base of the Los Molles Formation. It corresponds to the V Interval

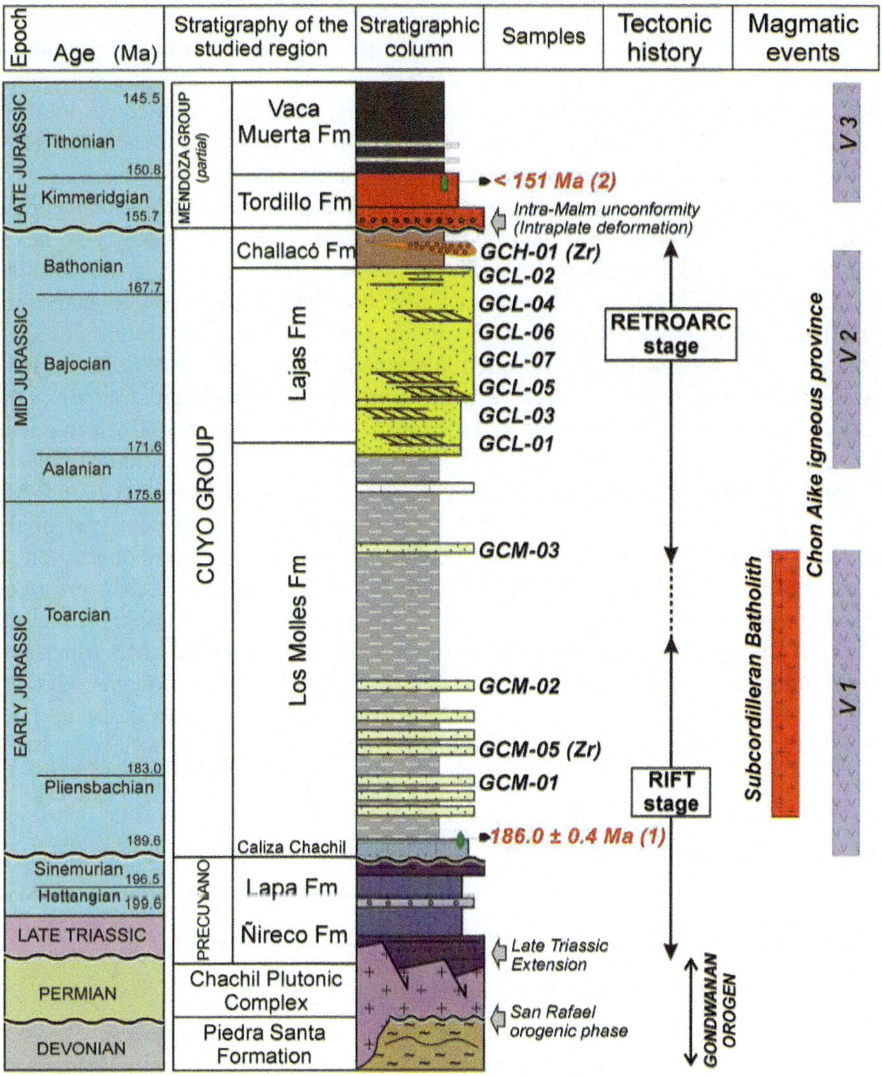

Fig. 3 Simplified stratigraphic column of the Sierra de Chacaico area with Paleozoic basement rocks and Late Triassic to Jurassic deposits (modified from Gulisano and Gutiérrez Pleimling 1994). Magmatic events linked to the Subcordilleran batholith (Rapela et al. 2005) and the Chon Aike Igneous Province (Kay et al. 1989; Pankhurst et al. 2000) are also shown. U-P zircon ages are from (1) Leanza et al. (2013) and (2) Naipauer et al. (2012)

defined by Gulisano y Gutiérrez Pleimling (1994) in the Arroyo Lapa profile. The sampled rock is a quartz-feldspathic sandstone, medium- grained, and yellowish-green color. It was sampled in a one meter thick-level interbedded within a thick pelitic succession. The sandstone body has pelitic intraclasts (10 cm) at its base. Other sandstone with synsedimentary deformation structure levels is also observed.

The sample GCH-01 (GPS: 39°21′05.4″S–70°20′24.4″W) corresponds to the base of the Challacó Formation. It was taken from a 2-m-thick fine conglomerate. The conglomerates, which have lenticular bed geometry, are interbedded within a thick succession of dark green and red pelites (Fig. 3). The sample corresponds to the Interval I of the Challacó Formation as defined by Gulisano and Gutiérrez Pleimling (1994) in their profile from the Charahuilla locality.

Detrital zircon separation was performed by standard techniques of concentration of heavy minerals in the Departamento de Ciencias Geológicas de la Universidad de Buenos Aires (Argentina). The detrital zircons from samples GCM-05 and GCH-01 were observed under binocular microscope to determine external morphological features (Fedo et al. 2003). Also, cathodoluminescence images were performed to define internal textures and to guide the location of the laser spot in the U-Pb analyzes. Zircons from the non-magnetic fraction were handpicked under the microscope and mounted with standards in a 1-inch-diameter epoxy puck and slightly grounded to expose the surface and keep as much material as possible for laser ablation analyses. U-Pb age determinations with LA-ICP-MS followed by Hf isotopic analyses by LA-MC-ICP-MS in an adjacent part of the crystal within the same cathodoluminescence-defined zone were conducted at Washington State University. The LA-ICP-MS U-Pb analyses were performed using a New Wave Nd:YAG UV 213-nm laser coupled to a Thermo-Finnigan Element 2 single collector, double-focusing, magnetic sector ICP-MS (operating procedures and parameters are similar to those of Chang et al. (2006)). LA-MC-ICP-MS HF zircon isotope analyses were conducted using the same laser coupled to a Thermo-Finnigan Neptune Multi-Collector ICP-MS following operation conditions in Gaschnig et al. (2011). Analytical data (U-Pb and Lu–Hf) are available in the Appendix (Tables A.2 and A.3).

5 Results

5.1 Sandstone Petrography

The samples from Los Molles Formation show the highest content of volcanic rock fragments (>30%), mostly andesites (Fig. 4a, b); the monocrystalline quartz is less than 30%, and plagioclases are close to 24%. Sedimentary rock fragments (including chert) are higher than 10%, and the polycrystalline quartz is close to 5%, suggesting a metamorphic source. The monocrystalline quartz ranges between 10 and 30%. Sandstones of the Lajas Formation show more than 30% of mono and polycrystalline quartz and low contents of volcanic rock fragments (< 14%). Basic/intermediate rocks are more abundant than acid rocks (Fig. 4c, d). The lithic rock fragments are close to 10%. The sample from the Challacó Formation shows a high content of acid volcanic rock fragments (30%) with a minor proportion of quartz (25%) and feldspars.

The ternary diagrams for the Los Molles samples (Dickinson et al. 1983) show a trend from transitional and dissected volcanic arc to a recycled orogen. On the other

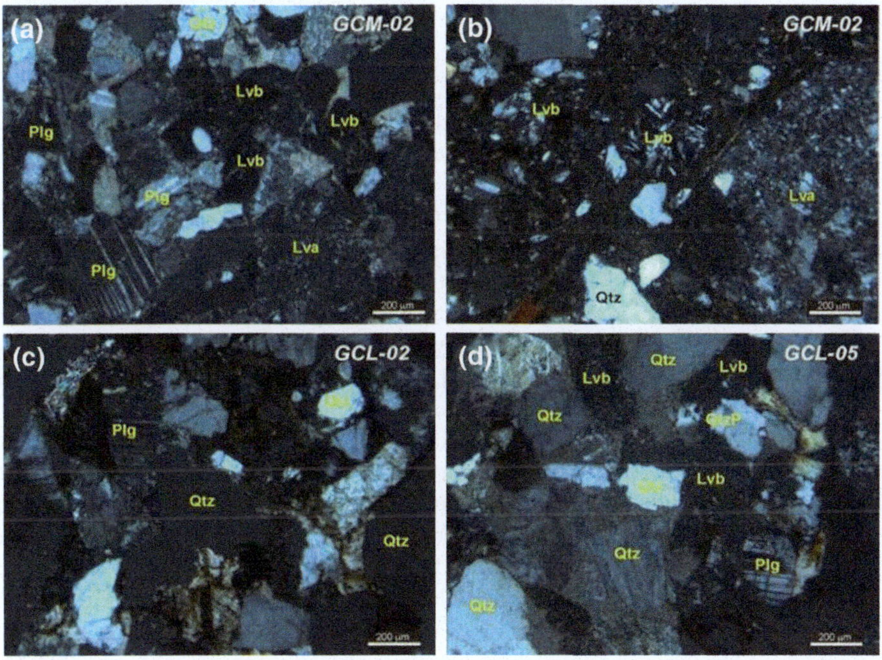

Fig. 4 Microphotographs of thin section from samples of the Charahuilla-Lapa profile (crossed nicols): **a–b** Los Molles Formation and **c–d** Lajas Formation. Qtz: quartz grains; QtzP: polycrystalline quartz; Plg: plagioclase; Lvb: basic volcanic lithic fragments; Lva: acidic volcanic lithic fragments

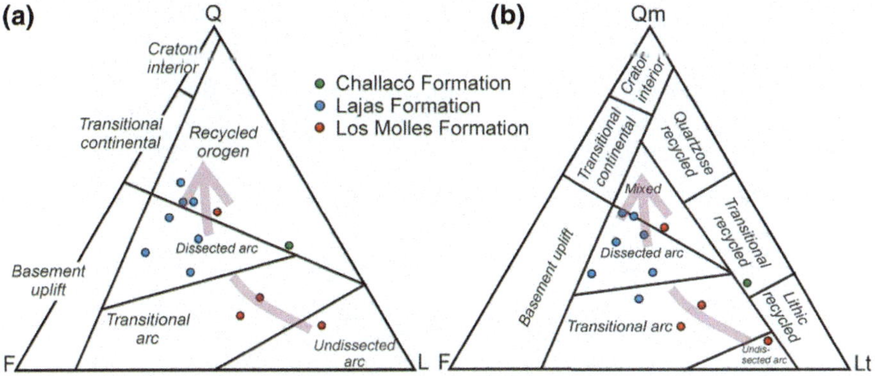

Fig. 5 Tectonic setting plots of the studied samples from Cuyo Group in Neuquén Basin, after Dickinson et al. (1983); **a** Q-F-L diagram and **b** Qm-F-Lt diagram. Q: total quartz grains; Qm: monocrystalline quartz grains; F: feldspar grains; L: lithic fragments; Lt: lithic grains including polycrystalline quartz

hand, sandstones of the Lajas Formation mainly plot in the dissected arc and recycled orogen fields. Finally, the sample from the Challacó Formation plots in the recycled orogen and transitional recycled fields (Fig. 5a, b).

5.2 Detrital Zircon Analysis

Most of the crystals observed in both samples are characterized by a prismatic habit with elongation between 3 and 4. They have preserved pyramidal faces and have idiomorphic to subidiomorphic forms. Only a few crystals have a rounded shape. The zircons generally are translucent or with yellowish color (Fig. 6a). A zircons population with a long prismatic habit (elongation >5) and idiomorphic form is remarkable, and many of them are broken but preserve their pyramidal faces (Fig. 5a). We interpret that this population has probably a volcanic origin. The internal texture observed in the CL images showed typically oscillatory zoning indicating an igneous origin for the analyzed zircons (Fig. 6b). Particularly, in the sample, CGH-01 crystals of garnet and rutile were also observed, suggesting a contribution of metamorphic basement rocks.

124 U-Pb ages were determined in detrital zircon of the GCH-01 sample from the Challacó Formation. The ages range between ca. 176 and 2620 Ma. The age distribution has a multimodal pattern with peaks of maximum frequency at ca. 182, 210, 240, and 267 Ma. The main represented zircon ages are Permian to Triassic (65%) and Early Jurassic (29%). Isolated ages also appear in the Early Paleozoic, Neoproterozoic, Mesoproterozoic, and Neoarchean (Fig. 7a). 124 U-Pb ages were also obtained in detrital zircons separated from the GCM-05 sample (Los Molles Formation). The ages are in a short time interval between ca. 175 and 284 Ma (Fig. 7b). The distribution of the data is clearly bimodal with two peaks of maximum frequency at ca. 192 Ma (59%) and 253 Ma (41%).

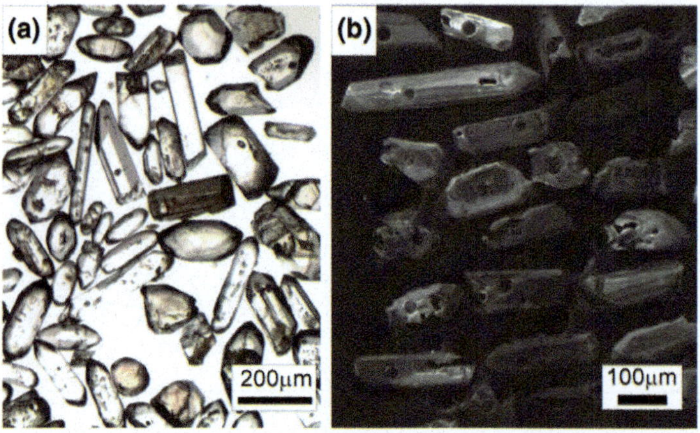

Fig. 6 Representative images of the detrital zircons from the studied samples: **a** binocular images of the zircons from Challacó Formation and **b** cathodoluminescence images of the zircons from Los Molles Formation

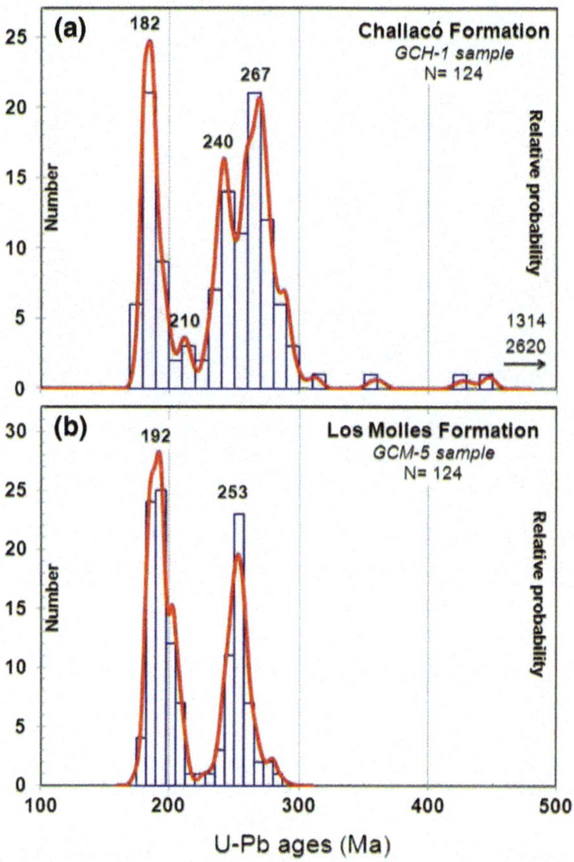

Fig. 7 Frequency histogram and relative probability plots of U-Pb ages on detrital zircons from **a** Challacó Formation and **b** Los Molles Formation

5.3 Maximum Depositional Age

The maximum depositional ages for each unit were calculated from the population of the youngest zircons analyzed (Fig. 8). The youngest significant detrital zircon age peak from sample GCM-05 (Los Molles Formation) at 192 Ma (Sinemurian) is significantly older than the biostratigraphic age of the unit. Therefore, a weighted mean age for all consecutively youngest grains whose 1σ errors overlap (N = 7) was calculated, yielding a value of 182.4 ± 2.3 Ma (Early Toarcian) (Fig. 8a), coherent with the stratigraphic position of the sample. The youngest significant peak from sample GCH-01 Challacó Formation is at ca. 182 Ma, and the weighted mean age calculated for the three youngest zircons is 177,2 ± 4,0 Ma (Toarciano) (Fig. 8b). The two values of the maximum depositional ages are older than the

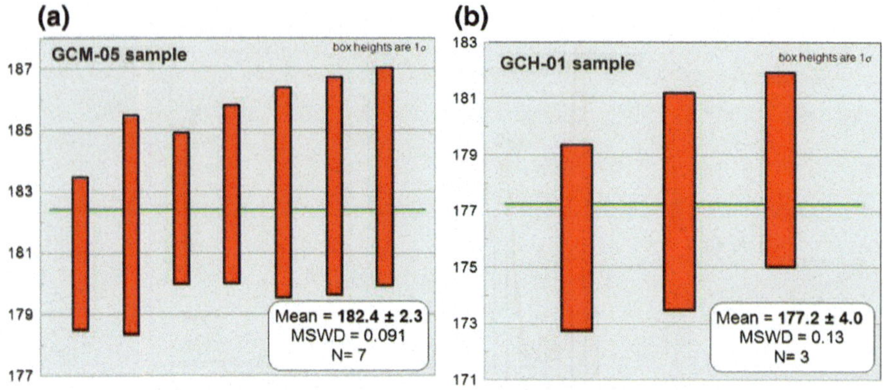

Fig. 8 The weighted mean ages calculated with the youngest zircon grains (1σ errors overlap) from the samples: **a** GCM-05 (Los Molles Formation) and **b** GCH-01 (Challacó Formation)

stratigraphic age of the Challacó Formation, classically assigned to the Bathonian based on its position above the Lajas Formation (Gulisano and Gutierrez Pleimling 1994).

5.4 *Hf Isotope Data*

In order to characterize and determinate potential source rocks of the detrital zircons grains from the Cuyo Group, Hf isotopes were analyzed for selected populations. The potential sources are related to the Choiyoi igneous province (ca. 290–240 Ma), the Precuyano volcanic rocks (ca. 220–190 Ma), and Lower Jurassic magmatic rocks (<190 Ma). A group of 15 detrital magmatic zircons of the Los Molles Formation (GCM-05) and 16 of the Challacó Formation (GCH-01) was analyzed. The most important feature is that all detrital zircons analyzed have negative εHf_t values (−0.7 to −15.5), indicating evolved magmatic sources with important crustal contribution; zircon Hf model ages of these population were calculated, have Early Neoproterozoic and Mesoproterozoic values (between 900 and 1500 Ma; see Table A.3 in the Appendix). The oldest zircon (ca. 447 Ma) has a low εHf_t of −8.9, and the Permian to Triassic ones (U-Pb ages between 283 and 236 Ma) show the negative values between −15.5 and −7.4 (with the exception of one data −0.7), suggesting that the magmatic source of these zircons have strong crustal contamination. The εHf_t values of the Late Triassic to Early Jurassic zircons (229–192 Ma) are also low (−8.1 to −2.1), and the values of the youngest zircons (186–175 Ma) are mostly in the range of −2,2 to −4,9 (with the exception of one data −9.0), also indicating a crustal origin for the magmatic source (Fig. 9).

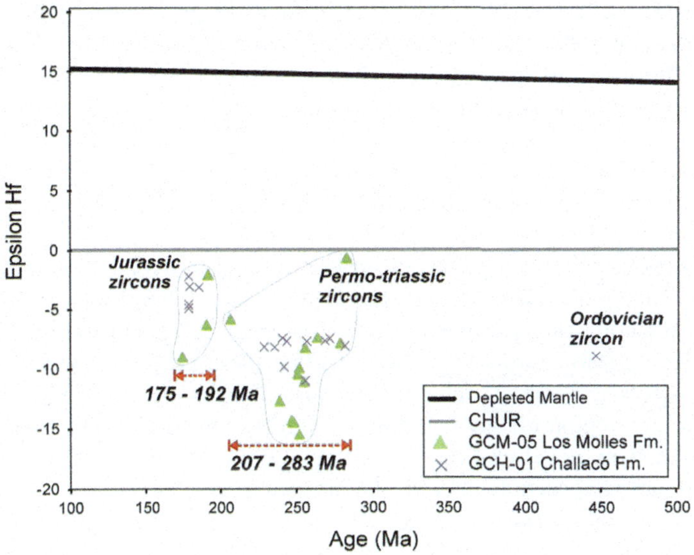

Fig. 9 Hf isotopic analysis on selected detrital zircons (samples GCM-05 and GCH-01)

6 Discussion and Conclusions

Our sandstone petrographic analysis of the Cuyo Group suggests important contributions from volcanic sources. At the base of the sequence (Los Molles and Lajas formations), andesitic lithics are dominant. Towards the upper Challacó Formation, they are progressively replaced by lithics of acidic composition. The upper part of the sequence also contains sedimentary lithic clasts and polycrystalline quartz grains which suggest a basement source. Ternary diagrams (Dickinson et al. 1983) suggest a classical evolution of the sediment source regions in a continental margin, from non-dissected to transitional and dissected volcanic fields, combined with contributions from a recycled orogen (Fig. 5). The morphology and internal structure of the detrital zircons indicate a dominant volcanic origin. U-Pb ages and Hf isotopes obtained in the analyzed zircons indicate a more complex configuration given by multiple igneous components.

According to the bimodal frequency diagram of the detrital zircon ages, obtained in the Los Molles Formation, two main sediment source regions can be distinguished. The older, with a maximum age peak at 253 Ma, overlaps several U-Pb zircon ages (TIMS and SHRIMP) obtained in igneous and metamorphic rocks from northern Patagonia (Llambías et al. 1984; Varela et al. 2005; Pankhurst et al. 2006, 2014; Naipauer and Ramos 2016). These rocks are exposed along the southern margin of the Neuquén Basin and have also been recorded in the subsurface of the basin (Schiuma and Llambías 2008). The second source was defined by the main Jurassic peak of zircon ages at ca. 192 Ma. This value has the same absolute U-Pb (SHRIMP) age at 191.7 ± 2.8 Ma obtained in tuff zircons of the Piedra del Águila

Formation (Spalletti et al. 2010). This unit crops out about 100 km south of the study area and integrates the Precuyano cycle with the Paso Flores and Sañicó formations (Fig. 1). The 1,500 m thick Sañicó Formation, composed of volcaniclastic and basaltic, andesitic, and rhyolitic volcanic rocks, is the most widespread in the area (D'Elia et al. 2015). Therefore, we can interpret that these units of the Precuyano cycle were exhumed at the time of deposition of the Los Molles Formation, acting as source rock that would explain the basic to acidic volcanic fragments recorded within the Jurassic zircon population (ca. 192 Ma).

The weighted mean age obtained for the youngest zircons at 182.4 ± 2.3 Ma indicates a maximum depositional age at the base of the Toarcian for the Los Molles Formation. This absolute age, very close to the biostratigraphic age of the unit and the crystals morphology, suggests that these zircons were delivered by a contemporaneous volcanic source at the time of sedimentation. According to its age, the source region of this volcanism could be compatible both with the intraplate magmatism developed in the extra-Andean Patagonia, e.g., the Marifil Formation and equivalents and with the arc magmatism (Fig. 1).

The detrital zircon ages of the Challacó Formation showed a multimodal frequency diagram in contrast from that defined in the Los Molles Formation. Maximum peaks are in the Permian, Triassic, and Early Jurassic, existing also isolated ages in the Early Paleozoic and Precambrian that were not present in the previous unit. Permian and Triassic detrital zircons probably come, as in the Los Molles Formation, from the North Patagonian Massif. Also, the oldest zircon ages of Early Paleozoic and Precambrian probably came from the basement rocks of the northeastern sector of the North Patagonia Massif (see Naipauer and Ramos 2016).

On the other hand, the better represented population in the Challacó Formation has detrital zircon ages around 182 Ma, similar to the youngest zircons of the Los Molles Formation. These zircons probably come from the same source rock defined by the Los Molles Formation, but not as a contemporary volcanism with the sedimentation, because the Challacó Formation is younger than 177 Ma. The zircon ages at ca. 177 Ma were interpreted as the maximum depositional age; however, the age of the Challacó Formation is much younger due to its stratigraphic position in the Bathonian. According to the crystal morphology, the source rock of these zircons was interpreted as volcanic. Due to their Early Jurassic age, it could be related either to the Andean volcanic arc or to the intraplate Chon Aike Igneous Province (Fig. 1).

The sediment source regions were determined with better precision through the Hf isotopic analysis and were a powerful tool to discriminate volcanic source rocks. Although there are only few Hf isotope data available on zircons of source rocks within the study area (Fanning et al. 2011; Hervé et al. 2013; Pankhurst et al. 2014), there are published Nd isotope analyses in the major basement rocks of Patagonia (Pankhurst et al. 2000; Lucassen et al. 2004; Rapela et al. 2005; Varela et al. 2005; Pankhurst et al. 2006; Martínez Dopico et al. 2011). Therefore, the εNdt parameters were also used to compare with the εHf_t values determinate in these works. The low values of εHf_t (between −15.5 and −0.7) analyzed in Permian and Triassic detrital zircons indicate a typical evolved source with strong crustal contribution. These data are in agreement with the negative values of εNd_t calculated in Permian granitoids from the North Patagonian Massif (Varela et al. 2005; Martinez Dopico et al. 2011), as well as for several Permian granitoids of Fanning et al. (2011), but the values are slightly higher than the values obtained in the detrital zircons of this work. However, εHf_t values between −11.4 and −7.0 from several igneous zircons from granite dated at 249 ± 2 Ma (U-Pb SHRIMP) were recently reported by Pankhurst et al. (2014). The values are equivalent to those determined in the detrital zircons from the Cuyo Group and support the interpretation that the source region of these zircons is related to the Gondwanic magmatism of the north of Patagonia.

The main objective of the provenance analysis of this work was determinate the source region of the volcanic material and the detrital zircons dated between 182 and 177 Ma. This range of ages is compatible with both the Andean volcanic arc and the Chon Aike Igneous Province. However, as the geochemical and isotopic compositions of the Lower Jurassic magmatic arc (Rapela et al. 2005; Parada et al. 2007; Vásquez et al. 2011) differ from the Jurassic Chon Aike Igneous Province (Pankhurst et al. 2000; Rapela et al. 2005), Hf isotopes were a powerful tool to discriminate this magmatic source (Fig. 1). The negative εHf_t values (about −4) of the analyzed grains indicated a crustal origin for the Jurassic zircons and are clearly compatible with the isotopic composition of Chon Aike Igneous Province, mainly with the Marifil Formation. A sedimentary and volcanic source region from the Andean magmatic arc is dismissed; isotope characteristics of the Subcordillerano batholith and the Cordillera de la Costa batholith have values of εNdt close to 0 (between +2.8 and −1.5) indicating an origin from a mantle source (Figs. 1 and 8). In addition, the high content of volcanic fragments of acidic composition determined in the petrography supports that the Marifil Formation supplied the volcanic materials. Therefore, the importance of Chon Aike volcanism as an area of sedimentary and volcanic contribution in the Neuquén Basin is highlighted.

On the other hand, the detrital zircon ages at ca. 182 Ma found in this study are similar to the U-Pb (TIMS) absolute ages between 182 and 186 Ma obtained in tuff zircons from the Chachil Limestone and Puesto Araya Formation (Mazzini et al. 2010; Leanza et al. 2013; Armella et al. 2016). The tuffs interlayered in these units of the base of the Cuyo Group are classically interpreted as coming from the Andean volcanic arc. However, these tuffs could have come from the Chon Aike volcanism as it was demonstrated by the Hf isotopes of the detrital zircons from Los Molles and Challacó formations.

The variations in the provenance patterns described in this work, between the base and the top of the Cuyo Group, allow us to make some final observations about the paleogeographic evolution of the Neuquén Basin. We interpret for Los Molles Formation, a source region formed by the combination of local basement uplifts composed of Permian igneous rocks (Choiyoi magmatism) and volcanic rocks from the Precuyano cycle. The relief and paleogeography of the basin were inherited from the rift setting, which agree with the seismic data of the Huincul High, where several authors identified normal faults affecting the sediments of Los Molles Formation (Silvestro and Zubiri 2008; Pángaro et al. 2009; Mosquera et al. 2011). An important change in the provenance pattern of the Cuyo Group was observed in the Challacó Formation. Lower Jurassic volcanic rocks and Lower Paleozoic to Precambrian basement rocks are the main sediment source regions, and this fact suggests the integration of more distal and ancient sources, linked with the North Patagonian Massif. In addition, the source rocks related to the magmatism of the Precuyano cycle decrease in proportion, possibly because they were eroded and/or buried by the retroarc sedimentation.

Acknowledgements Maximiliano Naipauer acknowledges the financial support of CONICET and ANPCyT PICT-2010–2099, Argentina. This is the contribution **R-235** of the Instituto de Estudios Andinos "Don Pablo Groeber" (CONICET-UBA).

Appendix

Table A.1 Petrographic data

Sample	Qtz	QtzP	Plg	FK	MTX/CTO	Lit Sed	Lvb	Lva	Mu	Mi	Total points
Los Molles Formation											
QCM1	91	11	44	11	74	41	14	7	4	3	300
QCM2	22	14	34	5	43	16	108	55	1	2	300
QCM3	31	7	71	13	46	8	83	26	14	1	300
QCM5	44	7	56	10	58	23	77	22	2	1	300
Lajas Formation											
QCL1	115	23	53	23	40	25	9	4	5	3	300
QCL2	101	22	55	21	42	22	20	10	5	2	300
QCL3	112	8	67	15	51	35	4	6	1	1	300
QCL4	50	17	74	21	57	52	17	4	3	5	300
QCL5	68	23	67	12	46	23	35	8	13	5	300
QCL6	95	16	80	15	47	28	10	5	1	3	300
QCL7	69	15	101	22	51	31	6	3	1	1	300
Challacó Formation											
QCH1	61	28	23	7	47	32	16	81	2	3	300

Qtz: quartz grains; QtzP: polycrystalline quartz; Plg: plagioclase; FK: K feldespar; MTX/CTO: matrix/cement

Lvb: basic volcanic lithic; Lva: acidic volcanic lithic; Mu: muscovite; Mi: microcline

Table A.2 U-Pb data

Sample	U	Th	238U	1 sigma	207Pb	1 sigma	206/238	1 sigma	207/206	1 sigma	Best	1 sigma
Name	ppm	U	206Pb	% error	206Pb	% error	age	abs err	age	abs err	age	abs err Ma
Los Molles Formation (sample GCM-05)												
GCM5-124	482	1.12	25.064	1.89%	0.0524	1.03%	252.2	4.7	303.5	23.3	252.2	4.7
GCM5-123	496	1.19	25.122	2.07%	0.0552	1.42%	251.6	5.1	422.3	31.4	251.6	5.1
GCM5-122	401	1.34	26.116	2.21%	0.0542	1.68%	242.2	5.3	378.6	37.4	242.2	5.3
GCM5-121	470	1.20	25.735	1.98%	0.0516	1.04%	245.8	4.8	269.8	23.7	245.8	4.8
GCM5-120	550	1.58	25.017	1.90%	0.053	1.06%	252.7	4.7	327.7	23.8	252.7	4.7
GCM5-119	504	1.10	23.844	1.88%	0.0523	1.05%	264.8	4.9	297.6	23.8	264.8	4.9
GCM5-118	683	1.70	33.132	1.93%	0.0508	1.20%	191.7	3.6	233.8	27.4	191.7	3.6
GCM5-117	212	2.67	31.790	1.98%	0.0507	1.61%	199.7	3.9	228.2	36.7	199.7	3.9
GCM5-116	741	0.79	24.934	1.84%	0.0534	0.93%	253.5	4.6	345.0	21.0	253.5	4.6
GCM5-115	903	0.58	26.460	2.07%	0.0509	1.19%	239.1	4.9	236.2	27.3	239.1	4.9
GCM5-114	1151	0.82	25.066	1.89%	0.0519	0.85%	252.2	4.7	282.4	19.4	252.2	4.7
GCM5-113	977	1.24	25.420	1.88%	0.0509	0.89%	248.7	4.6	235.6	20.4	248.7	4.6
GCM5-112	250	0.68	34.937	1.97%	0.0499	1.54%	181.9	3.5	192.5	35.4	181.9	3.5
GCM5-111	466	0.26	36.306	1.91%	0.049	1.09%	175.2	3.3	149.3	25.4	175.2	3.3
GCM5-110	387	1.00	25.093	1.87%	0.0518	1.04%	251.9	4.6	275.2	23.5	251.9	4.6
GCM5-109	523	2.54	25.605	1.90%	0.0524	1.05%	247.0	4.6	303.3	23.8	247.0	4.6
GCM5-108	350	1.13	24.783	1.91%	0.0572	1.15%	255.0	4.8	500.7	25.1	255.0	4.8
GCM5-107	565	0.66	33.012	1.95%	0.0532	1.13%	192.4	3.7	337.5	25.3	192.4	3.7
GCM5-106	531	0.96	22.209	2.35%	0.0556	1.21%	283.9	6.5	437.5	26.7	283.9	6.5
GCM5-105	1212	0.28	25.009	1.88%	0.0517	0.83%	252.7	4.7	273.5	18.9	252.7	4.7
GCM5-104	1391	1.26	24.597	1.88%	0.0517	0.78%	256.9	4.7	272.2	17.9	256.9	4.7

(continued)

Table A.2 (continued)

Sample	U	Th	238U	1 sigma	207Pb	1 sigma	206/238	1 sigma	207/206	1 sigma	Best	1 sigma
Name	ppm	U	206Pb	% error	206Pb	% error	age	abs err	age	abs err	age	abs err Ma
GCM5-103	925	1.40	25.162	2.00%	0.0574	1.10%	251.2	4.9	505.8	24.1	251.2	4.9
GCM5-102	86	1.17	30.550	2.14%	0.0504	2.46%	207.6	4.4	212.7	56.0	207.6	4.4
GCM5-101	608	0.27	22.593	1.90%	0.0524	0.96%	279.2	5.2	304.5	21.7	279.2	5.2
GCM5_100	205	1.56	33.422	2.29%	0.0503	2.87%	190.0	4.3	209.4	65.3	190.0	4.3
GCM5_99	286	0.81	32.893	1.28%	0.0500	1.81%	193.1	2.4	197.0	41.5	193.1	2.4
GCM5_98	181	2.25	30.278	1.36%	0.0535	1.93%	209.5	2.8	349.1	43.0	209.5	2.8
GCM5_97	253	1.98	31.317	1.27%	0.0518	1.85%	202.6	2.5	278.2	41.8	202.6	2.5
GCM5_96	1022	1.28	26.394	1.25%	0.0535	2.75%	239.7	2.9	351.9	60.9	239.7	2.9
GCM5_95	240	2.79	25.270	1.25%	0.0505	1.87%	250.2	3.1	216.5	42.7	250.2	3.1
GCM5_94	543	1.66	31.422	1.23%	0.0498	1.73%	202.0	2.4	188.0	39.7	202.0	2.4
GCM5_93	272	0.72	32.981	1.30%	0.0499	1.87%	192.6	2.5	192.2	42.9	192.6	2.5
GCM5_92	389	1.69	24.774	1.30%	0.0515	1.65%	255.1	3.2	262.5	37.5	255.1	3.2
GCM5_91	177	1.55	34.324	1.41%	0.0520	2.02%	185.1	2.6	287.0	45.5	185.1	2.6
GCM5_90	441	1.96	33.601	1.31%	0.0508	1.74%	189.1	2.4	229.7	39.6	189.1	2.4
GCM5_89	465	1.11	33.997	1.24%	0.0501	1.68%	186.9	2.3	201.2	38.7	186.9	2.3
GCM5_88	425	0.43	32.641	1.27%	0.0498	1.76%	194.5	2.4	184.2	40.5	194.5	2.4
GCM5_87	262	0.88	34.362	1.61%	0.0500	1.97%	184.9	2.9	194.3	45.1	184.9	2.9
GCM5_86	153	1.97	24.965	1.34%	0.0512	2.03%	253.2	3.3	248.1	46.0	253.2	3.3
GCM5_85	557	1.50	33.531	1.25%	0.0526	1.70%	189.4	2.3	313.7	38.2	189.4	2.3
GCM5_84	357	0.62	34.362	1.27%	0.0514	1.78%	184.9	2.3	260.3	40.3	184.9	2.3
GCM5_83	71	1.19	25.661	1.53%	0.0509	2.28%	246.4	3.7	238.2	51.7	246.4	3.7
GCM5_82	378	1.45	33.057	1.22%	0.0500	1.72%	192.1	2.3	196.9	39.4	192.1	2.3

(continued)

Table A.2 (continued)

Sample	U	Th	238U	1 sigma	207Pb	1 sigma	206/238	1 sigma	207/206	1 sigma	Best	1 sigma
Name	ppm	U	206Pb	% error	206Pb	% error	age	abs err	age	abs err	age	abs err Ma
GCM5_81	806	1.10	31.636	1.20%	0.0525	1.69%	200.6	2.4	305.5	38.1	200.6	2.4
GCM5_80	412	0.77	34.330	1.21%	0.0503	1.79%	185.1	2.2	207.5	41.0	185.1	2.2
GCM5_79	159	1.20	32.913	1.45%	0.0497	2.12%	192.9	2.8	182.4	48.7	192.9	2.8
GCM5_78	233	0.96	35.118	1.38%	0.0507	1.99%	181.0	2.5	227.2	45.3	181.0	2.5
GCM5_77	459	1.31	26.080	1.26%	0.0538	1.69%	242.6	3.0	361.2	37.6	242.6	3.0
GCM5_76	181	1.77	34.737	1.88%	0.0492	2.71%	183.0	3.4	158.9	62.2	183.0	3.4
GCM5_75	267	2.01	30.509	1.60%	0.0506	2.07%	207.9	3.3	223.1	47.3	207.9	3.3
GCM5_74	358	1.23	24.562	1.18%	0.0513	1.71%	257.3	3.0	255.3	38.8	257.3	3.0
GCM5_73	300	2.91	34.833	1.35%	0.0498	1.75%	182.5	2.4	183.6	40.2	182.5	2.4
GCM5_72	472	0.68	22.596	1.18%	0.0526	1.47%	279.2	3.2	313.4	33.1	279.2	3.2
GCM5_71	1010	0.86	33.527	1.19%	0.0543	1.41%	189.5	2.2	385.5	31.4	189.5	2.2
GCM5_70	196	1.14	24.450	1.31%	0.0518	1.67%	258.4	3.3	276.5	37.7	258.4	3.3
GCM5_69	171	1.71	25.181	1.32%	0.0554	1.83%	251.0	3.2	428.1	40.3	251.0	3.2
GCM5_68	190	0.95	33.992	1.31%	0.0499	1.91%	186.9	2.4	191.5	43.7	186.9	2.4
GCM5_67	222	0.68	32.557	1.26%	0.0491	1.77%	195.0	2.4	151.6	40.8	195.0	2.4
GCM5_66	57	2.39	28.009	1.91%	0.0551	2.99%	226.1	4.3	416.1	65.4	226.1	4.3
GCM5_65	1190	0.44	32.539	1.21%	0.0512	1.43%	195.1	2.3	248.9	32.5	195.1	2.3
GCM5_64	337	1.36	24.645	1.27%	0.0510	1.57%	256.4	3.2	240.7	35.8	256.4	3.2
GCM5_63	257	1.10	31.081	1.36%	0.0488	1.73%	204.1	2.7	135.9	40.2	204.1	2.7
GCM5_62	255	1.67	31.393	1.33%	0.0520	1.86%	202.1	2.6	284.9	42.1	202.1	2.6
GCM5_61	458	0.96	34.621	1.31%	0.0510	1.50%	183.6	2.4	241.5	34.3	183.6	2.4
GCM5_60	96	1.64	25.856	1.52%	0.0523	2.18%	244.6	3.6	299.9	49.0	244.6	3.6
GCM5_59	240	1.94	32.421	1.31%	0.0496	1.72%	195.8	2.5	174.4	39.7	195.8	2.5

(continued)

Table A.2 (continued)

Sample	U	Th	238U	1 sigma	207Pb	1 sigma	206/238	1 sigma	207/206	1 sigma	Best	1 sigma
Name	ppm	U	206Pb	% error	206Pb	% error	age	abs err	age	abs err	age	abs err Ma
GCM5_58	323	2.86	31.312	1.28%	0.0529	1.65%	202.7	2.6	324.5	37.0	202.7	2.6
GCM5_57	222	1.52	31.038	1.51%	0.0521	2.75%	204.4	3.0	288.9	61.5	204.4	3.0
GCM5_56	148	1.77	29.704	1.48%	0.0549	1.91%	213.4	3.1	410.0	42.1	213.4	3.1
GCM5_55	257	0.99	32.855	1.61%	0.0498	1.75%	193.3	3.1	184.6	40.3	193.3	3.1
GCM5_54	87	2.34	30.077	2.02%	0.0509	2.59%	210.8	4.2	234.6	58.7	210.8	4.2
GCM5_53	154	2.03	24.774	1.68%	0.0526	1.82%	255.1	4.2	312.8	40.9	255.1	4.2
GCM5_52	160	1.23	24.371	1.65%	0.0519	1.73%	259.2	4.2	282.1	39.2	259.2	4.2
GCM5_51	297	0.42	24.000	1.72%	0.0527	1.71%	263.2	4.4	313.8	38.4	263.2	4.4
GCM5_50	415	1.60	30.852	1.54%	0.0508	1.56%	205.6	3.1	230.4	35.7	205.6	3.1
GCM5_49	503	0.60	33.093	1.62%	0.0527	1.56%	191.9	3.1	315.5	35.2	191.9	3.1
GCM5_48	509	1.36	25.950	1.54%	0.0513	1.47%	243.7	3.7	255.1	33.4	243.7	3.7
GCM5_47	454	1.54	24.619	1.51%	0.0521	1.50%	256.7	3.8	289.2	33.8	256.7	3.8
GCM5_46	417	1.43	23.900	1.50%	0.0508	1.49%	264.2	3.9	232.9	34.1	264.2	3.9
GCM5_45	196	1.57	33.209	1.72%	0.0498	1.74%	191.3	3.2	183.8	40.1	191.3	3.2
GCM5_44	277	1.15	34.745	1.59%	0.0495	1.82%	182.9	2.9	171.8	42.0	182.9	2.9
GCM5_43	390	1.45	25.366	1.52%	0.0532	1.51%	249.3	3.7	338.5	33.7	249.3	3.7
GCM5_42	201	1.76	30.618	1.60%	0.0519	1.76%	207.2	3.3	281.3	39.7	207.2	3.3
GCM5_41	164	1.29	33.386	2.25%	0.0498	3.06%	190.3	4.2	184.6	69.7	190.3	4.2
GCM5_40	399	0.66	33.082	1.56%	0.0503	1.57%	192.0	2.9	210.3	35.9	192.0	2.9
GCM5_39	165	1.44	31.975	1.78%	0.0494	1.98%	198.5	3.5	164.9	45.7	198.5	3.5
GCM5_38	427	1.11	25.176	1.61%	0.0508	1.61%	251.1	4.0	229.5	36.7	251.1	4.0
GCM5_37	553	0.82	25.185	1.54%	0.0510	1.46%	251.0	3.8	243.1	33.3	251.0	3.8
GCM5_36	54	1.13	33.626	1.88%	0.0484	2.91%	188.9	3.5	120.6	67.2	188.9	3.5

(continued)

Table A.2 (continued)

Sample	U	Th	238U	1 sigma	207Pb	1 sigma	206/238	1 sigma	207/206	1 sigma	Best	1 sigma
Name	ppm	U	206Pb	% error	206Pb	% error	age	abs err	age	abs err	age	abs err Ma
GCM5_35	142	2.07	34.568	2.15%	0.0485	1.89%	183.8	3.9	125.5	44.0	183.8	3.9
GCM5_34	172	1.33	24.911	2.08%	0.0511	1.46%	253.7	5.2	244.7	33.2	253.7	5.2
GCM5_33	425	2.01	31.558	2.01%	0.0527	1.16%	201.1	4.0	314.0	26.2	201.1	4.0
GCM5_32	144	2.19	34.444	2.10%	0.0510	1.80%	184.5	3.8	240.8	41.0	184.5	3.8
GCM5_31	379	0.59	32.931	2.12%	0.0506	1.28%	192.8	4.0	221.0	29.2	192.8	4.0
GCM5_30	863	1.05	25.100	1.96%	0.0521	1.02%	251.8	4.8	287.7	23.3	251.8	4.8
GCM5_29	450	0.43	32.630	1.97%	0.0499	1.19%	194.6	3.8	188.7	27.4	194.6	3.8
GCM5_28	746	0.22	30.574	2.00%	0.0526	1.05%	207.5	4.1	312.9	23.6	207.5	4.1
GCM5_27	471	0.32	23.577	2.01%	0.0510	1.09%	267.8	5.3	240.7	24.9	267.8	5.3
GCM5_26	455	0.56	25.895	2.12%	0.0512	1.10%	244.3	5.1	249.5	25.2	244.3	5.1
GCM5_25	110	1.25	33.335	2.26%	0.0580	1.82%	190.5	4.2	530.3	39.3	190.5	4.2
GCM5_24	743	1.30	25.910	2.15%	0.0572	1.21%	244.1	5.2	498.4	26.5	244.1	5.2
GCM5_23	168	1.80	34.181	2.11%	0.0511	1.49%	185.9	3.9	246.8	33.9	185.9	3.9
GCM5_22	195	0.74	33.888	2.07%	0.0491	1.63%	187.5	3.8	152.4	37.8	187.5	3.8
GCM5_21	338	0.81	34.125	2.04%	0.0502	1.38%	186.2	3.7	203.8	31.8	186.2	3.7
GCM5_20	94	1.46	23.860	2.13%	0.0512	1.94%	264.7	5.5	248.5	44.1	264.7	5.5
GCM5_19	216	1.43	32.033	2.17%	0.0497	1.57%	198.2	4.2	182.3	36.2	198.2	4.2
GCM5_18	251	1.77	31.043	1.96%	0.0517	1.26%	204.4	3.9	274.2	28.6	204.4	3.9
GCM5_17	51	0.79	34.496	2.39%	0.0489	2.39%	184.2	4.3	145.2	55.1	184.2	4.3
GCM5_16	266	0.93	32.842	2.11%	0.0488	1.29%	193.4	4.0	137.5	29.9	193.4	4.0
GCM5_15	526	1.25	33.656	1.98%	0.0494	0.87%	188.7	3.7	165.3	20.3	188.7	3.7
GCM5_14	499	0.36	34.331	1.92%	0.0496	0.86%	185.1	3.5	178.5	20.0	185.1	3.5
GCM5_13	547	0.97	27.043	2.02%	0.0508	0.98%	234.1	4.6	233.4	22.4	234.1	4.6

(continued)

Table A.2 (continued)

Sample	U	Th	238U	1 sigma	207Pb	1 sigma	206/238	1 sigma	207/206	1 sigma	Best	1 sigma
Name	ppm	U	206Pb	% error	206Pb	% error	age	abs err	age	abs err	age	abs err Ma
GCM5_12	514	1.14	25.633	1.91%	0.0506	0.75%	246.7	4.6	223.5	17.2	246.7	4.6
GCM5_11	160	0.59	24.059	2.10%	0.0536	0.99%	262.5	5.4	353.7	22.2	262.5	5.4
GCM5_10	334	1.20	33.977	1.95%	0.0504	1.01%	187.0	3.6	215.5	23.3	187.0	3.6
GCM5_9	136	0.94	32.328	2.09%	0.0488	1.65%	196.4	4.0	138.9	38.4	196.4	4.0
GCM5_8	462	2.00	34.693	1.94%	0.0490	0.92%	183.2	3.5	146.3	21.3	183.2	3.5
GCM5_7	887	0.87	23.445	1.89%	0.0514	0.59%	269.3	5.0	258.4	13.6	269.3	5.0
GCM5_6	467	0.74	34.637	1.94%	0.0494	0.87%	183.5	3.5	165.4	20.1	183.5	3.5
GCM5_5	643	0.52	24.601	1.90%	0.0521	0.57%	256.9	4.8	289.3	13.1	256.9	4.8
GCM5_4	174	2.33	32.272	2.00%	0.0501	1.27%	196.7	3.9	197.4	29.2	196.7	3.9
GCM5_3	604	0.59	33.112	2.01%	0.0507	0.88%	191.8	3.8	226.2	20.1	191.8	3.8
GCM5_2	244	0.79	32.950	1.97%	0.0501	1.25%	192.7	3.7	199.3	28.7	192.7	3.7
GCM5_1	261	0.63	32.153	1.95%	0.0502	1.14%	197.4	3.8	202.8	26.1	197.4	3.8
Challacó Formation (sample GCH-01)												
GCH1_128	283.2	1.42	22.123	1.91%	0.0561	1.14%	285.0	5.3	455.6	25.2	285.0	5.3
GCH1_127	208.4	0.41	23.983	1.95%	0.0527	1.36%	263.3	5.0	317.2	30.7	263.3	5.0
GCH1_126	103.6	0.82	23.613	2.15%	0.0608	1.87%	267.4	5.6	632.8	39.8	267.4	5.6
GCH1_125	541.3	1.16	35.014	2.03%	0.0583	1.59%	181.5	3.6	541.8	34.4	181.5	3.6
GCH1_124	97.23	1.15	26.662	2.06%	0.0503	1.93%	237.4	4.8	208.0	44.1	237.4	4.8
GCH1_123	1138	0.61	34.717	1.86%	0.0513	0.90%	183.1	3.4	256.3	20.6	183.1	3.4
GCH1_122	428.2	0.90	34.411	1.89%	0.0508	1.16%	184.7	3.4	232.2	26.6	184.7	3.4
GCH1_121	51.39	1.65	26.137	2.68%	0.0563	2.88%	242.0	6.4	463.4	62.7	242.0	6.4
GCH1_120	407.3	0.81	23.997	2.10%	0.0521	1.20%	263.2	5.4	287.8	27.1	263.2	5.4

(continued)

Table A.2 (continued)

Sample	U	Th	238U	1 sigma	207Pb	1 sigma	206/238	1 sigma	207/206	1 sigma	Best	1 sigma
Name	ppm	U	206Pb	% error	206Pb	% error	age	abs err	age	abs err	age	abs err Ma
GCH1_119	1372	104.8	31.572	2.10%	0.0551	0.92%	201.0	4.1	414.6	20.5	201.0	4.1
GCH1_118	531.2	0.72	26.891	2.09%	0.0518	1.18%	235.4	4.8	277.0	26.9	235.4	4.8
GCH1_117	524.1	0.88	24.030	2.09%	0.0517	1.08%	262.8	5.4	273.3	24.5	262.8	5.4
GCH1_116	124.4	0.71	25.514	2.20%	0.0547	1.88%	247.8	5.4	398.1	41.6	247.8	5.4
GCH1_115	407.6	0.67	23.649	2.09%	0.0542	1.11%	267.0	5.5	380.3	24.8	267.0	5.5
GCH1_114	395.3	0.33	23.294	2.08%	0.0521	1.17%	271.0	5.5	291.3	26.5	271.0	5.5
GCH1_113	563.7	0.93	35.853	2.18%	0.0599	1.36%	177.3	3.8	600.4	29.3	177.3	3.8
GCH1_112	143.2	0.61	23.826	2.18%	0.0528	1.71%	265.0	5.7	320.5	38.5	265.0	5.7
GCH1_111	377.2	0.70	23.666	2.07%	0.0524	1.18%	266.8	5.4	302.1	26.6	266.8	5.4
GCH1_110	388.9	1.00	23.622	2.11%	0.052	1.17%	267.3	5.5	286.6	26.6	267.3	5.5
GCH1_109	1190	0.31	24.060	2.06%	0.0516	0.91%	262.5	5.3	268.8	20.6	262.5	5.3
GCH1_108	343	0.88	34.392	2.13%	0.0608	2.40%	184.8	3.9	633.1	50.9	184.8	3.9
GCH1_107	792.4	0.40	23.476	2.08%	0.0523	0.96%	268.9	5.5	299.7	21.7	268.9	5.5
GCH1_106	4752	34.20	30.266	2.07%	0.053	0.79%	209.6	4.3	330.4	17.9	209.6	4.3
GCH1_104	1228	0.83	34.336	2.07%	0.0543	1.20%	185.1	3.8	383.9	26.6	185.1	3.8
GCH1_103	213.1	0.73	22.289	2.11%	0.0506	1.46%	282.9	5.8	220.7	33.3	282.9	5.8
GCH1_102	392.9	0.47	25.202	2.09%	0.052	1.17%	250.8	5.2	284.4	26.5	250.8	5.2
GCH1_101	371.2	0.40	23.097	2.10%	0.0523	1.33%	273.2	5.6	297.9	30.0	273.2	5.6
GCH1_100	159.3	0.96	25.949	2.25%	0.05	1.56%	243.8	5.4	194.5	35.8	243.8	5.4
GCH1_99	418.7	0.34	22.704	2.09%	0.0521	1.12%	277.9	5.7	289.6	25.5	277.9	5.7
GCH1_98	275.1	1.23	32.729	2.14%	0.053	1.50%	194.0	4.1	330.3	33.6	194.0	4.1
GCH1_97	245.2	0.70	21.702	2.18%	0.0504	1.41%	290.4	6.2	214.9	32.3	290.4	6.2
GCH1_96	82.36	1.24	26.048	2.24%	0.0532	2.79%	242.8	5.3	337.8	62.0	242.8	5.3

(continued)

Table A.2 (continued)

Sample	U	Th	238U	1 sigma	207Pb	1 sigma	206/238	1 sigma	207/206	1 sigma	Best	1 sigma
Name	ppm	U	206Pb	% error	206Pb	% error	age	abs err	age	abs err	age	abs err Ma
GCH1_95	802.1	0.83	33.332	2.08%	0.0502	1.07%	190.6	3.9	205.0	24.7	190.6	3.9
GCH1_94	407.2	0.93	32.954	2.32%	0.0571	1.63%	192.7	4.4	494.6	35.5	192.7	4.4
GCH1_93	1249	0.23	22.743	2.06%	0.0583	0.99%	277.4	5.6	541.4	21.6	277.4	5.6
GCH1_92	613.1	0.96	26.662	2.12%	0.0516	1.06%	237.4	4.9	269.6	24.0	237.4	4.9
GCH1_91	584.6	0.61	32.619	2.11%	0.052	1.11%	194.7	4.0	284.9	25.2	194.7	4.0
GCH1_90	325.3	1.04	29.713	2.10%	0.0511	1.30%	213.4	4.4	243.8	29.8	213.4	4.4
GCH1_89	635.6	0.45	23.293	2.09%	0.053	0.98%	271.0	5.5	330.3	22.0	271.0	5.5
GCH1_88	949.9	1.52	33.372	2.08%	0.0541	1.08%	190.3	3.9	373.6	24.0	190.3	3.9
GCH1_87	605.5	0.30	21.734	2.08%	0.0522	0.97%	290.0	5.9	294.6	22.1	290.0	5.9
GCH1_86	1090	0.11	22.902	1.86%	0.0568	1.39%	275.5	5.0	482.8	30.4	275.5	5.0
GCH1_85	1011	0.80	33.170	1.82%	0.0511	1.01%	191.5	3.4	244.4	23.1	191.5	3.4
GCH1_84	955.3	1.08	34.287	1.78%	0.0504	1.08%	185.3	3.3	212.4	24.9	185.3	3.3
GCH1_83	461.6	0.54	26.198	1.82%	0.0502	1.16%	241.5	4.3	205.2	26.8	241.5	4.3
GCH1_82	290.2	0.55	22.901	1.81%	0.051	1.38%	275.5	4.9	242.8	31.4	275.5	4.9
GCH1_81	1261	0.37	23.844	2.13%	0.054	1.32%	264.8	5.5	370.7	29.4	264.8	5.5
GCH1_80	482.8	0.80	23.036	1.82%	0.0522	1.14%	273.9	4.9	294.0	25.8	273.9	4.9
GCH1_79	801.6	0.52	32.165	1.78%	0.0579	1.44%	197.4	3.5	524.6	31.3	197.4	3.5
GCH1_78	106.2	1.19	26.199	1.94%	0.0546	1.95%	241.5	4.6	394.1	43.1	241.5	4.6
GCH1_76	122.3	1.46	23.018	1.95%	0.0595	1.78%	274.1	5.2	584.6	38.2	274.1	5.2
GCH1_75	589.9	0.52	26.000	1.80%	0.051	1.09%	243.3	4.3	241.2	25.0	243.3	4.3
GCH1_74	314.3	0.64	24.330	1.82%	0.053	1.44%	259.7	4.6	330.3	32.4	259.7	4.6
GCH1_73	1397	0.42	5.669	1.79%	0.0849	0.90%	1047.2	17.3	1314.0	17.4	1314.0	17.4

(continued)

Table A.2 (continued)

Sample	U	Th	238U	1 sigma	207Pb	1 sigma	206/238	1 sigma	207/206	1 sigma	Best	1 sigma
Name	ppm	U	206Pb	% error	206Pb	% error	age	abs err	age	abs err	age	abs err Ma
GCH1_72	529.1	0.56	24.108	1.80%	0.0518	1.10%	262.0	4.6	276.0	24.9	262.0	4.6
GCH1_71	516.4	0.93	34.945	1.99%	0.0525	1.53%	181.9	3.6	307.5	34.4	181.9	3.6
GCH1_70	202	0.82	11.293	2.56%	0.0631	1.49%	547.0	13.4	710.9	31.3	547.0	13.4
GCH1_69	808	1.55	34.558	1.83%	0.059	1.31%	183.9	3.3	565.6	28.3	183.9	3.3
GCH1_68	751.1	0.62	33.962	1.87%	0.0505	1.19%	187.1	3.4	219.3	27.3	187.1	3.4
GCH1_67	1229	0.19	25.345	2.19%	0.052	1.36%	249.5	5.4	283.7	30.9	249.5	5.4
GCH1_66	301	1.65	29.971	1.85%	0.0505	1.44%	211.6	3.9	216.9	33.0	211.6	3.9
GCH1_65	791.6	0.68	23.493	1.79%	0.0525	1.04%	268.7	4.7	308.0	23.5	268.7	4.7
GCH1_64	282.9	1.13	33.846	1.88%	0.0587	1.67%	187.7	3.5	557.0	36.1	187.7	3.5
GCH1_63	1001	0.43	22.932	1.80%	0.0517	1.00%	275.2	4.9	274.2	22.7	275.2	4.9
GCH1_62	426.9	0.88	23.488	1.81%	0.0526	1.13%	268.8	4.8	312.2	25.5	268.8	4.8
GCH1_61	425.8	0.76	34.191	1.81%	0.0498	1.29%	185.8	3.3	187.1	29.8	185.8	3.3
GCH1_60	303.1	0.74	23.755	1.84%	0.0531	1.30%	265.8	4.8	331.4	29.1	265.8	4.8
GCH1_59	1126	0.48	25.072	1.83%	0.0524	1.16%	252.1	4.5	303.0	26.2	252.1	4.5
GCH1_58	1344	0.79	24.814	1.76%	0.0523	0.96%	254.7	4.4	297.9	21.8	254.7	4.4
GCH1_57	871.4	0.25	14.564	1.80%	0.0604	0.91%	428.1	7.4	616.3	19.5	428.1	7.4
GCH1_56	1235	0.08	20.179	1.78%	0.0531	0.93%	311.8	5.4	334.5	21.0	311.8	5.4
GCH1_55	283	0.95	32.644	1.93%	0.0504	1.33%	194.5	3.7	214.6	30.6	194.5	3.7
GCH1_54	333.7	0.67	34.077	1.94%	0.0533	1.24%	186.5	3.6	342.2	27.8	186.5	3.6
GCH1_53	469.8	0.38	35.626	1.94%	0.051	1.09%	178.5	3.4	243.0	25.0	178.5	3.4
GCH1_52	702.9	0.63	21.600	2.04%	0.06	1.13%	291.7	5.8	602.1	24.2	291.7	5.8
GCH1_51	134.5	1.39	27.656	2.01%	0.0553	1.45%	229.0	4.5	424.5	32.1	229.0	4.5
GCH1_50	150.1	0.81	21.410	2.01%	0.0524	1.40%	294.3	5.8	301.9	31.6	294.3	5.8

(continued)

Table A.2 (continued)

Sample	U	Th	238U	1 sigma	207Pb	1 sigma	206/238	1 sigma	207/206	1 sigma	Best	1 sigma
Name	ppm	U	206Pb	% error	206Pb	% error	age	abs err	age	abs err	age	abs err Ma
GCH1_49	573.1	1.01	33.911	1.92%	0.0531	1.10%	187.4	3.5	333.2	24.7	187.4	3.5
GCH1_48	280	0.60	26.973	1.92%	0.0517	1.26%	234.7	4.4	271.2	28.6	234.7	4.4
GCH1_47	842.4	0.27	17.445	1.87%	0.0542	0.87%	359.3	6.5	379.8	19.5	359.3	6.5
GCH1_46	273.3	0.44	34.272	1.94%	0.05	1.22%	185.4	3.5	194.3	28.2	185.4	3.5
GCH1_44	61.14	1.23	26.996	2.21%	0.0514	2.48%	234.5	5.1	259.4	56.0	234.5	5.1
GCH1_43	357.9	0.71	23.392	1.89%	0.0526	1.13%	269.9	5.0	313.0	25.6	269.9	5.0
GCH1_42	247.3	0.78	24.093	2.02%	0.054	1.33%	262.2	5.2	370.5	29.7	262.2	5.2
GCH1_41	186.9	1.24	34.079	1.97%	0.0498	1.60%	186.4	3.6	184.5	36.8	186.4	3.6
GCH1_40	65.18	0.94	29.305	2.55%	0.0563	3.31%	216.3	5.4	464.6	71.7	216.3	5.4
GCH1_39	157.9	0.81	34.971	1.99%	0.056	1.51%	181.7	3.6	453.4	33.2	181.7	3.6
GCH1_38	1236	0.39	26.087	1.85%	0.0517	0.88%	242.5	4.4	272.9	20.0	242.5	4.4
GCH1_37	452.6	0.62	34.880	1.93%	0.0502	1.18%	182.2	3.5	204.1	27.2	182.2	3.5
GCH1_36	330.4	0.82	24.348	1.89%	0.0517	1.09%	259.5	4.8	273.8	24.7	259.5	4.8
GCH1_35	455.7	0.72	24.946	1.91%	0.0533	1.20%	253.4	4.7	340.8	27.0	253.4	4.7
GCH1_34	365.4	0.35	3.188	2.91%	0.1765	0.81%	1758.6	44.7	2619.8	13.5	2619.8	13.5
GCH1_33	115.6	1.15	26.248	1.97%	0.0543	1.68%	241.0	4.7	384.6	37.4	241.0	4.7
GCH1_32	149.2	0.63	25.285	1.98%	0.0525	1.45%	250.0	4.9	307.3	32.6	250.0	4.9
GCH1_31	719.2	1.08	36.115	1.88%	0.0546	1.10%	176.1	3.3	393.9	24.4	176.1	3.3
GCH1_30	317.8	200.5	31.756	2.29%	0.0632	1.33%	199.9	4.5	713.5	27.9	199.9	4.5
GCH1_29	388.3	0.76	22.879	1.16%	0.0555	1.26%	275.8	3.1	431.3	27.9	275.8	3.1
GCH1_28	390.9	0.14	21.793	1.17%	0.0532	1.14%	289.2	3.3	335.5	25.6	289.2	3.3
GCH1_27	378.7	0.78	23.505	1.19%	0.0519	1.24%	268.6	3.1	282.0	28.1	268.6	3.1
GCH1_26	926.1	0.38	26.528	1.26%	0.0544	1.25%	238.5	2.9	385.6	27.8	238.5	2.9

(continued)

Table A.2 (continued)

Sample	U	Th	238U	1 sigma	207Pb	1 sigma	206/238	1 sigma	207/206	1 sigma	Best	1 sigma
Name	ppm	U	206Pb	% error	206Pb	% error	age	abs err	age	abs err	age	abs err Ma
GCH1_25	584.2	0.74	23.172	1.14%	0.0524	1.14%	272.4	3.0	300.9	25.9	272.4	3.0
GCH1_24	660.4	0.09	24.479	1.15%	0.0517	1.13%	258.1	2.9	271.1	25.6	258.1	2.9
GCH1_23	1114	0.74	34.776	1.17%	0.0508	1.05%	182.8	2.1	231.6	24.1	182.8	2.1
GCH1_22	87.95	1.23	25.431	1.44%	0.05	2.31%	248.6	3.5	196.5	52.7	248.6	3.5
GCH1_21	300.8	0.91	33.782	1.21%	0.0507	1.57%	188.1	2.2	227.6	35.9	188.1	2.2
GCH1_20	244.7	1.40	26.107	1.22%	0.0519	1.52%	242.3	2.9	282.1	34.4	242.3	2.9
GCH1_19	395.9	0.39	26.168	1.17%	0.0515	1.32%	241.8	2.8	265.2	30.0	241.8	2.8
GCH1_17	608	0.10	23.864	1.39%	0.0513	1.50%	264.6	3.6	253.5	34.1	264.6	3.6
GCH1_16	593.9	0.13	25.877	1.41%	0.055	1.38%	244.4	3.4	411.4	30.6	244.4	3.4
GCH1_15	11.61	0.18	24.839	3.25%	0.0608	4.72%	254.4	8.1	632.8	98.5	254.4	8.1
GCH1_14	633.8	0.40	21.806	1.16%	0.0526	1.16%	289.0	3.3	312.3	26.2	289.0	3.3
GCH1_13	479.2	0.62	22.332	1.16%	0.0523	1.20%	282.4	3.2	298.4	27.0	282.4	3.2
GCH1_12	527.5	0.36	24.271	1.21%	0.0522	1.19%	260.3	3.1	292.1	26.9	260.3	3.1
GCH1_11	50.34	1.05	27.633	1.75%	0.0535	2.99%	229.2	3.9	348.7	66.3	229.2	3.9
GCH1_10	521.6	0.51	24.525	1.16%	0.0521	1.16%	257.6	2.9	288.8	26.4	257.6	2.9
GCH1_9	125.8	1.29	26.729	1.44%	0.0505	1.94%	236.8	3.3	217.1	44.3	236.8	3.3
GCH1_8	1246	0.44	23.396	1.13%	0.0576	1.05%	269.8	3.0	515.6	22.9	269.8	3.0
GCH1_7	352.2	0.54	13.914	1.22%	0.0567	1.39%	447.4	5.3	480.8	30.3	447.4	5.3
GCH1_6	693.7	0.88	35.349	1.29%	0.0509	1.23%	179.8	2.3	236.8	28.2	179.8	2.3
GCH1_5	223.6	0.70	35.382	1.37%	0.0491	1.74%	179.7	2.4	152.2	40.3	179.7	2.4
GCH1_4	414.3	0.50	35.400	1.25%	0.0499	1.37%	179.6	2.2	188.3	31.5	179.6	2.2
GCH1_3	755.6	0.58	34.118	1.24%	0.0525	1.13%	186.2	2.3	308.8	25.4	186.2	2.3

(continued)

Table A.2 (continued)

Sample	U	Th	238U	1 sigma	207Pb	1 sigma	206/238	1 sigma	207/206	1 sigma	Best	1 sigma
Name	ppm	U	206Pb	% error	206Pb	% error	age	abs err	age	abs err	age	abs err Ma
GCH1_2	405.2	0.32	24.664	1.30%	0.0551	1.24%	256.2	3.3	417.0	27.4	256.2	3.3
GCH1_1	601.1	0.59	35.276	1.13%	0.0507	1.21%	180.2	2.0	228.7	27.7	180.2	2.0

Table A.3 Lu–Hf data

		TotalHfBeam	*AGE (U-Pb)*	*Initial*		
$^{180}Hf/^{177}Hf$	*2SE*	*(Volts)*	*Ma*	*eHf*	*2SE*	TDM_{ZIRC}
1.88704	0.00011	15.65	279	-7.8	1.1	1201
1.8869	0.00013	10.07	207	-5.8	1.6	1064
1.88691	0.000091	15.53	251	-10.4	0.9	1298
1.88693	0.00011	12.27	256	-8.2	1.1	1250
1.887023	0.000086	15.44	252	-9.9	0.9	1276
1.88693	0.00013	12.39	248	-14.5	1.6	1457
1.88704	0.0001	10.45	247	-14.3	1.1	1485
1.88684	0.00013	10.68	255	-11.1	1.9	1379
1.886928	0.000098	11.96	192	-2.1	1.2	930
1.88702	0.00016	13.47	283	-0.7	1.5	938
1.88686	0.00012	12.46	175	-9.0	1.5	1188
1.88696	0.00016	13.64	239	-12.7	1.8	1376
1.88696	0.00016	9.9	191	-6.3	1.7	1127
1.88685	0.00015	13.72	264	-7.4	1.2	1167
1.88697	0.00013	11.65	252	-15.5	1.7	1514
1.887	0.00011	10.78	180	-3.1	1.2	1001
1.8869	0.0001	15.37	256	-11.0	1.3	1318
1.88701	0.00013	11.3	186	-3.2	1.4	973
1.88707	0.00012	11.33	179	-4.9	1.3	1063
1.88686	0.00016	9.18	179	-2.2	1.5	904
1.88693	0.00012	10.17	179	-4.6	1.7	1046
1.88701	0.00015	13.75	447	-8.9	1.6	1420
1.88699	0.00014	15.61	269	-7.6	1.3	1198
1.88697	0.00015	12.04	236	-8.2	1.7	1194
1.88698	0.00017	11.78	257	-7.7	1.8	1199
1.887	0.00019	13.44	229	-8.1	2.3	1179
1.88707	0.00017	11.35	242	-9.8	1.6	1289
1.88706	0.00013	15.32	241	-7.4	1.4	1154
1.88696	0.00011	12.02	244	-7.7	1.3	1171
1.887042	0.000098	13.9	282	-8.1	1.1	1220
1.886983	0.000099	15.13	272	-7.4	1.0	1172

References

Al-Suwaidi AH, Angelozzi GN, Baudin F, Damborenea SE, Hesselbo SP, Jenkyns HC, Manceñido MO, Riccardi AC (2010) First record of the Early Toarcian Oceanic Anoxic Event from the Southern Hemisphere, Neuquén Basin, Argentina. Geol Soc of London 167:633–636

Armella C, Leanza HA, Corfu F (2016) Synsedimentary ash rains and paleoenvironmental conditions during the deposition of the Chachil Formation (Pliensbachian) at its type locality, Neuquén Basin, Argentina. J South Am Earth Sci 71:82–95

Benedini L, Gregori D, Strazzere L, Falco JI, Dristas JA (2014) Lower Pliensbachian caldera volcanism in high-obliquity rift systems in the western North Patagonian Massif, Argentina. J South Am Earth Sci 56:1–19

Burgess PM, Flint S, Johnson S (2000) Sequence stratigraphic interpretation of turbiditic strata: an example from Jurassic strata of the Neuquén basin, Argentina. Geol Soc Am Bull 112 (11):1650–1666

Chang Z, Vervoort JD, McClelland WC, Knaack C (2006) U-Pb dating of zircon by LA-ICP-MS. Geochem Geophy Geosy 7:1–14

Cúneo R, Ramezani J, Scasso R, Pol D, Escapa I, Zavattieri AM, Bowring SA (2013) High-precision U-Pb geochronology and a new chronostratigraphy for the Cañadón Asfalto Basin, Chubut, central Patagonia: implications for terrestrial faunal and floral evolution in Jurassic. Gondwana Res 24:1267–1275

De la Cruz R, Suárez M (1997) El Jurásico de la cuenca de Neuquén en Lonquimay, Chile: Formación Nacientes del Biobío (38–39°8). Rev Geol de Chile 24(1):3–24

D'Elia L, Bilmes A, Franzese JR, Veiga GD, Hernández M, Muravchik M (2015) Early evolution of the southern margin of the Neuquén Basin, Argentina: tectono-stratigraphic implications for rift evolution and exploration of hydrocarbon plays. J South Am Earth Sci 64:42–57

Dickinson WR, Beard LS, Brakenridge GR, Erjavec JL, Ferguson RC, Inman KF, Knepp RA, Lindberg FA, Ryberg PT (1983) Provenance of North American Phanerozoic sandstone in relation to tectonic setting. Geol Soc Am Bull 94:222–235

Féraud G, Alric V, Fornari M, Bertrand H, Haller M (1999) 40Ar/39Ar dating of the Jurassic volcanic province of Patagonia: migrating magmatism related to Gondwana break-up and subduction. Earth Planet Sci Lett 172:83–96

Fanning CM, Hervé F, Pankhurst RJ, Rapela CW, Kleiman LE, Yaxley GM, Castillo P (2011) Lu-Hf isotope evidence for the provenance of Permian detritus in accretionary complexes of western Patagonia and the northern Antarctic Peninsula region. J South Am Earth Sci 32:485–496

Franzese JR (1995) El Complejo Piedra Santa (Neuquén, Argentina): parte de un cinturón metamórfico neopaleozoico del Gondwana suroccidental. Rev Geol de Chile 22(2):193–202

Franzese JR, Spalletti LA (2001) Late Triassic-early Jurassic continental extension in southwestern Gondwana: tectonic segmentation and pre-break-up rifting. J South Am Earth Sci 14:257–270

Franzese JR, Pankhurst RJ, Rapela CW, Spalletti LA, Fanning M, Muravchik M (2002) Nuevas evidencias geocronológicas sobre el magmatismo Gondwánico en el noroeste del Macizo Nordpatagónico. 15 Congreso Geológico Argentino, Actas en CD, El Calafate

García Morabito E, Ramos VA (2012) Andean evolution of the Aluminé fold and thrust belt, Northern Patagonian Andes (38°30′–40°30′S). J South Am Earth Sci 38:13–30

Gaschnig RM, Vervoort JD, Lewis RS, Tikoff B (2011) Isotopic evolution of the Idaho batholith and Challis intrusive province, northern US Cordillera. J Petrol 52:2397–2429

Gulisano CA (1981) El ciclo cuyano en el norte de Neuquén y sur de Mendoza. XVIII Congreso Geológico Argentino (San Juan), Actas III, pp 573–592

Gulisano CA, Gutiérrez Pleimling AR (1994) Guía de campo El Jurásico de la Cuenca Neuquina, b) Provincia de Mendoza. Asociación Geológica Argentina, Serie E 3, 1–103, Buenos Aires

Gulisano CA, Gutiérrez Pleimling AR, Digregorio RE (1984) Esquema estratigráfico de la secuencia jurásica del oeste de la Provincia de Neuquén. IX Congreso Geológico Argentino, Actas 1:236–259

Hervé F, Calderón M, Fanning CM, Pankhurst RJ, Godoy E (2013) Provenance variations in the Late Paleozoic accretionary complex of central Chile as indicated by detrital zircons. Gondwana Res 23:1122–1135

Howell JA, Schwarz E, Spalletti LA, Veiga GD (2005) The Neuquén Basin: an overview. In: Veiga GD, Spalletti LA, Howell JA, Schwarz E (eds) The Neuquén Basin, Argentina: a Case Study in Sequence Stratigraphy and Basin Dynamics, vol 252. Geol Soc. Special Publications, London, pp 1–14

Kamo SL, Riccardi AC (2009) A new U-Pb zircon age for an ash layer at the Bathonian-Callovian boundary, Argentina. GFF 131:177–182

Kay SM, Ramos VA, Mpodozis C, Sruoga P (1989) Late Paleozoic to Jurassic silicic magmatism at the Gondwana margin: analogy to middle Proterozoic in North America? Geology 17:324–328

Leanza HA, Hugo C (1997) Hoja Geológica 3969-III. Picún Leufú. Provincias de Neuquén y Río Negro. Servicio Geológico Minero Argentino, Boletín 218, 135 pp., Buenos Aires

Leanza HA, Mazzini A, Corfu F, Llambías EJ, Svensen H, Planke S, Galland O (2013) The Chachil Limestone (Pliensbachian-earliest Toarcian) Neuquén Basin, Argentina: U-Pb age calibration and its significance on the Early Jurassic evolution of southwestern Gondwana. J South Am Earth Sci 42:171–185

Llambías EJ, Caminos R, Rapela CW (1984) Las plutonitas y vulcanitas del Ciclo Eruptivo Gondwánico. Noveno Congreso Geológico Argentino, Relatorio:85–117

Llambías EJ, Leanza HA, Carbone O (2007) Evolución tectono-magmática durante el Pérmico al Jurásico Temprano en la Cordillera del Viento (37°05′S-37°15′S): nuevas evidencias geológicas y geoquímicas del inicio de la cuenca Neuquina. Rev Asoc Geol Argentina 62(2):217–235

Lucassen F, Trumbull R, Franz G, Creixell C, Vásquez P, Romer RL, Figueroa O (2004) Distinguishing crustal recycling and juvenile additions at active continental margins: the Paleozoic to Recent compositional evolution of the Chilean Pacific margin (36-41° S). J South Am Earth Sci 17:103–119

Martínez Dopico C, Lopez de Luchi M, Rapalini AE, Kleinhanns IC (2011) Crustal segments in the North Patagonian Massif, Patagonia: an integrated perspective based on Sm/Nd isotope systematics. J South Am Earth Sci 31:324–341

Mazzini A, Svensen H, Leanza HA, Corfu F, Planke S (2010) Early Jurassic shale chemostratigraphy and U-Pb ages from the Neuquén Basin (Argentina): implications for the Toarcian Oceanic Anoxic Event. Earth Planet Sc Lett:297, 633e645

Mosquera A, Silvestro J, Ramos VA, Alarcón M, Zubiri M (2011) La estructura de la Dorsal de Huincul. In: Leanza HA, Arregui C, Carbone O, Danieli JC, Vallés JM (eds) Geología y recursos naturales de la provincia de Neuquén. Relatorio del XVIII Congreso Geológico Argentino, Buenos Aires, pp 385–397

Mpodozis C, Ramos VA (1989) The Andes of Chile and Argentina. In: Ericksen GE, Cañas MT, Reinemud JA (eds) Geology of the Andes and Its Relation to Hydrocarbon and Mineral Resources Circumpacific Council for Energy and Mineral Resources, 11. Earth Sci Series:59–90

Naipauer M, Ramos VA (2016) Changes in source areas at Neuquén Basin: Mesozoic evolution and tectonic setting based on U-Pb ages on zircons. In: Naipauer M, Sagripanti L, Ghiglione M, Orts D (eds) Folguera A. Growth of the Southern Andes, Springer, pp 33–61

Naipauer M, García Morabito E, Marques JC, Tunik V, Rojas Vera F, Vujovich GI, Pimentel MP, Ramos VA (2012) Intraplate Late Jurassic deformation and exhumation in western central Argentina: constraints from surface data and U-Pb detrital zircon ages. Tectonophysics 524–525:59–75

Oliveros V, Féraud G, Aguirre L, Fornari M, Morata D (2006) The Early Andean Magmatic Province (EAMP): 40Ar/39Ar dating on Mesozoic volcanic and plutonic rocks from the Coastal Cordillera, Northern Chile. J Volcanol Geotherm Res 157:311–330

Pángaro F, Pereira DM, Micucci E (2009) El sinrift de la Dorsal de Huincul, Cuenca Neuquina: evolución y control sobre la estratigrafía y estructura del área. Rev Asoc Geol Argentina 65 (2):265–277

Pankhurst RJ, Rapela CR (1995) Production of Jurassic rhyolite by anatexis of the lower crust of Patagonia. Earth Planet Sci Lett 134:23–36

Pankhurst RJ, Leat PT, Sruoga P, Rapela CW, Marquez M, Storey BC, RileyTR (1998) The Chon-Aike silicic igneous province of Patagonia and related rocks in West Antarctica: a silicic LIP. J Volcanol Geotherm Res 81:113–136

Pankhurst RJ, Riley TR, Fanning CM, Kelley SP (2000) Episodic silicic volcanism in Patagonia and the Antarctic Peninsula: chronology of magmatism associated with the break-up of Gondwana. J Petrol 41:605–625

Pankhurst R, Rapela C, Fanning C, Márquez M (2006) Gondwanide continental collision and the origin of Patagonia. Earth-Sci Rev 76:235–257

Pankhurst RJ, Rapela CW, López De Luchi MG, Rapalini AE, Fanning CM, Galindo C (2014) The Gondwana connections of northern Patagonia. Geol Soc of London 171(3):313–328

Parada MA, López-Escobar L, Oliveros V, Fuentes F, Morata D, Calderón M, Aguirre L, Féraud G, Espinoza F, Moreno H, Figueroa O, Muñoz Ravo J, Troncoso Vásquez R, Stern CR (2007) Andean magmatism. In: Moreno T, Gibson W (eds) The Geology of Chile, vol 4. Geol Soc, Special Publications, London, pp 115–146

Fedo CM, Sircombe KN, Rainbird, RH (2003) Detrital zircon analysis of the sedimentary record. In: Hanchar JM, Hoskin PWO (eds) Zircon. Rev Mineral Geochem 53:277–303

Ramos VA (1999) Las provincias geológicas del territorio argentino. In: Caminos R L (ed) Geología Argentina. Segemar Anales 29(3):41–96

Ramos VA, Folguera A (2005) Tectonic evolution of the Andes of Neuquén: Constraints derived from the magmatic arc and foreland deformation. In: Veiga GD, Spalletti LA, Howell JA, Schwarz E (eds) The Neuquén Basin: A case study in sequence stratigraphy and basin dynamics, vol 252, Geol Soc, Special Publications, London, pp 15-35

Rapela CW, Pankhurst RJ, Fanning CM, Hervé F (2005) Pacific subduction coeval with the Karoo mantle plume: the Early Jurassic Subcordilleran belt of northwestern Patagonia. Geol Soc Lond, Spec Publ 246:217–239

Rolando AP, Hartmann LA, Santos JOS, Fernandez RR, Etcheverry RO, Schalamuk IA, Mcnaughton NJ (2002) SHRIMP zircon U-Pb evidence for extended Mesozoic magmatism in the Patagonian batholith and assimilation of Archaean crustal components. J South Am Earth Sci 15:267–283

Sato AM, Llambías EJ, Basei MAS, Castro CE (2015) Three stages in the Late Paleozoic to Triassic magmatism of southwestern Gondwana, and the relationships with the volcanogenic events in coeval basins J South Am. Earth Sci 63:48–69

Schiuma M, Llambías EJ (2008) New ages and chemical analysis on Lower Jurassic volcanism close to the Huincul High, Neuquén. Rev Asoc Geol Argentina 63(4):644–652

Silvestro J, Zubiri M (2008) Convergencia oblicua: modelo estructural alternativo para la Dorsal Neuquina (39° S), Neuquén. Rev Asoc Geol Argentina 63(1):49–64

Spalletti LA, Franzese J, Morel E, D´Elia L, Zúñiga A, Fanning M (2010) Consideraciones acerca de la sedimentología, paleobotánica y geocronología de la Formación Piedra del Águila (Jurásico Inferior, Neuquén). Rev Asoc Geol Argentina 66(3):305–313

Suarez M, Márquez M (2007) A Toarcian retro-arc basin of Central Patagonia (Chubut), Argentina: Middle Jurassic closure, arc migration and tectonic setting. Andean Geol 34:63–80

Varela R, Basei MAS, Cingolani CA, Siga O Jr, Passarelli CR (2005) El basamento cristalino de los Andes Norpatagónicos en Argentina: geocronología e interpretación tectónica. Rev Geol Chile 32(2):167–187

Vásquez P, Glodny J, Franz G, Frei D, Romer RL (2011) Early Mesozoic Plutonism of the Cordillera de la Costa (34–37° S), Chile: constraints on the onset of the Andean Orogeny. J Geol 119(2):159–184

Vergani GD, Tankard AJ, Belotti HJ, Welsink HJ (1995) Tectonic evolution and paleogeography of the Neuquén Basin, Argentina. In: Tankard AJ, Suárez Soruco R, Welsink HJ (eds) Petroleum basins of South America, vol 62. Am Assoc Pet Geol Bull, Memoirs, pp 383–402

Volkheimer W (1973) Palinología estratigráfica del Jurásico de la sierra de Chacai Co y adyacencias (Cuenca Neuquina, Argentina). I. Estratigrafía de las Formaciones Sierra Chacai Co (Pliensbachiano), Los Molles (Toarciano), Cura Niyeu (Bajociano) y Las Lajas (Caloviano inferior). Ameghiniana 10:105–109

Weaver C (1931) Paleontology of the Jurassic and Cretaceous of west central Argentina. In: Memoir, vol. 1. University of Washington, Seattle, pp 1–469

Part IV
The Early Andean Phases in the Chilean-Argentinean Margin

The Jurassic Paleogeography of South America from Paleomagnetic Data

María Paula Iglesia Llanos

Abstract For decades, it has been interpreted that South America had remained stationary in similar present-day latitudes during most of the Mesozoic and the Cenozoic. More recent paleomagnetic data however, suggest a thoroughly different scenario for the Jurassic. In order to test the stationary vs. the dynamic—continent model, a refined Jurassic South American apparent polar wander path was constructed and compared with those from Eurasia, Africa, and North America. Similarities are such that a master path for Pangea is proposed. This path shows remarkable different polar positions during the Early Jurassic that are interpreted to have been caused by true polar wander. An absolute paleogeographical reconstruction of Pangea is presented. Paleolatitudes changes are very well sustained by paleoclimatic proxies from the southern and northern hemispheres.

Keywords Paleomagnetism · Jurassic · Paleogeography · Pangea
Paleoclimate

1 Introduction

The study of the Earth's magnetic field as recorded in the rocks provided an important key in reconstructing the history of plate motions. A paleomagnetic pole (PP) marks the past position of the rotation axis with respect to the continent of observation, and there can be only one rotation axis at any particular geologic time. So if two continents are placed in their proper relative positions for a particular geologic time, their paleomagnetic poles for that time must coincide (Butler 1992). Positions of the PPs derived from the various continents are variable in time, generally falling farther away from the present-day geographic pole the greater the

M. P. Iglesia Llanos (✉)
Instituto de Geociencias Básicas, Ambientales y Aplicadas, Departamento de Ciencias Geológicas, Facultad de Ciencias Exactas y Naturales, Universidad de Buenos Aires–Consejo Nacional de Investigaciones Científicas y Técnicas, Buenos Aires, Argentina
e-mail: mpiglesia@gl.fcen.uba.ar

A. Folguera et al. (eds.), *The Evolution of the Chilean-Argentinean Andes*,
Springer Earth System Sciences, https://doi.org/10.1007/978-3-319-67774-3_9

age of the rock. The curve drawn joining the paleomagnetic poles is called apparent polar wander (APW) path, and is usually shown on the present geographic grid. Therefore, if continents had a fixed relative position for a significant interval of geologic time, their APW paths must coincide (Butler 1992).

The concept of apparent polar wander paths was helpful in determining the speed, direction, and rotation of continents. These paths can be viewed as a series of arcuate tracks separated by sharp corners called "cusps" (May and Butler 1986). Each track of APW is considered to result from the continent riding on a lithospheric plate that rotated about a fixed Euler pole for an extended interval of geologic time (e.g. 50 my). Different tracks represent rotations about different Euler poles, and cusps represent times of reorganization of the lithospheric plate boundaries and resulting driving forces (Butler 1992).

Paleomagnetic data allow to reconstruct the relative motions of two continents. Paleolatitude can be derived from the inclination of the paleomagnetic direction, but paleolongitude remains indeterminate, since the Geocentric Axial Dipole model (GAD) has the Earth's axis of rotation as its axis of symmetry (Lanza and Meloni 2006).

Latitude is the main variable that affects climate and therefore allows correlating paleomagnetic data with paleoclimatic data, which can be deduced from the fact that certain types of rocks and many fossils have a geographic distribution limited to certain climatic regions. The comparison cannot be very quantitative, but qualitative concordance between the results of completely different methodologies is always a convincing argument (Lanza and Meloni 2006).

From the Late Carboniferous until the Middle Jurassic continents were assembled in a quasi-rigid supercontinent called Pangea, which occupied most of a hemisphere, while the rest of the Earth′s surface was made up of a large ocean called Panthalassa. For a long time, it has been taken for granted that during most of the Mesozoic and the end of the Paleozoic, South America remained more or less stationary in similar present-day latitudes, based on the fact that the oldest paleomagnetic poles clustered around the geographic pole (e.g. Valencio et al. 1983; Oviedo and Vilas 1984; Rapalini et al. 1993; Beck 1999). Subsequently, Iglesia Llanos (1997), Vizán (1998) and Iglesia Llanos et al. (2006) argued that particularly in the case of the Early Jurassic, paleomagnetic poles fell away from the present-day geographic pole. The resultant South American APW path was thus compared with those of other major continents that were also part of Pangea, such as Eurasia, North America and Africa. These paths were rotated to South American coordinates using the kinematic model that produced the best clustering of poles and consequently, a master APW path for Pangea was obtained (Iglesia Llanos and Prezzi 2013). The resultant APW path shows a well-defined cusp—the intersection of two tracks—at ~197 Ma, followed by a relatively long track until ~185 Ma (Early Jurassic). The likely phenomena of shift in the pole positions could have been mainly due to either (1) lithospheric motion or, (2) true polar wander (TPW) defined as the drift of the whole rigid Earth relative to the spin axis. This analysis was accomplished using Morgan's (1983) grid of hotspots (HS) that goes back to 200 Ma. In addition, because TPW produces rapid important latitudinal changes, "absolute" (paleolongitudinally–controlled) reconstructions of "crude" Pangea were performed, assuming that during the Jurassic HS were more or less fixed to the mantle. Finally, the paleolatitudinal changes were confronted with paleoclimatic proxies derived from biogeographical and geological data from both the northern and southern hemispheres (Iglesia Llanos and Prezzi 2013).

2 Paleomagnetic Data

The geodynamic model proposed by Iglesia Llanos et al. (2006), and Iglesia Llanos and Prezzi (2013) for South America, is fundamentally based on two Lower Jurassic PP obtained in the Neuquén basin (Iglesia Llanos et al. 2006). Studied sections have been dated on the basis of ammonites from the Andean Region, that are correlatable with the Tethyan Standard Zonation (Iglesia Llanos 1997; Iglesia Llanos et al. 2006; Iglesia Llanos and Prezzi 2013).

The first paleopole was obtained in southern Mendoza province along the Atuel River (Fig. 1a) in two localities that comprise a c. 2 km-thick section bearing ammonites of Late Triassic to Toarcian age, made up of transgressive deposits that become younger and coarser-grained to the east. The oldest sampled rocks are in Arroyo Malo (AM in Fig. 1a), where the sampled section is c. 1 km-thick and is conformed from base to top, by the Arroyo Malo, El Freno and El Cholo formations (Riccardi et al. 1988, 1991; Riccardi and Iglesia Llanos 1999; Riccardi et al. 2003). Ammonites in Arroyo Malo belong to the Andean Choristoceras to Semicostatum

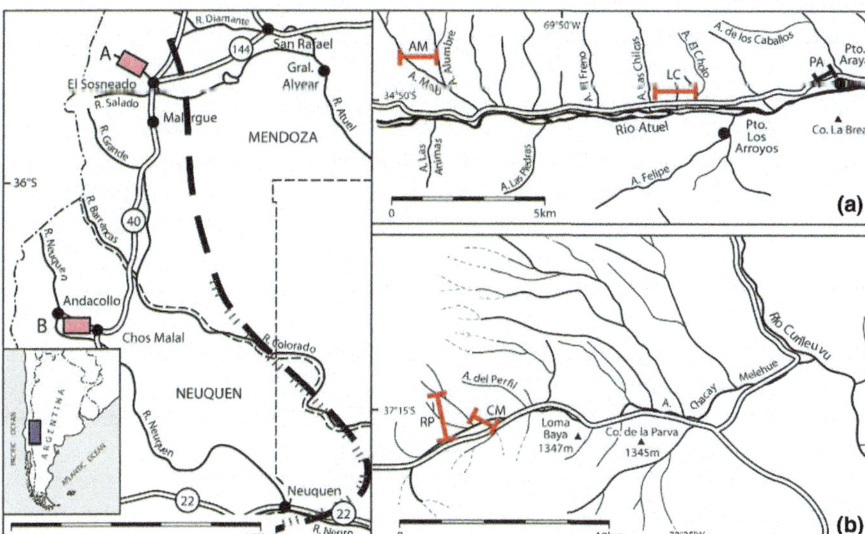

Fig. 1 Sketch map of the studied localities in the Neuquén Basin and detail **a** Arroyo Malo (AM), Las Chilcas (LC) and Puesto Araya (PA), **b** Rajapalo-Chacay Melehue (RP-CM) (Iglesia Llanos et al. 2006)

Zones that indicate Late Rhaetian to Early Sinemurian ages. Intercalated in the El Cholo Fm., there are at least three alkaline sills (Iglesia Llanos et al. 2006).

The other locality is Las Chilcas (LC in Fig. 1a), where the sampled section is 500 m thick and is made up of El Freno and El Cholo formations. Ammonites correspond to the Andean "Epophioceras" to Miltoceras Zones that indicate Early Sinemurian to Early Pliensbachian ages. A lamprophyric sill is intercalated in the El Cholo Fm. at c. 300 m from the base, marking the cessation of alkaline intrusions in the area. Thus, it is interpreted that these sills were intruded during the first stages of the rifting and were thus dated as Hettangian-Sinemurian (Iglesia Llanos et al. 2006). Sedimentary structures and ammonites in Arroyo Malo and Las Chilcas localities show that these sections have been subjected to compaction. Mean directions in Arroyo Malo and Las Chilcas pass several paleomagnetic field tests for paleomagnetic stability such as the fold, the reversal and baked contact tests, which prove the primary origin of the isolated magnetizations. The Hettangian-Sinemurian PP with compaction correction is located at 223° E, 51° S, $A_{95} = 6°$, $N = 25$ (Iglesia Llanos et al. 2006, 2008), and is assigned an absolute mean age of 197 Ma according to Ogg et al. (2016).

The other PP was derived from Pliensbachian-Toarcian sections cropping out at Puesto Araya (PA, Fig. 1a) and in northern Neuquén province in Rajapalo-Chacay Melehue (RP-CM, Fig. 1b). In Puesto Araya, the sampled section is c. 400 m-thick and is composed from base to top, of the El Freno and El Cholo formations. Ammonites correspond to the Andean Tropidoceras to Fanninoceras fannini Zones which indicate Early to Late Pliensbachian ages. In Rajapalo-Chacay Melehue (Fig. 1b), the sampled succession is c. 500 m-thick and is composed of the Lista Blanca and Los Molles formations. Ammonites belong to the Fanninoceras disciforme to Peronoceras largaense Zones thus recording Late Pliensbachian to Middle Toarcian ages. Mean directions calculated in Puesto Araya and Rajapalo-Chacay Melehue also pass several field tests such as the fold, reversal and conglomerate tests, proving that isolated magnetizations are primary as well. The PP calculated here is located at 67° E, 74° S, $A_{95} = 5°$, $N = 52$ (Iglesia Llanos et al. 2006, 2008), and is assigned a mean absolute age of 185 Ma according to Ogg et al. (2016).

These two PP were combined with other Jurassic paleopoles that passed strict reliability criteria to construct a more refined APW path of stable South America (Iglesia Llanos et al. 2006; Iglesia Llanos and Prezzi 2013). The resultant path is remarkably different from those previosly obtained by other authors, which showed PP clustered around the geographic pole. In the new path, there appears a distinctive W–E track between 215 and 197 Ma (Late Triassic to lowermost Jurassic), followed by a well-defined SE-NW track between 197 and 185 Ma (Sinemurian to Pliensbachian), clearly defining a cusp at ~197 Ma, and a subsequent standstill which might have persisted until the Late Jurassic.

In order to test whether this long APWP was also observed in other regions of Pangea, paleomagnetic data from the major lithospheric plates whose relative positions for this time are well-known such as Eurasia, North America, and Africa were compiled, and the APW paths constructed. PP were drawn, from the Global Palaeomagnetic Database v.4.6 2005 (http://www.tsrc.uwa.edu.au/data_bases) as well as from more recent literature (Table 1). Selection of PP was stringent and required to pass minimum reliability criteria such as: (i) coming from sections with constrained geological ages, (ii) unsuspected of being subjected to rotations around vertical axis, (iii) full demagnetization and vectorial analysis of paleomagnetic data and, (iv) paleomagnetic field tests and/or mineralogical studies to constrain magnetization ages.

The resultant APW paths from Eurasia, North America and Africa were subsequently rotated to South American present-day coordinates (Fig. 2). To achieve that, every classical paleoreconstruction available in the literature for Jurassic Pangea was tested as well as all likely combinations to bring continents to South America, looking for the one paleoreconstruction that provided the best clustering of poles. Thus, the best fit of poles was achieved (Table 2), by rotating Africa to South American coordinates using the palaeoreconstruction of Lawver and Scotese (1987). Eurasia was translated firstly to North America using Frei and Cox (1987), subsequently to Africa, and finally to South America using Lawver and Scotese (1987). North America on the other hand, was moved to Africa with the paleoreconstruction of Klitgord and Schouten (1986) and then to South America with Lawver and Scotese (1987).

By averaging non-overlapping poles with 5–10 Myr windows, a master path has been constructed for the Jurassic Pangea. The resultant curve shows (Fig. 2), when observed from South America, the same W–E track between during the Late Triassic to lowermost Jurassic, a SE–NW track between the Sinemurian and the Pliensbachian, and a cusp at ~197 Ma, followed by a standstill until the Late Jurassic. Such as in the case of South America, there appears a remarkable shift—*c*. 50°-in the pole positions between 197–185 Ma, showing minimum angular change of poles of approximately 4° Myr^{-1} (Iglesia Llanos and Prezzi 2013). The estimated minimum angular change of approximately 4° Myr^{-1} (0.4 m yr^{-1}) is consistent with Cretaceous true polar wander values predicted by Prévot et al. (2000) of 1– 5° Myr^{-1}, or the 3°–10° Myr^{-1} proposed by Sager and Koppers (2000) from data of the Pacific plate.

3 The Polar Shift

A noticeable difference in the pole positions could result from an Earth's magnetic field whose geometry does not fit with a dominant geocentric axial dipole, as it would occur when long-term, nondipole components prevail. If this is not so, the shift could be explained by: (1) lithospheric motion, or (2) true polar wander (TPW).

According to many authors, the Earth′s magnetic field was essentially geocentric and dipolar during the Phanerozoic (e.g. McElhinny and Brock 1975; Evans 1976; Livermore et al. 1983, 1984) (Fig. 3).

Table 1 Selected paleopoles from the GPDB (2005) and more recent studies (Iglesia Llanos and Prezzi 2013). A_{95}: 95% confidence interval. **1** in Torsvik et al. (2008), **2** in Kent and Irving (2010), **3** in Van der Voo and Torsvik (2004)

Code/Rock Unit	Mean age (MA)	Lat ° S	Long ° E	A_{95}	Reference
North America					
1. Passaic Fm.	210	57.6	310.6	7.0°	Kodama et al. (1994)
2. Newark Rutgers core *(1)*	215*	56.4	277.8	3.1°	Kent et al. (1995)
3. Chinle Fm. Upp. Shale *(1)*	218*	57.4	267.8	5.6°	Bazard and Butler (1991)
4. Newark Somerset Core *(1)*	213*	57.2	276.5	2.8°	Kent et al. (1995)
5. Newark Somerset Core *(1)*	210*	58.0	271.5	2.5°	Kent et al. (1995)
6. Redonda Fm. *(1)*	212*	57.7	259.1	4.2°	Reeve and Helsley (1972)
7. Newark Mart. *(1)*	206*	58.6	278.4	2.4°	Kent et al. (1995)
8. Culpeper Basin *(1)*	200*	65.5	253.1	4.8°	Kodama et al. (1994)
9. Wingate Sdt.	202	57.4	236.6	6.4°	Molina Garza et al. (2003)
10. Sugarloaf Fm.	200	57.7	261.3	9.1°	McEnroe and Brown (2000)
11. Holden Dlab.	197	60.1	260.5	5.1°	McEnroe and Brown (2000)
12. Chinle Red Mb.	198	51.3	251.7	8.8°	Molina Garza et al. (2003)
13. North Mount. Bas. 1 *(1)*	190*	66.4	251.9	10.7°	Hodych and Hayatsu (1988)
14. Copper Mt. Intr.	195	57.3	191.8	3.6°	Symons and Litalein (1984)
15. Pelham-Loudv. Dlab.	197	65.3	275.6	3.8°	McEnroe and Brown (2000)
16. Newark Supergr. Volo. *(1) (2)*	201*	68.0	268.6	3.9°	Prevot and McWilliams (1989)
17. Sil Nakya Volo. *(1)*	193	74.4	260.1	7.6°	Cohen et al. (1986)
18. Caraquet Dyke	190	74.1	294	8.5°	Seguin et al. (1981)
19. North Mt. Bas. 2	190	73.0	284	5.0°	Carmichael and Palmer (1958)
20. Kayenta Fm. 1 *(1)*	210*	61.9	254.4	6.8°	Steiner and Helsley (1974)
21. Kayenta Fm. 2 *(1)*	192*	59.0	246.6	3.3°	Bazard and Butler (1991)
22. Mulberry Wash	185	61.5	34.9	6.1°	Cohen et al. (1986)
23. Diabase Anticosti	183*	76.0	265	1°	Larochelle (1971)

(continued)

Table 1 (continued)

Code/Rock Unit	Mean age (MA)	Lat ° S	Long ° E	A_{95}	Reference
24. Tesin Crossing Pt.	181	84.5	69.4	8.8°	Harris et al. (1996)
25. White Mt. Volo. (2)	180	85.5	304.5	4.0°	Opdyke and Wensink (1966)
26. Picton Dyke	173	76.0	166	2.6°	Barnett et al. (1984)
27. Stump-Twin Fm.	170	64.0	342	8.0°	McWhinnie et al. (1990)
28. Summerville Fm.	159	67.0	289.8	4.3°	Steiner (1978)
Eurasia					
1. Sunnhordland dikes (3)	212	50.0	305	4.5°	Walderhaug (1993)
2. Volc. N Pyrenees 1	205	62.1	294.2	7.3°	Girdler (1968)
3. Pre-Azov	200	38.0	278.0	3.0°	Myhaylova et al. (1989)
4. Donbass	200	70.0	268.0	4.0°	Rusakov (1971)
5. Rhaetian Seds. (1)	208*	50.0	292.0	7.6°	Edel and Duringer (1997)
6. Hett.-Sinem. Limest (1)	201*	55.0	280.0	9.0°	Edel and Duringer (1997)
7. Paris Basin Seds. (1)	198*	51.3	285.0	3.1°	Yang et al. (1996)
8. Kerforne dikes (1) (2)	198*	61.3	258.8	7.5°	Sichler and Perrin (1993)
9. Main Caucasus	197	43.0	337.0	5.0°	Sintisin and Shevlyagin (1985)
10. Liassic Volcs. (1)	198*	64.9	323.6	6.7°	Girdler (1968)
11. N Caucasus Volc.	185	68.0	72.0	12.0°	Asanidze and Pechersky (1982)
12. E Donbass	190	85.0	41.0	11.0°	Myhaylova et al. (1989)
13. Liassic Seds. (1)	184*	76.9	314.7	3.0°	Hijab and Tarling (1982)
14. Thouars and Airvault Sections (1)	184*	70.5	276.3	11.8°	Galbrun et al. (1988)
15. Scania Basalts (1) (2)	179*	69.0	282	6.8°	Bylund and Halvorsen (1993)
16. Alsace Bajocian Seds. (1)	178*	63.0	300	6.0°	Kadzialko-Hofmokl et al. (1988)
17. Subtatric Nappe Seds. (1)	159*	71.7	312.2	4.0°	Kadzialko-Hofmokl and Kruczyk (1957)
18. Limestones. Krakow-Czest (1)	159*	72.3	330.4	7.3°	Kadzialko-Hofmokl and Kruczyk (1957)
19. Terres Notres (1)	158*	77.6	309.7	7.0°	Aubourg and Rochette (1992)

(continued)

Table 1 (continued)

Code/Rock Unit	Mean age (MA)	Lat ° S	Long ° E	A_{95}	Reference
20. Oxfordian Seds. (1)	157*	70.1	327	4.0°	Kruczyk and Kadzialko-Hofmokl (1988)
Africa					
1. U. Triassic Sed. (1)	221.5	54.9	43.3	11.5°	Ghorabi and Henry (1991)
2. Dolerites Morococo (1)	200	72.1	37.7	6.3°	Westphal et al. (1979)
3. Foum-Zguid	200	67.9	67.9	3.9°	Palenda-Ortas (2011)
4. Ighrem	200	78.4	58.2	6.7°	Palenda-Ortas (2011)
5. Zarzaitine Fm. (1)	206.5	70.9	55.1	2.6°	Kies et al. (1995)
6. Karroo Comb. (1)	198	65.4	75.1	9.5°	McElhinny and Jones (1965)
7. Freetown Cplx (1) (2)	193	82.9	32.7	6.2°	Hargraves et al. (1999)
8. Hank Volc. Mauri (1)	187*	60.4	52.0	4.1°	Dalrymple et al. (1975)
9. Hodn Volc. Mauri (1)	187*	71.4	60.2	6.1°	Hargraves et al. (1997)
10. Diabase dykes (1) (2)	185.5*	68.5	62.4	7.4°	Dairymple et al. (1975)
11. Karroo Ign. (2)	185	62.9	98.3	3.0°	Sichler et al. (1980)
12. Marangudzi Cplx. (1)	185	70.7	105.7	8.7°	Brock (1958)
13. Stormberg Lavas (1)	185	71.6	93.5	3.2°	Kosterov and Perrin (1995)
14. Lembobo Bas.	183	68.7	94.5	7.0°	Henthorn (1981)
South America					
1. Los Colorados Fm.	210	76.0	208.0	8.0°	Vizan et al. (2004)
2. Anari-Tapirapua Fm. (1) (2)	195.6 ± 0.4	65.5	250.0	3.5°	Montes Lauar et al. (1994)
3. Neuquen basin 1 (2)	197	51.0	223.0	6.0°	Iglesia Llanos et al. (2006)
4. Lepa-Osta Arena Fm. (2)	180–186	75.5	129.5	6.0°	Vizan et al. (1995)
5. Neuquen basin 2 (2)	182	74.0	67.0	5.0°	Iglesia Llanos et al. (2006)
6. Marifil Complex (2)	168–178	83.0	138.0	9.0°	Iglesia Llanos et al. (2003)
7. El Quemado Complex (N of 48° S) (1)	153–157	81.0	172.0	5.5°	Iglesia Llanos et al. (2003)

* Ages in Torsvik et al. (2008)

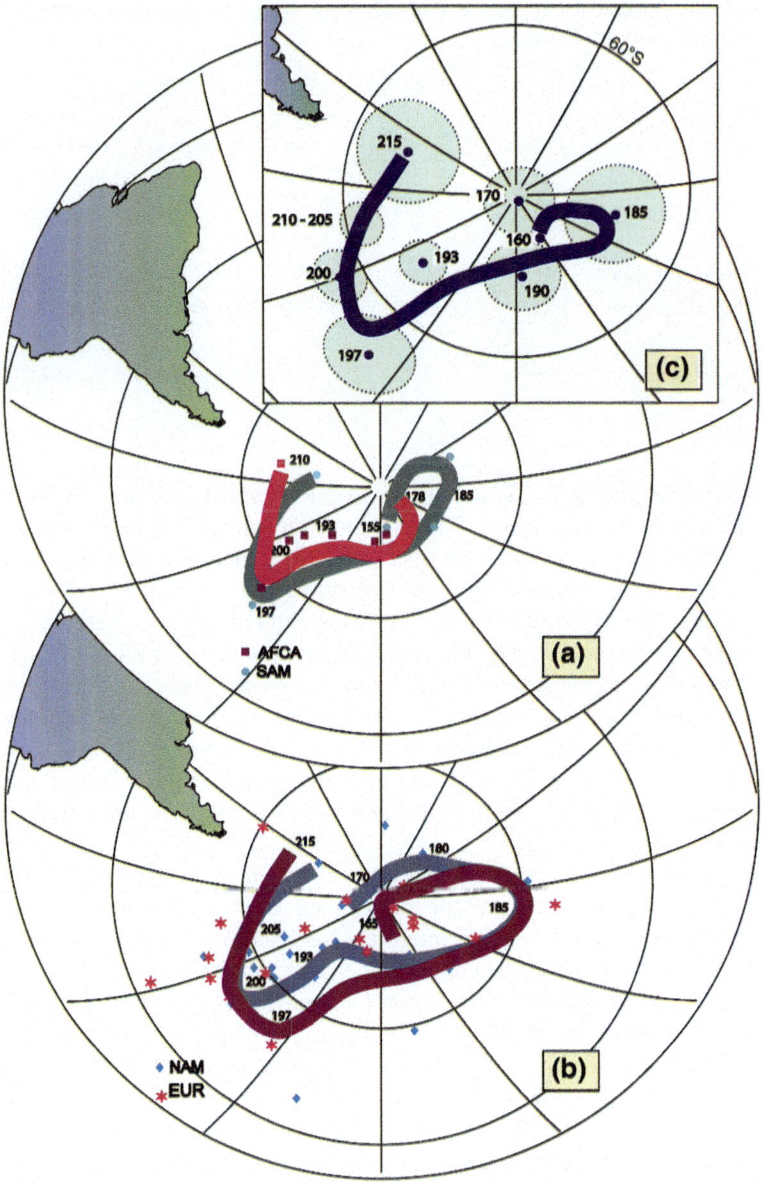

Fig. 2 Jurassic APW paths with mean ages for **a** Africa and South America, **b** North America and Eurasia. Africa, North America and Eurasia have been rotated to South American present-day coordinates using the kinematic parameters in Table 2. **c** Master APW path for Pangea with 95% confidence interval (from Iglesia Llanos and Prezzi 2013)

Table 2 Finite rotation parameters to translate North America, Eurasia and Africa to South American present-day coordinates

Plates	Lat ° S	Long ° E	Angle	Reference
Eurasia				
Eurasia–North America	88.0	−145.0	−42.0	Frei and Cox (1987)
North America–Africa	62.2	−15.9	78.8	Lawver and Scotese (1987)
Africa–South America	45.5	−32.2	−58.2	Lawver and Scotese (1987)
North America				
North America-Africa	67.0	347.7	75.5	Klitgord and Schouten (1986)
Africa-South America	45.5	−32.2	−58.2	Lawver and Scotese (1987)
Africa				
Africa-South America	45.5	−32.2	−58.2	Lawver and Scotese (1987)

Accordingly, during the Late Triassic and lowermost Jurassic, South America rotated (Fig. 8.3) 50° clockwise (CW). Did this movement link to lithospheric plate motion, South America would have overridden the trench as well as the Panthalassa plate, causing major compression in the overriding plate (see Chap. "Lower Jurassic to Early Paleogene Intraplate Contraction in Central Patagonia"). Yet, it is proposed that between ~200 and 90 Ma, the continent presented a retreating-type of margin which transferred important extension to the overriding plate (Royden 1993). Such extension caused the formation of major rifts in the southern Andes that developed mostly as backarc basins particularly in Peru, northern Chile and Argentina. However, recent work by Navarrete et al. (2016) has recognized important intraplate Early Jurassic deformation in Patagonia that could have been linked to this process.

An alternative option is to assign such large movements to the occurrence of true polar wander (TPW). TPW can be determined from displacement of hot spots relative to the spin axis, which are interpreted as fixed points relative to the mantle, once the lithospheric displacements have been accounted for (Table 3).

Hotspots are interpreted to be caused by upwellings from the deep mantle. Although they cannot be absolutely fixed in a convecting mantle, they move presumably much slower than plate motions, and reference frames with fixed or moving hotspots are both useful in distinguishing absolute plate motions and TPW (Steinberger and Torsvik 2008). TPW is a global event and thus, all APW paths from that particular time should portray consistent timing and direction. Therefore, a first approach to detect this phenomenon is to find resemblances in APW paths from different lithospheric plates at that time, and compare the direction of polar movement after removing the effects of plate motions. This is achieved by examining the APW paths in their hot spot reference frames (Sager and Koppers 2000). If TPW was much greater than 'continental drift', all plates on the same hemisphere should show the same sense of rotation (Steinberger and Torsvik 2008). Similarities in the APW paths (Fig. 2) were thus the first hint of TPW in the Early Jurassic.

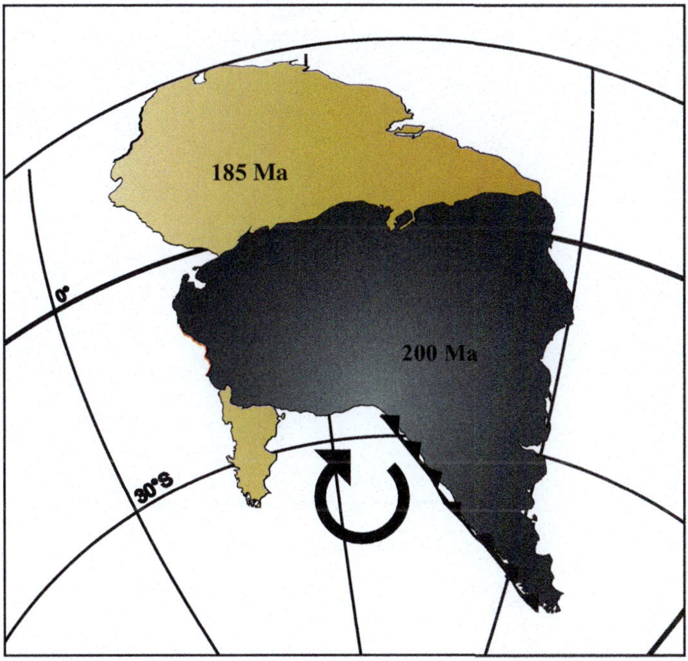

Fig. 3 Paleomagnetic data indicate that during the Early Jurassic South America rotated c. 50° clockwise (from Iglesia Llanos and Prezzi 2013)

Table 3 Morgan's (1983) grid of HS to compensate for the motion of South America respect to the Atlantic Ocean HS. Kinematic models in Table 2 (from Iglesia Llanos and Prezzi 2013)

Time (Ma)	Lat ° N	Long ° E	Ang
160	46.00	7.70	36.00
170	44.70	5.20	40.60
185	43.86	2.49	48.05
190	43.10	1.80	50.90
193	42.31	1.56	53.23
197	41.36	1.29	56.35
200	40.70	1.10	58.70
210–205	40.70	1.10	58.70
215	40.70	1.10	58.70

Another consequence of TPW is rapid regional latitudinal changes and thus, another straightforward way to assess TPW is through paleolatitudinal plots (Fig. 4).

To calculate TPW, continents were rotated to South American coordinates and positioned with respect to the mantle using Morgan's (1983) grid of HS, under the somewhat controversial assumption that HS remained more or less fixed to the

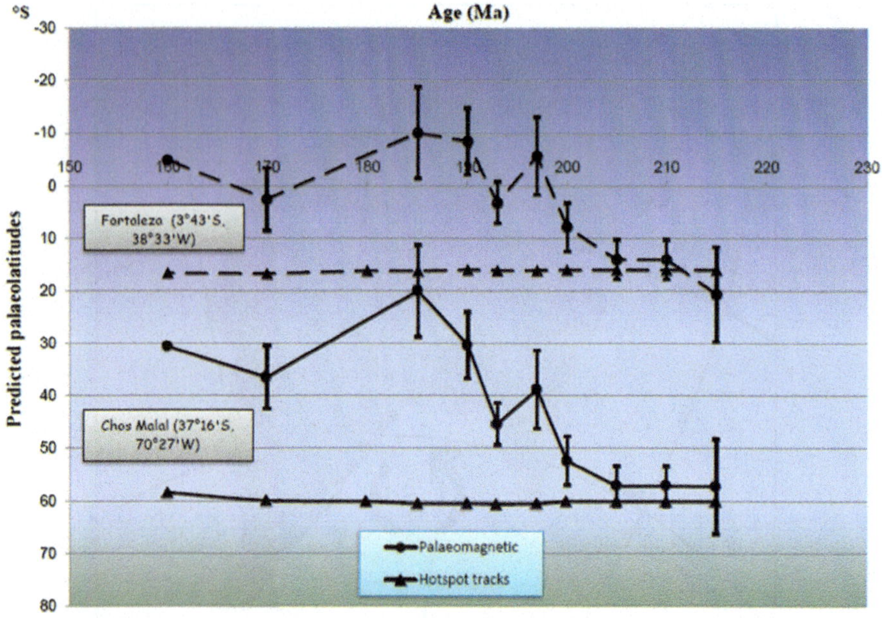

Fig. 4 Paleolatitudes plot supports the occurrence of TPW after 200 Ma. Two sets are shown for two South American localities, one from Brasil and the other from Argentina. One set of paleolatitudes was calculated from the master APW path -palaeomagnetic reference frame-, and the second set from Morgan's (1983) grid of HS (mantle reference frame). Had no TPW occurred, both curves should coincide (from Iglesia Llanos and Prezzi 2013)

mantle (Table 3). Two sets of paleolatitudes were calculated (Fig. 4), one from Argentina—Chos Malal—and the other from Brazil—Fortaleza-, that were derived from the master APW path (circles, Fig. 4) on the one hand, and from the array of HS (triangles, Fig. 4) on the other.

Assuming none or negligible drift of the hotspots frame, both paleolatitudinal sets should coincide if no TPW existed. However, results show that there is hardly any latitudinal change with respect to the HS, even though the continent was actually moving northward. Hence, TPW during the Early Jurassic has been interpreted (Iglesia Llanos and Prezzi 2013).

4 Paleogeography of Pangea

The paleomagnetic method can determine paleolatitudes and orientations of the continents, but due to the symmetry of the time-averaged geomagnetic field about the rotation axis, it cannot provide paleolongitudes. This is done by using the grid of HS. This implies that if the position of a continent such as South America can be determined with respect to the grid of HS at a certain time (Table 3), then plates can be put together within this frame, after applying the corresponding kinematic model (Table 2). Consequently, the combination of relative motions of the plates (Table 2), their motions relative to the hotspots (Table 3), together with paleomagnetic data marking the ancient position of the geomagnetic pole (PP) for a specific locality and time, paleolatitudinally and paleolongitudinally—controlled or "absolute" paleoreconstructions can be obtained.

To achieve absolute paleoreconstructions of Eurasia, Africa and North America, they were first translated to South American coordinates (Table 2), and then rotated once again to compensate for the motion of South America in relation to the Atlantic Ocean hotspots (Table 3). Rotations were performed using the GMAP software (Torsvik and Smethurst 1999). Figures 5 and 6 show changing positions and configurations of a crude Pangea in the Jurassic, in which only the former four major blocks were involved. Hence during the Late Triassic until the Sinemurian ($\sim$210–200 Ma), the supercontinent would have been at its southernmost position with continents rotated counter-clockwise (CCW) with respect to present-day orientation (Fig. 5). Then, Pangea moved northward at higher speeds (c. 4 Myr^{-1}) and rotated c. 50°CW. During the Pliensbachian ($\sim$185 Ma, Fig. 5), the supercontinent achieved its northernmost latitudes and by the end of the Early Jurassic, moved southward again until the Middle Jurassic when it located at similar present-day latitudes (Fig. 6).

5 Does Paleoecology Support Paleomagnetic Data?

Since latitude is one of the main variables that affects climate, it should be possible to correlate paleomagnetic and paleoclimatic data derived from fossils having a geographic distribution limited to certain climatic regions. Far from yielding a very quantitative correlation, a qualitative concordance between the results of completely different methodologies is a convincing argument (Lanza and Meloni 2006).

5.1 Marine Invertebrates

Bivalves are very sensitive to water temperatures. In the Southern Hemisphere during the Hettangian-Sinemurian ($\sim$200 Ma), the boundary between the South Pacific—high latitudes—and Tethyan—low latitudes—Realms was located (Fig. 5) in northern Chile (Damborenea 2001, 2002). According to the paleomagnetic data (Iglesia Llanos and Prezzi 2013), Pangea at this time was located at its southernmost position.

At the end of the Sinemurian and during the rest of the Early Jurassic, this boundary shifted southward until the Toarcian, when high-latitude bivalves became confined (Fig. 6) to the southernmost part of South America (Damborenea 2001, 2002). Paleomagnetic data at this time show that the supercontinent had moved

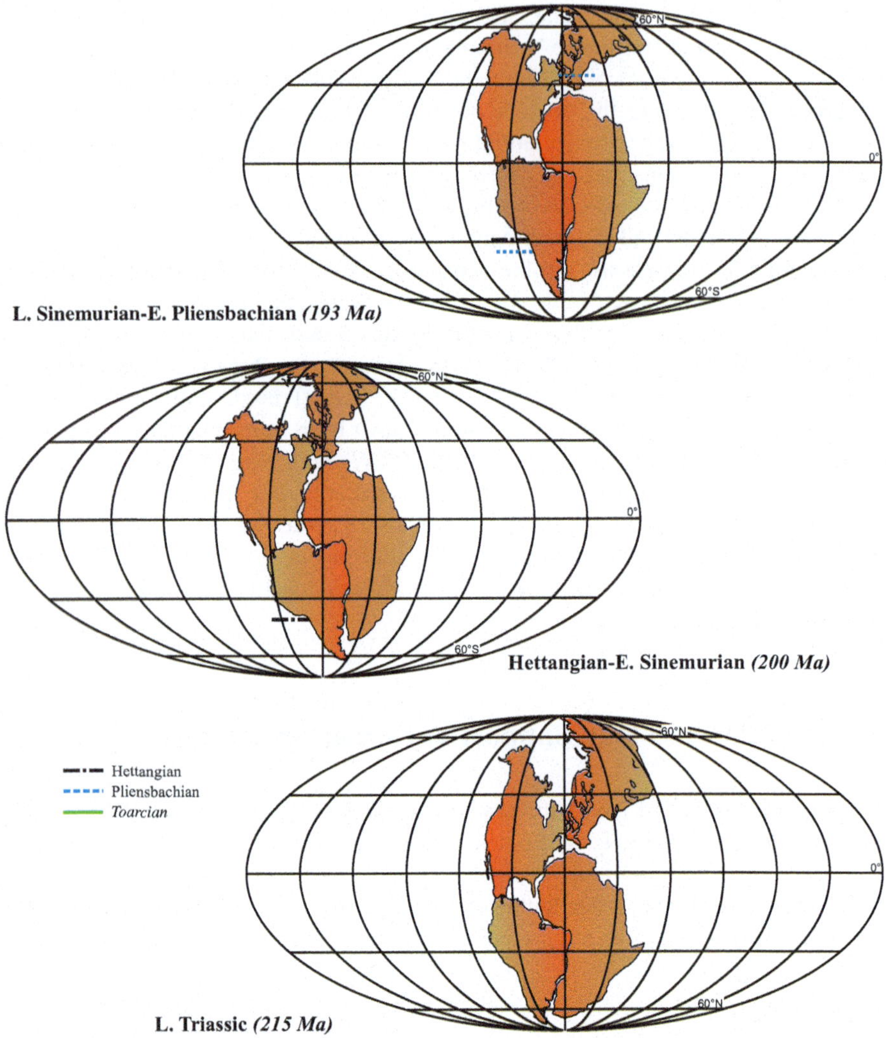

Fig. 5 "Absolute" paleogeographic reconstructions of Pangea for the Early Jurassic. During the Late Triassic-Early Sinemurian, Pangea was located at its southernmost position. The boundary—dashed line—between the South Pacific (high latitudes) and Tethyan (low latitudes) Realms of bivalves was in northern Chile. Then, Pangea moved northward and rotated CW, until the Late Pliensbachian-Early Toarcian when it reached its northernmost position (from Iglesia Llanos and Prezzi 2013)

significantly to the north. On the other hand, the first expansion of colonial corals (warm water temperatures) in west-central Argentina is recorded during the Pliensbachian. Meanwhile in the Northern Hemisphere, average water temperatures indicated a major drop—from 18 °C to 14 °C—during the Pliensbachian–Toarcian boundary (Rosales et al. 2004). Such drop was also recorded in the Arctic,

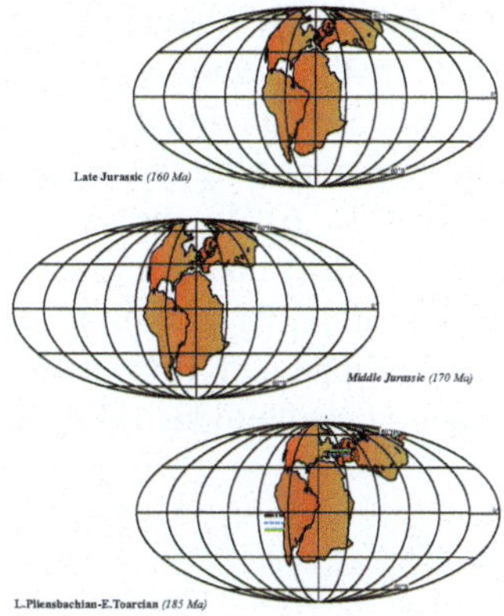

Fig. 6 "Absolute" paleogeographic reconstructions. The boundary between the high latitudes and low latitudes Realms of bivalves shifted toward the south during the late Early Jurassic until in the Early Toarcian, high-latitudes bivalves became confined to the southern extreme of the continent. By the end of the Early Jurassic, the continent ceased to rotate and started to move southward again, reaching similar present-day latitudes

and was the likely cause of the significant turnover recorded by Boreal marine invertebrates such as ammonites, ostracodes, foraminifers and particularly bivalves, during the Pliensbachian-Toarcian. The restoration in the biota in the late Early Toarcian was gradual: first, the Boreal-Arctic followed by the Tethyan taxa (Zakharov et al. 2006). Moreover, the Subtethyan and Low Boreal bivalves migrated southward from north of 68° N—present-day coordinates—in the terminal Pliensbachian, to 55° N at the beginning of the Toarcian (Figs. 8.5–8.6), to return to North Siberia in the Late Toarcian (Zakharov et al. 2006). In addition, ostracodes became extinct during the Late Pliensbachian—Early Toarcian (Arias 2009). Ammonites of the Tethyan Realm kept moving to the north during most of the Early Jurassic until the Late Pliensbachian, when the Boreal Realm started to shift southward replacing the Mediterranean fauna. Then by the Pliensbachian and Earliest Toarcian, ammonites were affected by a major extinction which affected particularly the Tethyan species, that disappeared and were replaced in the Early Toarcian (Macchioni and Cecca 2002). Precisely during this time, corresponding to the Pliensbachian-Early Toarcian, the continent was located at its northernmost position (Fig. 6).

5.2 *Paleoflora*

In the Southern Hemisphere, Volkeimer et al. (2008) reported that pollen *Classopollis*, typical of warm temperatures in the Neuquén Basin, showed abundance not greater than 60% during the Sinemurian-Early Pliensbachian. At this time, the Neuquén basin, presently at c. 37° S, was located at c. 50° S (Fig. 5).

However in the Late Pliensbachian-Early Toarcian, *Classopollis* presence increased to 91–99%. At this time, the first appearance of new group of Araucariaceae (*Callialasporites* spp.) also indicates a climatic amelioration (Volkheimer and Quattrocchio 1981). This time coincides with the recorded northernmost position of the Neuquén basin—c. 25° S—(Fig. 6). In the Northern Hemisphere, the wood *Xenoxylon* characteristic of wetter and cooler climate conditions was distributed widespread in Western Europe, attaining maximum occurrence during the Late Pliensbachian-Early Toarcian (Philippe and Thevenard 1996). The continent at this time was located at the highest palaeolatitudes (Fig. 6).

5.3 Geology

At the beginning of Jurassic, cool conditions apparently prevailed in the Southern Hemisphere. Such conclusion was proposed based on the absence of carbonates vs. the abundance of siliciclastic deposits (e.g. Spalletti et al. 1999, Gómez-Pérez 2003). By then, the continent was located at its southernmost position (Fig. 5). Conversely, by the Late Early–Middle Jurassic (Fig. 6), coal and well preserved flora indicate a humid climate (Volkeimer et al. 2008). Meanwhile in the Northern Hemisphere, a major change in the sedimentological record affected the Tethyan sections at the Pliensbachian–Toarcian boundary, leading the limestone-dominated deposits to be replaced by marls, as the result of a massive crisis in Europe that affected the carbonate platforms. This was assigned to an important cooling that occurred at the Pliensbachian-Toarcian boundary (Suan et al. 2006 and references therein). As already shown, during the Pliensbachian-Toarcian, Pangea was situated at its northernmost position (Fig. 6).

6 Conclusions

A master APW path for Jurassic Pangea was constructed from high-quality paleomagnetic poles derived from Eurasia, North America, Africa and South America. The path yields a distinctive W–E track between the Late Triassic to lowermost Jurassic, followed by a SE–NW track during the Sinemurian to the Pliensbachian, making up a noticeable cusp at ~197 Ma. From c. 197–c. 185 Ma in the Early Jurassic, position of poles changed drastically (near 50°), which imply a minimum angular change of poles of approximately 4° Ma^{-1}. According to APW path, Pangea was subjected to considerable CW rotation between the Sinemurian (~197 Ma) and Pliensbachian (~185 Ma), while it was moving northward. This motion is interpreted to be caused by true polar wander, rather than other type of phenomenon such as lithospheric motion or an artefact produced by long-lived nondipole components of the Earth's magnetic field. Absolute—paleolongitudinally and paleolatitudinally—controlled paleogeographic reconstructions reveal changes in latitudes that are consistent with paleoclimatic proxies derived from marine invertebrates, paleoflora, geological data, and water paleotemperatures derived from the Northern and Southern Hemispheres.

References

Arias C (2009) Extinction pattern of marine Ostracoda across the Pliensbachian-Toarcian boundary in the Cordillera Ibérica, NE Spain: causes and consequences. Geobios 42:1–15

Beck ME Jr (1999) Jurassic and Cretaceous apparent polar wander relative to South America: some tectonic implications. J Geophys Res 104:5063–5067

Butler RF (1992) Paleomagnetism: magnetic domains to geological terranes. Blackwell Scientific Publications, Oxford, UK, p 319

Damborenea SE (2001) Unidades paleobiogeográficas marinas jurásicas basadas sobre moluscos bivalvos: una visión desde el Hemisferio Sur. Anales Acad Nac Cs Ex Fís Nat 53:141–160

Damborenea SE (2002) Jurassic evolution of Southern Hemisphere marine palaeobiogeographic units based on benthonic bivalves. Geobios 35:51–71

Evans ME (1976) Test of the dipolar nature of the geomagnetic field throughout Phanerozoic time. Nat 262:676–677

Frei LS, Cox A (1987) Relative displacement between Eurasia and North America prior to the formation of oceanic crust in the North Atlantic. Tectonophysics 142:111–136

Gómez-Pérez I (2003) An early Jurassic deep-water stromatolitic bioherm related to possible methane seepage (Los Molles Formation, Neuquén, Argentina). Palaeogeogr. Palaeocl 201:21–49

Iglesia Llanos MP (1997) Magnetoestratigrafía y Paleomagnetismo del Jurásico Inferior marino de la Cuenca Neuquina, República Argentina, Ph.D. Thesis, Universidad de Buenos Aires, Buenos Aires, Argentina

Iglesia Llanos MP, Prezzi CB (2013) The role of true polar wander on the Jurassic palaeoclimate. Int J Earth Sci 102:745–759

Iglesia Llanos MP, Riccardi AC, Singer SE (2006) Palaeomagnetic study of Lower Jurassic marine strata from the Neuquén Basin, Argentina: a new Jurassic Apparent Polar Wander Path for South America. Earth Planet Sci Lett 252:379–397

Iglesia Llanos MP, Riccardi AC, Singer SE (2008) Reply to "A comment on Early Jurassic Palaeomagnetic study of Lower Jurassic marine strata from the Neuquén Basin, Argentina: a new Jurassic Apparent Polar Wander Path for South America. Earth Planet Sci Lett 265:316–319

Kent DV, Irving E (2010) Influence of inclination error in sedimentary rocks on the Triassic and Jurassic apparent pole wander path for North America and implications for Cordilleran tectonics. J Geophys Res 115. doi:https://doi.org/10.1029/2009JB007205

Klitgord KD, Schouten H (1986) Plate kinematics of the Central Atlantic. In: Vogt PR, Tulchoke BE (eds) The Geology of North America M, The western North Atlantic region. Geol Soc Am, New York, pp 351–378

Lanza R, Meloni A (2006). The earth's magnetism. An introduction for geologists Springer, Berlin, Heidelberg, New York, p 278

Lawver L, Scotese CR (1987) A revised reconstruction of Gondwanaland. In: McKenzie GD (ed.) Gondwana six: Structure, Tectonics and Geophysics. Am Geophys Union Monogr 40:17–23

Livermore RA, Vine FJ, Smith AG (1983) Plate motions and the geomagnetic field—I. Quaternary and late Tertiary. Geophys J Roy Astron Soc 73:153–171

Livermore RA, Vine FJ, Smith AG (1984) Plate motions and the geomagnetic field—II. Jurassic to tertiary. Geophys J Roy Astr Soc 79:939–961

Macchioni F, Cecca F (2002) Biodiversity and biogeography of middle-late liassic ammonoids: implications for the early Toarcian mass extinction. Geobios 35:165–175
May SR, Butler SR (1986) North American Jurassic Apparent polar wander: Implications for plate motions, paleogeography and Cordilleran tectonics. J Geophys Res 91:11519–11544
McElhinny MW, Brock A (1975) A new palaeomagnetic result from East Africa and estimates of the Mesozoic palaeoradius. Earth Planet Sci Lett 27:321–328
Morgan WJ (1983) Hotspot tracks and the early rifting of the Atlantic. Tectonophysics 94:123–139
Navarrete C, Gianni G, Echaurren A, Folguera A (2016) Episodic Jurassic intraplate compression during supercontinent breakup next to the Karoo LIP in Southwestern Gondwana, Central Patagonia. J Geodyn. doi:https://doi.org/10.1016/j.jog.2016.10.001
Ogg JG, Ogg G, Gradstein FM (2016) Jurassic. In Ogg JG, Ogg G, Gradstein FM (eds) A concise geologic time scale, Elsevier, pp 151–166
Oviedo E, Vilas JF (1984) Movimientos recurrentes en el Permo-Triásico entre el Gondwana Occidental y el Oriental. IX Congreso Geológico Argentino 3:97–114
Philippe M, Thevenard F (1996) Distribution and palaeoecology of the Mesozoic wood genus Xenoxylon: palaeoclimatological implications for the Jurassic of Western Europe. Rev Palaeobot Palyno 91:353–370
Prévot M, Mattern E, Camps P, Daignières M (2000) Evidence for a 20° tilting of the Earth's rotation axis 110 million years ago. Earth Planet Sci Lett 179:517–528
Rapalini AE, Abdeldayem AL, Tarling DH (1993) Intracontinental movements in Western Gondwanaland: a palaeomagnetic test. Tectonophysics 220:127–139
Riccardi AC, Iglesia Llanos MP (1999) Primer hallazgo de un amonite triásico en Argentina. Rev Asoc Geol Argent 54:298–300
Riccardi AC, Damborenea SE, Manceñido MO, Ballent SC (1988) Hettangiano y Sinemuriano marinos en la Argentina. IV Congreso Geológico Chileno, Santiago de Chile, pp C359–374
Riccardi AC, Damborenea SE, Manceñido MO, Ballent SC (1991) Hettangian and Sinemurian (Lower Jurassic) biostratigraphy of Argentina. J South Am Earth Sci 4:159–170
Riccardi AC, Damborenea SE, Manceñido MO, Iglesia Llanos MP (2003) The Triassic/Jurassic Boundary in the Andes of Argentina. Riv Ital Paleont Strat 110:69–76
Rosales I, Quesada S, Robles S (2004) Paleotemperature variations of Early Jurassic seawater recorded in geochemical trends of belemnites from the Basque–Cantabrian Basin, northern Spain. Palaeogeogr Palaeocl 203:253–275
Royden LH (1993) The tectonic expression slab pull at continental convergent boundaries. Tectonics 12:303–325
Sager WW, Koppers AAP (2000) Late cretaceous polar wander of the pacific plate: evidence of a rapid true polar wander event. Science 287:455–459
Spalletti LA, Franzese JR, MacDonald D, Gómez-Pérez I (1999) Palaeogeographic evolution of southern South America during the cretaceous. V Simposio sobre o Cretáceo do Brasil, Serra Negra, Brazil, pp 87–95
Steinberger B, Torsvik TH (2008) Absolute plate motions and true polar wander in the absence of hotspot tracks. Nature 452:620–624
Suan G, Pittet B, Mattioli E, Lécuyer C (2006) Palaeoclimatic and palaeoceanographic changes during the Pliensbachian-Toarcian (Early Jurassic): new results from stable isotope analyses of brachiopod shells. Volumina Jurassica 4:137–138
Torsvik TH, Smethurst MA (1999) Plate tectonic modelling: virtual reality with GMAP. Comput Geosci 25:395–402
Torsvik TH, Dietmar Müller R, Van der Voo R, Steinberger B, Gaina C (2008) Global plate motion frames: Toward a unified model. Rev Geoph 46(RG3004):1–44
Valencio DA, Vilas JF, Pacca IG (1983) The significance of the palaeomagnetism of Jurassic-Cretaceous rocks from South America: predrift movements, hairpins and Magnetostratigraphy. Geophys J R Astron Soc 73:135–151

Van der Voo R, Torsvik TH (2004) The quality of the European permo-triassic paleopoles and its impact on Pangea reconstructions. Timescales of the Paleomagnetic Field, Am Geophys Union Geophys Monograph Series 145:29–42

Vizán H (1998) Paleomagnetism of the Lower Jurassic Lepá and Osta Arena Formations, Argentine, Patagonia. J South Am Earth Sci 11:333–350

Volkheimer W, Quattrocchio M (1981) Distribución estratigráfica de los palinomorfos jurásicos y cretácicos en la faja andina y áreas adyacentes de América del Sur Austral con especial consideración en la Cuenca Neuquina. In: Volkheimer W, Musacchio EA (eds) Cuencas sedimentarias del Jurásico y Cretácico de América del Sur. Comité Sudamericano del Jurásico y Cretácico 2, Buenos Aires, pp 407–444

Volkheimer W, Rauhut OWM, Quattrocchio M, Martínez MA (2008) Jurassic Paleoclimates in Argentina, a review. Rev Asoc Geol Argent 63:549–556

Zakharov VA, Shurygin BN, Il'ina VI, Nikitenko BL (2006) Pliensbachian–Toarcian biotic turnover in North Siberia and the Arctic region. Stratigr Geol Correl 14, pp 399–417

Lower Jurassic to Early Paleogene Intraplate Contraction in Central Patagonia

César R. Navarrete, Guido M. Gianni, Andrés Echaurren and Andrés Folguera

Abstract Breakup and dispersion stages of Gondwana were ruled by crustal extension. In Patagonia, this regime was associated with the opening of extensional basins from the Jurassic onward, a process that was interrupted by the Andean Orogeny. New data generated from the hydrocarbon exploration allowed identifying Jurassic to Eocene contractional deformations, previously not registered in Central Patagonia. We summarize in this chapter evidence of five compressional events intercalated with the extensional regime that affected Central Patagonia from the Early Jurassic to the Paleogene. These events, denominated "C1," "C2," "C3," "C4," and "C5," acted diachronicronously producing tectonic inversion of the Jurassic–Cretaceous depocenters. The first three contractional pulses occurred during the Jurassic, while the two remaining were Late Lower Cretaceous and Early Paleogene. The origin of this compressive activity would be linked to different processes that comprehended from thermal weakening of the crust produced by expansion of the Karoo thermal anomaly in Mid- to Late Jurassic times; the southward continental drift since the Early Jurassic; the ridge push generated by the opening of Weddell Sea since Mid-Jurassic times; and two mid-ocean ridge collisions during the Cretaceous.

C. R. Navarrete (✉)
Departamento de Geología, Universidad Nacional de la Patagonia San Juan Bosco, Comodoro Rivadavia, Chubut, Argentina
e-mail: cesarnavarrete@live.com.ar

G. M. Gianni
Instituto Geofísico Sismológico Ing. Volponi (IGSV), Universidad de Nacional San Juan, San Juan, Argentina

G. M. Gianni
Consejo Nacional de Investigaciones Científicas y Técnicas (CONICET), Buenos Aires, Argentina

A. Echaurren
Instituto de Estudios Andinos Don Pablo Groeber, UBA—CONICET. Departamento de Ciencias Geológicas, FCEN, Universidad de Buenos Aires, Buenos Aires, Argentina

A. Folguera
Instituto de Estudios Andinos (IDEAN), Consejo Nacional de Investigaciones Científicas y Técnicas (CONICET), Universidad de Buenos Aires, Buenos Aires, Argentina

A. Folguera et al. (eds.), *The Evolution of the Chilean-Argentinean Andes*,
Springer Earth System Sciences, https://doi.org/10.1007/978-3-319-67774-3_10

Keywords Jurassic · Cretaceous · Within-plate contraction · South America absolute movement · Karoo thermal anomaly · Patagonia

1 Introduction

In this chapter, we review intraplate compressional zones in Patagonia in a context of generalized extensional deformation associated with Pangea (Gondwana) breakup. Even though mechanisms and timing of supercontinent breakup have been discussed in numerous studies, synchronous intraplate contraction is still matter of debate (Dietz and Holden 1970; Veevers 2004; Murphy and Nance 2013; Whalen et al. 2015). Initial dispersal of Pangea in Late Triassic times was associated with the emplacement of the Central Atlantic Magmatic Province (CAMP) LIP (200 Ma), while during Lower Jurassic times, the Karoo–Ferrar–Chon Aike LIP (183 Ma) affected southern Gondwana (Fig. 1) (Riley and Knight 2001; Jourdan et al. 2005).

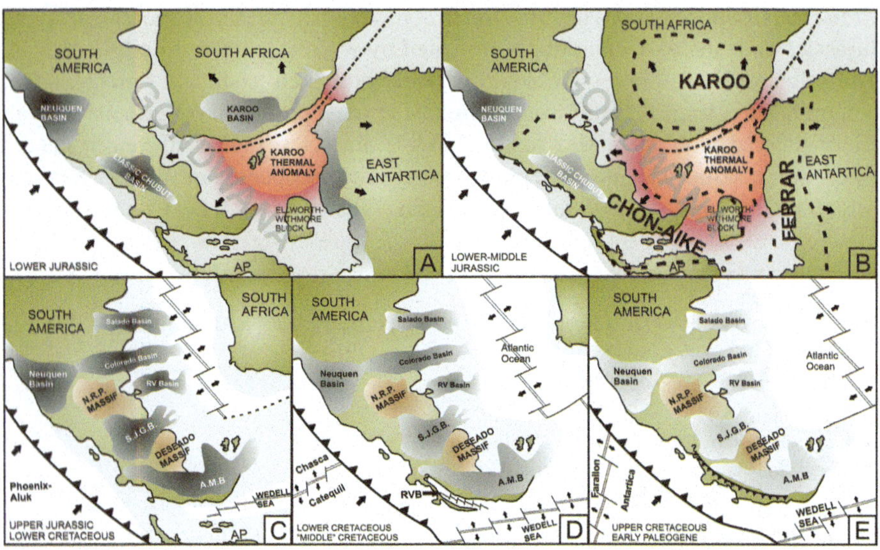

Fig. 1 a–e Patagonian region in the context of western Gondwana from Lower Jurassic to Paleogene with the development of Jurassic and Cretaceous basins (based on Storey et al. 1992; Elliot and Fleming 2000; Golonka and Bocharova 2000; Riley and Knight 2001; Jokat et al. 2003; Franke 2013; Rapela et al. 2005; Seton et al. 2012; Aguirre-Urreta et al. 2011; Maloney et al. 2013; Hastie et al. 2014; Frizon de Lamotte et al. 2015; Granot and Dyment 2015; Müller et al. 2016 and Ghiglione et al. 2016). The variable intensity of the colors inside the basins represents qualitatively the subsidence rate. AP.: Antarctic Peninsula; AMB: Austral and Magallanes Basins; NRP: North Patagonian; RVB: Rocas Verdes Basin; SJGB: San Jorge Gulf Basin

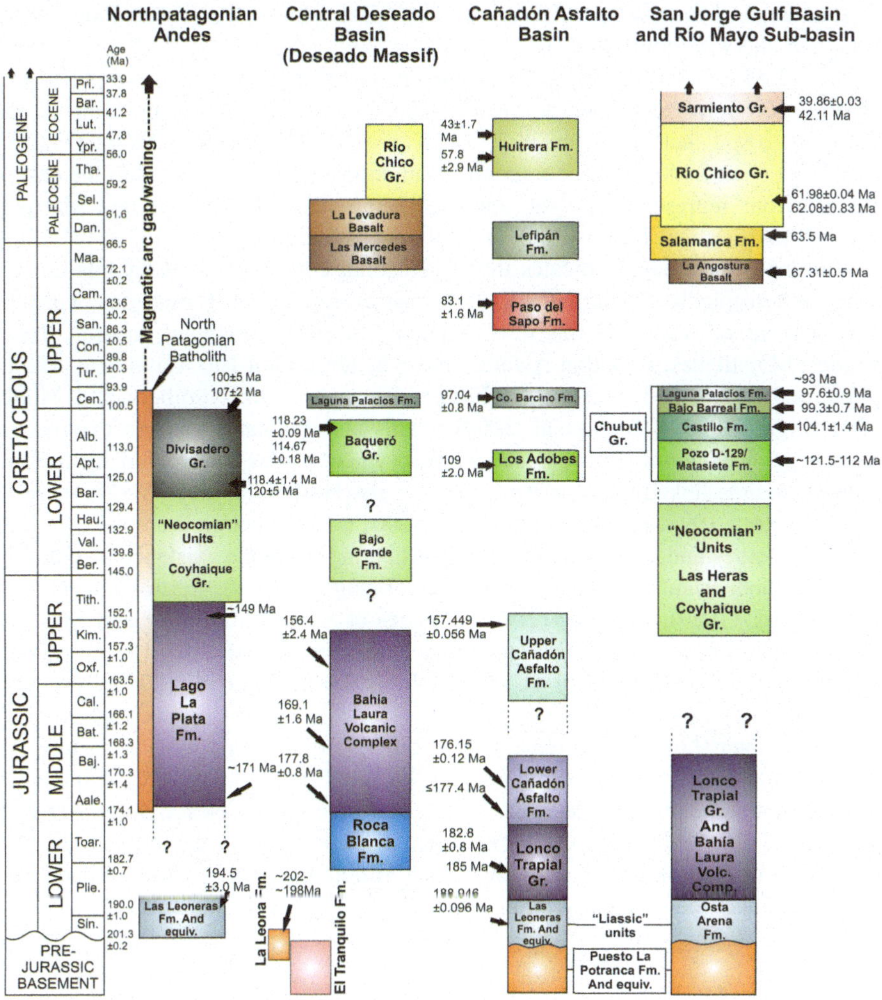

Fig. 2 Stratigraphic chart of Jurassic to Eocene units in Central Patagonia discussed in this work, based in: Pesce (1979), Rapela et al. (1988), Fitzgerald et al. (1990), Mazzoni et al. (1991), Varela et al. (1991), Pankhurst et al. (1993, 2000), Homovc and Constantini (2001), Rolando et al. (2002), Pankhurst et al. (2003), Blesa (2004), Rolando et al. (2004), Paredes et al. (2007), Raigemborn et al. (2010), Césari et al. (2011), Orts et al. (2012), Perez Loinaze et al. (2013), Cúneo et al. (2013), Dunn et al. (2013), Paredes et al. (2013), Clyde et al. (2014), Suárez et al. (2014), Navarrete et al. (2015), Navarro et al. (2015), Márquez et al. (2016), Echaurren et al. (2016)

Jokat et al. (2003) suggest that after LIP emplacement, the opening of the Weddell Sea began at 155 Ma, associated with extensional deformation described by Golonka and Bocharova (2000).

Discrete episodes of contractional deformation have been described as associated with accretion of parautochthonous blocks and closure of back arc basins in the surroundings of the CAMP LIP activity as well as other LIPs around Pangea (see Vaughan and Livermore 2005).

From Lower to Mid-Jurassic times, the crust of Central Patagonia stretched in a broad NNW-retroarc trough known as the Liassic Chubut Basin (Suárez and Márquez 2007) that coexisted with the Karoo LIP (Suárez and Márquez 2007; Pankhurst et al. 2000). Marine to continental deposits filled this basin to the east of the arc front represented by the sub-cordilleran batholith (Suárez and Márquez 2007; Mpodozis and Ramos 2008 and references therein).

Common knowledge indicates that extensional conditions in Patagonia continued uninterruptedly through the Late Jurassic and Lower Cretaceous with the development of several basins denominated Colorado, Neuquén, Austral, Rocas Verdes, Magallanes, Canadón Asfalto, and San Jorge Gulf-Río Mayo (Fig. 1C–E) (Aguirre Urreta and Ramos 1981; Fitzgerald et al. 1990; Nürnberg and Müller 1991; Figari 2005; Suárez et al. 2009b). This also indicates that the extensional regime persisted until Late Miocene times, when it was interrupted by a compressive regime linked with the Andean Orogeny (Homovc et al. 1995; Peroni et al. 1995; Chelotti 1997; Rodriguez and Littke 2001). Even though there were already classical studies and even recent works that had indicated the existence of previous compressional events in Patagonia such as Feruglio (1949), Lesta et al. (1980), Barcat et al. (1989), Márquez (2003), Paredes et al. (2006), Suárez and Márquez (2007), Allard et al. (2011), Márquez and Navarrete (2015), Allard et al. (2015), Reimer et al. (1996), Gianni et al. (2015a), Navarrete et al. (2015), previous assumptions persisted.

This chapter shows that extensional conditions in Patagonia were interrupted by five intraplate compressional events ("C1," "C2," "C3," "C4," and "C5") in Lower to Mid-Jurassic, Lower Cretaceous, and Lower Paleogene times. 2D and 3D seismic information distributed in a broad area of Central Patagonia allows a detailed inspection of Jurassic and Cretaceous strata at depth, showing regional unconformities produced contractually at the time of western Gondwana breakup.

2 Geological Framework

Initial stages of Gondwana breakup in Central Patagonia were characterized by a NNW-trending extensional basin known as Liassic Chubut Basin that affected Late Paleozoic low-grade metamorphic and magmatic rocks (Volkheimer 1964; Sesana 1968; Cortés 1988). This basin controlled a Pacific sea embayment that covered parts of Patagonia (Suárez and Márquez 2007) (Fig. 3).

However, the initial Lower Jurassic infill had started with fluvial successions interfingered with retroarc volcanic rocks [Puntudo Alto Formation (Robbiano 1971), Puesto Lizarralde Formation (Cortés 1988), Las Leoneras Formation (Nakayama 1973), Velázquez Formation (Turner 1975), El Córdoba Formation (Robbiano 1971)]. These were followed by sporadic shallow marine conditions and continental sections of the Osta Arena Formation (Herbst Herbst 1966; Malumián

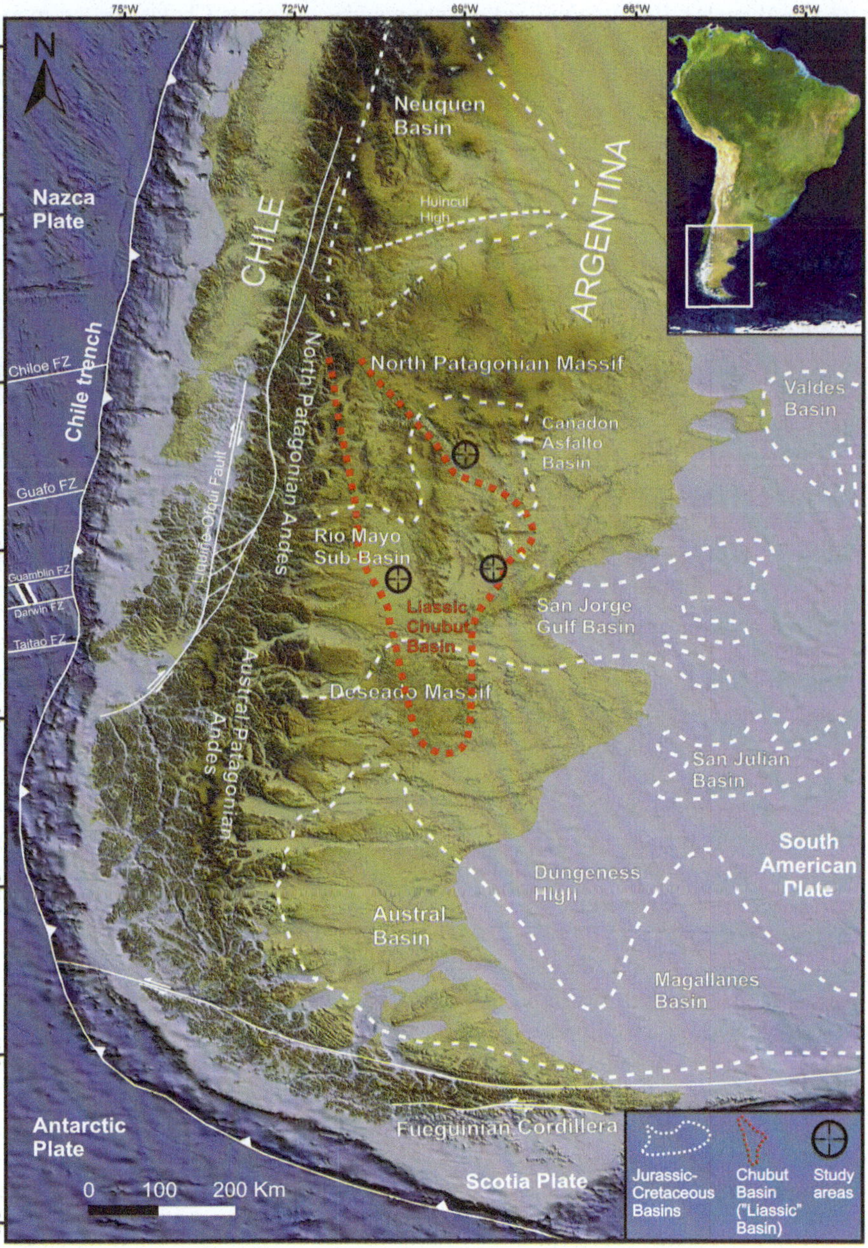

Fig. 3 Present tectonic setting of Patagonia with Jurassic and Cretaceous basins and the studied areas (see references in text) (taken from Navarrete et al. 2016)

and Ploszkiewicz 1976) composed of two members: Lomas Chatas for the marine fossiliferous facies (Malumián and Ploszkiewicz 1976) and Cerro Ferraroti for the continental sections (Ploszkiewicz 1987) and equivalents (Fernández Garrasino 1977). These units are recognized beneath thick packages of Upper Jurassic and Lower Cretaceous strata in Central Patagonia through two wells in the Río Mayo sub-basin and in the San Jorge Gulf Basin. To the south, these units crop out at 49° S over the Deseado Massif where the "Liassic" units are represented by acid pyroclastic rocks of Roca Blanca Formation (Herbst 1965, 1968).

The infill of the Liassic Chubut Basin was recently dated by Cúneo et al. (2013) in 188.946 ± 0.096 Ma (Pliensbachian) through high-precision geochronology (U–Pb CA-TIMS) in tuff beds of the Leoneras Formation (Fig. 2). This is further supported by a Sinemurian U–Pb age of 194.5 ± 3 Ma from detrital zircons obtained in "Liassic" marine deposits over the North Patagonian Andes (Orts et al. 2012) (Fig. 2). Overlaying these rocks, volcanic rocks of the Lonco Trapial Group have been dated in ~185–182 Ma (Zaffarana and Somoza 2012; Márquez et al. 2016). These geochronological data indicate that the Liassic Chubut Basin coexisted with the Chon Aike Magmatic Province to the east (Ferello and Lesta 1973; Franchi et al. 1989; Pankhurst et al. 1998; Suárez and Márquez 2007). This magmatic event developed through three eruptive stages youngling southwestward from 188–178 Ma (V1), 172–162 Ma (V2) to a last pulse in 157–153 Ma (V3) (Pankhurst et al. 2000).

Highly siliceous levels interfingered with Lower Jurassic strata of the Liassic Chubut Basin have been interpreted as crustal-derived components during the early stages of the Chon Aike (Karoo) LIPs (Cortés 1990; Pankhurst and Rapela 1995; Riley et al. 2001).

This extensional regime affecting Central Patagonia continued during the Lower–Middle Jurassic in the Cañadón Asfalto Basin (Figs. 2, 3) (Figari 2005; Fitzgerald et al. 1990), where U–Pb ages have yielded a Mid-Toarcian–Aalenian age (178.766 ± 0.092 Ma, ≤ 177.4 Ma, and 176.15 ± 0.12 Ma) to Oxfordian–Kimmeridgian ages (157.449 ± 0.056 and 157.387 ± 0.045 Ma) (Cúneo et al. 2013).

In Upper Jurassic to Lower Cretaceous times, rifting activity migrated to broad extensional systems crossing Central Patagonia with dominant NW to NNW strikes. These are gathered in the Río Mayo, Cañadón Asfalto, and San Jorge Gulf basins with marine and continental conditions (Fig. 2).

From the Aptian to Cenomanian, Central Patagonia was covered by continental deposits belonging to the Chubut Group (Lesta 1968) that started with lacustrine conditions of the Pozo D-129 Formation (Sciutto 1981), interfingered and overlaid by the fluvial Matasiete Formation (Lesta and Ferello 1972) with Aptian age (118.23 ± 0.09 and 114.67 ± 0.18) (Loinaze et al. and Césari et al. 2011) (Fig. 2). Later, a low sinuosity fluvial environment with increasingly higher pyroclastic components is established corresponding to the Castillo Formation (Lesta and Ferello 1972), passing to a meandering system of the Bajo Barreal Formation (Umazano et al. 2012) and paleosols in reworked pyroclastic horizons of the Laguna Palacios Formation (Sciutto 1981; Genise et al. 2002). Recently,

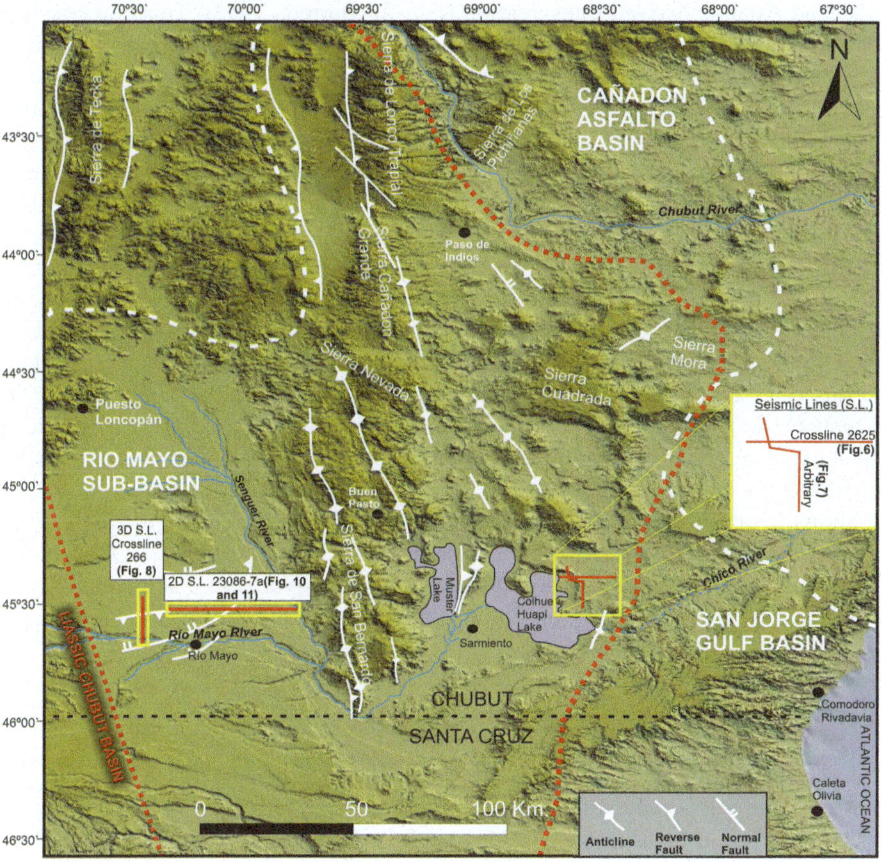

Fig. 4 Outline of the Lower Jurassic Chubut Basin and superimposed Upper Jurassic–Middle Cretaceous Río Mayo, San Jorge, and Cañadón Asfalto basins, with location of seismic information analyzed in this chapter: Colhué Huapi area: 3D Arbitrary, 3D Crossline 2625. Río Mayo sub-basin: 3D Crossline 266 and 2D Line 23086-7a

Suárez et al. (2009b, 2014) established an age of Albian to Cenomanian for the Castillo, Bajo Barreal, and Laguna Palacios formations through U–Pb in zircons (104.1 ± 1.4, 99.3 ± 0.7, 97.6 ± 0.9 Ma, respectively).

In the lower part of Paleogene, a first Atlantic marine transgression is registered on the southern edge of South America, represented in the San Jorge Gulf Basin by the Salamanca Formation (Lesta et al. 1980), recently dated in ca. 65.7–63.5 Ma (Danian) (Clyde et al. 2014) (Fig. 2).

From Eocene to Early Miocene times, a regressive cycle was associated with the deposition of the Sarmiento Group (Fig. 2) in Central Patagonia (Simpson 1941; Dunn et al. 2013), consisting of reworked tuffs in fluvial, aeolian, and lacustrine conditions (Mazzoni 1985). From Oligocene to Early Miocene, marine sediments of the Patagonia or Chenque Formation flooded the San Jorge Gulf Basin representing a second Atlantic derived transgression (Fig. 2) (Barreda 1992). This unit consisted of tuffaceous sandstones, cineritic levels, mudstones, and shales. During the middle Miocene, the Santa Cruz Formation (Fleagle et al. 1995) was deposited in fluvial and aeolian environments in response to a renewed uplift of the Patagonian Andes (Ramos 1999; Blisniuk et al. 2005).

Initial uplift of the Patagonian Andes and adjacent foreland zone has been located in the Late Early to Late Cretaceous as mentioned in Feruglio (1949), Barcat et al. (1989), Lesta et al. (1980), Homovc et al. (2011). More recently, Gianni et al. (2015a, b) and Navarrete et al. (2015) indicate that synorogenic deposition of the Chubut Group was associated with a broken foreland basin. An angular unconformity between the Neocomian deposits and the arc-related rocks of the Divisadero Group (120–100 Ma) in the Main Andes to the west has been related to this early uplift (Suárez and De la Cruz 2000; Folguera and Iannizzotto 2004; Suárez et al. 2009a, b). This Cretaceous contraction affecting Central Patagonia has been explained in a context of high plate convergence rates during the initial stages of formation of the South Atlantic Ocean (Somoza and Zaffarana 2008) and the shallowing of the subducted slab (Barcat et al. 1989), evidenced by an eastward arc migration at that time (Gianni et al. 2015a; see Chap. "Cretaceous Orogeny and Marine Transgression in the Southern Central and Northern Patagonian Andes: Aftermath of a Large-Scale Flat-Subduction Event?").

3 Jurassic Compressive Phase

Jurassic compressional events were identified in three sectors of Central Patagonia, the San Jorge Gulf Basin, the Rio Mayo sub-basin, and the Cañadón Asfalto Basin through 2D and 3D seismic information.

Seismic information and borehole data registered to the east of Colhue Huapi Lake, about 90 km westward to the city of Comodoro Rivadavia, allowed producing an isochronous map that shows extensional structures affecting the Late Paleozoic basement and delimiting "Liassic" depocenters (Fig. 5). This structural framework is characterized by two sets of structures, with NW and NE strikes (Fig. 5), controlling depocenters of the Lower Jurassic Osta Arena Formation, corresponding to a "Liassic" early synrift stage for the San Jorge Gulf Basin. Overlying these Liassic sections, volcanic successions of the Lonco Trapial Group, characterized by high contrasts of acoustic impedance, are controlled by the same structures determining a late synrift stage, limited to the central part of the analyzed area.

Interpretation of 3D seismic sections and 2D seismic lines across extensional structures controlling Early Jurassic depocenters provides evidence of inversion previous to deposition of Early Cretaceous sequences as depicted in Navarrete et al. (2016). For example, in an E-oriented seismic section across Fault "A" and "B," a marked angular unconformity between "Liassic" and "Neocomian" sections indicates inversion of the depocenters in Jurassic times (Fig. 6).

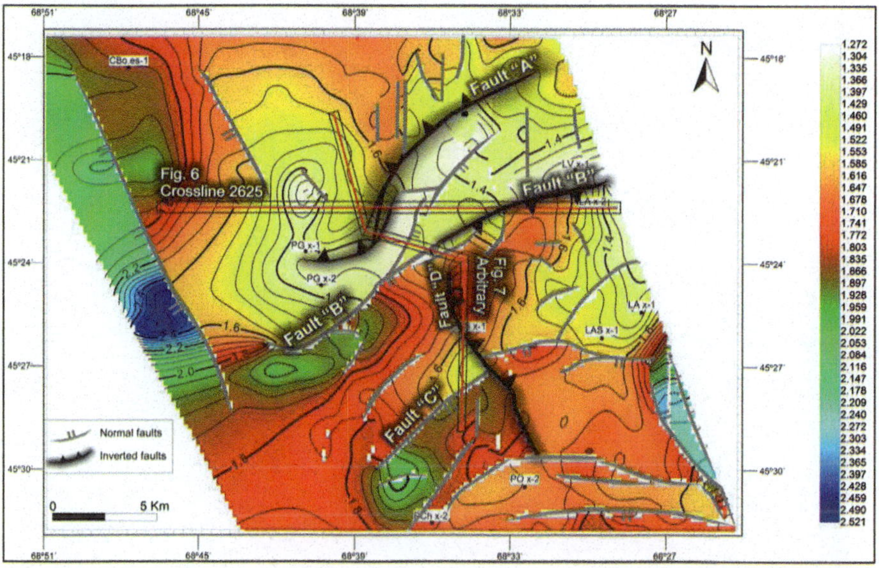

Fig. 5 Isochron map and interpretation of main structures controlling Lower to Middle Jurassic half-grabens in the western San Jorge Gulf Basin. Color scale is TWT (Two Way Travel Time) in msec (see location in Fig. 4)

Fault "B" controls a Jurassic depocenter that is compartmentalized in two smaller depocenters through a NNW-striking Fault "D" (Fig. 5). Seismic information shows that Fault "D" was also inverted previously to the deposition of "Neocomian" strata (Fig. 7). This implies that not only NE-oriented structures were affected by Jurassic contractional deformation but also the NNW ones (Fig. 5).

In the Río Mayo Sub-basin to the west, Suárez and Márquez (2007) had described a Mid-Jurassic orogenic compressive tectonic event associated with the deformation of "Liassic" rocks (Osta Arena Formation), before the deposition of Mid- to Upper Jurassic volcanic rocks assigned to Lago La Plata Formation [~171–149 Ma according to Blesa (2004) and Rolando et al. (2004), respectively (Fig. 2)]. 3D seismic data show at this site a strong angular unconformity between folded "Liassic" sections of Osta Arena Formation (drilled by the YPF.Ch.PRM. es-1 borehole) and tilted beds of the Lonco Trapial Group (Fig. 8). This unconformity is associated with the inversion of a "Liassic" ENE-trending depocenter, filled by the "Liassic" Osta Arena Formation, previously to the deposition of the Lower–Middle Jurassic volcanic rocks of the Lonco Trapial Group.

This compressive event associated with the inversion of Lower Jurassic depocenters at the Río Mayo sub-basin (Lower Jurassic inversion) was diachronous with timing of inversion of Lower Jurassic extensional structures described to the east in the San Jorge Gulf Basin (Colhué Huapí area) (Mid- to Upper Jurassic inversion).

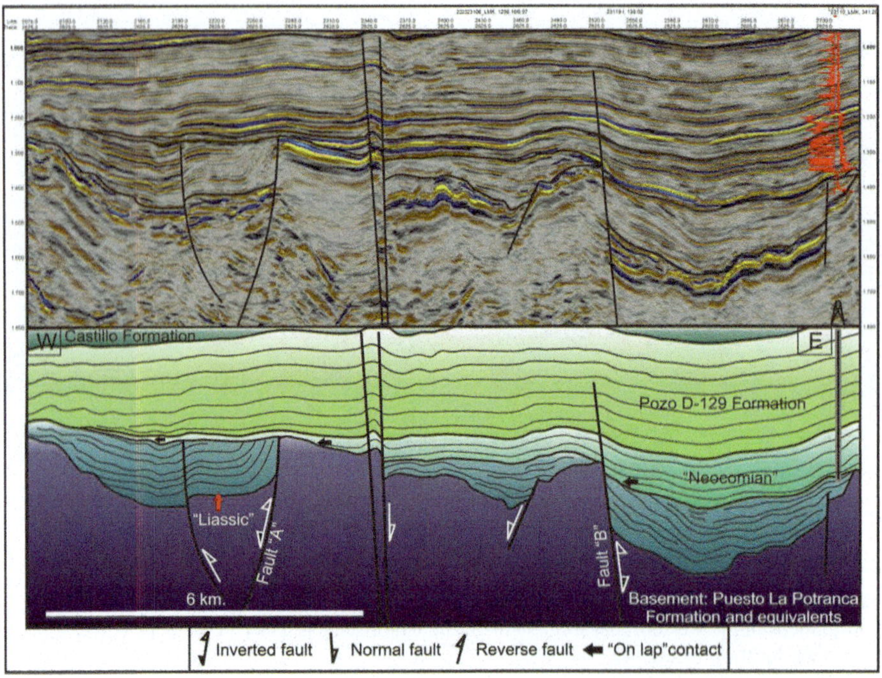

Fig. 6 Seismic section (Crossline 2625), where tectonic inversion of Fault "A" (left) and Fault "B" (right) is depicted, associated with folding and tilting of Lower Jurassic units, previously to the deposition of "Neocomian" deposits. See Figs. 4 and 5 for location in map view. Vertical scale is TWT (Two Way Travel Time) in msec (modified from Navarrete et al. 2016)

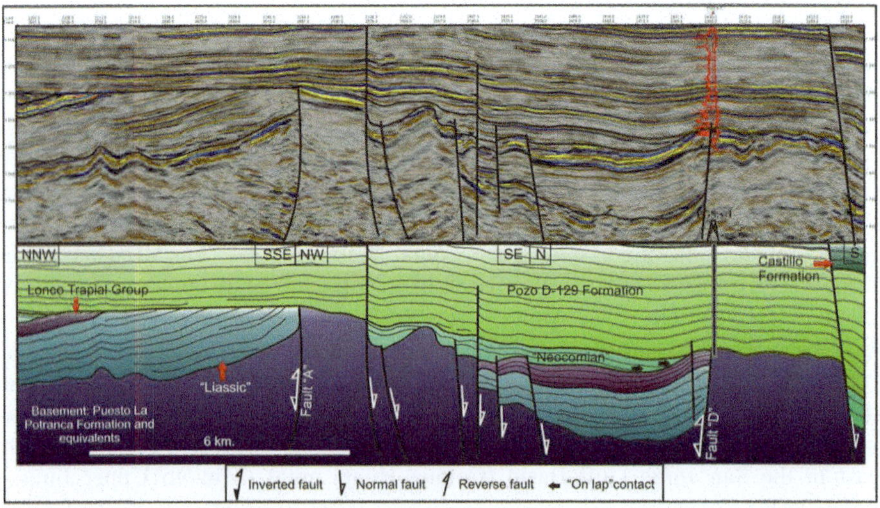

Fig. 7 Seismic section (Arbitrary) that shows selective inversion of Lower to Middle Jurassic half-grabens in the western San Jorge Gulf Basin. See Figs. 4 and 5 for location in map view. Vertical scale is TWT (Two Way Travel Time) in msec (modified from Navarrete et al. 2016)

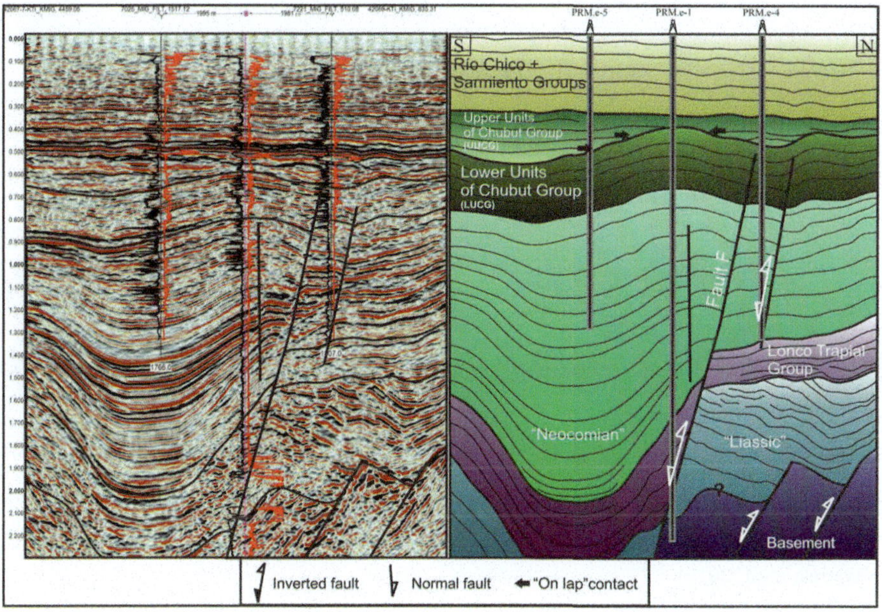

Fig. 8 Angular unconformity (below and to the right) between "Liassic" strata (Osta Arena Fm. —Sinemurian–Pliensbachian) and Pliensbachian–Toarcian volcanic sequences in the footwall of an inverted Lower Cretaceous half-graben of the Río Mayo sub-basin (see Fig. 4 for location in map view). Vertical scale is TWT (Two Way Travel Time) in msec (modified from Navarrete et al. 2016)

At the Cañadón Asfalto Basin, 2D seismic lines show three synrift superimposed episodes that span from the Lower Jurassic to Lower Cretaceous times (see Echaurren et al 2016 for a recent synthesis). Here, Jurassic extension was interrupted by a contractional event associated with clockwise block rotations and locally inversion of extensional structures during the Upper Jurassic to Lower Cretaceous (Somoza 1994; Geuna et al. 2000; Allard et al. 2011; Márquez and Navarrete 2011; Navarrete et al. 2016).

4 Cretaceous–Paleogene Compressive Phases

Cretaceous and Paleogene compressional events were also identified in the Rio Mayo sub-basin (Figs. 3, 4) through seismic information. The NE-Paso Rio Mayo Depocenter is limited by a normal fault (Fault E) roughly 50 km long (Fig. 9). Extensional activity of these structures began during the Lower Jurassic generating the space that accommodated thick sections of the Lonco Trapial Group that

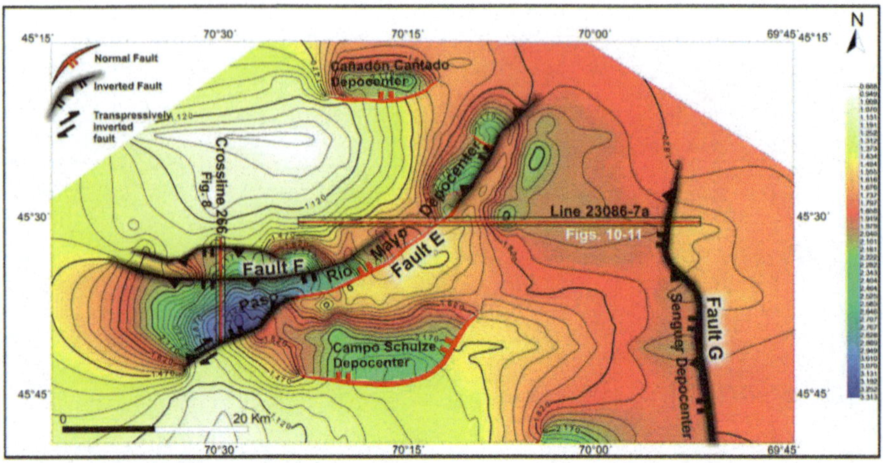

Fig. 9 Isochron map of the top prerift ("Liassic" + Paleozoic Basement) that shows four depocenters and location of seismic lines/sections (colors indicate depth in meters, see vertical scale to the right) (the location of the area is on Fig. 4). Color scale is TWT in msec (modified of Navarrete et al. 2015)

represents an early synrift stage (Figs. 8, 10). The maximum subsidence of this depocenter occurred during the deposition of the Paso de Río Mayo Formation ("Neocomian" unit), reaching up to 3,100 m (Figs. 8, 10). Then, the "Lower Units of Chubut Group" (LUCG) was deposited overlaying this units.

During the "Upper Units of Chubut Group" (UUCG) deposition, seismic line 23086-7a depicts tectonic inversion of an E–W fault (Fig. 10). This positive inversion generated a local depocenter where the UUCG accumulated syntectonically (Fig. 8, 10). Therefore, a synorogenic character of the UUCG is recognized through a strong internal angular unconformity and onlapping reflectors over the folded LUCG (Figs. 8, 10). On the other hand, the overlying Cenozoic units are preserved resting upon previous Mesozoic deposits without any deformation, constraining the age of the compressive deformation to the UUCG age.

To the east, an N-trending structure, recognized in a 2D seismic line (Figs. 9, 10), shows a similar kinematic history to the faults that delimit the Paso Río Mayo Depocenter. Here, the Lonco Trapial Group and Río Mayo Formation are characterized by synrift geometries (Fig. 10), with the LUCG exhibiting a tabular-shaped geometry. Contrastingly, the UUCG shows an internal unconformity and upper units onlapping tilted lower reflectors corresponding to the base of this group, which implies a syntectonic character (Fig. 10).

Maximum thicknesses of the Eocene Río Chico Group occur to the west of the inverted depocenter, indicating its syntectonic character, showing that the

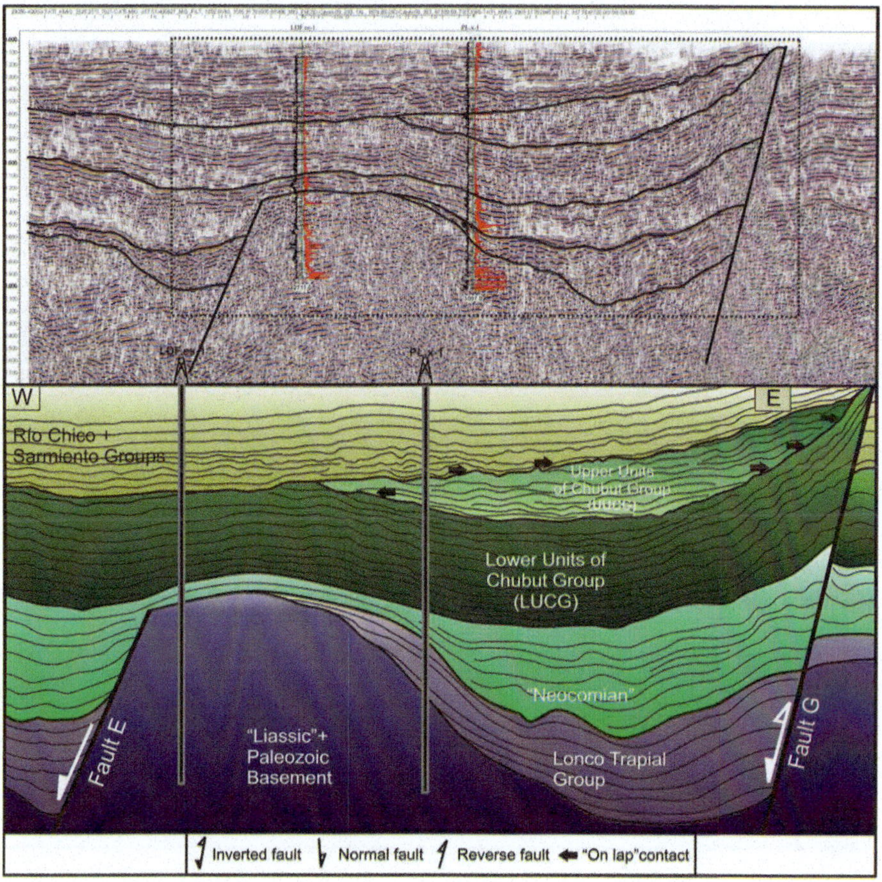

Fig. 10 2D seismic line 23086-7a associated with the Río Mayo and Río Senguer depocenters. An unconformity between the upper and lower terms of the Chubut Group is interpreted, associated with tectonic inversion. Two younger stages of tectonic inversion are interpreted that produced the wedging of reflectors from west to east in the Rio Chico Formation and the deformation of the upper Cenozoic units. Vertical scale is TWT in msec. (Modified from Navarrete et al. 2015)

Cretaceous orogenic event continued in Paleogene times, with a final stage during the Neogene (Peroni et al. 1995; Homovc et al. 1995). By flattening the same seismic line to the base of Sarmiento Group, this relation gets more evident (Fig. 11).

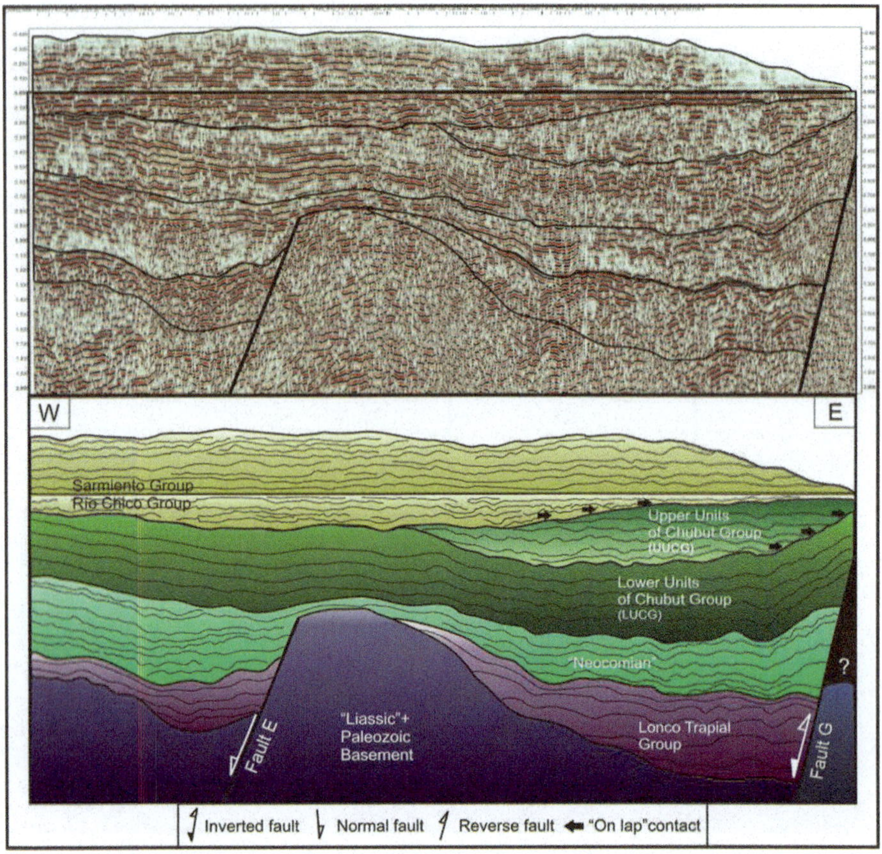

Fig. 11 2D seismic line 23086-7a flattened at the base of Sarmiento Group where the synorogenic character of the Río Chico Group and the Upper Units of Chubut Group (UUCG), produced by tectonic inversion of the Fault G, is noted. Vertical scale is TWT in msec (modified of Gianni et al. 2015a)

5 Timings of Intraplate Contractional Events in Patagonia

Seismic information analyzed in previous sections reveals that the extensional regime related to Gondwana breakup was temporally interrupted by intraplate contractional events in Central Patagonia. While in the Río Mayo sub-basin, a contractional intraplate deformation ("C1") took place in ~188–185 Ma, (Figs. 8, 12), in the San Jorge Gulf and Cañadón Asfalto basins a younger inversion event ("C2") is determined to the ~170–163 Ma time interval (Navarrete et al. 2016) (Fig. 12). Then, during ~157–130? Ma, an additional contractional intraplate event ("C3") may be distinguished based on the age of the folded Cañadón

Calcáreo Formation (equivalent to Upper Cañadón Asfalto Formation) and the flat-lying Chubut Group (Aptian–Cenomanian) (Volkheimer et al. Volkheimer et al. 2009; Allard et al. 2011; Márquez and Navarrete 2011; Ranalli et al. 2011) (Fig. 12). These Jurassic Compressive Phases keep strong similarities with the intraplate contractional events described for the southern Neuquén Basin to the north ($\sim 40°$ S; Fig. 3). There, the deformational pulses are linked to the uplift of an E- to ENE-oriented intraplate belt known as Huincul High (see Mosquera and Ramos 2006; Leanza 2009; Naipauer et al. 2012, for recent syntheses), where positive tectonic inversion and wrenching events are constrained to the Sinemurian to Callovian–Oxfordian (encompassing "C1" to "C2" events in this work) with late events in the Kimmeridgian and Valanginian ("C3") (Mosquera and Ramos 2006). Similarly, south of this feature, the Upper Triassic to Lower Jurassic Central Patagonian Batholith (42–44° S) is overlain by volcano-sedimentary rocks of 190–185 Ma, suggesting an episode of exhumation that exposed these plutonic rocks in Lower Jurassic times (Zaffarana et al. 2014).

On the other hand, the Cretaceous–Paleogene Compressive Phase began with the event denominated "C4" during the deposition of the upper units of the Chubut Group, likely from the Bajo Barreal Formation. Considering the Cenomanian to early Turonian age of the upper units of the Chubut Group ($\sim 100 \sim 93$ Ma, Suárez et al. 2014) (Fig. 2), the age of the "C4" event must to be restricted to the lapse $\sim 100 \sim 93$ (Fig. 12). This compressive event explains why the Maastrichtian Atlantic marine transgression corresponding to the Salamanca Formation did not overcome the central sector of the Patagonian, implying the existence of a latest

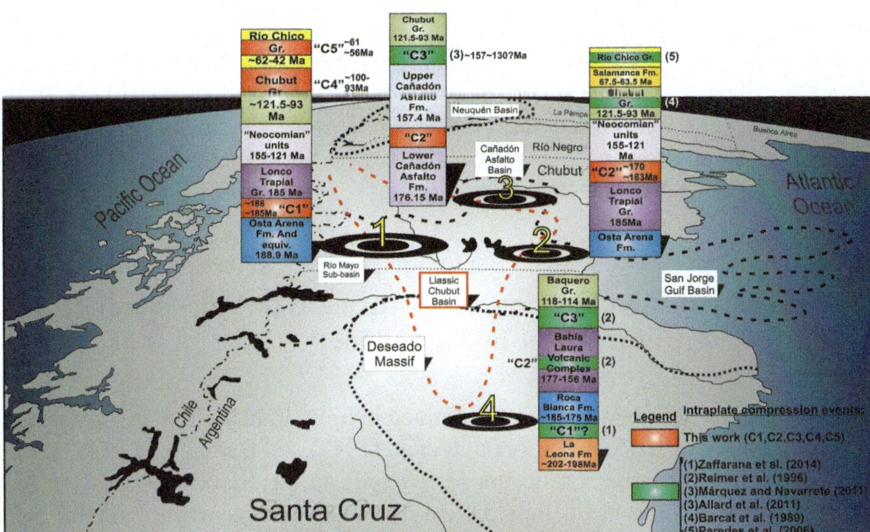

Fig. 12 Schematic representation showing the Jurassic to Paleocene compressive within-plate events in Patagonia interpreted in this chapter

Cretaceous topographic barrier to the west (see Navarrete et al. 2015) (see Chap. "Cretaceous Orogeny and Marine Transgression in the Southern Central and Northern Patagonian Andes: Aftermath of a Large-Scale Flat-Subduction Event?"). Similarly, a compressive event was defined for the Albian–Cenomanian to the north (~100 Ma ~similar with our "C4"), associated with synorogenic deposition of the Neuquén Group (Fennell et al. 2015).

Finally, a Paleogene compressive event is responsible for the synorogenic deposition of the Río Chico Group with a Paleocene (Danian–Selandian) to Mid-Eocene (Lutetian?) age (Raigemborn et al. 2010; Clyde et al. 2014), which circumscribes this "C5" event to ~61 to ~56? Ma (Fig. 12).

6 Intraplate Compressional Stress-Field Orientation in Patagonia

Jurassic contractional deformation inverted selectively faults striking in the ENE and NE directions in the Río Mayo sub-basin and in the San Jorge Gulf Basin (Figs. 4, 5). Therefore, Lower to Mid-Jurassic contractional deformation in Patagonia inverted normal faults striking close to ENE, NE, and WNW directions, implying a main stress ($\sigma 1$) disposed NNW or NNE (Fig. 13a). These paleostress determinations in Northern and Central Patagonia are compatible with stresses calculated for other intraplate belts in South America such as ~1,200-km-long Kimmeridgian Solimôes megashear that crosses the Amazonas Basin in an ENE sense (Caputo et al. 1991, 2014) (Fig. 14a).

On the other hand, contractional deformation beginning in the Late Lower Cretaceous selectively inverted previous extensional structures with E–W and N–S directions (Fig. 13b). Taking into account this, the largest horizontal stress ($\sigma 1$) since Late Early Cretaceous times should have been oriented approximately in the SW direction (Fig. 13b). In contrast, the Late Paleocene compression affected only the N–S-trending structures. Therefore, the largest horizontal stress ($\sigma 1$) is interpreted aligned in a SW direction (Fig. 13c).

7 Discussion: Possible Causes of Jurassic to Paleocene Intraplate Contraction in Patagonia

Jurassic deformation in Patagonia is explained in light of different hypotheses based on recent numerical models of Cloetingh et al. (2013) and Burov and Gerya (2014). Intraplate Jurassic contraction occurs peripherally to the Karoo–Ferrar LIP which would imply an E-oriented stress field, which is not consistent with the paleostress estimations that are close to an N direction (Figs. 13a, 14a).

Recent global plate reconstructions of Müller et al. (2016) describe sudden changes in plate motion of South America in Jurassic times that could be associated

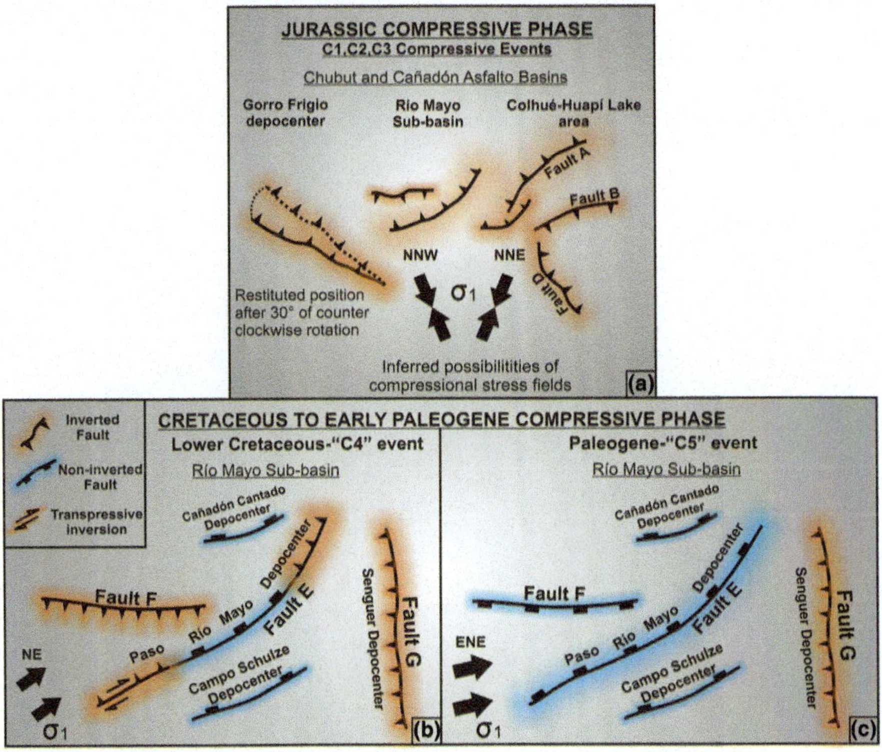

Fig. 13 Cartoon showing the structures that were partly inverted in Northern and Central Patagonia and inferred direction of the largest stress (σ1) for each time: **a** During the Jurassic Compressive Phase ("C1," "C2," and "C3" events) in Gorro Frigio, Río Mayo sub-basin, and Colhué Huapi Lake area (for more details, see Navarrete et al. 2016). **b** During the Lower Cretaceous compressive event ("C4") in Río Mayo sub-basin. **c** During the Paleogene compressive event ("C5") in Río Mayo sub-basin (for more details, see Navarrete et al. 2015)

with described intraplate contractional deformation (see also Chap. "The Jurassic Paleogeography of South America from Paleomagnetic Data"). After a motionless stage of South America between ~176 and 170 Ma, a rapid southward motion stage from ~170 to about 150 Ma started with rates of 8–9 cm/yr (Fig. 14b). This last path is compatible with the inferred compressional intraplate stress-field directions in Patagonia (Fig. 14a). An extra source of stress for the youngest "C3" (Upper Jurassic–lower Cretaceous) intraplate deformation could be associated with ridge-push forces linked to initial Weddell Sea opening that started by 155–145 Ma (Jokat et al. 2003) (Fig. 14c).

These intraplate contractional episodes overlapped in space and time with the Chon Aike Magmatic Province, following peaks of high magmatic fluxes V1 (188–178 Ma), V2 (172–162 Ma), and V3 (157–153 Ma) (Pankhurst et al. 2000) (Figs. 14a, c, 15a–f). Consequently, we speculate that Jurassic contractional events in Patagonia

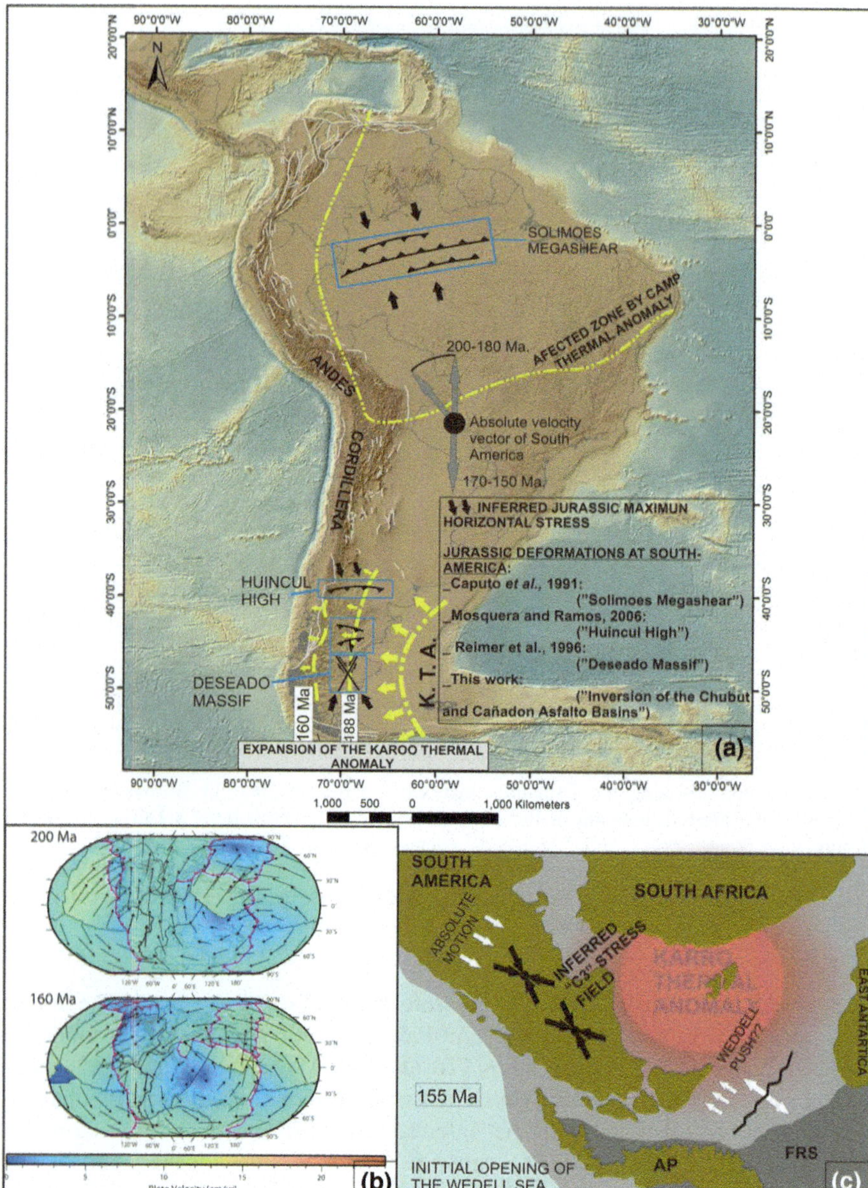

Fig. 14 **a** Jurassic intraplate compressional zones and associated maximum horizontal stresses. Note the spatial relation between intraplate structures and the areas influenced by LIP eruption in Jurassic times. LIP limits based on Pankhurst et al. (2000) and Coltice et al. (2008). **b** Global plate reconstructions from Müller et al. (2016) showing plate motion changes of South America at ~ 170–150 Ma from a NNE to NNW direction, to a southward motion at 8–9 cm/yr. **c** Weddell Sea opening at 155–145 Ma that could explain intraplate deformation "C3" associated with northward ridge-push forces (paleogeographic map taken from Jokat et al. 2003) and the thermal weakening by the Karoo thermal anomaly (KTA) (modified of Navarrete et al. 2016)

could have been related to Karoo plume-related magmatism (Chon Aike Magmatic Province), in a context of crustal thermo-mechanical weakening. Thus, a locally weakened continental lithosphere would enable accommodation of shortening under the imposed ~N–S directed compressional stress field (Figs. 14a).

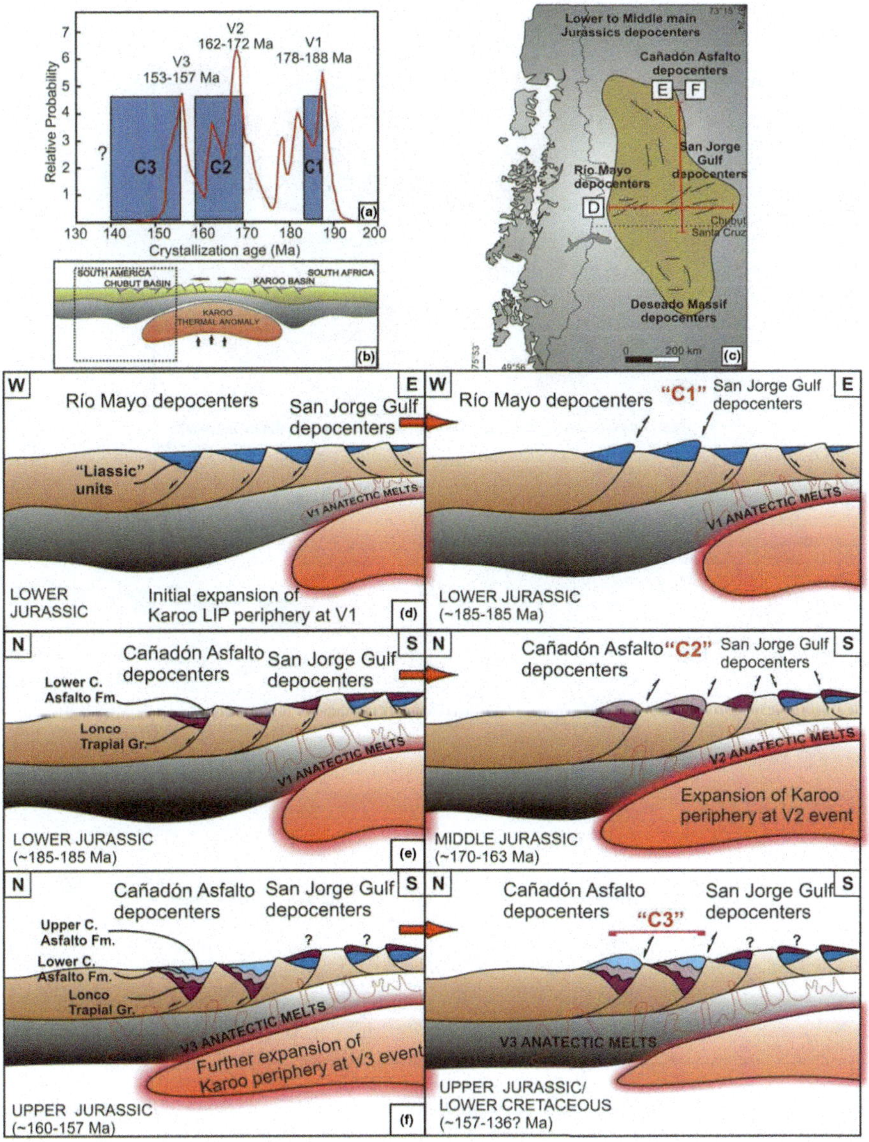

Fig. 15 **a** High flux magmatic events of the Chon Aike Magmatic Provinces linked to Karoo LIP activity of Pankhurst et al. (1998) and Jurassic intraplate compressional events discussed in this work

The Cretaceous–Paleogene Compressive Phases occurred in two stages, in the Cenomanian-early Turonian lapse (~100 ~93 Ma) and in Mid- to Upper Paleocene (~61~56 Ma). According to the global reconstruction by Seton et al.

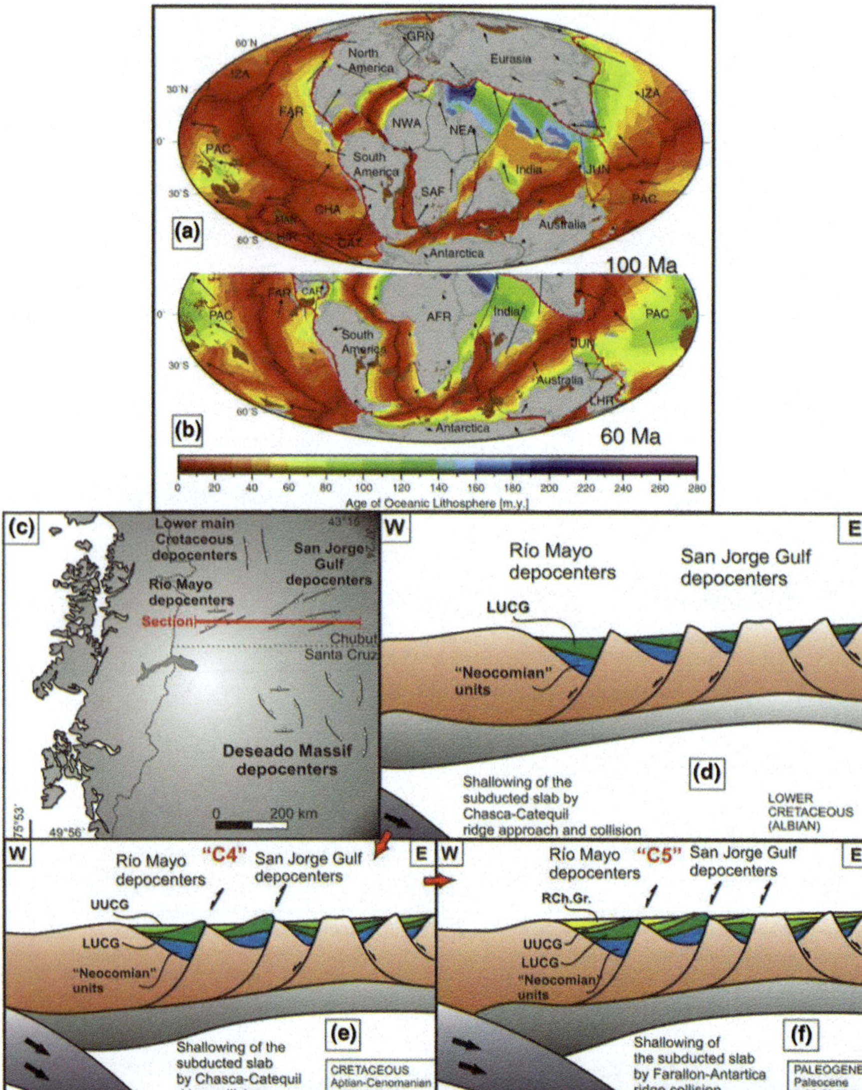

Fig. 16 **a** and **b** Global reconstruction from Seton et al. (2012), showing the Chasca–Catequil ridge collision against South America from Cenomanian to early Turonian (~100 ~93 Ma) **a** and the Farallon–Antarctica ridge collision from ~Santonian to Thanetian (~86~56 Ma). **C** Location of the analyzed area and Section D. **d** Synextensional deposition of Mid- to Upper Jurassic to "Neocomian" and Lower Units of Chubut Group (LUCG). **e** Cenomanian to early Turonian tectonic inversion of the Río Mayo depocenter (RMD), during the Upper Units of Chubut Group (UUCG) deposition ("C4" compressive event). **f** Paleocene tectonic inversion during the Río Chico Group (RCh. Gr.) deposition ("C5" compressive event)

(2012) (Fig. 16 a, b), subduction of two mid-ocean ridges would have occurred beneath Central Patagonia, first associated with the Chasca–Catequil ridge during the Late Lower Cretaceous and the beginning of the Upper Cretaceous ($\sim$120$\sim$95 Ma) (Fig. 16a) and a second associated with the Farallon–Antarctica ridge during the Upper Cretaceous–Paleocene ($\sim$86$\sim$56 Ma) (Fig. 16b). The collision of these two mid-ocean ridges could have been associated with a shallowing of the subducted slab inferred by a Cretaceous arc expansion explaining the broad development of the fold and thrust belt reaching the central sector of Patagonia (Gianni et al. 2015a, b) (see Chap. "Cretaceous Orogeny and Marine Transgression in the Southern Central and Northern Patagonian Andes: Aftermath of a Large-Scale Flat-Subduction Event?").

References

Aguirre Urreta MB, Ramos VA (1981) Estratigrafía y paleontología de la alta cuenca de río Roble. Cordillera Patagónica. VIII Congreso Geológico Argentino 101–138

Aguirre-Urreta B, Tunik M, Naipauer M, Pazos P, Ottone E, Fanning M, Ramos VA (2011) Malargüe Group (Maastrichtina-Danian) deposits in the Nuequén Andes, Argentina: implications for the onset of the first Atlantic transgression related to Western Gondwana break-up. Gondwana Res 19: 482–494. doi10.1016/j.gr.2010.06.008

Allard JO, Giacosa R, Paredes JM (2011) Relaciones estratigráficas entre la Formación los Adobes (Cretácico inferior) y su sustrato Jurásico: implicancias en la evolución tectónica de la cuenca de Cañadón Asfalto, Chubut, Argentina. In: Leanza HA, Franchini M, Impiccini A, Pettinari G, Sigismondi M, Pons J, Tunik M (eds) XVIII Congreso Geológico Argentino: 988–989

Allard J, Foix N, Rodriguez A, Sánchez F (2015) Evolución tectosedimentaria de la Cuenca del Golfo San Jorge en su margen occidental. XVI Reunión de tectónica, Argentina, pp 116–117

Barreda DV (1992) Muricingulisporis chenquensis, una nueva especie de espora de Pteridophyta del Terciario de Patagonia, Argentina. Ameghiniana 29:347–351

Barcat C, Cortiñas JS, Nevistic VA, Zucchi HE (1989) Cuenca Golfo San Jorge. In: Chebli GLS (ed) Cuencas Sedimentarias Argentinas. Serie Correlación Geológica 6:319–345

Blesa A (2004) Geology and mineralization of the Esquel area, Patagonia, Argentina. Msc. Thesis. Colorado School of Mines, Boulder

Blisniuk PM, Stern LA, Chamberlain CP, Idleman B, Zeitler PK (2005) Climatic and ecologic changes during Miocene surface uplift in the Southern Patagonian Andes. Earth Planet Sci Lett 230:125–142

Burov E, Gerya T (2014) Asymmetric three-dimensional topography over mantle plumes. Nature 513:85–103. https://doi.org/10.1038/nature13703

Caputo MV (1991) Solimões megashear: intraplate tectonics in northwestern Brazil. Geology 19:246–249

Caputo MV (2014) Juruá Orogeny: Brazil and Andean Countries. Braz J Geol 44:181–190

Césari SN, Limarino CO, Llorens M, Passalía MG, Perez Loinaze V, Vera EI (2011) High-precision late Aptian U/Pb age for the Punta del Barco Formation (Baqueró Group), Santa Cruz Province. Argent J South Am Earth Sci 31:426–431

Chelotti L (1997) Evolución tectónica de la Cuenca del Golfo San Jorge en el Cretácico y Terciario; algunas observaciones desde la interpretación sísmica. Bol Inf Petroleras 49:62–82

Cloetingh S, Burov E, Francois T (2013) Thermo-mechanical controls on intra-plate deformation and the role of plume-folding interactions in continental topography. Gondwana Res 24:815–837. https://doi.org/10.1016/j.gr.2012.11.012

Clyde W, Wilf P, Iglesias A, Slingerland R, Barnum T, Bijl P, Bralower T, Brinkhuis H, Comer E, Huber B, Ibañez-Mejia M, Jicha B, Krause J, Schueth J, Singer B, Raigemborn M, Schmitz M, Sluijs A, Zamaloa M (2014) New age constraints for the Salamanca Formation and lower Río Chico Group in the western San Jorge Basin, Patagonia, Argentina: Implications for Cretaceous-Paleogene extinction recovery and land mammal age correlations. Geol Soc Am Bull 126(3/4):289–306. https://doi.org/10.1130/B30915.1

Coltice N, Bertrand H, Rey PF, Jourdan F, Phillips BR, Ricard Y (2008) Global warming of the mantle beneath continents back to the Archaean. Gondwana Res 15:254–266. https://doi.org/10.1016/j.gr.2008.10.001

Cortés JM (1988) Descripción geológica de la Hoja 46d "Meseta del Canquel" (escala 1:200000), provincia del Chubut, PhD. Thesis, Universidad de Buenos Aires, Buenos Aires, Argentina

Cortés JM (1990) Estratigrafía de las sucesiones volcano-sedimentarias jurásicas del Chubut central, entre Paso de Indios y El Sombrero. Rev Asoc Geol Argentina 45:69–84

Cúneo R, Ramezani J, Scasso R, Pol D, Escapa I, Zavattieri AM, Bowring SA (2013) High-precision U-Pb geochronology and a new chronostratigraphy for the Cañadón Asfalto Basin, Chubut, central Patagonia: implications for terrestrial faunal and floral evolution in Jurassic. Gondwana Res 24:1267–1275

Dietz RS, Holden JC (1970) Reconstruction of Pangaea: breakup and dispersion of continents, Permian to present. J Geophys Res 75:4939–4956. https://doi.org/10.1029/JB075i026p04939

Dunn RE, Madden RH, Kohn MJ, Schmitz MD, Strömberg CAE, Carlini AA, Ré GH, Crowley J (2013) A new chronology for middle Eocene–early Miocene South American Land Mammal Ages: Geol. Soc. Am. Bull. 125: 539–555. https://doi.org/10.1130/B30660.1

Echaurren A, Folguera A, Gianni G, Orts D, Tassara A, Encinas A, Giménez M, Valencia V (2016) Tectonic evolution of the North Patagonian Andes (41°–44° S) through recognition of syntectonic strata. Tectonophysics 677–678:99–114. https://doi.org/10.1016/j.tecto.2016.04.009

Elliot DH, Fleming TH (2000) Weddell triple junction: the principal focus of Ferrar and Karoo magmatism during initial breakup of Gondwana. Geology 28:539–542. https://doi.org/10.1130/0091-7613(2000)28<539:WTJTPF>2.0.CO;2

Fennell L, Folguera A, Naipauer M, Gianni G, Rojas Vera E, Bottesi G, Ramos V (2015) Createaceous deformation of the southern Central Andes: synorogenic growth strata in the Neuquén Group (35°30′–37° S). Basin Res 1–22. https://doi.org/10.1111/bre.12135

Ferello R, Lesta P (1973) Acerca de la existencia de una dorsal interior en el sector central de la Serranía de San Bernardo (Chubut). V Congreso Geológico Argentino 19–26

Fernández Garrasino CA (1977) Contribución a la estratigrafía de la zona comprendida entre Estancia Ferraroti, Cerro Colorado y Cerrito Negro, Departamento de Tehuelches, provincia de Chubut. Argent Rev Asoc Geol Argent 32:130–144

Feruglio E (1949) Descripción Geológica de la Patagonia. Dirección General de Yacimientos Petrolíferos Fiscales, Editorial Coni

Figari E (2005) Evolución tectónica de la cuenca de Cañadón Asfalto (zona del Valle Medio del Rio Chubut). PhD. Thesis, Universidad Nacional de Buenos Aires. Buenos Aires, Argentina

Fitzgerald MG, Mitchum RM Jr, Uliana MA, Biddle KT (1990) Evolution of the San Jorge Basin. Argent Am Assoc Petr Geol Bull 74:879–920

Fleagle JG, Bown TM, Swisher III CC, Buckley G (1995) Ages of the Pinturas and Santa Cruz formations. VI Congreso Argentino de Paleontología y Bioestratigrafía 129–135

Folguera A, Iannizzotto NF (2004) The lagos La Plata and Fontana fold-and-thrust belt: long-lived orogenesis at the edge of western Patagonia. J South Am Earth Sci 16:541–566

Franchi M, Panza JL, de Barrio RR (1989) Depósitos triásicos y jurásicos de la Patagonia Extraandina. In: Chebli G, Spaletti L (eds) Cuencas Sedimentarias Argentinas. Universidad Nacional de Tucumán, Serie de Correlación Geológica, pp 347–378

Franke D (2013) Rifting lithosphere breakup and volcanism: comparison of magma-poor and volcanic rifted margins. Mar Petr Geol 43:63–87. https://doi.org/10.1016/j.marpetgeo.2012.11.003

Frizon de Lamotte D, Fourdan B, Leleu S, Leparmentier F, Clarens P (2015) Style of rifting and the stages of Pangea breakup. Tectonics 34:1009–1029. https://doi.org/10.1002/2014TC003760

Genise JF, Sciutto JC, Laza JH, González MG, Bellosi ES (2002) Fossil bee nests, coleopteral pupal chambers and tuffaceous paleosols from the Late Cretaceous Laguna Palacios Formation, Central Patagonia (Argentina). Palaeogeogr Palaeocl 177:215–235. https://doi.org/10.1016/S0031-0182(01)00333-9

Geuna SE, Somoza R, Vizán H, Figari E, Rinaldi CA (2000) Paleomagnetism of Jurassic and Cretaceous rocks in central Patagonia: a key to constrain the timing of rotations during the breakup of southwestern Gondwana? Earth Planet Sci Lett 181:145–160. https://doi.org/10.1016/S0012-821X(00)00198-9

Ghiglione M, Ramos V, Cuitiño J, Barberón V (2016) Growth of the southern Patagonian andes (46–53° S) and their relation to subduction processes. In: Folguera A, Naipauer M, Sagripanti L, Ghiglione M, Orts D, Giambiagi L (eds) Growth to the andes. Springer Earth Sys Sci 201–240

Gianni G, Navarrete C, Orts D, Tobal J, Folguera A, Giménez M (2015a) Patagonian broken foreland and related synorogenic rifting: the origin of the Chubut Group Basin. Tectonophysics 649:81–99. https://doi.org/10.1016/j.tecto.2015.03.006

Gianni G, Navarrete C, Folguera A (2015b) Synorogenic foreland rifts and transtensional basins: a review of Andean imprints on the evolution of the San Jorge Gulf, Salta Group and Taubaté Basins. J South Am Earth Sci 64:288–306. https://doi.org/10.1016/j.jsames.2015.08.004

Golonka J, Bocharova NY (2000) Hot spot activity and the break-up of Pangea. Palaeogeogr Palaeocl 161:49–69. https://doi.org/10.1016/S0031-0182(00)00117-6

Granot R, Dyment J (2015) The Cretaceous opening of the South Atlantic Ocean. Earth Planet Sci Lett 414:156–163. https://doi.org/10.1016/j.epsl.2015.01.015

Hastie WW, Watkeys MK, Aubourg C (2014) Magma flow in dyke swarms of the Karoo LIP: implications for the mantle plume hypothesis. Gondwana Res 25:736–755. https://doi.org/10.1016/j.gr.2013.08.010

Herbst R (1965) La flora fósil de la Formación Roca Blanca, provincia de Santa Cruz, Patagonia, vol 12. Opera Lilloana, Instituto Miguel Lillo, Tucumán, Argentina, pp 7–16

Herbst R (1966) La flora liásica del Grupo Pampa de Agnia, Chubut, Patagonia. Ameghiniana 4:337–349

Herbst R (1968) Las floras Liásicas Argentinas con consideraciones estratigráficas. III Jornadas Geológicas Argentinas, Comodoro Rivadavia, Argentina, pp 145–162

Homovc J, Conforto G, Lafourcade P, Chelotti L (1995) Fold Belt in the San Jorge Basin, Argentina: an example of Tectonic inversion. In: Buchanan J, Buchanan P (eds) Basin inversion. Special Publications, Geol. Soc, pp 235–248

Homovc JF, Constantini LA (2001) Hydrocarbon exploration potential within intraplate shear-related depocenters, Deseado and San Julián basins, southern Argentina. Am Assoc Petr Geol Bull 85:1795–1816

Homovc J, Navarrete C, Marshall P, Masquere S, Cerdan J (2011) Inversión tectónica intra-cretácica de la Subcuenca de Río Mayo, Chubut, Argentina. XVIII Congreso Geológico Argentino, 1418–1419

Jokat W, Boebel T, König M, Meyer U (2003) Timing and geometry of lower Gondwana breakup. J Geophys Res 108(B9):2428. https://doi.org/10.1029/2002JB001802

Jourdan F, Féraud G, Bertrand H, Kampunzu AB, Tshoso G, Watkeys MK, Le Gall B (2005) Karoo large igneous province: Brevity, origin, and relation to mass extinction questioned by new 40Ar/39Ar age data. Geology 33:745–748. https://doi.org/10.1130/G21632.1

Leanza HA (2009) Las principales discordancias del Mesozoico de la Cuenca Neuquina según observaciones de superficie. Rev Mus Arg Cienc Nat, n.s 11: 145–184. Buenos Aires

Lesta PJ (1968) Estratigrafía de la Cuenca del Golfo San Jorge. III Jornadas Geológicas Argentinas, Buenos Aires, pp 251–289

Lesta P, Ferello R (1972) Región Extraandina de Chubut y Norte de Santa Cruz. In: Leanza A (ed) Geología Regional Argentina, Academia Nacional de Ciencias: 602–687

Lesta P, Ferello R, Chebli G (1980) Chubut extraandino. In: Turner JC (ed) II Simposio de Geología Regional Argentina. Academia Nacional de Ciencias, Córdoba, pp 1307–1387

Perez Loinaze V, Vera E, Passalia M, Llorens M, Friedman R, Limarino C, Césari S (2013) High-precision U/Pb zircon age from the Anfiteatro de Ticó Formation: implications for the timing of the lower angiosperm diversification in Patagonia. J Soth Am Earth Sci 48:97–105. doi10.1016/j.jsames.2013.08.005

Maloney KT, Clarke GL, Klepeis KA, Quevedo L (2013) The late jurassic to present evolution of the Andean margin: drivers and the geological record. Tectonics 32:1049–1065. https://doi.org/10.1002/tect.20067

Malumián N, Ploszkiewicz JV (1976) El Liásico fosilífero de Loncopán, Departamento Tehuelches, Provincia del Chubut. República Argent Rev Asoc Geol Argent 31:279–280

Márquez MJ (2003) Las rocas volcano-sedimentarias liásicas en el cordón Esquel, Chubut, Argentina. I Simposio Argentino del Jurásico: 11–11

Márquez M, Navarrete C (2011) La tectónica compresiva pre-Aptiana-Albiana en la sierra de Pichiñanes, Chubut, Argentina. XVIII Congreso Geológico Argentino, 105–106

Márquez M, Navarrete C (2015) Deformaciones contraccionales previas al Cretácico Superior en la Patagonia Central Argentina. XVI Reunión de Tectónica, Argentina, pp 140–141

Márquez M, Zubia M, Giacosa R, Trevisol S,Fernández M (2016) Características geológicas y metalogenéticas del Depósito Navidad (Ag–Pb–Zn–C) Macizo Somún Curá, Chubut, Argentina. Instituto de Recursos Geológicos Mineros, Servicio Geológico Minero Argentino. Boletín N°40, 50 pp

Mazzoni MM (1985) La Formación Sarmiento y el volcanismo paleógeno. Rev Asoc Geol Argent 40:60–68

Mazzoni MM, Kawashita K, Harrison S, Aragón E (1991) Edades radimétricas eocenas en el borde occidental del Macizo Norpatagónico. Rev Asoc Geol Argent 46(1–2):150–158

Mosquera A, Ramos VA (2006) Intraplate deformation in the Neuquén Embayment. In: Kay SM, Ramos VA (eds) Evolution of an Andean margin: a tectonic and magmatic view from the Andes to the Neuquén Basin (35°–39° S lat). Geol Soc Am Spec Pap 407:97–123. https://doi.org/10.1130/2006.2407(05)

Mpodozis C, Ramos VA (2008) Tectónica jurásica en Argentina y Chile: extensión, subducción oblicua, rifting, deriva y colisiones? Rev Asoc Geol Argent 63:481–497

Müller DR, Seton M, Zahirovic S, Williams SE, Matthews KJ, Wright NM, Shephard GE, Maloney KT, Barnett-Moore N, Hosseinpour M, Bower DJ, Cannon J (2016) Ocean basin evolution and global-scale plate reorganization events since Pangea breakup. Annu Rev Earth Planet Sci 44:107–138. https://doi.org/10.1146/annurev-earth-060115-012211

Murphy JB, Nance RD (2013) Speculations on the mechanisms for the formation and breakup of supercontinents. Geosci Front 4: 85–194. doi: https://doi.org/10.1016/j.gsf.2012.07.005

Naipauer M, García E, Marques J, Tunik M, Rojas Vera E, Vujovich G, Pimentel M, Ramos V (2012) Intraplate Late Jurassic deformation and exhumation in western central Argentina: Constraints from surface data and U-Pb detrital zircon ages. Tectonophysics 524–525:59–75. https://doi.org/10.1016/j.tecto.2011.12.017

Nakayama C (1973) Sedimentitas prebajocianas en el extremo austral de la sierra de Taquetrén, Chubut, Argentina. V Congreso Geológico Argentino, 269–278

Navarrete CR, Gianni GM, Folguera A (2015) Tectonic inversión events in the western San Jorge Gulf Basin from seismic borehole and field data. J South Am Earth Sci 64:486–497. https://doi.org/10.1016/j.jsames.2015.09.012

Navarrete C, Gianni G, Echaurren A, Kingler F, Folguera A (2016) Episodic Jurassic to lower Cretaceous intraplate compression in Central Patagonia during Gondwana breakup. J Geodyn 102C:185–201. https://doi.org/10.1016/j.jog.2016.10.001

Navarro EL, Astini RA, Belousova E, Guler MV, Gehrels G (2015) Detrital zircon geochronology and provenance of the Chubut Group in the northeast of Patagonia. Argent J South Am Earth Sci 63:149–161

Nürnberg D, Müller DR (1991) The tectonic evolution of the South Atlantic from Late Jurassic to present. Tectonophysics 191:27–53. https://doi.org/10.1016/0040-1951(91)90231-G

Orts DL, Folguera A, Encinas A, Ramos M, Tobal J, Ramos VA (2012) Tectonic development of the North Patagonian Andes and their related Miocene foreland basin (41°30′–43° S). Tectonics 31(TC3012):1–24. https://doi.org/10.1029/2011TC003084

Pankhurst RJ, Rapela CW, Márquez MJ (1993) Geocronología y petrogénesis de los granitoides Jurásicos del noroeste del Macizo del Deseado. XII Congreso Geológico Argentina 4:134–141

Pankhurst RJ, Rapela CW (1995) Production of Jurassic rhyolite by anatexis of the lower crust of Patagonia. Earth Planet Sci Lett 134:23–36. https://doi.org/10.1016/0012-821X(95)00103-J

Pankhurst RJ, Leat PT, Sruoga P, Rapela CW, Márquez M, Storey BC, Riley TR (1998) The Chon Aike province of Patagonia and related rocks in West Antarctica: a silicic large igneous province. J Volc Geoth Res 81:113–136. https://doi.org/10.1016/S0377-0273(97)00070-X

Pankhurst RJ, Riley TR, Fanning CM, Kelley S (2000) Episodic silicic volcanism in Patagonia and the Antarctic Peninsula: Chronology of magmatism associated with the break-up of Gondwana. J Petrol 41:605–625. https://doi.org/10.1093/petrology/41.5.605

Pankhurst RJ, Hervé F, Fanning M, Suárez M (2003) Coeval plutonic and volcanic activity in the Patagonian Andes: the Patagonian Batholith and the Ibáñez and Divisadero formations, Aysén, southern Chile. X Congreso Geológico Chileno, Electronic files

Paredes JM, Azpiroz G, Foix N (2006) Tertiary tectonics and sedimentation in the Cerro Piedra Oil Field Golfo San Jorge basin, Argentina. IV Latin American Congress of Sedimentology, San Carlos de Bariloche, Argentina, 163

Paredes JM, Foix N, Colombo F, Nillni A, Allard JO, Marquillas R (2007) Volcanic and climatic control on fluvial style in a high energy system: the Lower Cretaceous Matasiete Formation, Golfo San Jorge basin, Argentina. Sed Geol 202:96–123

Paredes JM, Plazibat S, Crovetto C, Stein J, Cayo E, Schiuma A (2013) Fault kinematics and depocenter evolution of oil-bearing, continental successions of the Mina del Carmen Formation (Albian) in the Golfo San Jorge basin. Argent J South Am Earth Sci 46:63–79. https://doi.org/10.1016/j.jsames.2013.05.002

Peroni GO, Hegedus AG, Cerdan J, Legarreta L, Uliana MA (1995) Hydrocarbon Accumulation in an inverted segment of the Andean Foreland: San Bernardo Belt, Central Patagonia. In: Tankard AJ, Suárez R, Welsink HJ (eds) Petroleum basins of South America, 403–419

Pesce AH (1979) Estratigrafía de la Cordillera Patagónica entre los paralelos 43°30′ y 44° de latitud sur y sus áreas mineralizadas, provincia de Chubut. VII Congreso Geológico Argentino, 257–270

Ploszkiewicz J (1987) Descripción Geológica Hoja 47c, Apeleg. Dirección Nacional de Geología y Minería, Boletín 204, 95 pp

Raigemborn MS, Krause JM, Bellosi E, Matheos SD (2010) Redefinición estratigráfica del grupo Río Chico (Paleógeno Inferior), en el norte de la Cuenca del Golfo San Jorge. Chubut Rev Asoc Geol Argent 67:239–256

Ramos VA (1999) Los depósitos sinorogénicos terciarios de la Región Andina. Instituto de Geología y Recursos Minerales. Geología Argent Anales 29:651–682

Ranalli JN, Peroni GO, Boggetti DA, Manoni R (2011) Cuenca Cañadón Asfalto. Modelo tectosedimentario. VIII Congreso de Exploración y Desarrollo de Hidrocarburos, Simposio Cuencas Sedimentarias Argentinas: visión actual, 185–215

Rapela CW, Spalletti LA, Merodio JC, Aragón E (1988) Temporal evolution and spatial variation of early Tertiary volcanism in the Patagonian Andes (40 S–42 30′ S). J South Am Earth Sci 1:75–88

Rapela CW, Pankhurst RJ, Fanning CM, Hervé F (2005) Pacific subduction coeval with the Karoo mantle plume: the Lower Jurassic Subcordilleran belt of northwestern Patagonia. In: Vaughan APM, Leat PT, Pankhurst RJ (eds) Terrane processes at the margins of Gondwana. Geol Soc Lond Spec Publ 246:217–239

Reimer W, Miller H, Mehl H (1996) Mesozoic and Cenozoic palaeo-stress field of the South Patagonian Massif deduced from structural and remote sensing data. In: Storey BC, King EC, Livermore RA (eds) Weddell sea tectonics and Gondwana break-up. Geol Soc Lond Spec Publ 108:73–85. https://doi.org/10.1144/GSL.SP.1996.108.01.06

Riley TR, Leat PT, Pankhurst RJ, Harris C (2001) Origins of large volume rhyolitic volcanism in the Antarctic Peninsula and Patagonia by crustal melting. J Petrol 42:1043–1065. https://doi.org/10.1093/petrology/42.6.1043

Riley TR, Knight KB (2001) Age of pre-break-up Gondwana magmatism. Antarct Sci 13:99–110. https://doi.org/10.1017/S0954102001000177

Robbiano JA (1971) Contribución al conocimiento estratigráfico de la Sierra del Cerro Negro, Pampa de Agnia, provincia del Chubut. República Argent Rev Asoc Geol Argent 26:41–56

Rodriguez JF, Littke R (2001) Petroleum generation and accumulation in the Golfo San Jorge Basin, Argentina: a basin modeling study. Mar Pet Geol 18:995–1028

Rolando AP, Hartmann LA, Santos JOS, Fernandez RR, Etcheverry RO, Schalamuk IA, McNaughton NJ (2002) SHRIMP zircon U-Pb evidence for extended Mesozoic magmatism in the Patagonian Batholith and assimilation of Archean crustal components. J South Am Earth Sci 15:267–283

Rolando AP, Hartmann LA, Santos JO, Fernández RR, Etcheverry RO, Schalamuk IBA, McNaughton NJ (2004) SHRIMP UPb zircon dates from igneous rocks from Fontana Lake region, Patagonia: Implications for the age of magmatism, Mesozoic geological evolution and age of basement. Rev Asoc Geol Argent 59:671–684

Sesana F (1968) Rasgos petrológicos de la comarca de Río Chico, Río Negro. III Jornadas Geológicas Argentinas, Comodoro Rivadavia, pp 99–107

Seton M, Müller RD, Zahirovic S, Gaina C, Torsvik T, Shephard G, Talsma A, Gurnis M, Turner M, Maus S, Chandler M (2012) Global continental and ocean basin reconstructions since 200 Ma. Earth Sci Rev 113:212–270

Sciutto JC (1981) Geología del Codo del Río Senguer, Chubut, Argentina. VIII Congreso Geológico Argentino, 544–568

Simpson GG (1941) The Eogene of Patagonia. Am Mus Novit 1120:1–15

Somoza R (1994) South American reference pole for the mid-Cretaceous: further constraints in the interpretation of Andean paleomagnetic data. Geology 22:933–936. https://doi.org/10.1130/0091-7613(1994)-022<0933:SARPFT>2.3.CO;2

Somoza R, Zaffarana CB (2008) Mid Cretaceous polar standstill of South America, motion of the Atlantic hotspots and the birth of the Andean cordillera. Earth Planet Sci Lett 271:267–277

Storey BC, Alabaster T, Hole MJ, Pankhurst RJ, Wever E (1992) Role of subduction-plate boundary forces during the initial stages of Gondwana break-up: evidence from the proto-Pacific margin of Antartica. In: Storey BC, Alabaster T, Pankhurst RJ (eds) Magmatism and the causes of continental break-up. Geol Soc Spec Publ 68:149–163. https://doi.org/10.1144/GSL.SP.1992.068.01.10

Suárez M, De la Cruz R (2000) Tectonics in the eastern Patagonian Cordillera (45°30–47°30). J Geol Soc 15:995–1001

Suárez M, Márquez M (2007) A toarcian back-arc basin of Central Patagonia (Chubut), Argentina: Middle Jurassic closure, arc migration and tectonic setting. Rev Geol de Chile 34:63–79. https://doi.org/10.4067/S0716-02082007000100004

Suárez M, De La Cruz R, Bell M, Demant A (2009a) Cretaceous slab segmentation in southwestern Gondwana. Geol Mag 147(1):1–13

Suárez M, Márquez M, De La Cruz R, Fanning M (2009b) Aptian–Albian subaerial volcanic rocks in Central Patagonia: Divisadero and Chubut Groups. XII Congreso Geológico Chileno, 1–4

Suárez M, Márquez M, De La Cruz R, Navarrete C, Fanning M (2014) Cenomanian-? Lower Turonian minimum age of the Chubut Group, Argentina: SHRIMP U-Pb geochronology. J South Am Earth Sci 50:67–74. https://doi.org/10.1016/j.jsames.2013.10.008

Turner JCM (1975) Descripción Geológica de la Hoja 44d "Languiñeo", provincia de Chubut. Servicio Geológico Nacional (unpublished), Argentina

Umazano AM, Bellosi ES, Visconti G, Melchor RN (2012) Detecting allocyclic signals in volcaniclastic fluvial successions: facies, architecture and stacking pattern from the Cretaceous of Central Patagonia. Argent J South Am Earth Sci 40:94–115. https://doi.org/10.1016/j.jsames.2012.09.005

Varela R, Pezzuchi HD, Genini AD, Zubia MA (1991) Dataciones en el Jurásico de rocas magmáticas del nordeste del Macizo del Deseado. Santa Cruz Rev Asoc Geol Argent 46: 257–262

Vaughan APM, Livermore RA (2005) Episodicity of Mesozoic terrane accretion along the Pacific margin of Gondwana: implications for superplume–plate interactions. In: Vaughan APM, Leat PT, Pankhurst RJ (eds) Terrane processes at the margins of Gondwana. J Geol Soc Spec Publ 246:143–178. https://doi.org/10.1144/GSL.SP.2005.246.01.05

Veevers JJ (2004) Gondwanaland from 600.570 Ma assembly through 320 Ma merger in Pangea to 160 Ma breakup: supercontinental tectonics via stratigraphy and radiometric dating. Earth-Sci Rev 68:1–132. https://doi.org/10.1016/j.earscirev.2004.05.002

Volkheimer W (1964) Estratigrafía de la zona extraandina del Departamento de Cushamen (Chubut) entre los paralelos 42° y 42°30′y los meridianos 70° y 71°. Rev Asoc Geol Argent 19:85–107

Volkheimer W, Gallego O, Cabaleri N, Armella C, Narváez P, Silva Nieto D, Páez M (2009) Stratigraphy, palynology, and conchostracans of a Lower Cretaceous sequence at the Cañadón Calcáreo locality, Extra-Andean central Patagonia: age and palaeoenviroment significance. Cretac Res 30:270–282. https://doi.org/10.1016/j.cretres.2008.07.010

Whalen L, Gazel E, Vidito C, Puffer J, Bizimis M, Henika W, Caddick MJ (2015) Supercontinental inheritance and its influence on supercontinental breakup: The Central Atlantic Magmatic Province and the breakup of Pangea. Geochem Geophys Geosyst 16:3532–3554. https://doi.org/10.1002/2015GC005885

Zaffarana CB, Somoza R (2012) Palaeomagnetism and 40Ar/39Ar dating from Lower Jurassic rocks in Gastre, central Patagonia: further data to explore tectonomagmatic events associated with the break-up of Gondwana. J Geol Soc 169:371–379. https://doi.org/10.1144/0016-76492011-089

Zaffarana CB, Somoza R, López de Luchi M (2014) The late triassic central Patagonian batholith: magma hybridization, 40Ar/39Ar ages and thermobarometry. J South Am Earth Sci 55:94–122. https://doi.org/10.1016/j.jsames.2014.06.006

Mechanisms and Episodes of Deformation Along the Chilean–Pampean Flat-Slab Subduction Segment of the Central Andes in Northern Chile

Fernando Martínez, César Arriagada and Sebastián Bascuñán

Abstract The knowledge of the tectonic architecture, timing, and the mechanisms of deformation that affected the western slope of the Chilean–Pampean flat-slab subduction segment of the Central Andes in northern Chile are a key to understand the complete evolution of this Andean segment. In Chile, this segment is composed of two tectonic provinces: The Coastal and the Frontal Cordilleras. Traditionally, this broad intracontinental deformation zone that characterized this segment has been compared with the Rocky Mountains, in terms of structural styles and age of deformation, although the complex interaction between extensional Mesozoic structures and different styles of contractional structures suggest that this segment resulted from multiple episodes of deformation. We present the results of a study developed along the Coastal and Frontal Cordillera of northern Chile (27°–28° S), based on regional fieldwork, structural analysis, and geochronological dating of synorogenic deposits. Our results have revealed that the structure of this region consists of NNE-striking inverted and basement-involved contractional structures. The occurrence of these structural styles suggests that a hybrid tectonic mechanism dominated by tectonic inversion and basement-involving thrusting was responsible of its current configuration. On the other hand, the U–Pb ages determined in the synorogenic deposits exposed on the footwall of the main faults indicate that the Andean deformation could have initiated during the Late Cretaceous in the Coastal Cordillera associated with tectonic inversion and then this migrated to the east as basement-involved thrusting during the Paleocene–Miocene times.

Keywords Chilean–Pampean flat subduction zone · Cretaceous deformations
Cenozoic deformations · Tectonic inversion · Basement-involved structures
Northern Chile

F. Martínez (✉)
Facultad de Ingeniería y Ciencias Geológicas, Departamento de Geología,
Universidad Católica del Norte, Angamos Antofagasta 0610 Chile
e-mail: fernando.martinez@ucn.cl

C. Arriagada · S. Bascuñán
Departamento de Geología, Facultad de Ciencias Físicas y Matemáticas,
Universidad de Chile, Plaza Ercilla, Santiago 803, Chile
e-mail: cearriag.ca@gmail.com

A. Folguera et al. (eds.), *The Evolution of the Chilean-Argentinean Andes*,
Springer Earth System Sciences, https://doi.org/10.1007/978-3-319-67774-3_11

1 Introduction

The Chilean–Pampean flat-slab subduction segment of the Central Andes between 28° and 33° S (Baraganzi and Isacks 1976; Jordan et al. 1983; Isacks 1988) represents one of the two major flat-slab segments imaged below the continental margin of South America (Martinod Martinod et al. 2010) (Fig. 1). This Andean segment is characterized by a gap of active volcanism, a basement-involved deformation style and spread Late Cenozoic arc volcanism in the foreland area (Ramos et al. 2002). Between 27°–28° S, this segment consists of three morphotectonic units that include the Coastal and the Frontal Cordillera along the Chile–Argentina border, and the Precordillera and Sierras Pampeanas on the Argentinean side (Fig. 1). At these latitudes, the Frontal Cordillera, the Precordillera, and the Sierras Pampeanas reach elevations between 2 and 4 km and locally 6 km a.s.l, with intervening flat plains at less than 1 km a.s.l. (Jordan et al. 1983; Moscoso and Mpodozis 1988; Martínez et al. 2015b).

Traditionally, the Chilean–Pampean flat-slab subduction segment was considered as a modern analogue to the Rocky Mountain of North America in regards to its structure and age of deformation (Jordan and Allmendinger 1986; Allmendinger et al. 1997; Ingersoll 2012, among others), as it is fundamentally dominated by large and broad crystalline uplifted blocks (Jordan et al. 1983; Narr and Suppe 1994; Ramos 1999; Poblet and Lisle 2011) that resulted from Miocene horizontal compression generated by the shallowing of the Nazca Plate under the continental margin of South America. The knowledge of its tectonic history comes mainly from studies carried out along tectonic provinces and petroliferous basins located on the eastern side, such as the Sierras Pampeanas, the Tucumán Basin, the Métan Basin, the Santa Barbara System, among others. In these regions, previous studies using seismic information combined with field and oil well data have interpreted a thick-skinned structure composed of large and broad, basement-cored folds, and reverse faults intercalated with synorogenic basins related to the reactivation of ancient suture zones and normal faults, which have propagated as reverse faults accumulating around 150 km of minimum crustal shortening during Neogene times (Jordan et al. 1983; Coutand et al. 2001; Kley and Monaldi 2002; Heredia et al. 2002; Ramos et al. 2002; Cristallini et al. 2004; Carrapa et al. 2011; Iaffa et al. 2011).

On the other hand, recent studies developed in the western slope of northern Chile (Fig. 1), based mainly on geological observations, balanced cross sections and radiometric ages, have also documented a remarkable basement-involved deformation style (Moscoso and Mpodozis 1988; Amilibia et al. 2008; Peña et al. 2013; Martínez et al. 2013, 2015b, 2016). However, this has been mainly associated with inversion structures created from the shortening and partial closure of Lower Cretaceous and Jurassic extensional back-arc basins during Late Cretaceous–Paleocene and Miocene contractional pulses (Peña et al. 2013; Martínez et al. 2013, 2016). In contrast to previous interpretations made on the eastern Andean slope, the results obtained recently in northern Chile (Fig. 1) have opened a new debate about:

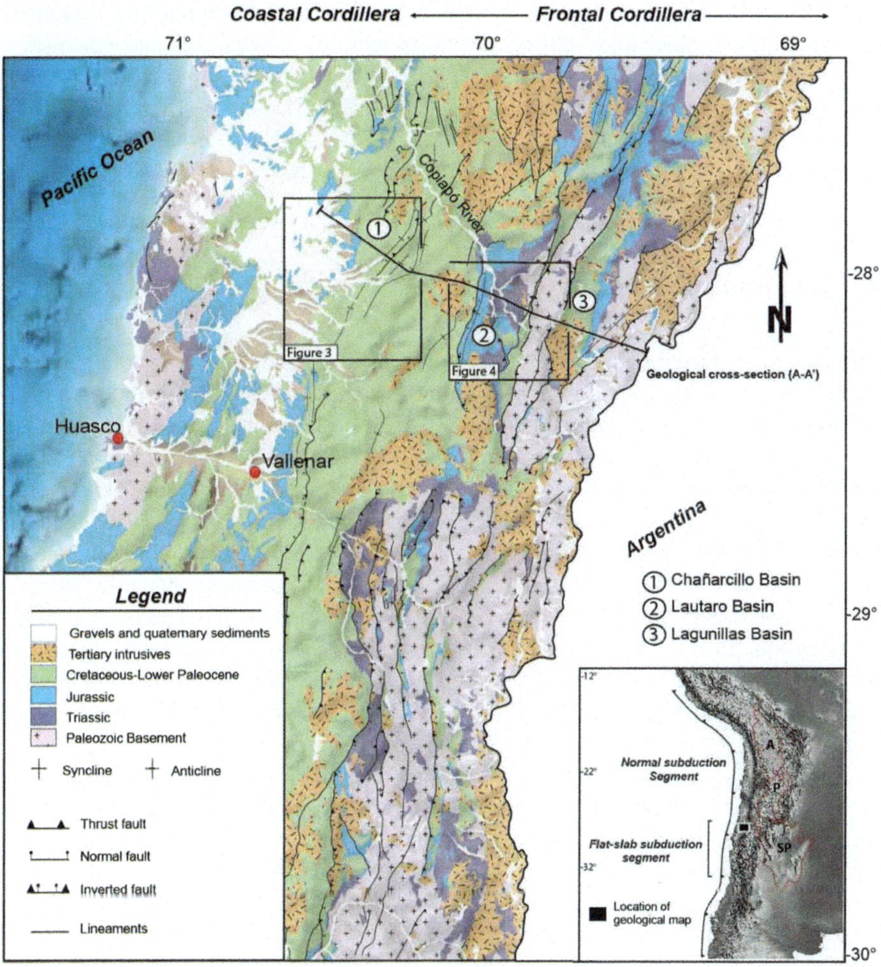

Fig. 1 Generalized geological map of the "Pampean" subduction segment of northern Chile between 28°–30° S, showing the spatial distribution of the main stratigraphic and tectonic units of the region

(a) What tectonic mechanisms are responsible for the current architecture of this Andean segment; (b) how the Mesozoic extensional systems previously established in this region have interacted with younger thrust systems during later Andean deformation; (c) what is the main tectonic control on the basement-involved deformation style; and (d) what are the main episodes of Andean deformation.

In order to discuss these topics, we present here the results of a recent study developed along the Coastal and Frontal Cordilleras of northern Chile (Fig. 1), which is based on regional geological mapping, field observations, structural analysis, and geochronological dating of synorogenic deposits. Finally, we compare these data with previous tectonic models proposed for the neighboring regions on the Argentinean side, highlighting the main tectonic mechanism and episodes of deformation that took place in these regions, showing that timing of development of the Chilean–Pampean flat subduction segment includes more than one phase of deformation and then that important sectors of the fold and thrust belt were constructed previously to the slab shallowing achieved in the last 17 Ma.

2 Regional Geological Context

Two tectonic provinces compose the present-day architecture of the Chilean–Pampean flat subduction segment in northern Chile: the Coastal Cordillera and the Frontal Cordillera, as illustrated in Fig. 1. The stratigraphic successions exposed along them comprise at least five major sequences (Fig. 2; Martínez et al. 2016). The oldest units correspond to Paleozoic and Lower Triassic rocks that include metasedimentary (Naranjo and Puig 1984; Grocott and Taylor 2002; Scheubert and Reutter 1992) and granitic plutonic complexes (245–250 Ma; Mpodozis and Ramos 1990; Martínez et al. 2015b; Del Rey et al. 2016, among others). They lie well-exposed both in the western Coastal Cordillera and the Frontal Cordillera (Fig. 1) and form the pre-rift basement of the Jurassic and Cretaceous extensional back-arc basins developed in this region after the breakup of Gondwana (Pindell and Dewey 1982; Mpodozis and Ramos 1990; Aguirre-Urreta 1993; Mpodozis and Ramos 1990, 2008; Ramos 2009; Del Rey et al. 2016).

These rocks are unconformably covered and flanked by a thick cover of volcanic and sedimentary sequences that vary laterally from the Coastal to the Frontal Cordillera, forming part of the infill of the Mesozoic, inverted Chañarcillo, Lautaro, and Lagunillas basins (Martínez et al. 2016; Figs. 1 and 2). These basins correspond to a series of NNE-oriented, former half-grabens that were created during the fragmentation of western Gondwana during the Late Triassic–Early Cretaceous periods (Aguirre-Urreta, 1993; Mpodozis and Ramos 1990, 2008; Martínez et al. 2016) (see Chap. "The Early Stages of the Magmatic Arc in the Southern Central Andes"). Within this cover, the oldest sequences correspond to around 2,100 m of Upper Triassic volcanic and sedimentary deposits (La Ternera Fm; Segerstrom 1960; Suárez and Bell 1992) that are mainly exposed along valleys and rivers cutting the Frontal Cordillera (Figs. 1 and 2). These deposits have been associated with a continental syn-rift sequence accumulated during early episodes of stretching and crustal extension, previous to the establishment of the Jurassic and Cretaceous back-arc basins in this region (Jensen 1976; Charrier 1979; Mpodozis and Cornejo 1997; Martínez et al. 2012). The Jurassic sequences can be divided into two sections, the westernmost deposits exposed along the eastern Coastal Cordillera which consist predominantly of Upper Jurassic volcanic successions (Punta del Cobre Fm; Fig. 2) related to intra-arc deposits (Lara and Godoy 1998; Marschik and Fontboté 2001), and the eastern, marine, and continental Jurassic deposits that crop out in the Frontal Cordillera, corresponding to intercalations of limestones, shale and volcano-sedimentary successions (Lautaro and Lagunillas formations;

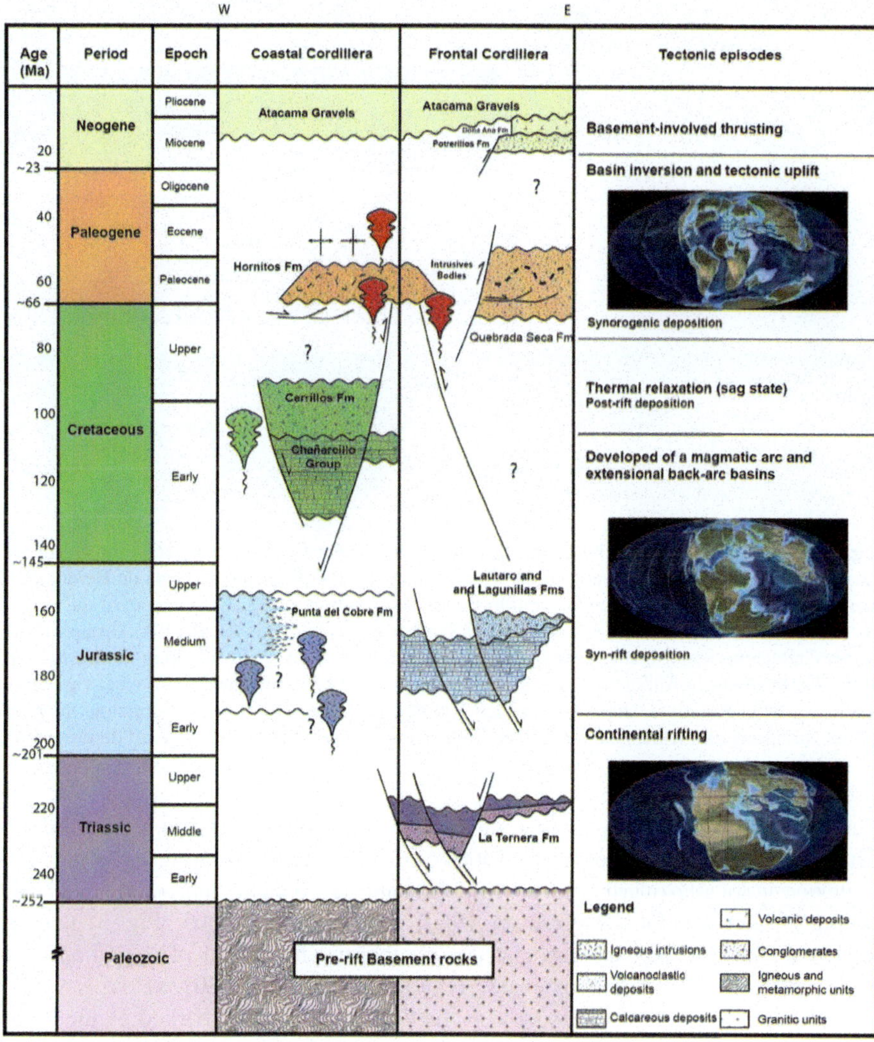

Fig. 2 Chronostratigraphic chart indicating the main sequences recognized on the western Andean slope of the northern Chilean–Pampean flat subduction zone and the tectonic stages proposed to the evolution of this Andean segment

Fig. 2), with remarkable variations of thickness, progressive unconformities, and internal normal faults, which have been interpreted as syn-rift deposits (Segerstrom 1960; Jensen 1976; Arévalo 2005; Martínez et al. 2012).

The Lower Cretaceous deposits in this region are exposed along the eastern Coastal Cordillera (Fig. 2). Here, they crop out as a NNE-oriented belt of almost 4,000 m of calcareous and siliciclastic rocks, forming a stratigraphic wedge internally affected by syn-extensional faults. These deposits have been interpreted as the

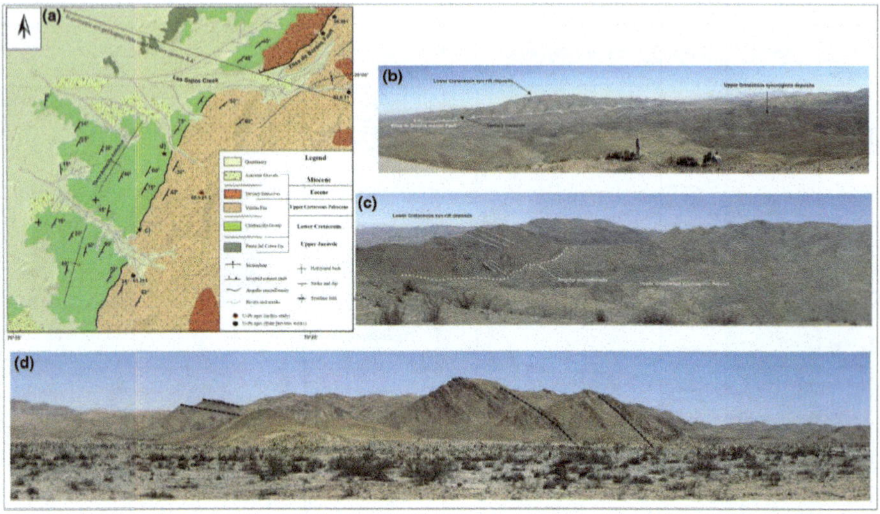

Fig. 3 **a** Geological map of the central region of the Lower Cretaceous Chañarcillo Basin (see location on Fig. 1), **b** frontal view of the Tierra Amarilla Anticline and the Elisa de Bordos Fault marking the eastern limit of the Chañarcillo Basin (see position in Fig. 3a), **c** aspect of the angular unconformity between the Lower Cretaceous syn-rift deposits of the Chañarcillo Group and the Upper Cretaceous synorogenic deposits of the Viñitas Formation observed along the frontal limb of the Tierra Amarilla Anticline (see positions on Fig. 3a), **d** panoramic and transversal view of the Tierra Amarilla Anticline, showing a typical geometry related to an inversion anticline. Note the lateral variations of thickness of the Lower Cretaceous (Chañarcillo Group) syn-rift deposits along the structure (see positions on Fig. 3a)

syn-rift deposits of the Chañarcillo Basin (Fig. 3) and they are called the Chañarcillo Group (Segerstrom 1960; Arévalo 1999; Mourgues 2004). This sequence is locally overlapped by Mid-Cretaceous volcanic and sedimentary successions (Cerrillos Fm; Segerstrom 1960; Arévalo 1994, 2005; Fig. 3) that are related to the sag deposits associated with thermal relaxation of the Chañarcillo Basin subsequent to crustal extension (Martínez et al. 2013, 2015a).

The transition from the Mesozoic to the Cenozoic cover is marked by the Upper Cretaceous and Paleocene sequences (Viñitas and Quebrada Seca formations; Figs. 3 and 4), which correspond to thick continental synorogenic deposits exposed along intramontane depressions in the eastern Coastal Cordillera (Viñitas Formation) and the eastern Frontal Cordillera (Quebrada Seca Formation; Figs. 3 and 4). They are composed of red sandstones, conglomerates, ignimbrites, and different volcanic flows with growth strata and progressive unconformities (Martínez et al. 2016), which in turn are intruded by kilometric, intrusive bodies (Figs. 3 and 4). The accumulation of these deposits has been related to regional contractional pulses associated with tectonic inversion and thrusting processes, as well as with the eastward displacement of the magmatic arc (Mpodozis and Ramos 1990; Cornejo et al. 1993; Martínez et al. 2013). Miocene continental and unconsolidated sediments (Atacama Gravels; Mortimer 1973) unconformably overlie the latter sequences.

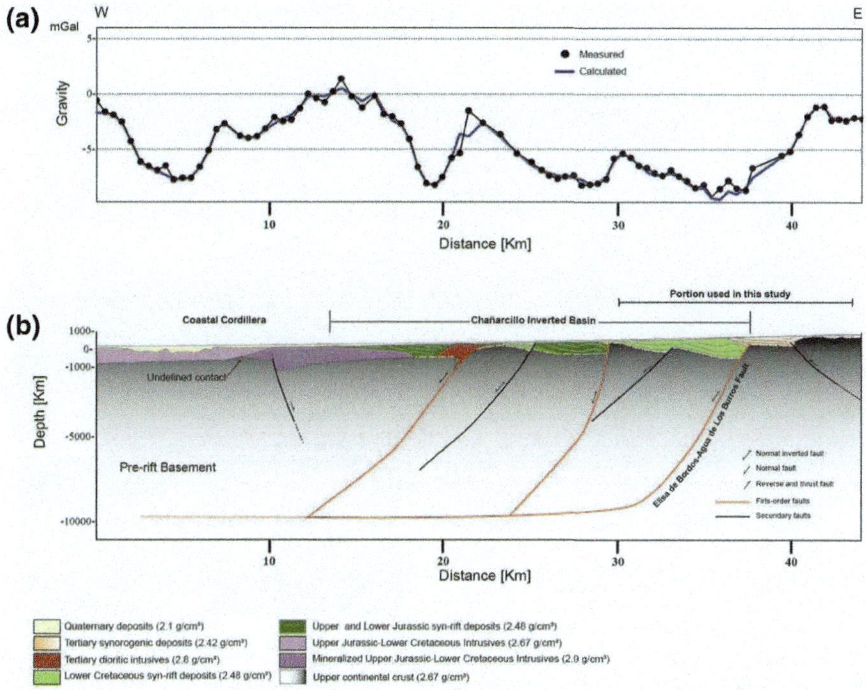

Fig. 4 **a** Gravity profile through the central section of the Chañarcillo Basin used in this study showing the Bouguer anomaly and the regional anomaly, **b** structural interpretation of the gravity profile supported by geological data (see position of this profile in Fig. 3) (modified from Martínez et al. 2015a)

3 Tectonic Framework and Mechanisms of Deformation

Two main structural styles have been recognized along this Andean segment: inversion and basement-involved structures (Jensen 1976; Moscoso and Mpodozis 1988; Arévalo 2005; Peña et al. 2013; Martínez et al. 2016). Their occurrence and distribution vary along the complete segment. Inversion structures are preferentially located at the edges of the Mesozoic basins forming the major limits between the Coastal and the Frontal Cordillera (Figs. 3 and 4), while basement-involved structures are specifically distributed along the Frontal Cordillera (Fig. 1). In order to document the mechanisms of deformation that took place in this region, we have constructed a regional cross section from the Coastal Cordillera to the Frontal Cordillera at 28° S (Figs. 1 and 5). We have constrained the surface structure by field work and geological maps (Moscoso et al. 2010; Peña et al. 2013) and the subsurface structure by gravimetric data (Martínez et al. 2015a). The cross section was constructed using Move software (@Midland Valley). Major detachments in this region are unknown, and they have been inferred by area-balanced methods,

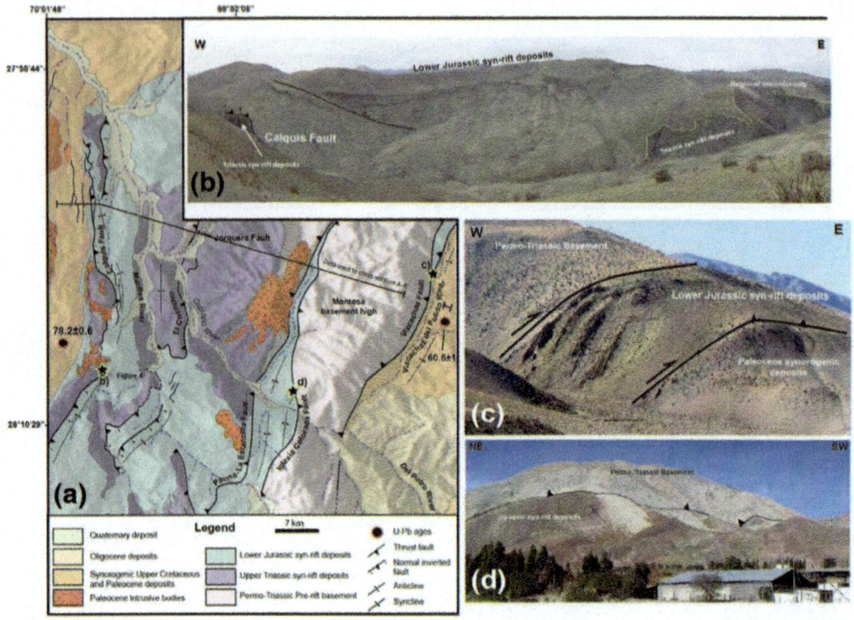

Fig. 5 **a** Geological map of the Lower Jurassic Lautaro Basin (see positions on Fig. 1), **b** panoramic view of the west-vergent inversion anticline recognized along the western edge (Calquis Fault) of the Lautaro Basin (see position on Fig. 4a), **c–d** structural expression of the basement-involved thrusts limiting the west **d** and east **c** sections of the Montosa Horst along the Frontal Cordillera (see location on Fig. 4a)

so this section is only an approximation held to illustrate the tectonic evolution of this region. Below, we describe the main characteristics of the observed deformation style.

3.1 Inversion Structures

The inversion structures correspond to two NNE-striking anticlines exposed along the eastern edge of the Chañarcillo Basin and the western edge of the Lautaro Basin, respectively (Figs. 3 and 4). These structures consist of large and asymmetrical anticlines with a long wavelength and arrowhead geometry (Figs. 3 and 4), similar to those reported from analogue models of inversion anticlines (Bonini et al. 2012 and works cited). Along the eastern edge of the Chañarcillo Basin (Fig. 3), the main structure is represented by the east-vergent Tierra Amarilla Anticline (Segerstrom 1960; Arévalo, 2005, among others), which involves the Jurassic–Lower Cretaceous syn-rift deposits and the Mid-Cretaceous post-rift deposits of the basin (Martínez et al. 2013; Fig. 3). The Tierra Amarilla Anticline shows along-strike variations in its geometry. For example, to the north of the basin, only the frontal limb is mostly exposed and the back limb is obscured by intrusive bodies, but to the southern region both limbs lie exposed. In the study area, the frontal limb and the hinge of the anticline form the most prominent relief along the eastern Coastal Cordillera (Fig. 3). Previous works have related the growth of this structure with partial reactivation of the NNE-striking Elisa de Bordos Fault (Fig. 3), which is interpreted as the master fault of the Chañarcillo Basin (Arévalo 2005; Martínez et al. 2013; Peña et al. 2013). Recent geophysical data have confirmed this interpretation, as well as the subsurface projection of these structures (Fig. 4; Martínez et al. 2015a).

The other inversion structure consists of a west-verging asymmetrical anticline located at the eastern edge of the Lautaro Basin (Fig. 5). This anticline is composed of a sub-horizontal back limb and an overturned frontal limb facing the NNE-striking Calquis Fault. This fault corresponds to the master fault of the Lautaro Basin (Fig. 4), mainly affecting the Triassic and Lower Jurassic syn-rift deposits (Fig. 5). Along this anticline, the Lower Jurassic marine syn-rift deposits are elevated over its regional level, shortened and buttressed against the footwall of the Calquis Fault (Fig. 5). Based on these geometric characteristics, previous studies (Martínez et al. 2012) have interpreted these as a result of the tectonic inversion and partial closure of the Lautaro Basin. Accordingly, we conclude that tectonic inversion is a fundamental process during the growth of this Andean segment, which took advantage of the former extensional Mesozoic structures established in this region. On the other hand, the inversion structures lie separated by a narrow NNE-striking belt composed of minor thin-skinned thrust (Fig. 6), associated with anticlines and synclines, which have been interpreted as secondary structures formed from the propagation of deformation into the footwall of both inverted faults (Elisa de Bordos and Calquis faults).

3.2 Basement-Involved Structures

Basement-involved structures are mostly located along the Frontal Cordillera, as illustrated in Fig. 1. Along this system, they consist of large, NNE-striking, doubly verging (west and east) reverse faults, and basement-cored anticlines (Fig. 1). These structures mainly affect the infill of the Lautaro (Fig. 5) and Lagunillas basins (Martínez et al. 2015b), and frequently place large, lower Triassic granitic blocks of the pre-rift basement of the basins over Mesozoic syn-rift and Cenozoic synorogenic deposits (Fig. 5). The faults predominantly form large east-verging ramps intercalated with west-verging backthrusts (Fig. 6), which have accumulated a minimum shortening of 18 km (Martínez et al. 2015b). On the other hand, the footwall faults generally form narrow and asymmetrical footwall synclines with

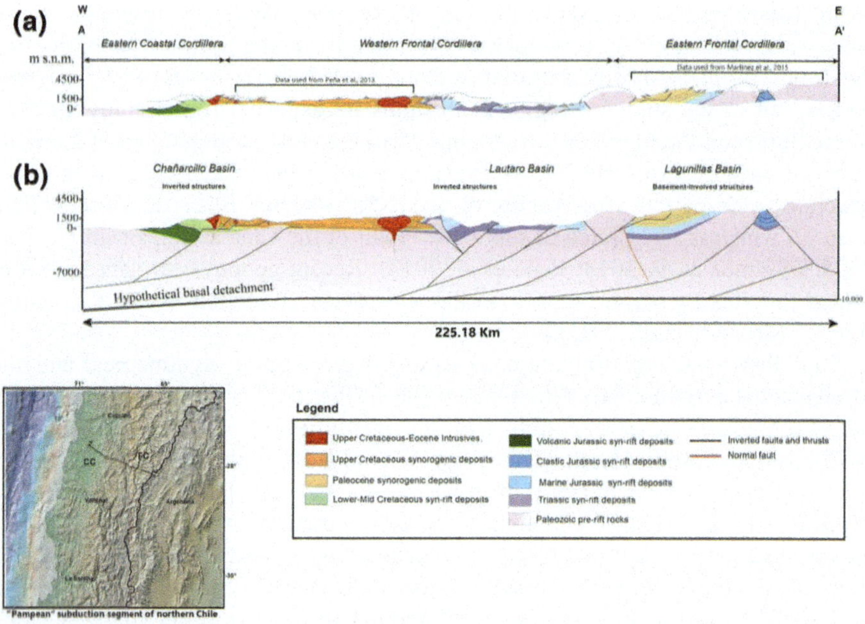

Fig. 6 **a** Geological data used to construct the regional cross section used in this study, **b** structural and semi-balanced cross section along the Chilean–Pampean flat subduction segment in northern Chile (see location of Fig. 1)

minor accommodation structures that involve synorogenic deposits. Previous works interpreted the basement-involved faults in this region as simple high-angle faults along which the basement blocks were uplifted and tilted (Jensen 1976; Godoy and Davidson 1976; Moscoso and Mpodozis 1988). However, recent studies based on balanced cross-sectional techniques and structural restoration have suggested that the geometry of the faults and folds are better related to moderate-angle faults and thrusts (Martínez et al. 2015b, 2016), which would have resulted from the faulting and folding of former basement paleohighs (e.g., Montosa Horst; Godoy and Davidson 1976) preserved in the inverted basins of this sector (Lautaro and Lagunillas basins).

Palinspastic restorations carried out along the Frontal Cordillera (Martínez et al. 2015b) have associated this structural style with large ramps that resulted from the shortening of previous extensional systems during different deformation phases that took place after the previous tectonic inversion events. If we consider that the previous extensional systems established in this region were composed of synthetic normal faults separated by basement highs (Martínez et al. 2016), then it is possible that some of these ramps could have truncated older, normal, or inverted faults (Fig. 6).

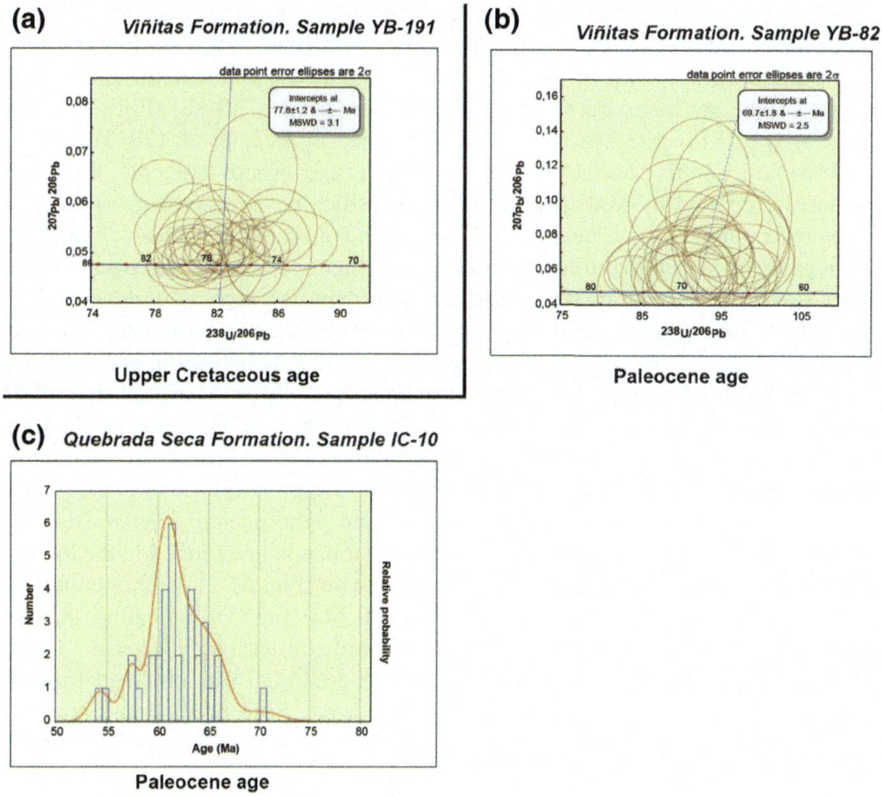

Fig. 7 **a–b** Concordia plots of LA-ICPMS U–Pb analyses of zircons from synorogenic deposits of the Viñitas Formation, **c** U–Pb frequency plot ages from detrital zircons of the Quebrada Seca Formation

4 Timing of Deformation

In order to obtain a constraint on the age of deformation in our study area, a series of U–Pb datings were performed using a LA-ICP Mass Spectrometer, which were then compared with previously reported ages. We obtained samples of detrital and igneous zircons from the continental red sandstone beds and ignimbrites belonging to synorogenic deposits (Viñitas and Quebrada Seca formations; Figs. 3, 5, and 7) exposed in the eastern Coastal Cordillera and the Frontal Cordillera, respectively. The samples were prepared applying the standard methods at the Geology Department of the University of Chile. The zircon selection was made by hand with a binocular microscope. The analytical work was carried out at the Laboratorio de Estudios Isotópicos of the Universidad Autonoma de Mexico (UNAM), using a M50 193 nm laser excimer connected to a Thermo Xii Series Quadrupole Mass Spectrometer, applying the technical details described in Solari et al. (2010). The mean ages were calculated using Isoplot v. 3.7 (Ludwing 2008).

Sedimentary and volcanic rocks of the middle and upper section of the Viñitas Formation, exposed over the angular unconformity formed in the frontal limb of the Tierra Amarilla Anticline in the eastern section of the Chañarcillo Basin, have reported ages that range between 65 and 68 Ma (sample YB-82) (Figs. 3 and 6). Older ages of 81,6 ± 1 Ma were also determined by Peña et al. (2013) from the basal section of this succession (Fig. 3). These age groups coincide with those previously reported by Maksaev et al. (2009), which have allowed us to place the tectonic inversion of the Chañarcillo Basin and the onset of the Andean deformation in the eastern Coastal Cordillera between the Upper Cretaceous–Paleogene ranges. Other stratigraphic features recognized in these deposits, such as internal progressive unconformities and growth strata, have confirmed their syn-kinematic character acquired during the interplay between tectonic uplift and continental accumulation. On the other hand, younger ages (Eocene; Peña et al. 2013; Martínez et al. 2016) obtained from syntectonic intrusive bodies located east of the Chañarcillo Basin have also suggested that the Andean deformation could have continued here during Eocene times.

Similar ages have been determined from the synorogenic deposits that rest unconformably over the frontal limb of the inversion anticline formed in the hanging wall block of the Calquis Fault in the Lautaro Basin (Fig. 5). Here, the ignimbrites and tuffs of the basal section have ages of 78 Ma (Fig. 5), indicating that this structure grew from the Upper Cretaceous onwards, at least in this sector. On the other hand, a younger Paleogene age of 60.5 Ma (IC-10) was obtained from the continental synorogenic successions of the Quebrada Seca Formation exposed to the east of the east-verging Viscachas Fault, in the Frontal Cordillera (Figs. 5 and 7). Here, Martínez et al. (2015b, 2016) also reported similar ages that range between 60 and 65 Ma. These ages have been frequently associated with propagation and uplift of the basement structures, which appear to be younger and subsequent to the creation of the inversion structures (Martínez et al. 2016; Rossel et al. 2016; Moscoso and Mpodozis 1988, among others). Lower Miocene ages have also been reported from basement-involved structures found in the easternmost portion of the Frontal Cordillera (Moscoso et al. 2010), which also confirm this interpretation. Based on the data reported here, we propose that the Andean deformation at these latitudes could have started during the Late Cretaceous, at least along the eastern Coastal Cordillera, and then migrated to the eastern regions in the Paleocene.

5 Discussion

5.1 The Chilean–Pampean Flat-Slab Segment of Northern Chile as a Result of a Hybrid Tectonic Mechanism

The structure of the Pampean–Chilean flat subduction segment of northern Chile is characterized by a complex interaction between inverted and basement-involved folds and faults (Fig. 6), or a hybrid mechanism. In this segment, the west and east variations of the structural styles are related to important differences of the previous structural configuration of the region. As recognized in other sectors of the Central Andes, the variations of the Andean structures appear to be influenced by the Mesozoic architecture inherent to former extensional tectonic episodes. In adjacent localities (e.g., Sierras Pampeanas, the Santa Barbará System, or the Salta Rift among others), the reactivation of previous normal faults and basement anisotropies is responsible for the final configuration of the Andean structures. In northern Chile, the position of inversion anticlines also allows the interpretation of this mechanism of deformation. Their doubly verging geometries reveal that they resulted from tectonic inversion of westward and eastward master faults separated by basement highs rather than by tectonic inversion of simple graben geometries (Figs. 6 and 8). Based on our field observations, we interpret that the crustal shortening in this region was initially accommodated by the reactivation of normal faults, such as those evidenced along the main edge of the Chañarcillo and Lautaro basins along the Coastal Cordillera and the Frontal Cordillera, respectively. Later shortening, however, was absorbed by large, basement-involved thrusting (Fig. 8).

On the other hand, the ages of the synorogenic deposits obtained here suggest that the Andean deformation in this segment started close to 78 Ma, with tectonic inversion of the Mesozoic basins indicating the occurrence of the Peruvian tectonic phase (Steinman 1929). Other younger, Paleocene ages determined from the synorogenic deposits exposed in the footwall of the basement-involved structures indicate: (a) an eastward migration of the Andean deformation front and (b) basement-involved thrusting occurring after the previous tectonic inversion episodes and during the K-T tectonic phase (Cornejo et al. 2003). Recent analyses of exhumation timings carried out in different regions of Argentina also indicate that the oldest episodes of Andean deformation occurred in the Late Cretaceous (Peruvian tectonic phase) (Balgord and Carrapa 2014; Folguera et al. 2015; Fennell et al. 2015, Echaurren et al. 2016).

In contrast to other regions, where the transition from tectonic inversion to basement-involved thrusting has been developed by shortcut faults, we have seen that the transition in this region could have been achieved by new thrusts that cut through basement highs and the hanging wall of the normal faults that dip away from the advancing thrusts (Figs. 6 and 8). Along the Andes, this mechanism of transition between inverted and basement-involved structures has been identified in the Cordillera de Carabaya in southern Peru (Pérez et al. 2016), in the Metán Basin in northwestern Argentina (Iaffa et al. 2011), in the Santa Barbara System (Kley and Monaldi 2002), the Eastern Cordillera of Colombia (Mora et al. 2009), as well as along other thrust belts located in the Alps, the Apennines, and the Atlas of Morocco (Welbon and Butler 1992; Scisciani et al. 2002; Burkhard et al. 2006, among others). From a point of view more speculative, this hybrid tectonic mechanism could be associated with a transition from tectonic inversion of half-grabens to the creation of basement-involved thrusts. Both geological processes are seen to have a broad wavelength, because both have been recognized in Chile and Argentina. In northern Chile, this hybrid mechanism was initially controlled by the coeval tectonic inversion of the ancient normal faults both in the

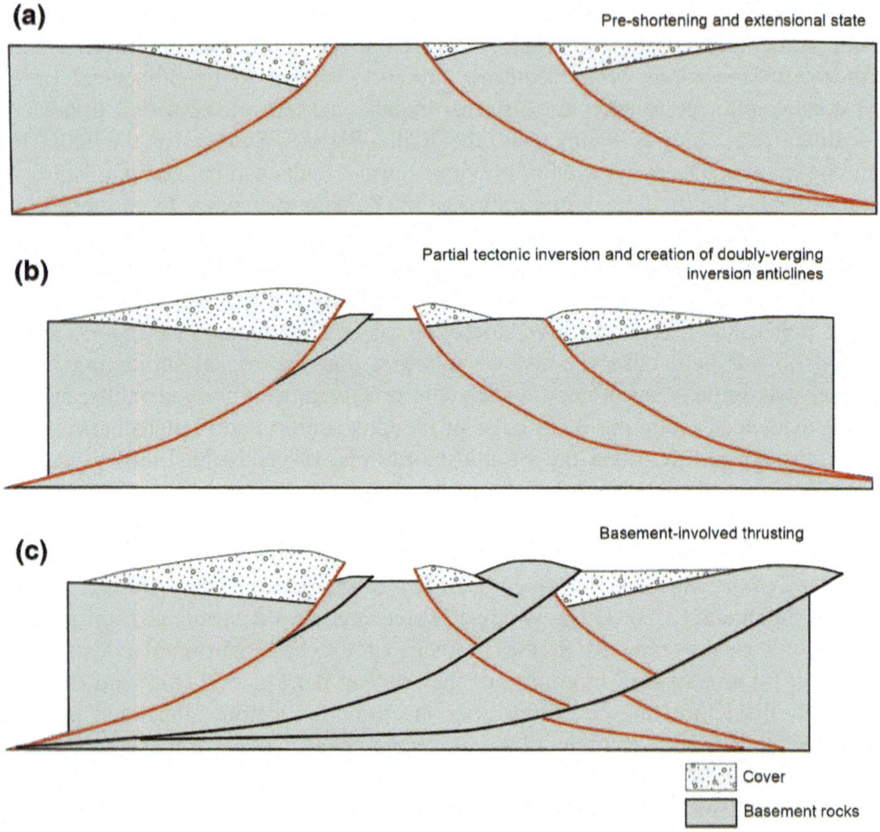

Fig. 8 Schematic cartoon illustrating the tectonic evolution of the Andean structures in the Chilean–Pampean flat subduction segment of northern Chile, as well as the interaction between the different tectonic mechanisms interpreted in this work

Coastal and Frontal Cordilleras; however, during the tectonic inversion is common that some faults are blocked by changes in their physics and therefore, new thrusts are created with a thick-skinned style taking advantage of a continental crust thermal and mechanically weakened. Similar interpretations have been proposed for the Malargüe Thrust and Fold Belt, where an interplay between inverted and basement-involved faults was also observed (Giambiagi et al. 2003; Mescua and Giambiagi 2012), previously established in a weak and thin continental crust.

Other, more regional factors that could have governed this mechanism are the convergence angle with respect to previous structural patterns and the magnitude of shortening; however, the last is most speculative, at least in this study area. Finally, our results indicate that the amplitude of the deformation, the degree of shortening, and the spatial development of contractional structures, as well as their kinematics, are fundamentally influenced by inherited structures in the basement and master normal faults. This was also interpreted by Ramos et al. (2002), who proposed that the contractional deformation along the Chilean–Pampean flat subduction segment is a product of the successive reactivation of weakness zones in the pre-rift basement of this region.

6 Conclusions

The Chilean–Pampean flat subduction segment of the Central Andes in northern Chile displays two main tectonic styles: inverted structures and basement-involved thrusts confined along the Coastal Cordillera and the Frontal Cordillera, respectively. The occurrence of both structural styles is associated with different tectonic mechanisms, which involve dip-slip reactivation of inherited extensional structures and anisotropies in the basement pre-rift units exposed along the region and basement-involved thrusting. The geological observations described here indicate that this structural interaction has resulted from a hybrid tectonic mechanism, which is marked by the transition from tectonic inversion to thrusting through inherited basement highs (Fig. 7). The first mechanism is mostly associated with the oldest episodes of Andean deformation that affected the continental margin during the Peruvian tectonic phase ($\sim$80 Ma), and the second is related to eastward and younger Paleocene tectonic episodes. We conclude that the tectonic evolution of this Andean segment is strongly influenced by the previous configuration of the pre-rift basement and the extensional basins that dominated this region during Mesozoic times, in terms of paleo-positions of normal faults and basement highs, among others. Finally, we suggest taking in consideration these results in order to analyze the tectonic context of northern Chile.

Acknowledgements This work has been supported by the research project "The crustal structure and timing deformation along the Chilean flat slab subduction segment (27° 29° S), Central Andes," funded by the Fondo Nacional de Desarrollo Científico y Tecnológico (Fondecyt), Chile (grant 3140557). We would like to thank Midland Valley Company for providing the academic license for the Move software used in this study. We also acknowledge Laura Giambiagi and Guido Gianni whose reviews greatly enriched earlier versions of this article. Finally, we thank Sergio Villagrán and Marco Vaccaris for their assistance in the field and J. Vargas and R. Valles (Geology Department, Universidad de Chile), for zircon sample preparation, as well as L. Solari for the U–Pb LA-ICP-MS analyses (LEI, UNAM).

References

Aguirre-Urreta B (1993) Neocomian ammonite biostratigraphy of the Andean basins of Argentina and Chile. Rev. Española de Paleontología 8:57–74

Allmendinger RW, Jordan TE, Kay SM, Isacks BL (1997) The evolution of the Altiplano-Puna Plateau of the central Andes. Annu Rev Earth Planet Sci 25:139–174

Amilibia A, Sàbat F, McClay KR, Muñoz JA, Roca E, Chong G (2008) The role of inherited tectono-sedimentary architecture in the development of the central Andean mountain belt: insights from the Cordillera de Domeyko. J Struct Geol 30:1520–1539

Arévalo C (1994) Mapa geológico de la Hoja Los Loros, Región de Atacama (1:100.000). Servicio Nacional de Geología y Minería, Documentos de, Trabajo No 6

Arévalo C (1999) The Coastal Cordillera–Precordillera boundary in the Copiapó area, northern Chile, and the structural setting of the Candelaria Cu–Au ore deposit. Unpublished Ph.D. Thesis. Kingston University, Kingston-upon-Thames, UK, p. 244

Arévalo C (2005) Carta los Loros, Región de Atacama, Carta Geológica Básica. Servicio Nacional de Geología y Minería, Santiago 92: 54

Balgord E, Carrapa B (2014) Basin evolution of Upper Cretaceous–Lower Cenozoic strata in the Malargüe fold-and-thrust belt: northern Neuquén Basin, Argentina. Basin Res: 1–24

Baraganzi M, Isacks BL (1976) Spatial distribution of earthquakes and subduction of the Nazca plate beneath South America. Geol 4:686–692

Bonini M, Sani F, Antonielli B (2012) Basin inversion and contractional reactivation of inherent normal faults: A review based on previous and new experimental models. Tectonophysics 522:55–88

Burkhard M, Caritg S, Helg U, Robert-Charrue C, Soulaimani A (2006) Tectonics of the Anti-Atlas of Morocco. CR Geosci 338:11–24

Carrapa B, Trimble JD, Stockli D (2011) Patterns and timing of exhumation and deformation in the Eastern Cordillera of NW Argentina revealed by (U–Th)/He thermochronology. Tectonics 30:1–30

Charrier R (1979) El Triásico en Chile y regiones adyacentes de Argentina: Una reconstrucción paleogeográfica y paleoclimática. Comunicaciones 26:1–47

Cornejo P, Mpodozis C, Kay S, Tomlinson A, Ramirez C (1993) Upper cretaceous lower eocene potasic volcanism in an extensional regime in the precordillera of Copiapó, Chile. In: Second IASG (Oxford). pp 347–350

Cornejo P, Matthews S, Pérez de Arce C (2003) The K-T compressive deformation event in northern Chile (24°–27°). In: 10° Congreso Geológico Chileno (Concepción), Chile

Coutand I, Cobbold P, Urreiztieta M, Gautier P, Chauvin A, Gapais D, Rossello E, Gamundi O (2001) Style and history of Andean deformation, Puna plateau, northwestern Argentina. Tectonics 20:210–234

Cristallini E, Comínguez A, Ramos V, Mercerat ED (2004) Basement double wedge thrusting in the northern Sierras Pampeanas of Argentina (27° S). Constraints from deep seismic reflection. In: McClay KR (ed) Thrust Tectonics and Hydrocarbon Systems. AAPG Mem 82:1–26

Del Rey A, Dejkart K, Arriagada C, Martínez F (2016) Resolving the paradigm of the late Paleozoic-Triassic Chilean magmatism: isotopic approach. Gondwana Res 37:172–181

Echaurren A, Folguera A, Gianni G, Orts D, Tassara A, Encinas A, Giménez M, Valencia V (2016) Tectonic evolution of the North Patagonian Andes (41°–44° S) through recognition of syntectonic strata. Tectonophysics 677–678:99–114

Fennell L, Folguera A, Naipauer M, Gianni G, Rojas Vera E, Bottesi G, Ramos V (2015) Cretaceous deformation of the southern Central Andes: synorogenic growth strata in the Neuquén Group (35°30'–37° S). Basin Res: 1–22. https://doi.org/10.1111/bre.12135

Folguera A, Bottesi G, Duddy I, Martín-Gonzalez Orts D, Sagripanti L, Vera, Rojas, Ramos VA (2015) Exhumation of the Neuquén Basin in southern Central Andes (Malargüe fold and thrust belt) from field data and low-temperature thermochronology. J S Am Earth Sci 1–18

Giambiagi L, Ramos V, Godoy E, Álvarez P, Orts D (2003) Cenozoic deformation and tectonic style of the Andes, between 33° and 34° south latitude. Tectonics 22(4):1–18. https://doi.org/10.1029/2001TC001354

Godoy E, Davidson JD (1976) Pilares en compresión de edad Mioceno superior en los Andes del Norte de Chile (22°–30° Latitud Sur). In: I Congreso Geológico Chileno. pp 87–103

Grocott J, Taylor G (2002) Magmatic arc fault systems, deformation partitioning and emplacement of granitic complexes in the Coastal Cordillera, north Chilean Andes (25°30′ to 27°00′S). J Geol Soc Lond 159:425–442

Heredia N, Rodríguez Fernández L, Gallastegui G, Busquest B, Colombo F (2002) Geological setting of the Argentine Frontal Cordillera in the flat-slab segment (30°00'-31°30'S latitude). J S Am Earth Sci 15:79–99

Iaffa D, Sàbat F, Muñoz JA, Mon R, Gutierrez AA (2011) The role of inherited structures in a foreland basin evolution. The Metán Basin in NW Argentina. J Struct Geol 33:1816–1828
Ingersoll R (2012) Tectonic of sedimentary basins, with revised nomenclature, in Tectonic of sedimentary basins recent advances. In: Bushy C and Azor A (ed). Wiley-Blackwell publication, pp 3–47
Isacks BL (1988) Uplift of the Central Andes Plateau and bending of the Bolivian Orocline. J Geophys Res 93:3211–3231
Jensen O (1976) Geología de las nacientes del río Copiapó, entre los 27°53′ y 28°20′ de latitud Sur, provincia de Atacama. Memoria de Titulo (Inédito), Universidad de Chile, Departamento de Geología, p 249
Jordan T, Allmendinger R (1986) The Sierras Pampeanas of Argentina: a modern analogue of Rocky Mountain foreland deformation. Am J Sci 286:737–764
Jordan T, Isacks BL, Allmendinger R, Brewer J, Ramos V, Ando C (1983) Andean tectonics related to geometry of subducted Nazca plate. Geol Soc Am Bull 94:341–361
Kley J, Monaldi C (2002) Tectonic inversión in the Santa Barbara System of the central Andean foreland thrust belt, northwestern Argentina. Tectonics 21:1–18
Lara L, Godoy E (1998) Mapa Geológico de la Hoja Quebrada Salitrosa, Región de Atacama. Servicio Nacional de Geología y Minería, Santiago, scale: 1:100.000
Ludwing KR (2008) Isoplot 3.6. 4. Berkley Geochronology Center Special, Publication
Maksaev V, Munizaga F, Valencia V, Barra F (2009) LA-ICP-MS zircón U-Pb geochronology to constrain the age of post-Neocomian continental deposits of the Cerrillos Formation, Atacama Region, northern Chile: tectonic and metallogenic implications. Andean Geol 36:264–287
Marschik R, Fontboté L (2001) The Candelaria-Punta del Cobre iron oxide Cu–Au (–Zn–Ag) deposits, Chile. Econ Geol 96:1799–1826
Martínez F, Arriagada C, Mpodozis C, Peña M (2012) The Lautaro Basin: a record of inversion tectonics in northern Chile. Andean Geol 39(2):258–278
Martínez F, Arriagada C, Peña M, Del Real I, Deckart K (2013) The structure of the Chañarcillo Basin: an example of tectonic inversion in the Atacama region, northern Chile. J S Am Earth Sci 42:1–16
Martínez F, Maksymowicz A, Ochoa H, Díaz D (2015a) Geometry of the inverted Cretaceous Chañarcillo Basin based on 2D gravity and field data-an approach to the structure of the western Central Andes of northern Chile. Solid-Earth 6:1–18
Martínez F, Arriagada C, Valdivia R, Deckart K, Peña M (2015b) Geometry and kinematics of the Andean thick-skinned thrust systems: insights from the Chilean Frontal Cordillera (28°–28.5° S), Central Andes. J S Am Earth Sci 34:307–324
Martínez F, Arriagada C, Peña M, Deckart K, Charrier R (2016) Tectonic styles and crustal shortening of the Central Andes "Pampean" flat-slab segment in northern Chile (27°–29° S). Tectonophysics 667:144–162
Martinod J, Husson L, Roperch P, Guillaume B, Espurt N (2010) Horizontal subduction zones, convergence velocity and the building of the Andes. Earth Planet Sci Lett 299:299–309
Mescua J, Giambiagi L (2012) Fault inversion vs. new thrust generation: A case study in the Malargüe fold-and-thrust belt, Andes of Argentina. J Struct Geol 31:51–63
Mora A, Gaona T, Kley J, Montoya D, Parra M, Quiroz LI, Reyes G, Strecker MR (2009) The role of inherited extensional fault segmentation and linkage in contractional orogenesis: a reconstruction of Lower Cretaceous inverted rift basins in the Eastern Cordillera of Colombia. Basin Res 21:111–137
Mortimer C (1973) The Cenozoic history of the southern Atacama Desert, Chile. J Geol Soc Lond 129:505–526
Moscoso R, Mpodozis C (1988) Estilos estructurales en el Norte Chico de Chile (28°–31° S), regiones de Atacama y Coquimbo. Rev Geol Chile 15:155–158
Moscoso R, Mpodozis C, Nassi C, Ribba L, Arévalo C (Compilador) (2010) Geología de la Hoja El Tránsito, Región de Atacama. Servicio Nacional de Geología y Minería de Chile, Serie Preliminar, 7, scale: 1:250.000, 3 anexos, Santiago

Mourgues FA (2004) Advances in ammonite biostratigraphy of the marine Atacama basin (Lower Cretaceous), northern Chile, and its relationship with the Neuquén basin, Argentina. J S Am Earth Sci 17:3–10
Mpodozis C, Cornejo P (1997) El rift Triásico- Sinemuriano de Sierra Exploradora, Cordillera de Domeyko (25°–26° S): Asociaciones de facies y reconstrucción tectónica. In: VIII Congreso Geológico Chileno 1:550–554
Mpodozis C, Ramos V (1990) The Andes of Chile and Argentina. In: Ericksen GE, Cañas Pinochet MT, Reinemud JA (ed). Geology of the Andes and its relation to hydrocarbon and mineral resources: Circumpacific Council for Energy and Mineral Resources. Earth Sci 11:59–90
Mpodozis C, Ramos VA (2008) Tectónica jurásica en Argentina y Chile: Extensión, Subducción Oblicua, Rifting, Deriva y Colisiones? Rev Geol Argent 63:479–495
Naranjo JA, Puig A (19849 Hojas Taltal y Chañaral, Regiones de Antofagasta y Atacama. Servicio Nacional de Geología y Minería, Santiago. pp 62–63
Narr W, Suppe J (1994) Kinematics of basement-involved compressive structures. Am J Sci 294:802–860
Peña M, Arriagada C, Martínez F, Becerra J (2013) Carta Geológica Yerbas Buenas-Tres Morros, Región de Atacama. Servicio Nacional de Geología y Minería, Santiago, scale: 1:100.000
Pérez N, Horton BK, Carlotto V (2016) Structural inheritance and selective reactivation in the central Andes: Cenozoic deformation guided by pre-Andean structures in southern Peru. Tectonophysics 671:264–280
Pindell J, Dewey J (1982) Permo-Triassic reconstruction of western Pangea and the evolution of the Gulf of Mexico/Caribbean region. Tectonics 1:179–211
Poblet J, Lisle RJ (2011) Kinematic evolution and structural styles of fold-and-thrust belts. In: Poblet J, Lisle RJ (ed) Kinematic evolution and structural styles of fold-and-thrust belts. Geol Soc Spec Pub 349:1–24
Ramos V (1999) El segmento de Subducción Subhorizontal de los Andes Centrales Argentino-Chilenos. Acta Geol Hisp 32(7):5–16
Ramos VA (2009) Anatomy and global context of the Andes: main geologic features and the Andean orogenic cycle. In: Kay SM, Ramos VA, Dickinson WR (ed) Backbone of the Americas: shallow subduction, plateau uplift, and ridge and terrane collision. Geol Soc Am Mem 204:31–65
Ramos VA, Cristallini EO, Pérez DJ (2002) The Pampean flat-slab of the Central Andes. J S Am Earth Sci 15:59–78
Rossel K, Aguilar G, Salazar E, Martinod J, Carretier S, Pinto L, Cabré A (2016) Chronology of Chilean Frontal Cordillera building from geochronological, stratigraphic and geomorphological data insights from Miocene intramontane-basin deposits. Basin Res. https://doi.org/10.1111/bre.12221
Scheubert E, Reutter KJ (1992) Magmatic arc tectonics in the Central Andes between 21° and 25° S. Tectonophysics 205:127–140
Scisciani V, Tavarnelli E, Calamita F (2002) The interaction of extensional and contractional deformations in the outer zones of the Central Apennines, Italy. J Struct Geol 24:1647–1658
Segerstrom K (1960) Cuadrángulo Quebrada Paipote, Provincia de Atacama, Carta Geológica de Chile. Instituto de Investigaciones Geológicas, Santiago de Chile. p 35
Solari LA, Gómez-Tuena A, Bernal JP, Pérez-Arvizu O, Tanner M (2010) U-Pb zircon geochronology by an integrated LAICPMS microanalytical work station: achievements in precision and accuracy. Geostand Geoanal Res 34(1):5–18
Steinman G (1929) Geologie von Peru. Carl Winters Universitats-Buchhandlung. p 448
Suárez M, Bell CM (1992) Triassic rift-related sedimentary basins in northern Chile (24°–29° S). J S Am Earth Sci 6:109–121
Welbon AI, Butler RWH (1992) Structural styles in thrust belts developed through rift basins: a view from the western Alps, in Structural and Tectonic Modelling and its Application to Petroleum Geology. In: Larsen RM, Brekke H, Larsen BT and Talleras E (ed) Spec publ Norw Petrol Soc:464–479

Cretaceous Orogeny and Marine Transgression in the Southern Central and Northern Patagonian Andes: Aftermath of a Large-Scale Flat-Subduction Event?

Guido M. Gianni, Andrés Echaurren, Lucas Fennell, César R. Navarrete, Paulo Quezada, Jonathan Tobal, Mario Giménez, Federico M. Dávila and Andrés Folguera

Abstract This review synthesizes the tectonomagmatic evolution of the southern Central and Northern Patagonian Andes between 35°30′S and 48° S with the aim to spotlight early contractional phases on Andean orogenic building and to analyze their potential driving processes. We examine early tectonic stages of the different fold and thrust belts that compose this Andean segment. Additionally, we study the magmatic arc behavior from a regional perspective as an indicator of potential past subduction configurations during critical tectonic stages of orogenic construction. This revision proposes the existence of a continuous large-scale flat-subduction with a similar size to the present-largest flat-slab setting on earth. This particular

G. M. Gianni (✉) · M. Giménez
Instituto Geofísico Sismológico Ing. Volponi (IGSV), Universidad de Nacional San Juan, San Juan, Argentina
e-mail: guidogianni22@gmail.com

G. M. Gianni · M. Giménez
Consejo Nacional de Investigaciones Científicas y Técnicas (CONICET), Buenos Aires, Argentina

A. Echaurren · L. Fennell · J. Tobal · A. Folguera
Universidad de Buenos Aires, Buenos Aires, Argentina
e-mail: andresfolguera2@yahoo.com.ar

C. R. Navarrete
Departamento de Geología, Universidad Nacional de la Patagonia San Juan Bosco, Comodoro Rivadavia, Chubut, Argentina

P. Quezada
Universidad Andres Bello, Santiago, Chile

F. M. Dávila
CICTERRA, CONICET—FCEFyN—Universidad Nacional de Córdoba, Córdoba, Argentina
e-mail: fmd.geodinamica@gmail.com

A. Folguera et al. (eds.), *The Evolution of the Chilean-Argentinean Andes*, Springer Earth System Sciences, https://doi.org/10.1007/978-3-319-67774-3_12

process would have initiated diachronically in late Early Cretaceous times and achieved full development in Late Cretaceous to earliest Paleocene, constructing a series of fold-thrust belts on the retroarc zone from 35°30′S to 48° S. Furthermore, dynamic subsidence focused at the edges of the slab flattening before re-steepening beneath the foreland zone may explain sudden paleogeographic changes in Maastrichtian–Danian times previously linked to continental tilting and orogenic loading during a high sea level global stage.

Keywords Andes · Fold and thrust belt · Foreland basin · Broken foreland Flat-slab · Dynamic subsidence

1 Introduction

In spite of many years of scientific research in the Andes, type locality of non-collisional orogenesis, several aspects regarding their evolution, deformational mechanisms, and timing of uplift remain unsolved. Even though most recent tectonic research is increasingly supporting a pre-Cenozoic initial growth of the Andean orogen (e.g., Soler and Bonhomme 1990; Sempere 1995; Fildani et al. 2003; Folguera and Iannizzotto 2004; Mpodozis et al. 2005; Jaimes and de Freitas 2006; Martin-Gombojav and Winkler 2008; Tunik et al. 2010, among others) (see Chaps. "Lower Jurassic to Early Paleogene intraplate contraction in Central Patagonia and Mechanisms and episodes of deformation along the Chilean-Pampean flat-slab subduction segment of the Central Andes in northern Chile"), contraction achieved during these stages is still underestimated in recent works assessing Andean evolution (e.g., Oncken et al. 2006; Husson et al. 2008; Barnes and Ehlers 2009; Maloney et al. 2013; Faccenna et al. 2013; Armijo et al. 2015).

Additionally, there exists a general disagreement when linking different proposed parameters driving non-collisional contraction and Andean evolution. This apparent lack of correlation between timing of the different orogenic stages and changes in plate kinematic parameters that rule upper plate stress state, such as upper plate motion and plate convergence, has been a matter of discussion through the last decade or so. For example, documented high convergence rate stages in Cenozoic times relate either to upper plate extension between 27° S and 44° S or contraction between 14° S and 25° S in the Andes (Pardo-Casas and Molnar 1987; Maloney et al. 2013; Jordan et al. 2001; Barnes and Ehlers 2009; Encinas et al. 2015). The same can be observed when analyzing the upper plate absolute velocity, considered an important factor controlling mountain building in subduction settings. Determinations of South American plate acceleration since ~30 Ma (Silver et al. 1998) correlate indistinctly with contractional as well as extensional events in different sectors of the Andean margin (Barnes and Elhers 2009; Armijo et al. 2015; Winocur et al. 2015; Pananont et al. 2004; Bechis et al. 2014). This observation precludes a straightforward link between absolute plate motion and Andean orogenesis despite the relatively good correlation shown in parametric studies of convergent settings worldwide (e.g., Heuret and Lallemand et al. 2005).

Alternatively, factors of local character such as shallow or flat-subduction acting on discrete margin segments have progressively taken a greater role on literature dealing with Andean evolution. Shallow subduction settings have been particularly helpful to explain cases where fold and thrust belt development took place in concert with magmatic arc expansions, which have been attributed to changes in the subduction angle (James and Sacks 1999; Gutscher et al. 2000; Ramos et al. 2002; Duerto et al. 2006; Ramos and Folguera 2009; Spagnuolo et al. 2012, Orts et al. 2012a, b; Horton and Fuentes 2016; Litvak et al. 2015). Interestingly, as previously noticed by Ramos and Folguera (2009), when present and past shallow/flat configurations are posted along the Andean subduction margin, it can be observed that these processes affected a vast portion of the Andean margin. Strikingly, in light of recent documentations of past subduction configurations (Duerto et al. 2006; Bissig et al. 2008; Espinoza et al. 2010; Schütte et al. 2010; Orts et al. 2012a, b; Guillot et al. 2011; García Morabito and Ramos 2012; Spagnuolo et al. 2012; Chiarabba et al. 2015; Gianni et al. 2015a; Echaurren et al. 2016a), it can be stated that practically the whole South American margin experienced this process at least once during Andean development (Fig. 1).

Some outstanding examples on the contribution of pre-Cenozoic stages to orogenic building and flat/shallow subduction processes in Andean uplift, come from studies carried on fold and thrust belts composing the southernmost sector of the Central Andes (Gansser 1973). In this study, we synthesize the latest geologic advances in the knowledge of the Malargüe, Chos Malal, Agrio, Aluminé and North Patagonian fold and thrust belts, as well as in the Patagonian broken foreland, to highlight the role of early orogenic stages and shallow subduction on the evolution of the Andes.

2 Cretaceous Contractional Deformation Stages

The following subsections constitute an update of different geological studies from extensive research done in the last decade in the Andes of southern South America (Fig. 1b). Particularly, we focus on a segment extending from the 35°30′S to 48° S that encompasses the Malargüe, Chos Malal, Agrio, Aluminé and North Patagonian fold and thrust belts; and the Patagonian broken foreland in the intraplate sector (Fig. 1b).

2.1 Late Early Cretaceous Deformation in the Malargüe Fold and Thrust Belt (FTB)

The Malargüe fold and thrust belt, as most of the sectors comprehended within the Mesozoic rift of the Neuquén basin, has been object of many classic studies developed in the last decades of the twentieth century due to its importance in the

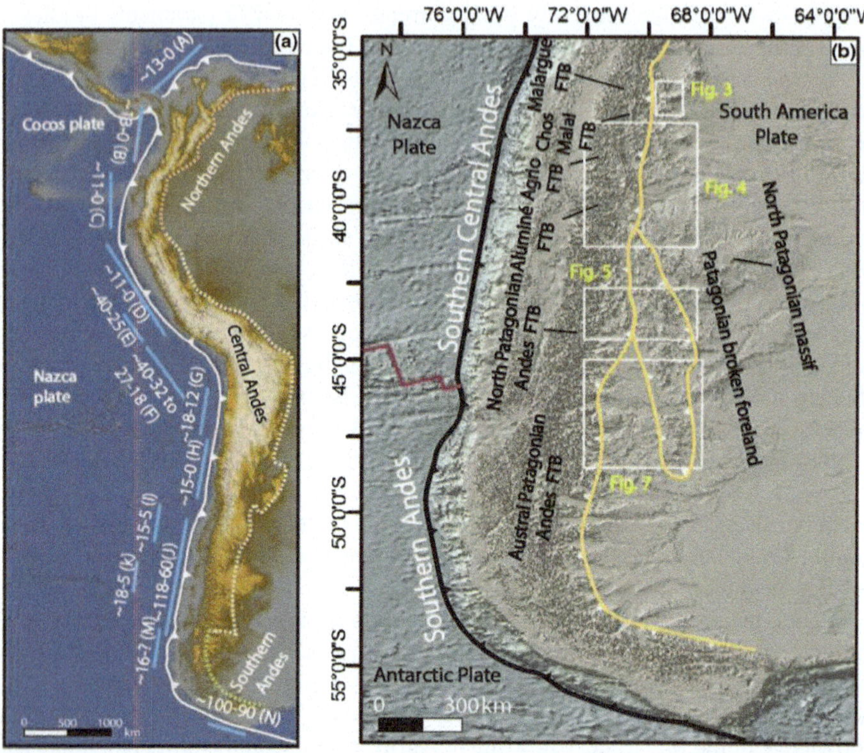

Fig. 1 **a** Current and past flat/shallow configurations proposed along the South American margin. References are: A Caribbean flat-slab, Duerto et al. (2006); B Bucaramanga flat-slab, Chiarabba et al. (2015); C Carnegie incipient flat-slab, Schütte et al. (2010); D Peruvian large flat-slab, Gutscher et al. (2000); E Southern Peru paleogeone flat-slab, Bissig et al. (2008); F Central Andes paleogene flat-slab, James and Sacks (1999); G Puna flat-slab, Kay et al. (1999); H Chilean/Pampean flat-slab, Ramos et al. (2002); I Payenia slab-shallowing, Kay et al. (2006); J Nalé large flat-slab proposed in this work composed of several segments from Ramos and Folguera (2005), García Morabito and Ramos (2012), Spagnuolo et al. (2012); Gianni et al. (2015a) and Echaurren et al. (2016a); K North Patagonian slab-shallowing, Orts et al. (2012a, b); M Northern Santa Cruz slab-shallowing, Espinoza et al. (2010); N Fueguian Andes Cretaceous flat-slab, Guillot et al. (2011). **b** DEM showing the Southern Central and the Andes with respective fold and thrust belts where Cretaceous orogeny has been documented. FTB: Fold and thrust belt

oil industry (Legarreta and Gulisano 1989; Uliana et al. 1989; Gulisano and Gutierrez Pliemling 1994; Manceda and Figueroa 1995; Vergani et al. 1995, among others). Since its definition by Kozlowski et al. (1993), there has been a general acceptance that its initial growth, as part of the Andean orogen, was during Miocene times (Silvestro et al. 2005; Giambiagi et al. 2008; Silvestro and Atencio 2009; Turienzo et al. 2012). These studies did not take in account older deformational phases, although the pioneer studies of Groeber (1946) had already detected in the area an angular unconformity developed in Late Cretaceous times. In light of new advances concerning the structural evolution of the Southern Andes (e.g., Pananont et al. 2004; Mpodozis et al. 2005; Ramos and Folguera 2005; Oncken et al. 2006), the geological evolution of this area had to be revised.

In this sense, the Late Cretaceous red beds of the Neuquén Group (Cazau and Uliana 1973) became the focus of numerous studies (Fig. 2). These were initially considered as part of the post-rift thermal subsidence in the Neuquén basin but, after the work of Tunik et al. (2010), its tectonic significance changed, becoming the first Andean foreland basin in Late Cretaceous times (for a summary, see Garrido 2010). In the Malargüe fold and thrust belt, numerous studies have arrived to similar conclusions based on sedimentological, magmatic, and seismic evidence (Galarza et al. 2009; Orts et al. 2012a, b; Spagnuolo et al. 2012; Mescua et al. 2013; Sánchez and Asurmendi 2014). Direct evidence of Cretaceous tectonic activity in individual structures of this belt has been recently described by Fennell et al. (2015). These authors identified contractional growth-strata in outcrops of the Neuquén Group in main structures of the Malargüe fold and thrust belt between 35°30′ and 37° S (Figs. 3a, b, and c). In line with this evidence, deformational synsedimentary structures were also described, interpreted by the authors as seismites, a feature that had already been documented elsewhere in the Neuquén Group by Sánchez et al. (2013).

Through the identification of growth-strata, Fennell et al. (2015) defined, with some degree of accuracy, the Late Cretaceous orogenic front, the general morphology of the orogen and the paleoflow of sediments between 35°30′S and 37° S. In order to represent the Late Cretaceous orogen, two schematic structural cross sections were performed by Fennell et al. (2015) (location in Fig. 3a). In the northern section (Fig. 3d), the Sierra Azul and the Ranquil Co anticlines were uplifted in Late Cretaceous times, defining a broad intermontane basin between them, where the Río Grande flows nowadays.

The first structure had its core exhumed and was actively eroded, while the second structure was located in the wedge top area (see DeCelles and Giles 1996). The Sierra Azul anticline was also the source of the sediments hosted in the Portezuelos Colorados syncline, structure active during Late Cretaceous times (Orts et al. 2012a, b). In the southern section (Fig. 3e), the Sierra de Cara Cura was uplifted, defining an intermontane basin between this structure and the Puntilla de Huincán, which has been reported as a growing structure due to thickness variations in the Neuquén Group (Galarza et al. 2009). This intermontane basin, where the Río Grande is flowing nowadays, was fed both from the Sierra de Cara Cura and Puntilla de Huincán structures. Toward the east, in the Bordo Alto del Payún, Sánchez and Asurmendi (2014) interpreted the geometry of a Late Cretaceous flexural depocenter that received sediments from the Sierra de Cara Cura at that time (Fig. 3e). As the activity of these structures decreased, a migration of the magmatic front took place in the area (Spagnuolo et al. 2012), characterized as a group of subvolcanic bodies and lava domes, aligned and parallel to the Late Cretaceous orogenic front, reaching the Sierra Azul and Puntilla de Huincán structures (Fig. 3a). Hence, the Late Cretaceous orogenic front was set 400–500 km east from the present trench, roughly reaching 69°30′W. The maximum depositional age was constrained through U-Pb dating on detrital zircons in synorogenic sequences, yielding an onset of deformation

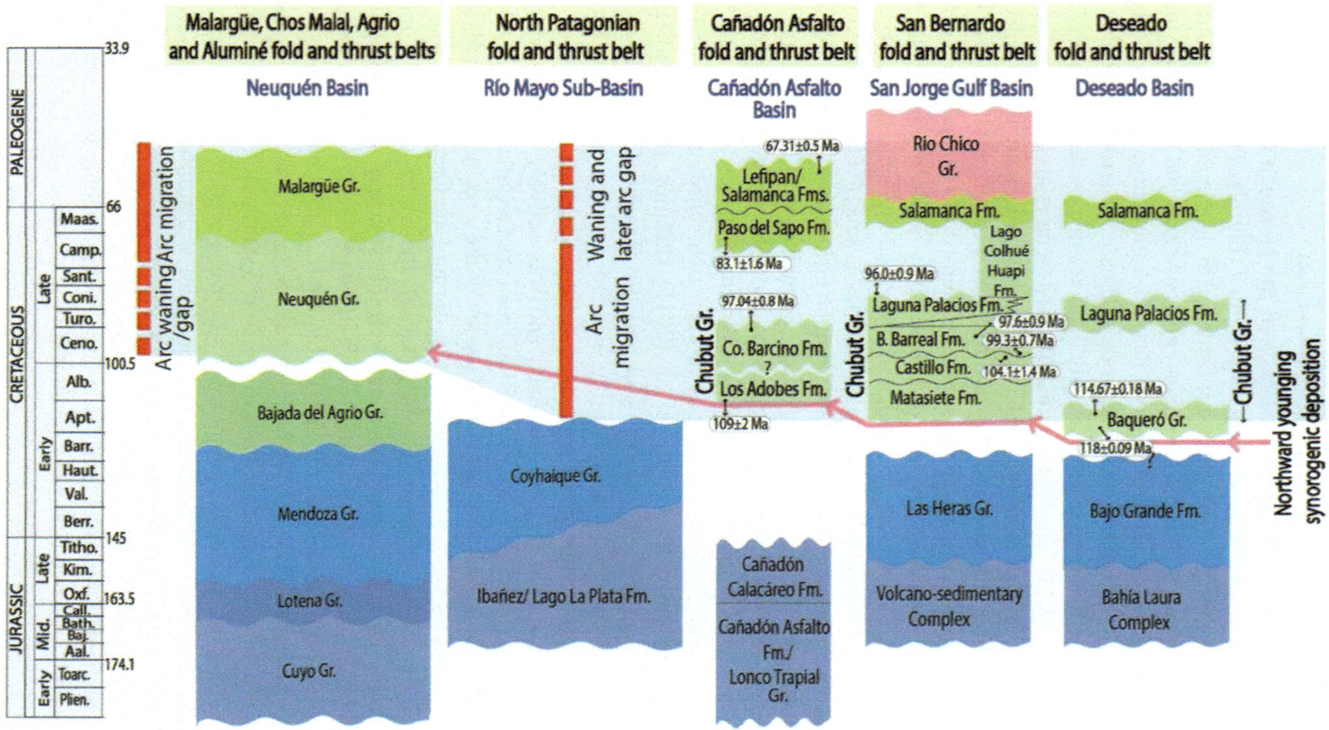

Fig. 2 Jurassic to Paleogene basin stratigraphy involved in the analyzed fold and thrust belts. Based on García Morabito and Ramos (2012), Fennell et al. (2015), Gianni et al. (2015a), Echaurren et al. (2016) and references cited in

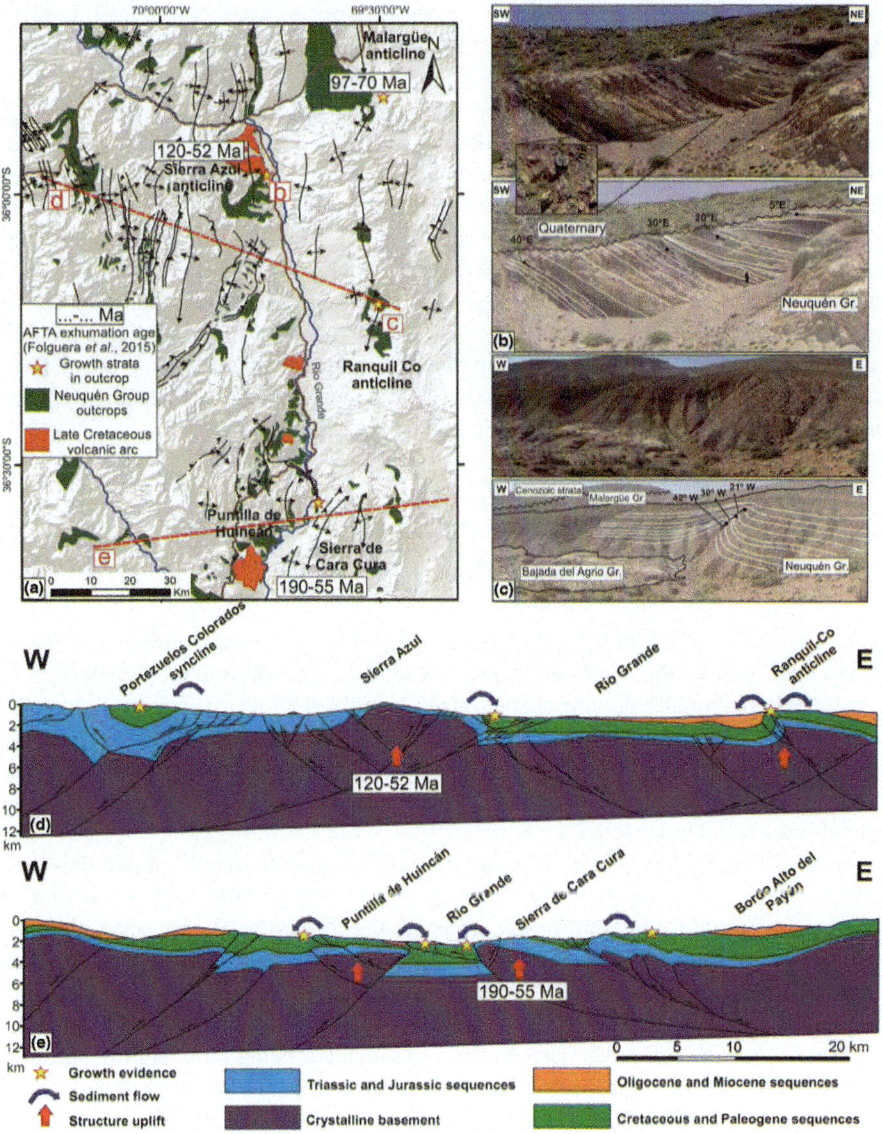

Fig. 3 **a** DEM of the study area in the Malargue FTB showing the location of the identified growth-strata in the Neuquén Group and apatite fission track exhumation ages (Folguera et al. 2015) associated with the initial uplift of main anticlines in the area. **b** Growth evidences in strata of the Neuquén Group, located in the eastern-frontal limb of the Sierra Azul anticline. In the inset, angular intraclasts of the Neuquén Group can be observed, implying synsedimentary deformation. **c** Growth-strata in the Ranquil Co anticline, where a maximum depositional age of 100 Ma for the Neuquén Group was obtained by U-Pb detrital zircon ages. **d** and **e** Schematic cross sections reflecting the dimensions of the Late Cretaceous orogen, along with the active structures identified through field growth evidences and other bibliographic mentions. Initial exhumation of the structures is constrained by apatite fission track analysis performed by Folguera et al. (2015). Location of cross sections is in Fig. 1a

in ~100 Ma (Fennell et al. 2015). To the North, between 35°30′S and 33° S, Mescua et al. (2013) had defined the Late Cretaceous orogenic front near the present international limit with Chile, but recent studies from the oil industry based on a 3-D seismic cube (Boll et al. 2014), indicate that the deformation reached 69°30′W, following the same structural trend defined in Fennell et al. (2015). Using a sedimentological and U-Pb detrital zircon analysis approach in a series of depocenters located at 35° S, Balgord and Carrapa (2016) recorded a 25–30 My hiatus between the Bajada del Agrio and Neuquén Groups (Fig. 2). According to these authors, this event represents the transition from post-rift thermal subsidence to forebulge erosion during initial flexural loading related to crustal shortening and uplift along the magmatic arc to the west by at least 97 ± 2 Ma. Finally, Folguera et al. (2015) supported field observations with apatite fission track analysis in representative structures of the Malargüe fold and thrust belt, documenting an exhumation episode in Late Cretaceous times (Figs. 3a, 3d and 3e).

2.2 *Late Cretaceous Deformation in the Agrio and Chos Malal Fold and Thrust Belts*

The Agrio and Chos Malal fold and thrust belts are located further east in the Argentinian side of the Andes and south of the Malargüe fold and thrust belt (Fig. 4). These deformational belts are characterized by mixed styles of deformation with a western thick-skinned area product of tectonic inversion of the Mesozoic Neuquén Basin and an eastern part dominated by thin-skinned structures (Rojas Vera et al. 2015, among others) (see Chap. "The Structure of the Southern Central Andes (Chos Malal Fold and Thrust Belt)").

Studies developed in the continental deposits of the Neuquén Group in these fold and thrust belts demonstrated the synorogenic character of this unit supporting the interpretation of a foreland basin stage in the Neuquén Basin since the Late Cretaceous (Cobbold and Rosello 2003; Zamora Valcarce et al. 2007, 2009; Tunik et al. 2010; Di Giulio et al. 2012, 2016). Particularly, throughout the entire Agrio fold and thrust belt (Fig. 1), an unconformity between the Neuquén Group and the Bajada del Agrio Group has been widely documented, being more significant in the inner part of the belt, and loosing expression toward the frontal structures (Groeber 1946; Ramos and Folguera 2005; Leanza 2009) (Fig. 2). Changes in thickness in the Neuquén Group were detected in subsurface by Cobbold and Rosello (2003), indicating Late Cretaceous activity along a thrust located in the frontal sector, defining it as a Late Cretaceous fold and thrust belt mildly reactivated in Miocene times (Ramos and Folguera 2005).

Additionally, Zamora Valcarce et al. (2006) dated a series of igneous rocks intruding folded structures in the area, constraining Late Cretaceous deformation between ~100 and ~73 Ma. Moreover, Zamora Valcarce et al. (2007) confirmed, based on paleomagnetic studies, that one of these igneous rocks intruded a previously deformed sequence, which was in turn deformed during Miocene times. Late Cretaceous contraction in the Agrio and Chos Malal fold and thrust belts has been

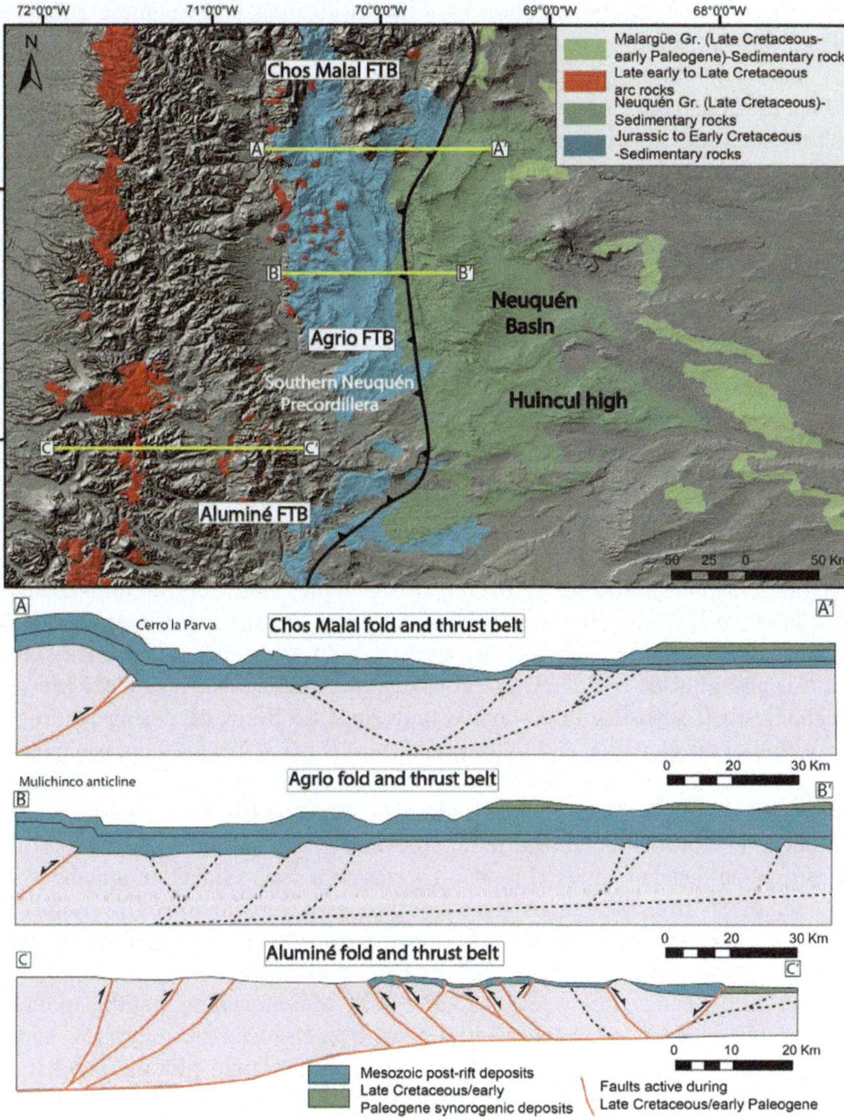

Fig. 4 DEM showing the location of the Chos Malal, Agrio and Aluminé fold and thrust belts. Cross sections reconstructed to the Late Cretaceous/Paleogene contractional stage are taken from García Morabito and Ramos (2012) and Rojas Vera et al. (2015)

recently confirmed through thermochronological studies of Rojas Vera et al. (2015). These authors published the results of a fission tracks analyses performed in the inner structures of these belts, which indicated a regional cooling/exhumation episode at ~70 Ma, younger than the one registered in the frontal structures of the Agrio fold and thrust belt at ca. ~100–95 Ma (Zamora Valcarce et al. 2009). These data supported a 6 km exhumation at ~69 Ma (Ramos and Folguera 2005), calculated from the cooling age of a Permian pluton (Kay and Copeland 2006).

Finally, Di Giulio et al. (2016) interpreted, through multi-proxy provenance data and lag times derived from apatite fission track analysis, a rapidly exhuming source within the Andes to the west in response to the Late Cretaceous deformational phase, coupled with the increasing contribution of material from the stable craton to the east as a consequence of the weak uplift and exhumation of the foreland, due to eastward migration of the forebulge, coeval with the migration of the volcanic arc (Spagnuolo et al. 2012).

2.3 Late Cretaceous Deformation in the Aluminé Fold and Thrust Belt

The Aluminé fold and thrust belt extends to the south between 38° S and 40°30′ S, as a latitudinal continuation of the Agrio fold and thrust belt, from the North Patagonian Andes in the west to a distal structured foreland sector known as the Neuquén Precordillera in the east (Fig. 4). In comparison to previous analyzed thrust belts, the Aluminé Fold and thrust belt has been significantly less studied. Nevertheless, most recent studies shed some light into its earliest evolutionary stages. This thick-skinned belt was also affected by Late Cretaceous contraction that deformed early Mesozoic rift and post-rift deposits of the southwestern Neuquén basin, triggering synorogenic deposition of the Neuquén and Malargüe Groups (Fig. 2) in a foreland basin (García Morabito and Ramos 2012). Notably, during the Late Cretaceous phase, deformation propagated distally to the foreland uplifting an intraplate belt known as the Southern Neuquén Precordillera whose relief prevented the Atlantic transgression in Maastrichtian–Danian times (Fig. 4). This eastern independent mountain system grew separately from the main Andean axis through a combination of inversion of the old rift systems and interaction with a pre-Andean belt which acted as a foreland obstacle. These deformed Mesozoic successions were covered by magmatic rocks with arc-affinities dated between ~75 and ~71 Ma, indicating that this mountain-building phase coexisted in space and time with an eastward arc-migration, and that this belt was mostly built in Late Cretaceous times (García Morabito and Ramos 2012). Finally, Miocene-Pliocene phase induced the reactivation of the internal and external sectors of the fold and thrust belt with minor propagation toward the foreland.

2.4 Late Early Cretaceous Contraction in the North Patagonian Fold-Thrust Belt (~42–44° S)

The North Patagonian Andes between 42°–44° S correspond to an orogenic segment characterized by a low topography (< 2000 m) at the cordilleran sector in the west and the "Patagonian broken foreland" to the east (Bilmes et al. 2013) (Fig. 5). In this segment of the Northern Patagonian Andes, the construction of the fold-thrust belt (Ramos and Cortés 1984) has been originally related to a Neogene stage of deformation (e.g., Giacosa et al. 2005).

The extra-Andean domain, with a wide geological record in a series of NNW-trending ranges forming the broken foreland, has also been interpreted as absorbing contraction since Neogene times after a protracted Mesozoic-Paleogene extensional regime (Figari 2005; Bilmes et al. 2013). However, recent studies have documented a late Early Cretaceous contractional stage that uplifted the North Patagonian fold and thrust belt near the plate margin and the Patagonian intraplate sector to the east (Orts et al. 2012a, b; Echaurren et al. 2016a).

The cordilleran sector is composed mainly by Mesozoic magmatic units, namely the North Patagonian Batholith and its cogenetic volcanic arc associations, the Middle-Late Jurassic Lago La Plata Formation and the mid-Cretaceous Divisadero Group (Fig. 2). The batholith has been primary formed by I-type, calc-alkaline granitoids of Jurassic-Cretaceous age with Neogene suites of less differentiated units (Pankhurst et al. 1992, 1999). The Mesozoic volcanic units consist of volcanic-volcanoclastic packages of rhyolithic ignimbrites, acid domes, and andesite-basalt lava flows (Lago La Plata Formation) and mainly rhyolithic lithic tuffs (Divisadero Group). Between these units, isolated sedimentary rocks of the shallow marine to delta sequences of the Neocomian Coyhaique Group (part of the Río Mayo Embayment, or Aysén Basin; Aguirre-Urreta and Ramos 1981; Suárez et al. 1996), are scattered and uplifted in the Andean eastern slope (e.g., Suárez et al. 2009a; Orts et al. 2012b).

Even though these deposits are better exposed to the south, at ~45–46° S, their presence in the proto-Andean margin and toward the foreland accounts for periods of widely developed backarc basin formation in Jurassic-Early Cretaceous times, being their synextensional character an evidence for contraction and mountain building taking place mainly by tectonic inversion.

In the cordilleran zone, the Situación Range exposes thick successions of intermediate to mafic volcanic rocks of the Lago La Plata Formation, covered by discontinuous acid volcanoclastic rocks of the Divisadero Group (Fig. 5b). The range is a north-plunging, double-vergent anticline, with an associated Neogene contractional reactivation. Echaurren et al. (2016a) identified wedges with synextensional geometries controlling deposition of the Lago la Plata Formation volcanic layers, as fault-bounded horizons present increasing dipping angles toward the base (Fig. 5b). Jurassic strata are unconformably underlying the Divisadero Group acid tuffs, as evidenced by a tight fault-propagated syncline (Fig. 5a). To the west, at the base of the Pirámides Range, steeply dipping horizons of the Coyhaique Group are separated gently dipping horizons of the Divisadero Group through an angular unconformity (Echaurren et al. 2016b). Further south, the Galeses Range exposes sections of the Lago La Plata Formation and the Coyhaique Group deformed into asymmetrical anticlines unconformably covered by the Divisadero Group that sealed the contractional episode (Echaurren et al. 2016b). The presence of this regional angular unconformity has been interpreted as caused by initial contraction

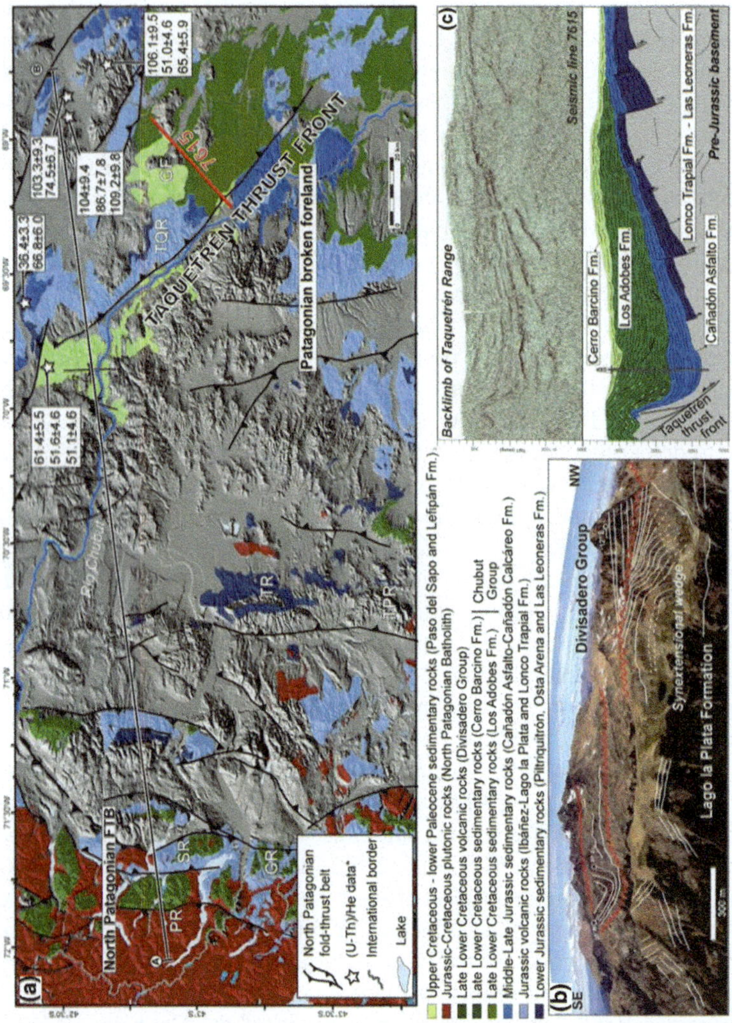
(a)
North Patagonian FTB
TAQUETRÉN THRUST FRONT
Patagonian broken foreland
7615
North Patagonian fold-thrust belt
(U-Th)/He data*
International border
Lake
Upper Cretaceous - lower Paleocene sedimentary rocks (Paso del Sapo and Lefipán Fm.)
Jurassic-Cretaceous plutonic rocks (North Patagonian Batholith)
Late Lower Cretaceous volcanic rocks (Divisadero Group)
Late Lower Cretaceous sedimentary rocks (Cerro Barcino Fm.)
Late Lower Cretaceous sedimentary rocks (Los Adobes Fm.)
Chubut Group
Middle-Late Jurassic sedimentary rocks (Cañadón Asfalto-Cañadón Calcáreo Fm.)
Jurassic volcanic rocks (Ibáñez-Lago la Plata and Lonco Trapial Fm.)
Lower Jurassic sedimentary rocks (Piltriquitrón, Osta Arena and Las Leoneras Fm.)
(b)
SE
NW
Divisadero Group
Lago la Plata Formation
Synextensional wedge
300 m
(c)
Backlimb of Taquetrén Range
Seismic line 7615
Cerro Barcino Fm.
Los Adobes Fm.
Cañadón Asfalto Fm.
Lonco Trapial Fm. - Las Leoneras Fm.
Pre-Jurassic basement
Taquetrén thrust front
(d)
NORTH PATAGONIAN ANDES
BROKEN FORELAND SYSTEM
CHUBUT RIVER MIDDLE VALLEY
CAÑADON ASFALTO BASIN
main, deeper detachment
shallow detachment
Depth (km)
50 km
Crustal basement
Cenozoic volcanic and sedimentary rocks

◂**Fig. 5** **a** DEM of the Patagonian Andes and the broken foreland sector showing the location of synorogenic deposits of the late Early Cretaceous Chubut Group, the Upper Cretaceous Paso del Sapo and Lefipan formations and Cretaceous arc-related volcanic and plutonic rocks (Echaurren et al. 2016a). Thermochronological (U-Th)/He data of the foreland region taken from Savignano et al. (2016). Abbreviations are: PR: Pirámides Range, SR: Situación Range, GR: Galeses Range, TR: Tecka Range, TPR: Tepuel Range, TQR: Tauqetrén Range, GF: Gorro Frigio area. **b** Partially inverted wedge-like depocenters of Mid to Late Jurassic sections of the Lago La Plata Formation and pre-Aptian contractional structures in the Situación Range, unconformably covered by Early Cretaceous volcanic rocks (Divisadero Group) (Echaurren et al. 2016a).**c** Subsurface growth-strata evidences in the Chubut Group (seismic line 7615 in a). **d** Balanced structural cross profile (Echaurren et al. 2016a)

of the North Patagonian fold and thrust belt (Orts et al. 2012a, b; Echaurren et al. 2016a, b). After this contractional event, in Late Cretaceous times (~100–80 Ma), during the major stage of emplacement of the North Patagonian Batholith, fission track ages indicate that the forearc region exhumed from depths of at least 10–12 km (Duhart and Adriasola 2008).

Within the episodic construction of the North Patagonian Batholith, its main peak of activity is constrained to the mid to Late Cretaceous, corresponding to plutonic suites of monzonites, granodiorites, and tonalities (Pankhurst et al. 1992, 1999). A notorious eastward expansion (~150 km) is registered for these intrusive units at ~43° S, where the granitoids in the Lago Aleusco area intrude Lower Jurassic plutonic rocks of the Subcordilleran Batholith and the Jurassic Lago La Plata Formation in the Tecka and Tepuel ranges. Radiometric dating indicates crystallization ages between ~90–75 Ma (Turner 1982), corresponding to slab-derived magmatism of I-type calc-alkaline granitoids (López de Luchi et al. 1992). Contemporaneous volcanic activity has also been identified in this foreland area, in the Don Juan Formation (Franchi and Page 1980). These andesitic packages have been correlated with upper sections of the Divisadero Group and recognized as unconformably underlying the within-plate volcanism of the Tres Picos Prieto Formation of ~80 Ma (Silva Nieto 2005), supporting a correlation with the deformational episode in the inner sector of the fold-thrust belt.

The extra-Andean domain is characterized by a series of ~N-trending belts exposing mainly Paleozoic and Mesozoic sedimentary and volcanic rocks, disrupted by the Paleocene-Eocene bimodal volcanic rocks of the Pilcaniyeu Belt or Ventana Formation (Rapela and Kay 1988; Aragón et al. 2011) and the NW inverted front of the Mesozoic Cañadón Asfalto Basin.

The Cañadón Asfalto Basin contains thick continental deposits of Jurassic sedimentary-volcanic rocks and Cretaceous sedimentary rocks deposited in an extensional system controlled by a complex arrangement of depocenters (Figari 2005). The Cretaceous infilling corresponds to the Chubut Group (late Early to Late Cretaceous), which has been separated in the area into two units, the basal Los Adobes Formation (Aptian) and the Cerro Barcino Formation (Albian) in the top, composed by continental and fluvial-lacustrine facies with variable volcaniclastic inputs (Fig. 2). Even though traditionally interpreted as sag deposits, or part of a synextensional reactivation of the basin (e.g., Figari 2005), the recent study of

Echaurren et al. (2016a) documented the presence of subsurface and surface contractional growth-strata highlighting their origin under a contractional regime (Figs. 5c and 6b). In Late Cretaceous–Early Paleocene times, an Atlantic transgression covered the eastern Patagonian platform forming an embayment in the Taquetrén range zone (Scasso et al. 2012). These sedimentary rocks, the continental Paso del Sapo Formation (Late Cretaceous) and marine Lefipán Formation (Danian), have also been interpreted as a sag sequence (Figari 2005). Nevertheless, field relations exhibit syncontractional growth-strata for both associations (Figs. 6a and 6d). The Paso del Sapo Formation is composed by continental sandstones and limestones with tuffs intercalations, which in the Gorro Frigio area are deformed into an overturned west-vergent fan of progressive unconformities thrust by the Los Adobes Formation (Echaurren et al. 2016a; Fig. 6a). A tuffaceous level was dated by the authors through U-Pb constraining the depositional age to ~83 Ma, with detrital material provided by the Chubut Group with an age peak of ~117 Ma. According to the authors, the Chubut Group age population probably suggests basin cannibalization during deposition of the Paso del Sapo Formation. Similar structural relationships are present to the north in Taquetrén range. In that area, the shallow marine Lefipán Formation is in fault contact with Early Jurassic deposits and preserves progressive unconformities. These evidences account for a contractional reactivation of the Cañadón Asfalto western edge that correlates with the recognized reactivation of the cordilleran area to the east. Recent, apatite (U-Th)/He ages in the broken foreland sector between 41°30′S–43° S, support previous tectonic interpretations evidencing a Cretaceous to Paleogene exhumation event (~110–50 Ma) (Savignano et al. 2016) (Fig. 5a).

2.5 *Late Early Cretaceous Uplift of the North Patagonian Andes Between 44°–46° S*

The North Patagonian Andes between ~44–46° S differ morphologically from the latter northern segment. At these latitudes, the cordilleran domain is clearly separated from the broken foreland area through a 250 km broad, practically undeformed sector (Fig. 7). In this segment of Central Patagonia, Jurassic to Cretaceous Western Gondwana breakup processes lead to the development of the continental San Jorge Gulf Basin in the intraplate sector and the marine to continental Río Mayo Sub-Basin or Aysén Basin near the continental margin in an intra-arc and inner retroarc position (Uliana et al. 1989; Aguirre-Urreta and Ramos 1981; Suárez et al. 2009a) (Figs. 2 and 7a). The Río Mayo Sub-basin and San Jorge Gulf Basin were progressively inverted during Andean orogeny constituting part of the Andes and broken foreland sectors, respectively. Particularly, the Río Mayo Sub-basin underwent a major reorganization during Aptian times reflected by the retreat of the Panthalasan Sea to the west and the appearance of subaereal subduction-related magmatism of the Divisadero Group and equivalent units, identified from

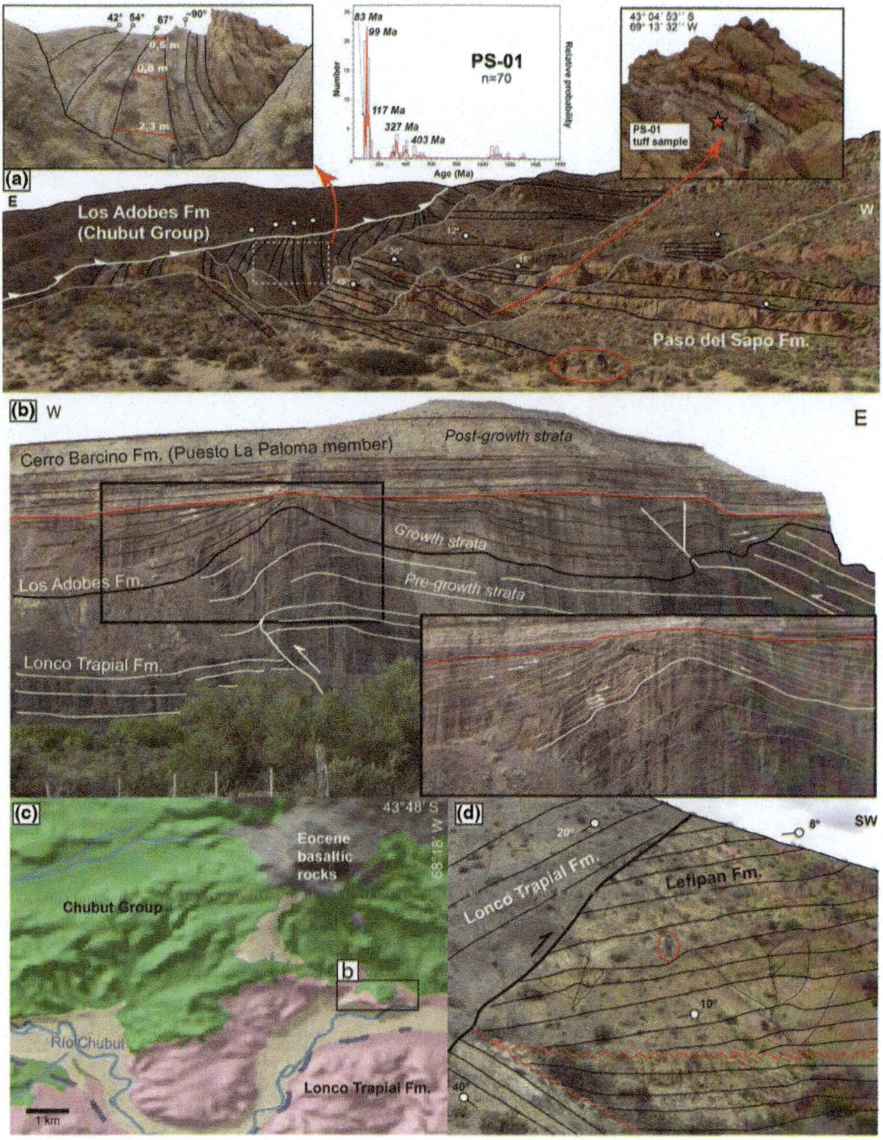

Fig. 6 **a** Contractional growth-strata in the Paso del Sapo Formation dated through U/Pb in ~83 Ma at the Taquetrén thrust-front. **b** and **c** Aptian-Albian angular unconformity and growth-strata in Los Adobes Formation at Los Altares location. **d** Contractional growth-strata in shallow marine Lefipán Formation (see locations in Fig. 5)

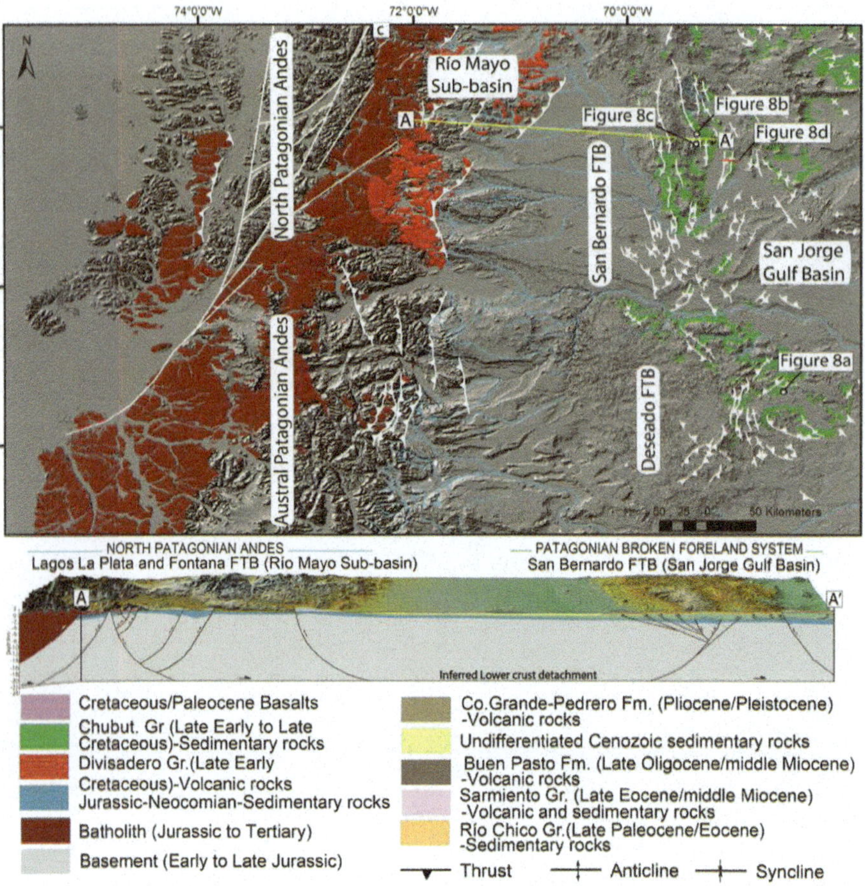

Fig. 7 DEM showing the southern sector of the North Patagonian Andes, the broken foreland sector (San Bernardo and Deseado fold belts) and a regional cross section from Gianni et al. (2017)

42° S–49° S (Suárez et al. 1996, 2007, 2009a and references therein; Suárez et al. 1996; Parada et al. 2001). U–Pb SHRIMP zircon ages in this unit range between ~118 and 102 Ma (Pankhurst et al. 2003; Suárez et al. 2009b).

The angular unconformity at the base of this unit is thought to reflect the earliest uplift episode of the North Patagonian Andes Ramos (1981). This orogenic event is well recorded in the east vergent Lagos La Plata and Fontana fold and thrust belts (Iannizzotto et al. 2004; Folguera and Iannizzotto 2004). Unlike in the northern sector of the North Patagonian Andes (previous section), a preserved stratigraphic record constrained through U/Pb zircon ages, allowed restricting this episode to the 121–118 Ma time interval (Aptian) (Ramos 1981; Iannizzotto et al. 2004; Folguera and Iannizzotto 2004; Suárez et al. 2009a).

An additional Late Cretaceous tectonic pulse has been interpreted based on the description of a gentle angular unconformity between Latest Cretaceous calc-alkaline volcanics and the Aptian Divisadero Group (Demant and Suárez 2007; Suárez et al. 2009a). This event is supported by regional observations such as out-of-sequence thrusting in the Patagonian batholith in the Lago La Plata-Fontana fold and thrust belt and more locally a K–Ar date of 105 Ma from cataclastic plutonic rocks (Iannizzotto et al. 2004; Folguera and Iannizzotto 2004; Suárez and De la Cruz 2001; Suárez et al. 2009a). Moreover, this tectonic pulse is coincident with the climax of tectonic emplacement of the Patagonian batholith (98 $\pm$ 4 Ma proposed by Ramos 1982). In concordance with above-mentioned structural observations, Tunik et al. (2004) interpreted an exposure of the volcanic arc roots over the cordillera in late Early Cretaceous times, based on a study of detrital modes in discrimination diagrams analyzing Cretaceous foreland deposits of the Chubut Group. East of the southern domain of the North Patagonian Andes, the Central sector of the Patagonian broken foreland uplifted as a product of tectonic inversion of Mesozoic rifts (Homovc et al. 1995; Peroni et al. 1995). This intraplate belt is characterized at these latitudes by the NNW-SSE-trending San Bernardo fold and thrust belt that exposes the western border of the San Jorge Gulf Basin and the Deseado fold belt to the south that deformed previous Permian to Jurassic extensional depocenters (Giacosa et al. 2010) (Fig. 7). As in the northern segment of the Patagonian broken foreland, the San Bernardo fold and thrust belt also presents extensive outcrops of the late Early Cretaceous to Late Cretaceous continental deposits Chubut Group. Here, these rocks are characterized by a high proportion of distal, ashfall deposits reworked in lacustrine and fluvial settings (Sciutto 1981). The high pyroclastic component of this unit has been associated with an increase in volcanic activity and a coeval eastward migration of the magmatic arc during its deposition (Barcat et al. 1989; Tunik et al. 2004). This succession is stratigraphically arranged as follows: The Pozo D-129/Matasiete, Castillo, Bajo Barreal, and Laguna Palacios Formations, ranging in age from Aptian to late Cenomanian–early Turonian (?) (121.5–98 Ma) based on fossil records and U/Pb zircon data (Fitzgerald et al. 1990; Suárez et al. 2014) (Fig. 2).

In the Deseado fold belt, Césari et al. (2011) and Perez Loinaze et al. (2013) constrained the age of a partially equivalent unit to the base of the Chubut Group, known as the Baqueró Group, through U/Pb detrital zircons to the $\sim$118–114 Ma time interval (Fig. 2). Surface and subsurface description of a regional angular unconformity along the Patagonian foreland region at the base of the Chubut Group has been commonly related to intraplate contraction and/or traspression (Clavijo 1986; Barcat et al. 1989; Homovc and Constantini 2001; Giacosa et al. 2010; Ranalli et al. 2011; Marquez and Navarrette 2011, among others) and has been directly linked with the initial growth of the North Patagonian Andes (Barcat et al. 1989; Ramos 1981; Iannizzotto et al. 2004; Folguera and Iannizzotto 2004; Suárez et al. 2009a) (Fig. 8a). More recently, in the northern part of the Austral Patagonian Andes, Ghiglione et al. (2013) have dated by U/Pb detrital zircons in synorogenic deposits of Río Belgrano and Río Tarde Formations, obtaining a maximum depositional age of $\sim$122 Ma and $\sim$118–111 Ma, respectively. These detrital zircon population yielded interesting results, instead of having a clear Andean provenance, they rather had come from the Patagonian foreland. Hence, the authors pointed out that these data could indicate a late Early Cretaceous post ~122 Ma uplift of different sectors of the foreland zone (the Deseado fold belt and the western edge of the North Patagonian Massif). Particularly, uplift of the San Bernardo fold and thrust belt is further supported by seismic surveys and surface detection of growth-strata that described a synorogenic character in the totality of the Chubut Group (Barcat et al. 1989; Navarrete et al. 2015; Gianni et al. 2015a, b) or at least the uppermost units (Allard et al. 2015) (Fig. 8b, c, and d) (see Chap. "Lower Jurassic to Early Paleogene intraplate contraction in Central Patagonia"). In the Deseado fold belt, a paleo-stress analysis carried out by Reimer et al. (1996) concluded that this area has been under compression since late Early Cretaceous times, producing diverse contractional structures that controlled different basaltic extrusion stages. Later identification of an Aptian angular unconformity of regional character by Giacosa et al. (2010) confirmed the beginning of contraction in this fold and thrust belt (Fig. 8a).

Across the foreland and offshore areas, the Chubut Group is capped by a regional angular unconformity, either by retroarc basalts or clastic deposits of Upper Cretaceous/Paleogene age, attesting for a major intraplate contractional event in latest Cretaceous times (Feruglio 1949; Lesta et al. 1980; Continanzia et al. 2011; Micucci et al. 2011; Navarrete et al. 2015).

3 Magmatic Arc Behavior During Early Andean Construction

Abnormal arc expansions along the Southern Central Andes from 35°30′S to 48° S have been broadly documented during early Andean orogenic stages from Cretaceous to Paleogene times (Barcat et al. 1989; Ramos and Folguera 2005; Kay and Copeland 2006; Suárez et al. 2009a; Folguera and Ramos 2011; Spagnuolo et al. 2012; García Morabito and Ramos 2012; Fennell et al. 2015; Gianni et al. 2015a; Echaurren et al. 2016a; Folguera et al. 2015). Noteworthy, authors have argued in favor of changes in subduction angle linked to slab-shallowing or flattening events as the most likely process to induce the observed arc expansions, as generally observed worldwide in subduction systems (e.g., Tatsumi and Eggins 1995). These studies have interpreted such episodes in terms of individual slab-shallowing or flattening processes that successfully explained deformation in discrete retroarc fold and thrust belts. Nevertheless, the possibility that a single large-scale slab flattening event could have been responsible for the identified Meso-Cenozoic arc shifting from 35°30 to 48° S has not been considered.

The magmatic arc between 35°30′S and 40°30′S underwent an eastward migration probably reflecting shallow subduction events in late Cretaceous to Eocene period, during the initial development of the Malargüe, Chos Malal, Agrio and Aluminé fold and thrust belts (Llambías and Rapela 1989; Franchini et al. 2003; Kay and Copeland 2006; Ramos and Folguera 2005; Zamora Valcarce et al. 2006;

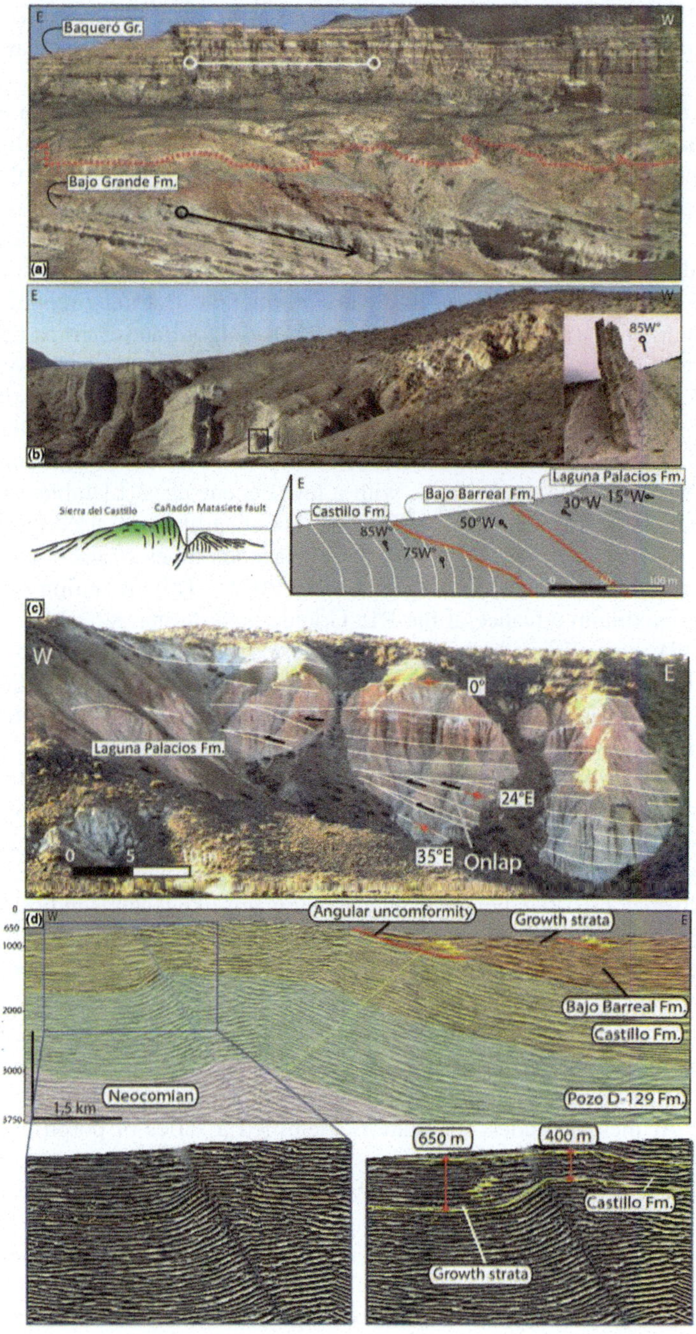

Fig. 8 **a** Bajo Grande unconformity evidencing an Aptian tectonic event in the Deseado fold belt. Surface **b** and **c** and subsurface (Sierra de Silva), **d** growth-strata in late Early to Late Cretaceous units of the Chubut Group in the San Bernardo fold belt. See locations in Fig. 7

García Morabito and Ramos 2012; Spagnuolo et al. 2012; Folguera et al. 2015, Rojas Vera et al. 2015, Fennell et al. 2015; Di Gulio et al. 2016; Horton and Fuentes 2016).

As reviewed in the work of Fennell et al. (2015), initial arc-migration at 35°30′S to 38° S latitudes is marked by a Late Cretaceous to Eocene arc gap on the Chilean side of the Andes (See Fig. 3.32; Charrier et al. 2007) (Fig. 9a). In this regard, Tunik et al. (2010), based on the analysis of the Mesozoic detrital zircons from the Neuquén Group, showed that there are two important peaks in the volcanic arc activity developed at 110 and 125 Ma, respectively, followed by a gradual decrease, interpreted as a waning in the arc production. Detrital zircon data shown by Di Giulio et al. (2012), Fennell et al. (2015), Balgord and Carrapa (2016), and Horton and Fuentes (2016) agree with this waning in arc activity between ~100 and ~85 Ma. Volcanic arc rocks in the Argentinean Andean slope have ages between ~85 and ~60 Ma (Domínguez et al. 1984; Munizaga et al. 1988; Linares and González 1990; Jordan et al. 2001; Franchini et al. 2003; Zamora Valcarce et al. 2006; Spagnuolo et al. 2012), with a few Eocene ages (Llambías and Rapela 1989; Cobbold and Rosello 2003) (Fig. 9a). This fact has been used to propose an eastward shifting of the arc-front in Late Cretaceous times (Ramos and Folguera 2005; Spagnuolo et al. 2012, Folguera et al. 2015). This migration would have reached a maximum advance at the Río Grande area during ~67 Ma (Spagnuolo et al. 2012), coevally with the uppermost terms of the Neuquén and Malargüe Groups (LLambías and Aragón 2011; Aguirre-Urreta et al. 2011). In this context, the Neuquén Group would be the synorogenic sedimentary unit associated with an early Andean orogen, produced by a shallow subduction regime prevailing from ~100 to ~60 Ma (Ramos and Folguera 2005; Fennell et al. 2015; Horton and Fuentes 2016). Main aspects of this magmatic belt between 35°30′S and 38° S have been summarized by Llambías and Rapela (1989), Franchini et al. (2003), Kay and Copeland (2006), Zamora Valcarce et al. (2006) and more recently by Spagnuolo et al. (2012). Chemical analyses show arc-like features indicated by high field strength elements (HFSE) depletion (La/Ta > 28; Ta/Hf < 0,15) and fluid mobile element enrichment (Ba/La > 20) (Kay and Copeland 2006; Zamora Valcarce et al. 2006).

A southward continuation of this magmatic belt has been proposed by García Morabito and Ramos (2012), based on an analysis of the igneous outcrops and a revision of previous and new radiometric data between 38°30′ and 40°30′S (Fig. 9a). At these latitudes, the authors identified a series of porphyric intrusive bodies of andesitic-rhyodacitic composition and associated extrusive series that were emplaced in close relation to NNW and NE-trending contractional structures. This spatial relation was interpreted as a syntectonic emplacement of igneous bodies or postectonic immediately after the Late Cretaceous contractional pulse (García Morabito and Ramos 2012). Geochemical data from outcrops of the Aluminé valley show typical volcanic arc signatures, indicated by trace elements ratios (La/Ta (33), Ba/La (20.35)), Nb negative anomalies, and a K, Rb, and Th enrichment with respect to N-MORB (Lagorio et al. 1998).

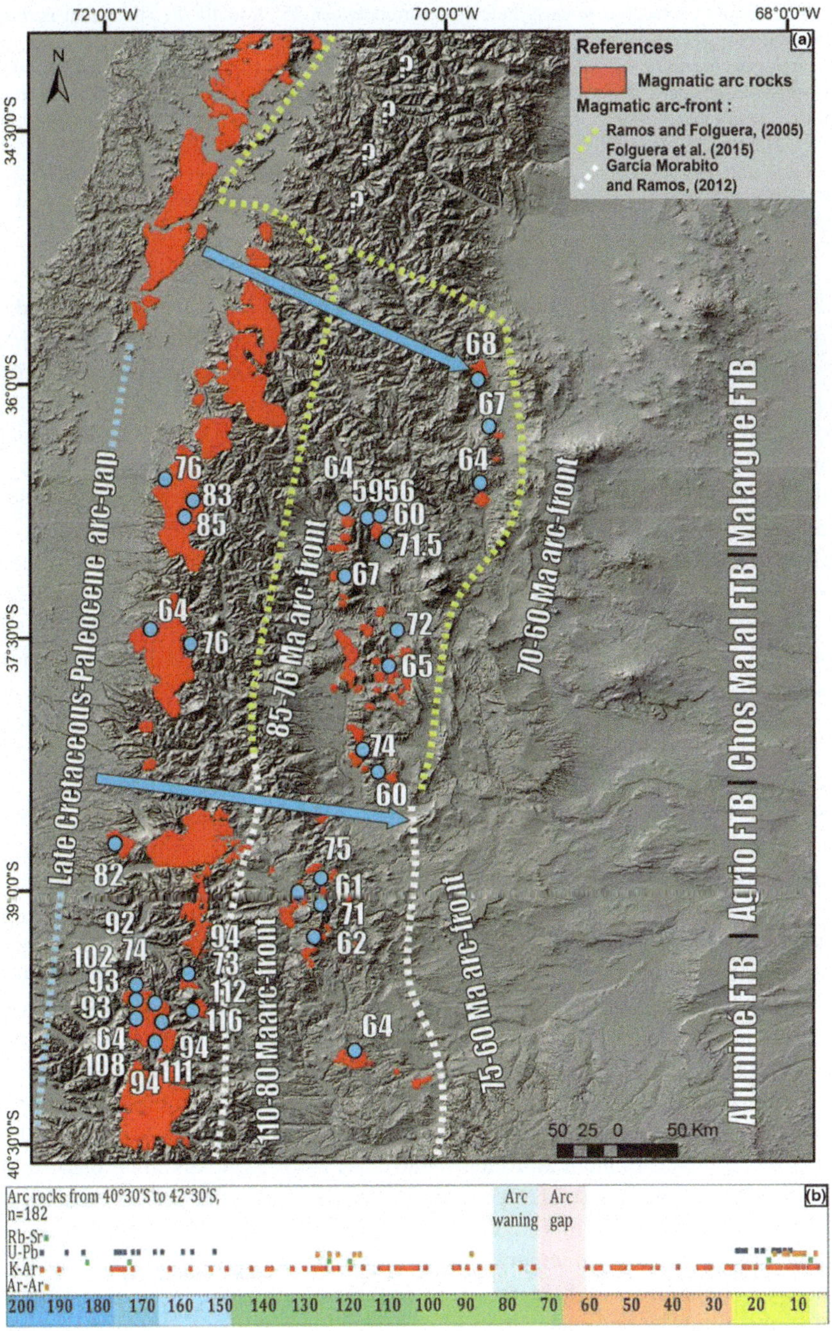
References
Magmatic arc rocks
Magmatic arc-front :
Ramos and Folguera, (2005)
Folguera et al. (2015)
García Morabito and Ramos, (2012)
Late Cretaceous-Paleocene arc-gap
85-76 Ma arc-front
70-60 Ma arc-front
110-80 Maarc-front
75-60 Ma arc-front
Aluminé FTB | Agrio FTB | Chos Malal FTB | Malargüe FTB
Arc rocks from 40°30'S to 42°30'S, n=182
Arc waning
Arc gap
Rb-Sr
U-Pb
K-Ar
Ar-Ar

◂**Fig. 9** **a** Middle Cretaceous to early Paleogene arc between 35° S to 40° S. The reader is referred to Folguera et al. (2015) and García Morabito and Ramos (2012) for complete references of the presented geochronological dataset. **b** Compiled radiometric dataset of arc rocks between 40°30′S to 42°30′S. References to the dataset are Adriasola et al. (2006); Aragón et al. (2011); Castro et al. (2011); Cazau et al. (1989); Ghiara et al. (1999); González Díaz (1979, 1982); Haller and Lapido (1982); Halpern et al. (1975); Linares and González (1990); Lizuaín (1979, 1980, 1981, 1987); Pankhurst and Hervé (1994); Pesce (1979); Rabassa (1978); Rapela et al (1983, 1987, 2005); SEGEMAR-JICA; Sepulveda and Viera (1980); Soechting (2001); Toubes and Spikerman (1973); Vattuone and Latorre (2004), Varela et al. (2005)

As a whole, arc-related outcrops from the Malargüe to the Aluminé fold and thrust belts between 35°30′ and 40°30′S indicate an eastward displaced volcanic arc-front of Late Cretaceous to early Paleogene age which occurred in the foreland region contemporaneously with the deposition of the Neuquén and Malargüe Groups, likely attesting a common slab-shallowing event at these latitudes (García Morabito and Ramos 2012; Fennell et al. 2015). Nevertheless, it is likely that shallow subduction extended even further to the south. In this regard, we compiled a radiometric dataset of arc rocks spanning from 40°30′ to 42° S (Fig. 9b). This analysis shows that magmatic activity in the North Patagonian Andes experienced a waning stage from 80–70 Ma and turned off from ~72 to 60 Ma, producing a ~10 Ma magmatic gap (Fig. 9b). Considering the proposed context of shallow subduction immediately to the north, we similarly interpret this arc behavior as produced by changes in the subduction angle. Notably, comparable arc gaps of ~15–5 Ma have been interpreted as related to slab-shallowing events elsewhere in the Andes (e.g., Haschke et al. 2002; Bissig et al. 2008).

Between 42° and 44°30′S, eastward shifting of the magmatic front since Cretaceous times has been interpreted as a progressive change from slab-shallowing in late Early Cretaceous to flat-subduction in Latest Cretaceous to early Paleocene (Echaurren et al. 2016a). These authors noticed that early contraction evidenced by an angular unconformity in the North Patagonian Andes was contemporaneous with eastward expansion of the arc rocks of the Divisadero Group at ~118 Ma, as also observed in the southern sector of this belt by Suárez and de la Cruz (2000) and Suárez et al. (2009a, b). Even though the magmatic suites of the North Patagonian Batholith are mainly concentrated in the North Patagonian Andes, satellite plutonic and volcanic units reached the foreland at ~90 Ma, evidencing an ~150 km eastward expansion of the magmatic activity (Fig. 10). These eastern bodies are calc-alkaline intermediate-to-acid volcanic rocks of the Don Juan Formation (K–Ar age of 91 ± 3 Ma; Franchi and Page 1980) and I-type, calc-alkaline, basic-to-acid plutonic suites of the Lago Aleusco area (López de Luchi et al. 1992). According to Echaurren et al. (2016a), radiometric K–Ar ages of ~90–75 Ma of this plutonism (Turner 1982) indicate a temporal association with retroarc deformation of the Taquetrén thrust front, where the ~83 Ma Paso del Sapo Formation deposited syntectonically. The end of this period of magmatic expansion is marked by an arc gap from ~74 to ~60 Ma that developed coetaneously to the Maastrichtian–early Paleocene synorogenic deposits of the Lefipán Formation deposited in an Atlantic

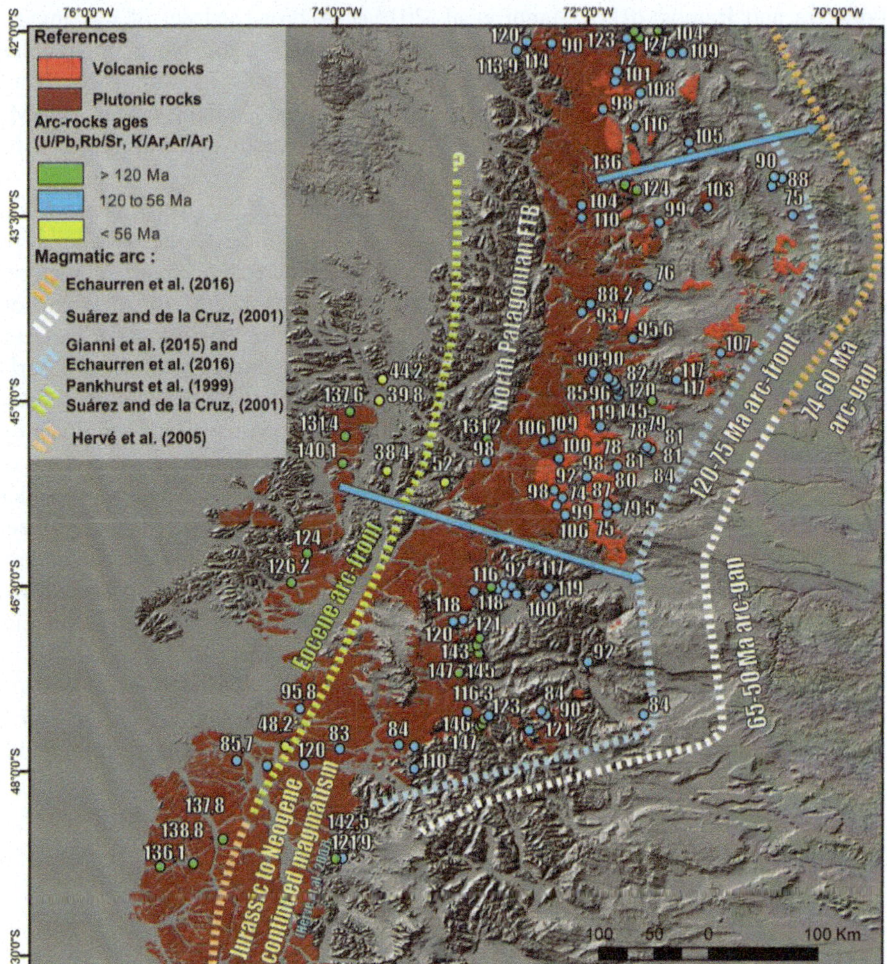

Fig. 10 Late early Cretaceous to early Paleogene arc between 42°30′ to 48° S. The reader is referred to Gianni et al. (2015a) and Echaurren et al. (2016a) for complete references of the presented geochronological dataset

embayment next to the eastern Andean front (Scasso et al. 2012; Echaurren et al. 2016a) (Fig. 2). Notably, this magmatic gap is strikingly similar in age to the gap described immediately to the north between 42°30′ and 40°30′S (Fig. 9b). According to Gianni et al. (2015a) and Echaurren et al. (2016a), the latter arc-front migration continued southward from 44°30′ to 48° S. Abnormal arc expansion at these latitudes has been described since pioneer works performed in the southern sector of North Patagonian Andes and its existence as well as its tectonothermal consequences have been analyzed in more recent studies (Barcat et al. 1989; Pankhurst et al. 1999; Suárez et al. 2009a; Suárez and de la Cruz 2000;

Folguera and Ramos 2011; Gianni et al. 2015a). A broad dataset of radiometric ages compiled for different arc-related units along-strike the North Patagonian Andes and broken foreland showed that eastward arc-migration took place from ~120 to 75 Ma, followed by a period of magmatic waning since ~74 Ma (Gianni et al. 2015a) that ultimately evolved into a magmatic gap stage from ~65 to 50 Ma (Suárez and de la Cruz 2000) (Fig. 10).

The arc units involved in this process have been grouped into several calc-alkaline units ranging from late Early Cretaceous silicic to mesosilicic rocks and late Cretaceous basaltic to dacitic rocks that share typical arc signatures (Pankhurst et al. 1999; Parada et al. 2001; Demant and Suárez 2007). Arc expansion at this time has been related to an initial shallow subduction configuration between late Early to Late Cretaceous and to flat-subduction from latest Cretaceous to Early Paleocene as reflected by the magmatic gap (Gianni et al. 2015a; Suárez and de la Cruz 2000). After this arc waning/gap period, the magmatic arc significantly retracted to the west and reappeared in Eocene times occupying the eastern flank of the Coastal Cordillera and the western flank of the North Patagonian Andes (Pankhurst et al. 1999, see Fig. 6) (Fig. 10). As initially suggested in the pioneer work of Barcat et al. (1989), late Early Cetaceous arc expansion was concomitant with the synorogenic deposition of the Chubut Group in the foreland area. Most recent studies echoed this view and demonstrated that arc-migration was coetaneous to broken foreland development (Gianni et al. 2015a; Navarrete et al. 2015). Therefore, in light of new data, it is now well established that regional changes in the subduction dynamics in late Early Cretaceous triggered uplift of the North Patagonian Andes fold and thrust belt, as well as in the Patagonian broken foreland.

4 Discussion

4.1 *Regional Arc-Migration Between 35°30 to 48° S: A ~1,350 Km Large-Scale Shallow Subduction Segment?*

As described in the previous section, arc-migration segments potentially linked to slab-shallowing/flattening events are not a local phenomenon. Instead they are latitudinally connected forming a continuous segment ~1,350 km long that could be conceived as a single process (Fig. 11a). If right, this process was evidently diachronic as reflected on the ages of initial expansion of the different arc segments (Figs. 9 and 10). In this sense, the first change in subduction angle may have taken place in the segment from 44°30′ to 48° S at ~120–118 Ma (Suárez et al. 2009a; Gianni et al. 2015a). Then, this episode seems to have generalized since 100–90 Ma, propagating northward and achieving its full development during the 75–60 Ma time interval (Figs. 9 and 10). It is worth noting that diachronism in arc shifting is apparently reflected in the beginning of synorogenic deposition in foreland as well as broken foreland basins related to the analyzed fold and thrust belts (Fig. 2). Nevertheless, at this time the configuration of the subduction system may have been complex as reflected by an eastwardly displaced arc north of the 40° 30'S and an arc gap to the south (Fig. 11a). Although complex, this situation is strikingly alike to the current situation of the Chilean/Pampean flat-slab (Álvarez et al. 2015. We speculate that similarly to the latter segment, it could be due to different slab configurations north and south of the 40°30′S latitude. A prevailing shallow angle to the north may have allowed arc magmatism production out of a preserved mantle wedge (Ramos and Folguera 2005), whereas a flat angle to the south extruded this feature during full plate horizontalization, canceling magmatic production (Gianni et al. 2015a; Echaurren et al. 2016a). The potential connection among some of the Southern Central Andes shallowing/flat-slab events was initially suspected by García Morabito and Ramos (2012). The regional analysis presented in this review expands this idea and supports the existence of a ~1,350 km long shallow to flat-subduction segment in Late Cretaceous–early Paleogene. This episode is here referred to as *Nalé large flat-slab* event (Fig. 11a). It is worth noting that this episode could have been linked to a subduction segment as large as the biggest flat-slab on earth, known as the Peruvian large flat-slab that runs ~1,400 km along the South American margin between 4° S and 15° S (Gutscher et al. 2000) (Fig. 11b). Such large-scale modifications in subduction settings are rare and poorly documented in the geologic record. In this regard, Gutscher et al. (2000) differentiated this type of processes from common small-scale slab-shallowings/flattenings and classified them as large flat-slab configurations.

Indeed, the Nalé event shares several features with other large flat-subduction events such as the Cretaceous to Paleogene Laramide event (USA) and the current Chilean/Pampean and Peruvian flat-slabs, where orogenic stages took place in concert with arc-front expansions followed, by stress transmission far inland developing intraplate contractional belts (Dickinson and Snyder 1978; Folguera and Ramos 2011; García Morabito and Ramos 2012; Gianni et al. 2015a; Echaurren et al. 2016a). The origin of current large flat-slabs has been explained by subduction of buoyant oceanic features on both sides of the flat-slab segments (Inca plateau and Nazca ridge, Peruvian and Juan Fernandez-Copiapó aseismic ridges, Chilean/Pampean) in order to maintain a large flat to shallow slab segment (Gutscher 2002; Álvarez et al. 2015). Using Seton et al. (2012) global plate reconstruction Gianni et al. (2015a), Fennell et al. (2015), and Echaurren et al. (2016a) showed that subduction of young oceanic lithosphere coincided in space and time with the described arc-magmatic expansions. Thus, these authors suggested that subduction of buoyant lithosphere in a context of known high convergence and absolute upper plate westward motion could have been the key to maintain the flat-slab configuration.

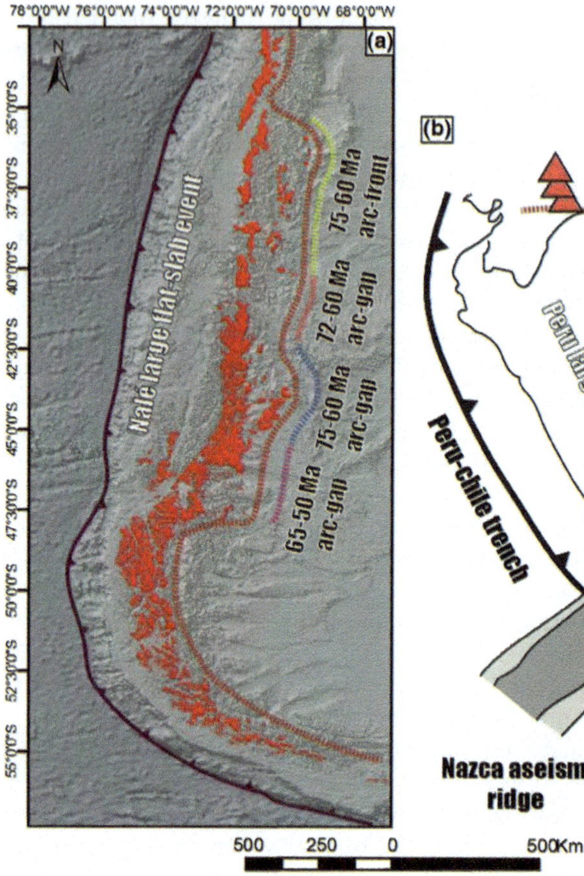

Fig. 11 **a** Distribution of Cretaceous to Paleocene arc-related outcrops and related arc gaps described in previous works and this study. **b** Image showing the Peruvian large flat-slab in order to be compared with the Nalé large flat-slab configuration proposed in this work

4.2 *The Nalé Large Flat-Slab Event and Its Bearings on Maastrichtian–Danian Paleogeography*

In Latest Cretaceous to Early Paleocene, dramatic paleogeographic changes took place in the Neuquén and Patagonian forelands during the full development of *Nalé large flat-slab*. At this moment, a restricted sector of the Southern Central Andes foreland was flooded by the first Atlantic-derived transgression since Gondwana breakup (Feruglio 1949; Aguirre-Urreta et al. 2011; Olivero and Medina 1994) (Fig. 12). South of 49° S, this transgression used a preexisting marine pathway through the Late Cretaceous foredeep of the Austral foreland basin of the Southern Andes (Fig. 12). On the other hand, between 48° and 35° S, the transgression formed an isolated flooded domain that occupied an extense intraplate area with some sectors reaching as far as the Andean foothills (Aguirre-Urreta et al. 2011; Scasso et al. 2012) (Fig. 12). In this sector, the transgression bordered the southern Patagonian broken foreland and entered axially through broken foreland depocenters in the northern sector forming the Paso del Sapo embayment (Scasso et al. 2012; Echaurren et al. 2016a). Maximum flooding achieved in Danian times caused interconexion among the Neuquén–Colorado, Golfo San Jorge, and Cañadón Asfalto basins as evidenced by common marine species and genera among them (Del Río 2015). Particularly, the ~83 Ma age obtained by Echaurren et al. (2016a) for the estuarine deposits of the Paso del Sapo Formation suggests that that flooding may have begun earlier, in Campanian times. The origin of the first Atlantic-derived transgression at studied latitudes has intrigued scientist since pioneer works in Patagonia (e.g., Weaver 1927; Feruglio 1949). The most accepted hypothesis invokes combined processes of Atlantic margin subsidence due to

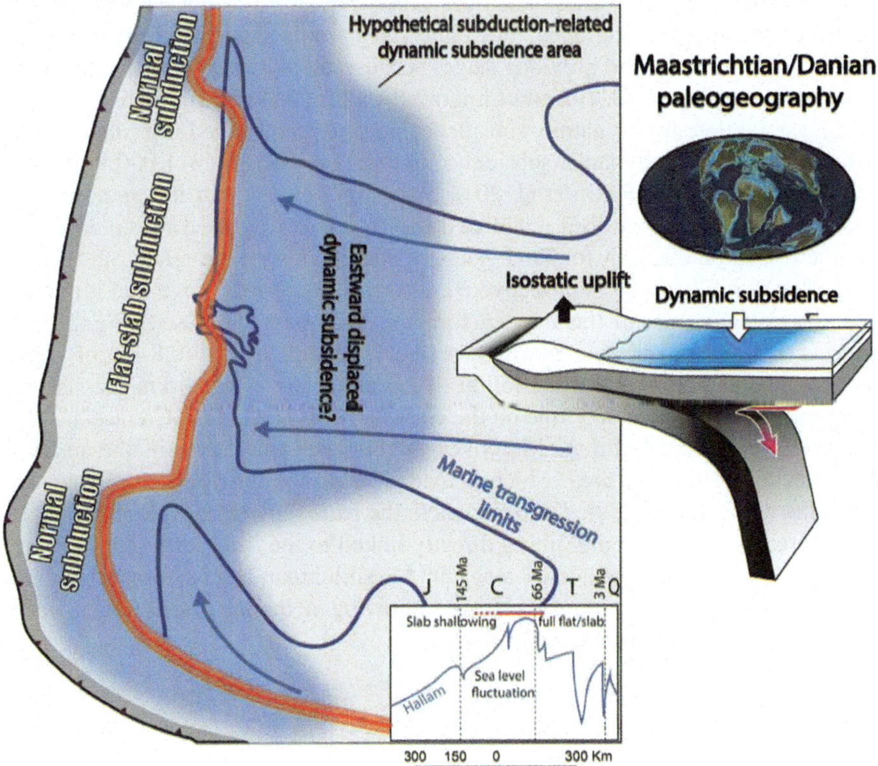

Fig. 12 Proposed model linking Late Cretaceous/Paleocene Atlantic transgressions in the Southern Central Andean foreland to *Nalé large flat-slab*. In this model, global high eustatic sea level (Hallam 1992) was concomitant with continental tilting driven by eastward shifted dynamic subsidence in the intraplate area and isostatic uplift in the margin, both consequences of flat-slab subduction

eastward continental tilting linked to Andean uplift in conjunction with a global eustatic high sea level in Maastrichtian–Danian times (e.g., Nullo and Combina 2011; Uliana and Bibble 1988; Aguirre-Urreta et al. 2011). However, in the complex Cretaceous–Paleocene tectonic context discussed in this work, additional factors may have been involved in the processes that lead to shallow marine flooding of the continental interior.

In this regard, the role of flat-subduction on driving regional continental tilting would be twofold. In the one hand, large flat-slab configurations produce forelandward shifting of subduction-related dynamic subsidence focus, producing broad areas of ~0.5–2 km of continental downwarping where the slab resumes its descent into the mantle (Dávila et al. 2010; Eakin et al. 2014). On the other hand, areas undergoing flat-subduct ion increase the continental margin buoyancy leading to an increment in vertical displacements of ~1–0.8 km above the subhorizontal slab (Dávila and Lithgow-Bertollini 2013). In this regard, changes in dynamic topography are null in those segments that remain at "normal" dip angles, and it is positive along flat-slabs. These positive values, however, do not denote a positive contribution on the topography, but rather they imply positive vertical displacements with respect to the previous stage, because the dynamic negative support is reduced or eliminated (Dávila and Lithgow-Bertollini 2013). Although speculative, using as a reference dynamic subsidence recently calculated for the Peruvian flat-slab that predicts dynamic subsidence of long wavelength (~1,000 km) and an amplitude of > 1 km (Eakin et al. 2014), we might expect that the proposed *Nalé flat-slab* potentially produced a similar subsiding area displaced to the east in the Southern Central Andean foreland. Thus, dynamic subsidence related to the *Nalé large flat-slab* where the slab recovered a normal angle may have had a probable position coincident with the flooded foreland zone here discussed (Fig. 12). We suggest that this process along with a predicted change of ~1–0.8 km of vertical motion (Dávila and Lithgow-Bertollini 2013) along the ~1,350 km flat-slab segment may have played a key role on the classically invoked latest Cretaceous–early Paleocene continental tilting. Moreover, it may elegantly explain the restricted character of the flooded area, which is evident from the distribution of paleoshore lines that coincided with the latitudes where the *Nalé flat-slab* developed (Fig. 12). Therefore, subtle continental tilting directly linked to the *Nalé large flat-slab event* during a high eustatic sea level stage in Maastrichtian–Danian times may have driven sudden paleogeographic changes allowing a marine transgression in the Southern Central Andean foreland (Fig. 12).

5 Conclusions

A synthesis of most recent tectonic studies in the Southern Central Andes presented in this work highlights the regional character of Cretaceous contraction. Initial orogenic building phases defined early growth stages in the Malargüe, Chos Malal, Agrio, Aluminé, and North Patagonian fold and thrust belts. Notably, part of the

Cretaceous shortening was absorbed far from the plate margin in the intraplate sector, giving place to an early expression of the Patagonian broken foreland. Most of the analyzed works identified a spatiotemporal relation between magmatic arc-front migrations or expansions and fold and thrust belt development. This phenomenon was unanimously related by different authors to transient changes in the subduction angle. In these works, distinct subduction segments undergoing slab-shallowing or flattening events were proposed to explain local contraction. However, the regional analysis here presented opens the possibility of the existence of a single process linked to a ~1,350 km long flat-slab here named *Nalé large flat-slab event*, which initiated diachronically in late Early Cretaceous and achieved full development in latest Cretaceous to earliest Paleocene. Noteworthy, diachronism in the slab-shallowing process was mirrored by a northward young trend in synorogenic deposition in foreland and broken foreland basins related to analyzed fold and thrust belts. Nevertheless, testing the feasibility of this proposal will require additional constraints on the geochronology and geochemistry of arc rocks along the 35°–48° S arc segment. The potential dimensions of this event allow us to classify it as a large flat-slab segment which could have been as long as the Peruvian flat-slab, the largest subhorizontal subduction configuration on earth. Additionally, continental tilting during high eustatic sea level often invoked to explain foreland marine flooding in Maastrichtian–Danian times may have been aided by *Nalé large flat-slab*, as these settings are expected to cause dynamic vertical motion of the continental margin and subsidence in the foreland area. Finally, this review underscores pre-Cenozoic contractional stages in Andean evolution and the importance of slab-shallowing or flattening as an effective driving process for non-collisional orogenesis.

Acknowledgements The authors want to recognize the support received by CONICET and the lively discussion with several members of the Laboratorio de Tectónica Andina and the Instituto Sismológico Ing. Fernando Volponi (IGSV) through the years. We are particularly grateful to Paola Ramírez, Vilma Jaramillo, and Conrado Gianni for fruitful discussions.

References

Adriasola AC, Thomson SN, Brix MR, Hervé F, Stöckhert B (2006) Postmagmatic cooling and late Cenozoic denudation of the North Patagonian Batholith in the Los Lagos region of Chile, 41°–42°15′S. Int J Earth Sci 95(3):504–528

Aguirre-Urreta B, Ramos, VA (1981) Estratigrafía y paleontología de la alta cuenca de río Roble, Cordillera Patagónica. In: VIII Congreso Geológico Argentino. Actas III. pp 101–138

Aguirre-Urreta B, Tunik M, Naipauer M, Pazos P, Ottone E, Fanning M, Ramos VA (2011) Malargüe Group (Maastrichtian–Danian) deposits in the Neuquén Andes, Argentina: Implications for the onset of the first Atlantic transgression related to Western Gondwana break-up. Gondwana Research, 19(2):482–494

Allard JO, Foix N, Rodríguez A, Sánchez F (2015) Evolución tectosedimentaria de la cuenca del golfo San Jorge en su margen occidental. Reunión de Tectónica (Río Negro). Electronic book

Álvarez O, Giménez M, Folguera A, Spagnotto S, Bustos E, Baez W, Braitenberg C (2015) New evidence about the subduction of the Copiapó ridge beneath South America, and its connection with the Chilean-Pampean flat slab, tracked by satellite GOCE and EGM2008 models. J Geod 91:65–88

Armijo R, Lacassin R, Coudurier-Curveur A, Carrizo D (2015) Coupled tectonic evolution of Andean orogeny and global climate. Earth Sci Rev 143:1–35

Aragón E, D' Eramo F, Castro A, Pinotti L, Brunelli D, Rabbia O, Rivalenti G,Varela R, Spakman W, Demartis M, Cavarozzi C, Aguilera Y, Mazzucchelli M, Ribot A (2011) Tectono-magmatic response to major convergence changes in the North Patagonian suprasubduction system; the Paleogene subduction–transcurrent plate margin transition. Tectonophysics 509:218–237

Balgord EA, Carrapa B (2016) Basin evolution of Upper Cretaceous-Lower Cenozoic strata in the Malargüe fold-and-thrust belt: northern Neuquén Basin, Argentina. Basin Res 28(2):183–206

Barcat C, Cortiñaz JS, Nevistic VA, Zucchi HE (1989) Cuenca Golfo San Jorge. In: Chebli LS (Ed) Cuencas Sedimentarias Argentinas, Serie Correlación Geológica 6 Tucumán, pp 319–345

Barnes JB, Ehlers TA (2009) End member models for Andean Plateau uplift. Earth-Sci. Rev. 97:(1–4)105–132. http://dx.doi.org/10.1016/j.earscirev.2009.08.003

Bechis F, Encinas A, Concheyro A, Litvak VD, Aguirre-Urreta B, Ramos VA (2014). New age constraints for the Cenozoic marine transgressions of northwestern Patagonia, Argentina (41–43 S): Paleogeographic and tectonic implications. J S Am Earth Sci 52:72–93

Bilmes A, D'Elia L, Franzese JR, Veiga GD, Hernández M (2013) Miocene block uplift and basin formation in the Patagonian foreland: The Gastre Basin, Argentina. Tectonophysics 601: 98–111

Bissig T, Ullrich TD, Tosdal RM, Friedman R, Ebert S (2008) The time-space distribution of Eocene to Miocene magmatism in the central Peruvian polymetallic province and its metallogenetic implications. J South Am Earth Sci 26(1):16–35

Boll A, Alonso J, Fuentes F, Vergara M, Laffitte G, Villar HJ (2014) Factores controlantes de las acumulaciones de hidrocarburos en el sector norte de la cuenca neuquina, entre los ríos Diamante y Salado, provincia de Mendoza, Argentina. Actas del IX Congreso de Exploración y Desarrollo de Hidrocarburos: trabajos técnicos 1:3–44

Castro A, Moreno-Ventas I, Fernández C, Vujovich G, Gallastegui G, Heredia N, Martino RD, Becchio R, Corretgé LG, Díaz-Alvarado J, Such P, García-Arias M, Liu DY (2011) Petrology and SHRIMP U-Pb zircon geochronology of Cordilleran granitoids of the Bariloche area. Argentina. J South Am Earth Sci 32(4):508–530

Cazau L, Uliana M (1973) El Cretácico Superior continental de la Cuenca Neuquina. V Congreso Geológico Argentino, Actas (Córdoba, Argentina) 3:131–163

Cazau L, Mancini D, Cangini J, Spalletti L (1989) Cuenca de Ñirihuau. In: Chebli G, Spalletti L (Eds.) Cuencas Sedimentarias Argentinas, Serie Correlación Geológica 6:299–318, Tucumá

Césari SN, Limarino CO, Llorens M, Passalia MG, Loinaze VP, Vera EI (2011) High-precision late Aptian Pb/U age for the Punta del Barco Formation (Baqueró Group), Santa Cruz Province, Argentina. J South Am Earth Sci 31:426–431

Charrier R, Pinto L, Rodriguez M (2007) Tectonostratigraphic evolution of the Andean orogen in Chile. In T Moreno and W Gibbons The Geology of Chile, Geol Soc Lond, pp 21–114

Chiarabba C, De Gori P, Faccenna C, Speranza F, Seccia D, Dionicio V, Prieto GA (2015) Subduction system and flat slab beneath the Eastern Cordillera of Colombia, Geochem. Geophys. Geosyst 16. https://doi.org/10.1002/2015GC006048

Clavijo R (1986) Estratigrafía del Cretácico Inferior en el sector occidental de la Cuenca del Golfo San Jorge. Boletín Inf. Pet. 9:15–32

Cobbold PR, Rosello EA (2003) Aptian to recent compressional deformation, foothills of the Neuquén Basin, Argentina. Mar Pet Geol 20:429–443

Continanzia J, Manceda R, Covellone GM, Gavarrino AS (2011) Cuencas de Rawson y Valdés : Síntesis del Conocimiento Exploratorio—Visión actual. In: Kozlowski E, Legarreta L, Boll A, Marshall PA (eds) Simposio Cuencas Argentinas Visión Actual: VIII Congreso de Exploración y Desarrollo de Hidrocarburos pp 47–64

Dávila F, Lithgow-Bertelloni C (2013) Dynamic topography in South America. J South Am Earth Sci 43:127–144

Dávila F, Lithgow-Bertelloni C, Giménez M (2010) Tectonic and dynamic controls on the topography and subsidence of the Argentine Pampas: the role of the flat slab. Earth Planet Sci Let 295(1):187–194

del Río CJ, Martínez SA (2015) Paleobiogeography of the Danian molluscan assemblages of Patagonia (Argentina). Palaeogeogr Palaeocl 417:274–292

DeCelles PG, Giles KA (1996) Foreland basin systems. Basin research 8(2):105–123

Demant A, Suárez M (2007) Geochronology and petrochemistry of Late Cretaceous- (?) Paleogene volcanic sequences from the eastern central Patagonian Cordillera (45°–45°40′S) Rev Geol Chile 34:3–21

Di Giulio A, Ronchi A, Sanfilippo A, Tiepolo M, Pimentel M, Ramos VA (2012) Detrital zircon provenance from the Neuquén Basin (south-central Andes): cretaceous geodynamic evolution and sedimentary response in a retroarc-foreland basin. Geology 40:559–562

Di Giulio A, Ronchi A, Sanfilippo A, Balgord EA, Carrapa B, Ramos VA (2016) Cretaceous evolution of the Andean margin between 36° S and 40° S latitude through a multi-proxy provenance analysis of Neuquén Basin strata (Argentina). Basin Res. https://doi.org/10.1111/bre.12176

Dickinson WR, Snyder WS (1978) Plate tectonics of the Laramide orogeny. Geol Soc Am Mem 151:355–366

Domínguez E, Aliotta G, Garrido M, Daniela JC, Ronconi N, Casé AM, Palacios M (1984) Los Maitenes-El Salvaje. Un sistema hidrotermal tipo porfírico. IX Congreso Geológico Argentino (Bariloche). Actas VII pp. 443–458

Duerto L, Escalona A, Mann P (2006) Deep structure of the Mérida Andes and Sierra de Perijá mountain fronts. Maracaibo Basin, Venezuela, AAPG Bull 90(4):505–528. https://doi.org/10.1306/10080505033

Duhart P, Adriasola AC (2008) New time-constraints on provenance, metamorphism and exhumation of the Bahía Mansa Metamorphic Complex on the Main Chiloé Island, south-central Chile. Rev Geol Chile 35:79–104

Eakin CM, Lithgow-Bertelloni C, Dávila FM (2014) Influence of Peruvian flat-subduction dynamics on the evolution of western Amazonia. Earth Planet Sci Let 404:250–260

Echaurren A, Folguera A, Gianni G, Orts D, Tassara A, Encinas A, Giménez M, Valencia V (2016a) Tectonic evolution of the North Patagonian Andes (41°–44° S) through recognition of syntectonic strata. Tectonophysics 677:99–114

Echaurren A, Oliveros V, Folguera A, Ibarra F, Creixell C, Lucassen F (2016b) Early andean tectonomagmatic stages in North Patagonia: Insights from field and geochemical data. J Geol Soc Lon. https://doi.org/10.1144/jgs2016-087

Encinas A, Folguera A, Oliveros V, De Girolamo Del Mauro L, Tapia F, Riffo R, Hervé F, Finger KL, Valencia VA, Gianni GM, Álvarez O (2015) Late Oligocene–Early Miocene submarine volcanism and deep marine sedimentation in an extensional basin of southern Chile: Implications on the tectonic development of the North Patagonian Andes. Geol Soc Am Bull 128:807–823

Espinoza F, Morata D, Polvé M, Lagabrielle Y, Maury R, de la Rupelle A, Guivel C, Cotten J, Bellon H, Suárez M (2010) Middle Miocene calc-alkaline volcanism in Central Patagonia (47° S): petrogenesis and implications for slab dynamics. Andean Geology 37(2):300–328

Faccenna C, Becker TW, Conrad CP, Husson L (2013) Mountain building and mantle dynamics. Tectonics 32(1):80–93. https://doi.org/10.1029/2012tc003176

Fennell LM, Folguera A, Naipauer M, Gianni G, Rojas Vera EA, Bottesi G, Ramos VA (2015) Cretaceous deformation of the Southern Central Andes: synorogenic growth strata in the Neuquén Group (35°30′–37° S). Basin Res. https://doi.org/10.1111/bre.12135

Feruglio E (1949) Descripción Geológica de la Patagonia. Dirección General de Yacimientos Petrolíferos Fiscales, Editorial Coni, p 334

Figari EG (2005) Evolución tectónica de la Cuenca de Cañadón Asfalto (Zona del valle medio del Río Chubut) Doctoral Thesis Universidad de Buenos Aires

Fildani A, Cope TD, Graham SA, Wooden JL (2003) Initiation of the Magallanes foreland basin: timing of the southernmost Patagonian Andes orogeny revised by detrital zircon provenance analysis. Geology 31:1081–1084

Fitzgerald MG, Mitchum RM Jr, Uliana MA, Biddle KT (1990) Evolution ofthe San Jorge Basin, Argentina. Am Assoc Pet Geol Bullettin 74:879–920

Folguera A, Iannizzotto NF (2004) The lagos La Plata and Fontana fold-and-thrust belt: long-lived orogenesis at the edge of western Patagonia. J South Am Earth Sci 16:541–566

Folguera A, Ramos VA (2011) Repeated eastward shifts of arc magmatism in the Southern Andes: a revision to the long-term, pattern of Andean uplift and magmatism. J South Am Earth Sci. https://doi.org/10.1016/j.jsames.2011.04.003

Folguera A, Bottesi G, Duddy I, Martín-González F, Orts D, Sagripanti L, Rojas Vera EA, Ramos VA (2015) Exhumation of the Neuquén Basin in the southern Central Andes (Malargüe fold and thrust belt) from field data and low-temperature thermochronology. J South Am Earth Sci 64:381–398

Franchi MR, Page R (1980) Los basaltos cretácicos y la evolución magmática del Chubut occidental. Rev. Asoc. Geol. Argent. 35:208–229

Franchini M, López-Escobar L, Schalamuk IBA, Meinert L (2003) Magmatic characteristics of the Paleocene Cerro Nevazón region and other Cretaceous to Early Tertiary cal-alcaline subvolcanic to plutonio units in the Neuquén Andes, Argentina. J South Am Earth Sci 16:399–421

Galarza BJ, Zamora Valcarce G, Folguera A, Bottesi GL (2009) Geología y Evolución tectónica del Frente Cordillerano a los 36°30′S: bloques de Yihuin-Huaca y Puntilla de Huincán. Mendoza. Rev As Geol Arg 65(1):170–191

Gansser A (1973) Facts and theories on the Andes: Journal of the Geol Soc Lond 129:93–131. https://doi.org/10.1144/gsjgs.129.2.0093

García Morabito E, Ramos VA (2012) Andean evolution of the Aluminé fold and thrust belt, Northern Patagonian Andes (38°300–40°300S). J South Am Earth Sci 38:13–30

Garrido AC (2010) Estratigrafía del Grupo Neuquén, Cretácico Superior de la Cuenca Neuquina (Argentina): nueva propuesta de ordenamiento litoestratigráfico. Rev Mus Argentino Cienc Nat 12(2):121–177

Ghiara MR, Haller MJ, Stanzione D, Barbieri M, Menditti I, Castorina F, Trudu C, Demichelis AH, Meister CM (1999) Calc-alkaline volcanic rocks from Cerro Ver, Patagonian Cordillera (43°10′S): geochemistry and geochronology. XIV Congreso Geológico Argentino, Salta 2:178–181

Ghiglione MC, Naipauer M, Sue C, Barberón V, Blampied JN, Ronda G, Ramos VA, Aguirre-Urreta BA, Valencia V (2013) Early Cretaceous (Aptian) uplift of Patagonia recorded on zircons detrital content from the northern Austral-Magallanes basin. In: int Geol Congress South Hemisph, Viña del Mar, pp 64–66

Gianni GM, Echaurren A, Folguera A, Likerman J, Encinas A, García HPA, Valencia VA (2017) Cenozoic intraplate tectonics in Central Patagonia: Record of main Andean phases in a weak upper plate. Tectonophysics.

Giacosa RE, Afonso JC, Heredia CN, Paredes J (2005) Tertiary tectonics of the sub-Andean region of the North Patagonian Andes, southern central Andes of Argentina (41–42 30′S). J South Am Earth Sci 20:157–170

Giacosa R, Zubia M, Sánchez M, Allard J (2010) Meso-Cenozoic tectonics of the southern Patagonian foreland: structural evolution and implications for Au–Ag veins in the eastern Deseado Region (Santa Cruz, Argentina). J. South Am. Earth Sci. 30:134–150

Giambiagi L, Bechis F, Garcia V, Clark AH (2008) Temporal and spatial relationships of thick- and thin-skinned deformation: a case study from the Malargüe fold-and-thrust belt, southern Central Andes. Tectonophysics 459:123–139

Gianni GM, Navarrete C, Orts D, Tobal J, Folguera A, Giménez M (2015a) Patagonian broken foreland and related synorogenic rifting: the origin of the Chubut Group Basin. Tectonophysics 649:81–99. https://doi.org/10.1016/j.tecto.2015.03.006

Gianni GM, Navarrete CG, Folguera A (2015b) Synorogenic foreland rifts and transtensional basins: A review of Andean imprints on the evolution of the San Jorge Gulf Salta Group and Taubaté Basins. J South Am Earth Sci 64:288–306
Gianni GM, Echaurren A, Folguera A, Likerman J, Encinas A, Dal Molin C,Valencia VA (2016) Cenozoic intraplate tectonics in Patagonia: Record of main Andean phases in a weak upper plate. Tectonics, in press
González Díaz E (1979) Estratigrafía del área de la Cordillera Patagónica entre los paralelos 40° 30′ y 41°00′ de latitud sur (provincia del Neuquén). VII Congreso Geológico Argentino, Neuquén, Actas 1:525–537
González Díaz E (1982) Zonación cronológica del plutonismo en los Andes Patagónicos Septentrionales entre los 40° y 42° sur: la migración de los ciclos intrusivos. Acta Geológica Lilloana 16(1):5–22
Groeber P (1946) Observaciones geológicas a lo largo del meridiano 70°. 2. Hojas Sosneado y Maipo. Rev Soc Geol Arg 2 (2):141–176
Guillot MG, Escayola M, Acevedo R (2011) Calc-alkaline rear-arc magmatism in the Fuegian Andes: Implications for the mid-cretaceous tectonomagmatic evolution of southernmost South America. J South Am Earth Sci 31(1):1–16
Gulisano CA, Gutiérrez Pleimling AR (1994) The Jurassic of the Neuquén Basin. Field Guide, Asoc Geol Arg Series E, p 111
Gutscher MA, Spakman W, Bijwaard H, Engdahl ER (2000) Geodynamic of flat subduction: seismicity and tomographic constraints from the Andean margin. Tectonics 19:814–833
Gutscher MA (2002) Andean subduction styles and their effect on thermal structure and interplate coupling. J S Am Earth Sci 15(1):3–10
Hallam A (1992) Phanerozoic sea-level changes. Columbia University Press
Haller MJ, Lech RR, Meister CM, Martínez O, Poma S, Viera RLM (2010) Descripción geológica de la Hoja Geológica 4373-IV/III, Trevelin, Provincia de Chubut. Serv Geol Min Arg, Instituto de Geología y Recursos Minerales, Boletín 322:55
Halpern M, Stipanicic PN, Toubes RO (1975) Geocronología (Rb/Sr) en los Andes Australes Argentinos. Rev Asoc Geol Arg 30(2):180–192
Haschke MR, Scheuber E, Günther A, Reutter KJ (2002) Evolutionary cycles during the Andean orogeny: repeated slab breakoff and flat subduction? Terra Nova 14(1):49–55
Homovc J, Conforto G, Lafourcade P, Chelotti L (1995) Fold Belt in the San Jorge Basin, Argentina: an Example of Tectonic Inversion. In: P Buchanan, J y Buchanan (eds), Basin Inversion. Geol Soc London Special Publication, pp 235–248
Homovc JF, Constantini L (2001) Hydrocarbon exploration potential within intraplate shear-related depocenters: Deseado and San Julián basins, southern Argentina. Am Assoc Pet Geol Bull 85:1795–1816
Horton BK, Fuentes F (2016) Sedimentary record of plate coupling and decoupling during growth of the Andes. Geology 44(8):647–650
Husson L, Conrad CP, Faccenna C (2008) Tethyan closure, Andean orogeny, and westward drift of the Pacific Basin. Earth Planet Sci Lett 271(1):303–310
Iannizzotto NF, Folguera A, Leal PR (2004) Control tectónico de las secuencias volcaniclásticas neocomianas y paleogeografía en la zona del Lago La Plata (45° S). Sector interno de la faja plegada y corrida de los lagos La Plata y Fontana. Rev Asoc Geol Argentina 59:655–670
Jaillard E, Bengtson P, Dhondt AV (2005) Late Cretaceous marine transgressions in Ecuador and northern Peru: a refined stratigraphic framework. J South Am Earth Sci 19:307–323
Jaimes E, de Freitas M (2006) An Albian-Cenomanian unconformity in the northern Andes: evidence and tectonic significance. J South Am Earth Sci 21:466–492
James DE, Sacks S (1999) Cenozoic formation of the Central Andes: a geophysical perspective. In: Skinner, B et al. (eds.) Geology and Mineral Deposits of Central Andes. Soc Ec Geol Special Publication 7:1–25
Jordan T, Burns W, Veiga R, Pángaro F, Copeland P, Kelley S, Mpodozis C (2001) Extension and basin formation in the Southern Andes caused by increased convergence rate: a Mid-Cenozoic trigger for the Andes. Tectonics 20(3):308–324

Kay SM, Copeland P (2006) Early to middle Miocene backarc magmas of the Neuquén Basin: geochemical consequences of slab shallowing and the westward drift of South America. In: Kay, SM, Ramos, VA (Eds.), Evolution of an Andean Margin: A Tectonic and Magmatic View from the Andes to the Neuquén Basin (35°–39° S lat.). Geol Soc Am Special Paper 407: 185–213

Kay SM, Mpodozis C, Coira B (1999) Neogene magmatism, tectonism, and mineral deposits of the Central Andes (22° S to 33° S). In: Skinner, B et al. (eds) Geology and Mineral Deposits of Central Andes. Soc Ec Geol, Special Publication 7:27–59

Kozlowski E, Manceda R, Ramos VA (1993) Estruc-tura. In: Ramos, VA (ed.) Geología y Recursos Naturales de Mendoza, Relatorio del XII Congreso Geológico Argentino y II Congreso de Exploración de Hidrocarburos 235–256

Lallemand S, Heuret A, Boutelier D (2005) On the relationships between slab dip, back-arc stress, upper plate absolute motion, and crustal nature in subduction zones. Geochemistry Geophysics Geosystems 6(9)

Lagorio S, Montenegro G, Massaferro, Vattuone ME (1998) Edad y geoquímica de las ignimbritas de Aluminé, provincia del Neuquén, Argentina. In: 10° Congreso Latinoamericano de Geología Económica (Buenos Aires), Actas 2:231–325

Leanza H (2009) Las principales discordancias del Mesozoico de la Cuenca Neuquina según observaciones de superficie. Rev del Mus. Argent. Ciencias Nat. 11(2):145–184

Legarreta L, Gulisano C (1989) Análisis estratigráfico secuencial de la Cuenca Neuquina (Triásico superior-Terciario inferior). In: Chebli G, Spalletti, L (eds) Cuencas Sedimentarias Argentinas, Correlación Geológica, serie 6:221–243

Lesta P, Ferello R, Chebli G (1980) Chubut extraandino. In: Turner JC (ed) II Simposio de Geología Regional Argentina. Academia Nacional de Ciencias, Córdoba, pp 1307–1387

Linares E, González RR (1990) Catálogo de edades radimétricas de la República Argentina 1957–1987. As Geol Arg, Publicaciones Especiales, Serie B, Didáctica y Complementaria 19:1–628

Litvak VD, Spagnuolo MG, Folguera A, Poma S, Jones RE, Ramos VA (2015) Late Cenozoic calc-alkaline volcanism over the Payenia shallow subduction zone, South-Central Andean back-arc (34°30′–37° S), Argentina. J South Am Earth Sci 64:365–380

Lizuaín A (1979) La edad de las sedimentitas del cerro Plataforma, provincia del Chubut. Rev Asoc Geol Arg 34(1):69–72

Lizuaín A (1980) Las formaciones Suprapaleozoicas y Jurásicas de laCordillera Patagónica. Provincias de Río Negro y Chubut. Rev Asoc Geol Arg 35(2):174–186

Lizuaín A (1981) Características y edad del plutonismo en los alrededores del lago Puelo. Provincia del Chubut. VIII Congreso Geológico Argentino, Neuquén, Actas 3:607–616

Lizuaín A (1987) El vulcanismo cretácico de la Cordillera Patagónica entre los lagos Puelo y Cholila, provincia del Chubut. X Congreso Geológico Argentino, San Miguel de Tucumán, Actas 4:213–21

LLambías EJ, Aragón E (2011) Volcanismo Paleógeno. In: Leanza, HA, Arregui O, Carbone O, Danieli JC, Vallés JM, Geología y recursos naturales de la Provincia del Neuquén Relatorio del XVIII Congreso Geológico Argentino 265–274

Llambías EJ, Rapela CW (1989) Las vulcanitas de Colipilli, Neuquén (37º S) y su relación con otras unidades paleógenas de la cordillera. As Geol Arg XLIV 1–4:224–236

López de Luchi MG, Spikermann JP, Strelin JA, Morelli J (1992) Geología y petrología de los plutones de la Tapera de Burgos, Arroyo el Rápido y Cerro Caquel, Departamento Languineo, Provincia del Chubut. Rev Asoc Geol Argent 47:87–98

Maloney KT, Clarke GL, Klepeis KA, Quevedo L (2013) The Late Jurassic to present evolution of the Andean margin: Drivers and the geological record, Tectonics 32. https://doi.org/10.1002/tect.20067

Manceda R, Figueroa D (1995) Inversion of the Mesozoic Neuquén rift in the Malargüe fold and thrust belt, Mendoza, Argentina. Tankard AJ, Suarez RS, Welsink HJ, Petroleum basins of South America, Am Assoc Petrol Geol Mem 62:369–382

Marquez M, Navarrette C (2011) La tectónica compresiva pre-Aptiana-Albiana en la sierra de Pichiñañez, Chubut, Argentina. In: XVIII Congreso Geológico Argentino, Neuquén, pp 105–106

Martin-Gombojav N, Winkler W (2008) Recycling of Proterozoic crust in the Andean Amazon foreland of Ecuador: implications for orogenic development of the Northern Andes. Terra Nova 20:22–31

Mescua JF, Giambiagi LB, Ramos VA (2013) Late Cretaceous Uplift in the Malargüe fold-and-thrust belt (35° S), southern Central Andes of Argentina and Chile. Andean Geology 40(1):102–116

Micucci EM, Continanzia J, Manceda R, Gavarrino AS (2011) Cuenca de San Julián: Síntesis del conocimiento exploratorio—visión actual. In: Kozlowski E, Legarreta L, Boll A, Marshall PA (eds), Simposio Cuencas Argentinas Visión Actual: VIII Congreso de Exploración y Desarrollo de Hidrocarburos, pp 17–46

Mpodozis C, Arriagada C, Basso M, Roperch P, Cobbold P, Reich M (2005) Late Mesozoic to Paleogene stratigraphy of the Salar de Atacama Basin, Antofagasta, Northern Chile: implications for the tectonic evolution of the Central Andes. Tectonophysics 399(1–4): 125–154

Munizaga F, Hervé F, Drake R, Pankhurst RJ, Brook M, Snelling N (1988) Geochronology of the Lake region of south-central Chile (39_e42_S): preliminary results. J South Am Earth Sci 1 (3):309–316

Navarrete CR, Gianni GM, Folguera A (2015) Tectonic inversion events in the western San Jorge Gulf Basin from seismic, borehole and field data. J South Am. Earth Sci. 64:486–497

Nullo F, Combina A (2011) Patagonian continental deposits (Cretaceous-Tertiary). Biological J Linn Soc 103(2):289–304

Olivero EB, Medina FA (1994) Sedimentología de la Formación Lefipán (Cretácico-Terciario) en el valle medio del Río Chubut. Rev Asoc Geol Arg 48:105–106

Oncken O, Hindle D, Kley J, Elger K, Victor P, Schemmann K, (2006) Deformation of the Central Andean upper plate system—Facts, fiction, and constraints for plateau models. In: Oncken O, Chong G, Franz G, Giese P, Götze HJ, Ramos VA, Strecker MR, Wigger P)The Andes, Springer Berlin Heidelberg 22:3–27

Orts DL, Folguera A, Giménez M, Ramos VA (2012a) Variable structural controls through time in the Southern Central Andes (~36°S). Andean Geology 39(2):220–224

Orts DL, Folguera A, Encinas A, Ramos M, Tobal J, Ramos VA (2012b) Tectonic development of the North Patagonian Andes and their related Miocene foreland basin (41°30′–43° S). Tectonics 31:1–24

Pananont P, Mpodozis C, Blanco N, Jordan TE, Brown LD (2004) Cenozoic evolution of the northwestern Salar de Atacama Basin, northern Chile. Tectonics 23(TC6007)

Pankhurst RJ, Hervé F (1994) Granitoid age distribution and emplacement control in the North Patagonian batholith in Aysen (44°–47° S). VII Congreso Geológico Chileno, Concepción, Actas 2:1409–1413

Pankhurst RJ, Hervé F, Rojas L, Cembrano J (1992) Magmatism and tectonics in continental Chiloé, Chile (42–42 30′S). Tectonophysics 205:283–294

Pankhurst RJ, Weaver SD, Hervé F, Larrondo P (1999) Mesozoic-Cenozoic evolution of the North Patagonian batholith in Aysen, southern Chile. J Geol. Soc. 156:673–694

Pankhurst RJ, Hervé F, Fanning M, Suárez M (2003) Coeval plutonic and volcanic activity in the Patagonian Andes: the Patagonian Batholith and the Ibáñez and Divisadero Formations, Aisén, southern Chile. In: X Congreso Geológico Chileno (ed Campos E). Electronic files

Pankhurst RJ, Rapela C, Fanning C, Márquez M (2006) Gondwanide continental collision and the origin of Patagonia. Earth Sci Rev 76:235–257

Parada MA, Lahsen A, Palacios C (2001) Ages and geochemistry of Mesozoic-Eocene back-arc volcanic rocks in the Aysén region of the Patagonian9 Andes. Rev Geol Chile 28:25–46

Pardo-Casas F, Molnar P (1987) Relative motion of the Nazca (Farallon) and South American plates since Late Cretaceous time. Tectonics 6(3):233–248

Perez Loinaze VS, Vera EI, Passalia MG, Llorens M, Friedman R, Limarino CO, Césari SN (2013) High-precision U-Pb zircon age from the Anfiteatro de Ticó Formation: implications for the timing of the early angiosperm diversification in Patagonia. J South Am Earth Sci 48:97–105
Peroni GO, Hegedus AG, Cerdan J, Legarreta L, Uliana MA (1995) Hydrocarbon Accumulation in an Inverted Segment of the Andean Foreland : San Bernardo Belt, Central Patagonia. In: Tankard AJ, Suárez R, Welsink HJ (eds), Petroleum Basins of South America, pp 403–419
Pesce A (1979) Estratigrafía de la Cordillera Patagónica entre los 43°30′ y 44° de latitud sur y sus áreas mineralizadas. VII Congreso Geológico Argentino, Buenos Aires 1:257–270
Rabassa J (1978) Estratigrafía de la región de Pilcaniyeu Comallo, Provincia de RíoNegro. VII Congreso Geológico Argentino, Neuquén, Actas 1:731–746
Ramos VA (1981) Descripción geológica de la hoja 47 ab Lago Fontana, Provincia de Chubut. Servicio Geológico Nacional. Bol., Buenos Aires 183,130
Ramos VA (1982) Geología de la región del lago Cardiel, provincia de Santa Cruz. Revista de la Asociación Geológica Argentina 37:23–49
Ramos VA (2008) Patagonia: a Paleozoic continent adrift? J South Am Earth Sci 26:235–251
Ramos VA, Cortés JM (1984) Estructura e interpretación tectónica, Geol. y Recur. Nat. la Prov. Río Negro, 1, 12
Ramos VA, Folguera A (2005) Tectonic evolution of the Andes of Neuquén: constraints derived from the magmatic arc and foreland deformation. Veiga GD, Spalletti LA, Howell JA, Schwarz E The Neuquén Basin, Argentina: a case Study in Sequence Stratigraphy and Basin Dynamics Geol Soc Lond, Special Publications 252:15–35
Ramos VA, Folguera A (2009) Andean flat slab subduction through time. In: Murphy, B (Ed.), Ancient Orogens and Modern Analogues. Geol Soc Lond, Special Publication 327:31–54
Ramos VA, Cristallini E, Pérez DJ (2002) The Pampean flat-slab of the Central Andes. J South Am Earth Sci 15:59–78
Ranalli JN, Peroni GO, Boggetti DA, Manoni R (2011) Cuenca Cañadón Asfalto. Modelo tectosedimentario. In: Kozlowski E, Legarreta L, Boll A, Marshall PA (eds), Simposio Cuencas Argentinas Visión Actual: VIII Congreso de Exploración y Desarrollo de Hidrocarburos pp 185–216
Rapela CW, Kay SM (1988) Late Paleozoic to recent magmatic evolution of northern Patagonia. Episodes 11:175–182
Rapela CW, Spalletti L, Merodio JC (1983) Evolución magmática y geotectónica de la Serie Andesitica andina (Paleoceno-Eoceno) en la cordillera norpatagónica. Rev Asoc Geol Arg 38 (3–4):469–484
Rapela C, Munizaga F, Dalla Salda L, Hervé F, Parada M, Cingolani C (1987) Nuevas edades K-Ar de los granitoides del sector nororiental de los Andes Patagónicos. X Congreso Geológico Argentino, Tucumán, Actas 4:18–20
Rapela CW, Pankhurst RJ, Fanning CM, Hervé F (2005) Pacific subduction coeval with the Karoo mantle plume: the Early Jurasssic Subcordilleran belt of northwestern Patagonia. Geol Soc Lond, Special Publications 246(1):217–239
Reimer W, Miller H, Mehl H (1996) Mesozoic and Cenozoic palaeo-stress fields of the South Patagonian Massif deduced from structural and remote sensing data. Geol Soc Lond, Special Publications 108(1):73–85
Rojas Vera E, Mescua J, Folguera A, Becker TP, Sagripanti L, Fennell L, Orts D, Ramos VA (2015) Evolution of the Chos Malal and Agrio fold and thrust belts, Andes of Neuquén: Insights from structural analysis and apatite fission track dating. J S Am Earth Sci 64:418–433
Sánchez ML, Asurmendi E (2014) Modelo de depósito de la Formación Cerro Lisandro: lóbulos de desembocadura y deltas de tipo Gilbert. Cretácico superior, región central de cuenca Neuquina, Argentina. Rev Mex Cien Geol, 31(2):141–162
Sánchez ML, Asurmendi E, Armas P (2013) Subgrupo Río Colorado (Grupo Neuquén): Registros de paleosismicidad en la cuenca de antepaís andina, Cuenca Neuquina, Provincias de Neuquén y Río Negro. Rev Asoc Geol Arg 70(1):96–114

Savignano E, Mazzoli S, Arce M, Franchini M, Gautheron C, Paolini M, Zattin M (2016) (Un) Coupled thrust belt-foreland deformation in the northern Patagonian Andes: new insights from the Esquel-Gastre sector (41°30′–43° S). Tectonics. In press

Scasso RA, Aberhan M, Ruiz L, Weidemeyer S, Medina FA, Kiessling W (2012) Integrated bio-and lithofacies analysis of coarse-grained, tide-dominated deltaic environments across the Cretaceous/Paleogene boundary in Patagonia. Argentina. Cretac. Res. 36:37–57

Schütte P, Chiaradia M, Beate B (2010) Geodynamic controls on Tertiary arc magmatism in Ecuador: constraints from U-Pb zircon geochronology of Oligocene-Miocene intrusions and regional age distribution trends. Tectonophysics 489(1):159–176

Sciutto JC, (1981) In: Geología del Codo del Río Senguerr, Chubut, Argentina. VIII Congreso Geológico Argentino, Actas 3. (San Luis).

SEGEMAR-JICA- MMAJ-JMEC 1999–2000–2001 Regional Survey for Mineral Ressources in the Southern Andes Area. Convenio de Cooperación Técnica entre el SEGEMAR, Counterpart for República Argentina y JICA, MMAJ y JMEC of Japan. SEGEMAR. Trabajo inédito. Buenos Aires

Sempere T 1995. Phanerozoic evolution of Bolivia and adjacent regions. In: Tankard, A.J., et al. (ed) Petroleum Basins of South America: Am Assoc Petr Geol, Memoir 62:207–230

Seton M, Müller RD, Zahirovic S, Gaina C, Torsvik T, Shephard G, Talsma A, Gurnis M, Turner M, Maus S, Chandler M (2012) Global continental and ocean basin reconstructions since 200 Ma. Earth Sci Rev 113:212–270

Sepúlveda EG, Viera RM (1980) Geología y área de alteración en el cerro Colorado y alrededores. Chubut noroccidental. Rev Asoc Geol Arg 35(2):195–202

Silva Nieto D (2005) Hoja Geológica 4369-III, Paso de Indios, provincia del Chubut Vol. 265. Instituto Geología y Recursos Minerales, Servicio Geológico Minero Argentino, Boletín, p. 72

Silver PG, Russo RM, Lithgow-Bertelloni C (1998) Coupling of South American and African Plate motion and Plate deformation. Science 279 (5347): 60–63. https://doi.org/10.1126/science.279.5347.60

Silvestro J, Atencio M (2009) La cuenca Cenozoica del Río Grande y Palauco: Edad, evolución y control estructural, faja plegada de Malargüe (36° S). Rev Asoc Geol Arg 65(1):154–169

Silvestro J, Kraemer P, Achilli F, Brinkworth W (2005) Evolución de las cuencas sinorogéncias de la Cordillera Principal entre 35°–36° S. Malargüe. Rev Asoc Geol Arg 60(4):627–643

Soechting W (2001) Tectonic control of epithermal gold mineralization in the Cordon de Esquel, Chubut. Argentina. Minera « El Desquite S.A. Esquel

Sole P, Bonhomme MG (1990) Relation of Magmatic Activity to Plate Dynamics in Central Peru from Late Cretaceous to Present. In Plutonism from Antarctica to Alaska; Special Paper 241; Kay, S.M

Spagnuolo MG, Folguera A, Litvak V, Rojas Vera EA, Ramos VA (2012) Late Cretaceous arc rocks in the Andean retroarc region at 36,5° S: Evidence supporting a Late Cretaceous slab shallowing. J South Am Earth Sci 38:44–56

Suarez M, De La Cruz R (2000) Tectonics in the eastern central Patagonian Cordillera (45 30′–47 30′ S). J Geol Soc 157(5):995–1001

Suárez M, De la Cruz R (2001) Jurassic to Miocene K-Ar dates from eastern central Patagonian Cordillera plutons, Chile (45–48S). Geol Mag 138:53–66

Suárez M, Márquez M (2007) A Toarcian retro-arc basin of Central Patagonia (Chubut), Argentina: middle Jurassic closure, arc migration and tectonic setting. Rev. Geol Chile 34(1). https://doi.org/10.4067/S0716-02082007000100004

Suárez M, De La Cruz R, Bell M (1996) Estratigrafía de la región de Coyhaique (45°–46° latitud S); Cordillera Patagónica, Chile. In: XIII Congreso Geológico Argentino and III Congreso de Exploración de Hidrocarburos (eds Ramos VA, Aguirre Urreta MB, Gulisano C, Kozlowski E, Limarino O, Méndez V, Mendía JE, Mozetic M, Ortiz A, Rapela CW, Riccard A, Robbiano JA, Turic MA, Vega VA, Irigoyen M), Buenos Aires, pp 575–90

Suárez M, De La Cruz R, Bell MC (2007) Geología del Area Ñireguao–Baño Nuevo, Región Aisén del General Carlos Ibáñez del Campo. Servicio Nacional de Geología y Minería, Carta Geológica de Chile, Serie Geología Básica 108:1–56. Santiago

Suárez M, De La Cruz R, Bell M, Demant A (2009a) Cretaceous slab segmentation in southwestern Gondwana. Geol Mag 147:193–205

Suárez M, Márquez M, De La Cruz R, Fanning M (2009b) Aptian-Albian subaerial volcanic rocks in central Patagonia : Divisadero and Chubut Groups. In: XII Congreso Geológico Chileno, pp 1–4

Suárez M, Márquez M, De La Cruz R, Navarrete C, Fanning M (2014) Cenomanian-? Early Turonian Minimum Age of the Chubut Group, Argentina: SHRIMP U-Pb geochronology. J South Am Earth Sci 50:64–74

Tatsumi T, Eggins S (1995) Subduction Zone Magmatism. Wiley-Blackwell

Toubes RO, Spikerman JP (1973) Algunas edades K/Ar y Rb/Sr de plutonitas de la Cordillera Patagónica entre los paralelos 40° y 44° de latitud sur. Rev Asoc Geol Arg 28(4):382–39

Tunik MA, Vietto ME, Sciutto JC (2004) Procedencia de areniscas del Grupo Chubut en el área central de la Sierra de San Bernardo. Análisis preliminar. Rev Asoc Geol Argentina 59: 601–606

Tunik M, Folguera A, Naipauer M, Pimentel M Ramos VA (2010) Early uplift and orogenic deformation in the Neuquén Basin: Constraints on the Andean uplift from U-Pb and Hf isotopic data of detrital zircons. Tectonophysics 489:258–273

Turienzo M, Dimieri L, Friscale C, Araujo V, Sánchez N (2012) Cenozoic structural evolution of the Argentinean Andes at 34°40′S: A close relationship between thick and thin-skinned deformation. Andean Geology 39(2):317–357

Turner JC (1982) Descripción geológica de la Hoja 44c, Tecka, Provincia del Chubut. Servicio Geológico Nacional, Boletín 180:1–92

Uliana MA, Biddle KT (1988) Mesozoic-Cenozoic paleogeographic and geodynamic evolution of southern South America. Rev Bra Geoc 18:172–190

Uliana MA, Biddle KT, Cerdan J (1989) Mesozoic extension and the formation of Argentine sedimentary basins. In: Tankard AJ, Balkwill HR (eds) Extensional Tectonics and Stratigraphy of the North Atlantic Margin pp 599–613

Varela R, Basei M, Cingolani C, Siga O Jr, Passarelli C (2005) El basamento cristalino de los Andes norpatagónicos en Argentina: geocronología e interpretación tectónica. Rev Geol Chile 32(2):167–187

Vattuone ME, Latorre CO (2004) Edades K/Ar al este del cerro Nahuel Pan, Chubut.Implicancias en la correlacion del Grupo Divisadero y del Choiyoi en el área. Rev Asoc Geol Arg 59(3): 510–513

Vergani GD, Tankard AJ, Belotti HJ, Welsink HJ (1995) Tectonic evolution and paleogeography of the Neuquén Basin, Argentina. In: Tankard AJ, Suárez Soruco R, Welsink HJ (Eds.) Petroleum Basins of South America. Am Assoc Petrol Geol, Memoirs, 62, pp 383–402

Weaver C (1927) The Roca Formation in Argentina. Am J Sci 13:417–434

Winocur DA, Litvak VD, Ramos VA (2015) Magmatic and tectonic evolution of the Oligocene Valle del Cura basin, main Andes of Argentina and Chile: evidence for generalized extension. Geol Soc Lond, Special Publications 399(1):109–130

Rapela, CW et al. (Eds.) Geol Soc Ame: Boulder, CO, USA, pp 173–192

Zamora Valcarce G, Zapata T, Del Pino D, Ansa A (2006) Structural evolution and magmatic characteristics of the Agrio Fol. and thrust belt. In: Kay SM, Ramos VA (Eds.), Evolution of an Andean Margin: A Tectonic and Magmatic View from the Andes to the Neuquén Basin (35°-39° lat). Geol Soc Am, Special Paper 407, pp 125–145

Zamora Valcarce G, Rapalini AE, Spagnuolo CM (2007) Reactivación de estructuras cretácicas durante la deformación miocena, faja plegada del Agrio. Neuquén. Rev Asoc Geol Arg 62 (2):299–308

Zamora Valcarce G, Zapata T, Ramos VA, Rodríguez F, Bernardo LM (2009) Evolución tectónica del frente andino en Neuquén. Rev Asoc Geol Arg 65(1):192–203

Tectonic Rotations Along the Western Central Andes

César Arriagada

Abstract Despite the evidence for protracted deformation, crustal shortening, and exhumation since at least 100 Ma along the Central Andes, the uplift of the Eastern Cordillera and Bolivian Orocline formation did not start until the Eocene–Oligocene. Moreover, the Central Andes also exhibit much younger recent surface uplift (e.g., 10 Ma) that would postdate significant shortening. Recent investigations were focused on the formation of the Bolivian Orocline by 2D Map-View Restoration of Non-plane Deformation experiments. Results from these 2D restorations support the hypothesis of the Paleogene formation of the Bolivian Orocline, due to differential shortening, concentrated in the Eastern Cordillera of Bolivia, Southern Peru, and northwestern Argentina. The "out of plane of cross-section" motion of material and the rotational components of deformation appear to be essential aspects for the formation of the Central Andes, although they are generally not included in models for orogenic systems evolution. One of the most remarkable results of the 2D restoration is the persistence of an "excess rotation" of 10–20° in northern Chile and 20–30° in Southern Peru which cannot be easily explained even if all the differential Paleogene shortening in the Eastern Cordillera and the Neogene shortening in the Sub-Andean zone were considered. Here, we discuss that a major proportion of rotation needs to be balanced in the forearc region by two major conjugate oblique shear zones (i.e., the Abancay Deflection and the Antofagasta–Calama Lineament). These shear zones are probably related to inherited lithospheric discontinuities and major changes in the magnitude of rotation likely occurring along these zones.

Keywords Oroclines · Paleomagnetism · Paleozoic terranes · Forearc
Shear zones · Paleogene deformation · Shortening gradients

C. Arriagada (✉)
Universidad de Chile, Santiago, Chile
e-mail: cearriag@cec.uchile.cl

A. Folguera et al. (eds.), *The Evolution of the Chilean-Argentinean Andes*,
Springer Earth System Sciences, https://doi.org/10.1007/978-3-319-67774-3_13

1 Introduction

The Andes are the type example of a mountain belt developed along an ocean–continent convergent margin, with geometry and style of Andean deformation very heterogeneous along strike (Fig. 1). The formation of the Andes, including their peculiar shape, the time lag since the initiation of subduction (see Chaps. "The Early Stages of the Magmatic Arc in the Southern Central Andes, Mechanisms and Episodes of Deformation Along the Chilean–Pampean Flat-Slab Subduction Segment of the Central Andes in Northern Chile", and "Cretaceous Orogeny and Marine Transgression in the Southern Central and Northern Patagonian Andes: Aftermath of a Large-Scale Flat-Subduction Event?"), and the activity of the different ranges are enigmatic features of one of Earth's major orogens and are not well understood. The main controversy focuses on how their arcuate a non-cylindrical structure, shortening, crustal thickening, topographic evolution, climate, and erosion are balanced across the entire orogen. Beyond some attempts to integrate geological information to understand the controlling mechanisms of the Central Andes evolution, how this orogen evolves and grows remain a matter of considerable debate. Although there is evidence for protracted deformation along the Central Andes, crustal shortening, and exhumation since at least 100 Ma (e.g., Carrapa and DeCelles 2015; Bascuñán et al. 2016), recent surface uplift as young as 10 Ma would postdate significant shortening (e.g., Garzione et al. 2008). Oroclines, the curvature of the volcanic arc and the continental margin's arcuate shape, are among the most conspicuous features along the Central Andes, separating different structural, magmatic, and basin domains with different timing of development.

2 Curvatures Along the Central Andes

Four large pronounced curvatures of tectonic trends by oroclinal bend are present along the topographic front of the Andes (i.e., Huancabamba, Abancay, Bolivian Orocline, and Patagonian Orocline, Fig. 1). One of the most remarkable tectonic features of the Central Andes is the curvature of the orogenic system and the South American continental margin at $\sim 20°$ S. The symmetry along the Bolivian Orocline correlates with the symmetric features of the Nazca plate (Fig. 1). Paleomagnetism is a useful tool for the correlation of tectonic rotations with arcuate shapes along the western margin of the Central Andes (see Beck 2004; Arriagada et al. 2000, 2003, 2006, 2013; Roperch et al. 2011; Somoza et al. 1999, 2015 and references therein). Isacks (1988) hypothesized that the Bolivian Orocline formed as a consequence of the geometrical constraints imposed by along strike changes in the magnitude of Neogene horizontal shortening along the Sub-Andean Fold and Thrust Belt. Later, Kley (1999) and McQuarrie (2002) reconstructed the evolution of the Bolivian Orocline and restored the original shape of the Andean margin before the onset of Late Cretaceous deformation.

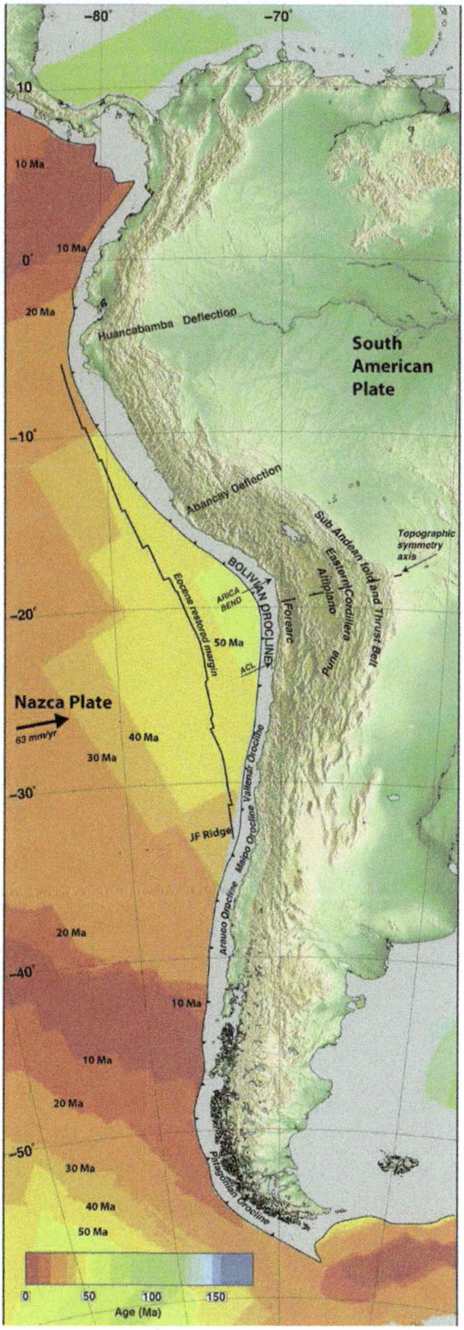

Fig. 1 Main tectonic features of the Andes with ages of oceanic crust

This was achieved by progressively removing the horizontal shortening components associated with successive deformation increments in the Sub-Andean zone (Neogene), Altiplano, Eastern Cordillera, and the modern forearc zone (Paleogene, Late Cretaceous). However, these reconstructions did not consider the effects of the paleomagnetically determined counterclockwise block rotations in Southern Peru, along with clockwise rotations in northern Chile that determine the Central Andean Rotation Pattern, CARP (Fig. 2).

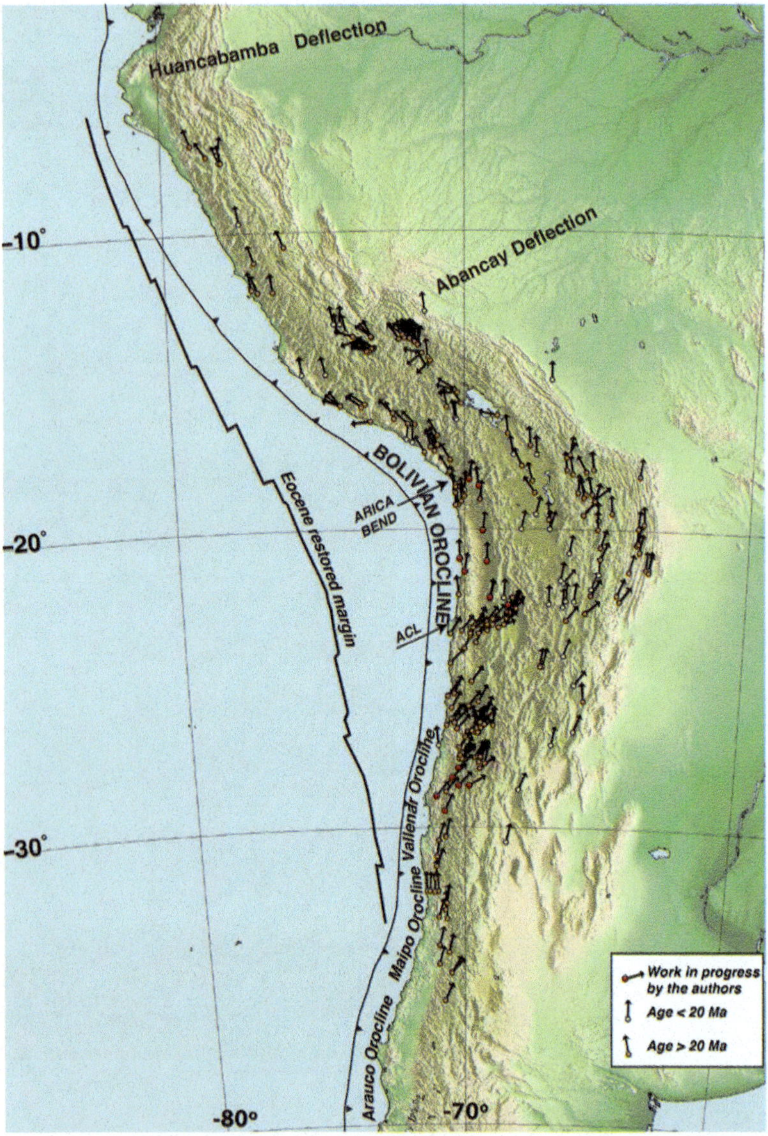

Fig. 2 Paleomagnetic database along the Central Andes for rocks <20 Ma, for rocks >20 Ma and work in progress by the authors (after Arriagada et al. 2008, 2013)

Although Isacks (1988) envisaged block rotations in the forearc, which were later supported, by numerous paleomagnetic studies (see Beck 2004; Arriagada et al. 2003, 2006; Roperch et al. 2011 and references therein), the magnitude of most of the forearc block rotations largely exceeded the predicted amounts in the Isacks model (Fig. 2). Somoza et al. (1999, 2015) and Arriagada et al. (2000, 2003, 2006) in northern Chile, and Roperch et al. (2011) in Southern Peru, showed that most of the block rotations in the forearc are pre-Early Miocene (Fig. 2), predating the onset of Neogene shortening in the Sub-Andean zone. Furthermore, the fact that no rock-age-dependent changes in the rotation magnitude occur in Mesozoic to Paleocene rocks in northern Chile which, in average exhibit the same amount of rotation (Fig. 2), led Arriagada et al. (2003, 2006) to hypothesize that most rotations were acquired between the Eocene–Oligocene. Assuming that rotations are essentially driven by oroclinal bending associated with shortening in the Eastern Cordillera and in the Sub-Andean belt, Arriagada et al. (2008) proposed that the total amount of shortening across the Central Andes is significantly greater than shortening estimated from balanced cross sections, and that a major phase of oroclinal bending associated with a broad Incaic phase of deformation affected extensive areas of the Central Andes between ca. 45 and 35 Ma. The rotations are, however, even larger than those driven by pure shortening, and rotations are likely enhanced locally through major deflections, like the Abancay deflection, where the main Andean morphotectonic features are sinistral (Roperch et al. 2011; Fig. 2). The spatial variation in the magnitude of rotations shows that the Central Andes region (Fig. 2) did not behave as a rigid block (Arriagada et al. 2008, 2013). In plan view, rotations seem to be correlated to the overall orientation of the margin, which is related to large blocks delimitated by major deflections, lineaments, or oroclines (Fig. 2) that can also be represented at crustal scale when the depth of the continental Moho is calculated (Tassara et al. 2006).

However, while the tectonic constraints of the plan view restoration of the deformation and rotations of Arriagada et al. (2008) are relatively accurate for the last 15 Ma, geological constraints on the amount of shortening are limited for Andean deformation older than 15 Ma. Müller et al. (2002), McQuarrie et al. (2005) and Horton (2005) highlighted the magnitude of Eocene–Oligocene deformation in the Eastern Cordillera, where shortening began at ca. 42 Ma by east-directed thrusting, continued by west-directed back-thrusting from late Eocene to late Oligocene and essentially terminated at early Miocene times, before the Andean deformation front finally jumped eastwards to the Sub-Andean zone (Müller et al. 2002; McQuarrie et al. 2005).

McQuarrie et al. (2005) indicated that almost 50% of the up to 262–326 km total horizontal shortening across the Central Andes in Southern Peru and Bolivia was accomplished by 40–20 Ma deformation in the Eastern Cordillera. The amount of shortening decreases progressively southwards, tracking changes in the upper crustal rheology from the weaker early Paleozoic sediments to the stronger, more crystalline Faja Eruptiva de la Puna Oriental and the northern Sierras Pampeanas (Allmendinger et al. 1997) (see Chaps. "Paleogeographic and Kinematic Constraints in the Tectonic Evolution of the Pre-Andean Basement Blocks", "The Pre-Andean Phases of Construction of the Southern Andes Basement in Neoproterozoic–Paleozoic Times" and "The Famatinian Orogen Along the Protomargin of Western Gondwana:

Evidence for a Nearly Continuous Ordovician Magmatic Arc Between Venezuela and Argentina"). Similarly, shortening also decreases northwards. A more recent map-view reconstruction of the eastern part of the Bolivian orocline between the Eastern Cordillera and Sub-Andean ranges (Eichelberger and McQuarrie 2015) suggests that deformation at the orocline core may be sufficient to account for the modern crustal thickness without significant additional shortening in the forearc.

The history of shortening and rotational strain presented for the Bolivian Orocline curvature (McQuarrie et al. 2005; Arriagada et al. 2008) was recently linked to phases of slab flattening and steepening which exerted a primary control on arc magmatism and regional crustal shortening (Martinod et al. 2010; O'Driscoll et al. 2012). Although flattened-slab promoted interplate friction, it may explain the increase of continental shortening in the Eastern Cordillera. Additionally, in a subduction corner-flow system such as the Andes, the presence of a deep lithospheric root of the upper plate may also play an important role because it restricts the supply of low-viscosity asthenosphere into the mantle wedge (O'Driscoll et al. 2012). According to O'Driscoll et al. (2012), flat-slab activity along the Central Andean margin during the late Eocene and Oligocene may have been caused by a combination of a cratonic root that inhibited drag basal forces triggered by trench roll back and subduction of an oceanic plateau. An abrupt transition to trench-normal subduction, after ca. 25 Ma, put an end to slab flattening, directing the subducted slab toward the more distal Sao Francisco Craton, and inhibiting slab suction (O'Driscoll et al. 2012). Sdrolias and Müller (2006) propose the subduction of an aseismic ridge along the Chilean margin during the Paleogene. The speculated ridge would have been active during the Paleocene (65–55 Ma), at the location of the present-day Juan Fernandez hot spot (Sdrolias and Müller 2006). Whereas during the Eocene the fossil ridge would have subducted with an open angle, sweeping the margin of Chile from south to north, the ridge would become more parallel to the margin of Southern Peru when it would have entered in subduction during the late Eocene–Oligocene (Martinod et al. 2010). This period is also characterized by a plate global reorganization followed by acceleration in the speed of subduction (Sdrolias and Müller 2006). Thus, ~50 Ma ago, this oceanic plateau would have migrated northward the subduction zone, reaching at 40 Ma ago the latitudes of the Central Andes, resulting in a horizontal subduction, producing tectonic rotations (Bolivian Orocline), and shortening of the Eastern Cordillera.

During the Neogene, the Juan Fernandez Ridge (JFR) (Fig. 1) swepts the Chilean trench from north to south (Yáñez et al. 2001; Arriagada et al. 2013). Between 18° and 30° S, the intersection point defined between the JFR and the subduction zone moved fairly quickly to the south between 22 and 10 Ma. However, since 10 Ma the JFR intersection has remained relatively stable at around 33° S (see Chap. "Structure and Tectonics of the Chilean Convergent Margin from Wide-Angle Seismic Studies: A Review"). In this scenario, about 10° of clockwise rotation associated with the formation of the Arica deflection would have increased by about 10° within the normal subduction segment south of Maipo Orocline (between 33° and 37° S, Fig. 2), while between 30° and 33° S about 7° to 10° of counterclockwise rotation could explain the current pattern of no rotations (Arriagada et al. 2013). Observed paleomagnetic data are consistent with a higher shortening around 33° S (130 to 150 km), which decreases steadily to its lowest value near 37° S (no more than 40 km) in its southern termination. Three-dimensional numerical subduction models provided another line of evidence regarding the role of the subducted plate to produce a concave margin during the late Eocene and Oligocene. Capitanio et al. (2011) found that the age of the downgoing lithosphere had a first-order control on the velocity of the subducting plate and on the topography of the overriding plate. The roll back of the subducted ocean lithosphere in the mantle drives traction toward the trench at the base of the upper plate (drag basal force), causing it to thicken. Thus, subduction of an older Nazca plate below the Central Andes could explain the locally thickened crust and consequently higher elevations. The present-day age gradient was not established until ~20 Ma (Sdrolias and Müller 2006) with older oceanic crust with an Eocene age facing the highest Altiplano plateau. At ~45–40 Ma ago, another strong along-trench age gradient formed between 30° and 45° S (Sdrolias and Müller 2006). The increased driving force in the north could explain the fast convergence through this period (Pardo Casas and Molnar 1987; Sdrolias and Müller 2006) and the initiation of the uplift of the Eastern Cordillera (Elger et al. 2005). This age gradient positioned ~70 my old lithosphere offshore from South America, which then became the oldest lithosphere subducted under the Central Andes since the early Cenozoic. In this sense, the initial development of the Central Andes and orocline formation in the Eocene could be directly linked to the subduction of Late Cretaceous ocean floor.

3 Paleomagnetic Domains in the Chilean Andes

Within the southern Central Andes four additional curvatures including important changes in the pattern of rotations have been discovered. From north to south these are (Fig. 2) the Antofagasta Calama Lineament (ACL), Vallenar, Maipo and Arauco oroclines. Collected and analyzed samples from numerous paleomagnetic sites (~200) along the forearc between 18° and 38° S confirm the hypothesis of oroclinal bending at these sites (Fig. 2). Results obtained between Arica and Antofagasta (Fig. 1) suggest that a dramatic change in the magnitude of rotation occurred, whereas no significant rotation has been detected in the Coastal Cordillera and Domeyko Cordillera between 18–22° S, a situation which is in sharp contrast with the large clockwise rotations (up to 65°) observed south of 22° S in the central part of the Antofagasta region (Arriagada et al. 2000, 2003, 2006; Figs. 2 and 3). Further south from Antofagasta–Calama to La Serena (30°S), the new dataset confirms the systematical occurrence of clockwise rotations (Figs. 2 and 3). New results from Mesozoic samples collected near Vallenar Orocline show rotation with a similar magnitude than other data from Antofagasta and Atacama regions. However, supplementary results obtained from Paleozoic to Jurassic intrusive rocks in Central Chile (32–34° S) and recalculated published data do not show evidences of rotation (Arriagada et al. 2013; Figs. 2 and 3). This fact implies that there should be a transition zone between two paleomagnetic domains, i.e., La Serena and Santiago.

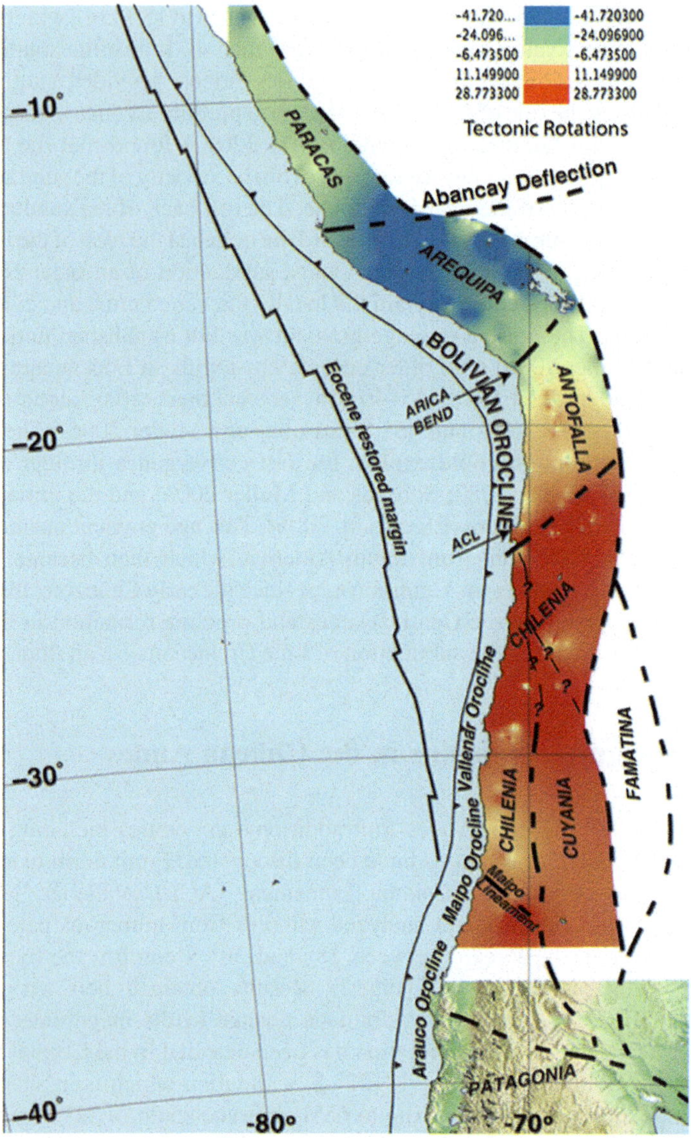

Fig. 3 Interrelationship between tectonic rotations from paleomagnetic data with proposed basement terranes (after Arriagada et al. 2008, 2013)

Antofagasta-Calama Lineament (ACL): Several NE–NNE striking faults and lineaments segment the forearc at the Antofagasta region (Arriagada et al. 2003). Here, the most important structural feature is a major lineament parallel to the Antofagasta–Calama road (ACL Lineament in Arriagada et al. 2003) (Figs. 2 and 3).

To the south-east of the ACL, there is a structurally uplifted block (Sierra del Buitre) consisting mainly of Mesozoic rocks. To explain the large clockwise rotations observed in the Antofagasta region (Fig. 2), Arriagada et al. (2003) suggested that differences in the magnitude of east-west shortening along the orogen, during the Incaic event, were accommodated along major oblique (NE, NNE) dextral transpressional shear zones (ACL in Fig. 2). When comparing our results obtained north of the ACL (Fig. 2) with the paleomagnetic data reported by Arriagada et al. (2003), immediately south of the ACL, this fault system (ACL) could delimitate major changes in the magnitude of differential rotations between two major tectonic domains (blocks). Gravity data (Götze and the Migra Group 1996) suggest that the eastern border of the Coastal Cordillera is significantly displaced to the east, north of the ACL. Additionally, the axis of the Cretaceous magmatic arc is also shifted toward the east near Calama.

Vallenar Orocline: Along the southern Atacama region, the analysis of the paleomagnetic data (Fig. 2) shows that the Vallenar Orocline, currently on the Chilean–Pampean flat-slab segment, separates two tectonic domains. While to the south a non-rotated domain coincides with a general N-trend of the main structural features, to the north, another domain is rotated in a clockwise sense characterized by faults bounding Paleozoic basement cored blocks which are trending NNE. Overall, the paleomagnetic vectors are parallel to the trend of the major faults, i.e., NNE to the north, NS-N10°E to the south, but a complex structure is preserved at the "hinge" or pivot zone between the two mayor rotation domains.

Maipo Orocline: The Andean margin at 33° S (Fig. 2) is characterized by another arcuate shape, known as the Maipo Orocline (Farías et al. 2008; Arriagada et al. 2013). Between 30° and 33° S, the topography, geologic units, and major structures strike N-, whereas south of 33° S they adopt a NNE-trend (Fig. 1). Whereas paleomagnetic analysis from Jurassic to Late Cretaceous rocks in the Chilean–Pampean flat-slab segment shows small to non-significant clockwise vertical-axis rotations, results obtained in Late Jurassic to Neogene rocks to the south, on the normal subduction segment, show systematically clockwise rotations of up to 40° (Arriagada et al. 2013). Paleomagnetic rotations are coeval with thrusting along the High Andes in Late Miocene times. Arriagada et al. (2013) propose a coupled Bolivian Orocline–Juan Fernández Ridge (JFR) model where a far-field component of clockwise rotation related to the formation of the Bolivian Orocline was first acquired by the margin between 31° and 33° S. The same area was later rotated by a slight counterclockwise component related to the subduction of the JFR over the last 10 Ma. South of 33° S the clockwise rotations related to the Bolivian Orocline component were amplified by the subduction of the JFR during the Miocene.

Arauco Orocline: The southern flank of the Maipo Orocline can be traced along strike to around 38° S. There, the Andean margin exposes another orogen bending, the Arauco "Orocline." However, presently, there are no detailed paleomagnetic studies to confirm the occurrence of these tectonic rotations.

4 Interrelation Between Tectonic Rotations and Proposed Basements Terranes

Additionally, in an attempt to illustrate a potential interrelation between tectonic rotations with similar magnitudes and defined large tectonic blocks related to hypothetic basement terranes (see Chaps. "Paleogeographic and Kinematic Constraints in the Tectonic Evolution of the Pre-Andean Basement Blocks", "The Pre-Andean Phases of Construction of the Southern Andes Basement in Neoproterozoic–Paleozoic Times", and "The Famatinian Orogen Along the Protomargin of Western Gondwana: Evidence for a Nearly Continuous Ordovician Magmatic Arc Between Venezuela and Argentina"), we have built a simple two-dimensional model (Fig. 3). The paleomagnetic database available for the Central Andes (rotations older than 20 Ma) was used to create a raster surface based on the tectonic rotation values (Fig. 3). The interpolation was created with the classic inverse distance weighting (IDW) method used for multivariable interpolation. Tectonic rotations along the margin of the Andes are usually interpreted in the frame of the Bolivian Orocline. However, equivalent magnitudes of tectonic rotations are arranged within large upper crustal blocks which are closely linked to the border of parautochthonous terranes derived from Gondwana, such as, Paracas, Arequipa, and Antofalla, and exotic terranes originated in Laurentia, such as Chilenia and Cuyania (Fig. 3). The available paleomagnetic data suggest that the major counterclockwise rotations are not found near the Arica bend but are closely related to the Abancay deflection (Arriagada et al. 2008; Roperch et al. 2011; Fig. 3).

The Abancay deflection is one of the most striking structural features in Southern Peru, which is clearly marked in the present-day topography (Fig. 1). There are several lines of geological evidence that indicate that the Abancay deflection corresponds to a long-lived paleogeographic structure, which has apparently been a major heterogeneity in the Peruvian Andes since the Paleozoic (see Sempere et al. 2002). Some of the main structures can be extended toward the coast separating the different lithospheric blocks, such as the Paracas and Arequipa terranes. The boundaries of these lithospheric blocks are interpreted as weakness zones along which more recent deformations could have been accommodated (Carlotto et al. 2008). The formation of the Abancay deflection and associated counterclockwise rotations (Fig. 2) predate the subduction of the Nazca ridge at these latitudes (Roperch et al. 2011) and may have resulted from processes associated with extensive crustal deformation in the eastern Altiplano and Eastern Cordillera (Fig. 1) coincident with a transition from normal to flat-slab subduction in late Eocene/early Oligocene times (James and Sacks 1999).

The Paracas terrane (Fig. 3) was interpreted either as a parautochthonous terrane, alike the Arequipa–Antofalla (Ramos and Basei 1997), or as an exotic terrane (see Chap. "The Famatinian Orogen Along the Protomargin of Western Gondwana: Evidence for a Nearly Continuous Ordovician Magmatic Arc Between Venezuela and Argentina"). The Paracas terrane collided against western Gondwana in Middle to Late Ordovician times associated with the Marañon orogeny during the

Famatinian cycle. Mamaní et al. (2007) indicate different Pb-isotopic composition in volcanos of the arc front north of the Arequipa domain delimiting the dimensions of this block. South of the Abancay deflection (Fig. 3), along the coastal region of Southern Peru and northern Chile, the Arequipa massif is exposed (Fig. 3). The southern boundary of the Arequipa–Antofalla terrane is located near the Arica bend. To the south between the Arica bend and the Antofagasta–Calama Lineament ACL, within the Antofalla Massif, rotations are not relevant.

However, the more significant clockwise rotations are found between the Antofagasta and Vallenar region within a N-trending block which is bounded by the NE ACL and the Vallenar Orocline (Fig. 3). Available information strongly suggests that the ACL corresponds to a major NE-trending strike-slip structure zone oblique to the orogen that was active during the Paleogene (Arriagada et al. 2003). The ACL seems to have controlled the emplacement of the 34–37 Ma porphyry copper deposits as well as rotated and displaced a late Eocene (41–44 Ma) stripe of the deposits (Palacios et al. 2007).

While the distribution of tectonic rotations of Peru is largely correlated to the series of proposed Paleozoic terranes, to the south the nature of the basement beneath the present-day forearc of northern Chile is not known, which precludes a direct correlation with the determined rotation pattern (Fig. 3). Preliminary paleomagnetic results show no significant rotations between the Arica bend and the ACL. This domain could be located within the Antofalla terrane (Fig. 3). However, large rotations observed between the ACL and the Vallenar Orocline could be placed both on Antofalla, Chilenia and even a part of the Cuyania terranes (Fig. 3).

5 Discussion

Differential Eocene–Oligocene tectonic shortening in the Eastern Cordillera was responsible for most of the bending of the Andean margin, the formation of the Bolivian Orocline and most of the block rotations associated to the Central Andean Rotation Pattern (CARP). However, the total rotation cannot be explained only by shortening in the Eastern Cordillera and needs to be linked to tectonic processes occurring in the forearc. We propose that an important component of the tectonic rotation along the forearc was accommodated by ductile deformation along two major conjugate oblique shear zones, i.e., the Abancay deflection and Antofagasta–Calama Lineament.

The Abancay deflection and Antoagasta–Calama Lineament may be related to major inherited lithospheric boundaries probably associated with the crustal structure determined from the accretion of basement terranes, which could be responsible for producing and delineating significant and abrupt changes in the magnitude of the CARP along the margin. Paleomagnetic rotations could be used as a tool to determine the boundaries between Antofalla, Cuyania, and Chilenia terranes in northern Chile whose boundaries still remain highly speculative.

References

Allmendinger RW, Jordan TE, Kay SM, Isacks BL (1997) The evolution of the Altiplano–Puna plateau of the Central Andes. Annu Rev Earth Planetary Sci 25:139–174

Arriagada C, Roperch P, Mpodozis C (2000) Clockwise block rotations along the eastern border of the Cordillera de Domeyko, northern Chile (22°45′–23°30′S). Tectonophysics 326:153–171

Arriagada C, Roperch P, Mpodozis C, Dupont-Nivet G, Cobbold PR, Chauvin A, Cortés J (2003) Paleogene clockwise tectonic rotations in the forearc of Central Andes, Antofagasta region, northern Chile. J Geophys Res 108(B1):2032. https://doi.org/10.1029/2001JB001598

Arriagada C, Roperch P, Mpodozis C, Fernandez R (2006) Paleomagnetism and tectonics of the southern Atacama Desert (25°–28° S) Northern Chile. Tectonics 25 (TC4001). https://doi.org/10.1029/2005TC001923

Arriagada C, Roperch P, Mpodozis C, Cobbold P (2008) Paleogene building of the Bolivian Orocline: Tectonic restoration of the central Andes in 2-D map view. Tectonics 27

Arriagada C, Ferrando R, Córdova L, Morata D, Roperch P (2013) The MaipoOrocline: A first scale structural feature in the Miocene to Recent geodynamic evolution in the central Chilean Andes. Andean Geology 40(3):419–437

Bascuñán S, Arriagada C, Le Roux J, Deckart K (2016) Unraveling the Peruvian Phase of the Central Andes: stratigraphy, sedimentology and geochronology of the Salar de Atacama Basin (22°30–23° S), northern Chile. Basin Res. https://doi.org/10.1111/bre.12114

Beck M (2004) The Central Andean rotation pattern: Another look. Geophys J Int 157

Capitanio FA, Faccenna C, Zlotnik S, Stegman DR (2011) Subduction dynamics and the origin of the Andean orogeny and the Bolivian orocline. Nature 480:83–86. https://doi.org/10.1038/nature10596

Carlotto V, Cárdenas J, Carlier G (2008) The lithosphere of Southern Peru: A result of the accretion of allochthonous blocks during the Mesoproterozoic. 7th International Symposium on Andean Geodynamics (ISAG), Extended Abstracts: 105–108

Carrapa B, DeCelles PG (2015) Regional exhumation and kinematic history of the central Andes in response to cyclical orogenic processes. In: DeCelles PG, Ducea MN, Carrapa B, Kapp (eds) Geol Soc Am Mem 212. https://doi.org/10.1130/2015.1212(11)

Eichelberger N, McQuarrie N (2015) Kinematic reconstruction of the Bolivian orocline. Geosphere 11(2):445–462

Elger K, Oncken O, Glodny J (2005) Plateau-style accumulation of deformation: Southern Altiplano. Tectonics 24 (TC4020). https://doi.org/10.1029/2004TC001675

Farías M, Charrier R, Carretier S, Martinod J, Fock A, Campbell D, Cáceres J, Comte D (2008). Late Miocene high and rapid surface uplift and its erosional response in the Andes of Central Chile (33°–35° S). Tectonics 27 (TC1005). https://doi.org/10.1029/2006TC002046

Garzione CN, Hoke GD, Libarkin JC, Withers S, MacFadden B, Eiler J, Ghosh P, Mulch A (2008) Rise of the Andes. Science 320:1304–1307. https://doi.org/10.1126/science.1148615

Götze H-J and the Migra Group (1996) Group updates the gravity data base in the central Andes (20–29° S). EOS Trans AGU (Available as http://www.agu.org/eos_elec/95189e.html)

Horton BK (2005) Revised deformation history of the central Andes: Inferences from Cenozoic foredeep and intermontane basins of the Eastern Cordillera, Bolivia. Tectonics 24

Isacks B (1988) Uplift of the Central Andean plateau and bending of the Bolivian orocline. J Geophys Res 93:3211–3231

James DE, Sacks S (1999) Cenozoic formation of the Central Andes: A geophysical perspective. S Soc Ec Geol, Special Publication 7: 1–25

Kley J (1999) Geologic and geometric constraints on a kinematic model of the Bolivian orocline. J South Am Earth Sci 12:221–235

Mamaní M, Tassara A, Wörner G (2007) Crustal domains in the Central Andes and their control on orogenic structures. 20th Colloquium Latin American Earth Sciences. Kiel, Germany: 27–28

Martinod J, Husson L, Roperch P, Guillaume B, Espurt N (2010) Horizontal subduction zones, convergence velocity and the building of the Andes. Earth Planet Sci Lett 299(2010):299–309

McQuarrie N (2002) Initial plate geometry, shortening variations and evolution of the Bolivian orocline. Geology 30(10):867–870
McQuarrie N, Horton B, Zandt G, Beck S, DeCelles P (2005) Lithospheric evolution of the Andean fold-thrust belt, Bolivia, and the origin of the Central Andean plateau. Tectonophysics 399:15–37
Müller JP, Kley J, Jacobshagen V (2002) Structure and cenozoic kinematics of the eastern Cordillera, southern Bolivia (21° S). Tectonics 21 (1037). https://doi.org/10.1029/2001TC001340
O'Driscoll LJ, Richards MA, Humphreys ED (2012) Nazca–South America interactions and the late Eocene–late Oligocene flat-slab episode in the central Andes. Tectonics 31 (TC2013)
Palacios C, Ramirez LE, Townley B, Solari M, Guerra N (2007) The role of the Antofagasta–Calama Lineament in ore deposit deformation in the Andes of northern Chile. Min Dep 42 (3):301–308
Pardo-Casas F, Molnar P (1987) Relative motions of the Nazca (Farallon) and South American plates since late Cretaceous time. Tectonics 6:233–248
Ramos VA, Basei MA (1997) Gondwanan, Perigondwanan, and exotic terranes of southern South America. In: South American Symposium on Isotope Geology, Extended Abstracts: 250–252
Roperch P, Carlotto V, Ruffet G, Fornari M (2011) Tectonic rotations and transcurrent deformation south of the Abancay deflection in the Andes of southern Peru. Tectonics 30 (TC2010)
Sdrolias M, Müller RD (2006) Controls on back-arc basin formation. Geochem Geophys Geosyst 7 (Q04016). https://doi.org/10.1029/2005GC001090
Sempere T, Carlier G, Soler P, Fornari M, Carlotto V, Jacay J, Arispe O, Neraudeau D, Cardenas J (2002) Late Permian—Middle Jurassic lithospheric thinning in Peru and Bolivia, and its bearing on Andean-age tectonics. Tectonophysics 345:153–181
Somoza R, Singer S, Tomlinson A (1999) Paleomagnetic study of upper Miocene rocks from northern Chile: Implications for the origin of late Miocene-Recent tectonic rotations in the southern Central Andes. J Geophys Res 104(B10):22923–22936
Somoza R, Tomlinson AJ, Zaffarana CB, Singer SE, Puigdomenech Negre CG, Raposo MI, Dilles JH (2015) Tectonic rotations and internal structure of Eocene plutons in Chuquicamata, northern Chile. Tectonophysics 654(18):113–130
Tassara A, Götze HJ, Schmidt S, Hackney R (2006) Three-dimensional density model of the Nazca plate and the Andean continental margin. J Geophys Res 111 (B09404)
Yáñez G, Ranero C, von Huene R, Díaz (2001) Magnetic anomaly interpretation across the southern central Andes (32°–34° S): The role of the Juan Fernández Ridge in the late tertiary evolution of the margin. J Geophys Res 106: 6325–6347

Paleogene Arc-Related Volcanism in the Southern Central Andes and North Patagonia (39°–41° S)

Sofía B. Iannelli, Lucía Fernández Paz, Vanesa D. Litvak, Rosemary E. Jones, Miguel E. Ramos, Andrés Folguera and Victor A. Ramos

Abstract The influence of tectonic processes in evolution of magmatic suites evaluated through their geochemical signature has always been a subject of debate. Late Paleocene arc volcanism in the Southern Central Andes, particularly in North Patagonia, can be used to infer a direct relationship between magmatic episodes and tectonic changes along the Andean margin. Eocene arc-related volcanism ($\sim$44 Ma) in the North Patagonian Andes shows evidence for limited influence of the subducting slab on the composition of arc magmas and they exhibit an alkaline tendency. By Oligocene times ($\sim$29 Ma), arc volcanic sequences in the Auca Pan depocenter show predominantly arc-like geochemical signatures and have been derived from a calc-alkaline source. However, a comparison with younger arc sequences (<28 Ma) in the region suggests that the magmatic source turned more tholeiitic in composition with a remarkable increase in the influence of slab-derived fluids, as seen in volcanic rocks from Cura Mallín and Abanico retro and intra-arc basins. It is proposed that the marked geochemical variations between these magmatic periods are related to the

S. B. Iannelli (✉) · L. Fernández Paz · V. D. Litvak · M. E. Ramos · A. Folguera
Instituto de Estudios Andinos (IDEAN), Consejo Nacional de Investigaciones Científicas y Técnicas (CONICET), Universidad de Buenos Aires, Buenos Aires, Argentina
e-mail: sofia.iannelli@hotmail.com.ar

V. D. Litvak
e-mail: vane_lit@yahoo.com.ar

R. E. Jones
Department of Earth Sciences, University of Oxford, South Parks Road, Oxford OX1 3AN, UK

V. A. Ramos
Instituto de Estudios Andinos "Don Pablo Groeber", Departamento de Ciencias Geológicas, FCEN, Universidad de Buenos Aires–CONICET, Buenos Aires, Argentina

A. Folguera et al. (eds.), *The Evolution of the Chilean-Argentinean Andes*, Springer Earth System Sciences, https://doi.org/10.1007/978-3-319-67774-3_14

tectonic changes associated with the breakup of the Farallon plate at ~28–26 Ma. The geochemical data from Eocene and Oligocene volcanic sequences provide further evidence for the strong link between tectonics and magmatism.

Keywords Arc-related rocks · Geochemical signature · Eocene Oligocene · Farallon plate

1 Introduction

The North Patagonian Andes are characterized by different magmatic episodes directly connected with changes in tectonic parameters. In particular, these parameters are associated with three contractional stages in the Late Cretaceous, Paleocene-early Eocene and middle to late Miocene (Cobbold et al. 1999; Charrier et al. 2007; Folguera and Ramos 2011; Navarrete et al. 2015; Gianni et al. 2015). These stages were separated by a widespread extensional stage during late Oligocene-early Miocene that extended from the arc into the retroarc zone (Encinas et al. 2015; Jordan et al. 2001; Muñoz et al. 2000; Orts et al. 2012). Thus, arc volcanism was strongly controlled by changes in deformational style.

During Paleocene to Miocene times, the North Patagonian Andes were affected by voluminous magmatic episodes, with a particular distribution in two different belts, the Pilcaniyeu and El Maitén Belts (Fig. 1) (Kay and Rapela 1987; Rapela et al. 1984, 1988). These belts show a clear north–south orientation, although they converge in the northernmost part of Patagonian Andes (Fig. 1). These volcanic episodes occurred partially diachronously with contrasting characteristics (e.g. Aragón et al. 2011, 2013; Bechis et al. 2014; Iannelli et al. 2017; Kay and Rapela 1987; Rapela et al. 1988; Wilf et al. 2010). The Pilcaniyeu Belt, with an age span of 60–42 Ma, is primarily located in the foreland zone and characterized by bimodal composition. The El Maitén Belt developed further west between 32 and 19 Ma and is basaltic to andesitic in composition (Aragón et al. 2011; Kay and Rapela 1987; Rapela et al. 1984, 1988).

The main focus of this work is to show the main contrasting petrological and geochronological features of these volcanic units where the belts converge in North Patagonia (Figs. 1 and 2). Furthermore, we present a regional comparison with spatial and temporal associated arc-related volcanism, to provide a more comprehensive review of the magmatic evolution of the North Patagonian Andes during the Paleogene. The primary studies on the lithological, stratigraphical and compositional features of the El Maitén and Pilcaniyeu Belts correspond to Turner (1973), González Bonorino and González Bonorino (1978), González Díaz (1979), Rapela et al. (1983, 1984, 1988), Kay and Rapela (1987). More recent work on the easternmost Pilcaniyeu magmatism is presented by Kay et al. (2007), Aragón et al. (2011, 2013); while stratigraphical, lithological and geochemical data on the westernmost sequences is detailed in recent studies by Bechis et al. (2014), Ramos et al. (2014), Litvak et al. (2014, 2016), Iannelli et al. (2016a, b, 2017), Fernández Paz et al. (2016a, b).

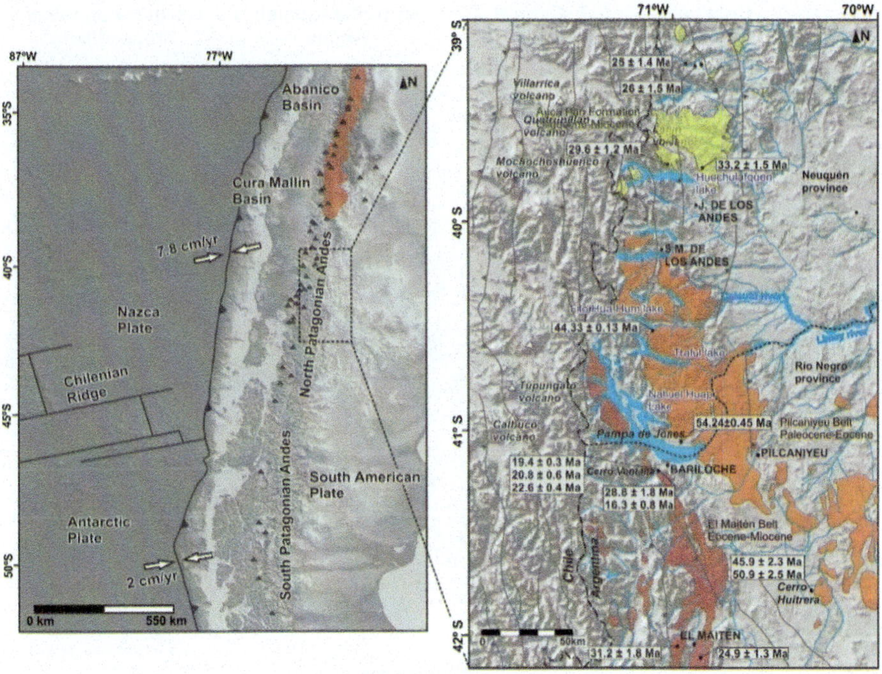

Fig. 1 Maps showing the tectonic setting of the North Patagonian Andes and the specific locations of the Paleogene to Neogene volcanic units. The present volcanic arc is indicated by triangles (López Escobar et al. 1991; Hickey et al. 1986; Hildreth and Morbath 1988; Stern 1991). Data on present-day plate convergence is taken from Pardo Casas and Molnar (1987) and Somoza (1998). Ages presented for the Cenozoic volcanism (represented by the Pilcaniyeu and El Maitén Magmatic Belts) are taken from Franzese et al. (2011), Ramos et al. (2014), Rapela et al. (1988), Iannelli et al. (2017), Wilf et al. (2010), Bechis et al. (2014), Mazonni et al. (1991). Based on Iannelli et al. (2017)

2 Geological and Tectonic Setting of Paleogene Volcanism in the North Patagonian Andes

The Andes at the studied latitudes are subdivided into a series of N-S morphostructural units (Fig. 2) (Orts et al. 2015; García Morabito and Ramos 2012; Giacosa et al. 2005). The westernmost unit is the Coastal Cordillera, which is formed of Permian–Triassic metamorphic rocks and segments of the Carboniferous batholith (Munizaga et al. 1988; Thomson and Hervé 2002) (Fig. 2). Immediately to the east, a continuous stretch of Paleogene to Neogene forearc depocenters define the Central Depression (Lavenu and Cembrano 1999) (Fig. 2). To the east of this, a series of east-vergent thrusts, which affect Paleozoic and Cenozoic units, are developed along the North Patagonian Andes (Figs. 1 and 2) (Ramos et al. 2014). Cenozoic units are composed of widespread volcanic and sedimentary sequences (Cucchi and Leanza 2005; Turner 1973; Franzese et al. 2011), which lie

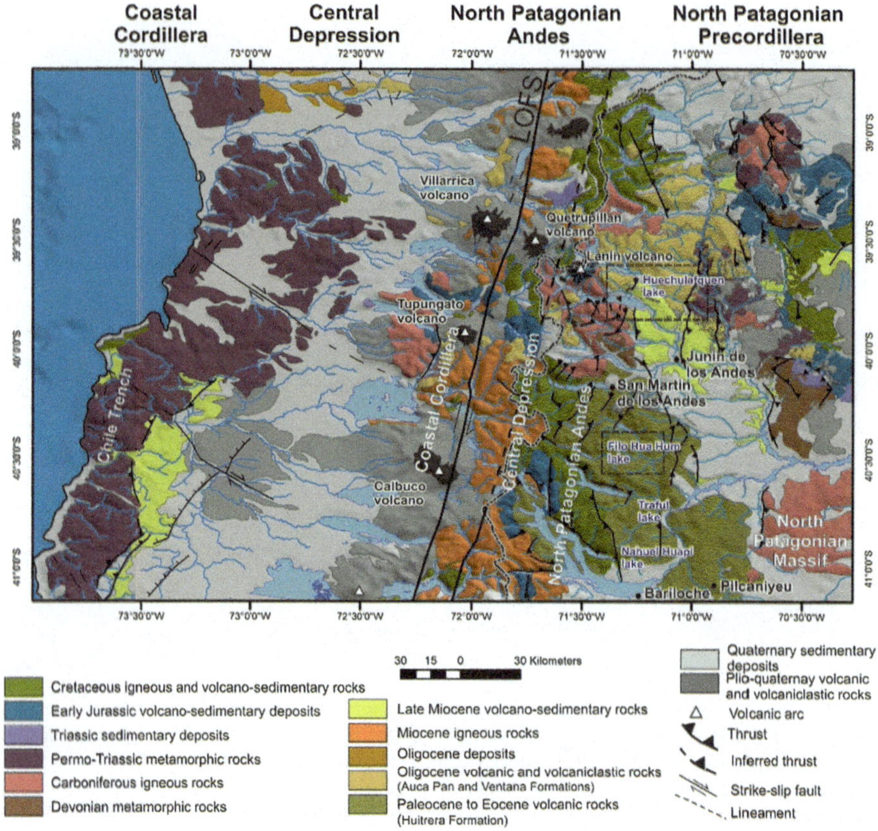

Fig. 2 Regional map of the area where the Pilcaniyeu and El Maitén Belts converge and the surroundings, showing the main morphostructural units and the outcropping stratigraphic formations (modified from Ramos et al. 2014). Black empty rectangles show the main areas where typical Eocene and Oligocene volcanic sequences outcrop near Filo Hua Hum and Huechulafquen Lakes, respectively

uncomformably over late Paleozoic intrusives and metamorphic rocks of the Colohuincul Formation (Turner 1973), and are locally covered by Quaternary retroarc basalts (Cucchi and Leanza 2005; Vattuone and Latorre 1998). Cenozoic volcanic sequences in the North Patagonian Andes are mainly exposed in two N-S belts between 39° and 40° S and have been included in the Auca Pan and Ventana Formations within El Maitén Belt, and Huitrera Formation, as part of Pilcaniyeu Belt (Bechis et al. 2014; González Bonorino 1979; Ramos 1982; Rapela et al. 1984, 1988; Turner 1973).

The Pilcaniyeu Belt is mainly composed of bimodal lava flows interbedded with silicic pyroclastic rocks, which have calc-alkaline K-rich compositions in the westernmost sequences within the belt (Rapela et al. 1984, 1988). The easternmost sequences consist of voluminous, siliceous calc-alkaline magmatism, which grades from alkaline to tholeiitic composition, but with a marked within-plate signature (Aragón et al. 2011, 2013). The age of the Pilcaniyeu Belt was constrained between the late Paleocene to Eocene, based on obtained whole-rock K–Ar ages ranging between 60 and 52 Ma (González Díaz 1979; Rapela et al. 1983, 1988) and Ar–Ar ages of 54.24 ± 0.45, 50.9 ± 2.5 and 45.9 ± 2.3 Ma (Mazzoni et al. 1991; Wilf et al. 2010) (Fig. 1). Development of this magmatism coincides with the oblique convergence of the Farallon plate with the South American plate (Rapela et al. 1988), and with the previous subduction of a spreading center around 42° S (Aluk-Farallon ridge; Cande and Leslie 1986).

The El Maitén Belt is composed of late Oligocene to Miocene volcanic rocks, which are primarily basaltic to andesitic in composition with mostly arc-like geochemical signatures (Fig. 1) (Kay and Rapela 1987; Rapela et al. 1984, 1988; Litvak et al. 2014; Iannelli et al. 2015, 2016a). The age of El Maitén Beltis is defined by a K–Ar ages ranging between 34 and 25 Ma (Cazau et al. 1989; Rapela et al. 1988), and younger U–Pb ages ranging between 23 and 19 Ma, which have been reported for interbedded tuffs within sequences cropping out in the type locality of the Ventana Formation (Bechis et al. 2014).

Both magmatic belts converge in the northern extreme of North Patagonia. Immediately to the north, late Oligocene to Miocene volcanism correlated with El Maitén Belt is represented by the Auca Pan Formation (Turner 1973) (Figs. 1 and 2). This volcanic unit is primarily composed of andesitic to basaltic rocks and locally occurring rhyolitic to dacitic sequences (Dalla Salda et al. 1981). Ages were mainly obtained in the surroundings of Huechulafquen Lake (K-Ar of ~33 Ma, Turner 1973), near Junín de los Andes (33.2 ± 1.5 Ma; Rapela et al. 1988), and near Aluminé city (25 ± 1.4 and 2 ± 1.5 Ma; Franzese et al. 2011) (Fig. 1). Recently, Ramos et al. (2014) presented a K-Ar age of 29.6 ± 1.2 Ma for a basaltic rock located on the northwest of Huechulafquen Lake (Fig. 1).

El Maitén Belt magmatism is associated with an arc-like setting (Fernández Paz et al. 2016a, b; Kay and Rapela 1987; Litvak et al. 2014, 2016; Rapela et al. 1984, 1988). The breakup of the Farallon plate at ~28–26 Ma (Cande and Leslie 1986; Somoza 1998) promoted a widespread extensional regime (Encinas et al. 2015; Franzese et al. 2011; Muñoz et al. 2000) that influenced the development of the arc-related volcanism associated with the El Maitén Belt (Kay et al. 2007; Litvak et al. 2014, 2016; Ramos et al. 2014). Further north of 40ºS, arc-related volcanism was also emplaced during this extensional regime, with volcanic and volcaniclastic deposits that comprise the infill of the Cura Mallín and Abanico basins (Fig. 1) (Folguera et al. 2010; Jordan et al. 2001; Kay et al. 2005, 2006; Radic et al. 2002; Rojas Vera et al. 2014).

The relatively orthogonal (N78° E) convergence between the Nazca and South American plates has prevailed since ~26 Ma (Fig. 1) (Pardo Casas and Molnar 1987; Somoza 1998). As a consequence of this tectonic configuration, an active volcanic arc became established over the axial part of the North Patagonian Andes. The current volcanic activity is represented by the volcanoes of Villarrica, Quetrupillán, Lanín, Calbuco and Osorno and minor monogenetic eruptive fields between ~39º–40° S (Deruelle 1982; López-Escobar et al. 1992, 1995; Lavenu and Cembrano 1999; Lara 2004) (Fig. 1).

3 Geology and Geochronology of the Eocene and Oligocene Volcanism Between 39° and 41° S

Eocene and Oligocene volcanism corresponding to the Pilcaniyeu and El Maitén Belts are exposed in the northernmost extreme of both belts: in the surroundings of the Filo Hua Hum Lake (40°30′S) and Huechulafquen Lake (39°45′S) (Figs. 1 and 2) representing contrasting features typical of these magmatic episodes.

The Eocene magmatism (Huitrera Formation, Fig. 2) is mainly composed of basaltic to rhyolitic lava flows interbedded with pyroclastic flows, with reported thicknesses of 150–300 m (Iannelli et al. 2017). The lava facies grade from rhyolites through basaltic to andesitic rocks. Basic-to-intermediate volcanic rocks show porphyritic textures with fresh plagioclase and partially altered mafic phases in an intersertal groundmass. In contrast, the rhyolitic facies contain plagioclase and amphibole phenocrysts in an alkaline feldspar, quartz-rich groundmass. Interbedded pyroclastic flows are composed of vitreous tuffs with subordinate feldspar and biotite crystal fragments. These pyroclastic facies grade between lithic tuffs and lapillis along typical stratigraphic profiles, with subordinate plagioclase and quartz crystal fragments in a vitreous matrix. Clast-supported conglomerate layers also appear interbedded with the volcanic facies and are composed of volcanic and pyroclastic clasts. The age of the sequence is constrained by an Ar–Ar age obtained from a basaltic lava flow, in the vicinity of Filo Hua Hum Lake, of 44.33 ± 0.13 Ma (Iannelli et al. 2017).

The Oligocene volcanic rocks (Auca Pan Formation, Fig. 2) are characterized by porphyritic lava facies interbedded with volcanic breccias and pyroclastic flows. The lava facies are mainly basalts and andesites composed of fresh plagioclase and clinopyroxene phenocrysts, and altered olivines, in an intersertal to intergranular groundmass. These facies are interbedded with volcanic breccias, which are comprised of basaltic lithic fragments, and are interpreted to be autoclastic lava deposits. Pyroclastic facies are composed of vitreous tuffs with predominant pumice fragments together with scarce biotite within a vitreous matrix (Ramos et al. 2014; Iannelli et al. 2017). The age of the section is constrained by the K–Ar age of 29.6 ± 1.2 Ma, which was obtained from a basaltic lava flow exposed on the northwest of Huechulafquen Lake (Ramos et al. 2014).

4 Geochemistry of Eocene and Oligocene Arc-Related Volcanism Between 39° and 41° S: Variable Slab Influence and Magmatic Source

According to major element geochemistry (Iannelli et al. 2017), the Eocene magmatism in North Patagonia (40°30′S) shows a continuous trend from basaltic to rhyolitic compositions (SiO_2: 53.97–75.17 wt.%), while the Oligocene volcanic rocks display a more restricted range in silica content (47.19–60.53 wt.% SiO_2), between subalkaline basaltic to andesitic compositions (Fig. 3a). Despite the disperse behavior of the Eocene magmatism, a more alkaline-like tendency is shown relative to the subalkaline trend of the Oligocene volcanism (Fig. 3a).

An increasingly arc-like geochemical signature is observed between the Eocene and the Oligocene. For example, the younger magmas have a La/Ta ratio >25, in contrast to the Eocene magmas, whose values primarily plot in the intraplate field (Fig. 3b). This is consistent with the typical arc-like anomalies shown in normalized multi-element diagrams, which are more pronounced for the Oligocene magmas (Iannelli et al. 2017). Despite these differences, no variation is observed in Ba/La ratios, possibly implying limited influence of slab-derived fluids during both magmatic periods (Eocene ratios: 7.54–26.79; Oligocene ratios: 13.38–20.76) (Fig. 3c). However, other trace elements ratios suggest relative increase in slab-derived fluids affecting the magma source region from the Eocene to Oligocene. For example, Ba/Nb (fluid mobile/fluid immobile) ratios are higher in Oligocene volcanic rocks (Ba/Nb: 38–127) than Eocene rocks (Ba/Nb: 13–66) (Fig. 3d). Thus, both the Eocene and Oligocene magmatic units are interpreted as arc-related retroarc sequences, with variable slab input, being more pronounced in the younger units.

Crustal contamination is considered not to be an important process in the evolution of the magmas, as suggested by Ba/La versus Th/La ratios. According to this, Eocene samples show low Ba/La and Th/La ratios, with no clear trend to crustal contamination (Fig. 3c). Instead, the Oligocene volcanic sequence shows a typical differentiation trend (Fig. 3c). Increasing melting degrees toward younger magmatism could be inferred due to decreasing Nb/Zr (Fig. 3d), as well as Ce/Pb and Nb/Yb ratios from Eocene to Oligocene sequences (Iannelli et al. 2017). Geochemical variations with time are also seen in Th/Hf versus Ta/Hf ratios, which are indicative of calc-alkaline versus enriched mantle sources. Higher Th/Hf values (0.37–3.53) and lower Ta/Hf ratios (0.05–0.17) for the Oligocene respect to the Eocene sequence (Th/Hf: 0.77–3.17; Ta/Hf: 0.17–0.41) suggest a more calc-alkaline mantle source in younger magmatism (Fig. 3e).

REE composition for both magmatic periods shows similar behavior, as demonstrated by La/Sm versus Sm/Yb ratios (Fig. 3f; Ramos et al. 2014; Iannelli et al. 2017), where Sm/Yb values lower than 2.5 would imply magmas equilibrating within a relatively normal crust ($\sim$30–40 km) (e.g., Kay and Kay 1991). Moreover, Sm/Yb together with Dy/Yb, and the slight depletion in intermediate REEs displayed in chondrite normalized plots, suggests magmatic equilibration under low-pressure conditions and the fractionation of residual pyroxene and amphibole (Ramos et al. 2014; Iannelli et al. 2017).

5 Paleogene to Neogene Regional Evolution of Arc-Magmatism in the North Patagonian Andes Constrained by Geochemical Features

Paleogene volcanism in the Southern Central Andes at North Patagonia (39–41° S) can be compared to spatially and temporally related sequences to the north and south in the Central Andes ($\sim$33°–42° S) (Fig. 1).

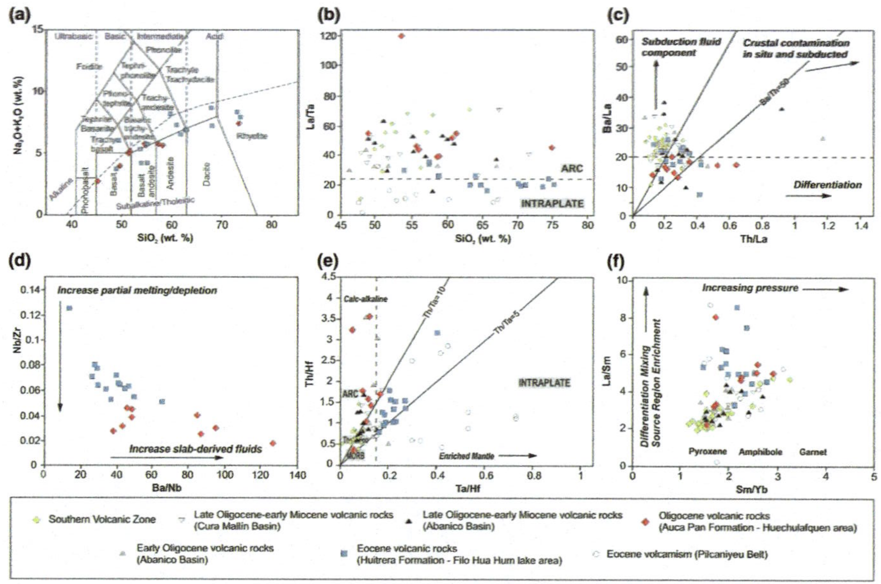

Fig. 3 **a** Major element classification (Total Alkali vs. Silica) of Eocene and Oligocene volcanic rocks (Le Maitre et al. 1989); **b** La/Ta ratios versus SiO_2 showing arc-like signatures in Oligocene rocks compared with intraplate-like signature of Eocene samples; **c** Ba/La versus Th/La suggests limited crustal contamination; **d** The lower Nb/Zr and higher Ba/Nb obtained for the Oligocene samples implies contributions from slab-derived fluids and higher degrees of partial melting, compared to the lesser degree of melting and more enriched signature obtained for the Eocene rocks; **e** A more calc-alkaline source characterized for Oligocene magmas on the basis of higher Th/Hf ratios relative to Eocene magmas, which are associated with a more enriched magmatic source. Data of Pilcaniyeu Belt is from Aragón et al. (2011), Iannelli et al. (2017) and Rapela et al. (1988); data from CuraMallín and Abanico basins are from Kay et al. (2006), Utgé et al. (2009) and Kay et al. (2005); data from Auca Pan depocenter is from Iannelli et al. (2017). Data from the present SVZ are from Deruelle (1982) and Lopez-Escobar et al. (1992, 1995)

Eocene magmatism (44 Ma) located in the western retroarc zone (Filo Hua Hum lake, 40º30′S) (Figs. 1 and 4a) is compared with the eastern and coeval magmatic sequences (~47–43 Ma), which are included as part of the Pilcaniyeu Belt (~60–42 Ma; Aragón et al. 2011; Rapela et al. 1988). As a result, ~W-E variations in Eocene magmatic activity are apparent along the belt (Figs. 1 and 4a).

During this time interval (47–43 Ma), an expected decrease in the arc-related signatures is observed from the western to the eastern retroarc zone. The decrease in La/Ta, Ba/La and Ba/Nb ratios from the western Eocene sequences in the High Andes toward the easternmost volcanic sequences shows the progression from arc-related signatures to a typical intraplate tendency further away from the Chile trench (Figs. 3b, c and 4a; Aragón et al. 2011; Iannelli et al. 2017; Rapela et al. 1988).

Ta/Hf ratios obtained for the western Eocene rocks compared to the eastern magmatism also suggest variable source compositions (Fig. 3e). The easternmost retroarc sequences are derived from an enriched source, typical of intraplate settings, while a more depleted source is evident for the western retroarc magmas, with scarce slab input (Fig. 4a). Higher Ce/Pb and lower Nb/Yb are also reported for the western units, implying an increase in slab fluid signal and higher melting degrees toward the Andean margin (Iannelli et al. 2017).

During Oligocene to early Miocene times, similar geochemical trends are defined between magmatism developed in Auca Pan depocenter (~29 Ma) and the Cura Mallín and Abanico basins along the Andean margin (~35–20 Ma; Iannelli et al. 2017; Kay et al. 2005, 2006; Muñoz et al. 2006; Ramos et al. 2014; Utgé et al. 2009)

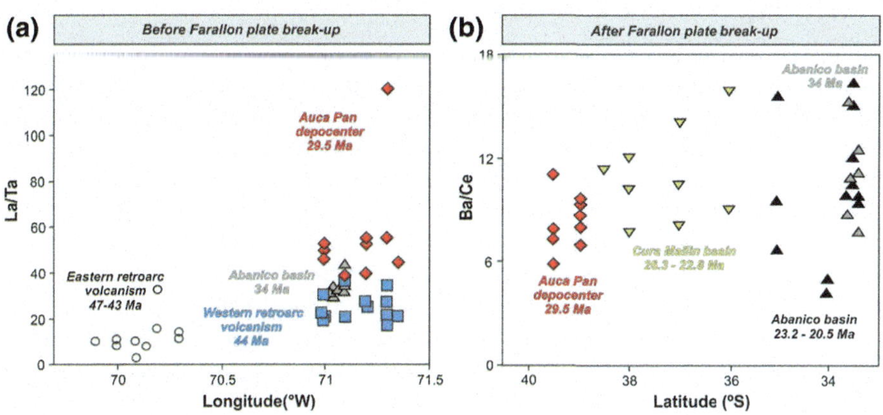

Fig. 4 **a** Increasing arc-like signatures in the High Andes are shown by progressively higher La/Ta from the Eocene to the Oligocene; **b** After the breakup of Farallon plate an increase in the influence of slab-derived fluids on the source of arc magmas is implied by the higher Ba/Ce ratios obtained for late Oligocene-early Miocene rocks compared to the volcanic rocks erupted just before (Auca Pan depocenter lavas). Geochemical data corresponds to Aragón et al. (2011), Iannelli et al. (2017), Kay et al. (2005, 2006), Muñoz et al. (2006), Utgé et al. (2009)

(Figs. 1 and 4b). All of them show arc-like signatures, as stated by La/Ta ratios over 25 and Ta/Hf ratios above 0.15 (Fig. 3b, e).

However, a more defined calc-alkaline source is seen for the oldest magmatism of this period, represented by Oligocene Auca Pan Formation (~29 Ma) and the lower sequences of Abanico basin (~34 Ma; Muñoz et al. 2006). The magmatic source seems to have evolved to a more tholeiitic composition with time, as suggested by the lower Th/Hf ratios obtained for the younger magmas of Cura Mallín and Abanico basins (~26–20 Ma; Kay et al. 2005, 2006) (Fig. 3e).

Increasing Ba/La and Ba/Ce ratios from the lower Abanico (~34 Ma) and Auca Pan (~29 Ma) Formations toward younger Abanico and Cura Mallín magmas (~26–20 Ma) would indicate an increase in the contribution from slab-derived fluids with time (Figs. 3c and 4b). The changes in slab-derived influence could be explained by variations in the composition of immobile elements from the subducted slab or related to a more anhydrous slab affecting older magmas (Goss et al. 2011).

Despite of the mentioned differences, no contrasting behavior is seen when considering Sm/Yb and La/Sm ratios for magmatic episodes during Oligocene to Miocene times. A garnet-free residual mineral assemblage is interpreted based on REE behavior, which could be associated with magmas equilibrating within a thin to normal crust (Fig. 3f; Burns et al. 2006; Iannelli et al. 2017; Jordan et al. 2001; Kay et al. 2006).

6 Arc-Related Magmatism and Its Implications on the Tectonic Evolution of Southern Central Andes During the Mid-Cenozoic

Eocene magmatism (~47–43 Ma) in North Patagonia was mainly developed in an intraplate setting in the easternmost foreland area, while minor arc-related magmas were developed to the west in retroarc basins (Aragón et al. 2011; Fernández Paz et al. 2016a, b; Iannelli et al. 2016b, 2017; Litvak et al. 2014, 2016; Rapela et al. 1984, 1988; Fig. 1). During Eocene times (~52 Ma), the ridge between the Aluk and Farrallon plates was subducted beneath the margin and convergence between the Farallon and South American plates changed to a less oblique angle (Cande and Leslie 1986). As a consequence, eastern intraplate magmas were associated with the development of a slab window (~47–43 Ma; Aragón et al. 2011; Muñoz et al. 2000) at ~42° S, while minor arc-like magmatism was developed to the west (~44 Ma; Iannelli et al. 2017) (Figs. 4a and 5a). This transition is clearly observed when comparing volcanic rocks from the eastern extreme of the Pilcaniyeu Belt (~47–43 Ma; Aragón et al. 2011) with those located in the former retroarc zone (~44 Ma). Eastern Eocene magmatism has been widely associated with a within-plate setting, with an enriched magmatic source and high degrees of decompression melting (Aragón et al. 2011); while further west, Eocene magmatism (~44 Ma) shows a slightly alkaline signature and limited influence of slab-derived fluids (Figs. 4a and 5a).

By late Oligocene times, a regional extensional regime prevailed in the North Patagonian Andes (Encinas et al. 2015; Jordan et al. 2001; Muñoz et al. 2000). This widespread extensional episode has been directly related to major plate reorganization in the southeast Pacific. An increase in convergence rates (from 6 cm/yr to 15 cm/yr) and a more orthogonal convergence occurred after ~28 Ma (Cande and Leslie 1986; Somoza 1998). These important changes in plate tectonics coincide with the breakup of the Farallon plate at ~28–26 Ma, and the starting orthogonal subduction regimen between the Nazca and the South America plates (Pardo Casas and Molnar 1987).

The extensional regime provoked the generation of intra-arc volcanic and sedimentary basins, such as the Auca Pan depocenter, and the Abanico and Cura Mallín basins to the north (Fig. 1) (Muñoz et al. 2000; Jordan et al. 2001; Muñoz et al. 2006; Ramos et al. 2014). Most of the volcanic record in these areas occurred after Farallon plate breakup, although the Auca Pan and Abanico basins do have volcanic exposures with older ages (Fig. 5b).

Marked geochemical differences are recognizable between volcanism developed before and after this major plate reorganization. Before Farallon plate breakup, the geochemistry of magmatic activity in the Abanico (~34 Ma; 33°30′S; Muñoz et al. 2006) and Auca Pan basins (~29 Ma; Ramos et al. 2014; Iannelli et al. 2016a, 2017) reflects the influence of slab-derived fluids and a calc-alkaline source. Volcanic rocks from the Auca Pan depocenter, which were emplaced in the stage immediately preceding Farallon plate breakup, show noticeable differences when comparing to volcanic deposits from immediately after. These latter magmatic episodes are exposed in the Cura Mallín basin (26.3 ± 1.5 to 22.8 ± 0.7 Ma; Jordan et al. 2001; Kay et al. 2006; Radic et al. 2002; Rojas Vera et al. 2010), and further north, in the younger units of the Abanico basin (~23–20 Ma; Hildreth and Morbath 1988; Jordan et al. 2001; Kay et al. 2005; Muñoz et al. 2006).

A more typical Andean-type magmatism was established as a result of the initiation of orthogonal convergence (Kay et al. 2006). Younger magmas from the Cura Mallín and Abanico basins show a change to a more tholeiitic magma source with an increase in slab fluid contribution, relative to the younger, early Oligocene, calc-alkaline magmatism from the retroarc Auca Pan depocenter (Iannelli et al. 2017; Kay et al. 2005; Ramos et al. 2014; Utgé et al. 2009) (Figs. 4b and 5c).

7 Conclusions

The major tectonic changes affecting the Andean margin during the Paleogene influenced the magmatic evolution of the margin and the petrogenesis of the resulting magmas, as evidenced by the variable geochemical compositions. During the Eocene (47–43 Ma; ~42° S) within-plate magmatism was developed in the

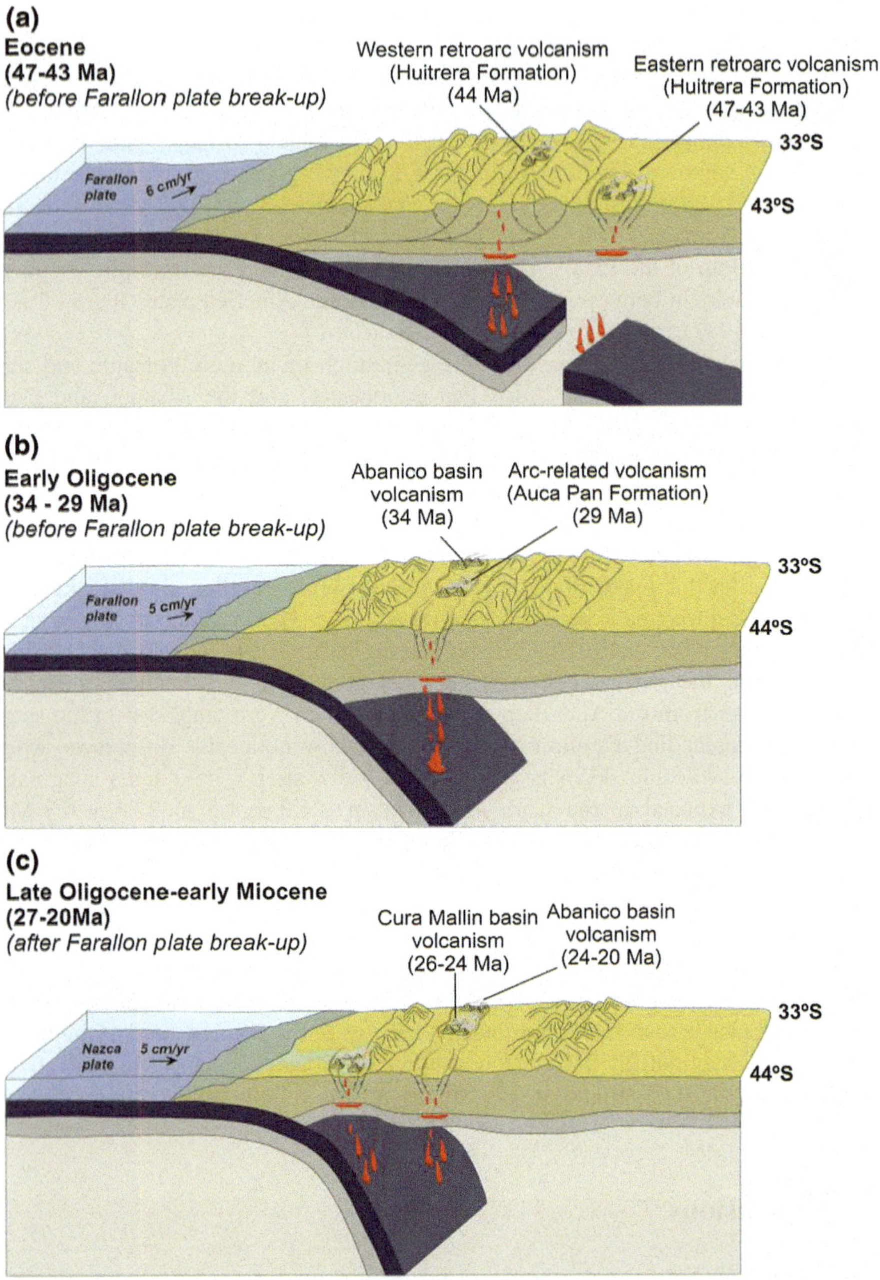
(a)
Eocene
(47-43 Ma)
(before Farallon plate break-up)
Western retroarc volcanism
(Huitrera Formation)
(44 Ma)
Eastern retroarc volcanism
(Huitrera Formation)
(47-43 Ma)
33ºS
43ºS
Farallon plate
6 cm/yr
(b)
Early Oligocene
(34 - 29 Ma)
(before Farallon plate break-up)
Abanico basin
volcanism
(34 Ma)
Arc-related volcanism
(Auca Pan Formation)
(29 Ma)
33ºS
44ºS
Farallon plate
5 cm/yr
(c)
Late Oligocene-early Miocene
(27-20Ma)
(after Farallon plate break-up)
Cura Mallin basin
volcanism
(26-24 Ma)
Abanico basin
volcanism
(24-20 Ma)
33ºS
44ºS
Nazca plate
5 cm/yr

◄**Fig. 5** Tectonic evolution of the North Patagonian Andes from Eocene (~47 Ma) to early Miocene times (~20 Ma). **a** Development of Pilcaniyeu magmatic belt during the subduction of Aluk/Farallon mid ocean ridge and the opening of a slab window. **b** Incipient extensional regime along the North Patagonian Andes, coeval with the emplacement of arc volcanism in the intra- and retroarc Auca Pan and older Abanico depocenters, immediately after Farallon plate breakup. **c** Subduction of the Nazca plate at a higher convergence rate and lower obliquity, provoking a westward arc retraction, crustal stretching, and formation of the Abanico and Cura Mallín basins (modified from Iannelli et al. 2017)

easternmost foreland, adjacent to the westernmost retroarc (44 Ma; 41°30′S), with alkaline-like magmas displaying limited slab-derived inputs.

By Oligocene times (34–29 Ma), a more calc-alkaline volcanism, with increased slab-derived influence, developed along the Abanico basin and the Auca Pan depocenter (39°45′S). Immediately after, a clear evolution to a more tholeiitic magmatic source, with a higher contribution from slab-derived fluids, is observed for the younger (26–20 Ma) arc-related rocks of the Cura Mallín and Abanico basins. Changes in the geochemical signatures of the Oligocene arc magmas correlate with, and appear to be controlled by, tectonic reconfigurations primarily related to the breakup of the Farallon plate. This provides evidence for a strong link between magma petrogenesis and changes in subduction zone parameters along the Andean margin.

References

Aragón E, D'Eramo F, Castro A, Pinotti L, Brunelli D, Rabbia O, Rivalenti G, Varela R, Spakman W, Demartis M, Cavarozzi CE, Aguilera YE, Mazzucchelli M, Ribot A (2011) Tectono-magmatic response to major convergence changes in the North Patagonian suprasubduction system; the Paleogene subduction–transcurrent plate margin transition. Tectonophysics 509:218–237

Aragón E, Pinotti L, D'Eramo F, Castro A, Rabbia O, Coniglio J, Demartis M, Hernando I, Cavarozzi CE, Aguilera YE (2013) The Farallon-Aluk ridge collision with South America: Implications for the geochemical changes of slab window magmas from fore to back-arc. Geosci Front 4:377–388

Bechis F, Encinas A, Concheyro A, Litvak VD, Aguirre Urreta B, Ramos VA (2014) New age constraints for the Cenozoic marine transgressions of northwestern Patagonia, Argentina (41°–43° S): paleogeographic and tectonic implications. J South Am Earth Sci 52:72–93

Burns WM, Jordan TE, Copeland P, Kelley SA (2006) The case for extensional tectonics in the Oligocene-Miocene Southern Andes as recorded in the Cura Mallín basin (36°–38° S). In: Kay SM, Ramos VA (eds) Evolution of an Andean margin: A tectonic and magmatic view from the Andes to the Neuquén Basin (35°–39° S lat), Geological Society of America, Special Paper 407, pp 163–184. https://doi.org/10.1130/2006.2407(08)

Cande SC, Leslie RB (1986) Late Cenozoic tectonics of the southern Chile trench. J Geophys Res: Solid Earth 91:471–496

Cazau L, Mancini D, Cangini J, Spalletti L (1989) Cuenca de Nirihuau. In: Chebli G, Sapalletti L (eds) Cuencas Sedimentarias Argentinas, Serie Correlación Geológica 6. Tucumán, Argentina, pp 299–318

Charrier R, Pinto L, Rodríguez M (2007) Tectonostratigraphic evolution of the Andean orogen in Chile. In: Moreno T, Gibbons W (eds) The geology of Chile. Geological Society, London, pp 21–114

Cobbold PR, Diraison M, Rossello EA (1999) Bitumen veins and Eocene transpression, Neuquén basin, Argentina. Tectonophysics 314:422–442

Cucchi R, Leanza H (2005) Hoja Geológica 3972-IV Junín de los Andes, provincia del Neuquén. Serv Geo Min Nac 357:1–102

Dalla Salla L, Leguizawjn M, Mazzoni M, Merodlo J, Rapela C, Spalletti L (1981) Características del vulcanismo paleógeno en la Cordillera Nordpatagónica entre las latitudes 39°30′ y 41°20′S. In: VIII° Congreso Geológico Argentino, San Luis, pp 629–657

Deruelle B (1982) Petrology of the Plio-Quaternary volcanism of the south-central and meridional Andes. J Volcanol Geoth Res 14:77–124

Encinas A, Folguera A, Oliveros V, Del Mauro LDG, Tapia F, Riffo R, Hervé F, Finger KL, Valencia V, Gianni G, Álvarez O (2015) Late Oligocene–early Miocene submarine volcanism and deep-marine sedimentation in an extensional basin of southern Chile: Implications for the tectonic development of the North Patagonian Andes. Geol Soc Am Bull. https://doi.org/10.1130/B31303.1

Fernández Paz L, Litvak VD, Echaurren A, Iannelli SB, Folguera A, Encinas A, Valencia V (2016a) Eocene magmatic evolution of North Patagonian Andes. In: I° Simposio de Tectónica Sudamericana, Santiago de Chile

Fernández Paz L, Iannelli SB, Litvak VD, Echaurren A, Folguera A (2016b) Petrogénesis del volcanismo de la Formación Ventana en el cordón Rivadavia, provincia de Chubut, Andes Nordpatagónicos. In: XII° Congreso de Mineralogía y Metalogénesis, Tucumán

Folguera A, Ramos VA (2011) Repeated eastward shifts of arc magmatism in the Southern Andes: A revision to the long-term pattern of Andean uplift and magmatism. J South Am Earth Sci 32:531–546

Folguera A, Rojas Vera E, Bottessi G, Zamora Valcarce G, Ramos VA (2010) The Loncopué Trough: a Cenozoic basin produced by extension in the southern Central Andes. J Geodyn 49:287–295

Franzese JR, D'Elia L, Bilmes A, Muravchik M, Hernández M (2011) Superposición de cuencas extensionales y contraccionales oligo-miocenas en el retroarco andino norpatagónico: la Cuenca de Aluminé, Neuquén, Argentina. Andean Geol 38:319–334

García Morabito E, Ramos VA (2012) Andean evolution of the Aluminé fold and thrust belt, Northern Patagonian Andes (38°30′–40°30′S). J South Am Earth Sci 38:13–30

Giacosa RE, Alfonso JC, Heredia CN, Paredes J (2005) Tertiary tectonics of the sub-Andean region of the North Patagonian Andes, southern central Andes of Argentina (41°–42°30′S). J South Am Earth Sci 20:157–170

Gianni G, Navarrete C, Orts D, Tobal J, Folguera A, Giménez M (2015) Patagonian broken foreland and related synorogenic rifting: the origin of the Chubut Group Basin. Tectonophysics 649:81–99

González Bonorino F (1979) Esquema de la evolución geológica de la Cordillera Norpatagónica. Rev Asoc Geol Argent 34:184–202

González Bonorino F, González Bonorino G (1978) Geología de la región de San Carlos de Bariloche. Rev Asoc Geol Argent 33:175–210

González Díaz E (1979) La edad de la Formación Ventana en el área al norte y al este del lago Nahuel Huapi. Rev Asoc Geol Argent 34:113–124

Goss AR, Kay SM, Mpodozis C (2011) The geochemistry of a dying continental arc: the Incapillo Caldera and Dome Complex of the southernmost Central Andean Volcanic Zone (28° S). Contrib Miner Petrol 161:101–128

Hickey RL, Frey FA, Gerlach DC, López-Escobar L (1986) Multiple sources for basaltic arc rocks from the southern volcanic zone of the Andes (34°41′S): trace element and isotopic evidence for contributions from subducted oceanic crust, mantle and continental crust. J Geophysical Research 91:5963–5983

Hildreth W, Morbath S (1988) Crustal contributions to arc magmatism in the Andes of central Chile. Contrib Miner Petrol 98:455–489

Iannelli S, Fernández Paz L, Ramos M, Litvak VD, Folguera A (2015) Caracterización de la Formación Auca Pan en los alrededores del lago Huechulafquen y cerro Auca Pan, provincia de Neuquén. III° Simposio de Petrología Ígnea y Metalogénesis Asociada. Actas, General Roca, pp 80–81

Iannelli SB, Fernández Paz L, Ramos M, Litvak VD, Folguera A (2016a) Geoquímica de la Formación Auca Pan en los alrededores del lago Huechulafquen, provincia de Neuquén. In: XII° Congreso de Mineralogía y Metalogénesis, Actas 159, Tucumán

Iannelli SB, Litvak VD, Fernández Paz L, Folguera A, Ramos ME (2016b) Late Paleogene arc-related volcanism in the North Patagonian Andes (39–41° S). In: I° Simposio de Tectónica Latinoamericana, Santiago de Chile

Iannelli SB, Litvak VD, Fernández Paz L, Folguera A, Ramos ME, Ramos V (2017) Evolution of Eocene to Oligocene arc-related volcanism in the North Patagonian Andes (39–41° S), prior to the break-up of the Farallon plate. Tectonophysics 696–697:70–87. https://doi.org/10.1016/j.tecto.2016.12.024

Jordan T, Burns W, Veiga R, Pángaro F, Copeland P, Kelley S, Mpodozis C (2001) Extension and basin formation in the Southern Andes caused by increased convergence rate: Amid-Cenozoic trigger for the Andes. Tectonics 20:308–324

Kay RW, Kay SM (1991) Creation and destruction of lower continental crust. Geol Rundsch 80:259–278

Kay SM, Rapela CW (1987) El volcanismo del Terciario inferior y medio en los Andes Norpatagónicos (40°–42°30′S): Fuente de origen de los magmas y su relación con variaciones en la oblicuidad de la zona de subducción. In: X° Congreso Geológico Argentino, Actas 4, Tucumán, pp 192–194

Kay SM, Godoy E, Kurtz A (2005) Episodic arc migration, crustal thickening, subduction erosion, and magmatism in the south-central Andes. Geol Soc Am 117:67–88

Kay SM, Burns M, Copeland P (2006) Upper Cretaceous to Holocene magmatism and evidence for transient Miocene shallowing of the Andean subduction zone under the northern Neuquén Basin. In: Kay SM, Ramos VA (eds) Evolution of an Andean margin: a tectonic and magmatic view from the Andes to the Neuquén Basin (35–39° S). Geological Society of America, Special Paper 407, pp 19–60

Kay SM, Ardolino AA, Gorring ML, Ramos VA (2007) The Somuncura large igneous province in Patagonia: interaction of a transient mantle thermal anomaly with a subducting slab. J Petrol 48:43–77

Lara LE (2004) Geología del Volcán Lanín, Región de La Araucanía. Servicio Nacional de Geología y Minería, Carta Geológica de Chile 87:18

Lavenu A, Cembrano J (1999) Compressional- and transpressional-stress pattern for Pliocene and Quaternary brittle deformation in fore arc and intra-arc zones (Andes of Central and Southern Chile). J Struct Geol 21:1669–1691

Le Maitre RW, Bateman P, Dudek A, Keller J, Lameyre J, Le Bas MJ, Sabine PA, Schmid R, Sorensen H, Streckeisen A, Woolley AR, Zanettin B (1989) A classification of igneous rocks and glossary of terms: recommendations of the international union of geological sciences subcommission on the systematics of igneous rocks. Blackwell Scientific Publications, Oxford

Litvak VD, Encinas A, Oliveros V, Bechis F, Folguera A, Ramos VA (2014) El volcanismo mioceno inferior vinculado a las ingresiones marinas en los Andes Nordpatagónicos. XIX° Congreso Geológico Argentino. Actas CD, Córdoba, pp 22–35

Litvak VD, Folguera A, Encinas A, Bechis F, Iannelli SB, Fernández Paz L, Echaurren A, Ramos V, Valencia V (2016) Middle to Late Cenozoic arc magmatism along the Patagonian Andean margin. In I° Simposio de Tectónica Latinoamericana, Santiago de Chile

López-Escobar L, Tagiri M, Vergara M (1991) Geochemical features of southern Andes Quaternary volcanic rocks between 41°50′ and 43°00′S. In: Harmon RS, Rapela CW (eds) Andean magmatism and its tectonic setting: Geological Society of America Special Paper 265, pp 45–56

López-Escobar L, Parada MA, Moreno H, Frey FA, Hickey-Vargas RL (1992) A contribution to the petrogenesis of Osorno and Calbuco volcanoes, Southern Andes (41°00′–41°30′S): comparative study. Andean Geol 19:211–226

López-Escobar L, Cembrano J, Moreno H (1995) Geochemistry and tectonics of the Chilean Southern Andes basaltic Quaternary volcanism (37–46° S). Andean Geol 22:219

Mazzoni MM, Kawashita K, Harrison S, Aragón E (1991) Edades radimétricas eocenas. Borde occidental del macizo Norpatagónico. Rev Asoc Geol Argent 46:150–158

Munizaga F, Hervé F, Drake R, Pankhurst RJ, Brook M, Snelling N (1988) Geochronology of the Lake Region of south-central Chile (39°42′S): preliminary results. J South Am Earth Sci 1:309–316

Muñoz J, Troncoso R, Duhart P, Crignola P, Farmer L, Stern CR (2000) The relation of the mid-Tertiary coastal magmatic belt in south-central Chile to the late Oligocene increase in plate convergence rate. Revista Geológica de Chile 27:177–203

Muñoz M, Fuentes F, Vergara M, Aguirre L, Olov Nyström J, Féraud G, Demant A (2006) Abanico East formation: petrology and geochemistry of volcanic rocks behind the Cenozoic arc front in the Andean Cordillera, central Chile (33°50′S). Revista Geológica de Chile 331:109–140

Navarrete CR, Gianni GM, Folguera A (2015) Tectonic inversion events in the western San Jorge Gulf Basin from seismic, borehole and field data. J South Am Earth Sci 64:486–497

Orts DL, Folguera A, Encinas A, Ramos M, Tobal J, Ramos VA (2012) Tectonic development of the North Patagonian Andes and their related Miocene foreland basin (41°30′–43° S). Tectonics 31:1–24

Orts DL, Folguera A, Giménez M, Ruiz F, Vera EAR, Klinger FL (2015) Cenozoic building and deformational processes in the North Patagonian Andes. J Geodyn 86:26–41

Pardo Casas F, Molnar P (1987) Relative motion of the Nazca (Farallon) and South American plates since late Cretaceous time. Tectonics 6:233–248

Radic JP, Rojas L, Carpinelli A, Zurita E (2002) Evolución tectónica de la cuenca terciaria de Cura-Mallín, región cordillerana chileno argentina (36°30′–39°00′S). In: XV° Congreso Geológico Argentino, Actas 3, Calafate, pp 233–237

Ramos VA (1982) Las ingresiones pacíficas del Terciario en el Norte de la Patagonia. III° Congreso Geológico Chileno. Actas I(A), Concepción, pp 262–288

Ramos ME, Folguera A, Fennell L, Giménez M, Litvak VD, Dzierma Y, Ramos VA (2014) Tectonic evolution of the North Patagonian Andes from field and gravity data (39–40° S). J South Am Earth Sci 51:59–75

Rapela CW, Spalletti LA, Merodio CJ (1983) Evolución magmática y geotectónica de la "Serie Andesítica" Andina (Paleoceno-Eoceno) en la Cordillera Nordpatagónica. Rev Asoc Geol Argent 38:469–484

Rapela CW, Spalletti L, Merodio J, Aragón E (1984) El vulcanismo paleoceno-eoceno de la provincia andino-patagónica. In: Ramos VA (ed) Geología y Recursos Naturales de la Provincia de Río Negro, Buenos Aires, pp 189–214

Rapela C, Spalletti L, Merodio J, Aragón E (1988) Temporal evolution and spatial variation of early Terciary volcanism in the Patagonian Andes (40° S–42°30′S). J South Am Earth Sci 1:75–88

Rojas Vera EA, Folguera A, Zamora Valcarce G, Giménez M, Ruiz F, Martínez P, Bottesi G, Ramos VA (2010) Neogene to Quaternary extensional reactivation of a fold and thrust belt: The Agrio belt in the Southern Central Andes and its relation to the Loncopué trough (38°–39° S). Tectonophysics 492:279–294

Rojas Vera EA, Selles D, Folguera A, Gímenez M, Ruíz F, Orts D, Zamora Valcarce G, Martínez P, Bechis F, Ramos VA (2014) The origin of the Loncopué Trough in the retroarc of the Southern Central Andes from field, geophysical and geochemical data. Tectonophysics 637:1–19

Somoza R (1998) Updated Nazca (Farallón)-South America relative motions during the last 40 My: implications for mountain building in the central Andean region. J South Am Earth Sci 11:211–215

Stern CR (1991) Role of subduction erosion in the generation of Andean magmas. Geology 19:78–81

Thomson SN, Hervé F (2002) New time constraints for the age of metamorphism at the ancestral Pacific Gondwana margin of southern Chile (42°52′S). Rev Geol Chile 29:1–16

Turner JCM (1973) Descripción geológica de la Hoja 37 a-b, Junín de los Andes, provincia del Neuquén. Serv Geo Min Nac, Boletín 138, Buenos Aires, pp 1–86

Utgé S, Folguera A, Litvak V, Ramos VA (2009) Geología del sector norte de la cuenca de Cura Mallín en las lagunas de Epulaufquen, Neuquén. Rev Asoc Geol Argent 64:231–248

Vattuone ME, Latorre CO (1998) Caracterización geoquímica y edad K/Ar de basaltos del Terciario superior de Aluminé, Neuquén. In: X° Congreso Latinoamericano de Geología, VI° Congreso Nacional de Geología Económica, Actas 2, Buenos Aires, pp 184–190

Wilf P, Singer BS, Zamaloa MC, Johnson KR, Cúneo NR (2010) Early Eocene 40Ar/39Ar age for the Pampa de Jones plant, frog, and insect biota (Huitrera Formation, Neuquén province, Patagonia, Argentina). Ameghiniana 47:207–216

Part V
The Late Andean Stages in the Chilean-Argentinean Margin

Mantle Influence on Andean and Pre-Andean Topography

Federico M. Dávila, Carolina Lithgow-Bertelloni, Federico Martina, Pilar Ávila, Julieta Nóbile, Gilda Collo, Miguel Ezpeleta, Horacio Canelo and Francisco Sánchez

Abstract Mantle convection can drive long-wavelength and low-amplitude topography, which can occur synchronously and superimposed to tectonics. The discrimination between these two topographic components, however, is difficult to assert. This is because there are still several uncertainties and debates in the geodynamic community, for example, the scales and rates of dynamic topography. Geological, geomorphological, geophysical measurements, and/or landscape analyses might assist to validate models. In this contribution, we provide new geological evidences along the Central and Patagonian Andes, which demonstrate that dynamic topography has been an important component on the South American landscape formation as well as in the ancient western Gondwana. Our examples in the Argentine Pampas show that dynamic topography is required to explain not only the basin subsidence but also the whole observed topography. We also suggest that the dynamic components in this region are much lower than numerical models (average dynamic subsidence rates of ~ 0.04 mm/yr—this work— which contrast with the ~ 0.1 mm/yr estimated in the US). We also propose two strategies to analyze ancient cases. The first requires of comparing a total elevation proxy, like the equilibrium lines (or ELA) in glaciated areas, with model topography derived from geochemical studies of mantle rocks. A second strategy was the analysis of the Triassic rifting evolution of western Argentina (post-rift sag deposits). Sag deposit thicknesses exceed 2 km, which do not correlate with the 100 m thick thermal calculated by rift subsidence modeling.

Keywords Dynamic uplift · Dynamic subsidence · Mantle · Flat subduction zone Glaciations · Triassic rifting

F. M. Dávila (✉) · F. Martina · P. Ávila · J. Nóbile · G. Collo · M. Ezpeleta
H. Canelo · F. Sánchez
CICTERRA, CONICET—FCEFyN—Universidad Nacional de Córdoba, Córdoba, Argentina
e-mail: fmdavila@unc.edu.ar

C. Lithgow-Bertelloni
University College London, London, England

A. Folguera et al. (eds.), *The Evolution of the Chilean-Argentinean Andes*, Springer Earth System Sciences, https://doi.org/10.1007/978-3-319-67774-3_15

1 Introduction

Mantle convection is considered the main driving mechanism of long-wavelength (thousands of kilometers) and low-amplitude surface uplift and subsidence (i.e., dynamic topography, Gurnis 1992). Examples of dynamic topography are the elevations of the southern part of Africa (Lithgow-Bertelloni and Silver 1998) and the Cretaceous subsidence phases of the Western Interior of the USA (Gurnis 1990). These topographies occur synchronously and superimposed to the tectonic deformation components (Fig. 1), which produce lower wavelength and higher amplitude topographies (i.e., tectonic topography). While tectonic topography responses to isostasy and flexure (Turcotte and Schubert 2002), dynamic topography is mainly driven by viscous mantle flow (cf. Richards and Hager 1984). Although both mechanisms evolve mostly together, crustal deformation is more frequently used to explain elevations and basins. Dynamic topography is only contemplated for those situations where the tectonic deformation approaches required of abnormal parameters to match observations and modeling (Liu and Nummedal 2004). The major problem in topographic analysis is the quantification of both components. In practice, the discrimination between deformation- and mantle-driven topographies requires of multidisciplinary approaches. In subducting margins, like the Andes, the crustal and lithospheric contribution on topography is strong and, therefore, the separation is tricky. The deviations between observed (basin subsidence) and modeled flexural curves might be one way to justify dynamic topography (e.g., Dávila et al. 2010). However, over or underestimation of the mechanical properties of the lithosphere might introduce errors and, consequently, under or overestimations of the remaining dynamic topography.

The development of sophisticated computational tools has been crucial to understand and quantify the deep Earth behavior (Hager and O'Connell 1979, 1981; Moresi and Gurnis 1996; Liu and Gurnis 2008; among many others). However, to Present day, scales and rates of dynamic topography are still strongly debated, and sometimes models differ in orders of magnitude between them depending on the physical model or parameters used (Fig. 2). Although geological, geomorphological, geophysical measurements, and/or landscape analyses have been considered strategic to validate mantle models (e.g., Lithgow-Bertelloni and Gurnis 1997; Gurnis et al. 2000; Roberts and White 2010; Roberts et al. 2012), a few works have proposed new observational features to correlate. Another point to consider is that most contributions are on the Late Cretaceous—Early Cenozoic Western Interior of the USA (Mitrovica et al. 1989; Gurnis 1990; Burgess et al. 1997; Spasojevic et al. 2009; Liu et al. 2011; among others). In the remaining continental areas, the studies are indeed scarce. Particularly, South America has provided some few examples (e.g., Dávila et al. 2010; Shephard et al. 2010, 2012; Dávila and Lithgow-Bertelloni 2013, 2015; Eakin et al. 2014; Flament et al. 2015), especially along flat subduction segments. There, the geodynamic settings have allowed modeling the mantle contribution to topography to explain the strong

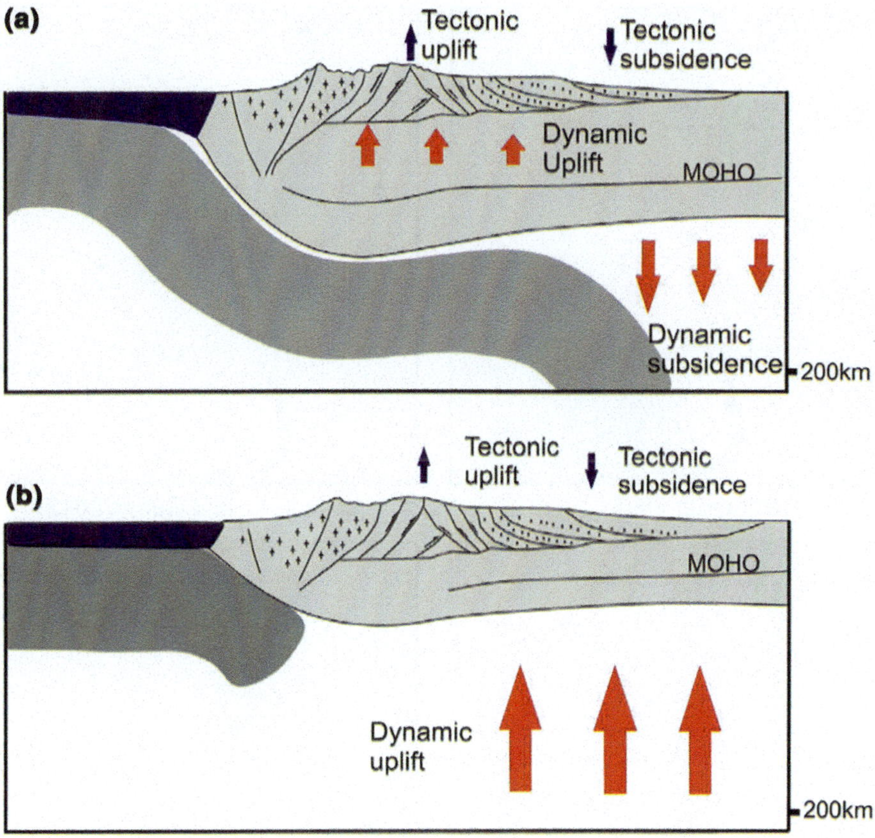

Fig. 1 Forces that contribute to the total (observed) topography formation. Note that crust and lithospheric mantle create topography by isostasy, whereas asthenospheric mantle flow affects the isostatic equilibrium by viscous shear forces

changes of topography in the Peruvian Amazon basin and Argentine Sierras Pampeanas foreland that do not respond to tectonic forces (see Dávila and Lithgow-Bertelloni 2013).

In this contribution, we provide new geological evidences along the Central and Patagonian Andes, which demonstrate that dynamic topography has been an important component on the South American landscape formation as well as in the ancient western Gondwana. After a brief chapter on theory and state of the art, we explore Neogene-Present examples in the Pampas and southernmost Patagonia and Present a model to estimate dynamic components in extinct Paleozoic and Mesozoic settings: the Mississippian of western Gondwana, dominated by high-elevation alpine-type glaciers, and the Triassic basins of western Sierras Pampeanas, subject to post-rifting thermal subsidence.

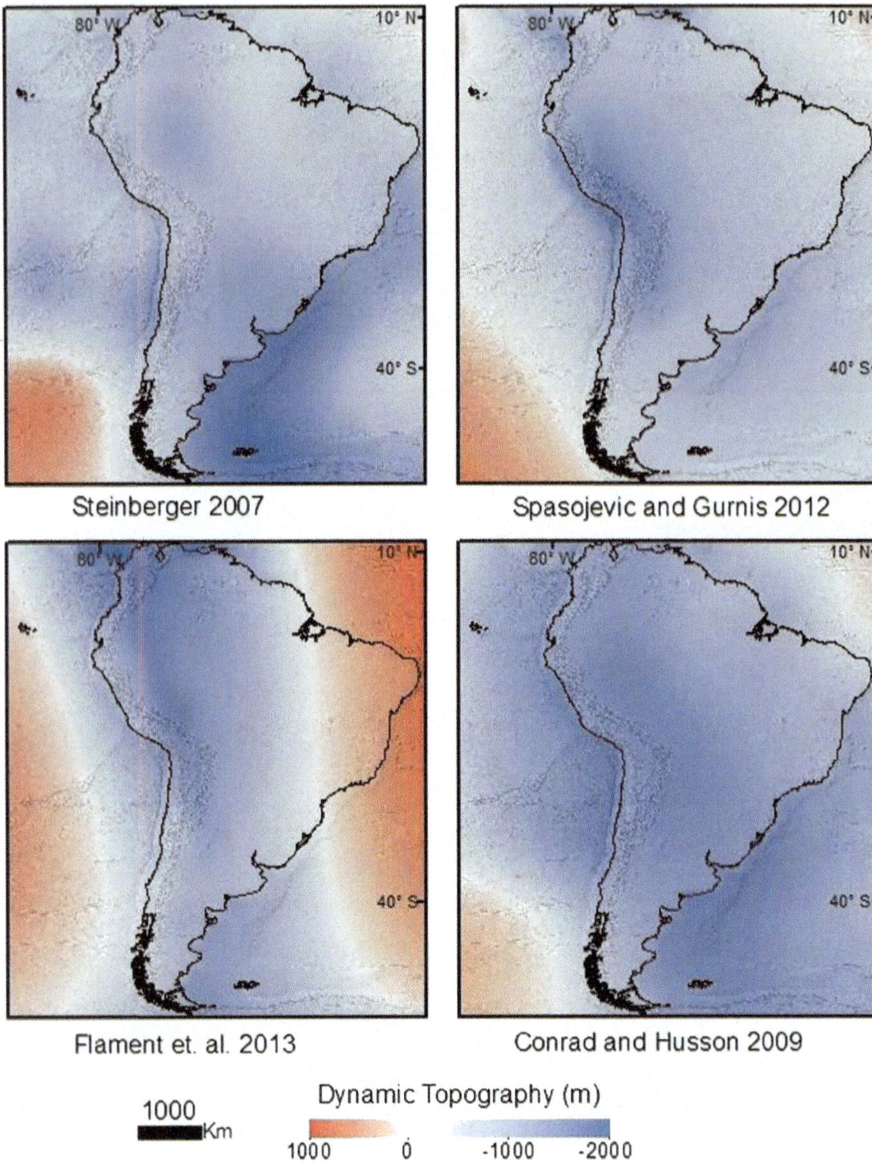

Fig. 2 Different dynamic topography models showing the strong differences in amplitude and distribution. Note that the models reproduce a large subsidence area that occupies most of South America

2 Theory and State-of-the-Art Along the Andes

We define dynamic topography (Richards and Hager 1984) as the net displacement of the Earth surface in response to mantle flow, caused by the vertical stresses arising from viscous flow in the asthenosphere. While this topographic amplitude (h) is inversely proportional to the density contrast (deltarho) across the boundary (lithosphere-air or lithosphere-ocean), it is directly proportional to the magnitude of the stresses (Trr) impinging on the surface (h = −Trr/deltarho*g).

Mantle flow models are based on solutions to the governing equations for the conservation of mass, momentum (non-inertial), and energy (advection and diffusion with sources from radioactive decay) together with constitutive relations for rheology abstracted from laboratory measurements and theoretical considerations. Such models have shown that the mantle behaves rigidly only on short time scales (e.g., propagation of seismic shear waves) and like a fluid (slow creeping flow) on long time scales (>10^4 years). Most heat is transported to the surface by large-scale solid-state convective flow, with typical overturn times of 100 My.

While the governing equations are known and numerical solutions abound (Hager and O'Connell 1981; Liu and Gurnis 2008; among others), the problem for the plate-mantle system lies in identifying the proper mantle and lithosphere rheology. Obtaining the appropriate law that captures both plate boundary behavior and internal dynamics and accounts for the large range of conditions in pressure, temperature, composition, and stress in the Earth has proven difficult (e.g., Liu and Gurnis 2008). It is possible to solve the equations numerically from first principles with a given rheological law to study the fluid dynamics of the system (e.g., Tackley et al. 1993). However, such models cannot, without knowledge of the initial conditions and all contingencies of Earth history, reproduce a Present-day Earth for direct comparison to geological observations. A recent strategy has been to impose a plate motion history (Han and Gurnis 1999; Liu and Gurnis 2008), combined with adjoint models and data assimilation and the latter can potentially allow for very sophisticated comparisons to observations (Liu et al. 2008; Shephard et al. 2010). Numerical computations, particularly adjoint models, are limited by the choice or availability of seismic tomographic models at a guess for the initial state of mantle structure. Moreover, although seismic resolution has improved, tomography cannot yet capture the large variations along strike in slab morphology and density structure Present, for example, in the Nazca slab. Alternatively to capture such variations and analyze the impact of slab morphology and density structure on dynamic topography, recent work (e.g., Dávila and Lithgow-Bertelloni 2013, 2015; Eakin et al. 2014) has taken a simpler approach. Rather than solving the coupled energy, momentum, and mass conservation equations numerically, we solve only for the conservation of mass and momentum analytically to an arbitrary resolution (Hager and O'Connell 1981). The key approximations are choosing a Newtonian rheology and assuming knowledge of the 3-D density heterogeneity in the mantle (Hager et al. 1985). These simpler models compare reconstructed history of subduction with Present-day structures of a slab (e.g., Nazca and Antarctica, Dávila

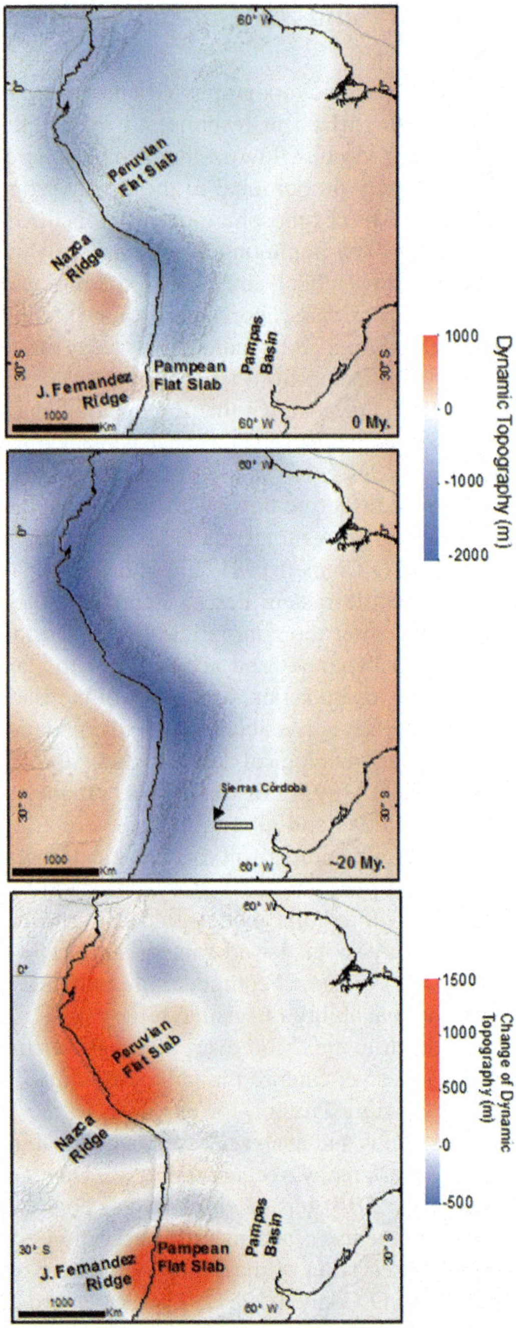

Fig. 3 The two upper maps show instantaneous dynamic topography at Present day and 20 My (cf Dávila and Lithgow-Bertelloni 2015). The lower map is the change of dynamic topography, highlighting the two flat-slab regions and connection with aseismic ridges, calculated by subtracting the Present-day component, based on the modern subduction (uppermost map), to an hypothetical early Miocene setting, when subduction was nearly homogeneous along strike, plunging ~30° E (middle map). The middle map also shows the location of the Sierras de Cordoba and location of the swath of Fig. 4

and Lithgow-Bertelloni 2015). These calculations of instantaneous viscous flow induced in the mantle by the presence of mass anomalies in different period of times allowed estimating the change of dynamic topography (Fig. 3).

The relationship between mantle dynamics and topography is direct, acting in parallel with other processes like tectonics and/or erosion. As a plate migrates over a mantle upwelling or downwelling, large vertical motions lead to the emergence or submergence, respectively (positive and negative dynamic topographies, Lithgow-Bertelloni and Gurnis 1997). Mantle forces would have caused tilting of entire continents (Mitrovica et al. 1989; Gurnis 1990) and influence on the shape of the ocean surface via its effect on the geoid. At continental long-wavelength scales, dynamic topography is the most important contributor to Earth's topography (Ricard et al. 2006) and drives the state of stress of the lithosphere in areas of long-lived subduction (such as Southeast Asia and the Americas, Lithgow-Bertelloni and Guynn 2004; Liu et al. 2008) as well as can create anomalous deep basins (Pysklywec and Mitrovica 1999; Burgess et al. 1997) or elevations (Espurt et al. 2007; Guillaume et al. 2013).

The subduction history of western Gondwana (Terra Australis orogen, cf. Cawood 2005) and South America (Andes) allows presuming a global-scale mantle downwelling cell and continental-scale dynamic subsidence (Lithgow-Bertelloni and Gurnis 1997; Steinberger 2007; Heine et al. 2008; Conrad and Husson 2009). Given the westward motion of South America since the late Cretaceous, long-wavelength dynamic uplift (Shephard et al. 2010) seems to have act synchronously with major crustal deformation ($\sim$20 my) along the Andean belt in the western part of the continent. This dynamic uplift, however, is not an absolute upward component of deformation, but rather it results from a reduction of the downwelling forces (Shephard et al. 2012), i.e., the change shows rising. Such evolution was proposed on the base of global tomography reconstructions (e.g., Grand 2002). However, as demonstrated by Dávila and Lithgow-Bertelloni (2013, 2015) and Eakin et al. (2014), continental-scale studies have not been able to capture the variations in morphology and density structure of subducting slabs. Moreover, the gradual changes in topography in South America would have not been as uniform as proposed (in fact they are marked by fluctuating episodes of uplift-subsidence, influenced by a changing tectonic-dynamic scenario, Ramos 2009). Based on local seismic tomography slab reconstructions (e.g., Gutscher 2002; Gilbert et al. 2006), recent contributions (Dávila and Lithgow-Bertelloni 2013, 2015; Eakin et al. 2014) have proposed a remarkable matching between slab geometry and dynamic topography. While horizontal slab segments are associated with dynamic uplift, plunging slabs are linked to dynamic subsidence (Dávila and Lithgow-Bertelloni 2013) (Fig. 3).

The geoid, residual topography, and basin subsidence have served traditionally to correlate numerical computing (Flament et al. 2013). Nevertheless, geoid data are too long wavelength (>5000 km, global scale) to capture regional variations of topography. Residual topography and basin subsidence, in turn, are strongly dependent on the lithosphere model and rheological parameters used (see Dávila and Lithgow-Bertelloni 2015).

The few reports of dynamic topography records are mostly on extinct cases (e.g., Mitrovica et al. 1989; Burgess et al. 1997). Spasojevic et al. (2008) using the difference between global eustasy and local sea level (calculated from borehole data) estimated a subrecent dynamic subsidence along the eastern passive coast of the USA. This component was associated with an extinct piece of the Farallon plate located below the eastern North American plate margin and derived from the subduction history of the western margin. In South America, the cases are in the Amazonas, Pampean flat-slab segment, southernmost Patagonia, and Argentine abyssal basin (Dávila et al. 2007, 2010; Shephard et al. 2012; Dávila and Lithgow-Bertelloni 2013, 2015; Eakin et al. 2014).

3 New Observational Cases in the Andes

3.1 Subsidence of the Argentine Pampas

The extensive and stable pericratonic forelands of South America, like the Argentine Pampas, are excellent analogue to analyze dynamic topography components. This region occurs >900 km inland from the trench, >700 km from the main Andean loads (mountain belt), and relief cannot be explained by faulting or bending. Therefore, sublithospheric forces are required to match observed and modeled topography (see more details in Dávila and Lithgow-Bertelloni 2013, 2015; Eakin et al. 2014).

Strategy We here combine stratigraphy and morphometric analysis to demonstrate that, at the leading edge of the Pampean flat slab (see location in Fig. 3), corner mantle flow conducted dynamic subsidence at thousands of kilometers from the Chilean trench and mountain belts. In this segment, the easternmost tectonic load is the Sierras de Cordoba (black arrow in Fig. 3), which contribute poorly in flexural subsidence (see more details in Dávila et al. 2010). We also demonstrate with this example that the maximum dynamic amplitude should be <<1 km, much less than previously proposed (e.g., Liu et al. 2008). The central Argentine modern foreland requires of dynamic topography to construct its landscape given that is tectonic and thermally inactive since the late Mesozoic. According to Dávila et al. (2010) and more recently Dávila and Lithgow-Bertelloni (2015), Mio-Pliocene slab flattening and eastward shifting of the mantle wedge would have been essential to create the Pampas basin.

Data In the Pampas, the total Cenozoic thickness is $\sim$500 m (Marengo 2006; Astini et al. 2014). This sedimentation, however, could be "geodynamically" divided into two phases, pre- and post-slab flattening (Dávila et al. 2010), which occurred at $\sim$7 Ma (Kay and Gordillo 1994). This stratigraphic division coincides in turn with the occurrence of a widespread marine horizon (the Paranaense bed, Ruskin et al. 2011). We consider in the Present analysis the Mio-Pliocene strata accumulated after the arrival of the flat slab (Fig. 4). Given that the Argentine plains are not at sea level, we also take into account the modern relief (to contemplate the whole topography) across the Pampas using a topographic swath extracted from a SRTM 90 m DEM.

Results Figure 4 shows a topographic swath curve integrating the maximum relief (100–50 m above sea level) extracted from a DEM across the Pampas (see location in Fig. 3), and the Upper Miocene-Pleistocene thicknesses (≤ 200 m), extracted from subsurface studies (Marengo 2006). Both curves account for ≤ 300 m of total topography across the Pampas (Fig. 4). We demonstrated that decompaction does not add significant thicknesses given that the unit considered and strata overlying it are very thin (see also Dávila et al. 2010).

Geology-Modeling Correlation The observations on the modern Pampas topography correlate remarkably well with our computed dynamic topography models using instantaneous computations (Dávila and Lithgow-Bertelloni 2013, 2015). Our models reproduce hundreds of meters of dynamic topography. Assuming the flat slab and eastward shifting of the mantle wedge at $\sim$7 Ma, an average dynamic subsidence rate of <0.04 mm/yr was estimated. This is an order of magnitude smaller than those estimated ($\sim$0.1 mm/yr) in the Western Interior basins of the USA (Lithgow-Bertelloni and Richard 1998; Liu et al. 2008; Spasojevic et al. 2009).

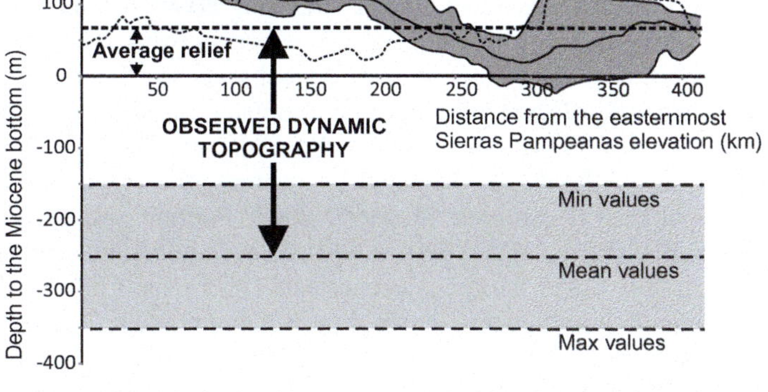

Fig. 4 Topographic swath curve integrating the maximum relief extracted from a DEM across the Pampas and the Upper Miocene–Pleistocene thicknesses extracted from subsurface studies (Marengo 2006). See location of the swath in Fig. 3

3.2 *Southern Andes Foreland Basin Exhumation and the Patagonia Plateau Uplift*

The Patagonian plateau (between the Southern Andes and Atlantic coast, Argentina, Fig. 5) is a plateau-like and low-elevation surface, no isostatically supported and located far away from the main Andean thrusts (see discussion in Dávila and Lithgow-Bertelloni 2013). This plateau was associated with mantle upwellings (Guillaume et al. 2009, 2010, 2013) driven by the development of an asthenospheric or slab window (Gorring et al. 1997; Russo et al. 2010).

Data The occurrence of dynamic topography in Patagonia is clearly evidenced by different stratigraphic relationships. Marine and estuarine strata from the Upper Oligocene–Lower Miocene (Centinela or Estancia 25 de Mayo Fms; Cuitiño and Scasso 2010), suggest the Patagonia was topographically near sea level. These are covered conformably by subhorizontal (no deformed) Middle–Upper Miocene continental alluvial beds (Santa Cruz Fm), which are interlayered by subhorizontal Upper Miocene (12-3 Ma) basalts (Gorring et al. 1997; Lagabrielle et al. 2004; Scalabrino et al. 2010). In addition to this significant paleoenvironmental change, from marine to alluvial during the Miocene, the flat Upper Miocene is observed at Present-day at relatively high elevations (between 500 and 1500 m asl) (Fig. 3). Another evidence of this uplift is the Patagonian cliff along the Atlantic coast, which shows the largest uplift respect to the rest of the SW Atlantic margin (Pedoja et al. 2011). The Patagonian extra-Andean foreland is not or poorly affected by thrusting, and crustal thicknesses are ~35 km (Fosdick et al. 2011), and, consequently, the recent elevations require of sublithospheric forces (Guillaume et al. 2009).

Model The mantle upwelling forces and related slab window have been associated with the subduction the Chile seismic ridge and kinematics of the Chilean Triple Junction (CTJ) during the middle–late Miocene (Breitsprecher and Thorkelson 2009). The seismic ridge subducts today below the Patagonia at ~47° SL. Plate reconstructions (e.g., Breitsprecher and Thorkelson 2009), however, suggest it began to affect the southernmost South American margin (what it is ~54° S today) in the early Miocene (~19 Ma). Our work, consequently, analyzes three key subduction stages: 19, 12, and 0 Ma, which correlate with the beginning, migration to the north, and Present position of the Chilean ridge, respectively. The northward shifting of the ridge conducted to the formation of a slab window as suggested by volcanic (Gorring et al. 1997, 2003; Gorring and Kay 2001) and seismic tomography (Russo et al. 2010) studies. The temporal and spatial Antarctic slab reconstruction is based on Breitsprecher and Thorkelson (2009), whereas the subduction dip angle interpretation is based on Boutonnet et al. (2010). Given that our model cannot simulate asthenospheric windows, i.e., slab breaks, we use a continuous subduction mesh but with neutral density contrast (no contrast) between slab and asthenosphere in slab window areas as well as where subduction suddenly

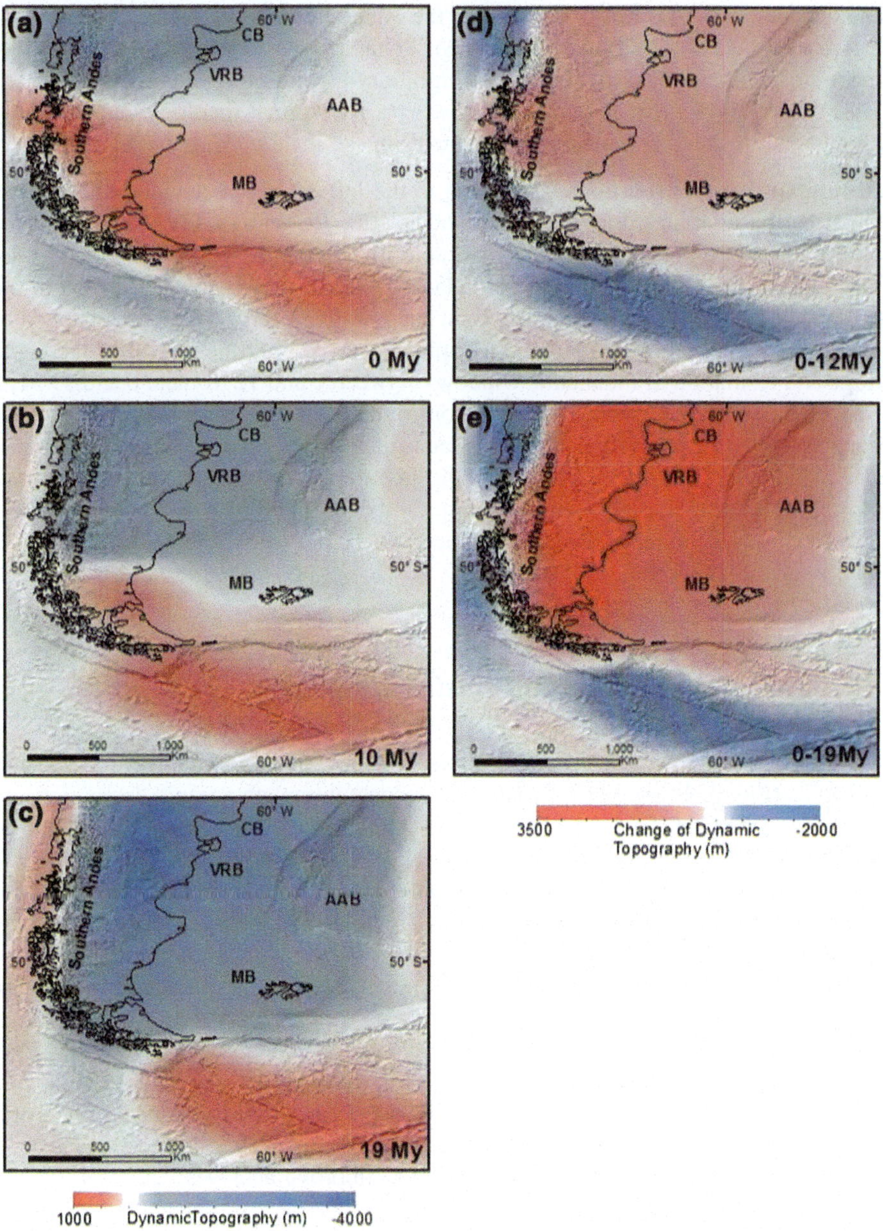

Fig. 5 a, b, c Instantaneous dynamic topography model in the southernmost South America (Patagonia) at 19, 12, and 0 Ma based on the plate reconstructions of Breitsprecher and Thorkelson (2009) and subduction geometry of Boutonnet et al. (2010). The change of dynamic topography is illustrated on the right (D: between 12 My and Present and E: between 19 My and Present), which evidences a widespread dynamic uplift in Patagonia and South Atlantics (in the Argentine abyssal basin AAB and different Argentine shelf basins from south to north, Malvinas basin: MB, Valdes-Rawson basins: VRB, and Colorado basin: CB)

ends up below ~100 km (Fig. 5) as in the Antarctic plate in south Patagonia (cf. Russo et al. 2010). As stated in the theory chapter, we solve the conservation of mass and momentum analytically to an arbitrary resolution (Hager and O'Connell 1981). The key approximations are choosing a Newtonian rheology and assuming knowledge of the 3-D density heterogeneity in the mantle (Hager et al. 1985).

Results and Interpretations The maps to the left on Fig. 5 depict instantaneous dynamic topographies during the key stages at 0, 12, and 19 My. Notice that the oldest stage is represented by a remarkable dynamic subsidence along the entire Patagonia, which explains the thick foreland basins formed before the arrival of the ridge subduction. The maps to the right on Fig. 5, in turn, show the changes of dynamic topography and the dynamic uplift occurred after the arrival of the Chile ridge and formation of the mantle window (between 0 and 12 My and 0 and 19 My).

Our model agrees with a change from dynamic subsidence to uplift contemporaneously with the arrival of the Chilean ridge and formation of the Patagonian slab windows. This also agrees with the geological evidences, among them the change from the marine Centinela Fm to the fluvial Santa Cruz Fm and the development of fluvial and marine terraces along the eastern Patagonian Atlantic margin (Guillaume et al. 2009).

3.3 The Argentine Abyssal Basin Subsidence

Tectonic Setting and Subsidence Timing The abyssal Argentine Basin in the SW Atlantics (Fig. 5) is one of the largest and deepest intraplate oceanic depressions on Earth (~4,000,000 km^2 in area, >6 km below sea level, and holds >4 km of sediments, cf. NGDC global sediment thickness grid, Divins 2008). The basaltic substrate of the basin (corrected by sediment loading) is ~1 km below mean global oceanic subsidence values (Hohertz and Carlson 1998). This subsidence anomaly matches in turn with a large negative free-air gravity anomaly (Winterbourne et al. 2009). This allows confirming the Argentine Basin is a world-class negative topographic anomaly.

Despite the different approaches and models (Hohertz and Carlson 1998; Steinberger 2007; Winterbourne et al. 2009; Shepard et al. 2012; Dávila and Lithgow-Bertelloni 2013), rigorous ages and subsidence timing of this basin are still unknown. Consequently, interpretations and derived modeling are too speculative. We Present new constraints below in order to improve analyses.

Basin Constraints The Argentine Basin rests today on oceanic basaltic floor that is < 130 Ma (Müller et al. 1997; Torsvik et al. 2009), indicating that the main subsidence should be younger. Oceanic half-space subsidence models, in turn, reproduce the main sinking phase in the firsts ~70 My before flattening (thermal instability, Parsons and McKenzie 1978). Therefore, one might presume the

Argentine abyssal basin should have formed within the Cenozoic. This agrees with the Husson et al. (2012) predictions, which interpreted a downwelling cell < 50 Ma in the SW Atlantics. Although these analyses allow approximating the subsidence time, no stratigraphic data support the modeling deductions (e.g., Shephard et al. 2013).

The leg 71 / site 513 report of the Deep Sea Drilling Project (DSDP) (47°34.99′ S–24°38.40′W, out of Fig. 5) provides two key data to analyze the subsidence history in the study area: (1) the calcite compensation depths, or CCD, and (2) the foraminiferal lysocline, or FL. While the CCD is the depth at which the dissolution flux balances the rain flux of calcite to the sediments, the lysocline is the greater depth at which dissolution occurs (Zeebe 2012). This DSDP report shows that the CCD and the planktonic FL were high in the early Oligocene (~34 Ma), the planktonic FL was low, and the benthonic FL high in the late Oligocene (~28 Ma), whereas the CCD was very low in the Neogene (< 23 Ma). Tyrrell and Zeebe (2004), in turn, demonstrated that between early Oligocene and Pliocene the global CCD would have placed between 4000 and 4500 m below sea level. The comparison of these results indicates that the subsidence anomaly should be younger than 23 Ma (i.e., early Miocene). This estimation correlates with the sedimentation rates reported by the DSDP studies, which show a remarkable increase from 4 to 11 m/My in the late Oligocene—early Miocene to 33–35 m/My in the late Miocene–Pliocene.

An alternative to constrain the abyssal basin subsidence is to analyze the Argentine shelf stratigraphic evolution since the Atlantic opening. In the southernmost shelf, along the Malvinas Basin (between 50° and 55° S) (Fig. 5), Baristeas et al. (2012) interpreted a major stratigraphic change at ~5–6 Ma, when the sedimentary sequences changed their arrangements from net vertical aggradation to basinward progradation (aggradation is represented by their U5a sequence, constrained between 42.5 and 5.5 Ma, whereas U5b represents progradation since 5.5 Ma). A 1000 km northward, in the Valdés-Rawson Basins (between 41° and 45° S, surrounding the main part of the abyssal Argentine Basin), Gruetzner et al. (2012) interpreted a paleotopographic change in the late Miocene based on seismic stratigraphy, represented by canyons between the shelf and the abyssal basin. Martinelli and Franzin (1996) also interpreted in this area a major sedimentation change in the late Miocene based on a borehole analysis (Tayra x-1) located 250 km offshore from the Patagonian coasts (Fig. 5). In this region, the upper Miocene–Pliocene rests mostly on Cretaceous, whereas Paleogene and lower Miocene are thin, condensed, or not developed. More northward, across the offshore Colorado Basin (Fig. 5) at 39°–40° S and bounding the Argentine Basin to the North, Loegering et al. (2013) also reported an increase in the sediment supply during the Oligo-Miocene (Barranca Final Formation), resulting in prograding sedimentary wedges, interpreted as an event of basin subsidence.

In synthesis, along the whole of the Argentine shelf (along the boundary of the Argentine Basin), the sedimentation history agrees with a regional basin level fall during the Miocene.

Subsidence in the Oceanic Floor, Uplift in the Continental Lithosphere Are the Patagonian and Argentine Basin anomalies contemporaneous and related? We demonstrated above that both occurred in the Miocene, and consequently, the oceanic subsidence would have been contemporaneous with the dynamic uplift in Patagonia (Guillaume et al. 2009) as well as with the dynamic-driven coastal rising along the SE margin of South America (Pedoja et al. 2011) (Fig. 5).

We compared our dynamic topography results in southern South America (Fig. 5) (based on plate reconstructions, Chile ridge kinematics, and subduction angles since the early Miocene, see above) with the Argentine Basin position. We can clearly observe that asthenospheric mantle did not influence on the subsidence anomaly in the SW Atlantic. Furthermore, recent seismic tomography (cf. Colli et al. 2013) shows no slab feature evidences (high-velocities) to justify a mass anomaly, which could have driven downwelling flows and dynamic subsidence. Therefore, the only possibility is to survey into the lithospheric mantle.

We propose alternatively that the Argentine abyssal Basin formed by a local oceanic floor obduction, which drove lithospheric thickening (similar to Garcia Castellanos et al. 2002; Bajolet et al. 2012). O'Reilly et al. (2009) interpreted thick continental roots along the eastern South America Plate across the Argentine shelf and abyssal Basins on the base of a world magnetic anomaly map and tomography model. This duplication on a thick lithosphere might have created an extra load and instability. Nevertheless, it is still uncertain the sharp change from uplift in Patagonia to the remarkable subsidence in the oceanic basin.

4 Fossil Cases

The influence of mantle forces on topography is difficult to demonstrate in modern and subrecent settings but much more for extinct ones. This is because records are mostly extinctive and variable to quantify some properties, reason by which these need to be totally deduced. Some works, however, have supported mantle-driven topographies based on correlations and extrapolations with similar scenarios. A better approximation, in contrast, is to make use of the isostatic analysis. When "observed" topography (total topography) does not match with modeled topography (based on a specific lithospheric structure), a residue generates. This residual topography has been attributed to mantle dynamics, especially to convective (asthenospheric) mantle activity. However, it might derive from the conductive (lithospheric) part of the mantle by changes of thicknesses and/or densities. Although ancient model topography can be proposed on the base of hypotheses and theoretical scenarios, a comparison with observed topographies requires of geological proxies. In the next two sections, we show two examples from western Argentina, which pretend to spark a discussion on mantle contribution on topography on extinct settings.

4.1 Dynamic Supports and Carboniferous Glaciations

A potential good proxy to estimate total (observed) topography is the occurrence of mountain glacier records. There is a well-studied relationship between climate (temperature and moisture), latitudinal position, and altitude. Then, if we know the geographic position and climate, we might calculate paleoelevations (Condom et al. 2007). On the other hand, if we can calculate a model elevation based on a lithospheric structure (thickness and internal temperature and composition), then we could compare observations and models to get a difference or residue (see further details in Dávila and Martina, in revision).

An enigmatic episode to analyze is the early Carboniferous glaciations occurred in western Gondwanaland. What triggered this mountain glacier formation? The main causes have been explained as changes of the paleolatitudinal position, global cooling, and/or mountain building, when continental areas would have uplifted over the paleo-equilibrium line altitude (or ELA). Most works, however, agreed with the tectonic-driven uplift glaciation (Isbell et al. 2012) related to the collisional tectonics (Ramos et al. 1986), slab flattening (Astini et al. 2005), or rifting (Astini et al. 2011). But the evidences for any of these hypotheses are not convincing and refutable. We propose an alternative Carboniferous geodynamic setting, where deep mantle forces would have conducted non-isostatic uplift (upwelling). Our proposal is based on the comparison between estimations of the total paleoelevations (using the ELA paleoaltimetry, following the Condom et al. 2007 method) with the theoretical paleoelevation from isostatic calculations (Lachenbruch and Morgan 1990). Granite geochemistry (Dahlquist et al. 2014, 2016) and thermochronology studies support our proposal (Jordan et al. 1989; Enkelmann et al. 2014). For paleo-ELA calculations, we compile paleogeographic and paleoclimate data from Gulbranson et al. (2010). Figure 6 shows the Carboniferous ELA (maximum and minimum) corrected by topographic evidences, which allow estimating a minimum total elevation >1 km across the glaciated areas. The isostatic elevations, in turn, require of a geochemical approach to estimate the Paleozoic lithospheric structure. Studies indicate that the different crust-mantle and lithosphere–asthenosphere boundary depths affect the rare earth elements (REE) composition of the magmas generated in the mantle. This is because the heavy REE partitioning depends strongly on pressure. We estimated crustal and lithospheric mantle thicknesses from Early Carboniferous basalts underlying the postglacial deposits, using the Mantle and Collins (2008) and Gibson and Geist (2010) approaches, respectively. The lithospheric structure provides an isostatic mean elevation of $\sim 0.16 \pm 0.7$ km (minimum 0.525 km below sea level and maximum 0.866 on sea level).

The comparison between total and hypothetical elevations indicates that the glaciated high elevations of southwestern Gondwana were strongly overcompensated by sublithospheric forces. Considering that the Carboniferous crustal heat flow would have been remarkably high (intrusive phase of A-type granites, Dahlquist et al. 2014, 2016) and rapid basement exhumation interpreted from thermochronology (Jordan et al. 1989; Dávila and Carter 2013; Enkelmann et al. 2014;

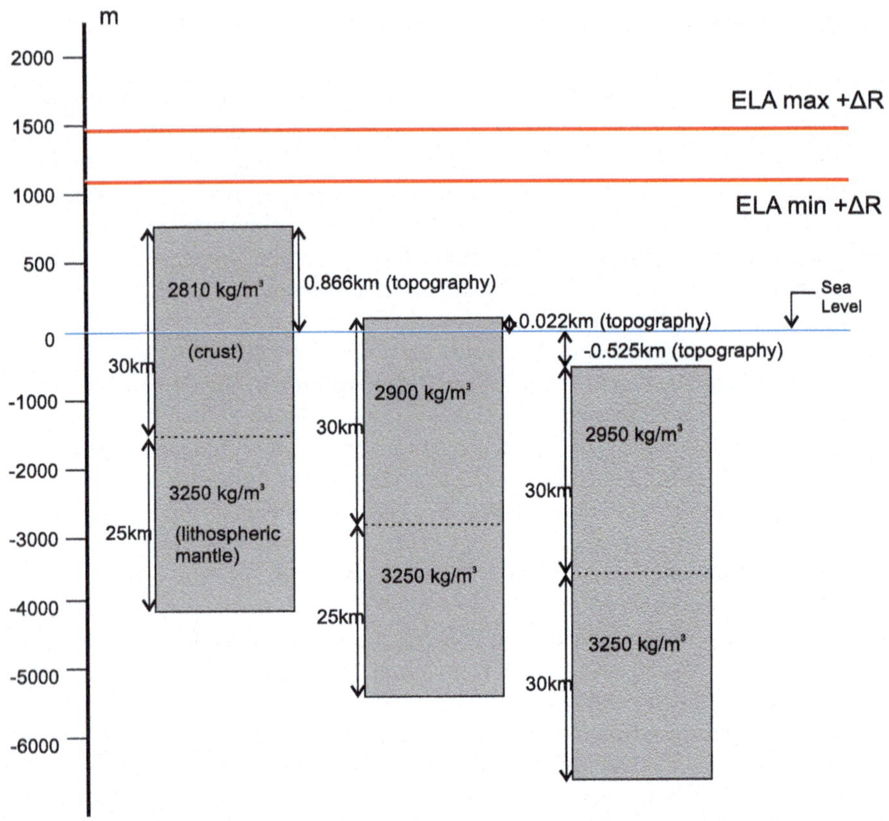

Fig. 6 Diagram comparing the paleoelevations estimated using paleo-ELA and preserved relives of 1.5 km (ΔR) estimated from Carboniferous mountain glaciers, with model elevation based on a paleo reconstruction of the lithosphere using early Carboniferous basalts (see text for further details and Dávila and Martina, in rev.)

Nóbile et al. 2015), we propose three potential scenarios to explain the non-isostatic uplift, driven by asthenospheric upwelling forces: slab rollback, slab windows, or mantle plumes. All of them are more consistent with shallow crustal extension rather than shortening (collisional tectonics) as proposed by Astini et al. (2011).

4.2 *Post-Rifting Subsidence of Triassic Basins in Western Argentina*

In South America, as stated above, Dávila and Lithgow-Bertelloni (2013) showed that the foreland basin systems require mantle forces, to be fully reproduced (especially thicknesses), of mantle forces. This hypothesis also might apply to the pre-Cenozoic rift basins developed along the western Gondwana. These rifts are formed during lithospheric stretching. From their structural and stratigraphic features, these basins can be divided in two main phases: a thick syn-rift stage (thicknesses over 5 km), where deposits accumulate within the normal fault system and a thermal-driven post-rift (cooling phase), where deposits overlap and exceed the spaces created by faulting. This thermal subsidence is very low-amplitude and, under special circumstances, long-lived (hundreds of millions of years, McKenzie 1978).

One of the major rift basin systems in western Sierras Pampeanas is the Talampaya-Ischigualasto-Ischichuca basins, developed between the early late Triassic (250 and 215? Ma) (see Chap. "The early stages of the magmatic arc in the Southern Central Andes"). These are asymmetric systems with thicker depocenters to the west. The basin evolution was divided into two syn-rift phases (occurred between 250 and ~222 ± 2.2 Ma, followed by a thick post-rift stage (Los Colorados Fm) (Milana and Alcober 1994). While the syn-rift phase accumulated approximately 5–6 km of continental sediments in ~30 My, the post-rift stage ~2 km in < 10-5 My. This suggests the rift phase accumulated at rates of ~0.2 mm/yr, and the sag thermal phase at rather similar rates of ~0.4–0.2 mm/yr. This evolution does not agree with theoretical rift subsiding curves, which show a strong asymptotic geometry in time, flattened during the thermal subsidence (Fig. 7). According to the McKenzie (1978) thermal subsidence model (assuming a lithospheric thinning from 60 to 55 km, Martina pers. com and stretching factor $\beta < 1.1$ between Permo-Triassic to late Triassic), the expected accommodation

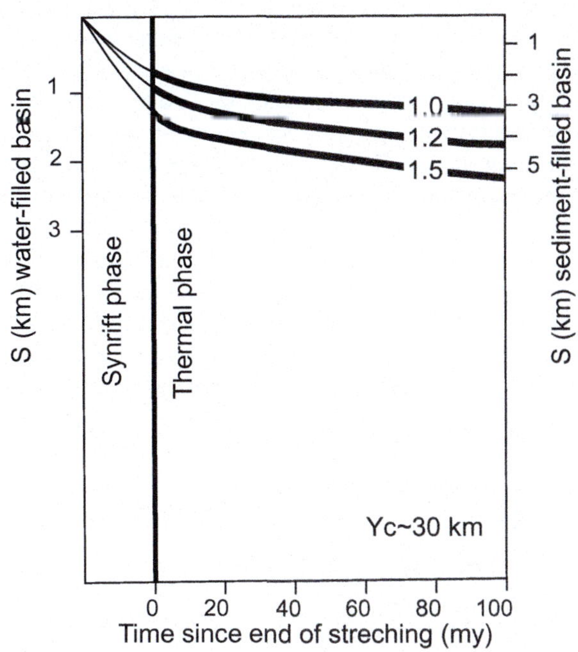

Fig. 7 Stretching subsidence model taken from McKenzie (1978). Note that the thermal phase, after the main rifting (when sag deposits occurs), reproduces a few hundred meters of stratigraphy. Considering a stretching factor between 1 and 1.2 (thickest black lines), the >2 km of post-rift deposits in Triassic basins of western Argentina are difficult to explain by this model alone (see text for further details)

space would be of a few hundreds of meters in tens or hundreds of millions years (Fig. 7). Although Los Colorados Formation shows a clear stratigraphic post-rift relationship, overlapping the rift shoulders for hundreds of kilometers, the thicknesses are much larger than those expected by thermal models (over 2 km). We propose, likewise in the foreland basins of South America, that thermal subsidence (at least) works together with sublithospheric forces, likely driven by mantle convection during the Triassic subduction that affected the western part of Gondwanaland (see Chap. "The early stages of the magmatic arc in the Southern Central Andes"). Future mantle modeling during the Paleozoic and Triassic in Gondwana will allow us to corroborate our hypotheses.

5 Concluding Remarks and Future Perspectives

We demonstrated in this contribution that geological/geophysical observations are needed for geodynamic modeling, especially to constrain physical parameters. Our examples in Argentina show that dynamic topography is required to explain the whole observed topography and that the amount of dynamic components is much lower than proposed in previous work. Average dynamic subsidence rates are ~0.04 mm/yr, which contrast remarkably with the ~0.1 mm/yr estimated in the US (Liu et al. 2008; Spasojevic et al. 2009). This drove us to think that the total topography was formed by an isostatic crustal component complemented by non-isostatic mantle forces, either uplift or subsidence. We also alert on the importance of the isostatic lithospheric mantle contribution on topography, especially when mismatching between crustal tectonics and dynamic topography arises. These three components, together to erosion, drive the observed landscapes.

We also propose two strategies to analyze ancient cases, estimating the residual topography or analyzing post-rifting strata. The first requires of comparing a total elevation proxy, like the ELA in glaciated areas, with model topography derived from geochemical studies of mantle rocks. The stratigraphy and thicknesses of Triassic sag deposits (post-rift) and correlation with thermal subsidence modeling allow calculating other contributions on basin preservation. This latter fossil case is particularly important given that no analogue modern examples are known.

These studies not only help us to understand the Earth dynamics and to quantify the different forces that affect on topography but also for economic purposes. Hydrocarbon source rock formations as well as maturation are strongly controlled by negative topography creation, or subsidence, which might drive marine incursions (source) to deep burial histories (temperature) or by uplift and exhumation. It is important to consider that in addition to classical subsidence or uplift episodes, a deep mantle component might amplify or retard the topographic amplitudes and wavelengths as well as the thermal state of the lithosphere.

Acknowledgements FONCyT, PUE 2016 CICTERRA CONICET, and SECyT-UNC (Argentina), the Royal Society and UCL (UK), and the Marie Curie Fellowship IIF Program (ANDYN Project, EU) supported our studies in South America.

References

Astini RA, Dávila FM, López Gamundí OR, Gómez F, Collo G, Ezpeleta M, Martina F, Ortiz A (2005) Cuencas de la región precordillerana. In: Chebli G, Spalletti L (eds) Frontera Exploratoria de la Argentina, Buenos Aires, pp 115–145

Astini RA, Dávila FM, Martina F (2011) La Formación Los Llantenes en la Precordillera de Jagüé (La Rioja) y la identificación de un episodio de rifting en la evolución de las cuencas del Paleozoico superior en el oeste argentine. Revista Geológica de Chile 38:245–267

Astini RA, Tauber AA, Marengo HG, Oviedo N, Del V (2014) Cubierta cenozoica (Paleógeno-Neógeno). In: Relatorio de la geología y recursos Naturales de la Provincia de Córdoba, Asociación Geológica Argentina, pp 539–591

Bajolet F, Galeano J, Funiciello F, Moroni M, Negredo AM, Faccenna C (2012) Continental delamination: insights from laboratory models. Geochemistry Geophysics Geosystems 13: Q02009. https://doi.org/10.1029/2011GC003896

Baristeas N, Anka Z, Di Primio R, Rodriguez JF, Marchal D, Domínguez F (2012) Distribution of hydrocarbon leakage indicators in the Malvinas Basin, offshore Argentine continental margin. Mar Geol 332:56–74

Boutonnet E, Arnaud N, Guivel C, Lagabrielle Y, Scalabrino B, Espinoza F (2010) Subduction of the South Chile active spreading ridge: A 17Ma to 3Ma magmatic record in central Patagonia (western edge of Meseta del Lago Buenos Aires, Argentina). J Volcanol Geoth Res 189 (3):319–339

Breitsprecher K, Thorkelson DJ (2009) Neogene kinematic evolution of the Nazca e Antarctice Phoenix slab windows beneath Patagonia and the Antarctic Peninsula. Tectonophysics 464, pp 10–20

Burgess PM, Gurnis M, Moresi L, (1997) Formation of sequences in the cratonic interior of North America by interaction between mantle, eustatic, and stratigraphic processes. Geol Soc Am Bull 109(12):1515–1535

Cawood PA (2005) Terra Australis Orogen: Rodinia breakup and development of the Pacific and Iapetus margins of Gondwana during the Neoproterozoic and Paleozoic. Earth-Sci Rev 69 (3):249–279

Colli L, Fichtner A, Bunge HP (2013) Full waveform tomography of the upper mantle in the South Atlantic region: Imaging a westward fluxing shallow asthenosphere? Tectonophysics 604: 26–40

Condom T, Coudrain A, Sicart JE, Théry S (2007) Computation of the space and time evolution of equilibrium-line altitudes on Andean glaciers (10°N–55° S). Global Planet Change 59(1):189–202

Conrad CP, Husson L (2009) Influence of dynamic topography on sea level and its rate of change. Lithosphere 1:110–120. https://doi.org/10.1130/L32.1

Cuitiño JI, Scasso RA (2010) Sedimentología y paleoambientes del Patagoniano y su transición a la Formación Santa Cruz al sur del Lago Argentino, Patagonia Austral. Rev Asoc Geol Argent 66(3):406–417

Dahlquist JA, Alasino PH, Bello C (2014) Devonian F-rich peraluminous A-type magmatism in the proto-Andean foreland (Sierras Pampeanas, Argentina): geochemical constraints and petrogenesis from the western-central region of the Achala batholith. Mineral Petrol 108 (3):391–417

Dahlquist JA, Verdecchia SO, Baldo EG, Basei MA, Alasino PH, Urán GA, Zandomeni PS (2016) Early Cambrian U-Pb zircon age and Hf-isotope data from the Guasayán pluton, Sierras

Pampeanas, Argentina: implications for the northwestern boundary of the Pampean arc. Andean Geol 43(1)
Dávila FM, Carter A (2013) Exhumation history of the Andean broken foreland revisited. Geology 41(4):443–446
Dávila FM, Lithgow-Bertelloni C (2013) Dynamic topography in South America. J South Am Earth Sci 43:127–144
Dávila FM, Lithgow-Bertelloni C (2015) Dynamic uplift during slab flattening. Earth Planet Sci Lett 425:34–43
Dávila FM, Lithgow-Bertelloni C, Giménez M (2010) Tectonic and dynamic controls on the topography and subsidence of the Argentine Pampas: The role of the flat slab. Earth Planet Sci Lett 295(1):187–194
Divins DL (2008) NGDC Total sediment thickness of the world's oceans and marginal seas. http://www.ngdc.noaa.gov/mgg/sedthick/sedthick.html
Eakin CM, Lithgow-Bertelloni C, Dávila FM (2014) Influence of Peruvian flat-subduction dynamics on the evolution of the Amazon basin. Earth Planet Sci Lett 404:250–260. https://doi.org/10.1016/j.epsl.2014.07.027
Enkelmann E, Ridgway KD, Carignano C, Linnemann U (2014) A thermochronometric view into an ancient landscape: Tectonic setting, development, and inversion of the Paleozoic eastern Paganzo basin, Argentina. Lithosphere 6(2):93–107
Espurt N, Baby P, Brusset S, Roddaz M, Hermoza W, Regard V, Antoine P-O, Salas-Gismondi R, Bolaños R (2007) How does the Nazca Ridge subduction influence the modern Amazonian foreland basin? Geology 35:515–518
Flament N, Gurnis M, Muller RD (2013) A review of observations and models of dynamic topography. Lithosphere 5(2):189–210. https://doi.org/10.1130/L245.1
Flament N, Gurnis M, Müller RD, Bower DJ, Husson L (2015) Influence of subduction history on South American topography. Earth Planet Sci Lett 430:9–18
Fosdick JC, Romans BW, Fildani A, Bernhardt A, Calderón M, Graham SA (2011) Kinematic evolution of the Patagonian retroarc fold-and-thrust belt and Magallanes foreland basin, Chile and Argentina, 51 30′ S. Geol Soc Am Bull 123(9–10):1679–1698
Flament N, Gurnis M, Müller RD, Bower DJ, Husson L (2015) Influence of subduction history on South American topography. Earth Planet Sci Lett 430:9–18
Garcia-Castellanos D, Fernandez M, Torné M (2002) Modeling the evolution of the Guadalquivir foreland basin (southern Spain). Tectonics 21(3)
Gibson SA, Geist D (2010) Geochemical and geophysical estimates of lithospheric thickness variation beneath Galápagos. Earth Planet Sci Lett 300(3):275–286
Gilbert H, Beck S, Zandt G (2006) Lithospheric and upper mantle structure of central Chile and Argentina. Geophys J Int 165(1):383–398
Gorring ML, Kay SM (2001) Mantle processes and sources of Neogene slab-window magmas in southern Patagonia. J Petrol 42(6):1067–1094
Gorring M, Kay S, Zeitler P, Ramos V, Rubiolo D, Fernandez M, Panza J (1997) Neogene Patagonian plateau lavas: continental magmas associated with ridge collision at the Chile Triple Junction. Tectonics 16(1):1–17
Gorring M, Singer B, Gowers J, Kay S (2003) Plio-Pleistocene basalts from the Meseta del Lago Buenos Aires, Argentina: evidence for asthenosphere lithosphere interactions during slab window magmatism. Chem Geol 193:215–235
Grand SP (2002) Mantle shear–wave tomography and the fate of subducted slabs. Philosophical Transactions of the Royal Society of London A: Mathematical, Physical and Engineering Sciences 360(1800):2475–2491
Gruetzner J, Uenzelmann-Neben G, Franke D (2012) Variations in sediment transport at the Central Argentine continental margin during the Cenozoic. Geochem Geophys Geosyst 13(10)
Guillaume B, Martinod J, Husson L, Roddaz M, Riquelme R (2009) Neogene uplift of central eastern Patagonia: dynamic response to active spreading ridge subduction? Tectonics 28 (TC2009). https://doi.org/10.1029/2008TC002324

Guillaume B, Moroni M, Funiciello F, Martinod J, Faccenna C (2010) Mantle flow and dynamic topography associated with slab window opening: Insights from laboratory models. Tectonophysics 496(1):83–98

Guillaume B, Gautheron C, Simon-Labric T, Martinod J, Roddaz M, Douville E (2013) Dynamic topography control on Patagonian relief evolution as inferred from low temperature thermochronology. Earth Planet Sci Lett 364:157–167

Gulbranson EL, Montañez IP, Schmitz MD, Limarino CO, Isbell JL, Marenssi SA, Crowley JL (2010) High-precision U-Pb calibration of Carboniferous glaciation and climate history, Paganzo Group, NW Argentina. Geol Soc Am Bulletin 122(9–10):1480–1498

Gurnis M (1990) Ridge spreading, subduction and sea level fluctuations. Science 250:970–972

Gurnis M (1992) Rapid continental subsidence following the initiation and evolution of subduction. Science 255(5051):1556–1558

Gurnis M, Mitrovica JX, Ritsema J, van Heijst H (2000) Constraining mantle density structure using geological evidence of surface uplift rates: The case of the African Superplume. Geochem Geophys Geosyst 1:1525–2027. https://doi.org/10.1029/1999GC000035

Gutscher MA (2002) Andean subduction styles and their effect on thermal structure and interplate coupling. J South Am Earth Sci 15(1):3–10

Hager BH, O'Connell RJ (1981) A simple model of plate dynamics and mantle convection. J Geophys Res 86(B6):4843–4867

Hager BH, O'Connell RJ (1979) Kinematic models of large-scale flow in the Earth's mantle. J Geophys Res 84:1031–1048

Hager BH, Clayton RW, Richards MA, Dziewonski AM, Comer RP (1985) Lower mantle heterogeneity, dynamic topography, and the geoid. Nature 313:541–545

Han L, Gurnis M (1999) How valid are dynamic models of subduction and convection when plate motions are prescribed? Physics of Earth and Planetary Interiors 110:235–246

Heine C, Müller RD, Steinberger B, Torsvikb TH (2008) Subsidence in intracontinental basins due to dynamic topography. Phys Earth Planet Inter 171:252–264

Hohertz WL, Carlson RL (1998) An independent test of thermal subsidence and asthenosphere flow beneath the Argentine Basin. Earth Planet Sci Lett 161:73–83

Husson L, Conrad CP, Faccenna C (2012) Plate motions, Andean orogeny, and volcanism above the South Atlantic convection cell. Earth Planet Sci Lett 317:126–135

Isbell JL, Henry LC, Gulbranson EL, Limarino CO, Fraiser ML, Koch ZJ, Dineen AA (2012) Glacial paradoxes during the late Paleozoic ice age: evaluating the equilibrium line altitude as a control on glaciation. Gondwana Res 22(1):1–19

Jordan TE, Zeitler P, Ramos VA, Gleadow AJW (1989) Thermochronometric data on the development of the basement peneplain in the Sierras Pampeanas, Argentina. J South Am Earth Sci 2:207–222

Kay SM, Gordillo CE (1994) Pocho volcanic rocks and the melting of depleted continental lithosphere above a shallowly dipping subduction zone in the Central Andes. Contrib Mineral Petr 117:25–44

Lachenbruch AH, Morgan P (1990) Continental extension, magmatism and elevation: Formal relations and rules of thumb. Tectonophysics 174:39–62. https://doi.org/10.1016/0040-1951(90)90383-J

Lagabrielle Y, Suárez M, Rossello EA, Hérail G, Martinod J, Régnier M, de la Cruz R (2004) Neogene to Quaternary tectonic evolution of the Patagonian Andes at the latitude of the Chile Triple Junction. Tectonophysics 385(1):211–241

Lithgow-Bertelloni C, Richards MA (1998) The dynamics of Cenozoic and Mesozoic plate motions. Rev Geophys 36: 27–78 Magoon, LB. y Dow, WG (eds) 1994. The petroleum system from source to trap. AAPG Memoir 60

Lithgow-Bertelloni C, Silver PG (1998) Dynamic topography, plate driving forces and the African superswell. Nature 395(6699):269–272

Lithgow-Bertelloni C, Gurnis M (1997) Cenozoic subsidence and uplift of continents from time-varying dynamic topography. Geology 25(8):735–738

Lithgow-Bertelloni C, Guynn JH (2004) Origin of the lithospheric stress field. J Geophys Res: Solid Earth 109(B1)

Liu L, Gurnis M (2008) Simultaneous inversion of mantle properties and initial conditions using an adjoint of mantle convection. J Geophys Res 113(B08405). https://doi.org/10.1029/2008JB005594

Liu S, Nummedal D (2004) Late Cretaceous subsidence in Wyoming: quantifying the dynamic component. Geology 32:397–400

Liu L, Spasojevi'c S, Gurnis M (2008) Reconstructing Farallon plate subduction beneath North America back to the Late Cretaceous. Science 322:934–938

Liu S, Nummedal D, Liu L (2011) Migration of dynamic subsidence across the Late Cretaceous United States Western Initerior Basin in response to Farallon plate subduction. Geology 39:555–558

Loegering MJ, Anka Z, Autin J, Di Primio R, Marchal D, Rodriguez JF, Vallejo E (2013) Tectonic evolution of the Colorado Basin, offshore Argentina, inferred from seismo-stratigraphy and depositional rates analysis. Tectonophysics 604:245–263

Mantle GW, Collins WJ (2008) Quantifying crustal thickness variations in evolving orogens: Correlation between arc basalt composition and Moho depth. Geology 36(1):87–90

Marengo HG (2006) Micropaleontología y estratigrafía del Mioceno marino de la Argentina: Las transgresiones de Laguna Paiva y del "Entrerriense-Paranense". M. S. thesis. Buenos Aires University, p 124. Unpublished

Martinelli RV, Franzin HJ (1996) Cuencas de Rawson y Península Valdés. In: Ramos VA, Turic MA (eds) Geología y recursos naturales de la Plataforma continental Argentina. Relatorio del XIII° Congreso Geológico Argentino y III° Congreso de Exploración de Hidrocarburos, Buenos Aires, pp 159–170

McKenzie D (1978) Some remarks on the development of sedimentary basins. Earth Planet Sci Lett 40(1):25–32

Milana JP, Alcober OA (1994) Modelo tectosedimentario de la cuenca triásica de Ischigualasto (San Juan, Argentina). Rev Asoc Geol Argent 49:217–235

Mitrovica JX, Beaumont C, Jarvis GT (1989) Tilting of the continental interior by the dynamical effects of subduction. Tectonics 8:1079–1094

Moresi L, Gurnis M (1996) Constraints on the lateral strength of slabs from three-dimensional dynamic flow models. Earth Planet Sci Lett 138(1):15–28

Müller RD, Roest WR, Royer JY, Gahagan LM, Sclater JG (1997) Digital isochrons of the world's ocean floor. J Geophys Res: Solid Earth 102(B2):3211–3214

Nóbile JC, Collo G, Dávila FM, Martina F, Wemmer K (2015) Successive reactivation of older structures under variable heat flow conditions evidenced by K-Ar fault gouge dating in Sierra de Ambato, northern Argentine broken foreland. J South Am Earth Sci 64:152–165

O'Reilly SY, Zhang M, Griffin WL, Begg G, Hronsky J (2009) Ultradeep continental roots and their oceanic remnants: A solution to the geochemical "mantle reservoir" problem? Lithos 112:1043–1054

Parsons B, McKenzie D (1978) Mantle convection and the thermal structure of the plates. J Geophysic Res 83(B9):4485–4496

Pedoja K, Husson L, Regard V, Cobbold PR, Ostanciaux E, Johnson ME, Kershaw S, Saillard M, Martinod J, Furgerot L, Weill P, Delcaillau B (2011) Relative sea-level fall since the last interglacial stage: are coasts uplifting worldwide? Earth Sci Rev. https://doi.org/10.1016/j.earscirev.2011.05.002

Pysklywec RN, Mitrovica JX (1999) The role of subduction-induced subsidence in the evolution of the Karoo Basin. J Geol 107:155–164

Ramos VA (2009) Anatomy and global context of the Andes: main geologic features and the Andean orogenic cycle. In: Kay SM, Ramos VA, Dickinson W (eds) Backbone of the Americas: Shallow subduction, Plateau uplift, and Ridge and Terrane collision. Geol Soc Am, Memoir 204, pp 31–65

Ramos VA, Jordan TE, Allmendinger RW, Mpodozis C, Kay SM, Cortés JM, Palma M (1986) Paleozoic terranes of the central Argentine-Chilean Andes. Tectonics 5(6):855–880

Ricard Y, Chambat F, Lithgow-Bertelloni C (2006) Gravity observations and 3D structure of the Earth. Comptes Rendus de l'Acad (c)mie des Sciences. Comptes Rendus Geoscience 338: 992–1001

Richards MA, Hager BH (1984) Geoid anomalies in a dynamic earth. J Geophys Res 89: 5987–6002

Roberts GG, White N (2010) Estimating uplift rate histories from river profiles using African examples. J Geophys Res: Solid Earth 115(B2)

Roberts GG, White NJ, Martin-Brandis GL, Crosby AG (2012) An uplift history of the Colorado Plateau and its surroundings from inverse modeling of longitudinal river profiles. Tectonics 31(4)

Ruskin BG, Dávila FM, Hoke GD, Jordan TE, Astini RA, Alonso R (2011) Stable isotope composition of middle Miocene carbonates of the Frontal Cordillera and Sierras Pampeanas: Did the Paranaense seaway flood western and central Argentina? Palaeogeogr Palaeoclimatol Palaeoecol 308(3):293–303

Russo RM, VanDecar JC, Comte D, Mocanu VI, Gallego A, Murdie RE (2010) Subduction of the Chile ridge: uppermantle structure and flow. GSA Today 20(9):4–10

Scalabrino B, Lagabrielle Y, Malavieille J, Dominguez S, Melnick D, Espinoza F, Rossello E (2010) A morphotectonic analysis of central Patagonian Cordillera: Negative inversion of the Andean belt over a buried spreading center? Tectonics 29(2)

Shephard GE, Liu L, Gurnis M, Muller RD (2010) Miocene drainage reversal of the Amazon River driven by plate-mantle interaction. Nat Geosci 3:870–875

Shephard GE, Liu L, Müller RD, Gurnis M (2012) Dynamic topography and anomalously negative residual depth of the Argentine Basin. Gondwana Res 22(2):658–663

Shephard GE, Müller RD, Seton M (2013) The tectonic evolution of the Arctic since Pangea breakup: Integrating constraints from surface geology and geophysics with mantle structure. Earth-Sci Rev 124:148–183

Spasojevic S, Liu L, Gurnis M, Muller RD (2008) The case for dynamic subsidence of the United States east coast since the Eocene. Geophys Res Lett 35, L08305.http://dx.doi.org/10.1029/2008GL033511

Spasojevic S, Liu L, Gurnis M (2009) Adjoint models of mantle convection with seismic, plate motion, and stratigraphic constraints: North America since the Late Cretaceous. Geochem Geophys Geosyst 10(Q05W02). https://doi.org/10.1029/2008GC002345

Steinberger B (2007) Effects of latent heat release at phase boundaries on flow in the Earth's mantle, phase boundary topography and dynamic topography at the Earth's surface. Phys Earth Planet Inter 164(1–2):2–20. https://doi.org/10.1016/j.pepi.2007.04.021

Tackley PJ, Stevenson DJ, Glatzmaier GA, Schubert G (1993) Effects of an endothermic phase transition at 670 km depth on a spherical model of convection in the Earth's mantle. Nature 361:699–704

Torsvik TH, Rousse S, Labails C, Smethurst MA (2009) A new scheme for the opening of the South Atlantic Ocean and the dissection of an Aptian salt basin. Geophys J Int 177 (3):1315–1333

Turcotte D, Schubert G (2002) Geodynamics. Cambridge University Press

Tyrrell T, Zeebe RE (2004) History of carbonate ion concentration over the last 100 million years. Geochim Cosmochim Acta 68(17):3521–3530

Winterbourne J, Crosby A, White NJ (2009) Depth, age and dynamic topography of oceanic lithosphere beneath heavily sedimented Atlantic margin. Earth Planet Sci Lett 287(1–2): 137–151. https://doi.org/10.1016/j.epsl.2009.08.019

Zeebe RE (2012) History of seawater carbonate chemistry, atmospheric CO2, and ocean acidification. Annu Rev Earth Planet Sci 40:141–165

Cenozoic Uplift and Exhumation of the Frontal Cordillera Between 30° and 35° S and the Influence of the Subduction Dynamics in the Flat Slab Subduction Context, South Central Andes

Ana C. Lossada, Laura Giambiagi, Gregory Hoke, José Mescua, Julieta Suriano and Manuela Mazzitelli

Abstract This review explores the timing of uplift and exhumation of the Frontal Cordillera range in the South Central Andes between 30° and 35° S summing up the available published evidence up to present. In this segment of the Andes, the Frontal Cordillera straddles a transition from the Chilean–Pampean flat-slab segment to a normal subduction segment and exhibits remarkable along-strike variations in the amount of horizontal shortening, onset of Miocene deformation, and orogenic width, while mean elevations remain steady. None of these variations seem to correspond, in time or space, to the shift in the subduction dynamics; instead, we propose that they represent the expression of inherited features in the continental crust such as the presence of post-Triassic basins or paleo-relief.

Keywords Frontal cordillera · Exhumation · Cenozoic contraction
South Central Andes · Flat slab

A. C. Lossada (✉) · L. Giambiagi · J. Mescua · M. Mazzitelli
Centro Regional de Investigaciones Científicas y Tecnológicas, IANIGLA, CCT Mendoza, Mendoza, Argentina
e-mail: alossada@mendoza-conicet.gob.ar

G. Hoke
Department of Earth Sciences, Syracuse University, Syracuse NY, USA

J. Suriano
IGeBA—FCEyN, UBA-CONICET, Buenos Aires, Argentina

A. Folguera et al. (eds.), *The Evolution of the Chilean-Argentinean Andes*,
Springer Earth System Sciences, https://doi.org/10.1007/978-3-319-67774-3_16

1 Introduction

The Andes of Argentina and Chile between 30° and 35° S are located in a transition zone where several major tectonic transitions occur (Fig. 1). These include a shift in the subduction geometry of the Nazca plate from flat subduction (c. 5–10° at 100 km depth) north of 33° S to normal subduction (~30°) to the south, the southern termination of the Precordillera and Frontal Cordillera morphostructural units, at 33° and 34°40′S, respectively and, finally, the interruption of the Central Depression (~33° S; Fig. 1). In this complex scenario, the Andes shows remarkable along-strike variations in the pattern of deformation, with changes in horizontal shortening, crustal thickening, and mean topographic elevation (Giambiagi et al. 2012). Several causes were proposed to explain the observed trends such as variations in subduction dynamics and age and geometry of the slab (Jordan et al. 1983; Ramos et al. 2004; Ramos 2010; Yáñez and Cembrano 2004), rheological and inherited heterogeneities of the South American plate (Ramos et al. 2004), lithospheric strength variations (Tassara and Yáñez, 2003; Oncken et al. 2006), and enhanced climate-related erosion (Lamb and Davis, 2003), but the degree of influence of each one remains unclear. Despite a north-to-south decrease in crustal shortening, and punctuated drop around 33° S, there is no corresponding change in average elevation or in orogenic width. Instead, changes in orogenic width and mean elevation are coincident with disappearance of the Frontal Cordillera at ~34.5° S. This observation suggests that other mechanisms, in addition to upper-crustal shortening, must be invoked to explain the present topography of the Andes. Classical crustal shortening models propose a diachronous north-to-south onset of late Cenozoic deformation of the Precordillera and the Frontal Cordillera (Ramos et al. 1996, 2002) in connection with the development of flat-slab conditions due to the southward migration of the subducting Juan Fernández aseismic ridge. At these latitudes, the Frontal Cordillera represents most of the high Andes, but attempts to study the pattern of uplift and exhumation, and the latitudinal variations of this range are scarce.

Several studies were carried out, mostly in the Precordillera domain, in attempt to discriminate between the traditional north-to-south diachronous versus synchronous onset of contraction (Giambiagi et al. 2012; Walcek and Hoke, 2012; Levina et al. 2014; Hoke et al. 2014; Suriano et al. 2017); however, there is still no clear consensus. In this review, we attempt to address the controversy related to the Cenozoic uplift and exhumation of the Frontal Cordillera between 30° and 35° S and, by extension, elucidate the feedback mechanisms between topographic uplift, crustal deformation, and subduction dynamics. Differentiating between these two models requires determining whether Neogene contraction in the Frontal Cordillera is largely controlled by a major change in the subduction dynamics, such as the initiation of the flat-slab subduction system, or standard subduction zone stress-building processes and variations in upper-plate crustal strength.

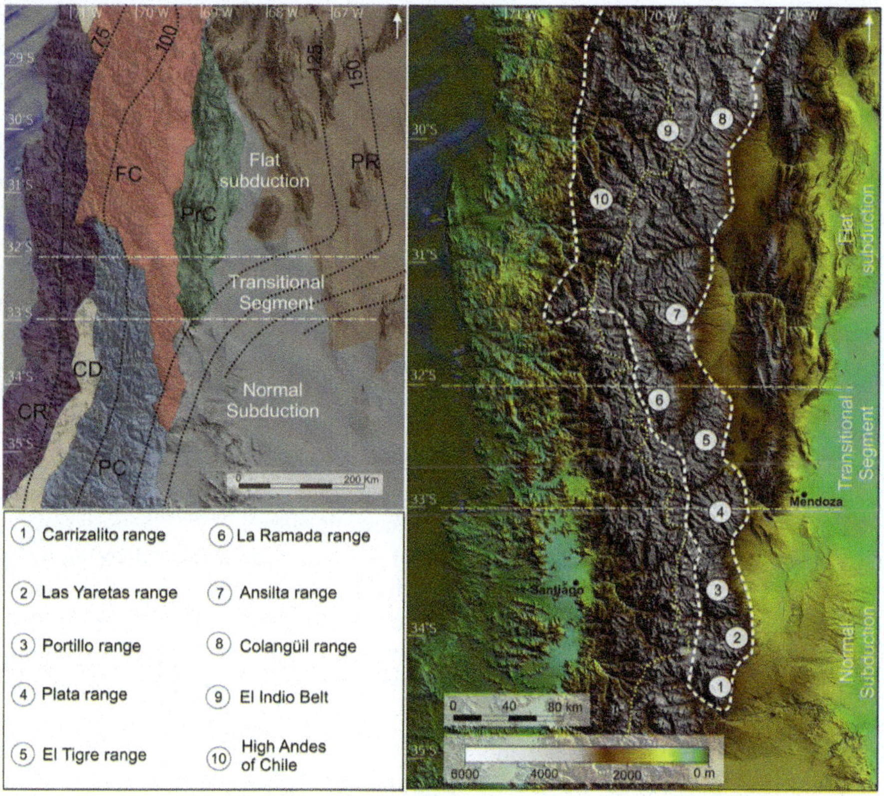

Fig. 1 Morphotectonic units of the Andes between 30° and 35° S in the flat-slab subduction context. Contours of the Wadati–Benioff zone are indicted (Cachill and Isacks 1992). CR: Coastal Range, CD: Central Depression, FC: Frontal Cordillera, PC: Principal Cordillera, PrC: Precordillera, and PR: Pampean Range. Location of the ranges in the Frontal Cordillera unit

2 Tectonic Setting

It is widely accepted that the growth of the Andes is the product of crustal shortening associated with the convergence of the Nazca and South America plates. Thus, it is easy to deduce that the current topography and crustal thickness of the Andes are the result of the uninterrupted subduction of the Pacific plates since at least the Jurassic (Mpodozis and Ramos, 1989; Oliveros et al. 2007; Rossel et al. 2013) (see Chap. "The early stages of the magmatic arc in the Southern Central Andes"). However, the main phase of construction of the Andes occurred during the Miocene times in association with high plate convergence rates between the Nazca and South American plates (Ramos et al. 1996; Somoza 1998; Pardo-Casas and Molnar 1987; Charrier et al. 2007). Here we focus on the development of the Andes in an area that spans the transition between the flat-slab and normal subduction segments, making it a perfect scenario to study the relative influence of different tectonic settings on the uplift of the Andean range. In the flat-slab segment (28°–33° S), high elevations in several tectonic provinces, lack of active volcanism, and the basement-involved foreland deformation characterize the mountain chain.

During the Miocene–Pliocene, deformation and uplift advanced to the foreland, together with arc-related magmatic activity (Kay et al. 1987), sequentially uplifting the Principal Cordillera (early to late Miocene), the Frontal Cordillera (middle to late Miocene), the Precordillera (late Miocene), and the Sierras Pampeanas (<5 Ma). The origin of the flattening of the slab has been linked with the subduction of the buoyant Juan Fernández aseismic ridge (Barazangi and Isacks 1976; Jordan et al. 1983; Yáñez et al. 2002), and traditional models of shortening have postulated a diachronous north-to-south onset of contraction along the Andes that followed the southward propagation of the ridge since ca. 18 Ma (Pilger 1984; Yáñez et al. 2002; Ramos et al. 2002). However, Manea et al. (2012)'s numerical modelings of the generation of the Chilean flat slab suggest that the origin of the flattening could be rather related to the trenchward motion of a thick continental lithosphere and trench rollback instead of subduction of a buoyant oceanic ridge.

The normal subduction segment, south of 33° S, is characterized by an active magmatic arc straddling the main drainage divide, the Tupungatito volcano marks the initiation of the Southern Volcanic Zone, and by the absence of the morphotectonic units of Precordillera and Sierras Pampeanas. The transition between both segments seems to occur as a smooth flexure rather than a sharp tear (Cahill and Isacks 1992; Yáñez et al. 2001).

The Frontal Cordillera is an 800km-long, thick-skinned (basement-involved) range located between the morphotectonic units of Precordillera to the east, and Principal and Coastal Cordilleras to the west (Fig. 1). Originally defined on the eastern slope of the Andes by Groeber (1938), the Frontal Cordillera extends from almost 27° to 34.5° S with many peaks in excess of 6 km and comprises the bulk of the high Andes at these latitudes. North of 31° S, the Frontal Cordillera straddles the drainage divide and is an important component of the western flank of the Andes (Mpodozis and Moscoso 1988).

The Frontal Cordillera is composed of several discrete longitudinally-elongated ranges, separated by transfer oblique zones. From north to south, it corresponds to the Guanta range, El Indio Belt, Colangüil, Ansilta, Santa Cruz, La Ramada-El Espinacito, El Tigre, El Plata, El Portillo, Las Yaretas, and El Carrizalito ranges (Fig. 1). Proterozoic metamorphic rocks, lower Paleozoic metasedimentary and metavolcanic rocks, upper Paleozoic marine deposits, Carboniferous to Permian granitoids, and Permian to Triassic volcanic rocks crop out along these ranges (Polanski 1964, 1972; Heredia et al. 2012). North of 33° S, the Frontal Cordillera is uplifted by a blind ramp, deeply seated in the basement; this structure is inferred to ramp up into the Precordillera fold-and-thrust belt (Allmendinger et al. 1990). South of 33° S, the Frontal Cordillera is bounded by emergent faults, such as the La Carrera fault system (Caminos 1965; Folguera et al. 2004; Casa et al. 2010) consisting of four N to NNE-trending reactivated Permian structures (Giambiagi et al. 2014b).

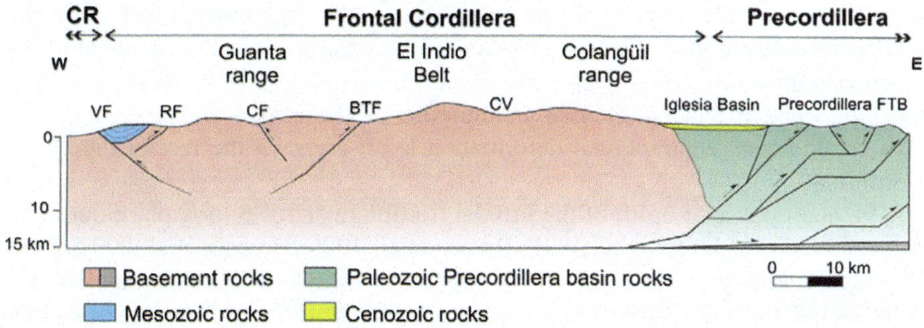

Fig. 2 Schematic cross-section of the El Indio transect (30° S), based on Allmendinger et al. (1999). CR: Coastal Range, VF: Vicuña Fault, RF: Rivadavia Fault, CF: Cuartitos Fault, BTF: Baños del Toro Fault, CV: Cura Valley, FTB: Fault and Thrust Belt

3 Cenozoic Uplift and Exhumation at 30° S: The El Indio Belt

At 30° S, the Frontal Cordillera achieves its maximum width of ~125 km and constitutes most of the Andes of Chile and Argentina. Several N-thrust fault-bounded ranges exist within the Frontal Cordillera. These main ranges are from west to east: the Guanta range which borders the Coastal Range through the Vicuña Thrust Fault; the high peaks of the El Indio Belt in the central portion of the range; and the Colangüil range limiting to the east with the Iglesia/Calingasta Valleys; and the Precordillera further east (Figs. 1 and 2). At this latitude, both the Central Depression in Chile and the Principal Cordillera are absent.

Structural (Pineda and Emparán 2006) and thermochronology (Cembrano et al. 2003; Rodríguez 2013; Lossada et al. 2015, 2017) evidence suggests an Eocene to early Oligocene deformational event affecting the westernmost sector of the Frontal Cordillera, linked to the activity of the Vicuña Fault, in the western slope (Fig. 2). In the eastern slope of the Frontal Cordillera, at the Colangüil range, Fosdick et al. (2015) reported a late Eocene—early Oligocene apatite fission track cooling age, interpreted as an initial uplift of this sector, followed by a main episode of uplift during the middle Miocene. Data from Lossada et al. (2017) also point toward late Eocene cooling ages for the Colangüil range. This pre-Neogene (~45–35 Ma) constructional event, referred to as the Incaic, is widely recognized to the North in the Puna and Altiplano domains (Coira et al. 1982; Isacks 1988; Benevides-Cáseres 1999, among others); however, widespread Paleogene deformation is absent at this latitude (30° S) and further south within the south Central Andes. Instead, at this latitude, available evidence suggests that deformation was concentrated along the eastern and western edges of the Frontal Cordillera (Lossada et al. 2017). Subsequent extension during Oligocene times resulted in the development of the Doña Ana intra-arc (Maksaev et al. 1984; Martin et al. 1995) and the Valle del Cura back-arc (Winocur et al. 2015) basins, which are nearly synchronous with the

Abanico basin (Charrier et al. 2002) to the south. This extensional event was associated with a low absolute velocity of the Nazca plate (Silver et al. 1998) together with a low convergence rate (Pardo-Casas and Molnar, 1987). At 30° S, the Oligocene extension focused in the core of the range, and no evidence of contemporaneous contractional deformation is observed in the rest of the Frontal Cordillera.

The main phase of uplift of the Frontal Cordillera at 30° S took place during the Miocene (Allmendinger et al. 1990; Ramos et al. 1990; Heredia et al. 2002; Bissig et al. 2001; Kay and Mpodozis 2002; Mpodozis and Cornejo 2012; Aguilar et al. 2013; Winocur et al. 2015), and it was focused in the El Indio Belt (Maksaev et al. 1984) and in the Argentinian slope (Fig. 1). Bissig et al. (2002) were the first to describe three discrete periods of deformation by identifying several paleosurface levels between Valle del Cura and the El Indio Belt straddling the international border, which they interpreted as uplifted features. Based on the geochronology of the overlying tuffs, they recognized three pulses of tectonic uplift affecting this portion of the Frontal Cordillera: early to middle Miocene (15–17 Ma), middle Miocene (12.5–14 Ma), and late Miocene (6–10 Ma). The oldest pulses are in good agreement with results from Giambiagi et al.'s (2014a, 2017) structural analysis carried on in the core of the range. They proposed that between 18 and 13 Ma, the zone was under a compressive regime, and deformation and uplift was concentrated near the international border, and then, the tectonic regime changed to strike-slip and deformation migrated eastward into the Precordillera (Giambiagi et al. 2014a, Suriano et al. 2017). Winocur et al. (2015) documented the same early to middle Miocene (from ~18 Ma) contractional episode by analyzing the synorogenic deposits and structures associated with the tectonic inversion of Oligocene extensional structures.

Several thermochronology studies were conducted across the range in an attempt to better constrain its Neogene exhumation history. In the core of the range, Lossada et al. (2017) performed AFT thermochronology and AHe dating in a ~2 km vertical transect, that constrained the onset of Cenozoic deformation to the late Eocene. In addition, 1-D thermal models of AFT and AHe systems from Rodríguez (2013) suggest that this sector was affected by an episode of accelerated exhumation at ~22–18 Ma. This timing of exhumation is coincident with the tectonic inversion of the Oligocene intra-arc basin and a period of high convergence rate (Charrier et al. 2007). As mentioned by Winocur (2010), some Oligocene extensional structures were inverted during the early Miocene, others during the late Miocene, while others were never reactivated and are still preserved as extensional. Rodríguez (2013) also reported a late Miocene (~7 Ma) episode of exhumation, near the crest of the range that correlates in time with the second pulse of tectonic inversion and with the surface uplift event described by Bissig et al. (2002). Using AFT thermochronology, Cembrano et al. (2003) constrained the main phase of exhumation in the same area to be slightly older, 15–10 Ma, which they correlated with the collision of the Juan Fernández ridge at ~30° (Yáñez et al., 2002). On the eastern Andean slope, thermal models of AFT and AHe thermochronology from Fosdick et al. (2015) and Lossada et al. (2017) show an initial uplift of the Colangüil range during late Eocene, but the main phase of exhumation is recorded between 10 and 8 Ma, with a Pliocene ($\sim$5 Ma) reactivation. Finally, Suriano et al.'s (2017) work in the intermontane basin of the western Precordillera indicates that the Frontal Cordillera must have been a positive relief by at least 13 Ma, in concordance with previous works (Allmendinger et al. 1990; Jordan et al. 1993). Collectively, the different lines of evidences point toward two spatially and temporally distinct phases of contractional deformation, horizontal shortening, uplift, and ultimate construction of topography of the Frontal Cordillera at 30° S. The first phase occurred during the Eocene–early Oligocene and affected mainly the westernmost sector, followed by a protracted episode during the Miocene. Miocene activity in the central and eastern sector can be subdivided into discrete early-, middle- and late-Miocene pulses. Localized extension occurred in the arc and back-arc sectors during the Oligocene. In the Bermejo foreland basin and related intermontane within and adjacent to the Precordillera, post-22 Ma foredeep deposits linked with the Miocene constructional event are well recognized. Traditionally, the Eocene–early Oligocene foredeep deposits were absent; however, recent detrital zircon U-Pb dating from Suriano et al. (2017) and Fosdick et al. (2017) of the pre-22 Ma red beds deposits interpreted as Permo-Triassic (Borello and Cuerda, 1968; Limarino et al. 2000) reveal an Eocene depositional age, likely related to this older pulse of tectonic uplift of Frontal Cordillera at 30° S. In order to better constrain this Eocene-early Oligocene phase, more work that describes the generation of orogenic relief and associated synorogenic sedimentation is necessary.

4 Cenozoic Uplift and Exhumation Between 31° and 32° S: The Ramada-Espinacito and Tigre Ranges

This sector of the Frontal Cordillera is composed of, from west to east, the Santa Cruz, Ramada-Espinacito, and the northern end of the Tigre ranges (Fig. 1). These ranges are bounded to the east by west-dipping high-angle faults and are segmented by transverse NNE-trending lineaments in an *en-échélon* pattern (Cristallini and Ramos 2000). The thick-skin structural style results in the tectonic inversion of late Triassic rift systems (Álvarez and Ramos 1999; Cristallini et al. 1995; Ramos et al. 1996; Cristallini and Ramos 2000). To the north and between the area described before, the Ansilta range extends, but unfortunately this sector remains practically unstudied with respect to the timing of exhumation and uplift. The La Ramada massif represents one of the highest non-volcanic ranges of the Andes, with the Cerro Mercedario (6,770 m.a.s.l.) belonging to the five highest mountains in the Americas.

The uplift sequence of these basement blocks is recorded in the strata of the Neogene Chinches Fm. (Mirré 1966) of the Manantiales foreland basin (Pérez 1995; Jordan et al. 1996; Iglesia Llanos 1995). The Manantiales basin is nested between different blocks of the Frontal Cordillera (Fig. 3).

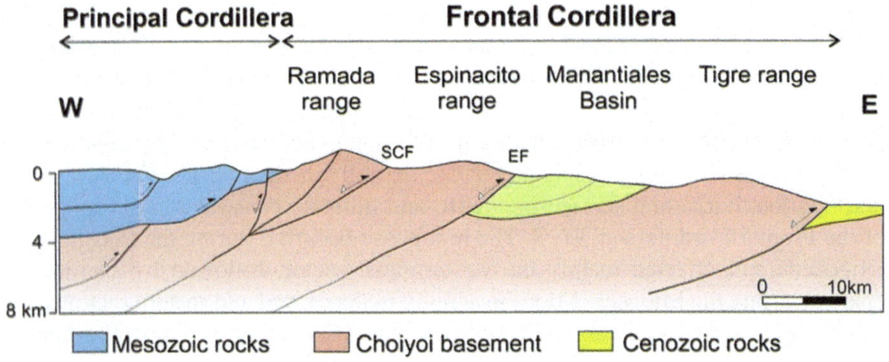

Fig. 3 Schematic cross section of the La Ramada transect (32° S), based on Ramos et al. (1996). SCF: Santa Cruz Fault, EF: Espinacito Fault

The unroofing studies carried out in this basin by Pérez (1995, 2001) and Jordan et al. (1996) allowed them to investigate and constrain the timing of uplift of the Frontal Cordillera blocks, together with the previous uplift (~20 Ma at this latitude) of the Principal Cordillera fold-and-thrust belt further west. Jordan et al.'s (1996) stratigraphic description and magnetic polarity stratigraphy recognized four cycles of coarsening upward facies and constrained the age of initiation of each one at 19, 15.7, 12.5, and 10.5 Ma. They interpreted the first cycle to be related with deformation in the Principal Cordillera, while the second and third episodes were the result of the uplift of the La Ramada (15.7 Ma) and the El Espinacito (12.5 Ma) ranges, respectively. The youngest cycle (10 Ma) is interpreted as reflecting a minor reactivation without the development of a new deformation front.

In contrast, Pérez (1995, 2001) divided the Manantiales basin deposits into 7 members (Tc0–Tc6) and correlated the uplift of the Frontal Cordillera basement blocks with the conglomerates and breccia of the upper member (Tc6). He restricted the episode of uplift and exhumation to between ~14 and 12 Ma, based on the ages of subhorizontal lavas (12.7 ± 0.6 and 10.7 ± 0.7 Ma) in the El Espinacito range and the inferred marine deposits linked with the Paranaense ingression (13.5 Ma, Pérez et al. 1996). Subsequently, deformation migrated to the east after 9 Ma, uplifting the Tigre range, tilting the strata of the Manantiales basin to the west, and isolating it from the Iglesia Basin. The last episode of deformation occurs during Plio-Pleistocene time as a reactivation and final uplift of the El Espinacito range, through an out-of-sequence thrust, evidenced by the strong angular unconformity between the late Miocene and younger deposits. The structural work done by Cristallini and Ramos (2000) is in agreement with Pérez's (1995, 2001) observations. Even though different authors have proposed different stratigraphic interpretations for the Manantiales basin record, they report reconcilable and consistent ages for the uplift of the Frontal Cordillera (Santa Cruz, La Ramada and El Espinacito ranges) between ~15 and 9 Ma.

Alarcón and Pinto (2015) performed a sediment geochemical provenance study of the Chinches Fm. based on sandstone petrography and whole-rock geochemistry (major and trace elements). Following the paleomagnetic reversal stratigraphy of Jordan et al. (1996), they defined three pulses of tectonic uplift: In the first one, at ca. 19 Ma, the main contribution were the intermediate rocks of Principal Cordillera; the second one, between 19 and 12.5 Ma, was characterized by a mixed source of acidic to intermediate rocks, while in the last stage, from 12.5 to 10 Ma, the provenance was markedly acidic. They interpreted the acidic source as corresponding to the volcanic Choiyoi Group in the Frontal Cordillera (El Espinacito range). This implies that the Frontal Cordillera uplift begun at ca. 19 Ma, substantially earlier than previously proposed (Pérez 2001; Jordan et al. 1996). Finally, recent geochronologic analyses constrained the age of the Chinches Fm. in 16.8 Ma (U-Pb on a lapilli tuff at the base of the succession, Mazzitelli et al. 2015), and the detrital zircon provenance analyses also indicated that Frontal Cordillera was already exposed and actively eroding by the early Miocene.

In the Precordillera, north of the Manantiales Basin, at 31° S, Levina et al. (2014) developed a depositional model based on sediment provenance analysis and U-Pb dating in some intermontane basins located within the thrust sheets along the San Juan River. Their findings constrain the onset of activity in Frontal Cordillera at this latitude to between 21 and 17 Ma, with the main phase of shortening between 17–12 Ma, as indicated by the shift in the sedimentation facies to a more energetic regime, and the increment in Permo-Triassic zircon contribution, consistent with the studies discussed above.

The new geochemical (Alarcón and Pinto 2015) and geochronologic (Mazzitelli et al. 2015; Levina et al. 2014) data available indicate pre-Cenozoic exposure of the Choiyoi Group or an older onset of uplift of the Frontal Cordillera blocks at 32° S ($\sim$17 Ma) that contrast traditional models from Jordan et al. (1996) and Pérez (2001).

5 Cenozoic Uplift and Exhumation at 33° S: El Plata Range

Between the flat and normal subduction segments, there is a transition zone (33°–33°30′S) where the subducted slab has a smooth yet abrupt bend to a normal subduction geometry (Cachill and Isacks 1992; Yáñez et al. 2001). As a consequence, arc magmatism is reestablished with the Tupungatito volcano (33°30′S), and the Precordillera and Sierras Pampeanas disappear, narrowing the Andes. The Frontal Cordillera is represented at this latitude by the El Plata range (Fig. 1), which forms a NNE-trending mountain chain more than 6,000 m high, where Paleozoic basement is intruded by Permo-Triassic igneous rocks of the Choiyoi Group and Cenozoic sedimentary cover is exposed (Polanski 1958; Caminos 1965; Heredia et al. 2012). The Tigre range stands to the north and with a similar orientation.

Pre-Andean deformation affecting the Late Devonian to early Carboniferous (Chanic orogeny) and late Carboniferous to Early Permian (San Rafael orogeny) lithologies that comprise part of the Plata range has been reported (Giambiagi et al. 2011, 2014b). Many of these older structures were reactivated during Andean orogeny and control the morphology of the range (Giambiagi et al. 2014b).

The La Carrera fault system (Caminos 1964, 1965; Polanski 1972; Cortés 1993; Folguera et al. 2004; Casa 2005) forms the eastern boundary of the Plata range (Fig. 4) and was responsible for its Cenozoic uplift, although these structures initially developed during the Permian San Rafael orogeny (Giambiagi et al. 2011; Giambiagi et al. 2014b). It comprises four NNE-trending, east-vergent, high-angle reverse faults, of which, the innermost two continue to the north affecting the El Tigre range. The El Plata range is also segmented by transverse NW-trending sinistral strike–slip faults that accommodate the displacement and were interpreted as inherited Paleozoic features (Giambiagi and Martínez 2008). Neotectonic activity has been documented on the La Carrera fault system (Fauqué et al. 2000; Folguera et al. 2004; Casa et al. 2010).

The uplift sequence of the different morphotectonic units at this latitude, the thin-skinned Aconcagua fold-and-thrust belt in the Principal Cordillera, and the uplifted basement block of the El Plata range in Frontal Cordillera (Fig. 4) are recorded in the Cenozoic synorogenic sediments that filled the Cacheuta Basin to the east of the La Carrera fault system (Ramos 1988). Between 9 to 6 Ma, sedimentation in the Cacheuta Basin is correlated with exhumation of the Frontal Cordillera to the west on the basis of conglomerate clast count, Ar/Ar dating, and magnetostratigraphic study (Irigoyen 1997; Irigoyen et al. 1999, 2000). This is in good agreement with the integrated progression of deformation and arc migration to the foreland proposed by Ramos et al. (2002) at the latitude of 33° S. AHe data

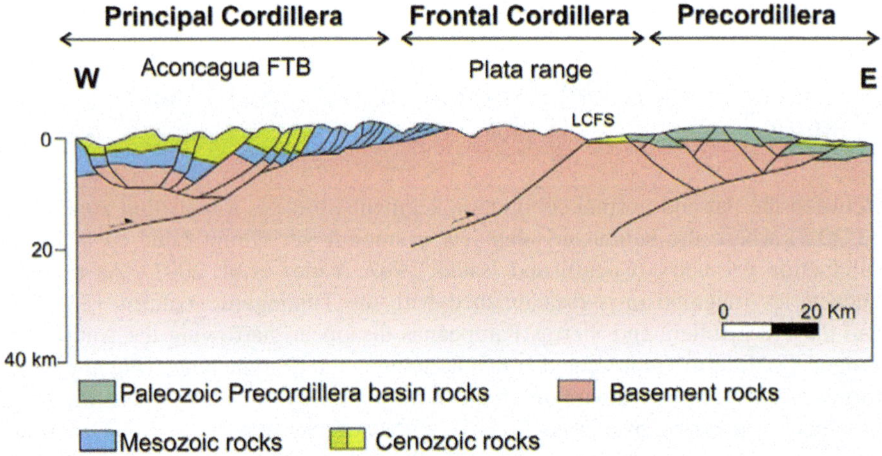

Fig. 4 Schematic cross section of the Aconcagua transect (33° S), based on Jara et al. (2015), Cegarra and Ramos (1996), and Giambiagi et al. (2011). LCFS: La Carrera Fault System

from a tributary of the Mendoza River in the El Plata range (Hoke et al. 2015) are complex, yet it is possible to identify accelerated exhumation at ~25 Ma until c. 10 Ma, with a subsequent rapid river incision and exhumation until present. Walcek and Hoke (2012) identified a similar ~10 Ma timing of uplift of the southern Precordillera based on geomorphic analysis and recognized significant preexisting relief in the region.

Recent sedimentologic, geochronologic (detrital U-Pb), and thermochronologic ((U-Th)/He) from the Cacheuta basin (Buelow et al. 2015) constrain orogenic exhumation and indicate a ~16 Ma uplift pulse of Frontal Cordillera, substantially older than previous estimates.

6 Cenozoic Uplift and Exhumation at 34° S: Portillo, Yaretas, and Carrizalito Ranges

Here, wholly within the normal subduction zone, the southern terminus of the Frontal Cordillera occurs at ~34.5° S, marking a major tectonic and physiographic transition. From north to south, the individual ranges are Portillo, Yaretas, and Carrizalito (Fig. 1). These ranges are bounded to the east by several east-vergent deeply seated faults of the Portillo fault system (Polanski 1958; Kozlowski et al. 1993; Giambiagi et al. 2003) in the north, and the Carrizalito fault (Baldi et al. 1984; Kozloswki et al. 1989; Nullo and Stephens 1993; Turienzo and Dimieri 2005) in the south (Fig. 5), which juxtaposes pre-Jurassic basement rocks over the middle Miocene to Quaternary sedimentary rocks deposited in the foreland basin (Polanski 1964; Giambiagi et al. 2003). At this latitude, the Frontal Cordillera presents its minimum width (~20 km), while the topographic mean elevation remains approximately invariant.

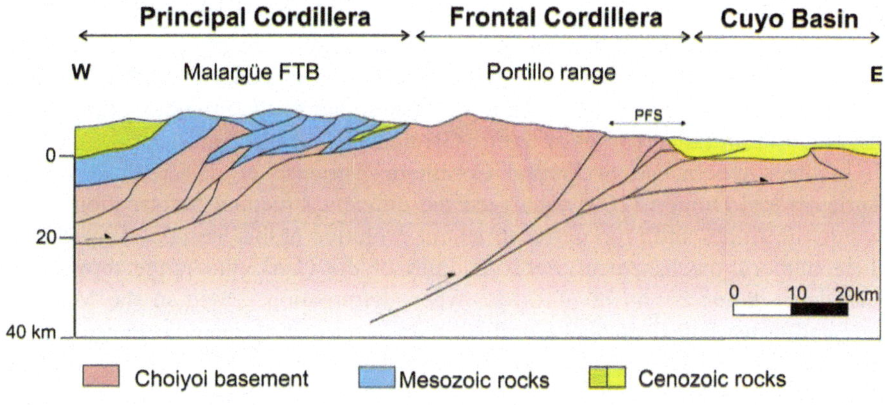

Fig. 5 Schematic cross section of the Maipo-Tunuyán transect (34° S), based on Giambiagi et al. (2015). PFS: Portillo Fault System

In the Portillo range, the northern region was affected by NNE-trending faults, whereas in the south, the structures have N to NNW trends related to late Paleozoic structures. The Yaretas range was uplifted along two NNW-trending high-angle faults. The Neogene Alto Tunuyán basin extends between the Frontal Cordillera and Principal Cordillera and contains synorogenic deposits associated with the uplift of these ranges (Fig. 5).

Giambiagi et al. (1999, 2001, and 2003) characterized the structural evolution of the principal morphostructural units at this latitude and established a sequence of deformation by performing clast counts and paleocurrent analysis in the Alto Tunuyán basin. Ages were assigned based on lithostratigraphic correlation with the Cuyo foreland basin, where radiometric and magnetostratigraphic data are available (Irigoyen et al. 2000). They proposed an upper late Miocene (9–6 Ma) uplift of the Portillo range and a Pliocene (6–2 Ma) uplift of the Yaretas range. Surface uplift estimates from latest Miocene to present in the Alto Tunuyán basin (Hoke et al. 2014) and in the western flank of the range (Farías et al. 2008) are consistent with this timing of deformation.

Recent provenance analysis based on petrography and geochemistry of sandstones (Porras et al. 2016) in the Alto Tunuyán basin identified an older signal of erosional products from the Frontal Cordillera between 15 and 11 Ma, interpreted as either (i) a peripheral bulge related to tectonic load in Principal Cordillera and developed in the actual position of the Frontal Cordillera, forming a high, erodible relief; or (ii) the results of recycling material from the Mesozoic sequences in the Principal Cordillera. Other explanation for this middle Miocene Frontal Cordillera signal is the proposed hypothesis of a proto-Frontal Cordillera local relief (Giambiagi et al. 2014b; Hoke et al. 2015), possibly resulted of the compressional San Rafael orogeny during late Paleozoic times. We will return to this idea in the discussion below.

The integrated Miocene to present kinematic model proposed by Giambiagi et al. (2015) for the Tunuyán-Maipo transect supports an upper late Miocene uplift of Frontal Cordillera at this latitude, coincidently with the achievement of a maximum crustal thickness of 50 km (Kay et al. 2005) and the need for the critical wedge to grow laterally in width instead of increasing in height, advancing to the east, and uplifting the Portillo range. Neotectonic activity is evidenced by shallow compressional focal mechanisms beneath the eastern part of the Frontal Cordillera and the Cuyo basin (Alvarado et al. 2007; García and Casa 2015).

In the southern end of Frontal Cordillera, Turienzo et al. (2010; 2012) and Turienzo and Dimieri (2005) analyzed the different structural deformational style within the Andes and proposed a structural evolution at this latitude. According to these authors, the basement-involved uplift of the Carrizalito range through the Carrizalito thrust begun at ~10 Ma, when deformation ceased in the Malargüe fold-and-thrust belt (Principal Cordillera), and ended in the middle Pleistocene as indicated by the age of the horizontal basalt that discordantly covers the structure (0.5 Ma, Barker et al. 2009). Neotectonic activity has also been documented in this area (Cortés and Srouga, 1998; Casa et al. 2011).

Taken together, the rise of the Portillo and the Carrizalito ranges would have occurred penecontemporaneously at around 10 Ma. The Yaretas range rose at 6 Ma, in concordance with a protracted deformation in the south and in the north. Thus, there is no systematic along-strike variation in the timing of deformation of these ranges, similar to that observed elsewhere in thick-skinned parts of the Andes (Strecker et al. 2009).

7 Along-Strike Variations in the Frontal Cordillera Range

As previously discussed, the South Central Andes exhibits remarkable along-strike variations in their structural and morphological characteristics, despite sharing a roughly common tectonic evolution. In this section, we identify and analyze the spatial and temporal patterns of deformation and mountain geometry of the Frontal Cordillera range since it is an important constituent to Andean orogeny between 30° and 35° S. We also discuss whether the observed variations are the result of the subduction geometry or if other causes need to be invoked.

7.1 Geometric Features of the Frontal Cordillera

Most of the topographic relief of the Frontal Cordillera is the result of crustal shortening and ensuing isostatic adjustments to the thickened crust. The presence of low-relief, relict landscape in the El Plata range suggests low rates of erosional exhumation and therefore nearly pure surface uplift, since the late middle Miocene (Hoke et al. 2015). For each latitude, we calculated the mean, maximum, and minimum elevation in the Frontal Cordillera range, in 20-km-wide swath profiles (Fig. 6a). Although maximum elevations reach about 6 km, the calculated mean topography of the Frontal Cordillera maintains a stable ~3.5 km elevation along different latitudes.

In contrast with the consistent mean elevation, the orogenic width of the Frontal Cordillera progressively decreases to the south, from ~140 km at 30° S to 0 km at 34°40′S where it is no longer present (Fig. 6b). This decrease in width is the consequence of a decrease in the amount of crustal blocks that comprise the range, with several blocks of both east and west-vergence in the north, to only one east-vergent block in the south. The eastern margin of the range is located at a fixed longitude, whereas the western margin is controlled by the extent of the Principal Cordillera (Fig. 1), which corresponds in turn to the presence of post-Triassic basins (the Neuquén Basin and the Abanico Basin) that favored the structural development of the thin-skinned fold-and-thrust belts.

Overall, range geometries summarized here not reflect any easily recognizable link (the exception of volcanism) to the dynamics of subducting slab; rather, they are correlated with the intrinsic characteristics of the continental lithosphere

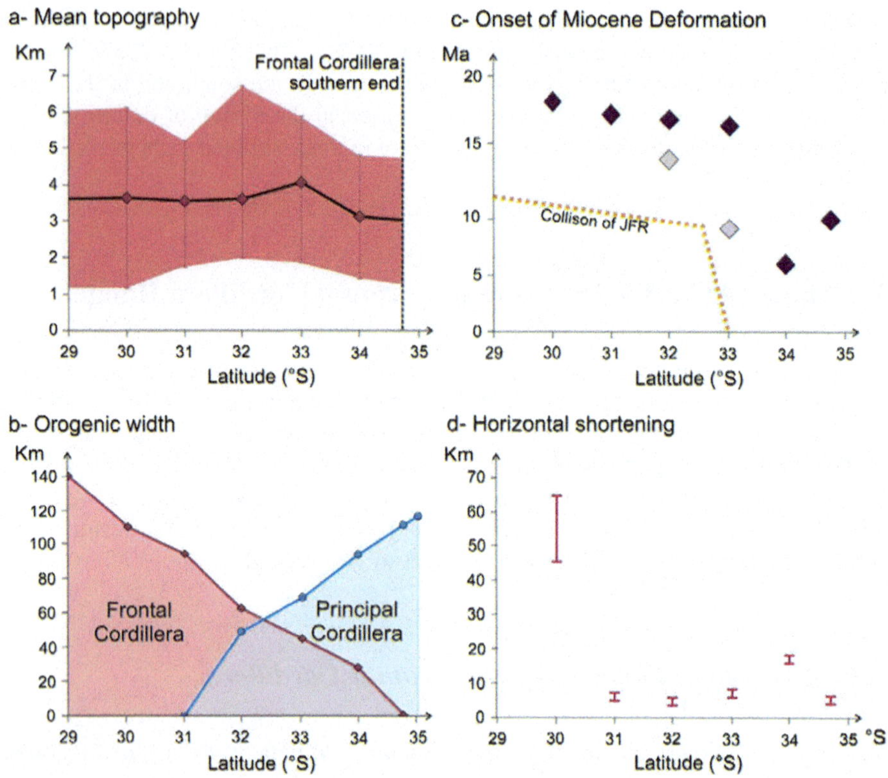

Fig. 6 Along-strike variations in the Frontal Cordillera range. **a** Minimum, mean, and maximum elevation of Frontal Cordillera range (20km-wide swath profiles). **b** Orogenic width of Frontal Cordillera and Principal Cordillera morphotectonic units. Note that the extension of the first is controlled by the development of the second one. **c** Onset of Miocene contraction in Frontal Cordillera (purple diamond) at each latitude (reference in text). At the latitudes of 32° and 33° S, pale purple diamonds indicate the traditional uplift dates against purple diamonds that indicate the recent datings (reference in text). Chronology of the collision of Juan Fernández ridge (JFR) is shown (orange path, modified from Yáñez et al. 2002). **d** Horizontal shortening data available for the Frontal Cordillera (reference in text)

inherited from late Paleozoic to late Mesozoic tectonic events, as the presence of Pre-Andean structures and thick sedimentary sequences (Mesozoic and Cenozoic basins).

7.2 *Structural Features of the Frontal Cordillera*

The Andean cordillera exhibits along-strike and intra-range variation in horizontal shortening, with an overall decrease toward the south (Ramos et al. 1996; Heredia et al. 2002; Mescua et al. 2016). The Frontal Cordillera acted as a rigid block accounting for little horizontal shortening during the Andean orogeny, while the majority of shortening of the Andes is considered to be accommodated in the fold-and-thrust belts of the Precordillera in the northern sector and Principal Cordillera toward the south. There are only a handful of studies that attempt to estimate the horizontal shortening in Frontal Cordillera, largely because the deep structure of the range is not well constrained, and the monotonous lithologies comprising most outcrops in the range are difficult to correlate across structures. Therefore, data on the amount of horizontal crustal shortening in the Frontal Cordillera are scarce, and estimates that do exist contain important caveats and assumptions made by each set of authors when modeling the structural evolution. To the north of the study area, at 30º S, Allmendinger et al. (1990) made a first-order estimation of total Cenozoic shortening in the Frontal Cordillera, obtaining ~45–65 km; however, how that shortening was partitioned in time between the Eocene to early Oligocene and the Miocene phases is unknown. At 31° S, Rodríguez Fernández et al. (1999) and Heredia et al. (2002) obtained a ~7 km (10%) of horizontal shortening calculated from their structural cross section. In the La Ramada cross section at 32° S, Cristallini and Ramos (2000) proposed tectonic inversion as the main mechanism responsible of block uplift, resulting in ~4 km of shortening. At 33° S, Ramos et al. (1996) calculated 8 km of shortening for the Frontal Cordillera in the Aconcagua cross section. In the southernmost part of Frontal Cordillera range, Giambiagi et al. (2015) estimated 17 km at ~34° S, and 5 km at 34°40′S, while Turienzo et al. (2012) arrived at a similar amount of horizontal shortening (4 km) for the Carrizalito range. The various estimates are summarized in Fig. 6d. More and better constrained shortening data are needed in order to model the contraction experienced in the Frontal Cordillera and the possible along-strike variations. Exhumation estimates gleaned through thermochronology may led to better constrains on the timing and magnitude of shortening of the Frontal Cordillera, allowing to discriminate between the exhumation events affecting the range, and within them, the different contractional phases that have acted.

The previous sections review the available evidence for the timing of exhumation and uplift of the Frontal Cordillera, from structural analysis, dating of synorogenic deposits and, in a less degree, thermochronology and paleoaltimetry estimates, mainly during the Miocene. Data referring to the onset of Miocene deformation in this range are summarized in Fig. 6c. If the new data from the Cacheuta (Buelow et al. 2015) and Manantiales (Mazzitelli et al. 2015, Alarcón and Pinto 2015) basins are incorporated, Miocene contraction proves to have been synchronous in the northern and central sectors (~18 Ma at 30° and ~16 at 33° S), with a later onset in the south (~10 Ma, Hoke et al. 2014). When comparing these data with the arrival of the Juan Fernández ridge at these latitudes (Fig. 6c, Yáñez et al. 2002), there appears to be no causal relationship; instead, deformation in the Frontal Cordillera predates the subduction of the aseismic ridge by ~5 Ma. Given the complexity of constrain accurately the time of uplift and deformation, the use of thermochronometers appears to be a powerful tool to get absolutely dating.

8 Discussions and Conclusion

In this chapter, we reviewed the data available concerning the timing of Cenozoic uplift and exhumation of the Frontal Cordillera range in the South Central Andes between 30° and 35° S and analyzed the along-strike variations in geometrical and structural features. Upper-crustal shortening, orogenic width, and the onset of Miocene deformation decrease to the south, while topography remains almost steady. Given the paucity of reliable shortening calculations for the Frontal Cordillera, more work is needed to better understand the structural evolution of the range and the linkages between shallow and deep structures that will allow an accurate characterization of what governs how the crust deforms, as well as the growth and maintenance of Andean topography. The orogenic width of the range decreases to the south, while the post-Triassic rift basins broaden. Ultimately, these rift basins favored the development of the fold-and-thrust belts in Principal Cordillera. Miocene deformation—the main pulse responsible for the Frontal Cordillera Cenozoic uplift—started first in the northern (~18 Ma at 30° S) and central (~16 Ma at 32° and 33° S) sectors and progressed to the south (10–5 Ma south of 33° S). Although the Frontal Cordillera lies within the transition from flat subduction to normal subduction segments, the observed variations cannot be correlated with this first-order change in the subduction dynamics, and other causes need to be invoked. Moreover, the onset of Miocene contraction predates the collision of the buoyant Juan Fernández ridge, which is widely considered the trigger for flattening of the subducting Nazca Plate. Thus, models that tie the onset of deformation with a Juan Fernández Ridge that induced flat-slab development (Pilger et al. 1984; Ramos et al. 1996, Yáñez et al. 2001, among others) are not applicable to the Frontal Cordillera (this work) nor Precordillera ranges (Giambiagi et al. 2012; Walcek and Hoke 2012; Levina et al. 2014; Suriano et al. 2017). As an alternative, we postulate that the diachronous onset of Miocene deformation observed in the Frontal Cordillera range is the product of the initial conditions set by the last major phase of deformation experienced (*i.e.,* related to inherited features of the continental crust). In this vain, we propose that contraction began earlier in the north, where Eocene constructive phase already built a substantial crustal root forcing the propagation of the deformation toward the foreland. Instead, in the south, where the Oligocene extension favored the development of the Abanico Basin, Miocene contraction would have needed to generate a crustal root thick enough for deformation to propagate into the Frontal Cordillera domain.

Receiver functions from passive source geophysics reveal a >60 km depth to the Moho in the north (at 31° S) which thins to ~55 km by 33.5° S (Gans et al. 2011), reinforcing the idea of pre-Miocene established root in the north. Moreover, between 31° and 32° S, receiver function geophysics from Ammirati et al. (2016) points toward similar depth to the Moho (>65 km) and the existence of eclogite facies beneath the Frontal Cordillera, based on the observation of a significant increase in shear-wave velocities. If the crust beneath Frontal Cordillera in the northern sector has already been through the transition to felsic granulite and/or ecoglite, this could explain why mean topography remains almost steady while crustal root decrease to the south, as suggested by Hoke et al. (2014).

Although the Miocene pulse is considered the main phase of construction of the Frontal Cordillera and responsible for its present configuration, there are several lines of evidence that point toward the importance of previous constructional events affecting different sectors of the range. These contractional events had resulted in the thickening of the crust and have the potential of influencing in the flattening of the slab; numerical models (Manea et al. 2012) showed that trenchward motion of thickened continental lithosphere increases the suction between ocean and continent and may result in flat subduction. In the northern sector (30° S), structural (Pineda and Emparán 2006) and thermochronometry studies (Cembrano et al. 2003; Rodríguez 2013; Lossada et al. 2015, 2017) emphasized the creation of Eocene relief. In the southern sector (33° S), recent provenance analyses in the Cacheuta Basin (Buelow et al. 2015), structural features (Giambiagi et al. 2014b), and thermochronometric constrains (Hoke et al. 2015) in the El Plata range suggest the existence of a paleo-relief in the Frontal Cordillera that could be a remnant of a Permian deformation phase. In order to better understand the influence of paleo-relief in the configuration of the Frontal Cordillera range as observed today and its relationship with the flat slab, if any, more estimates of paleoelevation, exhumation, and improved chronologic constraints on basin deposits are needed.

Taking into account the discussion above and regarding the question that promoted this review, we propose that inherited features of the continental lithosphere exert an important control on the tectonic characteristics and along-strike variations observed in the Frontal Cordillera today. Changes in subduction dynamics from normal to flat-slab configurations do not need to be invoked to explain these variations.

Acknowledgements Ana Lossada acknowledges the Tectonics Group—IANIGLA for the stimulating discussions. D. Yagupsky is thanked for a critical review of an early version of the manuscript, and the Chilean Geological and Mining Survey for the fieldwork facilities. This research was supported by a grant from the Argentine Agencia de Promoción Científica y Tecnológica (PICT 2011–1079) and an Argentine Presidential-Fulbright Fellowship in Science and Technology.

References

Aguilar G, Riquelme R, Martinod J, Darrozes J (2013) Rol del clima y la tectónica en la evolución geomorfológica de los Andes Semiáridos chilenos entre los 27–32° S. Andean Geol 40(1): 79–101

Alarcón P, Pinto L (2015) Neogene erosion of the Andean Cordillera in the flat-slab segment as indicated by petrography and whole-rock geochemestry from the Manatiales Foreland Basin (32°–32°30′S). Tectonophysics 639:1–22

Allmendinger RW, Figueroa D, Snyder D, Beer C, Mpodozis C, Isacks BL (1990) Foreland shortening and crustal balancing in the andes at 30° S latitude. Tectonics 9:789–809

Alvarado P, Beck S, Zandt G (2007) Crustal structure of the south-central Andes Cordillera and backarc region from regional waveform modeling.Geophys J Inter 170:858–875
Álvarez PP, Ramos VA (1999) The Mercedario rift system in the principal Cordillera of Argentina and Chile (32° SL). J South Am Earth Sci 12:17–31
Ammirati J-B, Pérez Luján S, Alvarado P, Beck S, Rocher S, Zandt G (2016) High-resolution images above the Pampean flat slab of Argentina (31–32° S) from local receiver functions: Implications on regional tectonics. Earth and Planet Sci Let 450:29–39
Baker S, Gosse J, McDonald E, Evenson E, Martínez O (2009) Quaternary history of the piedmont reach of Río Diamante, Argentina. J South Am Earth Sci 28:54–73
Baldi J, Ferrante R, Ferrante V, Martinez R (1984) Estructuras de bloques y su importancia petrolera en el ámbito Mendocino de la cuenca Neuquina. IX Cong Geol Arg, Bariloche 4:153–161
Barazangi M, Isacks BL (1976) Spatial distribution of earth quakes and subduction of the Nazca plate beneath South America. Geol 4:606–692
Benavides-Caceres V (1999) Orogenic evolution of the Peruvian Andes; the Andean Cycle. In: Skinner, BJ (ed.) Geology and ore deposits of the Central Andes. Society of Economic Geologists, pp 61–10
Bissig T, Lee JK, Clark AH, Heather KB (2001) The cenozoic history of volcanism and hydrothermal alteration in the central andean flat-slab region: new 40Ar-39Ar constraints from the El Indio–Pascua Au (-Ag, Cu) belt, 29° 20′–30° 30′ S. Int Geol Rev, 43(4):312–340
Bissig T, Clark AH, Lee JK, Hodgson CJ (2002) Miocene landscape evolution and geomorphologic controls on epithermal processes in the El Indio-Pascua Au-Ag-Cu belt. Chile and Argentina. Economic Geol 97(5):971–996
Borello A, Cuerda A (1968) Grupo Río Huaco (Triásico), San Juan. Com. Investig. Científicas la Prov. Buenos Aires 7:3–15
Buelow E, Suriano J (2015) Evolution of the Neogene Cacheuta Basin: a record of orogenic exhumation and basin inversion in the South Central Andes. GSA Metting
Cachill T, Isacks BL (1992) Seismicity and shape of the subducted Nazca plate. J Geophys Res 97:17503–17529
Caminos R (1964) Estratigrafía y tectónica del Espolón de la Carrera, Cordón del Plata, provincia de Mendoza. PhD Thesis, Universidad Nacional de Buenos Aires, p 145
Caminos R (1965) Geología de la vertiente oriental del Cordón del Plata, Cordillera Frontal de Mendoza. Rev Asoc Geol Arg 20(3):351–392
Casa AL (2005) Geología y neotectónica del piedemonte oriental del cordón del Plata en los alrededores de El Salto. Thesis degree, Universidad de Buenos Aires, 174 p
Casa AL, Borgnia MM, Cortés JM (2010) Evidencias de deformación pleistocena en el sistema de falla de La Carrera (32°40′–33°15′LS), Cordillera Frontal de Mendoza. Rev Asoc Geol Arg 67:91–104
Casa A, Cortés JM, Rapalini A (2011) Fallamiento activo y modifi cación del drenaje en el piedemonte del cordón del Carrizalito, Mendoza. XVIII Congreso Geológico Argentino (Neuquén) pp 712–713
Cembrano J, Zentilli M, Grist A, Yáñez G (2003) Nuevas edades de trazas de fisión para Chile Central (30°–34° S): Implicancias en el alzamiento y exhuma-ción de los Andes desde el Cretácico. X Congreso Geológico Chileno (Concepción)
Charrier R, Baeza O, Elgueta S, Flynn JJ, Gans P, Kay SM, Munoz N, Wyss AR, Zurita E (2002) Evidence for Cenozoic extensional basin development and tectonic inversion south of the flat-slab segment, southern Central Andes, Chile (33°–36° S). J South Am Earth Sci 15(1): 117–139
Charrier R, Pinto L, Rodríguez M (2007) Tectonostratigraphic evolution of the Andean Orogen in Chile. In: Moreno T, Gibbons W (eds) The Geology of Chile. Geological Society, London, pp 21–114
Coira B, Davidson J, Mpodozis C, Ramos V (1982) Tectonic and Magmatic Evolution of the Andes of Northern Argentina and Chile. Earth Sci Rev 18(3–4):303–332

Cortés JM (1993) El frente de la Cordillera Frontal y el extremo sur del valle de Uspallata, Mendoza. XII Cong Geol Arg (Mendoza), pp 168–178

Cortes JM, Sruoga P (1998) Zonas de fractura cuaternarias y volcanismo asociado en el piedemonte de la Cordillera Frontal (34°30′S), Argentina. X Congreso Latinoamericano de Geologia (Buenos Aires) 2:116–121

Cristallini EO, Ramos VA (2000) Thick-skinned and thin-skinned thrusting in the La Ramada fold and thrust belt: crustal evolution of the High Andes of San Juan, Argentina (32 SL). Tectonophysics 317(3):205–235

Cristallini E, Mosquera A, Ramos V (1995) Estructura de la Alta Cordillera de San Juan. Rev Asoc Geol Arg 49(1–2):165–183

Fauqué L, Cortés JM, Folguera A, Etcheverría M (2000) Avalanchas de rocas asociadas a neotectónica en el valle del río Mendoza, al sur de Uspallata. Rev Asoc Geol Arg 55:419–423

Folguera A, Etcheverria M, Pazos P, Giambiagi L, Cortés JM, Fauqué L, Fusari C, Rodríguez MF (2004) Descripción de la Hoja Geológica N° 3369–15 (Potrerillos). Carta Geologica de la República Argentina E. 1:100.000. Subsecretaría de Minería de la Nación, Dirección Nacional del Servicio Geológico, 262 p

Fosdick JC, Carrapa B, Ortíz G (2015) Faulting and erosion in the Argentine Precordillera during changes in subduction regime: reconciling bedrock cooling and detrital records. Earth Planet Sci Lett 432:73–83

Fosdick JC, Reat EJ, Carrapa B, Ortiz G, Alvarado PM (2017) Retroarc basin reorganization and aridification during paleogene uplift of the southern central andes. Tectonics, 36(3):493–514

Gans CR, Beck SL, Zandt G, Gilbert H, Alvarado P, Anderson M, Linkimer L (2011) Continental and oceanic crustal structure of the Pampean flat slab region, western Argentina, using receiver function analysis: new high-resolution results. Geophys J Int 186(1):45–58

García VH, Casa A (2015) Quaternary tectonics and seismic potential of the Andean retrowedge at 33°–34° S. In: Sepúlveda SA, Giambiagi LB, Moreiras SM, Pinto L, Tunik M, Hoke G, Farías M (eds.) Geodynamic Processes in the Andes of Central Chile and Argentina. Geol Soc, London, Sp Publ 399:311–327

Giambiagi LB (1999) Interpretación tectónica de los depósitos neógenos de la cuenca de antepaís del Alto Tunuyán, en la región del Río Palomares, cordillera principal de Mendoza. Rev Asoc Geol Argent 54:361–374

Giambiagi L, Martinez AN (2008) Permo-Triassic oblique ex-tension in the Uspallata-Potrerillos area, western Argentina. J South Am Earth Sci 26:252–260

Giambiagi L, Ramos V, Godoy E, Alvarez PP, Orts S (2003) Cenozoic deformation and tectonic style of the Andes, between 33° and 34° south latitude. Tectonics 22(4):1041

Giambiagi L, Mescua J, Bechis F, Martínez A, Folguera A (2011) Pre-Andean deformation of the Precordillera southern sector, Southern Central Andes. Geosphere 7:219–239

Giambiagi L, Mescua J, Bechis F, Tassara A, Hoke G (2012) Thrust belts of the Southern Central Andes: along strike variations in shortening, topography, crustal geometry and denudation. Geol Soc Am Bull 124(7–8):1339–1351

Giambiagi L, Álvarez PP, Godoy E, Polanco E (2014a) Modelo cinemático 3D e interpretación dinámica de estructuras en la región de El Elqui, Alta Cordillera de Chile Central (30° S). XIX Congreso Geológico Argentino, Córdoba, Argentina

Giambiagi L, Mescua J, Heredia N, Farías P, García Sansegundo J, Fernández C, Stier S, Pérez D, Bechis F, Moreiras SM, Lossada A (2014b) Reactivation of Paleozoic structures during Cenozoic deformation in the Cordón del Plata and Southern Precordillera ranges (Mendoza, Argentina). J Iberian Geol 40(2):309–320

Giambiagi L, Tassara A, Mescua J, Tunik M, Alvarez P, Godoy E, Hoke G, Pinto L, Spagnotto S, Porras H, Tapia F, Jara P, Bechis F, García V, Suriano J, Moreiras S, Pagano S (2015) Evolution of shallow and deep structures along the Maipo-Tunuyán transect (33°40º S): from the Pacific coast to the Andean foreland. In: Sepúlveda SA, Giambiagi LB, Moreiras SM, Pinto L, Tunik M, Hoke GD, Farías M (eds.) Geodynamic Processes in the Andes of Central Chile and Argentina. Geol Soc, Sp Publ 399:63–82

Giambiagi L, Álvarez PP, Creixell C, Mardonez D, Murillo I, Velásquez R, Lossada A, Suriano J, Mescua J, Barrionuevo M (2017) Cenozoic shift from compression to strike-slip stress regime in the high Andes at 30°S, during the shallowing of the slab: implications for the El Indio/Tambo mineral district. Tectonics

Groeber P (1938) Mineralogía y Geología. Espasa-Calpe, Argentina pp. 492

Heredia N, Fernández LR, Gallastegui G, Busquets P, Colombo F (2002) Geological setting of the argentine frontal cordillera in the flat-slab segment (30° 00′–31° 30′ S latitude). J S Am Earth Sci 15(1):79–99

Heredia N, Farías P, García Sansegundo J, Giambiagi L (2012) The basement of the Andean Frontal Cordillera in the Cordón del Plata (Mendoza, Argentina): geodynamic evolution. Andean Geol 39:242–257

Hoke G, Giambiagi L, Garzione C, Mahoney B, Strecker M (2014) Neogene paleoelevation of intermontane basins in a narrow, compressional mountain range, southern Central Andes of Argentina. Earth Planet Sci Lett 406:153–164

Hoke G, Graber N, Mescua J, Giambiagi L, Fitzgerald P, Metcalf J (2015) Near pure surface uplift of the Argentine Frontal Cordillera: insights from (U-Th)/He thermochronometry and geomorphic analysis. In: Sepúlveda et al (eds) Geodynamic Processes in the Andes of Central Chile and Argentina, Geolog Soc of London Sp Publ 339

Iglesia Llanos P (1995) Geología del area de Manantiales al este del cordón del Espinacito, Provincia de San Juan. Rev Asoc Geol Arg 50(1–4):195–211

Irigoyen MV (1997) Magnetic polarity stratigraphy and geochronological constraints on the sequence of thrusting in the Principal and Frontal cordilleras and the Precordillera of the Argentine Central Andes (33° S latitude). PhD thesis, Carleton University, Ottawa pp 392

Irigoyen MV, Villeneuve ME, Quigg F (1999) Calibration of a Neogene magnetostratigraphy by 40Ar-39Ar geochronology: The foreland-basin strata of northern Mendoza Province,Argentina. In Radiogenic age and isotopic studies: Report 12, Geological Survey of Canada, Current research, 1999-F: 27–41

Irigoyen MV, Buchan KL, Brown RL (2000) Magnetostratigraphy of Neogene Andean foreland-basin strata, lat 33° S, Mendoza province, Argentina. GSA Bull 112:803–816

Isacks B (1988) Uplift of the Central Andean Plateau and bending of the Bolvian Orocline. J Geophys Res 93:3211–3231

Jara P, Likerman J, Winocur D, Ghiglione M, Cristalini E, Pinto L, Charrier R (2015) Role of basin width variation in tectonic inversion: insight from analogue modeling and implications for the tectonic inversion of the Abanico Basin, 32°–34° S, Central Andes. In: Sepúlveda et al (eds) Geodynamic Processes in the Andes of Central Chile and Argentina, Geolog Soc of London Sp Publ 339: 83–107

Jordan T, Isacks BL, Allmendinger RW, Brewer JA, Ramos VA, Ando CJ (1983) Andean tectonics related to geometry of subducted Nazca Plate. Geol Soc Am Bull 94:341–361

Jordan TE, Allmendinger RW, Damanti JF, Drake RE (1993) Chronology of motion in a complete thrust belt: the precordillera, 30-31Â°S, andes mountains. J Geol 101(2):135–156

Jordan TE, Tamm V, Figueroa G, Flemings PB, Richards D, Tabbutt K, Cheatham T (1996) Development of the Miocene Manantiales foreland basin, Principal Cordillera, San Juan. Argentina. Andean Geol 23(1):43–79

Kay SM, Maksaev V, Moscoso R, Mpodozis C, Nasi C, Gordillo CE (1987) Tertiary Andean magmatism in Chile and Argentina between 28° S and 33° S: correlation of magmatic chemistry with a changing Benioff zone. J South Am Earth Sci 1(1):21–38

Kay S, Mpodozis C (2002) Magmatism as a probe to the neogene shallowing of the nazca plate beneath the modern chilean flat-slab. J South AM Earth Sci 15(1):39–57

Kay SM, Godoy E, Kurtz A (2005) Episodic arc migration, crustal thickening, subduction erosion, and magmatism in the south-central Andes. Geol Soc Am Bull 117:67–88

Kozloswki E, Cruz C, Condat P, Manceda R (1989) Interpretación del fallamiento de bajo ángulo en los sedimentos cretácicos del río Diamante. Pcia. de Mendoza. I Cong Nac Expl Hidrocarburos, Mar del Plata 2:675–688

Kozlowski E, Manceda R, Ramos V (1993) Estructura. In: Ramos V (ed.) Geologia y recusros naturales de Mendoza. XII Cong Geol Arg 1(18): 235–256

Lamb S, Davis P (2003) Cenozoic climate change as a possible cause for the rise of the Andes. Nature 425(6960):792–797

Levina M, Horton B, Fuentes F, Stockli D (2014) Cenozoic sedimentation and exhumation of the foreland basin system preserved in the Precordillera thrust belt (31–32° S), southern central Andes, Argentina. Tectonics 33:1659–1680

Limarino O, Net L, Gutierrez P, Barreda V, Caselli A, Ballent S (2000) Definicion litoestratigrafica de la Formacion Cienaga del Rio Huaco (Cretacico Superior), Precordillera central, San Juan, Argentina. Rev Asoc Geológica Arg 55:83–99

Lossada A, Mardonez D, Suriano J, Fitzgerald P, Hoke G, Mahoney B, Giambiagi L, Aragón E (2015) Uplift Sequence of the Main Morphostructural Units of the Southern Central Andes at 30° S: insights from a multidisciplinary approach. Am. Geophys. Union Meet T23A-2931

Lossada AC, Giambiagi L, Hoke GD, Fitzgerald PG, Creixell C, Murillo I, Mardonez D, Velásquez R, Suriano J (2017) Thermochronologic evidence for late eocene andean mountain building at 30° S. Tectonics

Maksaev V, Moscoso R, Mpodozis C, Nasi C (1984) Las unidades volcánicas y plutónicas del Cenozoico superior en la Alta Cordillera del Norte Chico (29°–31° S), Geología, alteración hidrotermal y mineralización. Rev Geol Chile 21:11–51

Manea VC, Pérez-gussinyé M, Manea M (2012) Chilean flat slab subduction controlled by overriding plate thickness and trench rollback. Geology 1:35–38

Martin M, Clavero J, Mpodozis C, Cutiño L (1995) Estudio geológico regional de la franja El Indio, Cordillera de Coquimbo. Informe registrado IR-95–6 Servicio Nacional de Geología y Minería, Chile, and Compañía Minera San José

Mazzitelli M, Mahoney B, Balgord E, Giambiagi L, Kimbrough D, Lossada A, Maccann C (2015) Evolution of the Manatiales Basin, San Juan, Argentina: constraining Miocene orogenic patterns in the South-Central Andes. GSA Meet 47(7):151

Mescua JF, Giambiagi L, Barrionuevo M, Tassara A, Mardonez D, Mazzitelli M, Lossada A (2016) Basement composition and basin geometry controls on upper-crustal deformation in the southern central andes (30–36° S). Geol Mag, 153(5–6):945–961

Mirré JC (1966) Geología del Valle del Río Los Patos (entre Barreal y Las Hornillas). Rev Asoc Geol Arg 21(4):211–231

Moscoso R, Mpodozis C (1988) Estilos estructurales en el norte chico de Chile (28–31° S), regiones de Atacama y Coquimbo. Andean Geol 15(2):151–166

Mpodozis C, Ramos V (1989) The Andes of Chile and Argentina. In: Ericksen GE, Cañas Pinochet MT, Reinemund JA (eds.) Geology of the Andes and Its Relation to Hydrocarbon and Mineral Resources. Circum-Pacific Council for Energy and Mineral Resources, Houston, Texas, pp 56–90

Mpodozis C, Cornejo P (2012) Cenozoic tectonics and porphyry copper systems of the Chilean Andes. Society Economic Geolog, Sp Publ 16:329–360

Nullo F, Stephens G (1993) Estructura y deformación terciaria en el área de las Aucas, sur de Mendoza. XII Cong Geol Arg, Mendoza 3:107–112

Oliveros V, Morata D, Aguirre L, Féraud G, Fornari G (2007) Jurassic to early cretaceous subduction-related magmatism in the Coastal Cordillera of northern Chile (18°30′–24° S): geochemistry and petrogenesis. Rev Geol Chile 34(2):209–232

Oncken O, Hindle D, Kley J, Elger K, Victor P, Schemann K (2006) Deformation of the Central Andean upper plate system—facts, fiction and constraints for plateau models. In: Oncken O, Chong G, Franz G, Giese P, Gotze HJ, Ramos VA, Strecker MR, Wigger P (eds) The Andes-active subduction orogeny. Springer, Berlin Heidelberg, pp 3–28

Pardo-Casas F, Molnar P (1987) Relative motion of the Nazca (Farallon) and South American plates since late cretaceous time. Tectonics 6(3):233–248

Pérez DJ (1995) Estudio geológico del cordón del Espinacito y regiones adyacentes, provincia de San Juan. (unpublished) PhD Thesis, Universidad de Buenos Aires

Pérez DJ (2001) Tectonic sand unroofing history of Neogene Manatiales foreland basin deposits, Cordillera Frontal (32°30′S), San Juan province, Argentina. J South Am Earth Sci 14:693–705
Pilger RH (1984) Cenozoic plate kinematics, subduction and magmatism: South American Andes. J Geol Soc 141(5):793–802
Pineda G, Emparán C (2006) Geología del área Vicuña-Pichasca, Región de Coquimbo. Carta Geológica de Chile 1:100.000. Servicio Nacional de Geología y Minería, Santiago, N°97 p 40
Polanski J (1958) El bloque varíscico de la Cordillera Frontal de Mendoza. Rev Asoc Geol Arg 12 (3):165–196
Polanski J (1964) Descripción geológica de la Hoja 25 a-b—Volcán de San José, provincia de Mendoza. Dirección Nacional de Geología y Minería, Buenos Aires 98:1–92
Polanski J (1972) Descripción geológica de la Hoja 24 a-b (Cerro Tupungato), provincia de Mendoza. Dirección Nacional de Geología y Minería, Buenos Aires, p 110
Porras H, Pinto L, Tunik M, Giambiagi L, Deckart K (2016) Provenance of the Alto Tunuyan Basin (33°40′S, Argentina) and its implications for the evolution of the Andean Range: insights from petrography, geochemistry and U-Pb LA-ICP-MS zircon ages. Tectonophysics (in press)
Ramos VA (1988) The tectonics of the Central Andes, 30° to 33° S latitude. In: Clark S, Burchfiel D (eds.) Processes in continental lithospheric deformation: Geological Soc Am, Sp Paper 218: 31–54
Ramos VA (2010) The tectonic regime along the Andes: present-day and mesozoic regimes. Geol Journal 45(1):2–25
Ramos V, Kay S, Page R, Munizaga F (1990) La ignimbrita Vacas Heladas y el cese del volcanismo en el Valle del Cura, provincia de San Juan. Rev Asoc Geol Arg 44:336–35
Ramos VA, Cegarra MI, Cristallini E (1996) Cenozoic tectonics of the High Andes of west-central Argentina (30°–36° S latitude). Tectonophys 259:185–200
Ramos VA, Cristallini E, Pérez D (2002) The Pampean flat-slab of the Central Andes. J South Am Earth Sci 15:59–78
Ramos VA, Zapata T, Cristallini E, Introcaso A (2004) The Andean thrust system: Latitudinal variations in structural styles and orogenic shortening. In: McClay KR (ed.) Thrust Tectonics and Hydrocarbon Systems. American Association of Petroleum Geologists Memoir 82, p. 30–50
Rodríguez MP (2013) Cenozoic uplift and exhumation above the southern part of the flat slab subduction segment of Chile (28.5–32° S). PhD Thesis (Unpublish), Universidad de Chile, Santiago p 220
Rossel P, Oliveros V, Ducea MN, Charrier R, Scaillet S, Retamal L, Figueroa O (2013) The Early andean subduction system as an analog to island arcs: evidence from across-arc geochemical variations in northern chile. Lithos 179:211–230
Silver PG, Russo RM, Lithgow-Bertelloni C (1998) Coupling of South American and African plate motion and plate deformation. Science 279:60–63
Somoza R (1998) Updated Nazca (Farallon)—South America relative motions during the last 40 My: implications for mountain building in the central Andean region. J South Am Earth Sci 11 (3):211–215
Strecker MR, Alonso R, Bookhagen B, Carrapa B, Coutand I, Hain MP, Hilley GE, Mortimer E, Schoenbohm L, Sobel ER (2009) Does the topographic distribution of the central Andean Puna Plateau result from climatic or geodynamic processes? Geology 37:643–646
Suriano J, Mardonez D, Mahoney B, Mescua J, Giambiagi L, Kimbrough D, Lossada A (2017) Uplift sequence of the Andes at 30° S: insights from sedimentology and U/Pb dating of synorogenic deposits. J South Am Earth Sci 75:11–34
Turienzo M (2010) Structural style of the Malargüe foldand—thrust belt at the Diamante River area (34°30′–34°50′S) and its linkage with the Cordillera Frontal, Andes of central Argentina. J South Am Earth Sci 29:537–556
Turienzo M, Dimieri L (2005) Geometric and kinematic model for basement-involved backthrusting at Diamante River, southern Andes, Mendoza province, Argentina. J South Am Earth Sci 19:111–125

Turienzo M, Dimieri L, Frisicale C, Araujo V, Sánchez N (2012) Cenozoic structural evolution of the Argentinean Andes at 34°40′S: a close relationship between thick and thin—skinned deformation. Andean Geol 39(2):317–357

Tassara A, Yáñez G (2003) Relación entre el espesor elástico de la litósfera y la segmentación tectónica del margen andino (15–47° S). Rev Geol Chile 32:159–186

Walcek AA, Hoke G (2012) Surface uplift and erosion of the southernmost Argentine Precordillera. Geomorphology 153–154:156–168

Winocur DA (2010) Geología y estructura del Valle del Cura y el sector central del Norte Chico, provincia de San Juan y IV Región de Coquimbo, Argentina y Chile. (unpublish) PhD thesis, Universidad de Buenos Aires

Winocur DA, Litvak VD, Ramos VA (2015) Magmatic and tectonic evolution of the Oligocene Valle del Cura Basin, Main Andes of Argentina and Chile: Evidence for generalized extension. Geol Soc, London 399:1–34

Yáñez G, Cembrano J (2004) Role of viscous plate coupling in the late tertiary Andean tectonics. J of Geophys Res 109 (B2)

Yáñez GA, Ranero CR, von Huene R, Díaz J (2001) Magnetic anomaly interpretation across the southern central Andes (32°–34° S): The role of the Juan Fernandez Ridge in the late tertiary evolution of the margin. J Geophys Res 106:6325–6345

Yáñez G, Cembrano J, Pardo M, Ranero C, Selles D (2002) The Challenger—Juan Fernandez—Maipo major tectonic transition of the Nazca-Andean subduction system at 33–34° S: geodynamic evidence and implications. J South AM Earth Sci 15:23–38

The Structure of the Southern Central Andes (Chos Malal Fold and Thrust Belt)

Martín Turienzo, Natalia Sánchez, Fernando Lebinson and Luis Dimieri

Abstract The Chos Malal fold and thrust belt, formed during the Andean orogeny, is characterized by the involvement of both the Paleozoic basement and Mesozoic strata of the Neuquén Basin into the deformation. Two detailed structural cross sections, built based on previous field mapping, new subsurface interpretations, and seismic and borehole data, allow characterizing the structural style of this orogenic belt. A close interaction between large thick-skinned structures (first order) and complex thin-skinned structures (second, third, and fourth order), related to the presence of multiple detachments in the sedimentary cover, is recognized. The largest thrusts form basement-involved duplex structures, with a lower detachment located at a depth of about 12–14 km and an upper detachment in the Jurassic evaporites of the Auquilco Formation. Displacement transmitted by these basement sheets in the inner zone of the Chos Malal fold and thrust belt produces a wide region of thin-skinned deformation, which contains second-order fault-bend folds that transfer deformation to the overlying Agrio Formation shales (Early Cretaceous) giving rise to third-order folds and thrusts involving this unit. In the outer zone, the basement-involved thrusts have less displacement and form monoclines and a complex thin-skinned deformation restricted to the deformation front, possibly caused by buttressing effect exerted by the overlying Miocene volcanic sequences. This impediment in forward deformation leads to an important out-of-sequence faulting, whose displacement is compensated by a passive-roof backthrust along the Cretaceous evaporites of the Bajada del Agrio Group forming a triangle zone. Second-order anticlines under this triangle zone, where the seismic data are of low quality, constitute important hydrocarbon oil fields such as El Porton and Filo Morado. Understanding the close relationship between the structures of different order cropping out in the inner zone of the Chos Malal fold and thrust belt is important to interpret the subsurface structures forming hydrocarbon oil fields in the outer zone as well as to identify other complex structures that may lead to new exploration opportunities. Restitution of the structural cross sections

M. Turienzo (✉) · N. Sánchez · F. Lebinson · L. Dimieri
Departamento de Geología, Instituto Geológico del Sur (INGEOSUR, UNS-CONICET), Universidad Nacional del Sur, Bahía Blanca, Argentina
e-mail: turienzo@uns.edu.ar

A. Folguera et al. (eds.), *The Evolution of the Chilean-Argentinean Andes*, Springer Earth System Sciences, https://doi.org/10.1007/978-3-319-67774-3_17

allowed calculating a tectonic shortening for this region in the order of 22–25 km (16–18%), higher than estimated by previous authors who generally simplified the thin-skinned deformation and considered the tectonic inversion of normal faults as the main mechanism of deformation in this orogen.

Keywords Neuquén Basin · Balanced cross sections · Thick-skinned structures

1 Introduction

The Chos Malal fold and thrust belt (FTB) is located in the northwestern of the Neuquén province, approximately between 36°50′ and 37°20′S, being part of the Andean orogenic belt, which deformed the western sector of the Neuquén Basin as a result of the subduction of a series of Pacific plates (Farallón, Nazca, etc.) beneath the South American plate (Fig. 1a, b). This orogenic belt is a classic site for the study of the Andean orogeny in the Neuquén Cordillera, due to the excellent quality of its outcrops and the profuse subsurface information derived from the large oil interests in the region. These conditions led to numerous works, in the past three decades, oriented to understand the structural evolution of this region and its relationship with the oil system of the area (Ploszkiewicz and Viñes 1987; Kozlowski et al. 1996, 1998; Cobbold et al. 1999; Zapata et al. 1999, 2001; Zamora Valcarce and Zapata 2005; Zamora Valcarce et al. 2006; Fantín et al. 2015; Gómez Omil et al. 2015; among others). Much of the previous structural studies are regional profiles, with an emphasis on subsurface structures adjacent to the deformation front where the areas of economic interest are located, and they are not fully detailed in the inner region where deformed Mesozoic rocks are exposed. There is not yet a widespread consensus about the structural style of this fold and thrust belt, the deformation mechanisms, the precise age of the main contractional pulses, and the tectonic shortenings (see Chaps. "Mechanisms and episodes of deformation along the Chilean-Pampean flat-slab subduction segment of the Central Andes in northern Chile", "Cretaceous orogeny and marine transgression in the Southern Central and Northern Patagonian Andes: Aftermath of a large-scale flat-subduction event?", and "Cenozoic Uplift and Exhumation of the Frontal Cordillera between 30° and 35° S and the Influence of the Subduction Dynamics in the Flat Slab Subduction Context, South Central Andes"). The Chos Malal FTB is a basement-involved orogenic belt, which is exposed at the Cordillera del Viento (Fig. 1c), and two ends structural models have been proposed to explain its origin. One model considers that the basement participates in the deformation by tectonic inversion of preexisting normal faults, inherited from the initial rifting stage of the basin, which would have been reactivated during subsequent compressive deformation (Booth and Coward 1996; Chaveau et al. 1996; Zapata et al. 1999; Zamora Valcarce and Zapata 2005; Zamora Valcarce et al. 2006; Rojas Vera et al. 2014, 2015a, b). The structural cross sections interpreted with this model generally show small shortenings, related to the high angle of the normal faults, and a sparse

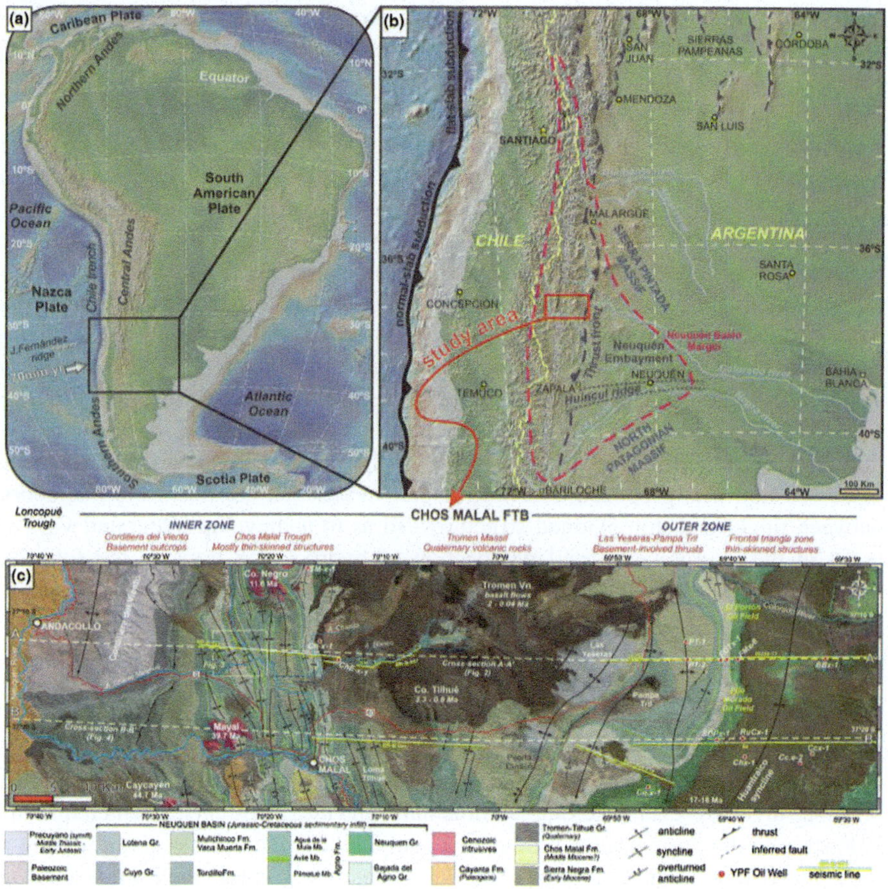

Fig. 1 **a, b** Location of the Chos Malal fold and thrust belt (FTB) in the western region of the Neuquén Basin, at the southern segment of the Central Andes ($\sim 37°$ S). **c** Regional geological map of the Chos Malal FTB (based on Groeber 1946; Zöllner and Amos 1973; Holmberg 1975; Gulisano and Gutiérrez Pleimling 1995; Llambías et al. 2007; and own field data). Detailed geology of the inner zone is simplified from Sánchez (2015) and Sánchez et al. (2015). Location of two balanced cross sections and seismic and well data used for the interpretations are shown

deformation in the sedimentary cover. The other model considers that the main thick-skinned structures are associated with thrusts caused by the Andean contraction and that these low-angle faults generate intense thin-skinned deformation of overlying sedimentary sequences, and therefore, the calculated tectonic shortenings are higher (Ploszkiewicz and Viñes 1987; Nocioni 1996; Sánchez et al. 2014, 2015). There is also no consensus on the style of deformation that affects Mesozoic sedimentary rocks, although most agree that the structures are complex and they have variable vergence. In some cases, the folds are interpreted as detachment and/or fault-propagation folds mainly due to their steeply dipping flanks (Kozlowski et al. 1996, 1998; Zapata et al. 2001; Zamora Valcarce and Zapata 2005), while in other cases, the presence of repeated units along multiple decollements in the sedimentary cover allows inferring fault-bend folding mechanisms and duplex structures (Ploszkiewicz and Viñes 1987; Nocioni 1996; Turienzo et al. 2014).

In this chapter, two regional structural sections, based on a detailed field mapping and interpretation of the available subsurface information, are presented in order to characterize the structural style of the Chos Malal FTB. Despite the great length of these cross sections, the thin-skinned deformation is analyzed in detail considering the complex structures of diverse orders that affect the Mesozoic rocks between the Cordillera del Viento and the Tromen volcano (Sánchez et al. 2014, 2015; Turienzo et al. 2014) and reinterpreting the structural configuration at the deformation front. Restitution of the structural cross sections allowed calculating accurately the tectonic shortenings for the entire Chos Malal FTB, which will serve as support for future contributions that analyze the tectonic evolution of this segment of the Andes. With the purpose of better understanding the development of this orogenic belt, the relative importance of the tectonic inversion mechanism is discussed and it is proposed a new structural model that considers the interaction between thick- and thin-skinned structures, so as to achieve an integrated view to assist in the interpretation and prospection of hydrocarbon traps in the fold and thrust belts of the Neuquén Basin.

2 Tectono-Stratigraphic Setting

The geological history of the Southern Central Andes in the north of the Neuquén province involves several tectonic, sedimentary, and magmatic cycles, which from the Paleozoic to the present have produced a complex morphotectonic configuration in the southwestern margin of Gondwana (Ramos and Folguera 2005; Ramos and Kay 2006; Ramos et al. 2011, 2014). The Paleozoic basement crops out in the core of the Cordillera del Viento anticline (Fig. 1c), and the oldest rocks are a thick Carboniferous volcano-sedimentary sequence of the Andacollo Group that has been strongly deformed during the Early Permian San Rafael orogeny (Zöllner and Amos 1973; Llambías et al. 2007; Danieli et al. 2011; Giacosa et al. 2014). These rocks were intruded by granitoids and unconformably covered by Permian mesosilicic volcanic sequences that form the Huingancó volcanic–plutonic complex, emplaced in an extensional environment (Llambías et al. 2007; Llambías and Sato 2011). Extensional tectonics continued during the Middle Triassic–Early Jurassic causing NW-oriented half-grabens with varying polarity, which controlled the accumulation of synrift sequences mainly composed of andesitic to rhyolitic volcanic rocks grouped within the Precuyano cycle (Gulisano 1981; Franzese and Spalletti 2001; Llambías et al. 2007; Carbone et al. 2011) (see Chap. "A Provenance Analysis from the Lower Jurassic Units of the Neuquén Basin. Volcanic Arc or Intraplate Magmatic Input?"). The synrift deposits crop out on the eastern flank of the Cordillera del Viento (Fig. 1c), showing a complex architecture and variable thickness controlled by normal faults (Sagripanti et al. 2014), but the geometry of the half-grabens is observed more clearly in seismic lines located in areas away from the deformation front where they often form important hydrocarbon deposits (Veiga et al. 1999; Pángaro et al. 2002; Barrionuevo et al. 2005; Cristallini et al. 2006, 2009; Carbone et al. 2011). Thermal subsidence in a backarc setting during the Jurassic and Cretaceous favored the alternate accumulation of marine and continental rocks in the Neuquén Basin (Legarreta and Uliana 1991; Vergani et al. 1995; Howell et al. 2005; Arregui et al. 2011a), bounded on the northeast by the Sierra Pintada massif and to the southeast by the North Patagonian massif (Fig. 1b) (see Chaps. "The Early Stages of the Magmatic Arc in the Southern Central Andes" and "A Provenance Analysis from the Lower Jurassic Units of the Neuquén Basin. Volcanic Arc or Intraplate Magmatic Input?"). The different cycles of sedimentation that filled the Neuquén Basin established by Groeber (1929, 1946) generally remain in use, and subsequent studies allowed adjusting the knowledge about the processes of sedimentation and subsidence by focusing on sequence stratigraphy (Legarreta and Gulisano 1989) and through the field recognition of unconformities (Leanza 2009). The onset of marine sedimentation from the Pacific Ocean in this part of the Neuquén Basin took place from the Pliensbachian–Toarcian with the accumulation of Cuyo Group (Gulisano and Gutiérrez Pleimling 1995; Arregui et al. 2011b; Leanza et al. 2013). The accumulation of this group was locally influenced by the Huincul ridge (Fig. 1b), an ~EW morphostructural element that divided the Neuquén Basin into two depocentres, whose tectonic activity at this stage was related to a vector of convergence from the NW (Mosquera et al. 2011) (see Chap. "Lower Jurassic to Early Paleogene Intraplate Contraction in Central Patagonia"). The second cycle of marine sedimentation corresponds to the Lotena Group (Callovian–Oxfordian) and is composed of sandstones, limestones, and evaporites of the Lotena, La Manga, and Auquilco Formations, respectively, the latter of great importance for the structural development of the fold and thrust belt because it acts as one of the major detachment levels in the region. The next cycle, the Mendoza Group, is undoubtedly one of the most important and best exposed in the Neuquén Basin (Fig. 1c). This group begins with continental red beds of the Tordillo Formation (Kimmeridgian) which are covered by black shales of the Vaca Muerta Formation (Tithonian–Early Valanginian), the main hydrocarbon source rock in the basin. In the Early Cretaceous, a sudden relative sea-level drop allowed the accumulation of Mulichinco Formation (Valanginian), while a new marine flooding led to a thick succession of black shales with interbedded limestones and sandstones that constitute the Agrio Formation (Late Valanginian–Early Barremian). Within this formation is the Avilé Member, formed by fluvial–aeolian sandstones with a variable thickness between 10 and 70 m, which besides being one of the main reservoirs of hydrocarbons is an excellent key bed that helps to reconstruct the tectonic structures in the fold and thrust belt. The overlying Huitrín Formation (Barremian–Aptian) reveals a noticeable regressive event during the Early Cretaceous and consists of fluvial–aeolian sandstones, limestones, and evaporites deposited in a shallow and hypersaline sea (Leanza 2003). The Rayoso Formation consists of fine grained sandstones, reddish mudstones, and evaporites, accumulated in a continental environment that marks the disconnection of the Neuquén Basin with the proto-Pacific Ocean, which together with the Huitrín Formation constitute the Bajada del Agrio Group. Above lies the Neuquén Group (Early Cenomanian–Middle Campanian), formed by thick continental clastic sequences that are considered accumulated in a foreland basin related to the beginning of the Andean orogeny since the Late Cretaceous (Cobbold and Rossello 2003; Ramos and Folguera 2005; Tunik et al. 2010; Di Giulio et al. 2015). Andean compression deformed the western sector of the Neuquén Basin, although diachronically creating the Malargüe, Chos Malal, Agrio, and Aluminé fold and thrust belts (Fig. 1b). There are many contributions that examine the Cenozoic tectonic evolution of the Central Andes evidencing complex relationships between deformation, magmatism, and sedimentation, with particularities along the different segments that compose this mountain chain (Ramos 1999, 2010; Ramos et al. 2004, 2014; Ramos and Kay 2006; Charrier et al. 2007, 2015; Cobbold et al. 2007; Maloney et al. 2013, among others). Magmatic activity during the development of the Andes was intense, with subvolcanic and volcanic rocks emplaced in an arc and backarc setting and closely connected with the tectonic evolution of the region. In the study area, Paleogene andesitic volcanic rocks of the Cayanta Formation are noteworthy, which unconformably cover the exhumed Paleozoic basement west of Andacollo, the Eocene, and Miocene subvolcanic bodies intruding the folded Mesozoic sediments and thick sequences of Miocene and Quaternary basaltic lavas cropping out in the Huantraico syncline and around the Tromen volcano, respectively (Fig. 1c). The ages of these rocks and their linkage with the tectonic structures forming the Chos Malal FTB will be detailed in Sect. 5 of this work (Structural Evolution).

3 Regional Structure of the Chos Malal Fold and Thrust Belt

The structural configuration of this segment of the Southern Central Andes is characterized by large-scale basement-involved folds located in the inner zone, as in the Cordillera del Viento where Paleozoic rocks crop out, and in the outer zone where participation of the basement is recognized through wells and seismic lines in the region of Las Yeseras–Pampa Tril (Fig. 1c). The first regional cross sections through the Chos Malal FTB illustrate this structural style (Ploszkiewicz and Viñes 1987; Kozlowski et al. 1996, 1998; Nocioni 1996; Booth and Coward 1996; Cobbold et al. 1999; Zapata et al. 1999) and, although they propose different deformation mechanisms to explain the thick-skinned structures, laid the foundation that supported the subsequent structural studies (Zamora Valcarce et al. 2006; Folguera et al. 2007; Galland et al. 2007; Rojas Vera et al. 2014, 2015a, b; Sagripanti et al. 2014, 2015a, b; Sánchez et al. 2014, 2015; Turienzo et al. 2014). The main basement-involved structures have a dominant eastward vergence, with subordinate development of backthrusts in the hangingwalls, and generate a complex thin-skinned deformation in their leading edges characterized by different-size fault-related folds with NS to NNW trend affecting the Mesozoic units of the

Neuquén Basin. All these structures were mapped in detail by means of successive fieldwork carried out regularly in the last decade on which about 1,000 control points were measured, each with multiple structural data and observations that for reasons of scale are not shown on the map of Fig. 1c. To comprehend the geometry in depth and establish the relationship between the thick- and thin-skinned structures of different orders, and with the aid of the available subsurface information, two structural cross sections were built throughout the Chos Malal FTB. These two EW cross sections, of more than 100 km long each (Fig. 1c), were restored by the line-length conservation method to ensure their geometric viability and to calculate the tectonic shortenings.

3.1 Cross Section A-A′ (~37°13′S)

This cross section has a length of 111.8 km and extends from around Andacollo to the eastern flank of the Huantraico syncline, east of the Filo Morado oil field (Fig. 2a). The Cordillera del Viento anticlinorium, which extends for more than 65 km with NS trend, is undoubtedly the most important structure of the Chos

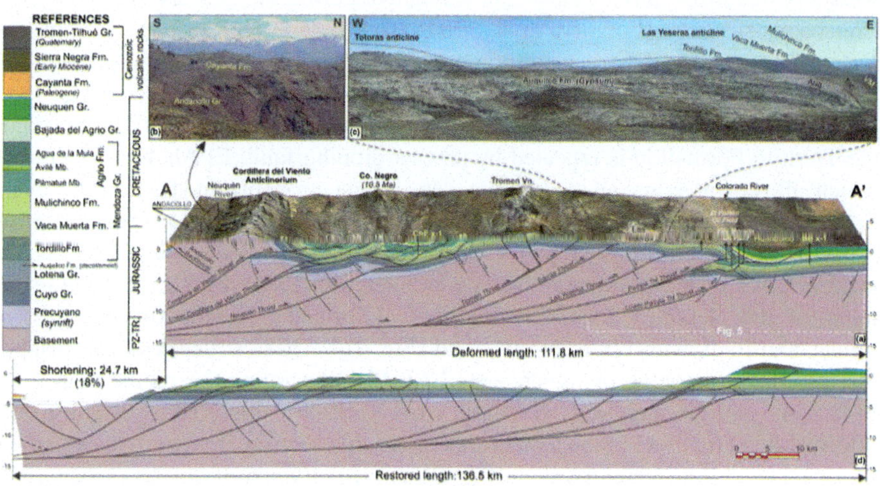

Fig. 2 **a** Structural cross section across the north segment of the Chos Malal FTB (see Fig. 1c for location), showing a complex interaction between thick- and thin-skinned structures. Numerous basement-involved thrusts cut preexisting normal faults and transmit the deformation to the cover rocks creating thrust-related folds of different orders involving the Mesozoic rocks of the Neuquén Basin. Interpretation of the inner zone is based on Sánchez et al. (2015). EA: El Alamito backthrust, CS: Cañada Seca syncline, CN: Cerro Negro anticline, LM: Las Máquinas anticline, AB: Arroyo Blanco anticline. **b** Panoramic picture to the west of Huinganco showing the unconformity between the Paleozoic Andacollo Group and the overlying Paleogene volcanics of the Cayanta Fm. **c** Panoramic photograph of the Totoras and Las Yeseras anticlines, to the east of Tromen volcano. **d** Restitution of the A-A′ cross section

Malal FTB. This anticlinorium, with variable width between 10 and 20 km, is a first-order structure that exposes the Paleozoic basement rocks (Zöllner and Amos 1973; Kozlowski et al. 1996, 1998). The western flank of this large fold was strongly exhumed, and Carboniferous rocks of the Andacollo Group were unconformably covered by Paleogene volcanic sequences of the Cayanta Formation (Fig. 2b). Such feature has led to propose a west-vergent fault system in depth, which some authors interpret as inverted normal faults (Booth and Coward 1996; Zapata et al. 1999; Cobbold and Rossello 2003). In the Cordillera del Viento, Sagripanti et al. (2014) recognized NW- and WNW-trending normal faults bounding Late Triassic depocenters, segmented by minor NE faults, which were only partially inverted by the Andean compression while NS contractional structures did not respect rifting architecture cutting through the depocenters.

Theoretically, the inversion of normal faults is complete when the prerift rocks reach the null point (Williams et al. 1989, Buiter and Pfiffner 2003, Yamada and McClay 2004). The mentioned unconformity involves several kilometers of basement uplift and erosion of the entire Mesozoic sequence. This feature cannot be accounted only by inversion of preexistent faults; therefore, a newly formed backthrust system is further proposed in the hangingwall of the Cordillera del Viento thrust in depth (Fig. 2a). The overlaying Paleogene volcanic rocks dip to the west along the entire west slope of the Cordillera del Viento, which imply that this west-dipping thrust into the basement was active during the Tertiary and it tilted the Cretaceous structure (Folguera et al. 2007). The eastern flank of the Cordillera del Viento forms a homoclinal east-dipping sequence ($\sim 30°$–$40°$) involving the whole Triassic–Cretaceous succession, interpreted as the forelimb of a large basement-involved fault-bend anticline (Kozlowski et al. 1996, 1998). In this flank, the Tordillo Formation is repeated by an east-dipping fault, El Alamito backthrust, which allows inferring an underlying detachment probably in the Jurassic evaporites of the Auquilco Formation at the top of the Lotena Group (Sánchez et al. 2015). Between the Cordillera del Viento and the western slope of Tromen volcano, there is a wide area dominated by thin-skinned structures in which the Mesozoic rocks of the Neuquén Basin are intensely deformed, showing diverse sizes of thrusts and folds with NS to NNW trend (Fig. 1c). These structures were mapped and interpreted in detail by Sánchez et al. (2015), illustrating the geometry in depth by means of four balanced structural cross sections supported with subsurface information. The thin-skinned structures along this section are mainly east-vergent thrust-related folds, which have a lower detachment at the roof of the Lotena Group, and the whole Mendoza Group is involved in the deformation (Fig. 2a). The most prominent of these second-order folds is the Las Máquinas anticline, located south of the Cerro Negro intrusive, which is the only one in the sector that exposes the rocks of the Vaca Muerta Formation. The important structural relief of this fold is provoked by a basement-involved structure at depth, recognized in the BR-B-025 seismic line and interpreted as a slightly inverted east-dipping normal fault cut by the west-dipping Neuquén thrust (Sánchez et al. 2015). When the ramping Neuquén thrust reaches the Lotena Group evaporites, it becomes a flat and its eastward displacement gives rise to several thrust-related anticlines involving the Mendoza

Group, which show less development and in some cases no surface expression as below the ChE x-1 well (Fig. 2a). To the west of the Las Máquinas anticline, another thrust involving the Mendoza Group originates the Cerro Negro anticline. Based on seismic data, it was interpreted that this thrust, in addition to connect a lower detachment in the Aquilco Formation with an upper detachment in the Agrio Formation, presents an intermediate detachment within the Vaca Muerta Formation creating a more complex folding geometry (Sánchez et al. 2015). In turn, this sector is affected by the Lower Cordillera del Viento thrust in depth, which despite having a moderate displacement contributes to the total uplift of the Cordillera del Viento (Fig. 2a). The Cañada Seca syncline is located between the Cerro Negro anticline and the El Alamito backthrust. The Bajada del Agrio Group crops out in the syncline core, while the Agrio Formation is exposed in the flanks, where the three members of this unit are intensely deformed. Sandstones of the Avilé Member located in the middle of thick sequences of black shales (Pilmatué and Agua de la Mula Members) constitute an excellent key bed that allows mapping in detail the NNW thrusts and folds affecting the Agrio Formation in the Chacay Melehue area (Fig. 3a). These are smaller-scale folds with shorter wavelengths than second-order anticlines involving the entire Mendoza Group and therefore are considered third-order structures. Layers of sandstones are faulted and tectonically repeated using both under- and overlying shales as detachment, which produce complex structures by combining imbricate thrusts and thrust-related folds whose geometry and kinematics were studied in detail by Turienzo et al. (2014). The Chacay Melehue anticline, located between the Cañada Seca syncline and the Curí Leuvú River, is the most notable and best exposed of these third-order folds (Fig. 3). This anticline has a clear eastward vergence to the north of the Chacay Melehue stream, with steeper dips in the forelimb, and the Avilé Member in its gently dipping backlimb is repeated tectonically (Fig. 3b). A striated slickenside measured in the fault zone contains down-dip lineations and steps that evidence a pure thrust displacement with eastward-directed sense of slip. The Avilé Member is also repeated in the forelimb of the anticline, cropping out parallel to the Curí Leuvú River (Fig. 3a), but no kinematic indicators were found. A probable interpretation is that the same thrust ramping in the backlimb duplicated the Avilé Member, and then, this repeated sequence was folded by a blind thrust in depth (Turienzo et al. 2014). Southward of the Chacay Melehue stream, the structure is even more complex, consisting of two exposed backthrusts that repeated the Avilé Member and a third inferred blind backthrust which caused a west-vergent anticline (Fig. 3c). The two exposed faults have slickensides lineations and steps that corroborate a top-west shear sense.

A kinematic model of these structures shows that initially from a basal detachment within the Pilmatué Member shales, the Chacay Melehue backthrust (1 in Fig. 3c) produces an imbrication of the Avilé Member, then a second backthrust (2 in Fig. 3c) completely duplicates this unit using as upper decollement the overlying shales, and finally, a blind backthrust (3 in Fig. 3c) forms the west-vergent anticline (Turienzo et al. 2014). The change in the fold vergence

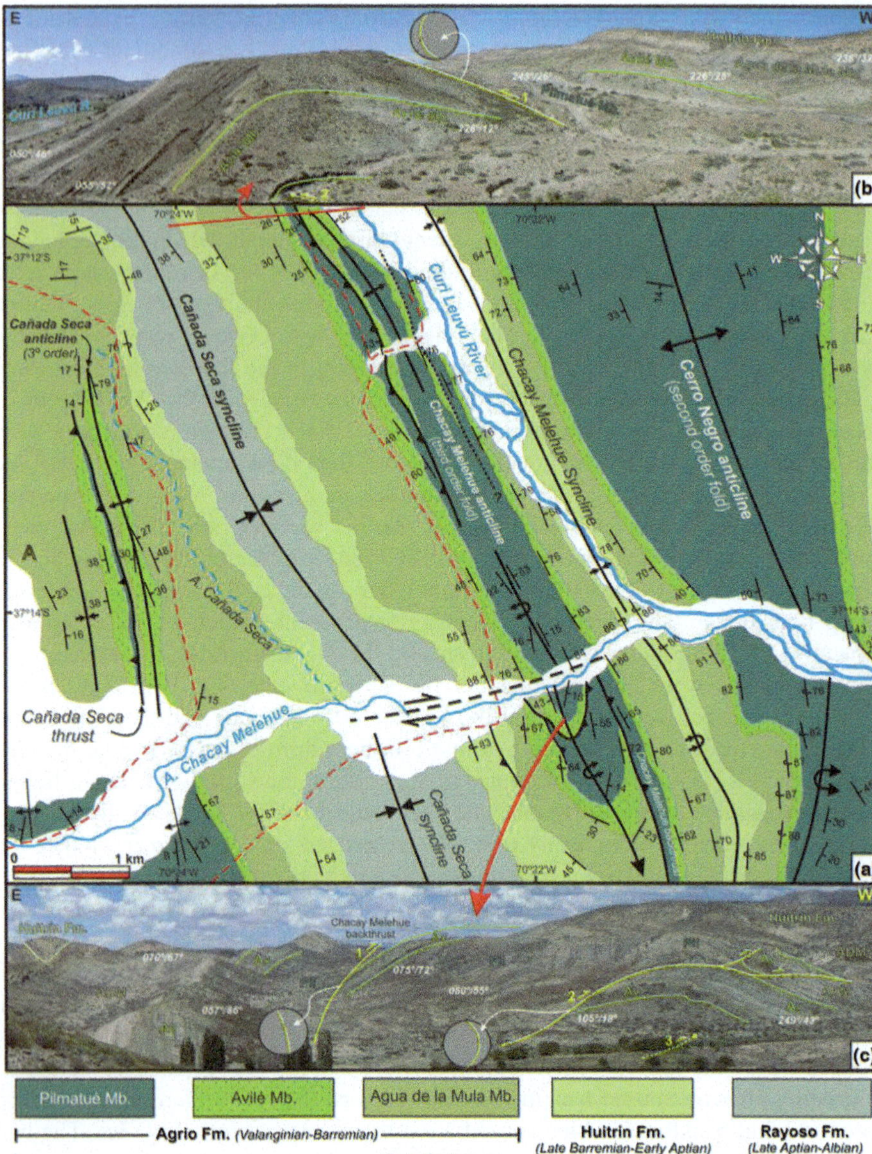

Fig. 3 **a** Detailed geological map of the second- and third-order thrusts and folds affecting the Agrio Formation in Chacay Melehue area (modified from Turienzo et al. 2014). **b** Southward view of the third-order Chacay Melehue anticline folding the Avilé Member, which is tectonically repeated in the backlimb. Note the eastward vergence of the fold. **c** Panoramic picture of the structures to the south of Arroyo Chacay Melehue, where the Avilé Member is deformed by a complex backthrust system. A tear fault interpreted along the creek allows explaining the change in the anticline vergence. Inserted plots show the measured fault planes

observed to the north and south of the Chacay Melehue stream is explained by means of an ENE dextral strike-slip fault, which acts as a tear fault (Fig. 3a).

The three major basement-involved thrusts in the inner zone are connected to a basal detachment interpreted at a depth of about 14 km, possibly corresponding to the brittle–ductile transition into the upper crust. These faults are also linked to an upper detachment in the sedimentary cover, in the evaporites of the Auquilco Formation at the top of the Lotena Group, forming basement-involved duplex structures (Fig. 2a). The structural style is different in the outer zone of the Chos Malal FTB, east of the Tromen volcano (Fig. 1c). Five west-dipping basement-involved thrusts are interpreted, most of which have very small displacements and generate incipient fault-propagation folding in the overlying sedimentary strata (Fig. 2a). The Tromen thrust is the westernmost structure, and although it shows little stratigraphic offset, related fault breccias have been observed on the surface (Galland et al. 2007). In the hangingwall of this thrust, an east-dipping normal fault with no evidence of tectonic inversion has been interpreted based on seismic information (Sánchez et al. 2015). Also in that sector, Sagripanti et al. (2015a, b) have interpreted backthrusts generating west-vergent folds that deform the Mesozoic strata in the Blanco stream, which together with the Tromen thrust form a thick-skinned pop-up structure (Fig. 2a) onto which the Quaternary volcanism of the Tromen volcano develops (Galland et al. 2007). The Yeseras and Totoras thrusts form wide anticlines, although with subhorizontal backlimbs defining basically monocline geometries, which are recognized in the field due to changes of dip in the Auquilco Formation (Fig. 2c). To the north of the cross section (Fig. 1c), Early Jurassic black shales of the Cuyo Group crop out in the core of the Totoras anticline (Stipanicic 1965; Holmberg 1975, Gulisano and Gutiérrez Pleimling 1995). Resistant layers of the Mendoza Group form a continuous and prominent crest in the east-dipping forelimb of the Yeseras anticline (Fig. 2c). The most significant thick-skinned structure in the outer zone of the Chos Malal FTB is undoubtedly the Pampa Tril anticline, a kilometer-scale doubly plunging anticline with NS trend, which exposes the Mulichinco Formation in its core (Fig. 1c). The PT-1 well was the first exploratory well (1946) drilling the Pampa Tril anticline, and although not hydrocarbons were found in this structure, it verified the huge vertical offset ($\sim$4 km) of the basement relative to its position in the adjacent foreland (Fig. 2a). Different models and hypotheses have been proposed in order to explain the structural configuration of this deformation front, which will be discussed in detail in Sect. 4 of this contribution. The structure observed in the field is relatively simple since the Pampa Tril anticline connects eastward with the broad Huantraico syncline, which contains a thick sequence of Miocene volcanic rocks (Fig. 1c). However, with the aid of seismic data and by comparison with the models established by Jones (1982) in the Rocky Mountain foothill of Alberta, Canada, Ploszkiewicz and Viñes (1987) proposed the presence of a triangle zone with thin-skinned structures in depth. Independently of the arguments for or against, this model led to the discovery of the Filo Morado oil field and the underlying folding structure was verified by numerous wells drilled in the area (Fig. 2a). The Filo Morado anticline is a second-order fold involving the entire

Mendoza Group. The forward displacement of this fold above an upper detachment in the evaporites of the Bajada del Agrio Group gives rise to an incipient thrust affecting the Neuquén Group in the eastern flank of the Huantraico syncline. A partially inverted west-dipping normal fault was interpreted seismically underneath the thin-skinned structures (Fig. 2a). The Lower Pampa Tril thrust cuts through the previous extensional structure, and its forward movement is solved by means of backthrusts in the Cuyo Group. The restoration of the A-A′ structural cross section evidences a tectonic shortening of 24.7 km (18%) for this segment of the Chos Malal FTB (Fig. 2d).

3.2 *Cross Section B-B′ (~37°20′S)*

This cross section has a length of 112.3 km, almost the same as the A-A′ section, and extends from the Neuquén River to the east of the Huantraico syncline (Fig. 4a). The Cordillera del Viento anticlinorium plunges to the south, and thus, at the latitude of the B-B′ cross section, it shows less structural relief, cropping out the Tordillo Formation unconformably covered by Paleogene volcanic rocks (Fig. 1c).

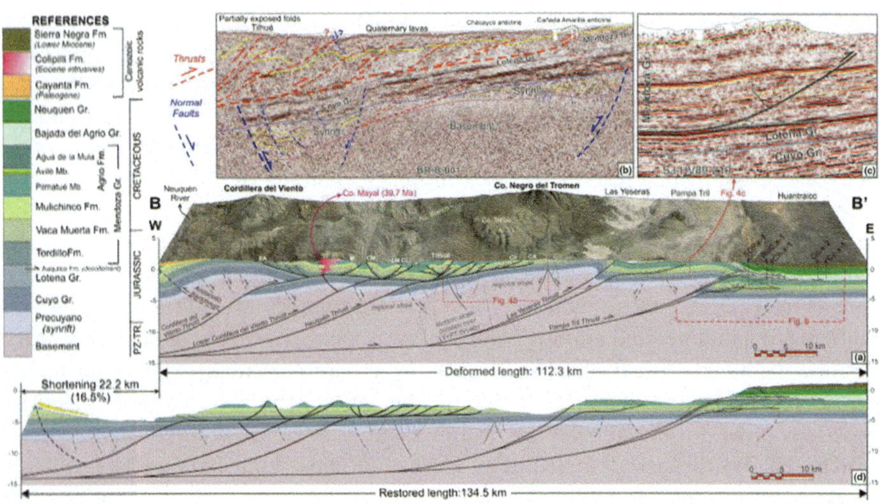

Fig. 4 **a** Structural cross section showing the structural style across the south segment of the Chos Malal FTB (see Fig. 1c for location). Interpretation of the inner zone is based on Sánchez et al. (2015). EA: El Alamito backthrust, M: Mayal anticline, CM: Chos Malal anticline, LM: Las Máquinas anticline, CL: Curí Leuvú anticline, Ch: Chacayco anticline, CA: Cañada Amarilla anticline. **b** Partial interpretation of a portion of the BR-B-001 seismic line (see location in Fig. 1c), where synrift geometries were recognized related to normal faults with no or very small evidences of positive tectonic inversion (modified from Sánchez et al. 2015). **c** Incipient east-vergent thrust-related anticline interpreted in the SJ-HV89-210 seismic line. **d** Restitution of the B-B′ cross section

The Cordillera del Viento anticlinorium has a gently dipping eastward forelimb, in which the El Alamito backthrust and other minor faults deform the Tordillo and Vaca Muerta Formations (Fig. 4a). The Eocene andesites of the Cerro Mayal were intruded as sills and laccoliths within the Agrio Formation shales in the core of a broad syncline (Cobbold and Rossello 2003). Numerous NNW- and NNE-trending folds involving the Mendoza Group were mapped from this syncline toward the foreland (Fig. 1c). All these thin-skinned structures were interpreted as second-order thrust-related anticlines; i.e., they have a basal detachment in evaporites of the Auquilco Formation at the top of the Lotena Group (Fig. 4a). Because no third-order structures requiring detachments within the Agrio Formation are observed, most of these anticlines were interpreted as fault-propagation folds. The Mayal anticline is a broad east-vergent fold, at the eastern slope of the intrusive, which has a poor surficial expression because the Agrio Formation shales are cropping out. A backthrust uplifting the Mulichinco Formation on the top of the Agrio Formation forms the west-vergent Chos Malal anticline (Fig. 4a). Evaporites of the Huitrín Formation are exposed in the core of a tight syncline, or triangle zone, formed between these two opposite-vergent anticlines. In this cross section, the east-vergent Las Máquinas and Curí Leuvú anticlines involve the Agrio Formation and they are very close together. As these anticlines plunge to the south, the Mulichinco Formation crops out north of the section and these folds are excellently exposed in a natural ~EW cut of the Curí Leuvú River (Sánchez et al. 2014, their Fig. 5). Together with the west-vergent Chos Malal anticline, these folds form a pop-up structure with a broad syncline between them containing the Agrio Formation (Fig. 4a). No seismic data are available in order to accurately interpret the subsurface structures in the western sector of the B-B′ cross section. The seismic line BR-B-001 located to the east (Fig. 1c) shows that the Paleozoic basement and the overlying Mesozoic sequence have slope to the west (Fig. 4b). If this basement-cover contact is extended to the west, it should be at a great depth generating more space than can be explained with the normal stratigraphic thickness of Mesozoic units. Thus, two basement slices were interpreted (Fig. 4a) in order to fill the missing space and by correlation with structures further north where seismic data show thick-skinned structures at depth (Sánchez et al. 2015). As in cross section A-A′, these basement-involved structures would be associated with the Lower Cordillera del Viento and Neuquén thrusts, and in their forward movement, they contribute to the thin-skinned deformation. Additionally, thrusts and backthrusts affecting basement rocks could act as favorable conduits for the emplacement of shallow intrusions as the Cerro Mayal (Fig. 4a), as it has been interpreted in different structures of the area (Galland et al. 2007; Gürer et al. 2015, Lebinson et al. 2015a) and of the Malargüe FTB (Dimieri 1992; Turienzo et al. 2012; Araujo et al. 2013). Two NNE-trending anticlines involving the Mendoza Group are exposed in the area of Loma Tilhué, east of Chos Malal (Fig. 1c). These anticlines have opposite vergence, and they form a large and continuous pop-up structure (Fig. 4a). The shales of Vaca Muerta Formation are exposed by means of a secondary thrust in the eastern side of the pop-up structure. The seismic line BR-B-001 allows to interpret three east-vergent thrust-related anticlines formed in

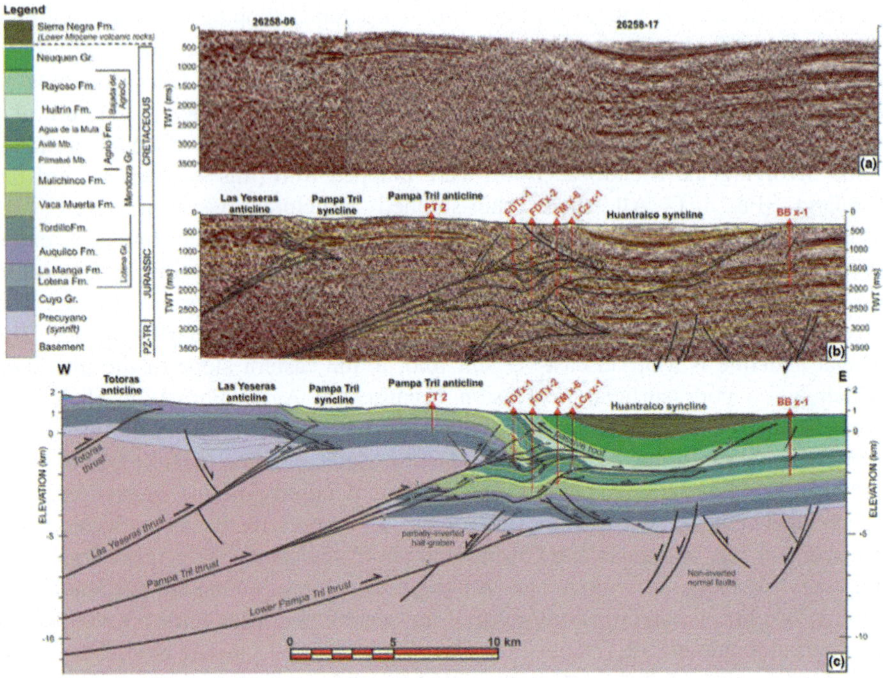

Fig. 5 **a, b** Uninterpreted and interpreted two-way travel time seismic lines in the mountain front of the Chos Malal FTB along the A-A′ cross section. **c** Reconstruction of the surface and subsurface structures evidencing a complex linkage between the basement-involved Pampa Tril thrust and the Filo Morado anticline and other minor structures involving the Mesozoic strata (see explanation on the text). The upper branch of the Pampa Tril thrust is connected with a major detachment in the evaporites of the Bajada del Agrio Group forming a triangle zone as it was proposed by early workers (Ploszkiewicz and Viñes 1987). Note that the underlying structures, a partially inverted normal fault and new formed thrusts and backthrusts, were not drilled, and thus, they could constitute future exploratory leads. Numbers indicate the faulting sequence

the footwall of the Tilhué pop-up (Fig. 4b). The westernmost of these subsurface structures crop out toward the south, involving the Mulichinco Formation, where it was interpreted as a fault-bend fold named Pampa Tilhué anticline (Lebinson et al. 2015b). Beneath these second-order structures, the seismic line shows a set of westward-thickening reflectors folded into a syncline, which are unconformably covered by flat and more continuous reflectors (Fig. 4b). The deep seismic reflections are interpreted as synrift deposits filling a half-graben associated with an east-dipping normal fault. A minimum reverse offset of overlying reflectors suggests this normal fault was slightly inverted, but most of its original extensional geometry is preserved. Minor synthetic and antithetic normal faults affect the reflectors that cover the synrift sequences, interpreted as the Jurassic Cuyo and Lotena Groups, which evidences that this younger faulting is not related to the Late Triassic rifting stage that originated the Neuquén Basin. This deformation including faulting and folding of layers located above synrift deposits is explained as resulting from differential subsidence over the half-grabens, as has been exemplified and modeled for half-grabens located in the northeastern of the Neuquén Basin (Cristallini et al. 2006, 2009). A package of strong subhorizontal reflectors on the top of the seismic line corresponds to Quaternary lava coming from the Tromen volcano, which locally are faulted and dipping to the west (Fig. 4b). Such geometry was interpreted as a rollover anticline associated with an east-dipping normal fault, which has led to propose the crustal collapse of this area during the Quaternary (Folguera et al. 2008). On the other hand, field structural data, analysis of borehole breakouts, and geomorphologic studies show contractional structures affecting Plio–Quaternary rocks under ~EW maximum horizontal stress (Galland et al. 2007; Guzmán et al. 2007; Messager et al. 2010). Alternatively, right-lateral strike-slip faulting has been proposed in order to explain the coexistence of compressive and distensive structures (Sagripanti et al. 2015b), which evidence a complex structural scenario. The Chacaico and Cañada Amarilla anticlines are two east-vergent second-order folds that involve the basal units of Mendoza Group to the south of the Cerro Tilhué (Fig. 4a). These anticlines can be recognized with difficulty on the seismic line, but they were mapped using field data and constitute the easternmost thin-skinned structures in the inner zone of the Chos Malal FTB (Fig. 1c). A set of folded and eastward-thickening reflectors, unconformably covered by planar and continuous reflectors at the eastern end of the seismic line BR-B-001, allow interpreting a half-graben related to a west-dipping normal fault (Fig. 4b). A small reverse offset at the top of synrift sequences evidences a very slight tectonic inversion, and this contraction produced minor backthrusts within the Cuyo Group.

The outer zone of the Chos Malal FTB along the B-B′ section is dominated by 2-km-scale thick-skinned structures that form the southward plunging Las Yeseras and Pampa Tril anticlines (Fig. 1c). The Las Yeseras anticline is an east-vergent asymmetric fold with a long, gently dipping backlimb and a short, steeply dipping forelimb (Fig. 4a). The vertical displacement of this structure is bigger than in the section A-A′ since the Lotena Group in the hanging wall of Las Yeseras thrust is at the same level as the Bajada del Agrio Group in the adjacent syncline. The anticline is a fault-propagation fold in which the thrust broke through the forelimb, east of Puerta Curacó, where the lower shales of the Agrio Formation are overturned and significantly thinned (Fig. 1c). A branch of the Las Yeseras thrust at depth connects with the detachment level on top of the Lotena Group, and this eastward displacement generates an incipient second-order anticline observed in the seismic line SJ-HV89-210 (Fig. 4c). As occurs in the north cross section, the Pampa Tril anticline is a broad asymmetric east-vergent fold, associated with a basement-involved thrust that splits into two branches (Fig. 4a). The upper thrust caused the high structural relief, and its large displacement is solved by backthrusting on the evaporites of the Bajada del Agrio Group forming a triangle zone. The lower branch of the Pampa Tril thrust connects with the detachment in the sedimentary cover and generates two second-order anticlines recognized seismically, which will be analyzed in detail in the next section of this paper. The restoration of the B-B′ structural cross section (Fig. 4d) evidences a tectonic shortening of 22.2 km (16.5%), slightly minor than shortening calculated from the A-A′ cross section. A smaller structural development of the basement-involved structures in the inner zone of the Chos Malal FTB at the southern region, expressed by the southward plunging of the Cordillera del Viento anticlinorium, could account for such differences in shortenings.

4 Related Thick- and Thin-Skinned Deformation at the Mountain Front: Implications for Oil Exploration

The exposed deformation front of the Chos Malal FCP is located in the region of Filo Morado, where Mesozoic sequences dipping to the east on the forelimb of the Pampa Tril anticline sink beneath Miocene volcanic rocks that fill the Huantraico syncline (Fig. 1c). As previously described, in the external sector of the regional structural cross sections, the surficial structure forms a large triangle zone that masks a complex thin-skinned deformation in depth (Figs. 2a and 4a). The sector is of particular interest because the structures affecting the Mesozoic sedimentary rocks are important hydrocarbon deposits such as El Porton and Filo Morado oil fields (Fig. 1c). The acquisition of 2D and 3D seismic data in the area, but fundamentally the hundreds of drilled wells for reservoir development, allowed insight into the geometry of some structures in depth (Zapata et al. 2001; Zamora Valcarce and Zapata 2005; Zamora Valcarce et al. 2006). The entrapment of hydrocarbons in El Porton and Filo Morado oil fields occurs in tight anticlines elongated in the NS direction, with closure in all four directions, which have been located in traditional sandy reservoirs as the Troncoso Inferior Member of the Huitrín Formation and the Avilé Member of the Agrio Formation (Selva et al. 2005). Additionally, the naturally fractured Vaca Muerta and Mulichinco Formations are potential reservoirs (Zapata et al. 2001; Fantín et al. 2015). Despite the economic relevance of these deposits and numerous studies in the area, there is still no consensus about the way in which the basement participates in deformation and its relationship with the structures developed in the sedimentary cover. The analysis of the abundant seismic and well data in the outer zone of the two regional structural cross sections (Fig. 1c) allowed to elaborate a new interpretation that characterize in detail the structural configuration of the deformation front of the Chos Malal FTB (Figs. 5 and 6). The first interpretations of this region show a wide basement sheet thrust on the sedimentary cover originating duplex structures that are connected toward the surface with a passive-roof duplex giving birth to a triangle zone (Ploszkiewicz and Viñes 1987; Ramos and Barbieri 1989; Nocioni 1996). Other authors linked this thin-skinned deformation to basement structures originated as fault-propagation folds, with subvertical forelimbs, which are further displaced over a detachment in the sedimentary cover (Viñes 1989; Kozlowski et al. 1998; Allmendinger et al. 2004). More recent studies, mainly supported by numerous well data, reveal that the

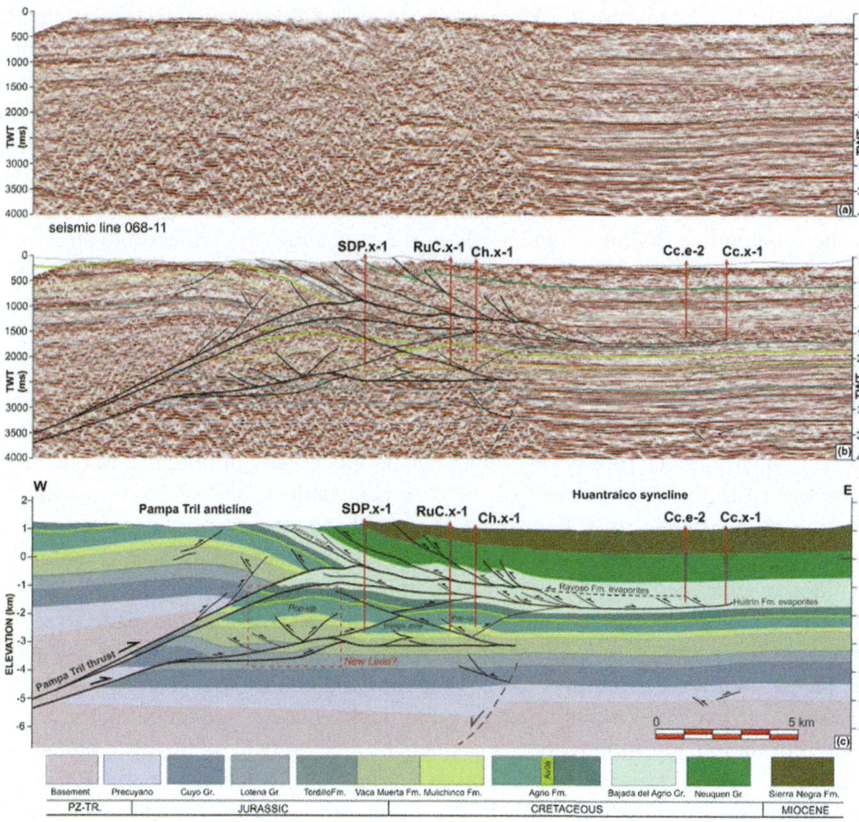

Fig. 6 **a, b** Uninterpreted and interpreted two-way travel time seismic line in the mountain front of the Chos Malal FTB along the B-B′ cross section. **c** Detailed reconstruction of the related thick- and thin-skinned structures. A lower branch of the Pampa Tril thrust produced thrust-related anticlines at depth, including an undrilled pop-up. The upper branch of the Pampa Tril thrust cut up section and is inserted in the evaporites of Bajada del Agrio Group forming a triangle zone

structure of El Portón–Filo Morado oil fields can be depicted as a tight anticline with very steep flanks, which is considered a combination of detachment and fault-propagation folding (Zapata et al. 2001; Zamora Valcarce and Zapata 2005; Zamora Valcarce et al. 2006).

Zamora Valcarce et al. (2006) considered that the anticline was formed partially in response to the positive inversion of previous normal faults and that this mechanism of tectonic reactivation also originated kilometric uplifts of basement that made Las Yeseras and Pampa Tril structures. Several geometric inconsistencies in Zamora Valcarce and Zapata (2005) and Zamora Valcarce et al. (2006) sections make the inversion tectonic model a not feasible one. The most notable feature in these sections is that normal faults interpreted as inverted show no significant displacements, and yet they are used to explain the uplift of approximately 4 km of basement in the hangingwall. The classic models of inversion tectonics clearly show that inversion is complete when the basement or the prerift rocks reach its null point, i.e., regains its regional position (Williams et al. 1989; Buiter and Pfiffner 2003; Yamada and McClay 2004). Moreover, the basement-cover contact at the forelimb of the Pampa Tril anticline shows tight folding, such as a large fault-propagation fold, which cannot be produced by displacement in a previous fault (Allmendinger et al. 2004).

The seismic line 26258-17 and wells in the area show the remarkable structural uplift of the basement in the Pampa Tril anticline with respect to its position in the foreland below the Huantraico syncline (Fig. 5a, b). The structural configuration of this area can be explained with a faulting sequence that involves branches of the Pampa Tril thrust (Fig. 5c). The first fault generates an incipient basement sheet and transmits the displacement along a detachment within Auquilco Formation and thus begins to form the second-order Filo Morado anticline. It is worthy to note that at this section latitude the fold is very small and adopts greater importance to the south where therefore the displacement of the basement fault is larger. The Filo Morado anticline is interpreted as a fault-bend fold, with an upper detachment on Agrio Formation so explaining the development of third-order structures formed under the Huantraico syncline (Fig. 5c). A second branch of Pampa Tril thrust cuts the previous detachment and becomes horizontal at the base of Agrio Formation generating a repetition of Avile Member which was detected in the FM x-6 well. The progress of these structures toward the foreland could have been hampered by the weight of overlying units, including the thick volcanic sequence of the Sierra Negra Formation, promoting out-of-sequence development of a third branch of Pampa Tril thrust. This sheet reaches evaporites of Bajada del Agrio Group, a detachment level of high efficiency in the region, allowing it to accumulate a large displacement which is compensated near the surface through a passive-roof back-thrust setting a triangle zone (Fig. 5c). The exploratory well FDT x-1 was conducted in 1990 seeking to define the structural configuration in depth, particularly to verify the presence of a low-angle thrust from a series of subhorizontal reflectors observed in the seismic line near the surface ($\sim$700 ms, Fig. 5b). The well crossed these horizons and found that the whole sequence continued with a significant dip eastward, which led to rule out the hypothesis of a thrust sheet. This added to an abnormal thickness for the Vaca Muerta Formation, greater than 1,300 m, resulted in models that interpret the eastern flank of the Pampa Tril anticline with very high dip (even reversed) to very deep levels (Viñes 1989; Kozlowski et al. 1998; Zapata et al. 2001; Allmendinger et al. 2004; Zamora Valcarce and Zapata 2005; Zamora Valcarce et al. 2006; Rojas Vera et al. 2014, 2015a, b; Sagripanti et al. 2015a, b). However, detailed analysis of the well FDT x-1 report shows that the dip of Vaca Muerta Formation remains relatively constant, considering the measurement done directly from borehole cores, and the thickness is increased by reverse faults that displace the dipping layers (Fig. 5c). On the other hand, the study of calcareous nannofossils from well samples determined that under the shales of the Vaca Muerta Formation there are also shales but from the upper section of the Agrio Formation, which allows to interpret between both Formations an important faulting that is used to locate the Pampa Tril thrust. At this level, it was found in well FDTx-2 an important change of dips that supports the presence of the thrust, because in the hangingwall layers dip to the east while in the footwall layers dip to the west being part of the western flank of the Filo Morado anticline (Fig. 5c). The study of calcareous nannofossils in samples from this well detected the tectonic repetition of the Agrio Formation Members (Angelozzi 1995). Beneath these structures, an antiform is observed whose reflectors are unconformably covered, and they are interpreted as synrift deposits in a half-graben associated with a partially inverted west-dipping normal fault (the basement did not reach its null point). This anticline also affects Vaca Muerta and Mulichinco Formations, and since there have been no drilling through Pampa Tril sheet reaching these deep levels, this zone could be a future exploratory lead. The continuity of the tectonic compression gives rise to the Lower Pampa Tril thrust, which cuts the hangingwall block of the normal fault, whose front sprouts backthrusts involving the basal units of the sedimentary sequence and deforming the overlying structure of Filo Morado (Fig. 5c).

In the deformation front of the Chos Malal FTB along the B-B′ cross section, the 068-11 seismic line also allows recognizing the remarkable structural uplift of the basement between Pampa Tril and the adjacent foreland (Fig. 6a). In this section, the throw associated with the Pampa Tril thrust is about 3 km (Fig. 6b), slightly lower than in the north, which is consistent with the southward plunge of the associated anticline observed in surface (Fig. 1c). In this interpretation, the upper branches of Pampa Tril thrust are connected to two detachment zones, recognized in the seismic lines by major deformation, which correspond to the evaporites of the Huitrín and Rayoso Formations, both of the Bajada del Agrio Group (Fig. 6b and c). Throughout both horizons, it is possible to see other minor backthrusts, with associated folding that even affect the Neuquén Group and the overlying Miocene volcanics. This shows a significant contraction above these detachments which together act as a passive-roof detachment accommodating the displacement of Pampa Tril sheet and configuring a triangle zone. The lower branch of the Pampa Tril thrust forms a small wedge that is inserted into the Lotena Group, and the displacement transmitted to the east generates thin-skinned structures whose configuration is interpreted by subsurface information (Fig. 6b). One of these structures forms a structural high with faults on both sides developing a pop-up structure that involves the Mendoza Group, which has not been drilled and therefore could constitute a new exploration lead at the front of the Chos Malal FTB. Recent studies in the Malargüe FTB, based on reprocessing of seismic data, field geology, and integration of geochemical and gravimetric data, have identified complex tectonic structures yet unevaluated opening a new exploratory stage in these highly deformed regions (Periale et al. 2015). The western pop-up structure is interpreted with two thrusts that are attached to a roof detachment in common, within the Vaca Muerta Formation, forming a duplex. The displacement transmitted to the foreland by means of this detachment creates an oriental pop-up structure, the Ruca Carmelo anticline, which was drilled by the RuC.x-1 well (Fig. 6c). While this structure has not developed an oil field due to the high water content extracted from the units, wells detected good evidence of oil and gas. This shows that at least partially the oil system functioned and thus with new models and more information is possible to increase the exploration potential of the area.

5 Structural Evolution of the Chos Malal FTB

The structural development of the Chos Malal FTB was controlled by the subduction of a series of Pacific plates in the western margin of Gondwana from the Late Cretaceous to the present (see Chap. "Cretaceous Orogeny and Marine Transgression in the Southern Central and Northern Patagonian Andes: Aftermath of a Large-Scale Flat-Subduction Event?"). The influence of changes of the relative convergence rate and the angle of the subducted plate in the magmatic processes and deformation of the upper plate along the Neuquén Andes have been widely analyzed and debated in recent years (Ramos and Barbieri 1989; Ramos 1999; Jordan et al. 2001; Cobbold and Rosello 2003; Ramos and Folguera 2005; Ramos and Kay 2006; Cobbold et al. 2008; Ramos et al. 2011; Rojas Vera et al. 2015a; Sagripanti et al. 2015b). Based on the structures studied through field and subsurface data and geochronological information available in the region, a new structural model for Chos Malal FTB is proposed (Fig. 7).

This model is primarily oriented to understand the sequential development of the structures that form this orogenic belt, leaving as a secondary matter the precise chronology of these events, in order to understand the most effective mechanisms that led to the construction of this fold and thrust belt. The model is based on the main structures interpreted in the A-A′ cross section and their restitution, taking into account the thickness of the units and the original length of the deformed layers (Fig. 2a, d). In the initial stage of the model, the configuration of this sector of the Neuquén Basin is built, where the Paleozoic basement affected by dominant east-dipping normal faults with related synrift deposits covered unconformably by Jurassic and Cretaceous sedimentary successions is depicted (Fig. 7a). A similar scheme showing an east-dipping extensional system in this area was interpreted by Folguera et al. (2007), which is supported by the exposed faults in the Cordillera del Viento (Sagripanti et al. 2014). While seismic data allowed detecting the presence of these extensional structures, in no case it can be completely seen the associated rollover anticline, and therefore, the methodologies to calculate the depth of detachment of the normal faults could not apply (e.g., Dula 1991). There is widespread consensus that the compressive deformation in the western sector of the Neuquén Basin began in the Late Cretaceous and that deposits of the Neuquén Group were accumulated in a foreland basin in response to the start of the Andean orogeny (Cobbold and Rossello 2003; Ramos and Folguera 2005; Tunik et al. 2010; Di Giulio et al. 2015; Fenell et al. 2015) (see Chap. "Cretaceous Orogeny and Marine Transgression in the Southern Central and Northern Patagonian Andes: Aftermath of a Large-Scale Flat-Subduction Event?"). Particularly in the Chos Malal FTB, a 69.09 ± 0.13 Ma ($^{40}Ar/^{39}Ar$) biotite cooling age from the Varvarco

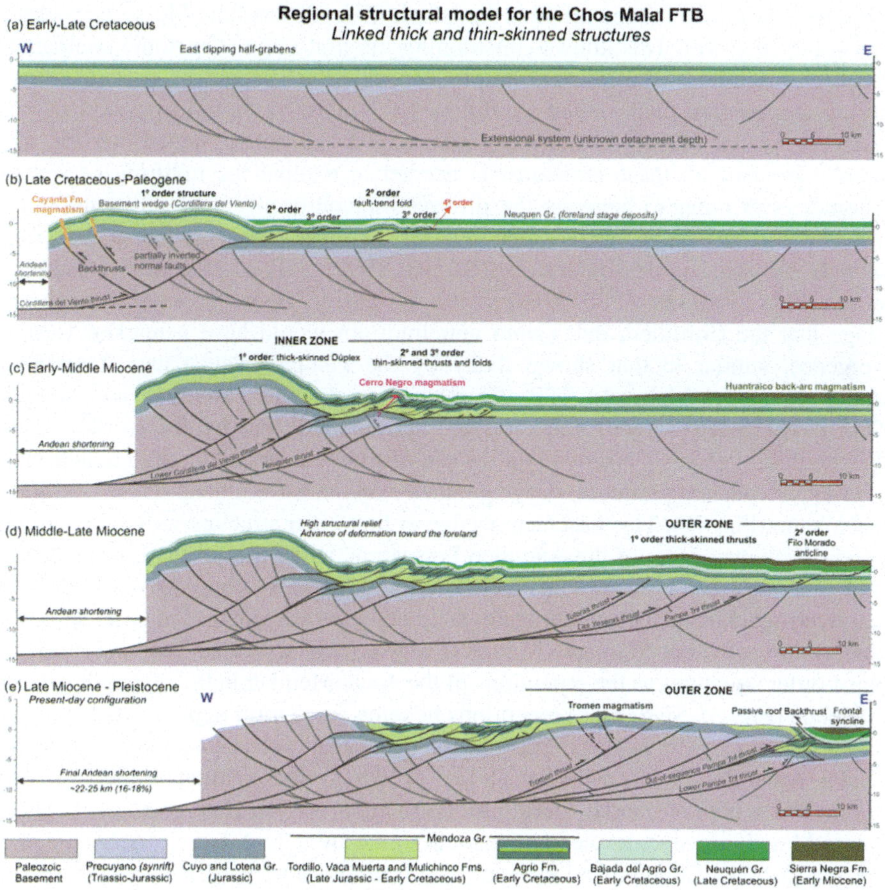

Fig. 7 Schematic structural evolution proposed for the Chos Malal FTB connecting thick- and thin-skinned structures. **a** Initial configuration prior to the Andean orogeny. **b** Late Cretaceous–Paleogene contraction produced a backthrust system that cut partially inverted normal faults in the western slope of the Cordillera del Viento. This west-vergent structure was intensely eroded and unconformably covered by Paleogene volcanic rocks. Further compression created the Cordillera del Viento thrust, which was inserted into the cover rocks producing second- and third-order thrust-related folds. **c** Neogene compression gave birth to new basement-involved thrusts at depth that in turn formed more thin-skinned structures in the inner zone of the Chos Malal FTB. **d** Propagation of the deformation toward the foreland produced new first-order thrusts into the basement rocks and subsequently different orders of folds in cover rocks. **e** Overburden in the mountain front provoked the out-of-sequence Pampa Tril thrust that is connected with an upper detachment to form a triangle zone

pluton was interpreted as an uplift age for the Cordillera del Viento (Kay et al. 2006). The volcanic rocks of the Cayanta Formation (Rapela and Llambías 1985), which have $^{40}Ar/^{39}Ar$ and fission track in zircon ages of 56.9 ± 1.1 and 48.9 ± 2.7 Ma, respectively (Jordan et al. 2001), unconformably cover the Paleozoic basement exhumed in the region of Andacollo (Fig. 2b) and the Mesozoic rocks to the south (Fig. 1c). This allows constraining the first stage of tectonic contraction between the Late Cretaceous and the Paleocene, which would have started structuring the western flank of the Cordillera del Viento by partial inversion of preexisting normal faults and the generation of a backthrust system (Fig. 7b). The combination of all these structures is needed to explain the uplifting of several kilometers in order to generate the observed unconformity, and additionally, the backthrusts could have acted as conduits for the Paleogene magmatism, a process already proposed in nearby regions (Dimieri 1992, Galland et al. 2007, Turienzo et al. 2012, Araujo et al. 2013, Gürer et al. 2015). It is possible to imagine at that stage that the Cordillera del Viento anticlinorium would have a marked western vergence, similar to that shown today by the Pampa Palauco and Ranquil-Co anticlines in the Malargüe FTB (Giambiagi et al. 2009; Silvestro and Atencio 2009). These retrovergent structures are subsequently cut by the Cordillera del Viento thrust, which is inserted into the sedimentary cover and through a fault-bend folding mechanism produces wedge geometry on the eastern flank of the Cordillera del Viento (Fig. 7b). This basement wedge or first-order structure moves toward the foreland on evaporites of the Auquilco Formation, at the roof of the Lotena Group, and generates second-order folds involving the Mendoza Group. Most of the latter are interpreted as fault-bend anticlines with an upper detachment in basal shales of the Agrio Formation (Pilmatué Member) whose transferred displacement produces third-order structures as the repetitions of the Avile Member in the region of Chacay Melehue (Fig. 3). Since these repetitions in some cases form duplex structures, part of the displacement is transmitted to an even shallower detachment within the shales Agua de la Mula Member and so producing deformation in the overlying levels that were considered fourth-order structures by Turienzo et al. (2014). The Cerro Mayal intrusive, located to the south of the study area, has a $^{40}Ar/^{39}Ar$ age of 39.7 ± 0.2 Ma (Cobbold and Rossello 2003) and intrudes and displaces the axes of the mentioned second-order folds (Sánchez et al. 2015). This allows constraining the age of the first stage of contraction toward the foreland in the inner zone of the Chos Malal FTB, which has occurred prior to Late Eocene.

The structural development continues with the Lower Cordillera del Viento and Neuquén thrusts (Fig. 7c), which also form basement wedges that are inserted into the sedimentary cover and cause the rest of the thin-skinned structures that characterize the inner zone of the Chos Malal FTB (Turienzo et al. 2014, Sánchez et al. 2015). North of the studied section, the middle Miocene volcanic sequences of the Charilehue Formation are folded (Folguera et al. 2007). The Cerro Negro intrusive complex consists of sills and N-S-striking dikes, with U-Pb ages of 11.63 ± 0.20 Ma and 11.55 ± 0.06 Ma, respectively, emplaced at the hinge zone of a large thin-skinned anticline that was interpreted as coeval with the tectonic development of the Chos Malal FTB (Gürer et al. 2015). The high structural relief achieved in the internal sector at this stage would favor the subsequent propagation of the deformation to the foreland, in view of the theory of Coulomb critical wedge (Davis et al. 1983), by generating new basement-involved thrusts in the outer zone (Fig. 7d). The Totoras and Las Yeseras thrusts caused low displacements, producing a stepwise rise of the basement and toward the surface fault-propagation folds with monoclinal geometry in the overlying Mesozoic strata. The Pampa Tril thrust reaches the detachment in the sedimentary cover, and it propagates toward the foreland forming the Filo Morado anticline and associated minor structures. The retroarc volcanic sequences in the Huantraico region, with $^{40}Ar/^{39}Ar$ ages of 17–18 Ma (Kay and Copeland 2006; Dyhr et al. 2013), are folded; moreover, some overlying clastic units are also folded suggesting a Late Miocene deformation (Garrido et al. 2012). The displacement of these structures toward the foreland was possibly hindered by the overburden exerted by these thick Tertiary sequences, so favouring the formation of an out-of-sequence branch of the Pampa Tril thrust (Fig. 7e). This new thrust uplifting the basement reaches the upper detachment within the evaporites of the Bajada del Agrio Group, and its forward displacement is compensated by a passive-roof backthrust. With the continuity of compression, the Lower Pampa Tril thrust is formed, which cuts a slightly inverted extensional structure, from whose front minor backthrusts emerge at the base of the sedimentary sequence. These structures in depth distort the original shape of the Filo Morado anticline, and they complete the structural configuration of the outer zone of the Chos Malal FTB (Fig. 7e). Neotectonic activity takes place by out-of-sequence faulting in the hangingwall of the Totoras and Las Yeseras thrusts, possibly with the development of backthrusts that control the deformation and magmatism in the region of Tromen volcano (Galland et al. 2007; Sagripanti et al. 2015a).

The final total shortening due to the structural development of the Chos Malal FTB considering the data of the two interpreted regional cross sections is in the order of 22–25 km (16–18%). Rojas Vera et al. (2015a) made a structural interpretation of this same section, extended from the western flank of the Cordillera del Viento anticlinorium to the Huantraico syncline, and obtained a shortening of 9.5 km ($\sim$10%). This substantial difference lies in two main factors. Rojas Vera et al. (2015a) considered that the main basement-involved structures in the Chos Malal FTB correspond to inverted normal faults, following the reconstruction of some previous authors (Zapata et al. 1999; Zamora Valcarce et al. 2006), and thereby, this implies significant uplifts and low shortenings. On the other hand, in their interpretation, thin-skinned deformation is represented by few fault-propagation anticlines, which are produced by faults that reach the surface branched from a deep detachment. Numerical and analogue models that analyze the inversion of normal faults clearly show that the total inversion of a half-graben occurs when the basement reaches the null point and the synrift infill is completely expelled, but if the compression continues thrusts are generated that meet the terms of the Coulomb criterion fracturing and they cut the previous structures (Williams et al. 1989; Buiter and Pfiffner 2003; Yamada and McClay 2004, among others). In addition to these concepts, all extensional structures recognized in the seismic lines (Fig. 4b) and in other sectors of the Chos Malal FTB (Sánchez et al. 2015) preserve their normal throw with little or no inversion. Although inversion of normal faults is a mechanism able to generate significant structures, and in many cases of great economic importance, these evidence show that it is not the main mechanism that has led to the construction of this orogen. Finally, detailed mapping in the area between the Cordillera del Viento and Tromen volcano (Sánchez et al. 2015) and available subsurface information allowed to recognize and characterize a complex thin-skinned deformation with fault-related folds with different wavelengths and affecting diverse geological units (Figs. 2a and 4a). The different geometry of the folds evidences the presence of multiple detachments in the sedimentary cover that can generate structures of second, third, and even fourth order (Turienzo et al. 2014), spatially and temporally related, which are directly linked to the first-order basement-involved structures (Fig. 7).

6 Conclusions

Balanced structural cross sections constructed throughout the Chos Malal FTB ($\sim 37°$ S), with the aid of detailed field mapping and subsurface information, have allowed to recognize a complex deformation pattern linking thick- and thin-skinned structures. This interaction between basement-involved structures and fault-related folds in the sedimentary cover is recognized in two discrete regions that lead to separate an inner zone from an outer one. The Cordillera del Viento structure is an extensive N-S-striking anticlinorium that exposes the Paleozoic basement and is the most important structure in the inner zone of the Chos Malal FTB. Its initial development is linked to the formation of a backthrust system on the western flank, cutting a series of partially inverted normal faults, which together originated a large west-vergent anticline. This structure underwent a major exhumation between the Late Cretaceous and the Paleocene, since Paleogene volcanic rocks lay unconformably above the basement in the north and above younger Mesozoic units to the south. Configuration of the Cordillera del Viento structure continued with the formation in depth of an important west-dipping thrust that was inserted into the sedimentary cover and produced a tectonic wedge on the eastern flank. Propagation toward the foreland of the first-order basement wedge produced complex thin-skinned structures (second, third, and fourth order) related to the presence of multiple detachments in the sedimentary cover. The main detachment on the cover rocks is interpreted interleaved in the Late Jurassic evaporites Auquilco Formation, atop the Lotena Group, because thin-skinned structures involving the overlying Tordillo Formation are exposed. This structural style is repeated and emphasized since the Miocene, where two new thrusts form basement slices that connect with the detachment on the cover and generate thin-skinned deformation that completes the structural development of the inner zone. Further tectonic compression provokes the expansion of the fold and thrust belt toward the foreland by means of new basement-involved thrusts in the external sector. The Las Yeseras and Totoras thrusts produced ample folds with monoclinic geometry. The Pampa Tril thrust instead is interpreted with ramifications that show a more complex deformation. Initially, it was connected to the detachment in the sedimentary cover, and the push of this first-order emerging structure gave rise to the Filo Morado anticline. Out-of-sequence faulting, possibly favoured by the overburden of Tertiary volcanic sequences in the frontal syncline, led to the uplift of a large basement slices which formed the Pampa Tril anticline. The large displacement of this structure is solved through passive-roof backthrust within the Cretaceous evaporites of the Bajada del Agrio Group, forming a triangle zone such as those proposed in some of the pioneering interpretations of the region. Detailed reinterpretation of subsurface information in the deformation front points to a complex structural configuration, whose identification and understanding open new exploratory expectations in the region. The restitution of sections, in addition to ensure the viability of the interpreted structures, allowed calculating a tectonic shortening of 22–25 km (16–18%) for the Chos Malal FTB. These shortening figures are substantially higher than those determined by other authors that infer the tectonic inversion of normal faults as the main mechanism of deformation and consider an underestimated thin-skinned deformation. This region of the Andes is characterized by excellent exposures, and the great diversity of structures observed highlights the importance of detailed study in the field as a fundamental tool for understanding complex subsurface structures. Thorough structural studies are essential for future tasks of hydrocarbon exploration in this region, and making balanced structural cross sections with accurate determinations of tectonic shortenings will provide support for regional models in order to analyze the tectonic evolution of this segment of the Southern Central Andes.

Acknowledgements The present studies were supported by several grants from CeCyT-UNS (24/H117), ANPCyT (Pict 0166) and CONICET (PIP 0390). Seismic and well data were provided by the Subsecretaría de Minería e Hidrocarburos de la Provincia del Neuquén. We kindly acknowledge the Municipalidad of Chos Malal and Andacollo for collaborating with our stay during the fieldwork. The authors thank Lucía Sagripanti and Emilio Rojas Vera for their reviews.

References

Allmendinger R, Zapata T, Manceda R, Dzelalija F (2004) Trishear kinematic modeling of structures, with examples from the Neuquén Basin, Argentina. In: McClay K (ed) Thrust tectonics and hydrocarbon systems, vol 82, American Association of Petroleum Geologists, Memoir, pp 356–371

Angelozzi G (1995) Nanofósiles calcáreos del pozo YPF Nq. FDT X-2 (Flanco del Tril). Boletín de Informaciones Petroleras 41:56–59

Araujo V, Dimieri L, Frisicale C, Turienzo M, Sánchez N (2013) Emplazamiento del intrusivo Laguna Amarga y su relación con las estructuras tectónicas andinas, sur de la provincia de Mendoza. Revista de la Asociación Geológica Argentina 70(1):457–469

Arregui C, Carbone O, Leanza HA (2011a) Contexto tectosedimentario. In: Leanza HA, Arregui C, Carbone O, Danieli JC, Vallés JM (eds) Geología y Recursos Naturales de la Provincia del Neuquén. Relatorio del XVIII Congreso Geológico Argentino, pp 29–36

Arregui C, Carbone O, Martínez R (2011b) El Grupo Cuyo (Jurásico Temprano-Medio) en la Cuenca Neuquina. In: Leanza HA, Arregui C, Carbone O, Danieli JC, Vallés JM (eds) Geología y Recursos Naturales de la Provincia del Neuquén. Relatorio del XVIII Congreso Geológico Argentino, pp 77–89

Barrionuevo M, Valenzuela M, Olea G, Gutiérrez Pleimling A (2005) Trampas características de las formaciones Triásico-Jurásicas en la plataforma oriental de la Cuenca Neuquina. In: Koslowski E, Vergani G, Boll A (eds) Las Trampas de Hidrocarburos en las Cuencas Productivas de Argentina. VI Congreso de Exploración y Desarrollo de Hidrocarburos, pp 209–224

Booth J, Coward M (1996) Basement faulting and inversion of the NW Neuquén Basin, Argentina. III international symposium on Andean Geodynamics, St. Malo (France), extended abstracts, pp 295–298

Buiter S, Pfiffner A (2003) Numerical models of the inversion of half-graben basins. Tectonics 22 (5). https://doi.org/10.1029/2002TC001417

Carbone O, Franzese J, Limeres M, Delpino D, Martínez R (2011) El Ciclo Precuyano (Triásico Tardío e Jurásico Temprano) en la Cuenca Neuquina. In: Leanza HA, Arregui C, Carbone O, Danieli JC, Vallés JM (eds) Geología y Recursos Naturales de la Provincia del Neuquén. Relatorio del XVIII Congreso Geológico Argentino, pp 63–76

Chauveau V, Niviere B, Cobbold P, Rossello E, Ballard J, Eichenseer H (1996) Structure of the Andean foothills, Chos Malal region, Neuquén Basin, Argentina. III international symposium on Andean Geodynamics, St. Malo (France), extended abstracts, pp 315–318

Charrier R, Pinto L, Rodríguez MP (2007) Tectono-stratigraphic evolution of the Andean orogen in Chile. In: Gibbons W, Moreno T (eds) Geology of Chile, Chapter 3. The Geological Society, London, Special Publication, pp 21–116

Charrier R, Ramos VA, Tapia F, Sagripanti L (2015) Tectono-stratigraphic evolution of the Andean Orogen between 31° and 37° S (Chile and Western Argentina). In: Sepúlveda SA, Giambiagi LB, Moreiras SM, Pinto L, Tunik M, Hoke GD, Farías M (eds) Geodynamic processes in the Andes of Central Chile and Argentina. Geological Society London, Special Publications 399, pp 13–61

Cobbold P, Rossello E (2003) Aptian to recent compressional deformation, foothills of the Neuquén Basin Argentina. Mar Pet Geol 20:429–443

Cobbold P, Diraison M, Rossello E (1999) Bitumen veins and Eocene Transpression, Neuquén Basin, Argentina. Tectonophysics 314:423–442

Cobbold P, Rossello E, Roperch P, Arriagada C, Gomez L, Lima C (2007) Distribution, timing, and causes of Andean deformation across South America. In: Ries AC, Butler RW, Graham RH (eds) Deformation of the continental crust "The legacy of Mike Coward". Geological Society, London, Special Publications, 272, 321–343

Cobbold P, Rosello E, Marquez F (2008) Where is the evidence for Oligocene rifting in the Andes? Is it in the Loncopué Basin of Argentina? In: VII international symposium on Andean Geodynamics, Nice (France), extended abstracts: 148–151

Cristallini EO, Bottesi G, Gavarrino A, Rodriguez L, Tomezzoli R, Comeron R (2006) Synrift geometry of the Neuquén Basin in northeastern Neuquén Province, Argentina. In: Kay SM, Ramos VA (eds) Evolution of an Andean margin: a tectonic and magmatic view from the Andes to the Neuquén Basin (35°–39° S lat). Geological Society of America, Special Paper 407, pp 147–162

Cristallini E, Tomezzoli R, Pando G, Gazzera C, Martínez J, Quiroga J, Buhler M, Bechis F, Barredo S, Zambrano O (2009) Controles precuyanos en la estructura de la Cuenca Neuquina. Rev Asoc Geol Argent 65:248–264

Danieli JC, Coppolecchia M, Elissondo M (2011) El Grupo Andacollo (Paleozoico Tardío). In: Leanza HA, Arregui C, Carbone O, Danieli JC, Vallés JM (eds) Geología y Recursos Naturales de la Provincia del Neuquén. Relatorio del XVIII Congreso Geológico Argentino, pp 49–52

Davis D, Suppe J, Dahlen F (1983) Mechanics of fold-and-thrust belts and accretionary wedges. J Geophys Res 88:1153–1172

Di Giulio A, Ronchi A, Sanfilippo A, Balgord E, Carrapa B, Ramos V (2015) Cretaceous evolution of the Andean margin between 36° S and 40° S latitude through a multi-proxy provenance analysis of Neuquén Basin strata (Argentina). Basin Research 1–21. https://doi.org/10.1111/bre.12176

Dimieri L (1992) Emplazamiento lacolítico a través de retrocorrimientos, Cerro Palao Mahuida, Bardas Blancas, Mendoza. Academia Nacional de Ciencias Exactas, Físicas y Naturales, Monografías 8:163–166

Dula W (1991) Geometric models of listric normal faults and rollover folds. Am Assoc Pet Geol 75(10):1609–1625

Dyhr CT, Holm PM, Llambias EJ, Scherstén A (2013) Subduction controls on Miocene back-arc lavas from Sierra de Huantraico and La Matancilla and new 40Ar/39Ar dating from the Mendoza region, Argentina. Lithos 179:67–83

Fantín J, Manceda R, Palacio B, López R, Mykietiuk K, Barberis R (2015) Caracterización de las fracturas naturales de las Formaciones Vaca Muerta y Mulichinco en la estructura de Filo Morado, Cuenca Neuquina, Argentina. IX Congreso de Exploración y Desarrollo de Hidrocarburos, pp 541–561

Fenell L, Folguera A, Naipauer M, Gianni G, Rojas Vera E, Bottesi G, Ramos V (2015) Cretaceous deformation of the Southern Central Andes: synorogenic growth strata in the Neuquén Group (35°30′–37° S). Basin Res. 1–22. https://doi.org/10.1111/bre.12135

Folguera A, Ramos VA, Zapata TR, Spagnuolo MG (2007) Andean evolution at the Guañacos and Chos Malal fold and thrust belts (36° 30′–37° S). J Geodyn 44:129–148

Folguera A, Bottesi G, Zapata T, Ramos VA (2008) Crustal collapse in the Andean backarc since 2 Ma: Tromen volcanic plateau, Southern Central Andes (36°40′–37°30′S). Tectonophysics 459:140–160

Franzese J, Spalletti L (2001) Late Triassic-Early Jurassic continental extension in southwestern Gondwana: tectonic segmentation and pre-break-up rifting. J South Am Earth Sci 14:257–270

Galland O, Hallot E, Cobbold P, Ruffet G, de Bremond d'Ars J (2007) Volcanism in a compressional Andean setting: a structural and geochronological study of Tromen volcano (Neuquén province, Argentina). Tectonics 26:1–24. TC4010. https://doi.org/10.1029/2006TC002011

Garrido A, Kramarz A, Forasiepi A, Bond M (2012) Estratigrafía, mamíferos fósiles y edad de las secuencias volcanosedimentarias eoceno-miocenas de la sierra de Huantraico-sierra Negra y cerro Villegas (provincia del Neuquén, Argentina). Andean Geol 39(3):482–510

Giacosa R, Allard J, Foix N, Heredia N (2014) Stratigraphy, structure and geodynamic evolution of the Paleozoic rocks in the Cordillera del Viento (37° S latitude, Andes of Neuquén, Argentina). J Iberian Geol 40:331–348. https://doi.org/10.5209/rev_JIGE.2014.v40.n2.45301

Giambiagi L, Ghiglione M, Cristallini E, Bottesi G (2009) Kinematic models of basement/cover interactions: insights from the Malargüe fold and thrust belt, Mendoza, Argentina. J Struct Geol 31:1443–1457

Gómez Omil R, Caniggia J, Borghi P (2015) La Formación Vaca Muerta en la faja plegada de Neuquén y Mendoza. Procesos que controlaron su madurez. IX Congreso de Exploración y Desarrollo de Hidrocarburos, Actas DVD, 71–96

Groeber P (1929) Líneas fundamentales de la geología del Neuquén, sur de Mendoza y regions adyacentes. Dirección General de Minas, Geología e Hidrología, vol 58. Buenos Aires, Publicación, p 110

Groeber P (1946) Observaciones geológicas a lo largo del meridiano 70°. 1. Hoja Chos Malal. Revista de la Sociedad Geológica Argentina 1(3): 177–208

Gulisano C (1981) El ciclo cuyano en el norte de Neuquén y sur de Mendoza. VIII Congreso Geológico Argentino (San Luis) Actas, vol 3, pp 573–592

Gulisano C, Gutiérrez Pleimling A (1995) Field Guide The Jurassic of the Neuquén Basin, a) Neuquén province. Asoc. Geológica Argentina Publicación Especial 158, 111 pp

Guzmán C, Cristallini E, Bottesi G (2007) Contemporary stress orientations in the Andean retroarc between 34° S and 39° S from borehole breakout analysis. Tectonics 26. http://dx.doi.org/10.1029/2006TC001958

Gürer D, Galland O, Corfu F, Leanza H, Sassier C (2015) Structure and evolution of volcanic plumbing systems in fold-and-thrust belts: a case study of the Cerro Negro de Tricao Malal, Neuquén Province, Argentina GSA Bulletin. https://doi.org/10.1130/B31341.1

Holmberg E (1975) Descripción geológica de la Hoja 32c, Buta Ranquil (Prov. Mendoza-Neuquén), Bull. 152, 71 pp. Serv. Geol. Minero Argent., Buenos Aires

Howell JA, Schwarz E, Spalletti LA, Veiga GD (2005) The Neuquén Basin: an overview. In: Veiga GD, Spalletti LA, Howell JA, Schwarz E (eds.) The Neuquén Basin, Argentina: a case study in sequence stratigraphy and basin dynamics, vol 352. Geological Society of London, London, Special Publications, pp 1–14

Jones PB (1982) Oil and gas beneath east-dipping underthrust faults in the Alberta Foothills, Canada. In: Powers RB (ed) Geologic studies of the Cordilleran thrust belt, Denver. Rocky Mountain Association of Geologists, pp 61–74

Jordan TE, Burns WM, Veiga R, Pángaro F, Copeland P, Kelley S, Mpodozis MC (2001) Extension and basin formation in the southern Andes caused by increased convergence rate. A mid-Cenozoic trigger for the Andes. Tectonics 20:308–324

Kay SM, Copeland P (2006) Early to middle Miocene back-arc magmas of the Neuquén Basin: geochemical consequences of slab shallowing and the westward drift of South America. In: Kay SM, Ramos VA (eds) Evolution of an Andean margin: a tectonic and magmatic view from the Andes to the Neuquén Basin (35°–39° S lat). Geol Soc of America, Special Paper, vol 407, pp 185–213

Kay S, Burns W, Copeland PC, Mancilla O (2006). Upper Cretaceous to Holocene magmatism and evidence for transient Miocene shallowing of the Andean subduction zone under the northern Neuquén Basin. In: Kay SM, Ramos VA (eds) Evolution of an Andean margin: a tectonic and magmatic view from the Andes to the Neuquén Basin (35°–39° S lat). Geol Soc of America, Special Paper, vol 407, pp 19–60

Kozlowski EE, Cruz CE, Sylwan CA (1996) Geología estructural de la zona de Chos Malal, Cuenca Neuquina, Argentina. XIII Congreso Geológico Argentino y III Congreso de Exploración de Hidrocarburos 1:15–26

Kozlowski EE, Cruz CE, Sylwan CA (1998) Modelo exploratorio en la faja corrida de la Cuenca Neuquina, Argentina. Boletín de Informaciones Petroleras 55:4–23

Leanza HA (2003) Las sedimentitas Huitrinianas y Rayosianas (Cretácico inferior) en el ámbito central y meridional de la Cuenca Neuquina, Argentina. Servicio Geológico Minero Argentino. Serie Contribuciones Técnicas, Geología 2:1–31

Leanza HA (2009) Las principales discordancias del Mesozoico de la Cuenca Neuquina según observaciones de superficie. Rev. del Museo Argentino de Ciencias Naturales, 11, 145–184

Leanza H, Mazzini A, Corfu F, Llambías E, Svensen H, Planke S, Galland O (2013) The Chachil Limestone (Pliensbachian-earliest Toarcian) Neuquén Basin, Argentina: U-Pb age calibration and its significance on the Early Jurassic evolution of southwestern Gondwana. J South Am Earth Sci 42:171–185

Lebinson F, Turienzo M, Araujo V, Sánchez N, Frisicale C, Dimieri L (2015) Control estructural en el emplazamiento de las rocas ígneas del cerro Caicayén, faja corrida y plegada del Agrio, Neuquén. XVI Reunión de Tectónica, Gral. Roca, Acta de Resúmenes, pp 138–139

Lebinson F, Turienzo M, Sánchez N, Araujo V, Dimieri L (2015b) Geometría y cinemática de las estructuras en el extremo septentrional de la faja plegada y corrida del Agrio, provincia de Neuquén. Revista de la Asociación Geológica Argentina 72(3):299–313

Legarreta L, Gulisano CA (1989) Análisis estratigráfico secuencial de la Cuenca Neuquina (Triásico superior-Terciario inferior, Argentina). In: Chebli G, Spalletti L (eds) Cuencas Sedimentarias Argentinas. Serie Correlación Geológica (6) Universidad Nacional de Tucumán, pp 221–243

Legarreta L, Uliana MA (1991) Jurassic/Cretaceous marine oscillations and geometry of a back-arc basin fill, central Argentine Andes. In: Mc Donald DIM (ed) Sedimentation, Tectonics and Eustacy. I.A.S, London, Special Publication 12, pp 429–450

Llambías EJ, Leanza HA, Carbone O (2007) Evolución tectono-magmática durante el Pérmico al Jurásico Temprano en la Cordillera del Viento (37°05′S–37°15′S): nuevas evidencias geológicas y geoquímicas del inicio de la cuenca Neuquina. Rev Asoc Geol Argent 62 (2):217–235

Llambías EJ, Sato AM (2011) Ciclo Gondwánico: la provincia magmática Choiyoi en Neuquén. In: Leanza HA, Arregui C, Carbone O, Danieli JC, Vallés JM (eds) Geología y Recursos Naturales de la Provincia del Neuquén. Relatorio del XVIII Congreso Geológico Argentino, pp 53–62

Maloney KT, Clarke GL, Klepeis KA, Quevedo L (2013) The Late Jurassic to present evolution of the Andean margin: drivers and the geological record. Tectonics 32:1049–1065. http://doi.org/10.1002/tect.20067

Messager G, Nivière B, Martinod J, Lacan P, Xavier J-P (2010) Geomorphic evidence for PlioQuaternary compression in the Andean foothillsof the southern Neuquén Basin, Argentina. Tectonics 29. https://doi.org/10.1029/2009TC002609

Mosquera A, Silvestro J, Ramos VA, Alarcón M, Zubiri M (2011) La estructura de la Dorsal de Huincul. In: Leanza HA, Arregui C, Carbone O, Danieli JC, Vallés JM (eds) Geología y Recursos Naturales de la Provincia del Neuquén. Relatorio del XVIII Congreso Geológico Argentino, pp 385–397

Nocioni A (1996) Estudio estructural de la Faja Plegada y Corrida de la Cuenca Neuquina—Surmendocina. XIII Congreso Geológico Argentino y III Congreso de Exploración de Hidrocarburos, Actas 2:353–372

Pángaro F, Corbera R, Carbone O, Hinterwimmer G (2002) Los Reservorios del Precuyano. In: Schiuma M, Hinterwimmer G, Vergani G (eds) Rocas Reservorio de las Cuencas Productivas de la Argentina. V Congreso de Exploración y Desarrollo de Hidrocarburos, Mar del Plata, pp 229–274

Periale S, Haring C, Olivieri G, Guerello R, Simonetto L (2015) Exploración en áreas productivas maduras. Malargüe: nueva etapa exploratoria. Cuenca Neuquina, Argentina. IX Congreso de Exploración y Desarrollo de Hidrocarburos. Actas DVD, 3–24

Ploszkiewicz V, Viñes R (1987) Filo Morado: Un descubrimiento exploratorio en cinturón plegado. Boletín de Informaciones Petroleras, Tercera Época IV (10): 97–102

Ramos VA (1999) Plate tectonic setting of the Andean Cordillera. Episodes 22(3):183–190

Ramos VA (2010) The tectonic regime along the Andes: Present-day and Mesozoic regimes. Geol J 45:2–25. https://doi.org/10.1002/gj.1193

Ramos V, Barbieri M (1989) El volcanismo Cenozoico de Huantraico: Edad y relaciones isotópicas iniciales, provincia del Neuquén. Rev Asoc Geol Argentina 43:210–223

Ramos VA, Folguera A (2005) Tectonic evolution of the Andes of Neuquén: constraints derived from the magmatic arc and foreland deformation. In: Veiga GD, Spalletti LA, Howell JA, Schwarz E (eds) The Neuquén Basin: a case study in sequence stratigraphy and basin dynamics, vol 252. Special Publications, Geol Soc London, pp 15–35

Ramos VA, Kay SM (2006) Overview of the evolution of the southern Central Andes of Mendoza and Neuquén (35°–39° S latitude). In: Kay SM, Ramos VA (eds) Evolution of an Andean margin: a tectonic and magmatic view from the Andes to the Neuquén Basin (35°–39° S Latitude). Geological Society of America, Special Paper, vol 407, pp 1–17

Ramos VA, Zapata T, Cristallini E, Introcaso A (2004) The Andean thrust system—latitudinal variations in structural styles and orogenic shortening. In: McClay KR (ed) Thrust tectonics and hydrocarbon systems, vol 82. American Association of Petroleum Geologists, Memoir, pp 30–50

Ramos VA, Mosquera A, Folguera A, García Morabito E (2011) Evolución tectónica de los Andes y del Engolfamiento Neuquino adyacente. In: Leanza HA, Arregui C, Carbone O, Danieli JC, Vallés JM (eds) Geología y Recursos Naturales de la Provincia de Neuquén. Relatorio del VXIII Congreso Geológico Argentino, Buenos Aires, pp 335–348

Ramos VA, Litvak VD, Folguera A, Spagnuolo M (2014) An Andean tectonic cycle: from crustal thickening to extension in a thin crust (34°–37°SL). Geosci Front 5:351–367. https://doi.org/10.1016/j.gsf.2013.12.009

Rapela C, Llambías E (1985) La secuencia andesítica terciaria de Andacollo, Neuquén, Argentina. IV Congreso Geológico Chileno, Antofagasta, Actas 4:458–488

Rojas Vera EA, Folguera A, Zamora Valcarce G, Bottesi G, Ramos VA (2014) Structure and development of the Andean system between 36° and 39° S. J Geodyn 73:34–52. https://doi.org/10.1016/j.jog.2013.09.001

Rojas Vera E, Mescua J, Folguera A, Becker TP, Sagripanti L, Fennell L, Orts D, Ramos VA (2015a) Evolution of the Chos Malal and Agrio fold and thrust belts, Andes of Neuquén: insights from structural analysis and apatite fission track dating. In: Folguera A, Alvarado P, Arriagada C, Ramos V (eds) Tectonics of the Argentine and Chilean Andes, J South Am Earth Sci, Special Issue 64 (2), 418–433

Rojas Vera E, Orts D, Folguera A, Zamora Valcarce G, Bottesi G, Fennell L, Chiachiarelli F, Ramos VA (2015b) The transitional zone between the Southern Central and Northern Patagonian Andes (36–39° S). In: Folguera A, Naipauer M, Sagripanti L, Ghiglione M, Orts D, Giambiagi L (eds) Growth of the Southern Andes. Springer ESS, pp 99–114

Sánchez N (2015) Evolución tectónica de las estructuras andinas en la región del río Neuquén (~37°20′ L.S), faja corrida y plegada de Chos Malal, provincia de Neuquén. PhD thesis (unpublished), Universidad Nacional del Sur, 220 pp, Bahía Blanca

Sánchez N, Turienzo M, Dimieri L, Araujo V (2014) Reconstrucción estructural a los 37°18′S, faja corrida y plegada de Chos Malal, Provincia de Neuquén. Revista de la Asociación Geológica Argentina 71(2):233–246

Sánchez N, Turienzo M, Lebinson F, Araujo V, Coutand I, Dimieri L (2015) Structural style of the Chos Malal fold-and-thrust belt, Neuquén Basin, Argentina: relationship between thick and thin-skinned tectonics. In: Folguera A, Alvarado P, Arriagada C, Ramos V (eds) Tectonics of the Argentine and Chilean Andes, Special Issue J South Am Earth Sci 64(2), 399–417

Sagripanti L, Folguera A, Giménez M, Rojas Vera EA, Fabiano JJ, Molnar N, Fennell L, Ramos VA (2014) Geometry of middle to late Triassic extensional deformation pattern in the Cordillera del Viento (Southern Central Andes): a combined field and geophysical study. J Iber Geol 40:349–366

Sagripanti L, Rojas Vera EA, Gianni GM, Folguera A, Harvey JE, Farías M, Ramos VA (2015a) Neotectonic reactivation of the western section of the Malargüe fold and thrust belt (Tromen volcanic plateau, Southern Central Andes). Geomorphology 232:164–181

Sagripanti L, Folguera A, Fennell L, Rojas Vera EA, Ramos VA (2015b) Progression of the deformation in the Southern Central Andes (37° S). In: Folguera A, Naipauer M, Sagripanti L, Ghiglione M, Orts D, Giambiagi L (eds) Growth of the Southern Andes. Springer ESS, pp 115–132

Selva G, Vittone J, Vergani G (2005) Trampas Estructurales en el Pie de Sierra de la Faja Plegada Neuquina. In: Kozlowski E, Vergani G, Boll A (eds) Las Trampas de Hidrocarburos en las Cuencas Productivas de Argentina. VI Congreso de Exploración y Desarrollo de Hidrocarburos, IAPG, pp 141–156

Silvestro J, Atencio M (2009) La cuenca cenozoica del Río Grande y Palauco: edad, evolución y control estructural, faja plegada de Malargüe. Rev Asoc Geol Argentina 65(1):154–169

Stipanicic P (1965) El Jurásico de la Vega de la Veranada (Neuquén), el Oxfordense y el diastrofismo diveseano (Agassiz-Yaila) en Argentina. Revista de la Asociación Geológica Argentina 20(4):403–478

Tunik M, Folguera A, Naipauer M, Pimentel MM, Ramos VA (2010) Early uplift and orogenic deformation in the Neuquén Basin: constraints on the Andean uplift from U-Pb and Hf isotopic data of detrital zircons. Tectonophysics 489(1–4):258–273

Turienzo M, Dimieri L, Frisicale C, Araujo V, Sánchez N (2012) Cenozoic structural evolution of the Argentinean Andes at 34°40′S: a close relationship between thick and thin-skinned deformation. Andean Geol 39(2):317–357

Turienzo M, Sánchez N, Dimieri L, Lebinson F, Araujo V. 2014. Tectonic repetitions of the Early Cretaceous Agrio Formation in the Chos Malal fold-and-thrust belt, Neuquén basin, Argentina: geometry, kinematics and structural implications for Andean building. J South Am Earth Sci 53: 1–19

Veiga R, Lara ME, Bruveris P (1999) Distribución de hidrocarburos sobre el margen externo en una cuenca de tras-arco. Ejemplos en la cuenca Neuquina, Argentina. Boletín de Informaciones Petroleras 60:142–164

Vergani G, Tankard AJ, Belotti HJ, Welsink HJ (1995) Tectonic Evolution and Paleogeography of the Neuquén Basin, Argentina. In: Tankard J, Suarez SR, Welsink J (eds) Petroleum Basin of South America: Am Assoc Petroleum Geologists, Memoir, vol 62, pp 383–402

Viñes R (1989) Interpretación de la estructura de Filo Morado. I Congreso Nacional de Exploración de Hidrocarburos (Argentina). Actas 2:1107–1124

Williams G, Powell C, Cooper M (1989) Geometry and kinematics of inversion tectonics. In: Cooper M, Williams G (eds) Inversion Tectonics. Geological Society of London, Special Publication 44, pp 3–15

Yamada Y, McClay K (2004) 3-D Analog modeling of inversion thrust structures. In: McClay KR (ed) Thrust tectonics and hydrocarbon systems, vol 82. American Asociation of Petroleum Geologists, pp 276–301

Zamora Valcarce G, Zapata T (2005) Estilo estructural del frente de la faja plegada Neuquina a los 37° S. VI Congreso de Exploración y Desarrollo de Hidrocarburos, Relatorio, CD room

Zamora Valcarce G, Zapata T, Ansa A, Selva G (2006) Three-dimensional structural modeling and its application for development of the El Portón field, Argentina. Am Assoc Petrol Geol Bull 90(3):307–319

Zapata T, Brissón I, Dzelalija F (1999) La Estructura de la faja plegada y corrida andina en relación con el control del basamento de la Cuenca Neuquina. Boletín de informaciones Petroleras, pp 112–121

Zapata T, Dzelalija F, Olivieri G (2001) Desarrollo de reservorios fracturados de la formación Mulichinco usando modelado estructural 3D: yacimiento Filo Morado, Cuenca Neuquina, Argentina. Boletín de Informaciones Petroleras 66:38–47

Zöllner W, Amos AJ (1973) Descripción geológica de la Hoja 32b, Chos Malal (Prov. Neuquén), Bull. 143, 91 pp. Serv. Geol Minero Argent, Buenos Aires

The Late Oligocene–Early Miocene Marine Transgression of Patagonia

Alfonso Encinas, Andrés Folguera, Florencia Bechis, Kenneth L. Finger, Patricio Zambrano, Felipe Pérez, Pablo Bernabé, Francisca Tapia, Ricardo Riffo, Luis Buatois, Darío Orts, Sven N. Nielsen, Victor V. Valencia, José Cuitiño, Verónica Oliveros, Lizet De Girolamo Del Mauro and Victor A. Ramos

Abstract The most important Cenozoic marine transgression in Patagonia occurred during the late Oligocene–early Miocene when marine waters of Pacific and Atlantic origin flooded most of southern South America including the present Patagonian Andes between ~41° and 47° S. The age, correlation, and tectonic setting of the different marine formations deposited during this period are debated. However, recent studies based principally on U–Pb geochronology and Sr isotope stratigraphy, indicate that all of these units had accumulated during the late Oligocene–early Miocene. The marine transgression flooded a vast part of southern South America and, according to paleontological data, probably allowed for the first time in the history of this area a transient connection between the Pacific and Atlantic oceans. Marine deposition started in the late Oligocene–earliest Miocene (~26–23 Ma) and was probably caused by a regional event of extension related to

A. Encinas (✉) · F. Pérez · P. Bernabé · F. Tapia · R. Riffo · V. Oliveros
L. De Girolamo Del Mauro
Departamento de Ciencias de la Tierra, Universidad de Concepción, Concepción, Chile
e-mail: aencinas@udec.cl

A. Folguera · V. A. Ramos
Instituto de Estudios Andinos "Don Pablo Groeber", Departamento de Ciencias Geológicas, FCEN, Universidad de Buenos Aires–CONICET, Buenos Aires, Argentina

F. Bechis
Instituto de Investigaciones en Diversidad Cultural y Procesos de Cambio (IIDyPCa), CONICET—Universidad Nacional de Río Negro, San Carlos de Bariloche, Argentina

K. L. Finger
University of California Museum of Paleontology, Berkeley, California, USA

P. Zambrano
Universidad Andres Bello, Facultad de Ingeniería, Geología, Concepción, Chile

L. Buatois
Department of Geological Sciences, University of Saskatchewan, Saskatoon, Canada

D. Orts
Instituto de Investigación en Paleobiología y Geología, Universidad Nacional de Río Negro–CONICET, General Roca, Argentina

A. Folguera et al. (eds.), *The Evolution of the Chilean-Argentinean Andes*, Springer Earth System Sciences, https://doi.org/10.1007/978-3-319-67774-3_18

major plate reorganization in the Southeast Pacific. Progressive extension and crustal thinning allowed a generalized marine flooding of Patagonia that reached its maximum extension at ~20 Ma. It was followed by a phase of compressive tectonics that started around 19–16 Ma and led to the growth of the Patagonian Andes. The youngest (~19–15 Ma) marine deposits that accumulated in the eastern Andean Cordillera and the extra-Andean regions are coeval with fluvial synorogenic deposits and probably had accumulated under a compressive regime.

Keywords Oligocene–Miocene transgression · Patagonia · Patagonian andes Extension · Orogenesis

1 Introduction

The Patagonia region in southern South America has been flooded by several marine transgressions throughout its history. During most of the Mesozoic, this area was characterized by a convergent margin dominated by extensional tectonics, which led to a low-relief volcanic arc and a series of interconnected back-arc basins that were periodically covered by marine transgressions of Pacific origin (Mpodozis and Ramos 1990) (see Chaps. "Lower Jurassic to Early Paleogene Intraplate Contraction in Central Patagonia" and "Cretaceous Orogeny and Marine Transgression in the Southern Central and Northern Patagonian Andes: Aftermath of a Large-Scale Flat-Subduction Event?"). By the late Early Cretaceous, a shift to a contractional tectonic regime was caused by the accelerated westward drift of South America that triggered the inversion of the Mesozoic back-arc basins and the initial uplift of the Andes (Mpodozis and Ramos 1990). From the latest Cretaceous onwards, a series of marine transgressions periodically covered part of Patagonia (Malumián 1999). Most of those transgressions had an Atlantic origin, as the orographic barrier imposed by a growing Andean Cordillera prevented the ingression of Pacific waters (Mpodozis and Ramos 1990).

The most extensive Cenozoic marine transgression in Patagonia occurred during the late Oligocene–early Miocene. During that period, marine waters of Pacific and Atlantic origin flooded most of southern South America and reached the present Patagonian Andes (Ramos 1982a, b; Elgueta et al. 2000; Malumián and Náñez 2011; Bechis et al. 2014; Encinas et al. 2016) (Fig. 1). Correlation of the marine deposits that accumulated during this period is difficult because they occur as small and discontinuous outcrops, principally in the Andean Cordillera (Fig. 1). This is due to several factors: (1) the dense vegetation that covers this area; (2) the high exhumation rates that affected the Patagonian Andes during the Late Cenozoic and

S. N. Nielsen
Instituto de Ciencias de la Tierra, Universidad Austral de Chile, Valdivia, Chile

V. V. Valencia
School of the Environment, Washington State University, Pullman, WA, USA

J. Cuitiño
Instituto Patagónico de Geología y Paleontología, CCT CENPAT-CONICET, Puerto Madryn, Chubut, Argentina

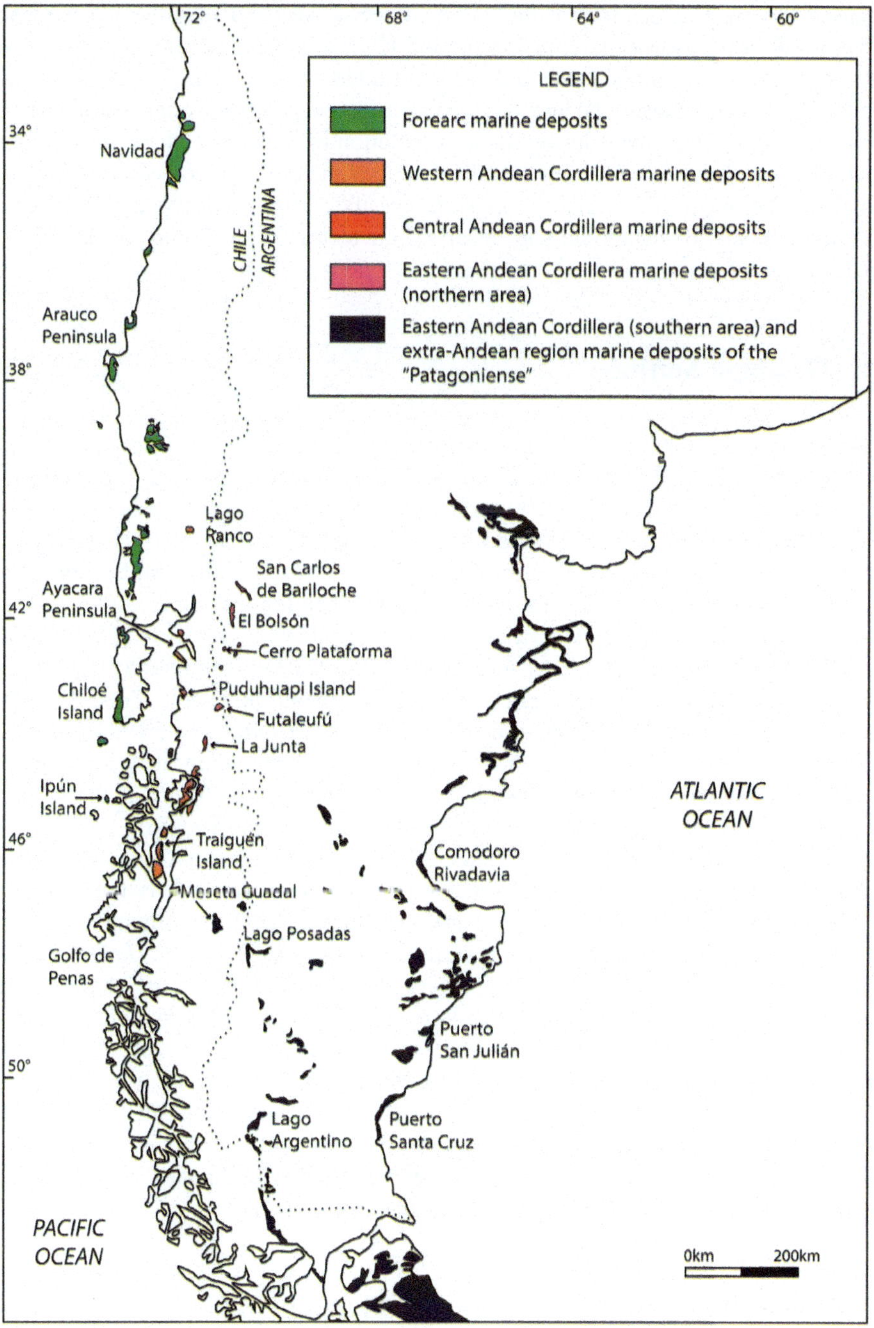

Fig. 1 Distribution of upper Oligocene–lower Miocene marine deposits in central and southern Chile and Argentina. Modified from Encinas et al. (2014)

caused the erosion of most of the Meso-Cenozoic volcano-sedimentary sequence (Thomson 2002; Adriasola et al. 2006); and (3) fossils from marine deposits of the Andean range are poorly preserved, which hinders their recognition at the species level. As a consequence, the age, tectonic setting, and possible connections with the Pacific or Atlantic oceans during the accumulation of these strata have been largely debated (see Encinas et al. 2016 and references therein). In this contribution, we analyze the most important antecedents on the late Oligocene–early Miocene marine deposits of Patagonia and discuss their tectono-sedimentary evolution and paleogeography.

2 Geologic Setting

The Patagonian Andes are located between ~38° and 47° S. Compared to the Central Andes to the north, this segment is characterized by lower elevation (1–2 km), narrower width (~300 km), reduced crustal thickness (~40 km), and less shortening (12–25 km) (Mpodozis and Ramos 1990; Hervé 1994; Orts et al. 2012). An active spreading center, the Chile Rise, is currently subducting at the latitude of the Taitao Peninsula (~46° S) and defines the Chile triple junction between the Nazca, South American, and Antarctic plates (Cande and Leslie 1986) (see Chap. Mantle Influence on Andean and Pre-Andean Topography). Also distinctive of the Patagonian Andes is the Liquiñe-Ofqui Fault Zone (LOFZ), a ~1000-km trench-parallel dextral strike-slip fault system (Hervé 1994). Five N–S-trending physiographic units characterize the Chilean margin at these latitudes. From west to east these are: (1) the offshore continental shelf; (2) the Coastal Cordillera; (3) the Central Depression, which is submerged south of ~42° S and gradually disappears toward the south; (4) the Main Andean Cordillera, which includes the highest peaks; and (5) the extra-Andean region, a mostly flat low-lying area that extends between the Main Andean Cordillera and the Atlantic coast.

The most extensive rocks in the study area are: (1) Paleozoic–Triassic metamorphic rocks, which crop out principally on the forearc (Hervé et al. 2003) but also in some parts of the Andean Cordillera and the western extra-Andean region (Ramos and Ghiglione 2008 and references therein); (2) Jurassic, Cretaceous, Eocene, and Miocene plutonic rocks of the Patagonian batholith (Hervé et al. 1985; Pankhurst et al. 1999); (3) Cenozoic sedimentary rocks that crop out in the Central Depression; (4) Mesozoic and Cenozoic sedimentary and volcanic rocks that cover the western and eastern flank of the Andes, and the extra-Andean region (Suárez and De La Cruz 2000; Ramos and Ghiglione 2008).

3 Late Oligocene–Early Miocene Marine Units

Late Oligocene–early Miocene marine strata extend across the forearc, the Andean Cordillera, and the extra-Andean region of Patagonia (Fig. 1). As previously noted, the age, correlation, and tectonic setting for the different marine formations defined in this vast area have been largely debated. In the following sections, we present the principal antecedents of the most important late Oligocene–early Miocene units that crop out in Patagonia.

3.1 The Forearc

3.1.1 The Navidad Formation and Equivalent Units

Late Oligocene–early Miocene marine strata are exposed at different localities in the forearc of central and southern Chile between ~34° and 45° S (Encinas et al. 2008, 2012; references therein) (Fig. 1). Their reference unit is the Navidad Formation exposed along the coast of central Chile at ~34° S (Cecioni 1980; Encinas et al. 2006). Strata showing similar lithofacies and invertebrate faunas crop out on the coast of the Arauco Peninsula (Ranquil Formation, ~37–38° S), in the Coastal Cordillera and

Fig. 2 Characteristic features from the Oligocene–Miocene marine units of the forearc. **a** Rhythmically interbedded sandstone and siltstone of the Lacuí Formation at Playa Chaumán, Chiloé Island, Chile. **b** Rhythmically interbedded sandstone and siltstone of the Ipún beds at Ipún Island, Chile. **c** Synsedimentary folds in sandstone and siltstone of the lower unit of the Grupo Chaicayán Sequence at Pan de Azúcar Island, Golfo de Penas, Chile. **d** shallow-marine sandstones of the upper unit of the Grupo Chaicayán Sequence at Hereford Island, Golfo de Penas, Chile

Central Depression between ~38–42° S (Santo Domingo Formation), on Chiloé Island (Lacuí Formation, ~42–43° S) (Fig. 2a), and on the Ipún, Lemo, and Stokes islands (Ipún beds, ~43–45° S) (Frassinetti 2001, 2004; Encinas et al. 2008, 2012; references therein) (Fig. 2b). These marine strata contain a diverse fossil biota that includes bivalves, gastropods, crabs, ostracodes, foraminifers, shark teeth, leaf impressions, and pollen (Finger et al. 2007; Kiel and Nielsen 2010; Encinas et al. 2008, 2012; references therein). Sedimentologic, ichnologic, and paleontologic studies suggest deposition at bathyal depths for part of the Navidad Formation and correlative units (Osorio and Elgueta 1990; Antinao 2000; Finger et al. 2007; Encinas et al. 2008, 2012).

Data based on the analysis of brittle deformation (Lavenú and Encinas 2005) and on the study of seismic lines (Elgueta et al. 2000, Jordan et al. 2001, Melnick and Echtler 2006) indicate the presence of normal faults and synextensional growth strata in these deposits. In contrast, Becerra et al. (2013) concluded that the marine strata of the Ranquil Formation on the Arauco Peninsula are syn-contractional (syn-inversional) based on the interpretation of an offshore seismic line. The Navidad Formation and correlative units are overlain by younger shallow-marine beds assigned to the middle Miocene–Pliocene (Martínez-Pardo 1976; Tavera 1979; Covacevich and Frassinetti 1990; Antinao et al. 2000; Encinas et al. 2006, 2012; Nielsen 2013).

The age of the Navidad Formation and equivalent units has been a matter of debate for decades (see Gutiérrez et al. 2013; Finger et al. 2013, and Le Roux et al. 2013; references therein). An early Miocene age for the Navidad Formation was derived from studies on mollusks, shark teeth, Sr isotopes, and radiometric (Ar/Ar and K/Ar) dating of volcanic clasts in this unit (Gutiérrez et al. 2013; Finger et al. 2013; references therein). A middle to late Miocene or early Pliocene age for this unit had been interpreted from the study of foraminifers, ostracodes, pollen, Sr isotopes, and Ar/Ar dating of volcanic clasts (Finger et al. 2013 and references therein). Finger (2013) reexamined planktic foraminifers ascribed to the late Miocene and early Pliocene (in Finger et al. 2007), whereupon they recognized that they were misidentified and reassigned the assemblages to the early Miocene (Burdigalian Stage, ~21–16 Ma). Fossils in the Navidad Formation and correlative units yielded $^{87}Sr/^{86}Sr$ ages of 24–16 Ma (Nielsen and Glodny 2009).

3.1.2 The Grupo Chaicayán Sequence

The region located around the Golfo de Penas (47° S) has the southernmost exposures of the Tertiary marine strata of the Chilean forearc (Forsythe and Nelson 1985) (Fig. 1). That area is where Forsythe et al. (1985) defined the Grupo Chaicayán Sequence, a Miocene marine succession that consists of sandstone, siltstone, and minor conglomerate. This unit overlies the metamorphic basement, but its upper contact is not exposed, although the marine beds are intruded by Pliocene porphyritic stocks and sills (Forsythe et al. 1985; Mpodozis et al. 1985).

Encinas et al. (2015) divided the Grupo Chaicayán Sequence into two units. The basal unit overlies the metamorphic basement and is a transgressive sequence that consists of a basal conglomerate, a shallow-marine massive sandstone, and a deep-water rhythmic turbidite of sandstone and siltstone (Encinas et al. 2015)

(Fig. 2c). Fossils occur locally in the basal unit and include bivalves, gastropods, scaphopods, brachiopods, solitary corals, echinoderms, and foraminifera (Encinas et al. 2015). The succession shows synextensional strata associated with normal faulting (Encinas et al. 2015).

The upper unit of the Grupo Chaicayán Sequence consists of sandstone, siltstone, and minor conglomerate deposited in a shallow-marine (shoreface and shelf) environment (Encinas et al. 2015) (Fig. 2d). This unit consists of horizontal strata and unconformably overlies the gently folded beds of the lower unit. The upper unit contains abundant bivalves, gastropods, echinoderms, crustaceans, and bryozoans (Forsythe et al. 1985; Covacevich and Frassinetti 1986; Stott and Webb 1989; Frassinetti 2006; Nielsen and Frassinetti 2007; Encinas et al. 2015).

Biostratigraphic studies have been carried out exclusively on the upper unit; hence, the age of the lower unit of the Grupo Chaicayán Sequence remains elusive. Encinas et al. (2015) dated detrital zircons from the lower unit with the U–Pb method and obtained a maximum age of 67 Ma (Maastrichtian), but the occurrence of the nautiloid *Aturia* sp. indicates a younger age, as it ranges Paleocene to Miocene (Kummel 1956). Encinas et al. (2015) correlated the lower unit with late Oligocene–early Miocene marine strata that crop out in the forearc of south-central Chile based on their similar facies. Forsythe et al. (1985) and Stott and Webb (1989) obtained a late Miocene age for the upper unit of the Grupo Chaicayán Sequence based on planktic foraminifera. Covacevich and Frassinetti (1986) compared the molluscan fauna of these strata with those of the Navidad and the overlying Licancheu and Rapel formations and suggested a middle to late Miocene age, but Nielsen and Frassinetti (2007) noted that, while the fauna is indeed similar, it is quite different on the species level in certain groups. Encinas et al. (2015) confirmed that age by U–Pb dating of detrital zircons, which yielded maximum ages of 16.8 Ma (Burdigalian) and 12.7 Ma (Serravalian) for the upper unit of the Grupo Chaicayán Sequence.

3.2 The Western Andean Cordillera

3.2.1 The Riñinahue Formation

Campos et al. (1998) described the Lago Ranco Formation as a volcano-sedimentary unit exposed along the western shores of Ranco Lake (~40° S) (Fig. 1). Bernabé (2014) recently redefined the marine deposits of this unit as the Riñinahue Formation, which is a ~120 m thick succession of fossiliferous sandstone, siltstone, conglomerate, and breccia (Fig. 3a).

The basal and upper contacts of this unit are covered. The Riñinahue Formation strata are highly deformed and overthrusted by late Oligocene–early Miocene volcanic rocks (Bernabé 2014). The bivalves, gastropods, brachiopods, echinoderms, and barnacles are scarce and preserved as molds (Covacevich 1996; Campos et al. 1998). The unit has been interpreted as a turbidite succession deposited in a deep-marine environment (Campos et al. 1998; Bernabé 2014). Covacevich (1996) assigned it to the Miocene based on the presence of *Chione* sp. and *Turritella*

Fig. 3 Characteristic features from the Oligocene–Miocene marine units of the western Andean Cordillera. **a** Rhythmically interbedded sandstone and siltstone of the Riñinahue Formation at Lago Ranco, Chile. **b** Rhythmically interbedded sandstone and siltstone of the Ayacara Formation at Isla Ica, near the Ayacara Peninsula, Chile. Note encircled person for scale (from Encinas et al. 2013). **c** Base of turbidite with flute casts of the Ayacara Formation at Isla Ica. Paleocurrent direction (157°) is indicated by the arrow (from Encinas et al. 2013). **d** Synsedimentary fold in sandstone and siltstone beds of the Puduhuapi Formation at Puduhuapi Island, near Chaitén, Chile (from Encinas et al. 2014). **e** Volcanic and sedimentary rocks from the Traiguén Formation at Isla Rojas, near Isla Traiguén, Chile. Sandstone (sd) and breccia (br) with basaltic clasts onlap lava flow deposits with well-developed pillows (pl). The three rocks are covered by a blocky lava flow (lv) (from Encinas et al. 2016). **f** Intraformation breccia with folded stratified block made of siltstone of the Traiguén Formation at Traiguén Island (from Encinas et al. 2016)

ambulacrum. However, *T. ambulacrum* is an Atlantic species (see Griffin and Nielsen 2008) that does not occur in Navidad-equivalent units. Given the poor preservation, it is difficult, if not impossible, to distinguish it from the coeval Pacific species, *T. pseudosuturalis*. Bernabé (2014) used the U–Pb method to date five samples of detrital zircons from the Lago Ranco Formation and obtained ages between ~21 and 19 Ma.

3.2.2 The Ayacara Formation

The Ayacara Formation (Levi et al. 1966) is exposed as small outcrops between Hornopirén and Península Ayacara (~42°–42°30′S) (Fig. 1). The most complete sections (~400 to 800 m thick) are those of Caleta Ayacara and Isla El Manzano (Levi et al. 1966; Encinas et al. 2013). Levi et al. (1966) divided the unit into two members and noted that it underlies Pleistocene(?) deposits, but its basal contact is not exposed. The Ayacara Formation consists of interbedded sandstone and siltstone with minor conglomerates and tuffs (Fig. 3b–c). Its fossils are scarce and poorly preserved, consisting of solitary corals (Solano 1978), echinoderms (Alarcón 1995), crustaceans (Feldmann et al. 2010), and foraminifera and silicoflagellates (Levi et al. 1966; Sernageomin 1995; Encinas et al. 2013). Strata are highly deformed and intruded by igneous dikes (Levi et al. 1966). Stratigraphic, sedimentologic, ichnologic, and foraminiferal studies by Encinas et al. (2013) indicate that the basal part of the Ayacara Formation accumulated in a shallow-marine, fan-delta system, whereas the upper part of this unit was deposited in a deep-marine turbidite system.

The Ayacara Formation was assigned to an Eocene–Miocene age by Levi et al. (1966) based on Martínez-Pardo's study of foraminifera and silicoflagellates. Martínez-Pardo (1965) ascribed a middle Miocene age to a foraminiferal sample from Isla El Manzano and that was supported by Fuenzalida (1979), who identified the planktic foraminifer *Orbulina universa*. Bourdillon (1994) criticized these studies because the foraminifera were identified only from petrographic thin sections, which in this case would have limited accuracy. Bourdillón (in Sernageomin 1995) ascribed foraminifera extracted from several Ayacara Formation samples to the middle Eocene–late Oligocene. Rojas et al. (1994) $^{40}Ar/^{39}Ar$-dated a tuff from the Ayacara Formation and obtained a middle Miocene age of 16.5 Ma. Encinas et al. (2013) dated five samples of detrital zircons with the U–Pb method and obtained a maximum depositional age in the Miocene, between 21.8 and 17.6 Ma. In the same study, they reported the occurrence of planktic foraminifera in the upper member of the Ayacara Formation that suggested a Langhian (early–middle Miocene) age, cautioning that their determination was based on poorly preserved specimens.

3.2.3 The Puduhuapi Formation

The Puduhuapi Formation is a marine unit defined by Levi et al. (1966) as a gently folded succession, ~35 m thick, of interbedded sandstone, siltstone, and minor conglomerate that crops out in the northwestern part of the Puduhuapi Island, approximately 7 km west of Chaitén (~43° S) (Fig. 1). The basal and upper contacts of the Puduhuapi Formation are not exposed. The unit crops out in the proximity of andesites of unknown age and their stratigraphic relationship with the Puduhuapi Formation has not been discerned. The occurrence of several andesitic dikes and sills intruding the marine unit suggest a genetic relationship with the volcanic rocks. The occurrence of turbidites and slumps, benthic foraminifera indicative of bathyal depths, and abundant planktic foraminifera indicate that the Puduhuapi Formation was deposited in a deep, offshore environment (Encinas et al. 2014) (Fig. 3d). Levi et al. (1966) assigned the Puduhuapi Formation to the Eocene–Miocene interval upon correlating it with the Ayacara Formation. Encinas et al. (2014) dated two samples of detrital zircons from this unit with the U–Pb method and obtained maximum depositional ages of ~23 Ma.

3.2.4 The Traiguén Formation

The Traiguén Formation is a marine volcano-sedimentary unit that crops out on the forearc and western part of the Andean Cordillera in the Aysén region between ~44° and 46° S (Espinoza and Fuenzalida 1971; Hervé et al. 1995) (Fig. 1). It was defined by Espinoza and Fuenzalida (1971), who designated Traiguén Island as its type locality. The Traiguén Formation consists of interbedded pillow basalts, tuffs, breccias, sandstones, and shales (Espinoza and Fuenzalida 1971; Hervé et al. 1995; Encinas et al. 2016) (Fig. 3e–f). It overlies the Paleozoic–Triassic metamorphic basement at Magdalena Island, but its upper contact is not exposed (Hervé et al. 1995). The stratigraphic thickness of this formation has not been determined because its exposures are limited to small coastal outcrops. The Traiguén Formation is intruded by numerous basic igneous dikes and locally by plutonic bodies of gabbroic to granodioritic composition (Hervé et al. 1995). Outcrops of this unit on Magdalena Island show greenschist- and amphibolite-grade metamorphism (Hervé et al. 1995; Silva 2003). The presence of pillow lava, turbidites, slump deposits, and bathyal benthic foraminifers in the Traiguén Formation indicates deposition in a deep-marine environment (Encinas et al. 2016). The occurrence of syndepositional normal faults in this unit indicates basin subsidence related to extensional tectonics (Encinas et al. 2016). In accordance, geochemical and petrographic analyzes suggest that the basalts of this formation represent the activity of subduction-related volcanism over a thinned crust (Encinas et al. 2016) (see Chap. "Neogene Growth of the Patagonian Andes").

The age of the Traiguén Formation has been a matter of debate. Céspedes (1975) suggested a Miocene age for the Traiguén Formation based on planktic foraminifera. Hervé et al. (1995) dated sedimentary and volcanic rocks of this unit on

Magdalena Island with the Rb–Sr method and obtained two imprecise ages of 20 ± 28 Ma and 20 ± 26 Ma, noting that they most likely correspond to their metamorphism. Hervé et al. (2001) dated detrital zircons from a sandstone sampled on Traiguén Island and obtained a 26.0 Ma U–Pb SHRIMP maximum age for this unit. Silva et al. (2003) reported that the basal Traiguén Formation could be as old as Eocene based on the K/Ar ages of 38–15 Ma for a suite of dikes that they considered to be contemporaneous with the Traiguén Formation's volcanic rocks. Encinas et al. (2016) applied the U–Pb method to date detrital zircons from five sandstones and one tuff of this unit and obtained four maximum ages between ~26 and 23 Ma (latest Oligocene–earliest Miocene).

3.3 The Central Andean Cordillera

3.3.1 The Vargas Formation

The Vargas Formation is a marine unit that crops out along the west bank of the Palena River, near the town of La Junta and close to the Liquiñe-Ofqui Fault Zone trace (~44° S) (Fig. 1). It is composed of fossiliferous black shales and minor sandstones. This unit was first studied by Steffen (1944), who referred to it as the "Esquistos Arcillosos (Clayey Shales) Formation." Subsequently, Bobenrieth et al. (1983) included these strata in the La Junta Formation. Urbina (2001) redefined the marine deposits of this unit as the Vargas Formation and separated it from the continental La Junta Formation. Outcrops of the Vargas Formation are small and discontinuous and its basal and upper contacts are not exposed. Their strata show significant deformation and are in fault contact with Cretaceous granitoids (Urbina 2001). This unit contains poorly preserved gastropods, bivalves, echinoderms, and planktic foraminifers. The poor preservation of the Vargas Formation strata hinders a refined sedimentologic interpretation, but the predominance of black siltstone and the relative abundance of planktic foraminifers suggest deposition on the outer continental shelf or the continental slope (Encinas et al. 2014).

Urbina (2001) assigned an Eocene age for the Vargas Formation based on the similarity of its molluscs with those of the La Cascada Formation that Thiele et al. (1978) and Castillo (1983) referred to that epoch. Encinas et al. (2014) dated two samples of detrital zircons from this unit by the U–Pb method and obtained two maximum depositional ages of ~39 Ma and 84 Ma. However, these authors tentatively identified the gastropod *Turritella ambulacrum* that indicates a late Oligocene–early Miocene age. Encinas et al. (2014) also dated detrital zircons from alluvial deposits of the La Junta Formation that crop out on the opposite shore of the Palena River and obtained a maximum depositional age of ~26 Ma. They noted that this unit could have been deposited during the initial stage of subsidence that affected this region prior to the marine transgression over this area.

3.4 The Eastern Andean Cordillera

We have separated the late Oligocene–early Miocene marine deposits of the eastern flank of the Patagonian Andes in two different areas. The northern area between ~41 and 43° S includes the Ventana, Río Foyel, Cerro Plataforma, and La Cascada formations, which most authors have attributed to a marine transgression of Pacific origin. The southern area between ~47 and 50° S encompasses the Guadal, El Chacay, and the Estancia 25 de Mayo formations, which have been affiliated with a marine transgression from the Atlantic.

The northern area (41–43° S)

3.4.1 The Ventana Formation

The Ventana Formation is a volcano-sedimentary unit defined by González Bonorino and González Bonorino (1978) that crops out in the Maitén volcanic belt, Argentina (41–43° S). This unit is dominated by volcanic rocks of andesitic to dacitic composition, with subordinate rhyolites, basalts, and tuffs, and minor continental, and marine sedimentary rocks (Rapela et al. 1988). At Cerro Ventana and Cerro Ñireco (~41° S), near San Carlos de Bariloche (Fig. 1), sandstones and siltstones bearing marine fossils are interbedded with the volcanic rocks of the Ventana Formation forming a succession ~500 m thick (Bechis et al. 2014) (Fig. 4a). The occurrence of synextensional growth strata (Orts et al. 2012; Echaurren et al. 2016) and geochemical data that indicate a relatively thinned crust (Rapela et al. 1988; Litvak et al. 2014) suggest an extensional setting for the Ventana Formation (Bechis et al. 2014) (see Chap. "Paleogene Arc-Related Volcanism in the Southern Central Andes and North Patagonia (39°–41° S))".

Several K–Ar and U–Pb ages of the Ventana Formation along the Maitén belt indicate an early Oligocene to early Miocene age, with most data ranging between 34 and 21 Ma (see Bechis et al. 2014 and references therein). Bechis et al. (2014) U–Pb dated zircons from two tuffs and one fossiliferous sandstone of the volcano-sedimentary succession at Cerro Ventana and Cerro Ñireco and obtained ages between ~23 and 20 Ma.

3.4.2 The Río Foyel Formation

The Río Foyel Formation, defined by Pöthe de Baldis (1984), is a marine unit that crops out in the El Bolsón area (~42° S) (Bechis et al. 2014) (Fig. 1). This unit comprises more than 600 m of black shales, sandstones, and minor limestones, and contains a rich marine invertebrate fauna (Bechis et al. 2014 and references therein) (Fig. 4b). At Rio Azul, a basal conglomerate of the Río Foyel Formation overlies Jurassic volcanic rocks and Mesozoic plutonic rocks of the Patagonian Batholith (Diez and Zubia 1981). The upper contact is usually covered and appears to be

Fig. 4 Characteristic features from the Oligocene–Miocene marine units of the eastern Andean Cordillera, northern area. **a** Volcanic and marine sedimentary rocks of the upper section of the Ventana Formation at Cerro Ñireco near Bariloche, Argentina. Basalts (Bs). Tuffs (Tf). Fossiliferous conglomerates and sandstones (Fcs). **b** Siltstones, sandstones, and limestones of the Río Foyel Formation at the banks of the homonymous river, Argentina. **c** Shallow-marine sandstones of the lower section of the Troncoso Formation at the Gendarmería Río Villegas, Argentina. **d** Shallow-marine sandstone, siltstone and minor conglomerate of the Cerro Plataforma Formation at Cerro Plataforma, Argentina. From base to top Late Cretaceous granitoids (Kg), Cerro Plataforma Formation (Cp) and middle Miocene volcano-sedimentary rocks (Vr) (modified from Orts et al. 2012). **e** Shallow-marine sandstones of the La Cascada Formation in a roadcut near Río Azul, southwest of Futaleufú, Chile. **f** Steinkern of "*Trochus* aff. *T. laevis,*" a fossil gastropod with Pacific affinities from the La Cascada Formation. Scale bar = 36 mm (from Encinas et al. 2014)

transitional with the Troncoso Formation (Bechis et al. 2014). The latter unit forms a ~750-m-thick succession, with a basal section consisting of green sandstones bearing marine invertebrates (Asensio et al. 2005; Bechis et al. 2014) (Fig. 4c). The upper section of the Troncoso Formation is a continental succession comprised of tuffs, tuffaceous pelites, sandstones, and conglomerates (Asensio et al. 2005), and shows synextensional growth strata according to Bechis et al. (2014). The Troncoso Formation has an erosional, low-angle, unconformable contact with the overlying coarse conglomerates of the Salto del Macho Formation (Asensio et al. 2005; Bechis et al. 2014). The Río Foyel Formation was interpreted as having been deposited in a restricted littoral marine environment (Bertels 1980; Barreda et al. 2003), but the occurrence of benthic foraminifera assigned to the genus *Transversigerina* suggest upper to middle bathyal depths (Malumián et al. 1984).

Regarding the origin of the marine transgression that deposited the Río Foyel Formation, Bertels (1980) favored an Atlantic origin based on her study of foraminifera whereas Feldmann et al. (2011) and Ramos (1982b) ascribed a Pacific origin based on its crustaceans and paleogeography, respectively.

The Río Foyel Formation has been ascribed to the Eocene on the basis of molluscs (Chiesa and Camacho 2001), to the Eocene–Oligocene based on fungal spores (Martínez et al. 2016), to the early Oligocene based on Sr isotopes (Griffin et al. 2004) and K/Ar minimum ages of crosscutting dikes (Giacosa et al. 2001), to the early–middle Oligocene to early Miocene(?) based on foraminifera (Bertels 1980), and to the late Oligocene–early Miocene based on palynomorphs (Barreda et al. 2003) and foraminifera (Malumián et al. 2008). U–Pb analysis of detrital zircons yielded maximum ages of 21.9 Ma for a sandstone of the Río Foyel Formation and 21.4 Ma for a sandstone in the basal, marine unit of the Troncoso Formation (Bechis et al. 2014). A crystalline tuff intercalated in continental strata from the upper unit of the Troncoso Formation yielded an age of 16.6 Ma. An early Miocene age for the Río Foyel Formation is also supported by the recognition of early to middle Miocene (Aquitanian–Langhian) calcareous nannofossils in this unit (Bechis et al. 2014).

3.4.3 The Cerro Plataforma Formation

The Cerro Plataforma Formation is a fossiliferous marine unit composed of ~300 m of sandstone, siltstone, and minor conglomerate exposed on a syncline at the Cerro Plataforma south of Lago Puelo, Argentina (42°30′S) (Lizuaín 1979; Griffin et al. 2002; Orts et al. 2012) (Fig. 1). The Cerro Plataforma Formation unconformably overlies late Early Cretaceous volcanic rocks of the Divisadero Group and Late Cretaceous granitoids and conformably underlies a middle Miocene volcano-sedimentary succesion (Lizuaín 1979) (Fig. 4d). The unit shows onlap relationships on both flanks of the Cerro Plataforma syncline, which has been interpreted by Orts et al. (2012) as evidence of contraction contemporaneous with sedimentation. The Cerro Plataforma Formation was deposited in a shallow-marine environment (Orts et al. 2012) and contains fossil molds of planktic foraminifers, bivalves, gastropods, and echinoderms (Griffin et al. 2002; Orts et al. 2012).

Griffin et al. (2002) correlated the molluscan fauna of this unit with those of the early Miocene Monte León and Chenque formations exposed along the east coast of Patagonia and which were deposited by an Atlantic transgression. Orts et al. (2012) dated two samples of detrital zircons from sandstone beds of the basal and upper part of the Cerro Plataforma Formation. The basal sample yielded an age of ~112 Ma and the upper sample yielded an age of 18.3 Ma. The overlying volcanic succession was K/Ar-dated by Lizuaín (1983) at 15 Ma, indicating the minimum age for the marine unit.

3.4.4 The La Cascada Formation

The La Cascada Formation is a ~130-m-thick succession of fossiliferous sandstone, siltstone, and conglomerate deposited in a shallow-marine environment (Encinas et al. 2014) (Fig. 4e). This unit was defined by Thiele et al. (1978) and is exposed as discontinuous outcrops near Futaleufú, Chile (43°20′S) (Fig. 1). It unconformably overlies Lower Jurassic volcanic rocks and Cretaceous granitoids but its upper contact is not exposed (Thiele et al. 1978). The La Cascada Formation contains poorly preserved bivalves and gastropods (Thiele et al. 1978; Castillo 1983; Encinas et al. 2014) (Fig. 4f). That molluscan fauna appears to have both Atlantic and Pacific affinities (Fig. 4f), but those with the Atlantic affinity dominate (Encinas et al. 2014). Thiele et al. (1978) assigned the La Cascada Formation to the Eocene because they considered that its lithology and paleontology were more similar to the Arauco succession than to the Navidad Formation. However, subsequent studies on the molluscs by Rubilar et al. (2001, 2002) and Encinas et al. (2014) concluded that the La Cascada Formation is of late Oligocene–early Miocene age. That agrees with the maximum age of 17.7 Ma that Encinas et al. (2014) obtained from one of two zircon samples from sandstones in the La Cascada Formation.

The southern area (47–50° S)

Late Oligocene–early Miocene shallow-marine deposits accumulated in several basins of the eastern Andean Cordillera and in the extra-Andean region of Patagonia due to a major Atlantic transgression (Cuitiño et al. 2015a, b). These strata were first documented by Darwin (1846), who referred to them as the Patagonian Tertiary Formation. Subsequently, other authors employed the term “Patagoniano” (Patagonian) or “Patagoniense” for these deposits (see Cutiño et al. 2015 and references therein) (Fig. 1). Better-studied units in the eastern Andes are the Guadal, El Chacay, and Estancia 25 de Mayo formations.

3.4.5 The Guadal Formation

The Guadal Formation is a fossiliferous marine unit with a stratigraphic thickness of 110 m and consisting of sandstone, siltstone, and limestone (De la Cruz and Suárez 2006). It was originally defined by Heim (1940) as the Mesa Guadal Formation and later renamed by Niemeyer (1975) as the Guadal Formation. The formation crops out on the flanks of an uplifted syncline at Meseta Guadal and has more limited exposures in the area to the east and south of Chile Chico (southern Chile, ~47° S) (Niemeyer 1975; De la Cruz and Suárez 2006) (Fig. 1). At Meseta Guadal, the Guadal Formation is paraconformable with the underlying lower Eocene fluvial deposits of the San José Formation (Fig. 5a) and transitional with the overlying early Miocene Santa Cruz Formation (De la Cruz and Suárez 2006). In the area south of Chile Chico, the Guadal Formation conformably(?) overlies the Paleocene–Eocene Basaltos Inferiores de la Meseta de Chile Chico and conformably(?) underlies the Basaltos Superiores de la Meseta de Chile Chico or the Santa Cruz Formation. The Guadal Formation contains a rich paleobiota that includes bivalves, gastropods, brachiopods, echinoderms, corals, bryozoans, sharks, whales, and leaves (Frassinetti and Covacevich 1999; Flynn et al. 2002; De la Cruz and Suárez 2006). This unit reflects a transgressive–regressive cycle in which the depositional environment began as estuarine, deepened to shelfal, and shoaled back to estuarine (Flint et al. 1994). Frassinetti and Covacevich (1999) interpreted the fauna of the Guadal Formation as late Oligocene–early Miocene and of Atlantic origin after correlating it with that of the Monte León Formation of Argentina. Their cited age is supported by a ~19 Ma U–Pb date obtained from detrital zircons in a sandstone from the upper part of the Guadal Formation (Suárez et al. 2015). The Guadal Formation represents the westernmost reach of the Atlantic "Patagonian" transgression (Niemeyer 1975; Ramos 1982a, b; Malumián et al. 1999; Cuitiño et al. 2015a).

3.4.6 The El Chacay Formation

The El Chacay Formation crops out in the Lago Posadas-Meseta Belgrano region (47°30′S) and corresponds to the southern extension of the Guadal Formation in Argentina (Fig. 1). In succession, this unit has been referred to as the Patagonia Formation (Ramos 1979), Centinela Formation (Riccardi and Rolleri 1980; Ramos 1982a), and El Chacay Formation (Chiesa and Camacho 1995). It unconformably overlies the Eocene Posadas Basalt and shows a transitional contact with the overlying early Miocene Santa Cruz Formation (Cutiño et al. 2015) (Fig. 5b). The El Chacay Formation is a 130m-thick succession composed of fossiliferous sandstone and mudstone (Cuitiño et al. 2015a). The strata indicate a general transgressive–regressive cycle with tide-dominated shallow-marine facies at the base that grade upward to offshore deeper-water mudstones, which in turn grade to estuarine facies (Cuitiño et al. 2015a). Ramos (1982a), based on the study of foraminifers from the El Chacay Formation by Malumián (1978), assigned a late–early Oligocene age to this unit. Parras et al. (2008) obtained two Sr ages of 25 and 26.3 Ma for the lower part of the El Chacay Formation. Cuitiño et al. (2015a) employed the same method to date 13 oyster shell samples collected from the base to the top of this unit and obtained younger ages in the range of 20.3–18.1 My.

Fig. 5 Characteristic features from the shallow-marine Oligocene–Miocene "Patagonian" deposits of the eastern Andean Cordillera and the extra-Andean region. **a** Sandstone strata of the Guadal Formation (Gu) at Pampa Castillo. The unit overlies the San José Formation (Sj). **b** Sandstone strata of the El Chacay Fm (Ch) at Lago Posadas. From base to top Rio Tarde Formation (Rt), Posadas Basalt (Pb), El Chacay Formation (Ch), and Santa Cruz Formation (Sc). **c** Sandstone strata of the Estancia 25 de Mayo Formation south of Lago Argentino. From base to top Estancia 25 de Mayo Formation (E25) and Santa Cruz Formation (Sc). **d** Coastal cliff exposing the upper part of the San Julián Formation (Sj) and the lowermost beds of the overlying Monte León Formation (Ml) at Playa la Mina, north of San Julián, Argentina. **e** Coastal cliffs exposing the upper part of the Monte León Formation (Ml) and the overlying Santa Cruz Formation (Sc). Monte León National Park, Argentina. **f** Sandstones and siltstones of the Chenque Formation at its type locality. This outcrop shows the lower and middle parts of the unit. Cerro Chenque in Comodoro Rivadavia, Argentina

3.4.7 The Estancia 25 de Mayo Formation

The Estancia 25 de Mayo crops out at Lago Argentino (50°30'S), Argentina, and in the adjacent Sierra Baguales, Chile (Cuitiño et al. 2012; Bostelmann et al. 2015) (Fig. 1). Furque and Camacho (1972) named the unit as the Centinela Formation, unaware that another unit in northern Argentina already had that name. Cuitiño and Scasso (2010) later renamed this unit, deposited during the Oligo–Miocene "Patagonian" transgression, as the Estancia 25 de Mayo Formation. It overlies the late Oligocene Río Leona Formation and transitions into the overlying early Miocene Santa Cruz Formation (Cuitiño et al. 2012; Bostelmann et al. 2015) (Fig. 5c). The Estancia 25 de Mayo Formation is a ~180 m-thick succession of fossiliferous sandstone and siltstone deposited in shallow-marine and estuarine environments (Cuitiño et al. 2012). Its fauna comprises corals, brachiopods, mollusks, echinoderms, arthropods, and annelids (Bostelmann et al. 2015). Numerous workers have provided chronologic and biostratigraphic age data for, and correlations of, the Estancia 25 de Mayo Formation. Parras et al. (2008) obtained two $^{87}Sr/^{86}Sr$ ages of 21.2 and 22.8 Ma from the lower part of this unit. Barreda et al. (2009) reported an early Miocene age for the Estancia 25 de Mayo Formation based on palynologic data. Casadío et al. (2000a) obtained a $^{40}Ar/^{39}Ar$ age of 46 Ma (Eocene) from a tuff bed but Casadío et al. (2000b) rejected that result shortly thereafter on the basis of paleontologic and stratigraphic evidence for a late Oligocene–early Miocene age, which Casadío et al. (2001) then confirmed with $^{87}Sr/^{86}Sr$ ages determined from oyster shells. Griffin et al. (2014) correlated the molluscan fossils of the Estancia 25 de Mayo Formation with those of the late Oligocene–early Miocene Monte León Formation. Cuitiño et al. (2012) obtained an U–Pb age of 19.14 Ma for a tuff bed in the lower part of the Estancia 25 de Mayo Formation, which is within the range of the $^{87}Sr/^{86}Sr$ ages they obtained from oysters (20.05 Ma in the lower part to 19.1 Ma in the upper part of this unit).

3.5 The Extra-Andean Region

In the Austral Basin to the south, the better-studied units deposited by the "Patagonian" transgression are represented by the San Julián and Monte León formations (Bertels 1970), and the Chenque Formation for those in Golfo San Jorge Basin to the north (Bellosi 1990). The Monte León and Chenque formations are correlative and both transitionally underlie the early to middle Miocene fluvial deposits of the Santa Cruz Formation (Cuitiño et al. 2015b). The age of the shallow-marine "Patagonian" deposits in the extra-Andean region has been largely debated. We cite below some of the studies on the age of these units; more are reported by Parras et al. (2012) and Cutiño et al. (2015).

The type locality of the San Julián Formation is located near Puerto San Julián (Figs. 1 and 5d). This unit has been assigned to the middle to late Eocene based on molluscs (Camacho 1985), to the late Eocene–Oligocene based on foraminifera

(Bertels 1975; Malumián and Náñez 1989), and to the Oligocene based on palynomorphs (Barreda 1997). More recently, Parras et al. (2008, 2012) obtained $^{87}Sr/^{86}Sr$ ages between 23.8 and 25.9 Ma (latest Oligocene) for this unit.

The type locality of the Monte León Formation is located at the mouth of the Río Santa Cruz, near Puerto Santa Cruz (Figs. 1 and 5e). This unit has been correlated with the late Oligocene (Bertels 1970) and late Oligocene–early Miocene (Malumián et al. 1999) based on foraminifera, and the early Miocene based on molluscs (del Río 2004). Parras et al. (2012) obtained $^{87}Sr/^{86}Sr$ ages of 22.1–17.9 Ma (early Miocene) for the entire formation.

The type locality of the Chenque Formation is located at Cerro Chenque, near Comodoro Rivadavia (Figs. 1 and 5f). This unit was assigned to the early to middle Miocene based on palynomorphs (Barreda and Palamarczuk 2000) and to the late Oligocene–early Miocene based on foraminifers (Malumián 2002) and molluscs (del Río 2004). K–Ar radiometric ages in the interval of 30–25 Ma (late Eocene–Oligocene) were obtained by Linares (1979), Riggi (1979), and Bellosi (1990). Recently, Cuitiño et al. (2015b) obtained 13 $^{87}Sr/^{86}Sr$ ages between 19.6 and 15.3 Ma (early to middle Miocene) for the Chenque Formation in the Comodoro Rivadavia section.

4 Discussion

4.1 Age and Correlations

It is evident that the main problem in correlating the Cenozoic marine deposits that extend from the Pacific to the Atlantic coasts of Patagonia is the uncertainty of their ages that is primarily manifested in the incongruent interpretations from those who have worked on the same units. In many cases, these authors based their conclusions on different kinds of paleontologic or isotopic data. As noted in the previous section, forearc deposits of the Navidad Formation and correlative units have been assigned either to the early Miocene or to the middle to late Miocene/earlyPliocene, whereas marine strata from the western Andean Cordillera (Lago Ranco, Ayacara, Puduhuapi, and Traiguén formations), central Andean Cordillera (Vargas Formation), eastern Andean Cordillera (Ventana, Río Foyel, Cerro Plataforma, La Cascada, Guadal, El Chacay, and Estancia 25 de Mayo formations), and extra-Andean region (Sand Julián, Monte León, and Chenque formations) have been assigned to the Eocene, Oligocene, or Miocene.

Recent studies based principally on U–Pb geochronology and Sr isotope stratigraphy, however, consistently indicate late Oligocene–early Miocene ages for all of these units. $^{87}Sr/^{86}Sr$ ages from the forearc deposits of the Navidad Formation and correlative units range 24–16 Ma (Nielsen and Glodny 2009). The Patagonian deposits of the extra-Andean region yielded $^{87}Sr/^{86}Sr$ ages of 25.9 and 23.8 Ma for the basal San Julián Formation (Parras et al. 2008, 2012), $^{87}Sr/^{86}Sr$ ages between

22.1 and 17.9 Ma for the overlying Monte León Formation (Parras et al. 2012), and $^{87}Sr/^{86}Sr$ ages of 19.6–15.3 Ma for the correlative Chenque Formation (Cuitiño et al. 2015b). The oldest deposits in the Patagonian Andes are those obtained by the U–Pb method for the Traiguén (~26–23 Ma), Ventana (~23–20 Ma), and Puduhuapi (~23 Ma) formations (Bechis et al. 2014; Encinas et al. 2014, 2016). U–Pb and Sr dates for the Riñinahue, Ayacara, Río Foyel, Troncoso, Cerro Plataforma, La Cascada, Guadal, El Chacay, and the Estancia 25 de Mayo formations are all in the early Miocene range of ~21–18 Ma. Thus, it appears that the marine transgression commenced in between 26 and 23 Ma with a significant incursion between ~21 and 18 Ma that flooded most of Patagonia. In agreement with this notion, Cutiño et al. (2015) indicated that the eustatic peak of the "Patagonian" Atlantic transgression that reached the eastern Andean Cordillera occurred at ~20 Ma and was followed by fluvial deposition that began around 19–18 Ma.

4.2 Tectonic Setting

The tectonic setting for the marine transgression that flooded most of Patagonia during the late Oligocene–early Miocene is debated. One reason for the discord is that most studies have focused on local areas and few have formulated an explanation for the tectono-sedimentary evolution of the entire region. Data based on the analysis of brittle deformation (Lavenu and Encinas 2005) and on the studies of seismic lines (Elgueta et al. 2000; Jordan et al. 2001; Melnick and Echtler 2006) and outcrops (Encinas et al. 2015) of the Navidad Formation and correlative units in the forearc indicate the presence of normal faults and synextensional growth strata. Elgueta et al. (2000) attributed them to extensional tectonics concurrent with a global rise in sea-level, but Melnick and Echtler (2006), Encinas et al. (2008), and Encinas et al. (2012) related them to basal subduction erosion. In contrast, Becerra et al. (2013) concluded that the marine strata of the Ranquil Formation are syn-contractional (syn-inversional) based on their interpretation of an offshore seismic line taken south of the Arauco Peninsula. We question that interpretation because Becerra et al. (2013) noted the presence of "pervasive extensional faulting" that they viewed as superimposed on the contractional tectonics and "for which it is not possible to attribute an origin." On the other hand, Forsythe and Nelson (1985) had hypothesized that the Cenozoic Golfo de Penas basin formed by a pull-apart mechanism in response to the northward displacement of the Chiloé block along the Liquiñe-Ofqui Fault Zone.

The tectonic context for the marine deposits in the Patagonian Andes is also disputed. Bechis et al. (2014) attributed deposition of the Ventana and Río Foyel formations to progressive crustal thinning related to extensional tectonics. A similar origin was proposed by Encinas et al. (2016) for the volcano-sedimentary marine strata of the Traiguén Formation. In contrast, Hervé et al. (1995) suggested that this unit formed in pull-apart or asymmetric extensional basins related to the dextral motion of the nearby strike-slip Liquiñe-Ofqui Fault Zone, similar to the scenario that Hervé (1994) envisioned for the Ayacara Formation. Orts et al. (2012) reported the presence of progressive unconformities in the marine strata of the Cerro Plataforma (~42° S), which they related to contractional tectonics. The "Patagonian" marine deposits exposed in the eastern Andean Cordillera between 47 and 49° S were considered to be synorogenic by Ramos (1982a, 1989) because they pinch out to the west, which was interpreted as evidence of a mountain chain. Frassinetti and Covacevich (1999) also assumed the existence of a cordillera that blocked the connection between the Pacific and Atlantic oceans in the Golfo de Penas and Meseta Guadal basins (~47° S) because of the invertebrate fossils of Atlantic origin that are present in the Guadal Formation. In the extra-Andean region, Giacosa et al. (2004) reported normal faults and synextensional strata in the Chenque Formation at Comodoro Rivadavia (46° S). In contrast, Paredes et al. (2006) described synorogenic strata in that unit based on their analysis of seismic lines from the Sierra de San Bernardo (46° S).

In the discussion of the tectonosedimentary evolution of the late Oligocene–early Miocene marine successions of Patagonia, we must consider the following:

(1) Their marine strata are contemporaneous and their deposition therefore is probably related to the same cause.
(2) The presence of deep-marine successions up to 1500 m thick in the forearc and western Andean Cordillera and 600 m in the eastern Andean Cordillera (Elgueta et al. 2000; Finger et al. 2007; Encinas et al. 2008, 2012, 2013; Bechis et al. 2014) indicates that subsidence was significant. This implies that tectonics, not global sea-level changes, was the predominant cause of the marine transgression.
(3) The marine strata accumulated in a huge area that extends between the Pacific and the Atlantic coasts of Patagonia. Therefore, the subsidence mechanism must have been due to a regional event, not local displacement along the Liquiñe-Ofqui Fault Zone or other local tectonic discontinuities.
(4) Subsidence driven by subduction erosion typically occurs in forearc regions (Clift and Vannucchi 2004), so this mechanism is an unlikely cause (at least as the only acting force) because late Oligocene–early Miocene marine deposits also extend across the Andean Cordillera and the extra-Andean region.
(5) The occurrence of late Oligocene–early Miocene marine deposits in the western, central, and eastern parts of the present Patagonian Andes between ~41 and 47° S indicates that this mountain range likely did not yet exist. This precludes flexural subsidence as the principal driving mechanism for the marine transgression, at least for the cited latitudinal interval.

As previously noted, late Oligocene–early Miocene marine deposition commenced at ~24 Ma in the forearc (Nielsen and Glodny 2009) and at ~26 Ma in the extra-Andean region (Parras et al. 2008). The oldest marine strata deposited in the Patagonian Andes during that period are those of the Ventana (~23–20 Ma), Traiguén (~26–23 Ma), and Puduhaupi formations (~23 Ma) that are interbedded with basalts or cross-cut by abundant andesitic dikes (Bechis et al. 2014; Encinas et al. 2014, 2016). Geochemical and structural evidence indicates that the Ventana and Traiguén formations were deposited on a thinned crust in an extensional tectonic regime (Kay and Copeland 2006; Bechis et al. 2014; Litvak et al. 2014; Encinas et al. 2016) (see Chap. "Paleogene Arc-Related Volcanism in the Southern Central Andes and North Patagonia (39°–41° S))". In agreement with the proposals of Bechis et al. (2014) and Encinas et al. (2016), we consider that the deposition of these volcano-sedimentary units is a consequence of the late Oligocene–early Miocene extensional episode that was described by Muñoz et al. (2000) and Jordan et al. (2001). This event resulted in a series of extensional basins that extended from the present Chilean coast to the retroarc in Argentina between ~33 and 46° S and filled with thick successions of volcanics, continental and marine sediments. Muñoz et al. (2000) and Jordan et al. (2001) related this episode to a major plate reorganization in the Southeast Pacific that resulted in a shift from slower, more oblique South America-Farallón convergence to more rapid, near-normal South America-Nazca convergence at ca. 28–26 Ma (see Pardo-Casas and Molnar 1987; Somoza 1998). Muñoz et al. (2000) proposed that the change in subduction geometry and the increased trench-normal convergence rate induced a transient period of vigorous asthenospheric-wedge circulation that triggered slab rollback of the subducting Nazca plate, which resulted in regional extension and widespread volcanism. Progressive extension and crustal thinning led to marine flooding of Patagonia that reached its maximum extension at ~20 Ma. It followed a phase of compressive tectonics that started at 19–16 Ma and resulted in the emersion, uplift, and deformation of the late Oligocene–early Miocene marine strata and the growth of the Patagonian Andes (Bechis et al. 2014; Cuitiño et al. 2016). Considering that fluvial synorogenic sedimentation of the Santa Cruz Formation and equivalent units started at ~19 Ma in the eastern flank of the southern Patagonian Andes (Cutiño et al. 2016), it is possible that deposition in the eastern Andean Cordillera of the Cerro Plataforma and La Cascada formations (Orts et al. 2012; Encinas et al. 2014) at ~18 Ma, and deposition in the extra-Andean region of the upper part of the Chenque Formation at 19.6–15.3 Ma (Cuitiño et al. 2015b) and the Monte León Formation at 22.1–17.9 Ma (Parras et al. 2012) occurred during the initial stages of Andean growth, which would explain the presence of the synorogenic strata described for some of these units (e.g., Paredes et al. 2006; Orts et al. 2012). It is probable that synorogenic marine deposition concurrent with fluvial sedimentation took place in areas of greater subsidence, as suggested by Cuitiño et al. (2015b).

The aforementioned antecedents indicate that extensional tectonics is the most plausible explanation to account for the widespread marine transgression that reached the Patagonian Andes during the late Oligocene–early Miocene. As noted before, the occurrence of marine deposits of this age in the western, central, and eastern parts of the present Andean Cordillera between approximately 41 and 47° S indicates that this mountain range likely did not yet exist, which precludes flexural subsidence as the principal driving mechanism for the marine transgression. However, several authors propose that near-continuous Andean uplift has been occurring in the area located south of ~48° S since the Late Cretaceous (e.g., Biddle et al. 1986; Fosdick et al. 2011). In addition, late Oligocene–early Miocene marine deposits occur exclusively in the eastern flank of the Andes and the extra-Andean region in this region. This has led to several authors (e.g., Ramos 1982a, 1989; Parras et al. 2012; Cuitiño et al. 2012, 2015a) to propose that the "Patagonian" deposits in this area accumulated in a foreland basin. We consider unlikely that coeval marine strata were deposited in different tectonic contexts north and south of ~47° S, but additional research is needed to confirm this.

4.3 *Pacific–Atlantic Connection*

Another point of contention is the possibility of a transient connection between the Pacific and Atlantic in the Patagonian Andes during the late Oligocene–early Miocene. Marine strata deposited in the forearc during this period are characterized by a Pacific fauna that have no species in common with the Atlantic faunas of Argentina but share several elements with southern Peru (DeVries and Frassinetti 2003; Finger et al. 2007; Kiel and Nielsen 2010). However, Nielsen (2005) described a gastropod of Atlantic origin, *Perissodonta ameghinoi*, in Miocene marine deposits located near the Pacific coast in the proximity of Valdivia (~40° S). The Riñinahue, Ayacara, Puduhuapi, and Traiguén formations that crop out on the western flank of the Andean Cordillera contain only poorly preserved fossils that defy certain identification. Nevertheless, their proximity to the late Oligocene–early Miocene forearc marine deposits indicates that they are likely to be of Pacific origin.

Ramos (1982b) proposed a Pacific origin for the Río Foyel Formation and equivalent units exposed on the eastern Andean Cordillera (~41–43° S) ~70 km east of the Ayacara Formation outcrops and ~250 km north of the "Patagonian" deposits of the Guadal Formation and equivalent units. He based this notion on the occurrence of Oligocene to middle Miocene continental successions north, south, and east of the Río Foyel Formation, which left the Pacific to the west as the only possible source. On the other hand, contradictory interpretations arise from paleontologic studies. Bertels (1980), and Griffin et al. (2002) respectively correlated the foraminifers of the Río Foyel Formation and the molluscs from the Cerro Plataforma Formation with similar faunas of the "Patagoniense" of Argentina. Instead, Feldmann et al. (2011) proposed a Pacific origin for the Río Foyel Formation because its crustaceans resemble those in the Navidad Formation and equivalent units in the Chilean forearc. Recently, Encinas et al. (2014) reported the likely presence of both Pacific and Atlantic molluscan taxa in late Oligocene–early Miocene marine strata of the La Cascada Formation (43–44° S). Farther south, at ~47°–50° S, the Guadal, El Chacay, and the Estancia 25 de Mayo formations contain invertebrates that were correlated with those of the Monte León Formation on the Atlantic coast of Argentina (Frassinetti and Covacevich 1999; Griffin et al. 2014; Cuitiño et al. 2015a).

The Late Cenozoic erosion of most of the Meso-Cenozoic volcano-sedimentary veneer of the Patagonian Andes and the relatively poor preservation of fossils in many of the upper Oligocene–lower Miocene marine deposits of this area preclude any reliable paleogeographic reconstruction that could resolve the issue of seaway connections in this region. Paleontologic data, however, suggest that such a connection probably existed, although its path remains debatable. The presence of upper Oligocene–lower Miocene marine deposits of the Puduhuapi, Vargas, and La Cascada formations at 43–44° S on the western, central, and eastern part of the Patagonian Andes, respectively, and the probable occurrence of invertebrate taxa of Atlantic and Pacific origin in the La Cascada Formation, indicate that this area likely had a transient connection between the two oceans (Encinas et al. 2014). On the other hand, Uliana and Biddle (1988) and Bechis et al. (2014) suggested a Pacific–Atlantic connection near ~46° S, where upper Oligocene–upper Miocene marine deposits of Pacific origin that also crop out on the Chilean forearc in the Golfo the Penas and Traiguén areas (Stott and Webb 1989; Hervé et al. 2001; Encinas et al. 2016) are closest to the coeval deposits of Atlantic origin that compose the Guadal Formation in the eastern part of the Andean Cordillera (Ramos 1982a, b; Frassinetti and Covacevich 1999). However, this proposition is not supported by paleontologic studies as no mixed faunas of Atlantic and Pacific origin have been reported in any of these marine deposits on either side of the Andes. Yet, the fact that paleontologic studies on the Golfo the Penas area were principally conducted on middle–late Miocene faunas of the upper Grupo Chaicayán Sequence and not on the late Oligocene–early Miocene(?) lower part of this unit (Encinas et al. 2015 and references therein) leaves certain possibilities that need to be addressed to solve this dispute.

5 Conclusions

It appears that recent data from U–Pb geochronology and Sr isotope stratigraphy that indicate ages in the late Oligocene–early Miocene interval have resolved the dispute over the age of the marine deposits extending between the Pacific and Atlantic coasts of Patagonia. The marine transgression flooded a vast part of southern South America and, according to paleontological studies, probably enabled the first transient connection in the area between the Pacific and Atlantic oceans. Marine deposition started in the late Oligocene–earliest Miocene (~26–23 Ma). The oldest marine strata in the Patagonian Andes are interbedded with basaltic rocks and show geochemical and structural evidence of a thinned crust and an extensional tectonic regime. Progressive extension and crustal thinning allowed a generalized marine flooding of Patagonia that reached its maximum extension at ~20 Ma. We ascribe marine deposition to a regional event of extension and widespread volcanism related to a major plate reorganization in the Southeast Pacific at ~28–26 Ma. This episode was followed by a phase of compressive tectonics that started around 19–16 Ma and resulted in the emersion, uplift, and deformation of the late Oligocene–early Miocene marine strata as well as the growth of the Patagonian Andes. The youngest (~19–15 Ma) marine strata that accumulated in the eastern Andean Cordillera and the extra-Andean regions are coeval with fluvial synorogenic deposits and likely accumulated under a compressive regime.

Acknowledgements AE was funded by Fondecyt projects 1010691, 3060051, 11080115, 1090165, 1110914, and 1151146 of Conicyt. FB was funded by Universidad Nacional de Río Negro project PI 40-B-159, ANPCyT projects PICT 2010–2051 and PICT 2014-2240. SN was funded by Fondecyt project 1150664.

References

Adriasola AC, Thomson SN, Brix MR, Hervé F, Stöckhert B (2006) Postmagmatic cooling and late Cenozoic denudation of the North Patagonian Batholith in the Los Lagos region of Chile, 41°–42°15′S. Int J Earth Sci 95:501–528

Alarcón B (1995) Geología del área comprendida entre los 41º45′–42º05′ latitud Sur y 72º25′–72º50′ longitud Oeste, Provincias de Llanquihue y Palena, X Región. Undergraduate Thesis, Departamento de Geología, Universidad de Chile, p 64

Antinao JL, Duhart P, Clayton J, Elgueta S, McDonough M (2000) Area de Ancud-Maullín, Región de Los Lagos, mapa geológico no. 17: Sernageomin, 1:100.000 scale, 1 sheet

Asensio M, Zavala C, Arcuri M (2005) Los sedimentos terciarios del río Foyel, provincia de Río Negro, Argentina. XVI Congreso Geológico Argentino, La Plata. Actas en CD-ROM

Barreda VD (1997) Palinoestratigrafía de la Formación San Julián en el área de Playa La Mina (provincia de Santa Cruz), Oligoceno de la cuenca Austral. Ameghiniana 34(3):283–294

Barreda VD, Palamarczuk S (2000) Palinomorfos continentales y marinos de la Formación Monte León en su área tipo, provincia de Santa Cruz, Argentina. Ameghiniana 37(1):3–12

Barreda V, García V, Quattrocchio ME, Volkheimer W (2003) Edad y paleoambiente de la Formación Río Foyel, cuenca Ñirihuau, provincia de Río Negro, Argentina. XII Simposio Argentino de Paleobotánica y Palinología, Abstracts, pp 8–9

Barreda VD, Palazzesi L, Marenssi S (2009) Palynological record of the Paleogene Río Leona Formation (southernmost South America): Stratigraphical and paleoenvironmental implications. Rev Palaeobot Palynol 151(1–4):22–33

Becerra J, Contreras-Reyes E, Arriagada C (2013) Seismic structure and tectonics of the southern Arauco Basin, south-central Chile (~38° S). Tectonophysics 592:53–66

Bechis F, Encinas A, Concheyro A, Litvak VD, Aguirre-Urreta B, Ramos VA (2014) New age constraints for the Cenozoic marine transgressions of Northwestern Patagonia, Argentina (41°–43° S): Paleogeographic and tectonic implications. J S Am Earth Sci 52:72–93

Bellosi ES (1990) Formación Chenque: Registro de la transgresión patagoniana (Terciario medio) de la cuenca de San Jorge. Argentina. In XI Congreso Geológico Argentino, Actas (2):57–60

Bernabé P (2014) Edad y ambiente de sedimentación de la Formación Riñinahue, Lago Ranco, región de Los Ríos, Chile. Undergraduate Thesis, Departamento de Ciencias de la Tierra, Universidad de Concepción, p 100

Bertels A (1970) Sobre el "Piso Patagoniano" y la representación de la época del Oligoceno en Patagonia Austral, República Argentina. Revista de la Asociación Geológica Argentina 25:496–501

Bertels A (1975) Bioestratigrafía del Paleógeno en la República Argentina. Rev Esp Micropaleontol 7:429–450

Bertels A (1980) Foraminíferos (Protozoa) y ostrácodos (Arthropoda) de las "Lutitas de Río Foyel" (Oligoceno) de la cuenca de Ñirihuau, Provincia de Río Negro, República Argentina. Ameghiniana 17:49–52

Bobenrieth L, Díaz F, Davidson J, Portigliati C (1983) Complemento Mapa Metalogénico XI Región, Sector Norte Continental, comprendido entre 45° latitud sur y el límite con la X Región. Unpublished report, Sernageomín, Santiago, Chile, p 154

Bostelmann E, Ugalde R, Oyarzún JL, Griffin M (2015) The Patagonian transgression, early Miocene (Burdigalian age), in Aysén and Magallanes. XIV Congreso Geológico Chileno (ST-4):825–828

Bourdillon C (1994) Estudio de los niveles bioestratigráficos de la Formación Ayacara en Chile, dentro del contexto del estudio de Recursos mineros. SERNAGEOMIN-BRGM, Internal Report, p 9

Camacho HH (1985) Presencia del genero Parynomya Olsson, 1928 (Mollusca, Bivalvia) en la Formacion San Julián (Eoceno) de la provincia de Santa Cruz, Argentina. Boletín de la Academia Nacional de Ciencias de Córdoba 56:119–125

Campos A, Moreno H, Muñoz J, Antinao J, Clayton J, Martin M (1998) Mapa Geológico No. 8, Area de Futrono-Lago Ranco, Región de los lagos. SERNAGEOMIN, Escala 1:1.000.000

Cande SC, Leslie RB (1986) Late Cenozoic tectonics of the southern Chile trench. J Geophys Res 91:471–496

Casadío S, Feldmann R, Foland K (2000a) 40Ar/39Ar age and oxygen isotope temperature of the Centinela Formation, southwestern Argentina: An Eocene age for crustacean-rich "Patagonian" beds. J S Am Earth Sci 13:123–132

Casadío S, Guerstein GR, Marenssi S, Santillana S, Feldmann R, Parras A, Montalvo C (2000b) Evidencias para una edad oligocena de la Formación Centinela, suroeste de Santa Cruz, Argentina. Ameghiniana 37(4):71R

Casadío S, Parras A, Marenssi S, Griffin M (2001) Edades 87Sr/86Sr de Crassostrea? hatcheri (Ortmann) (Bivalvia, Ostreoidea) en el "Patagoniano" de Santa Cruz, Argentina. Ameghiniana 38(4):30R

Castillo JC (1983) Geología del sector occidental de la comuna de Futaleufú, provincia de Palena, Décima Región de Los Lagos, Chile. Memoria de Título (inédito), Departamento de Geología de la Universidad de Chile, p 137

Cecioni G (1980) Darwin's Navidad embayment, Santiago region, Chile, as a model of the southeastern Pacific shelf. J Pet Geol 2–3:309–321

Céspedes S (1975) El Terciario de la Isla Traiguén y Golfo de Tres Montes, Provincia de Aysén. Empresa Nacional del Petróleo (ENAP – Chile) unpublished, p 5

Chiesa JO, Camacho HH (1995) Litoestratigrafía del Paleógeno marino en el noroeste de la provincia de Santa Cruz, Argentina. Monografías de la Academia Nacional de Ciencias Exactas, Físicas y Naturales de Buenos Aires 11(I):9–15

Chiesa JO, Camacho HH (2001) Invertebrados marinos eocenos de la parte inferior de la Formación Río Foyel, Provincia de Río Negro, Argentina. Revista Española de Paleontología 16:299–316

Clift P, Vannucchi P (2004) Controls on tectonic accretion versus erosion in subduction zones: Implications for the origin and recycling of the continental crust. Rev Geophys 42: RG2001. https://doi.org/10.1029/2003RG000127

Covacevich V (1996) Muestreos paleontológicos en el Teciario marino de la costa sur del lago Ranco. Unpublished report, Sernageomín, p 4

Covacevich V, Frassinetti D (1986) El género Cancellaria en el Mioceno de Chile, con descripción de cuatro especies nuevas (Gastropoda: Cancellariidae). Revista Geológica de Chile 28–29: 33–67

Covacevich V, Frassinetti D (1990) La Fauna de Lo Abarca: Hito biocronoestratigráfico y paleoclimático en el Terciario Superior marino de Chile Central. II Simposio sobre el Terciario de Chile, Concepción, pp 51–71

Cuitiño JI, Scasso RA (2010) Sedimentología y paleoambientes del Patagoniano y su transición a la Formación Santa Cruz al sur del Lago Argentino, Patagonia Austral. Revista de la Asociación Geológica Argentina 66(3):406–417

Cuitiño JI, Pimentel MM, Ventura Santos R, Scasso RA (2012) High resolution isotopic ages for the early Miocene "Patagoniense" transgression in southwest Patagonia: Stratigraphic implications. J S Am Earth Sci 38:110–122
Cuitiño JI, Ventura Santos R, Alonso Muruaga PJ (2015a) Sr-stratigraphy and sedimentary evolution of early Miocene marine foreland deposits in the northern Austral (Magallanes) Basin, Argentina. Andean Geology 42:364–385
Cuitiño JI, Scasso RA, Ventura Santos R, Mancini LH (2015b) Sr ages for the Chenque Formation in the Comodoro Rivadavia region (Golfo San Jorge Basin, Argentina): stratigraphic implications. Lat Am J Sedimentology Basin Anal 22(1):3–12
Cuitiño JI, Fernicola JC, Kohn MJ, Trayler R, Naipauer M, Bargo MS, Kay R, Vizcaíno SF (2016) U–Pb geochronology of the Santa Cruz Formation (early Miocene) at the Río Bote and Río Santa Cruz (southernmost Patagonia, Argentina): Implications for the correlation of fossil vertebrate localities. J S Am Earth Sci 70:198–210
Darwin C (1846) Geological observations on South America. Being the third part of the geology of the Voyage of the Beagle, Under the command of Capt. Fitzroy, R.N. during the years 1832 to 1836. Smith Elder and Co Ed., London, p 279
De la Cruz R, Suárez M (2006) Geología del área Puerto Guadal-Puerto Sánchez, Región Aisén del General Carlos Ibañez del Campo. Servicio Nacional de Geología y Minería, Carta Geológica de Chile, Serie Geológica Básica 95:58 p, 1 mapa escala 1:100.000. Santiago
del Río C (2004) Tertiary marine molluscan assemblages of eastern Patagonia (Argentina): A biostratigraphic analysis. J Paleontol 78(6):1097–1122
Diez OM, Zubia MA (1981) Sinopsis estratigráfica de la región de "El Bolsón", provincia de Río Negro. Revista de la Asociación Geológica Argentina 36(1):19–28
Echaurren A, Folguera A, Gianni G, Orts D, Tassara A, Encinas A, Gimenez M, Valencia V (2016) Tectonic evolution of the North Patagonian Andes (41°–44° S) through recognition of syntectonic strata. Tectonophysics 677–678:99–114
Elgueta S, McDonough M, Le Roux J, Urqueta E, Duhart P (2000) Estratigrafía y sedimentología de las cuencas terciarias de la Región de Los Lagos (39–41°30′S). Boletín de la Subdirección Nacional de Geología 57:1–50
Encinas A, Le Roux JP, Buatois LA, Nielsen SN, Finger KL, Fourtanier E, Lavenu A (2006) Nuevo esquema estratigráfico para los depósitos marinos mio-pliocenos del área de Navidad (33°00′–34°30′S), Chile central. Revista Geológica de Chile 33(2):221–246
Encinas A, Finger KL, Nielsen SN, Lavenu A, Buatois LA, Peterson DE, Le Roux JP (2008) Rapid and major coastal subsidence during the late Miocene in south-central Chile. J S Am Earth Sci 25:157–175
Encinas A, Finger KL, Buatois LA, Peterson DE (2012) Major forearc subsidence and deep-marine Miocene sedimentation in the present Coastal Cordillera and longitudinal depression of south-central Chile (38°30′S–41°45′S). Geol Soc Am 124(7–8):1262–1277
Encinas A, Zambrano PA, Finger KL, Valencia V, Buatois LA, Duhart P (2013) Implications of deep-marine Miocene deposits on the evolution of the North Patagonian Andes. J Geol 121:215–238
Encinas A, Pérez F, Nielsen SN, Finger KL, Valencia V, Duhart P (2014) Geochronologic and paleontologic evidence for a Pacific–Atlantic connection during the late Oligocene–early Miocene in the Patagonian Andes (43–44°S). J S Am Earth Sci 55:1–18
Encinas A, Folguera A, Oliveros V, De Girolamo Del Mauro L, Tapia F, Riffo R, Hervé F, Finger KL, Valencia VA, Gianni G, Álvarez O (2015) Late Oligocene–early Miocene submarine volcanism and deep-marine sedimentation in an extensional basin of southern Chile. Implications on the tectonic development of the North Patagonian Andes. XIV Congreso geológico Chileno, Coquimbo, AT1-ST4, 347
Encinas A, Folguera A, Oliveros V, De Girolamo Del Mauro L, Tapia F, Riffo R, Hervé F, Finger KL, Valencia VA, Gianni G, Álvarez O (2016) Late Oligocene–early Miocene submarine volcanism and deep-marine sedimentation in an extensional basin of southern Chile: Implications on the tectonic development of the North Patagonian Andes. Geol Soc Am Bull 128(5–6):807–823

Espinoza W, Fuenzalida P (1971) Geología de las Hojas Isla Rivero, Puerto Aysén y Balmaceda entre los paralelos 45° y 46° de latitud sur. Instituto de Investigaciones Geológicas Convenio Instituto CORFO Aysén, p 50

Feldmann RM, Schweitzer CE, Encinas A (2010) Neogene decapod Crustacea from southern Chile. Ann Carnegie Mus 78(4):337–366

Feldmann RM, Schweitzer CE, Casadío S, Griffin M (2011) New Miocene Decapoda (Thalassinidea; Brachyura) from Tierra del Fuego, Argentina: Paleobiogeographic implications. Ann Carnegie Mus 79:91–123

Finger KL (2013) Miocene foraminifera from the south-central coast of Chile. Micropaleontology 59:341–493

Finger KL, Nielsen SN, DeVries TJ, Encinas A, Peterson DE (2007) Paleontologic evidence for sedimentary displacement in Neogene forearc basins of central Chile. Palaios 22:3–16

Finger KL, Encinas A, Nielsen SN (2013). Comment on 'Evidence for an early-middle Miocene age of the Navidad Formation (central Chile): Paleontological, paleoclimatic and tectonic implications' of Gutiérrez et al. (2013, Andean Geology 40 (1): 66–78). Andean Geology 40:570–578

Flint SS, Prior DJ, Agar SM, Turner P (1994) Stratigraphic and structural evolution of the Tertiary Cosmelli Basin and its relationship to the Chile triple junction. J Geol Soc 151:251–268

Flynn JJ, Novacek MJ, Dodson HE, Frassinetti D, McKenna MC, Norell MA, Sears KE, Swisher CC, Wyss AR (2002) A new fossil mammal assemblage from the southern Chilean Andes: Implications for geology, geochronology, and tectonics. J S Am Earth Sci 15:285–302

Forsythe R, Nelson E (1985) Geological manifestations of ridge collision: Evidence from the Golfo de Penas-Taitao basin, southern Chile. Tectonics 4:477–495

Forsythe RD, Olsson RK, Johnson C, Nelson EP (1985) Stratigraphic and micropaleontologic observations from the Golfo de Penas-Taitao basin, Southern Chile. Revista Geológica de Chile 25–26:3–12

Frassinetti D (2001) Moluscos Bivalvos y Gastrópodos del Mioceno marino de isla Stokes, sur de Chile. Boletín del Museo Nacional de Historia Natural, Chile 50:73–90

Frassinetti D (2004) Moluscos fósiles del Mioceno marino de Isla Ipún, sur de Chile. Boletín del Museo Nacional de Historia Natural, Chile 53:71–83

Frassinetti D (2006) Moluscos fósiles del Mioceno marino de islas Crosslet y Hereford (Golfo Tres Montes, Aisén, Chile). Boletín del Museo Nacional de Historia Natural, Chile 55:61–74

Frassinetti D, Covacevich V (1999) Invertebrados fósiles marinos de la Formación Guadal (Oligoceno Superior–Mioceno Inferior) en Pampa Castillo, Región de Aysén, Chile. Boletín del Servicio Nacional de Geología y Minería de Chile 51:1–96

Fuenzalida JL (1979) Estudio geológico preliminar de Península Huequi, X Región. Undergraduate Thesis, Santiago, Departamento de Geología, Universidad de Chile, p 158

Furque G, Camacho HH (1972) El Cretácico Superior y terciario de la región austral del Lago Argentino (Provincia de Santa Cruz). Actas Jornadas Geológicas Argentinas, Buenos Aires, pp 61–75

Giacosa H, Heredia N, Césari O, Zubia M, González R, Faroux A (2001) Descripción geológica de la Hoja 4172-IV, San Carlos de Bariloche, Provincias de Río Negro y Neuquén. Servicio Geológico Minero Argentino. Instituto de Geología y Recursos Minerales. Boletín 279: p 61

Giacosa RE, Paredes JM, Nillni AM, Ledesma M, Colombo F (2004) Fallas normales de alto ángulo en el Neógeno del margen Atlántico de la Cuenca del Golfo San Jorge (46ºS–67º30′O, Patagonia Argentina). Boletín Geológico y Minero 115(3):537–550

González Bonorino F, González Bonorino G (1978) Geología de la región de San Carlos de Bariloche: un estudio de las formaciones terciarias del Grupo Nahuel Huapi. Revista de la Asociación Geológica Argentina 33(3):175–210

Griffin M, Pérez LM, Muravchik M (2002) Moluscos terciarios del Cerro Plataforma, en el noroeste de Chubut. VIII Congreso Argentino de Paleontología y Bioestratigrafía, Corrientes. Abstracts, p 101

Griffin M, Casadío S, Parras A, Feldmann R, Schweitzer C (2004) 87Sr/86Sr early Oligocene age for the Río Foyel formation, Río Negro, Argentina. Ameghiniana 41(4): Suplementos Resúmenes 13

Griffin M, Ugalde R, Genta Iturrería S, Bostelmann E, Parras A, Oyarzún JL (2014) Fossil invertebrates of the "Patagonian" beds in Sierra Baguales, Última Esperanza Province, Magallanes, Chile. Congreso Uruguayo de Zoología, Montevideo, Uruguay. Actas III I: 108–109

Gutiérrez N, Hinojosa LF, Le Roux JP, Pedroza V (2013) Evidence for an early-Middle Miocene age of the Navidad Formation (central Chile): Paleontological, paleoclimatic and tectonic implications. Andean Geol 40:66–78

Heim A (1940) Geological observations in the Patagonian Cordillera. Eclogae Geol Helv 33: 25–51

Hervé F, Kawashita K, Parada MA, Brook M, Pankhurst R, Snelling N, Drake R (1985) Granitoids of the coast range of central Chile, geochronology and geological setting. Comunicaciones 35:105–108

Hervé F (1994) The southern Andes between 39° and 44° S latitude: The geological signature of a transpressive tectonic regime related to a magmatic arc. In: Reutter KJ, Scheuber E, Wigger P (eds) Tectonics of the southern central Andes. Springer-Verlag, Berlin, pp 243–248

Hervé F, Pankhurst RJ, Drake R, Beck ME (1995) Pillow metabasalts in a mid-Tertiary extensional basin adjacent to the Liquiñe-Ofqui fault zone: the Isla Magdalena area, Aysén, Chile. J S Am Earth Sci 8(1):33–46

Hervé F, Sanhueza A, Silva C, Pankhurst RJ, Fanning MC, Campbell H, Crundwell M (2001) A Neogene age for Traiguén Formation, Aysén, Chile, as revealed by shrimp U–Pb dating of detrital zircons. III Simposio Sudamericano de Geología Isotópica, Pucón, Chile, pp 570–574

Hervé F, Fanning CM, Pankhurst RJ (2003) Detrital zircon age patterns and provenance of the metamorphic complexes of southern Chile. J S Am Earth Sci 16(1):33–46

Jordan T, Matthew W, Veiga R, Pángaro F, Copeland P, Kelley S, Mpodozis C (2001) Extension and basin formation in the southern Andes caused by increased convergence rate: a mid-Cenozoic trigger for the Andes. Tectonics 20:308–324

Kay SM, Copeland (2006) Early to middle Miocene backarc magmas of the Neuquén Basin: Geochemical consequences of slab shallowing and the westward drift of South America. Geol Soc Am, Special Papers 407:185–213

Kiel S, Nielsen SN (2010) Quaternary origin of the inverse latitudinal diversity gradient among southern Chilean mollusks. Geology 38:955–958

Kummel B (1956) Post-Triassic nautiloid genera. Bulletin of the Museum of Comparative Zoology Harvard 114:324–494

Lavenu A, Encinas A (2005) Deformación frágil de los depósitos neógenos de la cuenca de Navidad (Cordillera de la Costa, 34° S, Chile central). Revista Geológica de Chile 32(2): 229–248

Le Roux JP, Gutiérrez NM, Hinojosa LF, Pedroza V, Becerra J (2013) Reply to comment of Finger et al. (2013) on: 'Evidence for an early-middle Miocene age of the Navidad Formation (central Chile): Paleontological, paleoclimatic and tectonic implications' of Gutiérrez et al. (2013, Andean Geology 40 (1):66–78). Andean Geology 40(3):580–588

Levi B, Aguilar A, Fuenzalida R (1966) Reconocimiento geológico en las provincias de Llanquihue y Chiloé. Santiago, Instituto de Investigaciones Geológicas, Boletín 19:45 p

Linares E (1979) Catálogo de edades radimétricas determinadas para la República Argentina y Catálogo de edades radimétricas realizadas por INGEIS y sin publicar. Revista de la Asociación Geológica Argentina, Special Publication 6(B):32

Litvak VD, Encinas A, Oliveros V, Bechis F, Folguera A, Ramos VA (2014) El volcanismo Mioceno inferior vinculado a las ingresiones marinas en los Andes Nordpatagónicos. XIX Congreso Geológico Argentino, Córdoba, pp S22–S35

Lizuaín A (1979) La edad de las sedimentitas del cerro Plataforma, provincia del Chubut. Revista de la Asociación Geológica Argentina 34(1):69–72

Lizuaín A (1983) Geología de la Cordillera Patagónica entre las localidades de Lago Puelo y Leleque. Unpublished PhD thesis, Universidad de Buenos Aires, p 128
Malumián N (1978) Estudio micropaleontológico de muestras provenientes de las Hojas 55 a y b. Servicion Geológico Nacional, internal report, Buenos Aires
Malumián N (2002) El terciario marino. Sus relaciones con el eustatismo. In: Haller MJ (ed) Geología y recursos naturales de Santa Cruz. Relatorio del XV Congreso Geológico Argentino:237–244
Malumián N, Echevarría A, Martínez Macchiavello JC, Náñez C (1984) Microfósiles. En: Geología y Recursos Naturales de la Provincia de Río Negro, Ed. V.A. Ramos. Relatorio IX Congreso Geológico Argentino, San Carlos de Bariloche, Capítulo II-7:485–526
Malumián N, Náñez C (1989) Asociaciones de foraminíferos del Terciario medio de cuenca Austral: sus relaciones con eventos eustáticos globales. Revista de la Asociación Geológica Argentina 43(2):257–264
Malumián N, Ardolino AA, Franchi M, Remesal M, Salani F (1999) La sedimentación y el volcanismo terciarios en la Patagonia Extraandina. In: Caminos R (ed) Geología Argentina. Instituto de Geología y Recursos Minerales, Anales 29(18):557–612
Malumián N, Asensio MA, Cornou ME, Martínez MA, Quattrocchio ME (2008) Formación Río Foyel: La transgresión pacífica en la Cordillera Patagónica. XVII Congreso Geológico Argentino, Jujuy, Actas 2:861–862
Malumián N, Náñez C (2011) The late Cretaceous-Cenozoic transgressions in Patagonia and the Fuegian Andes: foraminifera, palaeoecology, and palaeogeography. Biol J Lin Soc 103:269–288
Martínez-Pardo R (1965) Edad de la Formación Ayacara en Caleta El Manzano e Isla El Manzano. Soc Geol Chile, Resúmenes:19–22
Martínez Pardo R (1976) Hallazgo de Sphaeroidinella dehiscens dehiscens (Parker and Jones) en el Plioceno de Arauco: su significado para la reinterpretación del Neógeno Superior en Chile. I Congreso Geológico Chileno, Actas: C125–C142
Martínez M, Bianchinotti MV, Saxena RK, Cornou ME, Quattrocchio ME (2016). Fungal spores from the Palaeogene El Foyel Group of Ñirihuau basin, Argentina. Papers in Palaeontology, 2016:1–20
Melnick D, Echtler HP (2006) Inversion of forearc basins in south-central Chile caused by rapid glacial age trench fill. Geology 34:709–712
Mpodozis C, Hervé M, Nasi C, Soffia J, Forsythe R, Nelson E (1985) El magmatismo Plioceno de península Tres Montes y su relación con la evolución del Punto Triple de Chile Austral. Revista Geológica de Chile 25–26:13–28
Mpodozis C, Ramos V (1990) The Andes of Chile and Argentina. In: Ericksen G, Cañas M, Reinemund J (eds) Geology of the Andes and its relation to hydrocarbon and mineral resources: Circum-Pacific Council for Energy and Mineral Resources. Earth Science Series 11:59–88
Muñoz J, Troncoso R, Duhart P, Crignola P, Farmer L, Stern CR (2000) The relation of the mid-Tertiary coastal magmatic belt in south-central Chile to the late Oligocene increase in plate convergence rate. Revista Geológica de Chile 27:177–203
Nielsen SN (2005) Cenozoic Strombidae, Aporrhaidae, and Struthiolariidae (Gastropoda, Stromboidea) from Chile: Their significance to biogeography of faunas and climate of the south-east Pacific. J Paleontol 79:1120–1130
Nielsen SN (2013) A new Pliocene mollusk fauna from Mejillones, northern Chile. Paläontologische Zeitschrift 87:33–66
Nielsen SN, Frassinetti D (2007) The Neogene Volutidae (Gastropoda: Neogastropoda) from the Pacific coast of Chile. J Paleontol 81(1):82–102
Nielsen SN, Glodny J (2009) Early Miocene subtropical water temperatures in the southeast Pacific. Palaeogeogr Palaeoclimatol Palaeoecol 280:480–488
Niemeyer H (1975) Geología de la región comprendida entre el Lago General Carrera y el Río Chacabuco, Provincia de Aisén, Chile. Undergraduate Thesis, Universidad de Chile, p 309

Orts DL, Folguera A, Encinas A, Ramos M, Tobal J, Ramos VA (2012) Tectonic development of the North Patagonian Andes and their related Miocene foreland basin (41°30′–43° S). Tectonics 31:TC3012

Osorio R, Elgueta S (1990) Evolución paleobatimétrica de la Cuenca Labranza documentada por Foraminíferos. II Simposio sobre el Terciario de Chile, Concepción, p 225–233

Pankhurst RJ, Weaver S, Hervé F, Larrondo P (1999) Mesozoic–Cenozoic evolution of the North Patagonian Batholith in Aysén, Southern Chile. J Geol Soc London 156:673–694

Pardo-Casas F, Molnar P (1987) Relative motion of the Nazca (Farallón) and South American plates since late Cretaceous time. Tectonics 6(3):233–248

Paredes JM, Azpiroz G, Foix M (2006) Tertiary tectonics and sedimentation in the Cerro Piedra Oíl Field (Golfo San Jorge Basin), Argentina. IV Congreso Latinoamericano de Scdimentología y XI Reunión Argentina de Scdimentología, San Carlos de Bariloche, Argentina, p 163

Parras A, Griffin M, Feldmann R, Casadío S, Schweitzer C, Marenssi S (2008) Correlation of marine beds based on Sr- and Ar-date determinations and faunal affinities across the Paleogene/Neogene boundary in southern Patagonia, Argentina. J S Am Earth Sci 26:204–216

Parras A, Dix GR, Griffin M (2012) Sr-isotope chronostratigraphy of Paleogene/Neogene marine deposits: Austral Basin, southern Patagonia (Argentina). J S Am Earth Sci 37:122–135

Pöthe de Baldis D (1984) Microfloras fósiles cenozoicas. En: Geología y Recursos Naturales de la Provincia de Río Negro, Ed. VA Ramos. Relatorio IX Congreso Geológico Argentino, San Carlos de Bariloche, Capítulo II(4):393–412

Ramos VA (1979) Tectónica de la región del Río y Lago Belgrano, Cordillera Patagónica, Argentina. II Congreso Geológico Chileno, Actas: B1–B32

Ramos VA (1982a) Geología de la región del Lago Cardiel, provincia de Santa Cruz. Revista de la Asociación Geológica Argentina 37(1):23–49

Ramos VA (1982b) Las ingresiones pacíficas del Terciario en el norte de la Patagonia (Argentina). III Congreso Geológico Chileno, Actas, A262–A288

Ramos VA (1989) Andean Foothills Structures in Northern Magallanes Basin, Argentina. The American Association of Petroleum Geologists Bulletin 73(7):887–903

Ramos VA, Ghiglione MC (2008) Tectonic evolution of the Patagonian Andes. In: Rabassa J (ed) Late Cenozoic of Patagonia and Tierra del Fuego: Developments in Quaternary Sciences 11. Ushuaia, Elsevier, p 57–71

Rapela CW, Spalletti LA, Merodio JC, Aragón E (1988) Temporal evolution and spatial variation of early Tertiary volcanism in the Patagonian Andes (40°S 42°30′S). J S Am Earth Sci 1. 75–88

Riccardi AC, Rolleri EO (1980) Cordillera Patagónica Austral. II Simposio de Geología Regional Argentina. Academia Nacional de Ciencias II:1173–1306

Riggi JC (1979) Nomenclatura, categoría litoestratigráfica y correlación de la Formación Patagonia en la costa atlántica. Revista de la Asociación Geológica Argentina 34:243–248

Rojas C, Beck ME, Burmester RF, Cembrano J, Hervé F (1994) Paleomagnetism of the Mid-Tertiary Ayacara Formation, southern Chile: Counterclockwise rotation in a dextral shear zone. J S Am Earth Sci 7:45–56

Rubilar A, Pérez E, Frassinetti D (2001). Fauna fósil recolectada al sur de Río Azul, oeste de Río Futaleufú, X Región. Sernageomin, Informe Paleontológico 2001–07:5 p

Rubilar A, Pérez E, Frassinetti D (2002) Invertebrados marinos cenozoicos recolectados al suroeste de Futaleufú (X Región), suroeste de Palena (XI Región) y al norte de la Provincia de Río Negro (Argentina). Sernageomin, Informe Paleontológico 2002–08:8 p

Sernageomin (1995) Carta metalogénica décima región sur, Chile. SERNAGEOMIN-BRGM, Internal Report IR-95-05, 10 vol, Santiago

Silva C (2003) Ambiente geotectónico de erupción y metamorfismo de metabasaltos almohadillados de los Andes Norpatagónicos (42°–46° S), Chile. Master Thesis, Departamento de Geología, Universidad de Chile, p 148

Silva C, Herrera C, Hervé F (2003) Petrogénesis de lavas y diques básicos de la Formación Traiguén, Región de Aysén (43°30′–46° S), Chile. X Congreso Geológico Chileno, Actas en CD-rom:11 p

Solano A (1978) Geología del sector costero de Chiloé continental entre los 41°50′ y 42°10′ de latitud sur. Undergraduate Thesis. Santiago, Departamento de. Geología, Universidad de Chile:123 p

Somoza R (1998) Updated Nazca (Farallon)-South America relative motions during the last 40 My: Implications for mountains building in the central Andean region. J S Am Earth Sci 11 (3):211–215

Steffen H (1944) Patagonia Occidental; las cordilleras patagónicas y sus regiones circundantes. Ediciones de la Universidad de Chile. Santiago, Chile:586 p

Stott LD, Webb PN (1989) The Neogloboquadrina continuosa last appearance datum level in the South Pacific. Micropaleontology 35:63–71

Suárez MR, De La Cruz R (2000) Tectonics in the eastern central Patagonian Cordillera (45°30S–47°30S). J Geol Soc London 157:995–1001

Suárez M, de la Cruz R, Etchart H, Márquez M, Fanning M (2015) Síntesis de la Cronología Magmática Meso-Cenozoica de Patagonia Central, Aysén, Chile: edades U–Pb SHRIMP. XIV Congreso Geológico Chileno, ST-4:789–792

Tavera J (1979) Estratigrafía y paleontología de la Formación Navidad, Provincia de Colchagua, Chile (lat 30°50′–34° S). Boletín del Museo Nacional de Historia Natural de Chile 36 (5–10), 176 p

Thiele R, Castillo JC, Hein R, Romero G, Ulloa M (1978) Geología del sector fronterizo de Chiloé continental entre los 43°00′ y 43°15′ L.S., Chile (Comunas de Futaleufú y Palena). VII Congreso Geológico Argentino, Neuquén, Abstracts 1: 577–591

Thomson N (2002) Late Cenozoic geomorphic and tectonic evolution of the Patagonian Andes between latitudes 42º S and 46º S: An appraisal based on fission-track results from the transpressional intra-arc Liquiñe-Ofqui fault zone. Geol Soc Am Bull 114:1159–1173

Uliana MA, Biddle KT (1988) Mesozoic–Cenozoic paleogeographic and geodynamic evolution of southern South America. Rev Bras Geociências 18(2):172–190

Urbina O (2001) Geología de la cordillera norpatagónica en el área del río Palena, XI Región de Aysén, Chile. Undergraduate Thesis (unpublished), Departamento de Geología, Universidad de Chile, p 112

Neogene Growth of the Patagonian Andes

Andrés Folguera, Guido M. Gianni, Alfonso Encinas, Orlando Álvarez, Darío Orts, Andrés Echaurren, Vanesa D. Litvak, César R. Navarrete, Daniel Sellés, Jonathan Tobal, Miguel E. Ramos, Lucas Fennell, Lucía Fernández Paz, Mario Giménez, Patricia Martínez, Francisco Ruiz and Sofía B. Iannelli

Abstract After a Late Cretaceous to Paleocene stage of mountain building, the North Patagonian Andes were extensionally reactivated leading to a period of crustal attenuation. The result was the marine Traiguén Basin characterized by submarine volcanism and deep-marine sedimentation over a quasi-oceanic basement floor that spread between 27 and 22 Ma and closed by 20 Ma, age of syndeformational granitoids that cut the basin infill. As a result of basin closure, accretion of the Upper Triassic metamorphic Chonos Archipelago took place against the Chilean margin,

A. Folguera (✉) · V. D. Litvak · J. Tobal · M. E. Ramos · L. Fennell
L. Fernández Paz · S. B. Iannelli
Instituto de Estudios Andinos (IDEAN), Consejo Nacional de Investigaciones Científicas y Técnicas (CONICET), Universidad de Buenos Aires, Buenos Aires, Argentina
e-mail: andresfolguera2@yahoo.com.ar

L. Fernández Paz
e-mail: luciafp@gl.fcen.uba.ar

S. B. Iannelli
e-mail: sofia.iannelli@hotmail.com.ar

G. M. Gianni · O. Álvarez · M. Giménez · P. Martínez · F. Ruiz
Instituto Geofísico Sismológico Ing. Volponi (IGSV), Universidad de Nacional San Juan, San Juan, Argentina
e-mail: orlando_a_p@yahoo.com.ar

G. M. Gianni · O. Álvarez · M. Giménez
Consejo Nacional de Investigaciones Científicas y Técnicas (CONICET), Buenos Aires, Argentina

A. Encinas
Departamento de Ciencias de la Tierra, Universidad de Concepción, Casilla 160-C, Concepción, Chile
e-mail: aencinas@udec.cl

D. Orts
Instituto de Investigación en Paleobiología y Geología, Universidad Nacional de Río Negro—CONICET, Av. J.A. Roca 1242. CP 8332, General Roca, Argentina
e-mail: dlorts@yahoo.com

A. Folguera et al. (eds.), *The Evolution of the Chilean-Argentinean Andes*,
Springer Earth System Sciences, https://doi.org/10.1007/978-3-319-67774-3_19

overthrusting a stripe of high-density (mafic) rocks on the upper crust, traced by gravity data through the Chonos Archipiélago. After this, contractional deformation had a rapid propagation between 19 and 14.8 Ma rebuilding the Patagonian Andes and producing a wide broken foreland zone. This rapid advance of the deformational front, registered in synorogenic sedimentation, was accompanied at the latitudes of the North Patagonian Andes by an expansion of the arc magmatism between 19 and 14 Ma, suggesting a change in the subduction geometry at that time. Then a sudden retraction of the contractional activity took place around 13.5–11.3 Ma, accompanied by a retraction of magmatism and an extensional reactivation of the Andean zone that controlled retroarc volcanism up to 7.3–(4.6?) Ma. This particular evolution is explained by a shallow subduction regime in the northernmost Patagonian Andes, probably facilitated by the presence of the North Patagonian massif lithospheric anchor that would have blocked drag basal forces creating low-pressure conditions for slab shallowing. Contrastingly, to the south, the accretion of the Chonos Archipelago explains rapid propagation of the deformation across the retroarc zone. These processes occurred at the time of rather orthogonal to the margin convergence between Nazca and South American plates after a long period of high oblique convergence. Finally, convergence deceleration in the last 10 My could have led to extensional relaxation of the orogen.

Keywords Patagonia · Fold and thrust belt · Foreland basin · Subduction accretion · Shallow subduction

1 Introduction

The Andes are unevenly explored. Much of their information comes from the Central Andes of Bolivia, southern Perú, and northern Argentina and Chile where orogenic processes show a strong linkage with isostatic readjustments associated with delamination of the lower lithosphere and climatic factors. In comparison, the Patagonian Andes lack integrated analyses that could allow discussing their present morphology, structure, and magmatism.

C. R. Navarrete
Departamento de Geología, Universidad Nacional de la Patagonia San Juan Bosco, Comodoro Rivadavia, Chubut, Argentina

A. Echaurren
Instituto de Estudios Andinos Don Pablo Groeber, UBA—CONICET. Departamento de Ciencias Geológicas, FCEN, Universidad de Buenos Aires, Buenos Aires, Argentina

D. Sellés
Aurum Consultores, Santiago de Chile, Chile

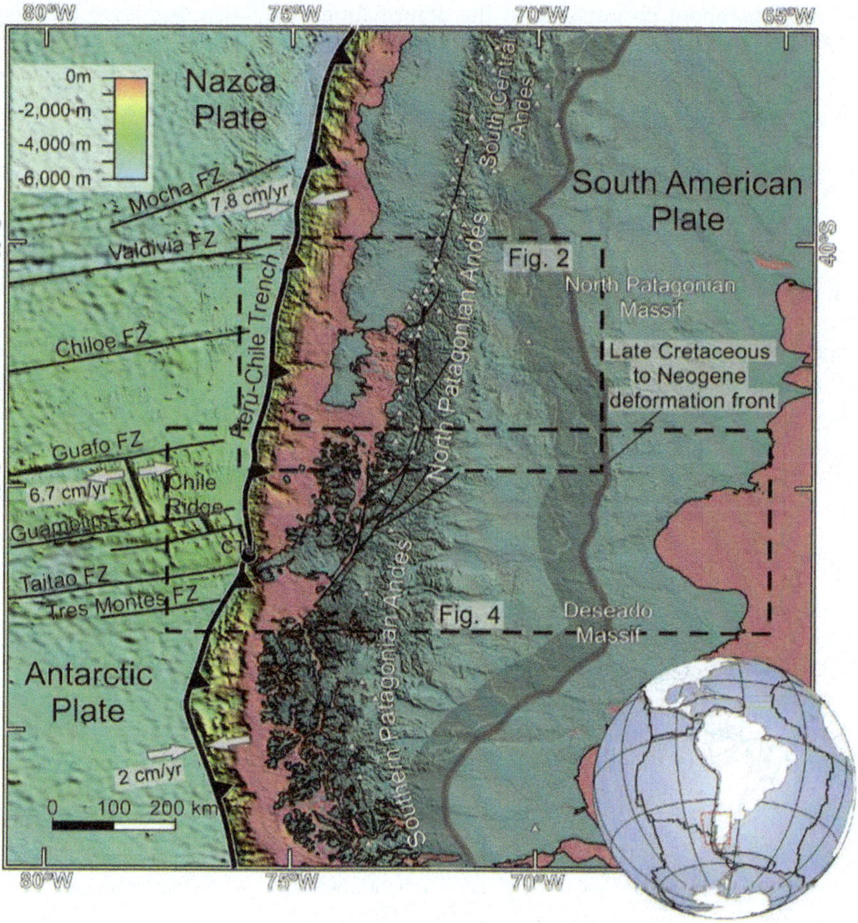

Fig. 1 Southern Central and Patagonian Andes. Note the anomalous foreland development of the fold and thrust belt in the northern Patagonian Andes (taken from Orts et al. 2012). Boxes indicate the two analyzed subareas corresponding to Figs. 2 and 4

This chapter constitutes a review centered on the Neogene evolution of the North Patagonian Andes, a period of time and a place with particular conditions that explain magmatic and deformational patterns achieved in the last 20 My. The deformational front in the North Patagonian Andes has a wider eastward development, which indicates relatively strong orogenic processes (Fig. 1). This deformational front is recognized on the basis of description of synorogenic strata (Giacosa and Heredia 2004; Ramos et al. 2011, 2015; Bilmes et al. 2013).

Radiometric ages of these strata grouped in the Ñirihuau and Collón Curá Formations have been scarce and mainly K/Ar during years (e.g., Marshall et al. 1977; Mazzoni and Benvenuto 1990) until U/Pb ages appeared. Since then, detailed descriptions about progression of the deformation have been discussed by Bechis et al. (2014), Orts et al. (2012, 2015), Tobal et al. (2015), Ramos et al. (2015), Gianni et al. (2015a, b) and Echaurren et al. (2016).

2 Geology of the Northern Patagonian Andes (~40°–46° S)

The present configuration of the Patagonian Andes is related to the subduction of the Nazca and Antarctic plates beneath the South American plate. The three plates presently interact at 46°30′S in the Aysen triple junction that separates two morphostructural domains along the Patagonian Andes: The Austral and North Patagonian Andes (Fig. 1). This triple junction has migrated northward along the trench in the last 14 Ma (Cande and Leslie 1986) producing an arc gap associated with the development of a slab window that separates the Austral volcanic zone from the South volcanic zone (Ramos and Kay 1992) (see Chap. "Mantle Influence on Andean and Pre-Andean Topography").

The Andes between 40° and 45° S comprise from west to east four physiographic units (Fig. 2): (i) the Coastal Cordillera, characterized by polyphase deformation since the Mesozoic and the development of extension during the late Eocene–Oligocene associated with the intra-arc Traiguén marine Basin (Hervé et al. 1995; Encinas et al. 2016) (see Chap. "The Late Oligocene–Early Miocene Marine Transgression of Patagonia"). (ii) The North Patagonian Andes separated from the Coastal Cordillera through a submerged Central Valley and associated with the dextral strike-slip Liquiñe-Ofqui fault zone (LOFZ) of Neogene to Quaternary age

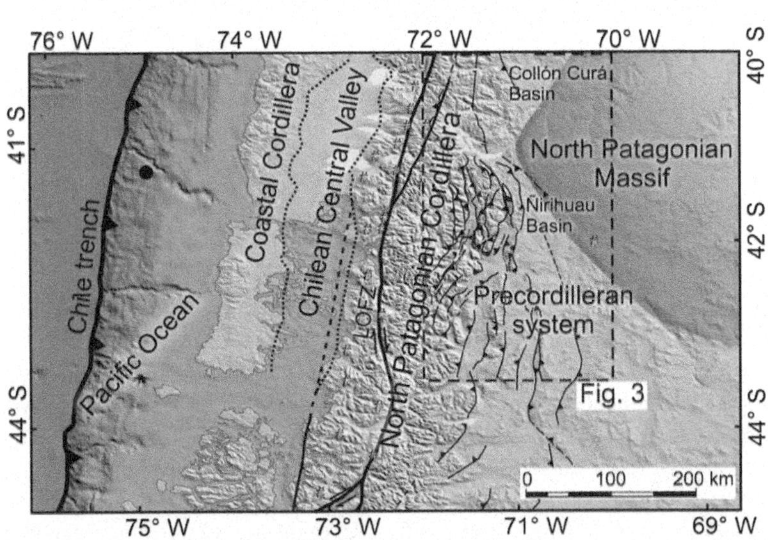

Fig. 2 Main structures at the northern analyzed area. Note how the fold and thrust belt narrows at the latitudes where the North Patagonian Massif is wider and closer to the subduction margin implying a strong rheology that inhibits propagation of the orogenic front to the foreland area (modified from Orts et al. 2015)

(Lavenu and Cembrano 1999, among others). This system constitutes the latest reactivation of a Mesozoic ductile fault system (Cembrano et al. 1996) that controls the emplacement of the Quaternary volcanic arc and affects a Mesozoic to Cenozoic Patagonian batholith. iii) To the east, the Patagonian broken foreland system, also known at these latitudes as the Bernárdides system, produced by the tectonic inversion of an Early Jurassic rift system (Feruglio 1947; Peroni et al. 1995) (see Chap. "Cretaceous Orogeny and Marine Transgression in the Southern Central and Northern Patagonian Andes: Aftermath of a Large-scale Flat-subduction Event?").

The pre-Andean tectonic evolution of northern Patagonia is associated with Paleozoic metamorphic and igneous rocks, deformed by the Gondwanan and Tabarín orogenies in late Carboniferous–middle Permian and late Permian–Triassic times, respectively (García-Sansegundo et al. 2009; Von Gosen 2009; Von Gosen and Loske 2004) (see Chap. "The Pre-Andean Phases of Construction of the Southern Andes Basement in Neoproterozoic–Paleozoic Times"). In Late Triassic times, metaturbidites of the Chonos complex (Hervé and Fanning 2001) (Fig. 1) were metamorphosed in high-pressure, low-temperature conditions associated with an early subduction event near the Triassic–Jurassic boundary (see Hervé et al. 1998). Early Jurassic times are characterized by the emplacement of I-type granitoids forming the NNW-trending Subcordilleran batholith, associated with marine/continental depocenters (Gordon and Orts 1993; Rapela et al. 2005; Vicente 2005) (see Chap. "A Provenance Analysis from the Lower Jurassic Units of the Neuquén Basin. Volcanic Arc or Intraplate Magmatic Input?"). Rift expansion took place in Mid Jurassic times with NW- and WNW-oriented depocenters associated with the eruption of the Karoo large igneous Province whose western products reached eastern Patagonia (Pankhurst et al. 1999; Rapela et al. 2005) (see Chap. "Lower Jurassic to Early Paleogene Intraplate Contraction in Central Patagonia").

After this, a NW-trending Neocomian marine to continental intra-arc to back-arc basin, known as the Río Mayo Embayment/Sub-basin (Ramos 1981) or Aysén Basin (Suárez et al. 1996), affected the western sector of northern Patagonia.

In the eastern parts of the North Patagonian Andes, volcanic rocks of the Divisadero Group (Ramos 1981) unconformably overlay Jurassic successions and older intrusive rocks. These volcanic successions with calc-alkaline affinities have Early Cretaceous ages (Rapela et al. 1987; Suárez and de La Cruz 2001). Their outcrops are found between 41° and 48° S and are emplaced over an angular unconformity that implies an Early Cretaceous orogenic event (Ramos 1981; Suárez and De La Cruz 2001; Gianni et al. 2015a, b) (see Chap. "Cretaceous Orogeny and Marine Transgression in the Southern Central and Northern Patagonian Andes: Aftermath of a Large-scale Flat-subduction Event?").

In this region, two Cenozoic magmatic belts were formed, related to extension (Giacosa and Heredia 2004): i) the eastern Pilcaniyeu belt (60–42 Ma, Huitrera Formation) and ii) the western El Maitén belt (33–23 Ma, Ventana Formation) (Rapela et al. 1988) (see Chap."Paleogene Arc-Related Volcanism in the Southern Central Andes and North Patagonia (39°–41° S)"). The Paleocene to Eocene Pilcaniyeu magmatic belt is characterized by bimodal volcanic rocks from predominant rhyolitic ignimbrite facies to subordinate andesites and basalts (Rapela et al. 1988). To the east, the Oligocene–early Miocene El Maitén magmatic belt is characterized by andesites and dacitic ignimbrites with subordinated basaltic lava flows and rhyolitic ignimbrites (Rapela et al. 1988). Volcanic depocenters show wedge-like geometries recognized from field observations associated with normal faults (Orts et al. 2012). This extensional stage ends with a marine ingression in the early Miocene from both the Pacific and the Atlantic interpreted as the product of a sag phase associated with the Paleogene extension (see Chap. "The Late Oligocene–Early Miocene Marine Transgression of Patagonia"). At this stage, Oligocene to Miocene volcanic plateaus occupy broad sectors of the Patagonian foreland in two main units: the Somuncura plateau in the north and the Meseta Cuadrada plateau in the south (Fig. 4) (Kay et al. 2006).

The Late Cenozoic sedimentary infill is accommodated in depocenters on the eastern slope of the Andes. These sedimentary successions correspond to the early to mid Miocene Ñirihuau and the late Miocene Collón Curá Formations (Fig. 3). These Miocene deposits are characterized by volcanoclastic and fluvial successions that onlap the flanks of the Cenozoic structures being interpreted as synorogenic units (Ramos et al. 2011; Orts et al. 2012; Bilmes et al. 2013).

Pliocene–Pleistocene basalts unconformably cover the Miocene successions (Fig. 3) (Haller et al. 2009; Mena et al. 2006; Pecskay et al. 2007). Their chemistry indicates a within-plate setting that differs from previous Andean arc-magmatic rocks (Massaferro et al. 2006).

3 Cenozoic Tectonic Evolution of the Patagonian Andes at the Latitudes of the Triple Junction Point

The Traiguén basin infill is exposed through a N-trend, between the submerged Coastal Cordillera represented by Los Chonos Archipelago and the North Patagonian Andes (43°45′–47° S) (Fig. 4) (Hervé et al. 1995, 2001) (see Chap. "The Late Oligocene–Early Miocene Marine Transgression of Patagonia"). This unit consists of alkali-subduction related pillow basalts interbedded with sandstones, shales, and breccias with facies and benthic foraminifers indicative of a deep-marine environment (Hervé et al. 1995; Encinas et al. 2016). The volcanic rocks have been metamorphosed to prehnite-pumpellyite facies in the Traiguén Island and to greenschist and amphibolites facies, typical of ocean-floor-type metamorphism (Hervé et al. 1995). Detrital zircons from this unit dated by U/Pb yielded a maximum depositional age between 26 and 23 Ma (late Oligocene–early Miocene) (Encinas et al. 2015). The Traiguén basin has been related to an extensional regime that affected the plate margin from 43° to 47° S, in harmony with the regional stress field that affected a vast portion of the Andean crust between 33° and 43° S in late Oligocene to early Miocene times (Muñoz et al. 2000; Jordan et al. 2001).

The closure of the Traiguén basin took place during the intrusion of 20–15 Ma syndeformational granitoids (Fig. 4) (Hervé et al. 1995). Synorogenic deposits are present at the subsurface sector of the Rio Mayo Sub-basin and at surface exposed at the Meseta de Chalía in the eastern Andean topographic break (Fig. 4), correlative to the Santa Cruz and Río Frías Formations deposited in high-energy fluvial

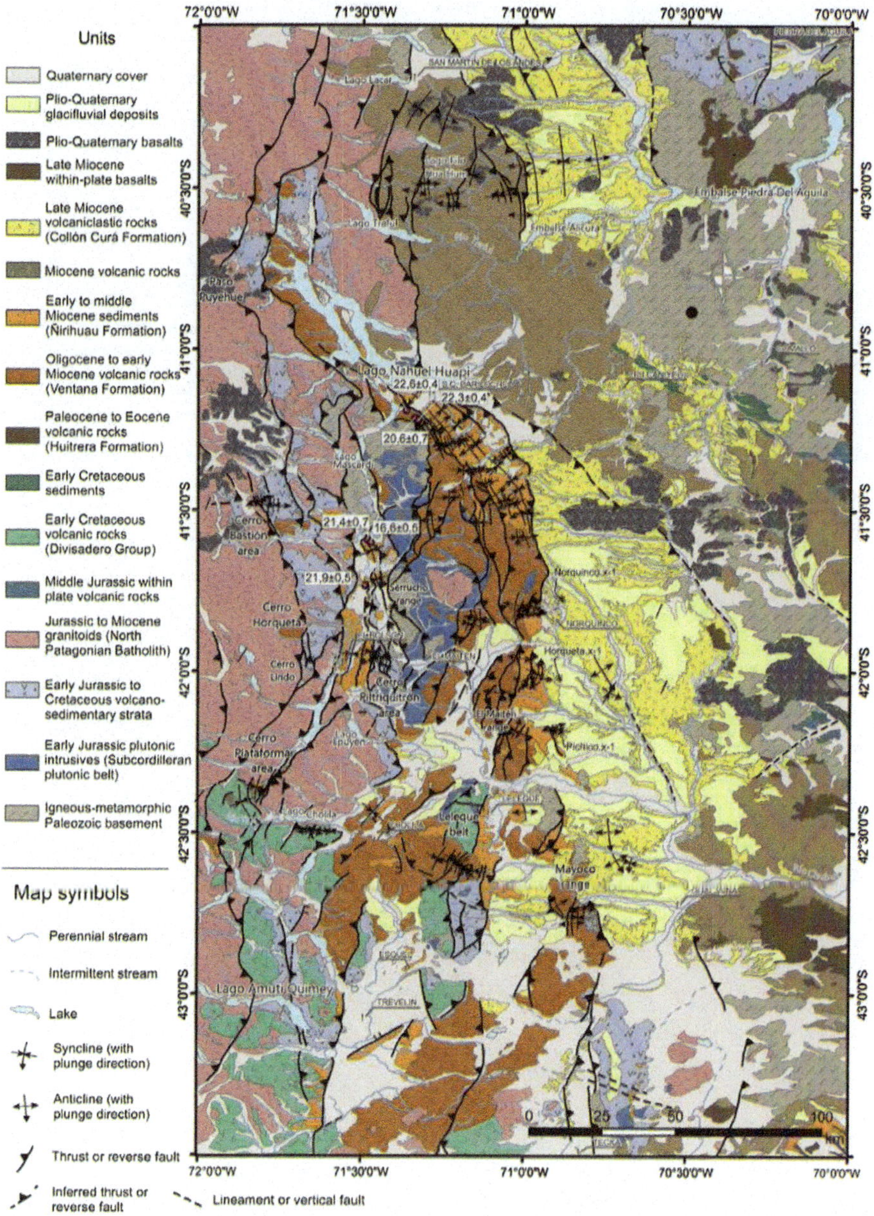

Fig. 3 Detail of the northern Patagonian Andes and western foreland basin (modified from Orts et al. 2015). Note that the foreland zone is characterized by the eruption of late Miocene to Quaternary within-plate volcanic rocks

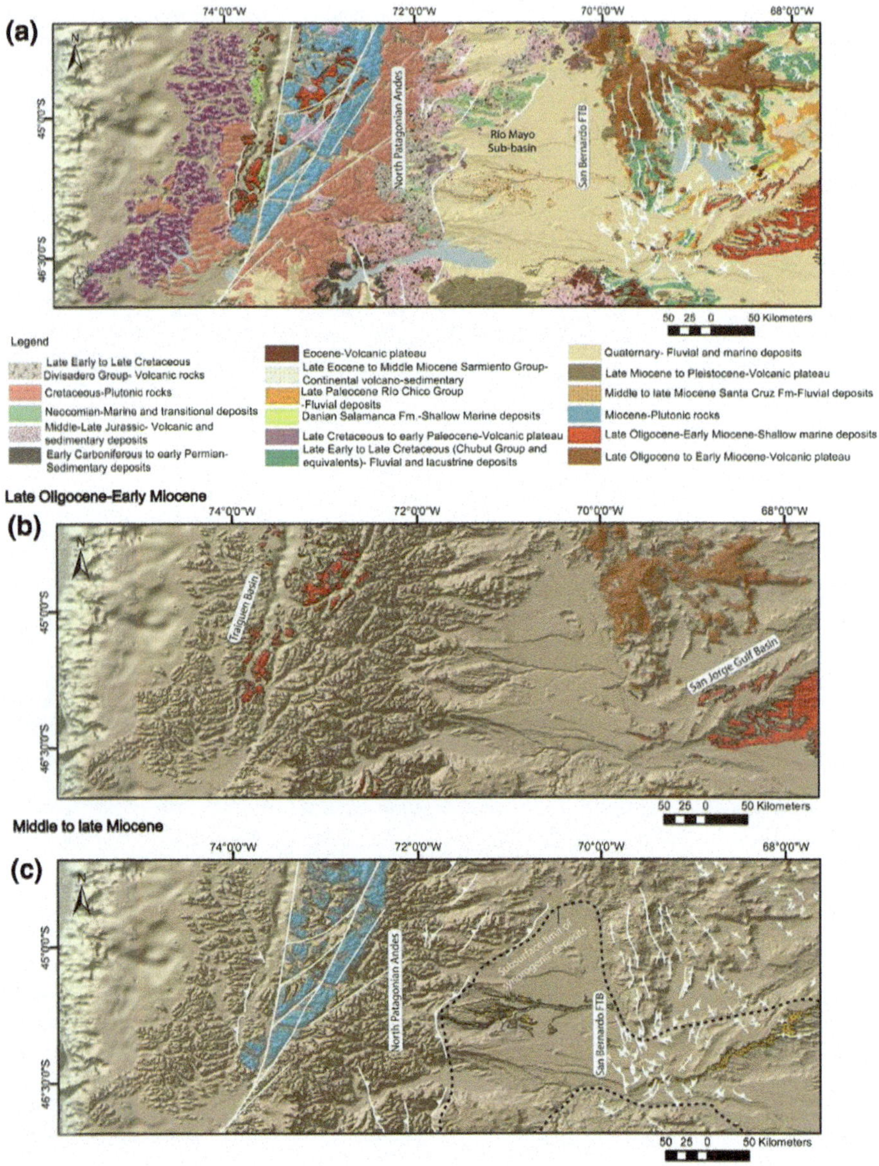

Fig. 4 **a** Geology of the southern analyzed zone in the northern Patagonian Andes at the latitudes of the triple junction between the Nazca, South America, and Antarctic plates. Note that Miocene plateau lavas are contractionally deformed at the foreland zone denoting the high amplitude of the Neogene fold and thrust belt at these latitudes. **b** Late Oligocene to early Miocene magmatic arc and sedimentary rocks corresponding to the Traiguén Basin and coetaneous retroarc volcanic products. **c** Miocene arc plutons in the present arc zone and coetaneous contractional deformation affecting the retroarc zone associated with synorogenic sedimentation. Taken from Gianni et al. (2015a)

systems dated between ~18 and 15 Ma (Marshall and Salinas 1990; Flynn et al. 2002). This time interval is expanded by Blisniuk et al. (2005) to between ~20 to 14 Ma based on Ar/Ar dating of intercalated tuffs in Santa Cruz Formation based on fission track data; Thomson (2002) identified a late Miocene–Pliocene period of enhanced cooling and denudation related to dextral transpression at the axial Andean sector initiated between ~16 and 10 Ma that roughly coincides with timing of Traiguén basin closure.

4 Crustal and Lithospheric Structure of the North Patagonian Andes from EGM2008 Data and Elastic Thicknesses

Crustal structure in the Chonos Archipelago at the westernmost part of the analyzed zone is poorly known. Seismic information is lacking as well as magnetotelluric surveys. However, gravity data and particularly models that combine the broad coverage of the satellite gravity surveys and the precision of terrestrial data such as the EGM2008 provide an unprecedented view of the density structure in the upper and lower crust. In this study, the vertical gravity gradient (Tzz) is calculated in terms of the spherical harmonic coefficients (Janak and Sprlak 2006) up to degree/order $N = 2159$ for the earth gravity field model EGM2008 (Pavlis et al. 2008, 2012), on a regular grid of 0.05° grid cell size. This allows obtaining a resolution of approximately $\lambda/2 \approx 2\pi R/Nmax \approx 9$ km with R being the mean Earth radius and Nmax the maximum degree and order of the harmonic expansion. The Tzz is obtained as the second derivative of the disturbing potential in the radial direction and is expressed in Eötvös (10^{-4} mGal/m). The Tzz represents a better theoretical resolution than the gravity vector for some geophysical features (Li 2001) and allows delineating the location of an anomalous mass with better detail and accuracy than the gravity anomaly itself (Braitenberg et al. 2011), particularly for shallower structures (Alvarez et al. 2012).

The potential generated by the topography was calculated from the digital elevation model (ETOPO1, Amante and Eakins 2009), using the software Tesseroids (Uieda et al. 2010), with 7,000 m as calculation height, to ensure that all values are above the topography. The Tzz was reduced for the gravity effect of topography by means of spherical prisms of constant density (see Grombein et al. 2013), to contemplate the Earth's curvature (Uieda et al. 2010). This calculation with spherical prisms is used since a planar approximation induces a considerable error (Braitenberg et al. 2011; Alvarez et al. 2013; Bouman et al. 2013). A standard density of 2.67 g cm^{-3} was used for masses above sea level and a density of 1.03 g cm^{-3} for the sea water.

The vertical gravity gradient delineated from the EGM2008 model shows a central area characterized by high values between two areas of low values, corresponding to the Chonos Archipiélago to the west and the mainland coastal zone to the east (Fig. 5). These areas are characterized by a Triassic metamorphic basement corresponding to the Chonos Archipiélago in the west and a suite of Mesozoic to

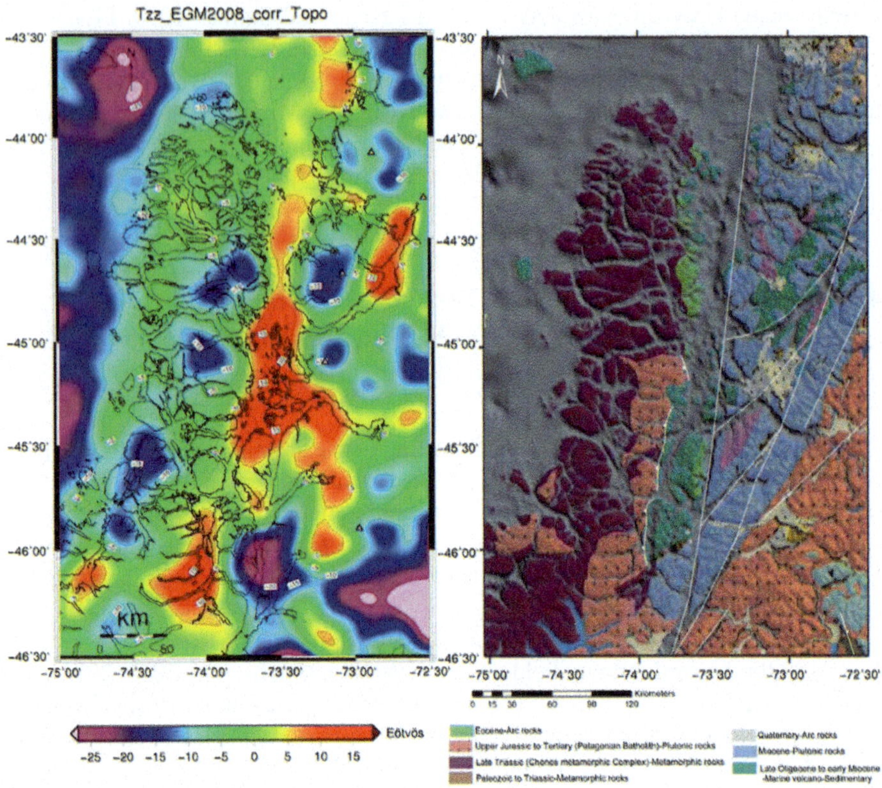

Fig. 5 Left) Vertical gravity gradient (Tzz) of the EGM2008 model constructed from an integration of terrestrial and GRACE satellite data in the arc and forearc zones in the Chonos Archipiélago. Note the stripe of high values comprehended between the Chonos Archipiélago to the west and the continent that corresponds to the exposures of the Traiguén Basin. This is interpreted as derived from mafic rocks that are the basement of the late Oligocene sedimentary successions (see Encinas et al. 2016). Right) Geological map of the arc and forearc region of the northern Patagonian Andes. The positive anomaly is interpreted as a Neogene quasi-ophiolite derived from the closure of the Traiguén intra-arc basin that led to the accretion to the parautochthonous Chonos Archipelago

Miocene plutons in the east that have lower gravity values than the rocks corresponding to the Traiguén and nearby islands. The high gravity values are explained by the presence of mafic rocks that crop out in this area. The shape of the anomaly elongated in a N-direction is interpreted as the product of basin closure during the early Miocene (~20–15 Ma) that produced a triangular zone delimited by an eastward thrust that uplifted Triassic rocks over the Traiguén basin infill and another with a west vergence that exhumed the Patagonian batholith (Figs. 5 and 6).

In order to study variable rheology through the study area, the elastic thickness (Te) was estimated based on the Bouguer anomaly. We calculated the flexural strength of the lithosphere using the convolution approach (Braitenberg et al. 2002; Wienecke 2006) and a newly derived analytical solution for the fourth order

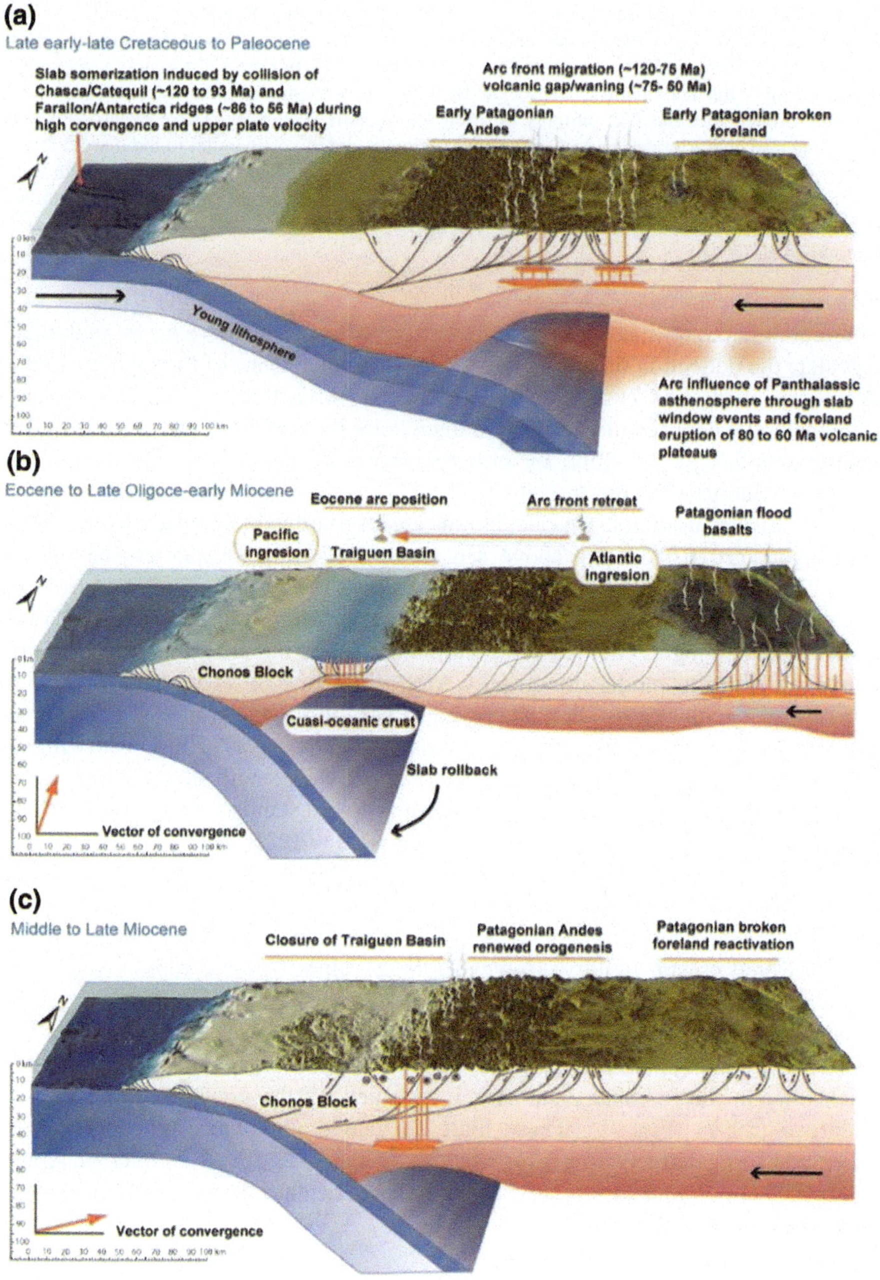
(a)
Late early-late Cretaceous to Paleocene
Slab somerization induced by collision of Chasca/Catequil (~120 to 93 Ma) and Farallon/Antarctica ridges (~86 to 56 Ma) during high corvengence and upper plate velocity
Arc front migration (~120-75 Ma) volcanic gap/waning (~75- 50 Ma)
Early Patagonian Andes
Early Patagonian broken foreland
Young lithosphere
Arc influence of Panthalassic asthenosphere through slab window events and foreland eruption of 80 to 60 Ma volcanic plateaus
(b)
Eocene to Late Oligoce-early Miocene
Eocene arc position
Arc front retreat
Patagonian flood basalts
Pacific ingresion
Traiguen Basin
Atlantic ingresion
Chonos Block
Cuasi-oceanic crust
Slab rollback
Vector of convergence
(c)
Middle to Late Miocene
Closure of Traiguen Basin
Patagonian Andes renewed orogenesis
Patagonian broken foreland reactivation
Chonos Block
Vector of convergence

◂**Fig. 6** Cartoons exemplifying the tectonic evolution of the southern analyzed area in the North Patagonian Andes: **a** Late early–late Cretaceous to Paleocene orogenic event that initially uplifted the North Patagonian Andes and broke the foreland area. **b** Late Oligocene extension and development of the Traiguén Basin associated with a highly attenuated-quasi-oceanic crust. To the east, the broken foreland was partially covered by volcanic sequences. These events took place under a period characterized by low upper plate velocity and almost parallel to the margin relative convergence. **c** Early Miocene accretion of the Chonos Archipiélago, in a context of high trench normal convergence and upper plate velocity, dated from postdeformational granitoids emplaced in the Traguén deformed basin filling (Hervé et al. 1995). Modified from Gianni et al. (2015a) and Encinas et al. (2016)

differential equation that describes the flexure of a thin plate (e.g., to analytically calculate the deflection of a thin plate for any irregular shape of the topography, see Wienecke et al. (2007) and references therein for a more detailed discussion). This method calculates the flexure parameters by the best fit of the observed crust–mantle interface (e.g., Moho by gravity inversion) and a crust–mantle interface computed due to a flexure model.

The densities used in these calculations are standard values already used in Rojas Vera et al. (2010) (density above sea level: 2.67 g/cm^3; upper crustal density: 2.7 g/cm^3; lower crustal density: 2.9 g/cm^3; upper mantle density: 3.3 g/cm^3). The Bouguer anomaly was inverted considering a normal crustal thickness of 35 km.

We obtained a gravimetric Moho undulation following the same steps described above. In order to model this gravity Moho in terms of an isostatic model, the effective elastic thickness (Te) is allowed to vary in the range of $1 < \text{Te} < 50$ km. The elastic thickness is constant over moving windows of 50 km × 50 km size. The window size was chosen by evaluating the wavelength of the main visible, geological structures. The flexure was calculated with standard values of the Young modulus (1011 N/m^2) and a normal Poisson ratio of 0.25.

Computed Te (elastic thickness) (Fig. 7) shows that areas of Pliocene–Quaternary within-plate magmatism are coincident with areas of low values. These can be explained by recent (less than 5 Ma) magmatic emplacement that would have lowered crustal rigidity. However, these values increase rapidly toward the foreland area where maximum Te is achieved. This maximum Te zone corresponds to the North Patagonian massif and its continuation to the north in the Neuquén platform area (Fig. 7). Here, higher Te values are coincident with lithospheric thickening in the foreland zone as it is revealed by the gravimetric modeling of Tašárová (2007) (Fig. 7b). These high Te values and higher lithospheric roots correspond to relatively fast Vp velocities in the tomographic model of Aragón et al. (2011) pointing to a relatively colder continental interior beneath the North Patagonian massif (Fig. 7c).

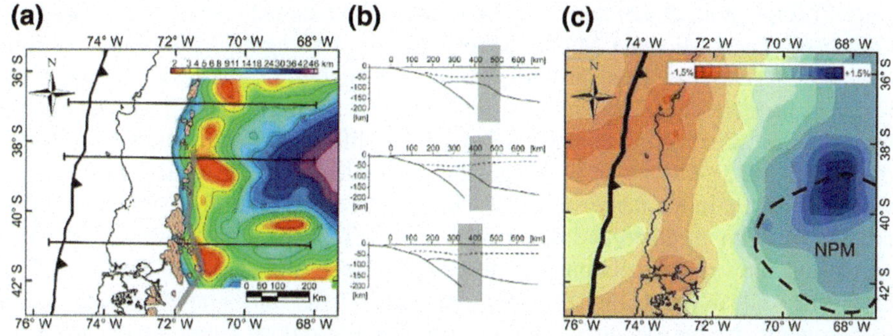

Fig. 7 **a** Elastic thicknesses computed from gravity data (see text for further details) of the North Patagonian Andes and the transition zone to the Southern Central Andes. Miocene arc rocks expanded over the eastern Andes are represented as a reference. **b** Gravity-modeled lithospheric cross-sections from Tašárová (2007) and transitional area from the narrow lithosphere beneath the arc to a thicker lithosphere in the foreland cratonic zone (thin dashed line is the Moho, while thin continuous line is the lithosphere/asthenosphere boundary). Note how the asthenospheric wedge thins toward the south, reaching its narrowest configuration near 42° S coinciding with the Miocene arc migration zone. **c** 300 km depth slice of the tomographic model obtained by Aragón et al. (2011). Note the relatively high Te values coincident with fast Vp velocities in the foreland area at the site of Miocene arc expansion

5 Tectonic Evolution of the Northernmost Patagonian Fold and Thrust Belt from Identification and Dating of Synorogenic Strata

Recent works focalized on the description and identification of synorogenic strata in the North Patagonian Andes provided a framework to analyze how deformation has progressed through time. A series of deformational stages can be separated affecting discrete zones of western North Patagonian Andes showing a complex picture in which contractional deformation would have progressed to the foreland zone in the early Miocene and then retracted at the time of extensional relaxation of the fold and thrust belt in late Miocene times.

5.1 First Orogenic Growth Stage (19–11 Ma): Uplift of the North Patagonian Cordillera, Initial Development of the Eastern Broken Foreland System, Out-of-Sequence Contractional Deformation

The oldest synorogenic sedimentation in the North Patagonian region is accumulated next to the axial Main Andes in the El Bolsón depocenter and other minor depocenters developed over the western Andean slope such as the Cerro Plataforma

(Fig. 8) (Orts et al. 2012) and over the Chilean Andean side (see Chap. "The Late Oligocene–Early Miocene Marine Transgression of Patagonia"). In these sections over the Argentinean side, progressive unconformities in marine to littoral sedimentary facies associated with a frontal syncline indicate that the Patagoniense transgression coexisted with an early Neogene uplift of the North Patagonian Andes (Orts et al. 2012). U/Pb ages in detrital zircons indicate that the axial Andean sector was constructed in the lapse ~19–17 Ma (Orts et al. 2012) with a source area first coming from the foreland and then progressively from the Main Andes in the west, which indicates disconnection with the source areas in the east at the time of an emergent Andean relief (Fig. 8) (Orts et al. 2012). After this, a rapid propagation of the orogenic wedge occurred, whose frontal easternmost sector associates with growth structures in the Gastre sector dated in 14.8 Ma (Ar/Ar; Bilmes et al. 2013) over the western edge of the North Patagonian Massif corresponding to an eastern broken foreland (Fig. 8).

Maximum U/Pb ages in detrital zircons associated with sedimentary sequences of the Ñirihuau Formation with progressive unconformities in the eastern flank of the El Maitén range have yielded 13.5–11.3 Ma, indicating that this sector, located in the western broken foreland (Precordillera), grew in an out-of-sequence order respect to the frontal sectors of the Gastre region (Ramos et al. 2011, 2015). Thus, a 1,600 m thick synorogenic depocenter is developed as a consequence of the exhumation of intermediate sectors of the North Patagonian fold and thrust belt, leaving to the east a broad sector structured in the previous early Miocene stages.

5.2 Extensional Reactivation of the Axial Zone of the North Patagonian Andes (14.8–4.6 Ma)

The axial Andean zone suffered an extensional relaxation since 14.8–7.3 Ma in the northernmost Patagonian Andes near ~42° S. This deformation is established from the age of synextensional volcanic successions with wedge-like geometries associated with normal faults that affected Mesozoic sequences, in Cerro Silvia (14.8 Ma U/Pb) and El Bastión (7.3–4.6 Ma) ranges near El Bolsón locality (Tobal et al. 2015). These Miocene ignimbrites and minor lava flows associated with caldera-type activity have a geochemical imprint that indicates direct extraction from the mantle with a common source with the arc front (Tobal et al. 2015). These synextensional volcanic activity indicates an extensional relaxation of discrete sectors of the axial zone of the North Patagonian fold and thrust belt in the 14.8–7.3 (4.6?) Ma interval (Fig. 9).

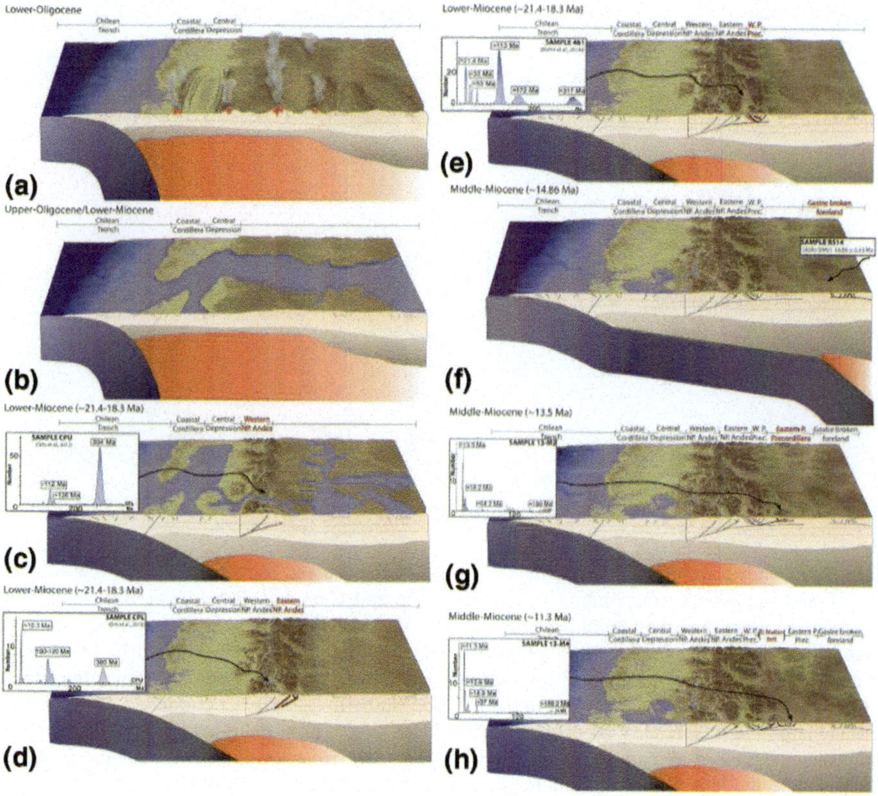

Fig. 8 Cartoons exemplifying the tectonic evolution of the northern analyzed area and source areas through time: **a** Eocene to early Miocene synrift stage; **b** Atlantic and Pacific connection during the early stages of the Patagoniense sedimentation in the early Miocene; **c** early uplift of the North Patagonian Andes in the late early Miocene with a marine foreland basin; **c** and **d** early synorogenic strata in the western fold and thrust belt that start to be fed from the foreland zone and then from the Andes (modified from Ramos et al. 2015, with data also taken from Orts et al. 2012 and Bechis et al. 2014); **e** rapid expansion of the northernmost Patagonian fold and thrust belt to the foreland zone at 14.8 Ma; **f**, **g** and **h** retraction of the orogenic front uplifting the eastern Precordillera between 13.5 and 11.3 Ma

5.3 Foreland Crustal Attenuation, Heat Flow, and Within-Plate Magmatism (Last 5 Ma) Between 41° and 44° S

Within-plate magmatism emplaced in the last 5 My in the foreland zone represented by monogenetic basaltic fields to the east of the Miocene orogenic front (Massaferro et al. 2006; Pécskay et al. 2007). This zone coincides with an area of lower crustal attenuation determined from inversion of gravity data and high heat flow determined from magnetic data (Fig. 9) (Orts et al. 2015). These volcanic

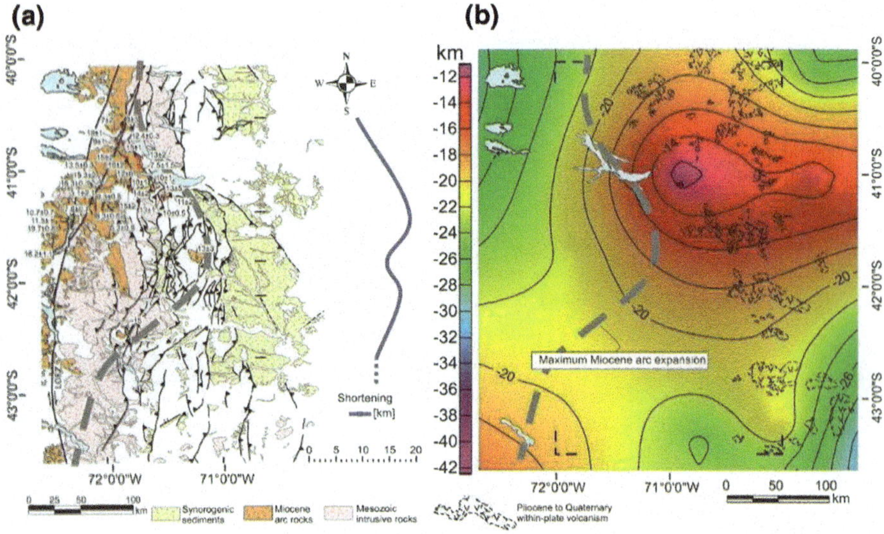

Fig. 9 **a** Main contractional structures developed in the arc and retroarc zones between 40° and 44° S and outcrops corresponding to the Mesozoic and Cenozoic Patagonian Batholith. Thick dashed line represents the line of maximum eastward expansion of the arc series during the mid to late Miocene. Note that maximum aerial development of the Neogene foreland basin coincides with the maximum expansion of the arc front and maximum shortening measured on retroarc structures (Orts et al. 2012, 2015). **b** Depth to Curie isotherm (~580 °C) computed from magnetic data (Maus et al. 2007; see Orts et al. 2015 for details). Note that shallower Curie isotherm, which indicates the area of maximum heat flow, coincides with retroarc within-plate eruptions and more locally with the latitudes of maximum expansion of the Miocene arc series

occurrences are interpreted as associated with lithospheric stretching affecting the foreland zone at the time of extensional relaxation of the highest Andes to the west (Orts et al. 2015).

6 Late Oligocene to Late Miocene Crustal Evolution from the Analysis of Arc-Related Rocks in the Transition Zone from the Central to Patagonian Andes

Oligocene to late Miocene arc-related volcanic rocks along the Andean margin, from 29° to 40° S, comprise different magmatic associations that evolved under variable compressive extensional regimes through the crust. Contractional tectonics could eventually be associated with increase of crustal thickness within the context of these arc-related products evolution. Thus, geochemical signature of the volcanic rocks, particularly trace elements behavior, could reflect fractionated residual mineral assemblages, from pyroxene to amphibole to garnet, as evidence from an increase in pressure conditions through magmas evolution, usually related to deeper crustal storage (Fig. 8) (e.g., Kay et al. 1991; Bissig et al. 2003; Litvak et al. 2007).

Selected arc-related suites from Southern Central to North Patagonian Andes show evidence for variable pressure conditions through magmas evolution since Oligocene to late Miocene–early Pliocene times, as reflected by their trace elements signature (Fig. 10). From north to south, representative Oligocene to early Miocene arc volcanic rocks include Doña Ana Group cropping out in the high Andes Cordillera over the present Pampean flat-segment (~29°–30° S), the Cura Mallín Formation magmatism nearby the Payenia shallow subduction segment (~36°–38° S) and the Ventana Formation (El Maitén Belt) volcanic rocks in the North Patagonian Andes (~40°–42° SL) (Kay et al. 1991; Bissig et al. 2003; Litvak et al. 2007; Utgé et al. 2009; Kay et al. 2006, 2007; Aragón et al. 2011) (see Chaps. "Paleogene Arc-Related Volcanism in the Southern Central Andes and North Patagonia (39°–41° S) and The Late Oligocene–Early Miocene Marine Transgression of Patagonia"). Younger exposures from Middle to Late Miocene age in the Chilean–Pampean flat slab correspond mainly to andesitic volcanism from Cerro de las Tórtolas Formation and pyroclastic deposits of Vacas Heladas Ignimbrites (Kay et al. 1991; Litvak et al. 2007; Bissig et al. 2003) (see Chap. "The Late Paleogene to Neogene Volcanic Arc in the Southern Central Andes (28°–37° S)"), while the easternmost expressions of arc-related volcanism associated with the Payenia shallow subduction zone comprises the Middle Miocene to Pliocene volcanic rocks from the present far retroarc zone over San Rafael Block (Litvak et al. 2015). Middle to Late Miocene Cerros Bastión and Silvia sequences constitute arc-related products on the North Patagonian Andes (Tobal et al. 2015).

Overall, Oligocene to early Miocene selected volcanic associations have a typical pattern characterized by low Sm/Yb ratios, which reflect the fractionation of pyroxene and amphibole as residual mineral assemblages all through the Central to Patagonian Andes (Fig. 10). These magmas would have equilibrated under similar conditions, in a relatively attenuated to normal crust (30–35 km), through the entire Oligocene to early Miocene times associated with the prevailing extensional setting (Kay et al. 1991, 2006; Bissig et al. 2003; Litvak et al. 2007; Jones et al. 2016; Utgé et al. 2009; Ramos et al. 2015).

Middle to late Miocene volcanic rocks, however, show a more variable behavior. Arc-related volcanism associated with the shallowing of the Payenia segment (36–38° S) shows low-pressure residual mineral assemblages, while arc rocks from the Pampean flat-slab to the north (27–33° S) show a distinguished increase in Sm/Yb ratios, reflecting garnet as a residual mineral assemblage in magmas equilibration site within the crust (Kay et al. 1991, 2006; Bissig et al. 2003; Litvak et al. 2007; 2015) (see Chap. "The Late Paleogene to Neogene Volcanic Arc in the Southern Central Andes (28°–37° S)"). Compressive tectonics that lead to the shallowing of both Pampean and Payenia slab segments have remarkable differences since in the

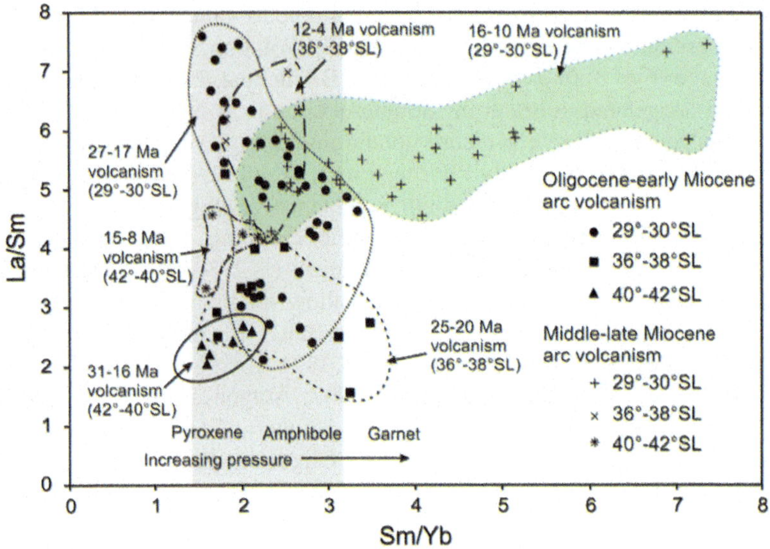

Fig. 10 Light rare earth elements (La/Sm) vs. heavy rare elements (Sm/Yb) for Oligocene–Miocene Andean arc volcanism: data from 29° to 30° are from Kay et al. (1987, 1991), Bissig et al. (2003) and Litvak et al. (2007); data from 36° to 38° belong to Kay et al. (2006), Utgé et al. (2009), and Litvak et al. (2015); values from 40° to 42° correspond to Kay et al. (2007) and Aragón et al. (2011). Note that Oligocene–early Miocene North Patagonian Andes (40°–42°) magmas evolved in equilibrium with low-pressure residual mineral assemblages, as well as Oligocene–early Miocene arc magmas in the southern Central Andes. Middle to Late Miocene/early Pliocene rocks of the Southern Central Andes (36°–38°) and North Patagonian ones also show low Sm/Yb ratios, in contrast with coetaneous magmas of the Pampean flat-slab, with garnet as residual mineral phase

northern segment, an increase in contractional deformation resulted in an increase of crustal thickness from ~30 to 60 km. In the particular case of the North Patagonian Andes, Middle to Upper Miocene arc-related volcanic rocks show Sm/Yb ratios within the range of the coetaneous Payenia shallow subduction volcanic rocks (Fig. 10). Thus, Patagonian magmatism would have also evolved with low-pressure residual mineral assemblages within a context of normal to attenuated crust, in contrast to the Pampean flat-slab magmas, indicating that Miocene contraction did not lead to substantial crustal thickening. However, compositional changes are registered in Miocene times with an increase in explosive facies when comparing Bastión and Silvia sequences with the Serie Andesítica (El Maitén Belt) products (Tobal et al. 2015).

7 A Brief Discussion About the Deformational Phases that Affected the North Patagonian Andes in the Last 20 My

Through the precedent sections we have described an evolutionary picture that characterizes the North Patagonian Andes in which: i) the orogenic wedge by 20–19 Ma comprehended the structure exhumed on the western Chilean slope, associated with a submarine and narrow foreland basin whose remnants lie cannibalized in the axial Andean sector; ii) a strong expansion of the orogenic wedge at 15 Ma occurred when the foreland fragmented into a series of intermontane depocenters reaching an amplitude close to 200 km; iii) an out-of-sequence growth stage at 13.5–11.3 Ma that was associated with the structuration of the western broken foreland (Precordillera of Chubut); iv) an extensional relaxation of the axial parts of the North Patagonian Andes that started more or less synchronously at 14.8 Ma and lasted until 7.3 Ma and probably 4.6 Ma; v) late propagation of the extensional deformation toward the foreland zone controlled the emplacement of Pliocene to Quaternary monogenetic basaltic fields.

During this time Rapela et al. (1987) and Rapela and Kay (1988) described an eastward expansion of Miocene magmatism over the Argentinean Andean slope, represented by isolated plutons intruding the Mesozoic batholiths and the Paleozoic basement. Shortenings calculated from construction of balanced cross-sections (Orts et al. 2015) show that maximum values coincide with the latitudes of Miocene arc expansion (Fig. 9). Additionally, the broadest synorogenic depocenters associated with the Ñirihuau and Collón Curá Formations were developed at the site of magmatic expansion. Radiometric data indicate that arc expansion occurred approximately at 19–14 Ma from the western Andean slope to the Eastern Andean foothills. After this expansión, magmatism reinstalled on the western Andean slope by 13 Ma (Orts et al. 2015). This scheme coincides with the deformational stages described for this sector: i) while the arc expanded between 19 and 14 Ma over the eastern Andean slope, the orogenic wedge broadened fragmenting the eastern foreland area; ii) when the arc retracted at 13 Ma, the High Andes relaxed extensionally (Tobal et al. 2015); iii) finally, the foreland zone was extensionally attenuated controlling the emplacement of mantle-derived products associated with decompression and high heat flows at the area of maximum arc expansion (Orts et al. 2015).

This stage of Miocene arc expansion occurred after a major plate reorganization in which the Farallón plate splits into two plates and the relative convergence between the resulting approaching plates became more orthogonal to the margin (Candie and Leslie 1986; Pardo-Casas and Molnar 1987; Somoza and Ghidella 2005) (Fig. 11). This change is also associated with an increase in the relative convergence rate that explains the fold and thrust belt development and foreland basin formation 6 million years latter described in this Chapter. During the 33–21 Ma time interval, volcanic and sedimentary sections are characterized by wedge-like geometries (Ventana and lower Ñirihuau Formations) depicted in

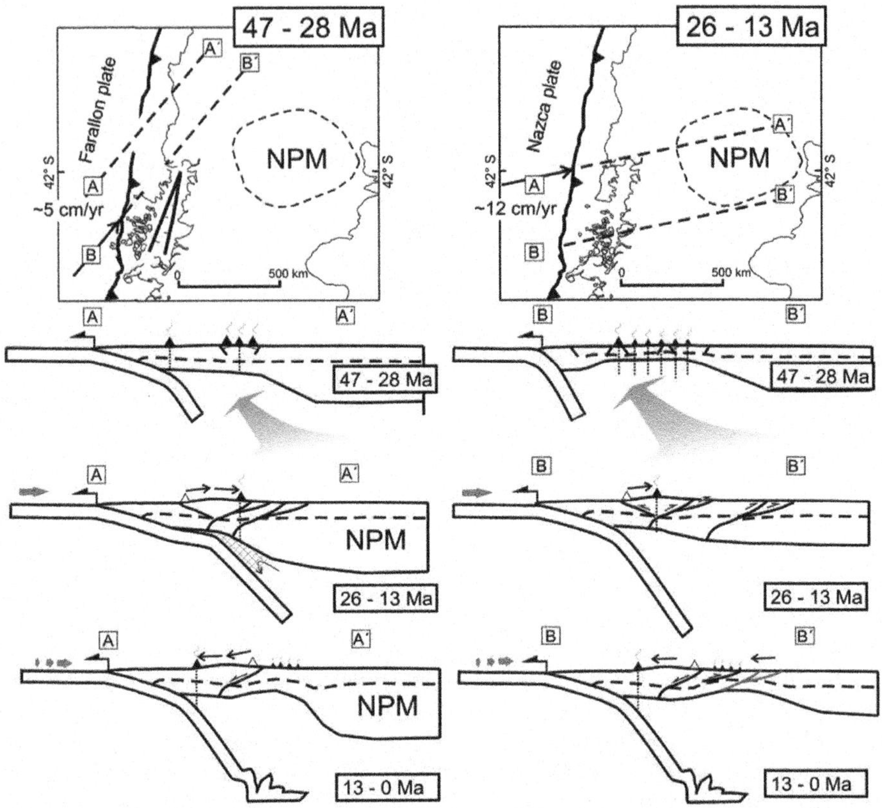

Fig. 11 Schematic cartoons showing two key stages during interaction between converging plates at the Chilean trench, before and after the major plate reorganization event that occurred between 28 and 26 Ma (Cande and Leslie 1986). After 26 Ma, a more frontal convergence between the Nazca and South American plates, after a period of high obliquity, coincides with a foreland expansion of the Miocene arc rocks at the time of fold and thrust belt development in the northern analyzed area (A-A′), although to the south arc expansion is limited and accretional processes derived from closure of the Traiguén basin dominate (B-B′). Note that this stage is coincident with the formation of the Ñirihuau and San Bernardo fold and thrust belts stacked over the western and southern North Patagonian massif (NPM), respectively. Convergence velocities at 42° S are taken from Somoza and Ghidella (2005)

seismic lines and in the field showing the existence of an extensional setting (Orts et al. 2015) (see Chap. "The Late Oligocene–Early Miocene Marine Transgression of Patagonia"). This stage coincides with a highly oblique convergence setting between the Farallón and South American plates (Fig. 11a).

After the 26 Ma plate reorganization stage (see Chap. "Paleogene Arc-Related Volcanism in the Southern Central Andes and North Patagonia (39°–41° S)"), the arc expanded toward the foreland area at the site of orogenic construction and lithospheric flexion associated with orogenic loading. At this site in the last 5 Ma, crustal stretching, as inferred from gravity inversion (Fig. 6a), and associated within-plate basalts indicate a drastic change in the tectonic conditions. This situation in which arc migration (19–14 Ma) coincides with the area of maximum orogenic shortening associated with foreland basin development (Fig. 11), followed by arc retraction since 13 Ma and injection of within-plate volcanic rocks in the last 5 My, could be explained by changes in the geometry of the subducted slab (Fig. 11). It follows that, as proposed previously in Orts et al. (2012), Miocene arc expansion could be connected to a shallow subduction setting that explains the differential shortening and synorogenic sedimentation magnitudes achieved in the retroarc area.

Shallow to flat subduction settings have been associated with a variety of processes associated with the subduction of anomalously buoyant oceanic crust related to aseismic oceanic ridges and plateaus originated in hot spots (see Gutscher et al. 2000, for a detailed revision) and the approximation of mid-ocean ridges (Ramírez de Arellano et al. 2012). However, more recently, new hypotheses have addressed the characteristics of the upper plate, particularly the existence of cold thick lithospheres that could act as barriers to the trenchward asthenospheric influx imposed by the rollback trench velocity (Manea et al. 2012) (see Chap. "Tectonic Rotations Along the Western Central Andes"). In Manea et al. (2012) they hypothesize that thick continental lithospheres in cold cratonic blocks could produce a low-pressure setting in the asthenospheric wedge that could allow the flattening of the subducted oceanic lithosphere when the upper plate advances toward the trench. Recently, O'Driscoll et al. (2012) have shown for the Central Andes that shallow subduction configurations were favored when subduction directed toward a close to the margin ancient craton. Particularly, these authors have explained the Eocene shallow subduction configuration in the Andes of southern Perú and Bolivia (James and Sacks 1999; Kay and Coira 2009) as produced by the collision of an oceanic plateau when subduction was directed toward the NE beneath the Amazonian craton.

Based on this, an analogous situation could be proposed for the North Patagonian segment in order to explain the described magmatic and basin development in the last ~20 Ma. Shallow subduction could be responsible for the registered magmatic expansion during 19–14 Ma when relatively high and normal to the trench convergence between Nazca and South American plates was achieved (Fig. 11b) (Somoza and Ghidella 2005) at the time of increased upper plate westward motion (Silver et al. 1998).

The proximity of a thick cratonic lithosphere to the South American plate margin represented by the North Patagonian massif would have favored subducted slab suction after 26 Ma when convergence rotated to the east. This situation requires deep upper plate lithospheres associated with low heat fluxes that are revealed at these latitudes by lithospheric density models, P-wave tomographies, and Te determinations (Fig. 11; Tašárová 2007, Aragón et al. 2011, and this work).

Thus, contractional deformation and synorogenic basin development in the retroarc area, contemporaneous to arc shifting, could be linked to a shallow subduction configuration favored by the deep roots of the North Patagonian massif following Manea et al.'s (2013) general hypothesis (Fig. 11).

This tectonic setting differs from the present subduction configuration, where the angle of penetration of the oceanic slab is around 30° E (Lange et al. 2007). Since late Miocene to Quaternary times, the arc retracted to its present position in the Cordilleran axis (Lara et al. 2001) probably reflecting a change (steepening) in the slab geometry. During this time within-plate volcanism emplaced in the retroarc area probably at the site of crustal stretching. A decrease in plate convergence since late Miocene times (Müller et al. 2008) could explain this steepening revealed by arc retraction after the proposed shallow subduction event.

8 Conclusions

The North Patagonian Andes are characterized by a strong propagation of the contractional deformation in the past 15 Ma into the foreland area. However, this event involved contrasting processes: on one hand, the closure of a back-arc basin in the south at the latitudes of the triple junction point and on the other hand a shallow subduction regime in the north. This shallow subduction could have been facilitated by the presence of the North Patagonian Massif at these latitudes in the foreland zone. Both processes have not thickened the crust considerably, although produced compositional variations in the arc-related magmas at the time when the deformational front advanced associated with synorogenic sedimentation. These compositional changes from dominant lava facies in late Oligocene–early Miocene times to caldera-type explosive facies in the mid to late Miocene are potentially linked to foreland underthrusting and forearc tectonic erosion during shallow subduction and Chonos block accretion that could have modified the asthenospheric wedge adding siliceous low-temperature melting materials. Mild crustal thickening and rapid propagation of the deformational front have occurred when plate convergence became more orthogonal after a period of high oblique convergence. After mild crustal thickening, extension relaxed discrete zones of the North Patagonian orogenic belt controlling within-plate and arc-related magmatism that emplaced at the time of retraction of the orogenic front and out-of-sequence thrusting development.

Acknowledgements We acknowledge PICT-2012-1490, PIP 2015–2017, 2015–2017. UBACYT, and Proyectos Fondecyt 11080115, 1110914, 1151146, for funding.

References

Alvarez O, Gimenez ME, Braitenberg C, Folguera A (2012) GOCE satellite derived gravity and gravity gradient corrected for topographic Effect in the South Central Andes region. Geophys J Int 190:941–959. https://doi.org/10.1111/j.1365-246X.2012.05556.x

Alvarez O, Gimenez ME, Braitenberg C (2013) Nueva metodología para el cálculo del efecto topográfico para la corrección de datos satelitales. Rev Asoc Geol Argent 70:422–429

Amante C, Eakins BW (2009) ETOPO1 1 Arc-Minute Global Relief Model: Procedures, Data Sources and Analysis. NOAA Technical Memorandum NESDIS NGDC-24. National Geophysical Data Center. Biblioteca Digital ILCE 1, p 19

Aragón E, Castro A, Díaz-Alvarado J, Liu DY (2011) The North Patagonian batholith at Paso Puyehue (Argentina-Chile). SHRIMP ages and compositional features. J South Am Earth Sci 32:547–554

Bechis F, Encinas A, Concheyro A, Litvak V, Aguirre-Urreta B, Ramos VA (2014) New age constraints for the Cenozoic marine transgressions of northwestern Patagonia, Argentina (41°–43° S): Paleogeographic and tectonic implications. J South Am Earth Sci 52:72–93

Bilmes A, D'Elia L, Franzese J, Veiga G, Hernández M (2013) Miocene block uplift and basin formation in the Patagonian foreland: The Gastre Basin, Argentina. Tectonophysics 601: 98–111

Bissig T, Clark AH, Lee JKW, von Quadt, (2003) Petrogenetic and metallogenetic responses to miocene slab flattenig: New constrains from the El Indio-Pascua Au-Ag-Cu Belt, Chile/ Argentina. Mineralium Deposita 38: 844–862

Cande SC, Leslie RB (1986) Late Cenozoic tectonics of the Southern Chile Trench. J Geophys Res 91:471–496

Bouman J, Ebbing J, Fuchs M (2013) Reference frame transformation of satellite gravity gradients and topographic mass reduction. J Geophys Res: Solid Earth 118:759–774. https://doi.org/10.1029/2012JB009747

Braitenberg C, Ebbing J, Gotze H (2002) Inverse modelling of elastic thickness by convolution method the Eastern Alps as a case example. Earth Planet Sci Lett 202:387–404

Braitenberg C, Mariani P, Ebbing J, Sprlak M (2011) The enigmatic Chad lineament revisited with global gravity and gravity gradient fields. In: van Hinsbergen DJJ, Buiter S, Torsvik TH, Gaina C, Webb S (eds) The Formation and Evolution of Africa: A Synopsis of 3.8 Ga of Earth History. Geological Society of London 357, London, pp 329–343

Cembrano J, Hervé F, Lavenu A (1996) The Liquiñe Ofqui fault zone: a long-lived intra-arc fault system in southern Chile. Tectonophysics 259:55–66

Echaurren A, Folguera A, Gianni G, Orts D, Tassara A, Encinas A, Giménez M, Valencia V (2016) Tectonic evolution of the North Patagonian Andes (41°–44° S) through recognition of syntectonic strata. Tectonophysics 677:99–114. https://doi.org/10.1016/j.tecto.2016.04.009

Encinas A, Folguera A, Oliveros V, del Mauro L, Tapia F, Riffo R, Hervé F, Finger K, Valencia V, Gianni G, Álvarez O (2015) Late Oligocene-Early Miocene submarine volcanism and deep-marine sedimentation in an extensional basin of southern Chile. Implications on the tectonic development of the North Patagonian Andes. Geol Soc Am Bulletin 128:807–823

Espinoza F, Morata D, Polve M, Lagabrielle Y, Maury R, De la Rupelle A, Guivel C, Cotten J, Bellon H, Suarez M (2010) Volcanismo calcoalcalino durante el Mioceno Medio en Patogonia Central (47° S): petrogenesis e implicaciones en la dinamica de placas. Andean Geol 37: 300–328

Feruglio E (1947) Descripción Geológica de la Hoja 40b, San Carlos de Bariloche, Río Negro, Carta Geológico-Económica de la República Argentina, escala 1:200,000. Dirección Nacional de Geología y Minería, Buenos Aires, Argentina

Flynn J, Novacek M, Dodson H, Frassinetti D, Mc Kenna M, Norell M, Sears K, Swisher C, Wuss A (2002) A new fossil mammal assemblage from the southern Chilean Andes: implications for geology, geochronology, and tectonics. J South Am Earth Sci 15:285–302

García-Sansegundo J, Farias P, Gallastegui G, Giacosa RE, Heredia N (2009) Structure and metamorphism of the Gondwanan basement in the Bariloche region (North Patagonian Argentine Andes). Int J Earth Sci 98:1599–1608

Giacosa R, Heredia N (2004) Structure of the North Patagonian thick-skinned fold-and-thrust belt, southern central Andes, Argentina (41°–42° S). J South Am Earth Sci 18:61–72

Gianni G, Navarrete C, Orts D, Tobal J, Folguera A, Giménez M (2015a) Patagonian broken foreland and related impactogene: The origin of the Chubut Basin. Tectonophysics 649:81–99

Gianni G, Folguera A, Navarrete C, Encinas A, Echaurren A (2015b) The North Patagonian orogen: Meso-Cenozoic evolution from the Andes to the foreland area. In: Folguera A, Naipauer M, Sagripanti L, Ghiglione M, Orts D, Giambiagi L (eds) Growth of the Southern Andes. Springer International Publishing, pp 173–200

Gordon A, Ort MH (1993) Edad y correlación del plutonismo subcordillerano en la provincias de Río Negro y Chubut (41°–42°30′ L.S.). In: Ramos VA (ed) Geología y Recursos Naturales de la Provincia de Mendoza. Mendoza, Relatorio del XII Congreso Geológico Argentino, pp 120–127

Grombein T, Heck B, Seitz K (2013) Optimized formulas for the gravitational field of a tesseroid. J Geod 87:645–600

Gutscher M-A, Spakman W, Bijwaard H, Engdahl ER (2000) Geodynamics of flat subduction: Seismicity and tomographic constraints from the Andean margin. Tectonics 19. https://doi.org/10.1029/1999TC001152

Haller M, Pécskay Z, Németh K, Gméling K, Massaferro GI, Meister CM, Nullo FE (2009) Preliminary K-Ar geochronology of Neogene back arc volcanism in Northern Patagonia, Argentina. III International Maar Conference. IAVCEI, Malargüe, pp 40–41

Hervé F (1998) Late Triassic rocks in the subduction complex of Aysén, southern Chile. Event Stratigraphy of Gondwana: J Afr Earth Sci 10:224

Hervé F, Fanning CM (2001) Late Triassic detrital zircons in meta-turbidites of the Chonos Metamorphic Complex, southern Chile. Rev Geol Chile 28:91–104

Hervé F, Pankhurst RJ, Drake R, Beck ME (1995) Pillow metabasalts in a mid-Tertiary extensional basin adjacent to the Liquiñe-Ofqui fault zone: the Isla Magdalena area, Aysén, Chile. J South Am Earth Sci 8:33–46

Hervé F, Sanhueza A, Silva C, Pankhurst RJ, Fanning MC, Campbell H, Crundwell M (2001) A Neogene age for Traiguén Formation, Aysén, Chile, as revealed by shrimp U-Pb dating of detrital zircons. In: III Simposio Sudamericano de Geología Isotópica, Pucón, pp 570–574

James D, Sacks IE (1999) Cenozoic formation of the Central Andes: a geophysical perspective. In: Skinner BJ (ed) Geology and Ore Deposits of the Central Andes. Soc Econ Geol: 1–26

Janak J, Sprlak M (2006) New Software for Gravity Field Modelling Using Spherical Armonic. Geodetic and Cartographic Horizon 52:1–8

Jordan T, Burns W, Veiga R, Pángaro F, Copeland P, Kelley S, Mpodozis C (2001) Extension and basin formation in the Southern Andes caused by increased convergence rate: Amid-Cenozoic trigger for the Andes. Tectonics 20:308–324

Kay SM, Coira BL (2009) Shallowing and steepening subduction zones, continental lithospheric loss, magmatism, and crustal flow under the Central Andean Altiplano-Puna Plateau. In: Kay SM, Ramos VA, Dickinson WR (eds) Backbone of the Americas: Shallow Subduction, Plateau Uplift, and Ridge and Terrane Collision. Geol Soc Am Memoirs 204, Boulder, pp 229–259

Kay SM, Maksaev VA, Moscoso R, Mpodozis C, Nasi C (1987) Probing the evolving Andean lithosphere: Mid-Late Tertiary Magmatism in Chile (29°–30°30′) over the modern zone of subhorizontal subduction. J Geophys Res 92:6173–6189

Kay SM, Mpodozis C, Ramos VA, Munizaga F (1991) Magma source variations for mid late Tertiary magmatic rocks associated with shallowing zone and thickening crust in the central Andes (28° to 33° S). In: Harmon RS, Rapela CW (eds) Andean magmatism and its tectonic setting. Geological Society of America, Special Paper 265: 113–137

Kay SM, Mpodozis C, Coira B (1999) Neogene magmatism, tectonism and mineral deposits of the Central Andes (22º–23º S Latitude). In: Skinner BJ (ed) Geology and Ore Deposits of the Central Andes. Society of Economic Geologists, Special Publication 7, pp 27–59

Kay SM, Burns M, Copeland P (2006) Upper Cretaceous to Holocene Magmatism over the Neuquén basin: Evidence for transient shallowing of the subduction zone under the Neuquén Andes (36° S to 38° S latitude). In: Kay SM, Ramos VA (eds) Late Cretaceous to Recent magmatism and tectonism of the Southern Andean margin at the latitude of the Neuquen basin (36–39º S). Geological Society of America, Special Paper 407, pp 19–60

Kay SM, Ardolino AA, Gorring ML, Ramos VA (2007) The Somuncura Large Igneous Province in Patagonia: Interaction of transcient mantle thermal anomaly with a subduction slab. J Petrol 48:43–77

Lange D, Rietbrock A, Haberland C, Bataille K, Dahm T, Tilmann F, Flüh ER (2007) Seismicity and geometry of the south Chilean subduction zone (41.5° S–43.5° S): Implications for controlling parameters. Geophys Res Lett 34:L06311. https://doi.org/10.1029/2006GL029190

Lara L, Rodriguez C, Moreno H, Pérez de Arce C (2001) Geocronología K-Ar y geoquímica del volcanismo plioceno superior-pleistoceno de los Andes del sur (39–42° S). Rev Geol Chile 28:67–90

Lavenu A, Cembrano J (1999) Compressional- and transpressional-stress pattern for Pliocene and Quaternary brittle deformation in fore arc and intra-arc zones (Andes of Central and Southern Chile). J Struct Geol 21:1669–1691

Li X (2001) Vertical resolution: gravity versus vertical gravity gradient. Lead Edge 20:901–904

Litvak VD, Poma S, Kay SM (2007) Paleogene and Neogene magmatism in the Valle del Cura region: a new perspective on the evolution of the Pampean flat slab, San Juan province, Argentina. J South Am Earth Sci 24:117–137

Litvak VD, Spagnuolo MG, Folguera A, Poma S, Jones R, Ramos VA (2015) Late Cenozoic calc-alkaline volcanism over the Payenia shallow subduction zone, South-Central Andean back-arc (34°30–3'7S), Argentina. J South Am Earth Sci 64:365–380

Manea VC, Pérez-Gussinyé M, Manea M (2012) Chilean flat slab subduction controlled by overriding plate thickness and trench rollback. Geology 40:35–38. https://doi.org/10.1130/G32543.1

Marshall L, Salinas P (1990) Stratigraphy of the Río Frías Formation (Miocene), along the Alto Río Cisnes, Aisén, Chile. Rev Geol Chile 17:57–87

Marshall LG, Pascual R, Curtis GH, Drake RE (1977) South american geochronology: radiometric time scale for middle to late tertiary mammal-bearing horizons in patagonia. Science 195:1325–8. https://doi.org/10.1126/science.195.4284.1325

Massaferro G, Haller M, D'Orazio M, Alric V (2006) Sub-recent volcanism in Northern Patagonia: A tectonomagmatic approach. J Volcanol Geoth Res 155:227–243

Maus S, Sazonova T, Hemant K, Fairhead JD, Ravat D (2007) Na-tional Geophysical Data Center candidate for the World Digital Magnetic Anomaly Map. Geochem Geophys Geosyst 8: Q06017. https://doi.org/10.1029/2007GC001643

Mazzoni M, Benvenuto A (1990) Radiometric ages of Tertiary ignimbrites and the Collón Curá Formation northwestern Patagonia. In: XI Congreso Geológico Argentino, San Juan, pp 87–90

Mena M, Ré GH, Haller MJ, Singer SE, Vilas JF (2006) Paleomagnetism of the late Cenozoic basalts from northern Patagonia. Earth Planets Sp 58:1273–1281

Müller RD, Sdrolias M, Gaina C, Roest WR (2008) Age, spreading rates, and spreading asymmetry of the world's ocean crust. Geochem Geophys Geosyst 9:1–19. https://doi.org/10.1029/2007GC001743

Muñoz J, Troncoso R, Duhart P, Crignola P, Farmer L, Stern CR (2000) The relation of the mid-Tertiary coastal magmatic belt in south-central Chile to the late Oligocene increase in plate convergence rate. Rev Geol Chile 27:177–203

Orts D, Folguera A, Encinas A, Ramos M, Tobal J, Ramos VA (2012) Tectonic development of the North Patagonian Andes and their related Miocene foreland basin (41°30′–43° S). Tectonics 31: TC3012. https://doi.org/10.1029/2011TC003084

Orts D, Folguera A, Gimenez M, Ruiz F, Rojas Vera E, Lince Klinger F (2015) Cenozoic building and deformational processes in the North Patagonian Andes. J Geodyn 86:26–41

O'Driscoll LJ, Richards MA, Humphreys ED (2012) Nazca-South America interactions and the late Eocene-late Oligocene flat-slab episode in the central Andes. Tectonics 31:TC2013. https://doi.org/10.1029/2011TC003036

Pankhurst RJ, Weaver SD, Hervé F, Larrondo P (1999) Mesozoic-Cenozoic evolution of the North Patagonian Batholith in Aysen, southern Chile. J Geol Soc London 156:673–694

Pardo-Casas F, Molnar P (1987) Relative motion of the Nazca (Farallón) and South American plates since Late Cretaceous time. Tectonics 6:233–248

Pavlis NK, Holmes SA, Kenyon SC, Factor JK (2008) An Earth Gravitational Model to Degree 2160: EGM2008. EGU General Assembly, Vienna, pp 13–18

Pavlis NK, Holmes SA, Kenyon SC, Factor JK (2012) The development and evaluation of the Earth Gravitational Model 2008 (EGM2008). J Geophys Res 117:B04406. https://doi.org/10.1029/2011JB008916

Pécskay Z, Haller M, Németh K (2007) Preliminary K/Ar geochronology of the Crater Basalt volcanic field (CBVF), northern Patagonia. Rev Asoc Geol Argent 62:25–29

Peroni GO, Hegedus AG, Cerdan J, Legarreta L, Uliana MA, Laffite G (1995) Hydrocarbon Accumulation in an Inverted Segment of the Andean Foreland: San Bernardo Belt, Central Patagonia. In: Tankard AJ, Suárez R, Welsink HJ (eds) Petroleum Basins of South America. Am Assoc Petrol Geol Memoir 62, pp 403–419

Ramírez de Arellano C, Putlitz B, Müntener O, Ovtcharova M (2012) High precision U/Pb zircon dating of the Chaltén Plutonic Complex (Cerro Fitz Roy, Patagonia) and its relationship to arc migration in the southernmost Andes. Tectonics 31: TC4009. https://doi.org/10.1029/2011TC003048

Ramos VA (1981). Descripción geológica de la hoja 47ab, Lago Fontana: provincia del Chubut. Buenos Aires

Ramos VA, Kay SM (1992) Southern Patagonian plateau basalts and deformation: Backarc testimony of ridge collisions. Tectonophysics 205:261–282. https://doi.org/10.1016/0040-1951(92)90430-E

Ramos M, Orts D, Calatayud F, Pazos P, Folguera A, Ramos VA (2011) Estructura, estratigrafía y evolución tectónica de la cuenca de Ñirihuao en las nacientes del río Cuyamen (Chubut, Argentina). Rev Asoc Geol Argent 68:210–224

Ramos M, Tobal J, Sagripanti L, Folguera A, Orts D, Giménez M, Ramos V (2015) The North Patagonian orogenic front and foreland evolution: Ñirihuau-Ñorquinco Depocenter (~42° S). J South Am Earth Sci 64:467–485

Rapela C, Kay S (1988) Late Paleozoic to Recent magmatic evolution of northern Patagonia. Episodes 11:175–182

Rapela C, Munizaga F, Dalla Salda L, Hervé F, Parada M, Cingolani C (1987) Nuevas edades potasio-argón de los granitoides del sector nororiental de los Andes Patagónicos. X Congreso Geológico Argentino, Tucumán, Actas 1:18–20

Rapela C, Spalletti L, Merodio JC, Aragón E (1988) Temporal evolution and spatial variation of early tertiary volcanism in the Patagonian Andes (40° S–42°30′S). J South Am Earth Sci 1:75–88. https://doi.org/10.1016/0895-9811(88)90017-X

Rapela CW, Pankhurst RJ, Fanning CM, Herve F (2005) Pacific subduction coeval with the Karoo mantle plume: the Early Jurasssic Subcordilleran belt of northwestern Patagonia. Geological Society of London, Special Publication 246:217–239. https://doi.org/10.1144/GSL.SP.2005.246.01.07

Rojas Vera EA, Folguera A, Valcarce GZ, Giménez M, Ruiz F, Martínez P, Bottesi G, Ramos V (2010) Neogene to Quaternary extensional reactivation of a fold and thrust belt: The Agrio belt in the Southern Central Andes and its relation to the Loncopué trough (38°–39° S). Tectonophysics 492:279–294

Silver PG, Russo RM, Lithgow-Bertelloni C (1998) Coupling of South American and African Plate Motion and Plate Deformation. Science 279:60–63. https://doi.org/10.1126/science.279.5347.60

Somoza R, Ghidella ME (2005) Convergencia en el margen occidental de América del sur durante el Cenozoico: Subducción de las placas de Nazca, Farallón y Aluk. Rev la Asoc Geol Argent 60:797–809

Suárez M, De La Cruz R (2001) Jurassic to Miocene K-Ar dates from eastern central Patagonian Cordillera plutons, Chile (45°–48° S). Geol Mag 138:53–66

Suárez M, De La Cruz R, Bell CM (1996). Estratigrafía de la región de Coyhaique (Latitud 45o-46oS), Cordillera Patagónica, Chile. In: XIII Congreso Geológico Argentino and III Congreso de Exploración de Hidrocarburos, Capital Federal, pp 575–590

Tašárová ZA (2007) Towards understanding the lithospheric structure of the southern Chilean subduction zone (36° S-42° S) and its role in the gravity field. Geophys J Int 170:995–1014. https://doi.org/10.1111/j.1365-246X.2007.03466.x

Thomson S (2002) Late Cenozoic geomorphic and tectonic evolution of the Patagonian Andes between latitudes 42° S and 46° S: An appraisal based on fission-track results from the transpressional intra-arc Liquiñe-Ofqui fault zone. Geol Soc Am Bulletin 114:1159–1173

Tobal J, Folguera A, Naipauer M, Sellés D, Likerman J, Boedo F, Gimenez M, Ramos VA (2015) Late Miocene extensional relaxation of the North Patagonian Andes (41°30′–42° S). Tectonophysics. https://doi.org/10.1016/j.tecto.2015.06.032

Uieda L, Ussami N, Braitenberg CF (2010) Computation of the gravity gradient tensor due to topographic masses using tesseroids. In: Eos Transactions AGU, 91, Meeting of the Americas supplement, Abstract G22A–04

Utgé S, Folguera A, Litvak VD, Ramos VA (2009) Geología del sector norte de la Cuenca de Cura Mallín: Zona de las lagunas de Epulaufquen (36°40'–50'S, 71°–71°10'W). Rev Asoc Geol Argent 64:230–247

Vicente JC (2005) Dynamic paleogeography of the Jurassic Andean Basin: pattern of transgression and localisation of main straits through the magmatic arc. Rev Asoc Geol Argent 60:221–250

Von Gosen W (2009) Stages of Late Palaeozoic deformation and intrusive activity in the western part of the North Patagonian Massif (southern Argentina) and their geotectonic implications. Geol Mag 146:48–71. https://doi.org/10.1017/S0016756808005311

Von Gosen W, Loske W (2004) Tectonic history of the Calcatapul Formation, Chubut province, Argentina, and the "Gastre fault system". J South Am Earth Sci 18:73–88

Wienecke S (2006). A new analytical solution for the calculation of flexural rigidity: significance and applications. PhD thesis, Freie Universität Berlin

Wienecke S, Braitenberg C, Goetze H (2007) A new analytical solution estimating the flexural rigidity in the Central Andes. Geophys J Int 169:789–794

The Late Paleogene to Neogene Volcanic Arc in the Southern Central Andes (28°–37° S)

Vanesa D. Litvak, Stella Poma, Rosemary E. Jones, Lucía Fernández Paz, Sofía B. Iannelli, Mauro Spagnuolo, Linda A. Kirstein, Andrés Folguera and Victor A. Ramos

Abstract Evolution of arc magmatism along the Southern Central Andes (~28°–37° S) is strongly controlled by changes in the geometry of the downgoing slab (e.g., slab dip angle). This is particularly evident in the present-day Chilean-Pampean flat-slab and the late Miocene Payenia shallow subduction segments. Typical Andean-type volcanism was established from the late Oligocene to late Miocene in the high Andes (29°30″–30°30″S), with arc-related calc-alkaline volcanism having geochemical signatures that reflect changes in the residual mineral assemblages related to increased crustal thickness (>50 km). The increase in crustal thickness resulted from increased compression along the Southern Central Andean margin due to the subduction of the Juan Fernández Ridge, and consequent shallowing of the downgoing slab during Mid to Late Miocene. Associated with the decrease in the slab dip angle, the volcanic front migrated to the east. Further south, magmas developed across the present-day Payenia back-arc region (35°–37° S) show an increase in slab-derived components in the middle Miocene to early

V. D. Litvak (✉) · L. Fernández Paz · S. B. Iannelli · M. Spagnuolo · A. Folguera
Instituto de Estudios Andinos (IDEAN), Consejo Nacional de Investigaciones Científicas y Técnicas (CONICET), Universidad de Buenos Aires, Buenos Aires, Argentina
e-mail: vane@gl.fcen.uba.ar

S. Poma
Instituto de Geociencias Básicas y Aplicadas de Buenos Aires (IGEBA), Universidad de Buenos Aires-Conicet, Buenos Aires, Argentina

R. E. Jones
Department of Earth Sciences, University of Oxford, South Parks Road, OX1 3AN Oxford, United Kingdom

L. A. Kirstein
School of GeoSciences, University of Edinburgh, The King's Buildings, James Hutton Road, EH9 3FE Edinburgh, United Kingdom

V. A. Ramos
Instituto de Estudios Andinos "Don Pablo Groeber", Departamento de Ciencias Geológicas, FCEN, Universidad de Buenos Aires–CONICET, Buenos Aires, Argentina

A. Folguera et al. (eds.), *The Evolution of the Chilean-Argentinean Andes*, Springer Earth System Sciences, https://doi.org/10.1007/978-3-319-67774-3_20

Pliocene times, which also suggests a progressive shallowing of the subducting slab at these latitudes. However, trace element ratios indicate a low-to-intermediate pressure residual mineral assemblage and no significant increase in crustal thickness is apparent, unlike further north in the Chilean-Pampean flat-slab segment. Although flat-slab geometry still prevails in this latter segment, re-steepening of the slab during early Pliocene times (~5–3 Ma) promoted an increase of arc and back-arc magmatism at these more southerly latitudes of the Southern Central Andes. A dynamic link between slab geometry, geochemistry, and volcanic activity is therefore observed in the Southern Central Andes.

Keywords Payenia shallow subduction · Pampean flat-slab · Crustal thickness Geochemistry · Arc magmatism

1 Introduction

Topographic variability along the length of the Andean mountain chain can be linked to the changes in the geodynamic driving forces, which correlate with the development of arc magmatism. The geometry of the subducting slab (e.g., slab dip angle) through time is one of the major factors controlling the development, intensity, migration, and/or expansion of volcanic arcs (Kay et al. 1991, 1999; Ramos and Folguera 2005; Folguera et al. 2011; Litvak et al. 2007).

The geodynamic setting of the Southern Central Andean margin (~28°–37° S) has changed significantly through the Cenozoic. In the late Oligocene (~25 Ma), the Farallón Plate broke up into the Nazca and Cocos Plates (see Chap. "Paleogene Arc-Related Volcanism in the Southern Central Andes and North Patagonia (39°–41° S)"), which resulted in a change from oblique to orthogonal convergence (e.g. Pardo Casas and Molnar 1987; Barckhausen et al. 2008). Typical Andean-type arc volcanism was initiated with a ~30° subduction angle. However, in the Southern Central Andes, the intersection of the Juan Fernández Ridge with the continental margin in the Miocene resulted in the shallowing of the angle at which the Nazca Plate subducts beneath the South American Plate. There are two regions affected by Cenozoic shallow and/or flat subduction in the Southern Central Andes: the present-day Chilean-Pampean flat-slab segment and the late Miocene Payenia shallow subduction segment (Kay et al. 1991, 1999, 2006a, b; Kay and Mpodozis 2002; Litvak et al. 2007, 2015; Jones et al. 2016; Fig. 1). Andean-type subduction and arc volcanism were developed along both of these segments in the late Oligocene. By Miocene times, a shallow and hence a more compressive subduction regime commenced, leading to the expansion of the arc activity toward the retro-arc and an increase of deformation into the Argentinean foreland. The main difference in the evolution of the volcanic arc within these segments is that the crust increased in thickness from ~30 km to >50 km in the Chilean-Pampean flat-slab segment during the latter part of the Cenozoic (Allmenndinger et al. 1990; Fromm et al. 2004; McGlashan et al. 2008; Chulick et al. 2013). Moreover, while the flat-slab

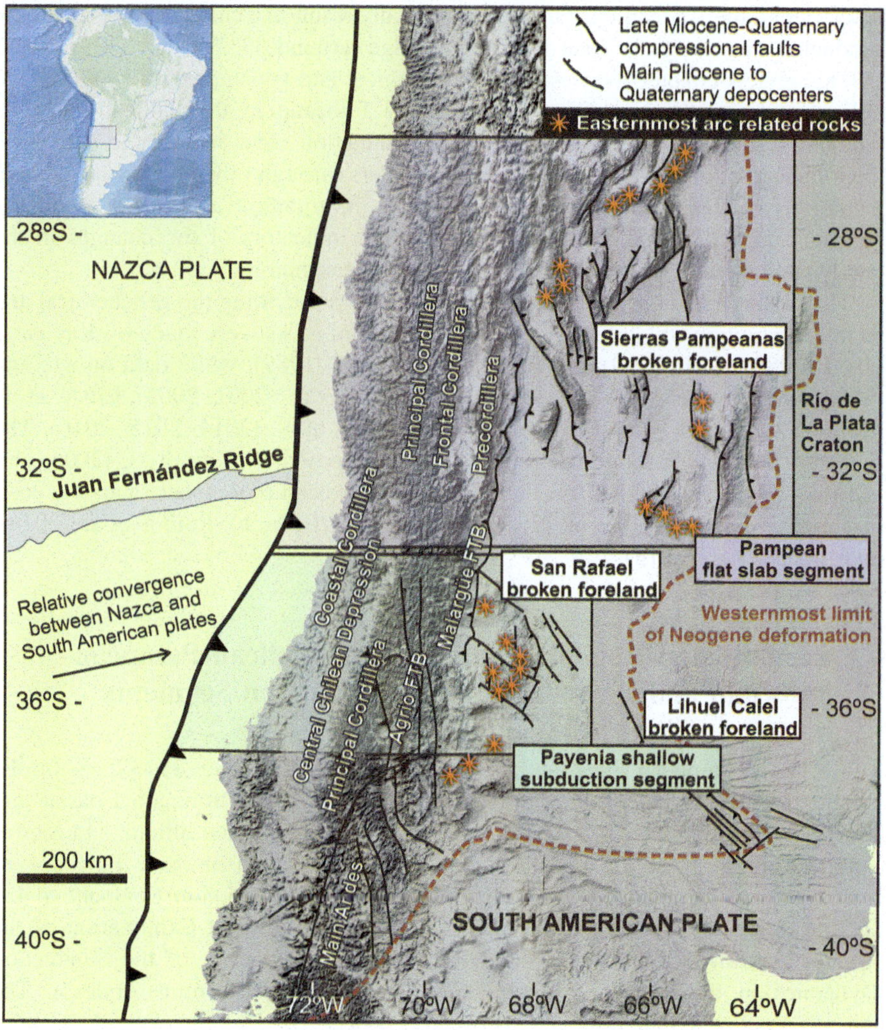

Fig. 1 Location and main morphotectonic features of the present-day Chilean-Pampean flat-slab segment and the partially contemporaneous late Miocene Payenia shallow subduction zone. The compressive regimes associated with both flat and shallow segments are responsible for the uplifting of the Sierras Pampeanas and San Rafael Block, respectively, in the foreland area. Location of arc-related centers after Ramos et al. (2002) (modified from Litvak et al. 2015)

geometry is still ongoing in the Chilean-Pampean flat-slab segment, the shallow subduction regime in the Payenia segment prevailed only until the beginning of the Pliocene ($\sim$5–3 Ma), when the subducting slab returned to a normal angle (Kay et al. 2006a, b; Ramos et al. 2014; Litvak et al. 2015). Given the spatial configuration and the timing at which the angle of subduction changed of both Chilean-Pampean and Payenia segments, the collision of the Juan Fernández Ridge is likely to explain the continuity and progressive shallowing throughout both segments during the Miocene. The variable influence of the ridge collision from

N to S along the Andean margin could eventually result in a change in the dip of the subducted slab away the Juan Fernández Ridge (around 33° S).

This review chapter focuses on the lithological and geochemical features of arc volcanism from the late Oligocene to early Pliocene in the Chilean-Pampean flat-slab segment and the Payenia shallow subduction zone, in order to determine the influence of changes in subduction parameters through time on arc magmatism. Variations in geochemical signatures provide information about the conditions beneath active volcanic arcs and thus are good indicators of the changes in the geodynamic setting of the Southern Central Andean margin.

The first and main scientific contributions to understanding the geochemical and tectonic evolution of Chilean-Pampean flat-slab arc rocks were made by Kay et al. (1987, 1988, 1991, 1999) and Ramos et al. (1987, 1989), while data on age and petrological features were later reported by Bissig et al. (2001, 2003), Litvak et al. (2007), Litvak and Poma (2010, 2014), and Jones et al. (2014, 2015, 2016). The Payenia shallow subduction regime was first proposed by Kay et al. (2006a, b), primarily based on studies on the Chachahuén volcanic complex, and was later discussed by Spagnuolo et al. (2012), Dyhr et al. (2013a, b), Ramos et al. (2014), and Litvak et al. (2015).

2 Tectonic and Regional Setting of the Chilean-Pampean Flat-Slab and Payenia Shallow Subduction Segments

The Chilean-Pampean flat-slab segment, located between ~28–32° S in the Southern Central Andes, is comprised of, in a W–E direction, the Coastal and Principal Cordilleras in Chile and the Frontal Cordillera, Precordillera, and Sierras Pampeanas of Argentina (Fig. 1) (see Chaps. "Mechanisms and Episodes of Deformation Along the Chilean–Pampean Flat-Slab Subduction Segment of the Central Andes in Northern Chile" and "Cenozoic Uplift and Exhumation of the Frontal Cordillera Between 30° and 35° S and the Influence of the Subduction Dynamics in the Flat Slab Subduction Context, South Central Andes"). The Payenia shallow subduction zone, between ~34–37° S, includes part of the Central Depression in Chile and the Cordillera Principal of Chile and Argentina, together with a lower mountain system to the east known as the San Rafael Block (Fig. 1). The San Rafael Block lies in a similar longitudinal position to the Precordillera in the present-day Pampean flat-slab segment (Fig. 1). However, its morphology, which corresponds to a dismembered peneplain that emerges in the foreland region, and the mechanics of its exhumation, makes the San Rafael Block similar to the Sierras Pampeanas, which is part of a broken foreland, located to the north of the San Rafael Block and to the east of the Precordillera. Formation of the Sierras Pampeanas and the San Rafael Block has been linked to compressive regimes, resulting from the shallowing of the subducting Nazca Plate in Miocene times and the reactivation of high-angle basement faults during Miocene to Quaternary times (Jordan et al. 1983; Ramos et al. 2002).

Since the Triassic, the Andean margin has been magmatically active at various intervals, producing a series of volcanic arcs in a setting which developed from being more extensional in the initial stages to increasingly compressive after the Late Cretaceous and then from the late Oligocene (e.g., Ramos et al. 2002; Charrier et al. 2007) (see Chaps. "The Early Stages of the Magmatic Arc in the Southern Central Andes", "A Provenance Analysis from the Lower Jurassic Units of the Neuquén Basin. Volcanic Arc or Intraplate Magmatic Input?", "Lower Jurassic to Early Paleogene Intraplate Contraction in Central Patagonia", "Mechanisms and Episodes of Deformation Along the Chilean–Pampean Flat-Slab Subduction Segment of the Central Andes in Northern Chile", "Cretaceous Orogeny and Marine Transgression in the Southern Central and Northern Patagonian Andes: Aftermath of a Large-Scale Flat-Subduction Event?", and "Paleogene Arc-Related Volcanism in the Southern Central Andes and North Patagonia (39°–41° S)"). The breakup of the Farallón Plate (~25 Ma) into the Nazca and Cocos caused an increase in convergence rates from ~8 cm/yr to ~15 cm/yr and a change from oblique to orthogonal convergence (Pardo Casas and Molnar 1987; Somoza 1998; Somoza and Ghidella 2012). In the Chilean-Pampean flat-slab segment, Andean-type volcanism was developed under this new plate configuration. One of the key areas for understanding the evolution of the arc is the El Indio–Valle del Cura belt (~29–30° S; Fig. 2), where late Oligocene–early Miocene main volcanic arc was formed of dacitic to andesitic pyroclastic and lava flows located in the present-day Chile, with associated volcanoclastic deposits primarily located in retro-arc extensional basins (Figs. 1 and 2) (Maksaev et al. 1984; Kay et al. 1991; Litvak et al. 2007; Winocur et al. 2015).

During mid to late Miocene times, arc magmatism (at 29°–30°S) was strongly influenced by the arrival of the Juan Fernández Ridge and the shallowing of the subducted slab (Kay and Mpodozis 2002, Pilger 1981, 1984, Yañez et al. 2002, 2001) (see Chaps. "Structure and Tectonics of the Chilean Convergent Margin from Wide-Angle Seismic Studies: A Review" and "The Peru–Chile Margin from Global Gravity Field Derivatives"). As a result, the volcanic arc front migrated into the present-day Argentinean Frontal Cordillera, while arc magmatism broadened and expanded to the east, as far as ~700 km away from the present trench (Kay and Gordillo 1994; Poma et al. 2005, 2014). The shallow subduction geometry caused a reduction in the magma volume erupted within the main arc, with cessation of arc magmatism by late Miocene (~6 Ma) times in the Chilean-Pampean flat-slab region (Ramos et al. 1989; Kay et al. 1991; Bissig et al. 2001; Litvak et al. 2007).

The Payenia shallow subduction segment is located immediately to the south of the Chilean-Pampean flat-slab (Fig. 1). This segment is characterized by widespread volcanic sequences that cover most of the retro-arc region, over 550 km east of the trench in the Cenozoic (Bermúdez 1991; Bermúdez et al. 1993; Nullo et al. 2002; Kay et al. 2006a, b; Spagnuolo et al. 2012; Dyhr et al. 2013a, b). As summarized by Dyhr et al. (2013a), the evolution of the mid-to-late Cenozoic volcanism in the

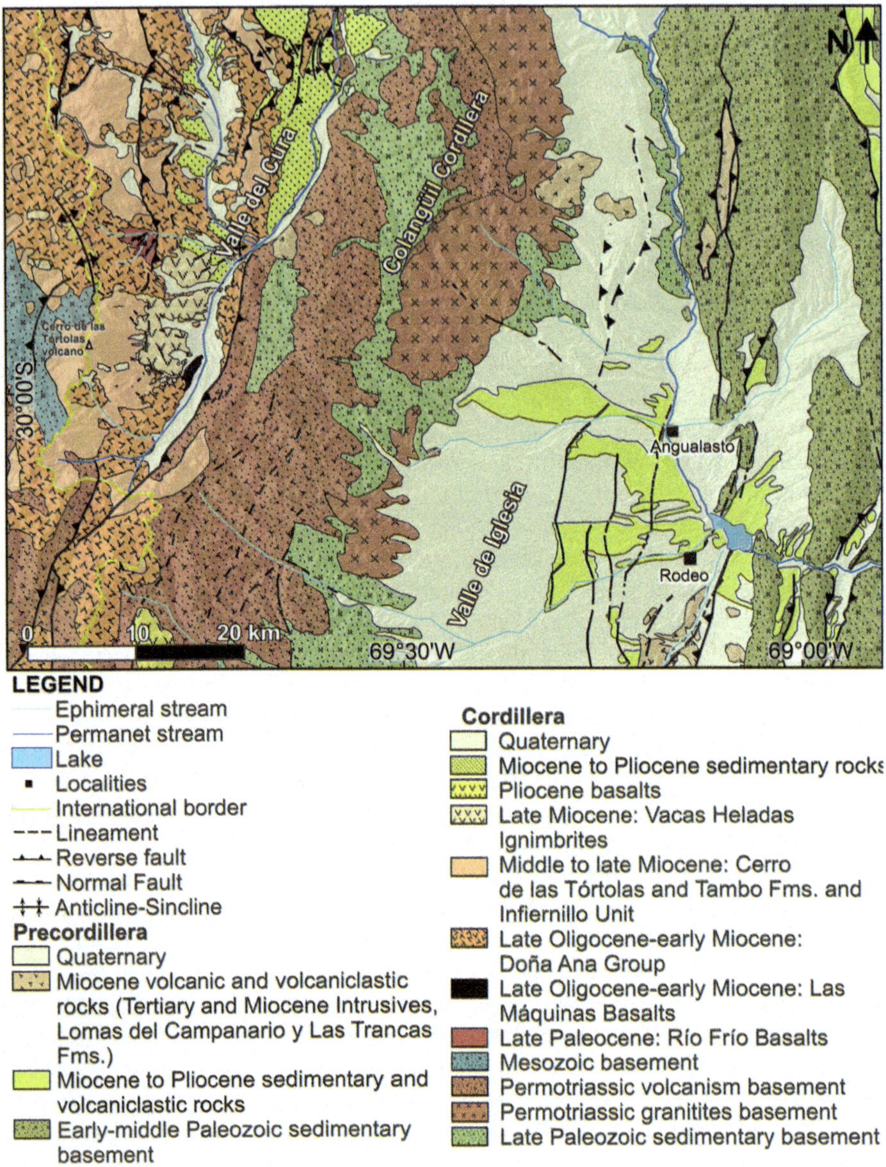

Fig. 2 Geological setting and distribution of arc-like sequences between 29°–30° S across the Frontal Cordillera and Precordillera, from the Valle del Cura to the Iglesia Basin, over the present-day Chilean-Pampean flat-slab segment (based on Litvak et al. 2007; Alonso et al. 2011; Litvak and Poma 2014)

retro-arc of the Payenia region, can be subdivided into three main episodes. These are differentiated on the basis of their age, composition, and tectonic setting: (i) an early Miocene episode of alkaline volcanism (Kay and Copeland 2006; Kay et al. 2006b;

Dyhr et al. 2013b); (ii) A middle Miocene-to-early Pliocene 'arc-like' episode (Nullo et al. 2002; Kay et al. 2006a, b; Spagnuolo et al. 2012; Dyhr et al. 2013a; Litvak et al. 2015); and (iii) An episode of late Pliocene-Recent volcanism, with alkali basalts displaying typical back-arc compositions and variable slab input (Bermúdez et al. 1993; Bertotto et al. 2009; Gudnason et al. 2012; Søager et al. 2013; Hernando et al. 2014). The second volcanic episode is key to understanding the development of the Payenia shallow subduction regime which is thought to have occurred between 18 and 3.5 Ma, with the plate at its shallowest angle between ~7.3 and 3.5 Ma when the easternmost arc-related sequences erupted over the present-day Payenia retro-arc zone (Kay et al. 2006a, b; Litvak et al. 2015).

The Andes between 34° and 38° S are formed of two east-verging mountain systems: firstly, the Malargüe fold and thrust belt and its continuation to the south into the Agrio fold and thrust belt (see Chaps. "Cenozoic Uplift and Exhumation of the Frontal Cordillera Between 30° and 35° S and the Influence of the Subduction Dynamics in the Flat Slab Subduction Context, South Central Andes" and "The Structure of the Southern Central Andes (Chos Malal Fold and Thrust Belt)") and secondly, the San Rafael Block (Fig. 1). Arc-related volcanism was particularly developed at these latitudes within retro-arc and intra-arc basins, such as the Abanico and Cura Mallin Basins, that were developed during the Late Oligocene to Early Miocene (see Chap. "Paleogene Arc-Related Volcanism in the Southern Central Andes and North Patagonia (39°–41° S)"), as the westernmost sections of the Malargüe and Agrio fold and thrust belts experienced an episode of extension. These basins contain volcanic and volcaniclastic sequences of over 3000 m in thickness, derived from activity in the main volcanic arc (e.g., Suárez and Emparan 1995; Charrier et al. 2002; Radic et al. 2002; Burns et al. 2006; Kay et al. 2006a, b; Muñoz et al. 2006). Extension in late Oligocene–early Miocene times was likely related to roll back of the subducting Nazca Plate caused by the deceleration of the South American Plate (see Chap. "The Late Oligocene–Early Miocene Marine Transgression of Patagonia). Within this setting, retro-arc alkaline, volcanic successions of late Oligocene–early Miocene age accumulated to the east, in the region of the southern section of the present-day Payenia retro-arc (Ramos and Barbieri 1989; Kay et al. 2006a, b; Dyhr et al. 2013b).

The middle Miocene-to-early Pliocene calc-alkaline volcanic sequences covered the distal foreland deposits within the present-day Payenia retro-arc (Fig. 3) (Kay et al. 2006a, b; Ramos et al. 2014; Litvak et al. 2015). Steepening of the subducting Nazca Plate (~5–3 Ma) in these region resulted in an extensional regime across the foreland, and the emplacement of Quaternary, mafic volcanic sequences that comprise one of the major volcanic provinces in the Andean back-arc (Groeber 1946; Muñoz Bravo et al. 1989; Bermúdez et al. 1993; Ramos and Folguera 2005, 2011; Folguera et al. 2009; Llambías et al. 2010; Gudnason et al. 2012; Søager et al. 2013; Hernando et al. 2014)

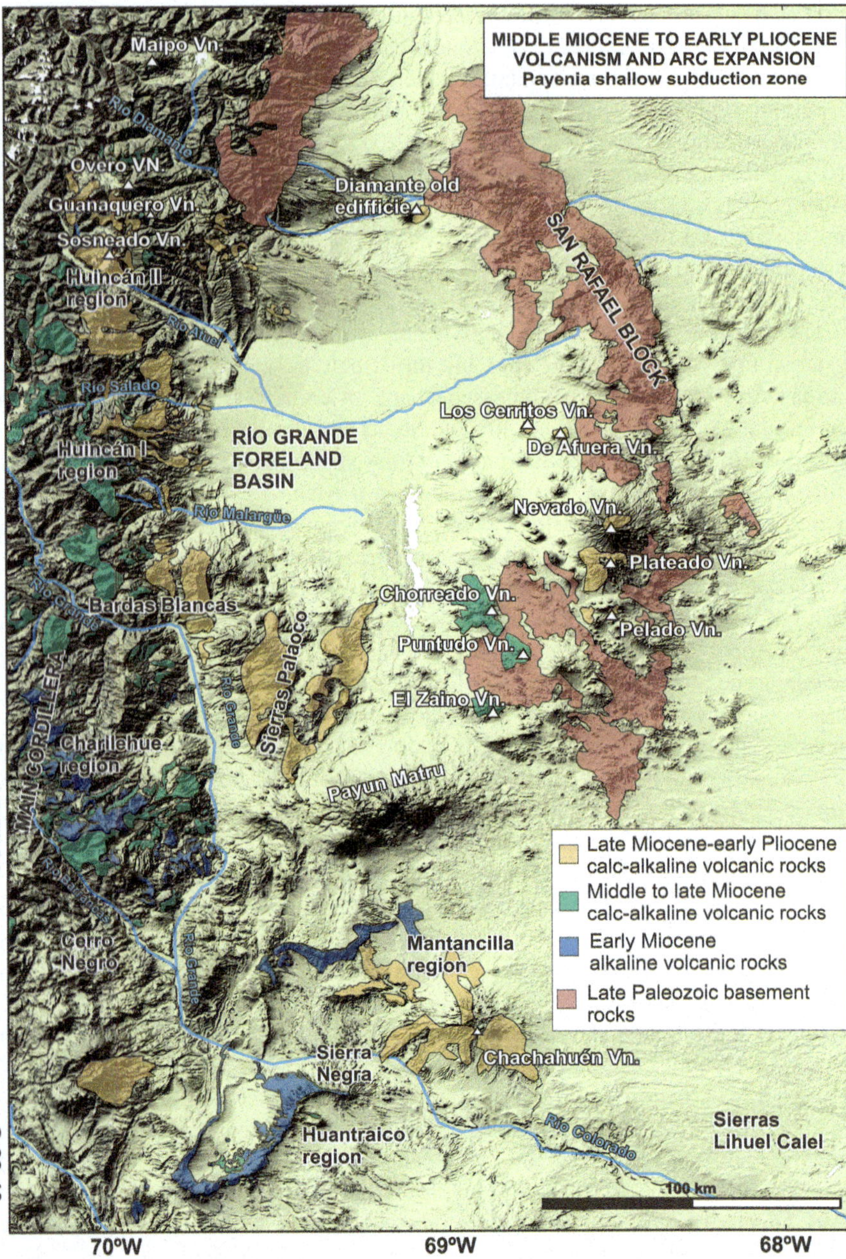
MIDDLE MIOCENE TO EARLY PLIOCENE
VOLCANISM AND ARC EXPANSION
Payenia shallow subduction zone
Maipo Vn.
Río Diamante
Overo VN.
Guanaquero Vn.
Sosneado Vn.
Huincán II region
Río Atuel
Diamante old edifficie
SAN RAFAEL BLOCK
Río Salado
RÍO GRANDE FORELAND BASIN
Los Cerritos Vn.
De Afuera Vn.
Huincán I region
Río Malargüe
Nevado Vn.
Plateado Vn.
Bardas Blancas
Río Grande
Chorreado Vn.
Pelado Vn.
Sierras Palaoco
Puntudo Vn.
El Zaino Vn.
MAIN CORDILLERA
Charilehue region
Payun Matru
Late Miocene-early Pliocene calc-alkaline volcanic rocks
Middle to late Miocene calc-alkaline volcanic rocks
Early Miocene alkaline volcanic rocks
Late Paleozoic basement rocks
Cerro Negro
Mantancilla region
Sierra Negra
Chachahuén Vn.
Huantraico region
Río Colorado
Sierras Lihuel Calel
100 km
34°30'S
35°30'S
36°30'S
37°30'S
70°W
69°W
68°W

◂**Fig. 3** Distribution of mesosilicic volcanic rocks from the eastern Main Andes to the San Rafael Block in the foreland area showing the eastward arc influence from middle Miocene to early Pliocene in the back-arc area. Modified from Litvak et al. (2015); (age compilation from Nullo et al. (1993, 1999, 2002), Ramos and Barbieri (1989), Ostera et al. (1999), Cobbold and Rosello (2003), Giambiagi et al. (2005), Kay and Copeland (2006), Kay et al. (2006a, b), Folguera et al. (2009), Spagnuolo et al. (2012), Dyhr et al. (2013a, b) and Ramos et al. (2014))

3 Age, Distribution, and Lithological Features of Cenozoic Arc-Related Volcanism in Southern Central Andes

3.1 The Chilean-Pampean Flat-Slab Segment: Valle Del Cura and Iglesia Basins

Arc-related volcanic units are distributed between ~29°30′–30°15′S in the Valle del Cura Basin, which forms part of the Cordillera Frontal in the high Andes of Argentina, and in the Valle de Iglesia Basin, which lies in the most westerly part of the Precordillera (Fig. 1). These two basins are divided by the Colangüil Cordillera (as part of the Cordillera Frontal), which is mostly comprised of Late Paleozoic and Permo-Triassic basement (Fig. 2).

The Valle del Cura Basin constitutes a mineralized hydrothermal system that, together with the El Indio belt in Chile, forms a north-south-trending block limited by high-angle reverse faults associated with both Au–Ag–Cu-rich epithermal and porphyry-type mineral deposits (e.g., Martin et al. 1995; Bissig et al. 2001, 2015; Litvak et al. 2007). Arc-related early Cenozoic volcanic rocks in the high Andes are mainly restricted to the Chilean Principal Cordillera with epizonal granodioritic plutons and extrusive basaltic to rhyolitic lavas and tuffs of Paleocene to Eocene age (Cogotí Supergroup and Los Equinos Formation, respectively, Parada et al. 1988; Pineda and Emparan 2006; Charrier et al. 2007; Jones et al. 2016). Paleocene volcanism was also developed to the east, in present-day Argentina, as volumetrically restricted, back-arc alkaline mafic volcanism (Río Frío Basalts), for which a K–Ar whole rock age of 55.9 ± 1.9 Ma has been obtained (Litvak and Page 2002; Litvak and Poma 2010) (Fig. 2). Overall, arc activity was reduced between ~39 and 26 Ma in the Southern Central Andes; however, some minor arc magmatism occurred, as represented by small dioritic and granodioritic subvolcanic bodies with reported K–Ar ages of 31.1 ± 1.2 and 39.5 ± 1.3 and Ar–Ar of 30.0 ± 1.9 and 35.9 ± 1.2 Ma (Bocatoma Unit; Nasi et al. 1990; Martin et al. 1995; Bissig et al. 2001).

Extensive arc magmatism was developed in the Southern Central Andes in the Late Oligocene (~25 Ma) as a consequence of the tectonic reconfiguration of the South American margin. This volcanism comprises the Doña Ana Group sequences, which crop out in both the Chilean and Argentinean side of the Andes representing arc and retro-arc-volcanic activity, respectively (Fig. 2). The older member of the Doña Ana Group, the Tilito Formation, includes rhyolitic and dacitic lavas and ignimbrites associated with minor basalts and andesites with reported ages from 27 to 23 Ma, based K–Ar, Ar–Ar, and U–Pb age dating; the younger member, the Escabroso Formation, is mainly composed of basaltic to andesitic lavas interbedded with volcanic agglomerates and breccias with reported ages between 21 and 18 Ma (Maksaev et al. 1984; Nasi et al. 1990; Martin et al. 1995; Bissig et al. 2001; Litvak 2009; Winocur et al. 2015; Jones et al. 2015, 2016). A major unconformity separates both formations as the result of a period of deformation around 20 Ma, with a hiatus of volcanic activity of 3 Myr (Martin et al. 1995).

The late Oligocene arc sequences have been affected by varying degrees of phyllic to advanced argillic hydrothermal alteration, which overall comprises biotite-bearing dacitic lavas and crystalline and vitreous tuffs associated with the Cerro Doña Ana eruptive center. Early Miocene lavas are pyroxene-bearing porphyritic basaltic andesites and andesites, occasionally with propylitic alteration. The Doña Ana Group sequences also expanded toward the former retro-arc, cropping out in the eastern sector of the present Valle del Cura Basin (Fig. 2), where they include pyroclastic and volcaniclastic deposits interbedded with clastic facies interpreted as river systems and alluvial fan deposits with associated ash-fall deposits due to the proximity of the magmatic vents (Limarino et al. 1999; Litvak 2009; Winocur et al. 2015). Minor alkaline basalt with slab-derived input (Las Máquinas Basalts), exposed in volcanic necks in the Valle del Cura region, is interpreted as back-arc lavas associated with the late Oligocene–early Miocene arc activity (Ramos et al. 1989; Kay et al. 1991; Litvak and Poma 2010).

Arc magmatism by middle to late Miocene times is represented by the Cerro de las Tórtolas Formation, which is distributed along the Valle del Cura (Fig. 2), with the Cerro de las Tórtolas and Vacas Heladas volcanoes on the Argentinean-Chilean border preserved as partially eroded eruptive centers. Two main sections have been recognized in this formation: (i) an andesitic to basaltic andesitic lower section, which is primarily comprised of porphyritic pyroxene to amphibole-bearing andesites within plagioclase-rich groundmass, with reported K–Ar, Ar–Ar, and zircon U/Pb ages between 17 and 14 Ma; and (ii) An andesitic to dacitic upper section, with mostly porphyritic amphibole-bearing rocks within hyalopilitic to trydimite-rich groundmass, with ages ranging from 13 to 10 Ma (Maksaev et al. 1984; Ramos et al. 1989; Kay et al. 1991, 1999; Martin et al. 1995; Bissig et al. 2001; Litvak et al. 2007; Litvak and Poma 2014; Jones et al. 2015, 2016). The lower section is widespread and distributed across the Chilean-Argentinean border, particularly on the eastern side of the high Andes between 29° to 30° S, whereas the upper section is restricted to the highest altitudes, proximal to the Cerro de las Tórtolas volcano (Fig. 2). The Miocene lavas are generally subhorizontal, although in some cases they have been tilted by late Miocene reverse faulting. In contrast to the late Oligocene–early Miocene arc rocks, the middle-to-late Miocene andesitic lavas are not particularly affected by hydrothermal alteration and appear fresh, preserving their primary textures and mineral assemblage. Further to the west, on the Chilean side of the high Andes, a series of high-K calc-alkaline, shallow level, intermediate intrusives (Infiernillo Unit, 18–15 Ma) have been interpreted as the subvolcanic equivalent of the Cerro de Las Tórtolas Formation (Kay et al. 1987, 1988, 1991; Bissig et al. 2001; Jones et al. 2016).

A limited number of exposures of dacitic crystalline tuffs with reported ages between 12 to 11 Ma (Tambo Formation, after Martin et al. 1995; also known as Vacas Heladas Formation—see Litvak et al. (2007) for further discussion on stratigraphy) crop out in the southern Valle del Cura, but extend mostly to the north, near the Veladero ore deposit. These tuffs show distinctive petrographic features, such as amphibole and titanite crystal fragments, and have been included as part of the middle-to-late Miocene volcanic sequences.

Late Miocene volcanism is volumetrically restricted in the area, as a consequence of the shallowing of the Nazca Plate since ~20–18 Ma. The Vacas Heladas Ignimbrites are comprised of homogenous crystalline dacitic tuffs, derived from surge and pyroclastic flows and ash deposits, and crop out on the eastern slope of the Vacas Heladas and Cerro de las Tórtolas volcanoes (Fig. 2), unconformably overlying Oligocene–Miocene sequences. These rocks represent the last significant volcanic activity in the region of the Valle del Cura on the Chilean-Pampean flat-slab segment, with reported K–Ar, Ar–Ar, and zircon U–Pb ages between 5.5 ± 0.1 Ma to 6.2 ± 0.19, and have been correlated with the Vallecito Formation in Chile (Ramos et al. 1989; Bissig et al. 2003; Jones et al. 2016).

Late Paleogene arc-related volcanism expanded into the foreland area associated with shallowing of the subduction regime. Andesitic to rhyolitic proximal block-and-ash pyroclastic flow deposits, ignimbrites, tuffs, and minor dacitic lava flows, which crop out as N–S thin belts in the western Precordillera (Fig. 2), have been identified as coeval with the Tilito Formation (Las Trancas Formation, Poma et al. 2005, 2014). A zircon U–Pb age of 22.6 ± 0.33 Ma, obtained from a rhyolite, confirms the beginning of arc-related activity in the Precodillera by late Oligocene–early Miocene times (Jones et al. 2016). Minor granitic to granodioritic intrusive bodies have also been recognized along the eastern slope of the Colangüil Cordillera, cropping out around 30° S in the Valle del Iglesia (Miocene Intrusives). K–Ar ages between 22 to 13 Ma (Llambías et al. 1990; Cardó et al. 2007) and recently produced U–Pb zircon age (22.2 ± 0.23 to 20.43 ± 0.31 Ma) assign this magmatism to the early Miocene (Jones et al. 2016).

Further south (~30°45′S), in the Valle de Iglesia, pyroclastic rocks and dacitic subvolcanic bodies were also associated with arc volcanism during the Miocene. They comprise subvolcanic porphyritic amphibole-bearing andesites, ignimbrites, block-and-ash deposits, and minor lavas interbedded with epiclastic continental deposits (Tertiary Intrusives unit and Lomas del Campanario Formation, respectively; Cardó and Díaz 1999; Poma et al. 2014). The age of the intrusive equivalents was determined by K–Ar age dating and ranges between 18.3 ± 2.5 Ma and 17.5 ± 5 Ma (Cardó and Díaz 1999; Leveratto 1976), with one younger age of 8.8 ± 0.3 Ma reported (Wetten 2005). Recently reported U–Pb zircon ages suggest an age between 11.7 ± 0.21 to 9.4 ± 0.18 Ma (Jones et al. 2016), so overall age for this magmatism comprises 18 to 9 Ma.

3.2 The Payenia Shallow Subduction Segment: Calc-Alkaline Magmatism in the Back-Arc Region

Late Oligocene–early Miocene main arc activity in the Payenia shallow subduction zone is represented by the Cura Mallín Formation, which developed as part of an intra-arc basin with pyroclastic and volcaniclastic deposits and minor basaltic andesitic lavas; reported Ar–Ar ages range between 24.6 ± 1.8 and 22.8 ± 0.7 (Jordan et al. 2001; Kay et al. 2006a), although younger K–Ar ages up to 11 Ma had been previously reported (Suárez and Emparan 1995). Volcanic rocks, dykes, and intrusive bodies cropping out on the Chilean slope are also considered as part of the Miocene main volcanic arc activity (Trapa-Trapa Formation; Niemeyer and Muñoz 1983), with reported K–Ar ages from 19 to 12 Ma (Niemeyer and Muñoz 1983; Muñoz and Niemeyer 1984; Suárez and Emparan 1995). To the east, on the Argentine side, west of the Cordillera del Viento, early-to-late Miocene arc-related volcanism is represented by andesitc to dacitic lavas and agglomerates, andesitic dykes and stocks, and pyroclastic rocks to ash deposits, with an overall age range from 14 to 5 Ma, according to reported K–Ar and Ar–Ar ages. This includes several geological units including the Trapa-Trapa, Cajón Negro, Quebrada Honda, and Pichi Neuquén Formations (see Kay et al. 2006a and reference therein) (see Chap. "The Structure of the Southern Central Andes (Chos Malal Fold and Thrust Belt)").

However, changes in the geodynamic conditions of the Andean margin through time, and the associated consequences on arc volcanism, are particularly reflected across the present-day Payenia retro-arc. Calc-alkaline products which crop out in the retro-arc region represent the eastward expansion of arc volcanism and have ages ranging from 20 to 4 Ma in the eastern foreland (Fig. 3) (Nullo et al. 2002; Baldauf 1997; Giambiagi et al. 2005, 2008; Kay and Copeland 2006; Kay et al. 2006a, b; Sruoga et al. 2008; Spagnuolo et al. 2012; Litvak et al. 2015; Dyhr et al. 2013a, b; Ramos et al. 2014).

Early Miocene volcanism in the present-day Payenia back-arc is associated with alkaline magmatism. The lavas from Sierra de Huantraico (Fig. 3), in the Neuquén Basin, provide the first evidence for the influence of the slab on a retro-arc position at around 20 Ma (Kay and Copeland 2006; Kay et al. 2006b; Dyhr et al. 2013b).

Middle Miocene-to-early Pliocene calc-alkaline mesosilicic volcanic sequences are exposed from the near back-arc to the far back-arc along the studied latitudes, between the eastern Malargüe fold and thrust belt in the west, and the San Rafael Block in the east (Fig. 3). These volcanic sequences can be divided into two main volcanic stages in Neogene times, based on their age distribution. The first stage, from the Middle to late Miocene (15 to 10 Ma), includes the Huincán I and part of the Huincán II andesites, Cerro Negro andesite, Charilehue volcanic rocks, and eruptive centers located nearby the San Rafael Block (Fig. 3) (Nullo et al. 2002; Kay and Copeland 2006; Kay et al. 2006a; Spagnuolo et al. 2012; Ramos et al. 2014; Litvak et al. 2015). These latter volcanic centers are mostly preserved as highly eroded stratovolcanoes, whose lava flows, pyroclastic deposits, and subvolcanic bodies were originally named the Cortaderas Formation (Delpino and Bermúdez 1985; Llambías et al. 2010). They comprise fresh pyroxene to amphibole-bearing basalts and basaltic andesites to andesites, with plagioclase-rich groundmass, with interstitial tridymite commonly occurring in the more silica-rich compositions.

The second volcanic stage of calc-alkaline volcanism in the back-arc includes late Miocene-to-early Pliocene (8 to 3.5 Ma) magmas represented by the younger Huincán II andesites, Palaoco volcanism, Chachahuén volcanic complex, and younger San Rafael volcanic rocks (Nullo et al. 2002; Kay et al. 2006b; Dyhr et al. 2013a; Ramos et al. 2014; Litvak et al. 2015). These younger calc-alkaline andesites from the San Rafael volcanic associations are related to the easternmost volcanic centers in the area, such as the Nevado-Plateado volcanic complex, which also include alkaline trachytes, trachyandesites, and basalts, grouped as the Nevado Formation by Bermúdez (1991), and considered to be Pliocene in age (Quidelleur et al. 2009; Ramos et al. 2014). The Chachahuén volcanic complex includes an older sequence (Vizcachas Group, from 7.3 to 6.8 Ma) with dacitic dikes along with andesitic to dacitic ignimbrites and lava flows that display intraplate chemical affinities (Kay et al. 2006b). However, arc-like rocks comprise the younger Early Chachahuén unit (6.8 to 6.4 Ma) with intermediate to mafic lavas and dykes, while the Late Chachahuén sequence is composed of andesitic to dacitic pyroclastic rocks overlain by basaltic to basaltic andesite lava flows, and hornblende-rich andesites (6.3 to 4.9 Ma) (Kay et al. 2006b). The andesites and dacites of the San Rafael Block volcanism share common petrographical characteristics with the Early–Late Chachahuén volcanic complex rocks (Kay et al. 2006a, b; Litvak et al. 2015).

4 Geochemistry of Late Oligocene-to-Late Miocene Arc Magmas Influenced by the Chilean-Pampean Flat-Slab Regime

Overall, the geochemical features for the late Oligocene–late Miocene volcanic rocks cropping out in the high Andes (29°30″–30°30″S) indicate a calc-alkaline affinity with ‘arc-like’ signatures. Variations in trace element ratios and isotopic signatures indicate changes in residual mineral assemblages over time, which have been linked with crustal thickening due to the shallowing of the downgoing slab (Kay et al. 1991, 1999; Kay and Mpodozis 2002; Bissig et al. 2003; Litvak et al. 2007; Jones et al. 2015, 2016).

The late Oligocene–early Miocene Doña Ana Group is comprised of subalkaline rocks, with a continuous SiO_2 wt.% content ranging from basaltic to rhyolitic compositions. Despite large differences in silica content, all of these volcanic rocks show an arc-type trace element pattern on mantle and chondrite normalized multi-element diagrams, e.g., La/Ta > 25, as well as the reported Ba/La and Ba/Ta ratios for the unit (Fig. 4). REE patterns and ratios (La/Yb, La/Sm, and Sm/Yb)

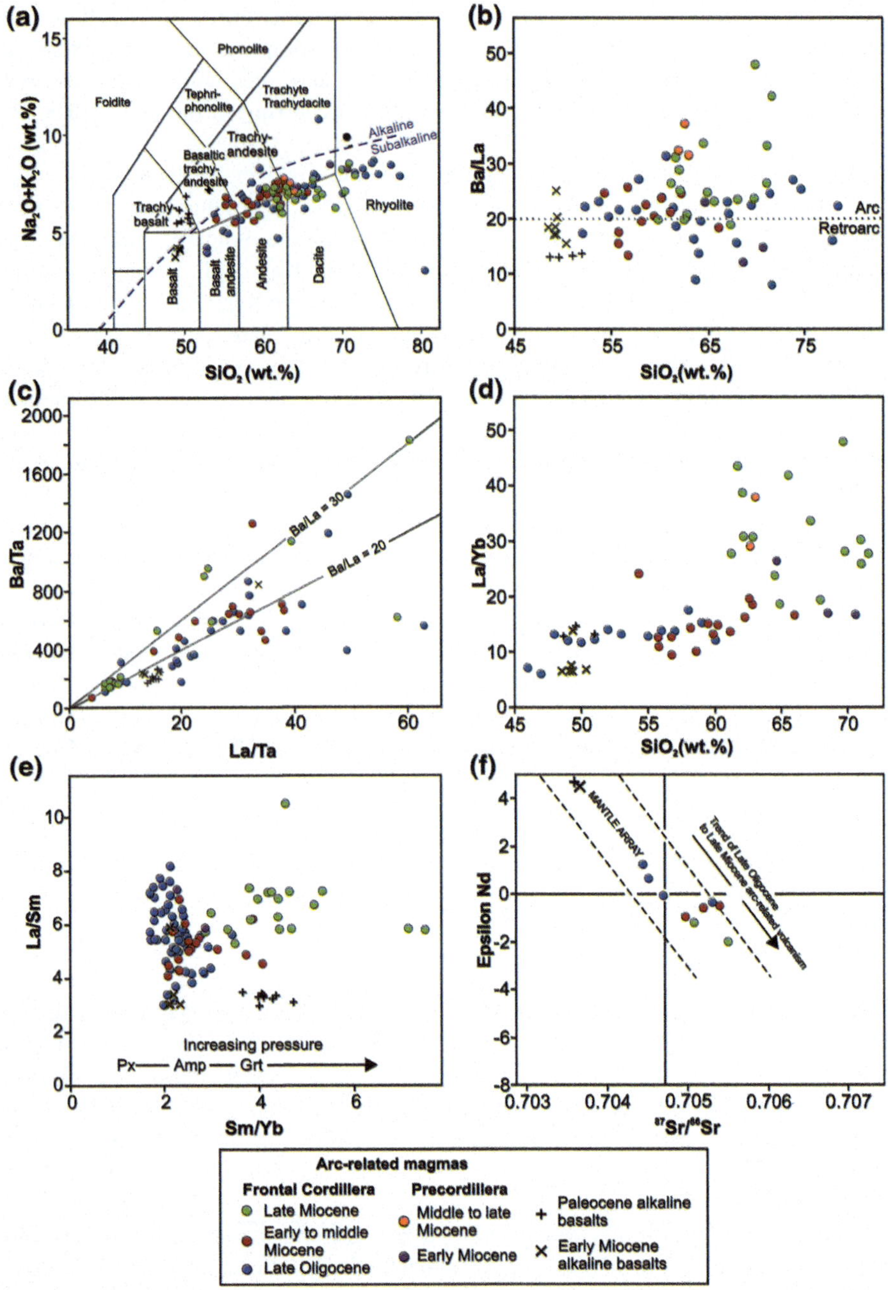
(a)
Foidite
Phonolite
Tephri-phonolite
Trachyte Trachydacite
Trachy-andesite
Basaltic trachy-andesite
Trachy-basalt
Basalt
Basaltic andesite
Andesite
Dacite
Rhyolite
Alkaline
Subalkaline
Na2O+K2O (wt.%)
SiO2 (wt.%)
(b)
Ba/La
Arc
Retroarc
SiO2(wt.%)
(c)
Ba/Ta
Ba/La = 30
Ba/La = 20
La/Ta
(d)
La/Yb
SiO2(wt.%)
(e)
La/Sm
Increasing pressure
Px — Amp — Grt
Sm/Yb
(f)
Epsilon Nd
MANTLE ARRAY
Trend of Late Oligocene to Late Miocene arc-related volcanism
87Sr/86Sr
Arc-related magmas
Frontal Cordillera
Late Miocene
Early to middle Miocene
Late Oligocene
Precordillera
Middle to late Miocene
Early Miocene
Paleocene alkaline basalts
Early Miocene alkaline basalts

◂**Fig. 4** Geochemical features of late Oligocene-to-late Miocene magmas over the Chilean-Pampean flat-slab segment, from the main arc (Cordillera Frontal) to its expansion to the east (western Precordillera). **a** Chemical classification according to total alkali versus silica content (Le Maitre et al. 1989); **b** arc-like signature given by Ba/La versus SiO_2; **c** Ba/Ta versus La/Ta show variable slab input for the Oligo-Miocene sequences; **d** REE behavior reflected in La/Yb versus SiO_2; **e** variable HREE signature evidenced from the La/Sm versus Sm/Yb ratios; **f** isotopic signatures that show an increase of crustal component toward the late Miocene arc magmas. Compiled data came from Bissig et al. (2003); Jones et al. (2016); Kay et al. (1987, 1988, 1991, 1999); Litvak et al. (2007), Litvak and Poma (2010) and Ramos et al. (1989)

suggest that this arc magmatism has variably equilibrated with a low-to-medium pressure residual mineral assemblage, including plagioclase and pyroxene (Kay et al. 1991, 1999; Bissig et al. 2003; Litvak et al. 2007; Jones et al. 2016).

Similar compositions, although less evolved and with lower Ba/Ta ratios, have been reported for the more eastern volcanic sequences, interbedded with epiclastic and volcanoclastic facies as part of the Doña Ana Group retro-arc basin infill (Litvak et al. 2007; Winocur et al. 2015; Jones et al. 2016). The partially contemporaneous early Miocene mafic volcanism (Las Máquinas Basalts) shows a more alkaline-like tendency and scarce slab input, with the least enrichment in the LILE, consistent with emplacement in a back-arc setting. Geochemical modeling suggests that fractional crystallization of these mafic magmas can produce the less evolved sequences of the Tillito Formation (early Miocene) that also crop out in a more easterly position, further away from the Chile trench. However, the more evolved magmas located in the main volcanic arc also require the assimilation of the Permo-Triassic basement to produce their compositions or, alternatively, they could be derived from a different mantle source with a greater influence from subducting components (Jones et al. 2016).

Evidence for crustal contributions in the late Oligocene-to-early Miocene arc magmas is given by their gradual increase in $^{87}Sr/^{86}Sr$ ratios and decrease in Nd isotope ratios, reflecting an increase in radiogenic crustal contribution with time. This is in contrast to alkaline-like Paleocene and early Miocene basalts, which have an isotopically depleted nature (Fig. 4) (Kay et al. 1991; Litvak and Poma 2010; Litvak y Page 2010; Kay and Abbruzzi 1996). Specific evidence for the assimilation of the Permo-Triassic basement by the Early Miocene silicic magmas (Tilito Formation) located in the main arc on Chilean slope comes from inherited zircon core ages (158.0 ± 2.4, 241 ± 2.7 and 388.1 ± 5.3 Ma; Jones et al. 2016) and is consistent with the relatively high $^{87}Sr/^{86}Sr$ ratios of the Doña Ana Group volcanic rocks (Kay and Abbruzzi 1996). However, the more intermediate and mafic units of the Tilito Formation, located further to the east, show no inherited zircon providing evidence for a more limited interaction with the local basement (Jones et al. 2015, 2016).

The first expression of early Miocene arc-related magmatism within the Precordillera is within the Valle de Iglesia (Miocene Intrusives) (Fig. 4). These magmatic rocks are more evolved and enriched in incompatible trace elements. Relatively high Nb/Zr ratios could be indicative of a relatively small degree of partial melting, consistent with their generation further away from the trench and therefore over a more dehydrated slab (Jones et al. 2016).

Arc-related volcanic activity was widespread during the early Miocene and throughout middle to late Miocene times in the location of the present Argentinean-Chilean border (Cerro de las Tórtolas and Tambo Formations) (Fig. 2). These lavas have calc-alkaline andesitic to dacitic compositions (Fig. 4) with typical arc-type trace element patterns (Kay et al. 1987, 1991, 1999; Ramos et al. 1989; Otamendi et al. 1994; Bissig et al. 2003; Litvak et al. 2007; Jones et al. 2016).

Overall, the reported La/Ta, Ba/Ta, Ba/La for these Miocene magmas are indicative of mantle sources influenced by the slab (Fig. 4). An increase in the 'arc-like' signature (e.g., high Ba/La, Fig. 4) is seen in middle-to-late Miocene magmas, which have higher fluid-mobile/immobile incompatible element ratios (e.g., Ba/Nb and U/Nb) suggesting a higher degree of partial melting of the mantle wedge due to an increase in fluids derived from the subducting slab, when compared to the late Oligocene magmas (Litvak et al. 2007; Jones et al. 2016).

Normalized REE patterns show evidence for a variable residual mineral assemblage throughout the Miocene arc magma evolution, as also seen in the La/Sm versus Sm/Yb plots (Fig. 4). La/Yb values are generally between 5 and 20; however, a main break in La/Yb ratio evolution is seen in the intermediate middle-to-late Miocene arc rocks, with the late Miocene showing a remarkable increase in La/Yb ratios. The trace element composition of the early Miocene-to-middle Miocene lavas infers that the residual assemblage was dominated by amphibole and pyroxene, whereas by the late Miocene it was dominated by garnet. Mass balance calculations and melting models have been used to conclude that the addition of garnet to the fractionation of middle-to-late Miocene magmas could generate the steep and variable REE patterns (Kay et al. 1987, 1991).

Overall, the late Miocene arc rocks display adakitic signatures (Drummond and Defant 1990), given by their $SiO_2 > 56\%$, $Al_2O_3 > 15\%$, and Sr > 400 ppm, together with low values of Y (<18 ppm) and high La/Yb ratios (>20) (Litvak et al. 2007; Kay and Mpodozis 2002; Jones et al. 2016). Although defining geochemical signatures and the origin of adakites remains controversial, there is consensus about the origin of this signal in the Chilean-Pampean flat-slab magmas ($\sim 30°$ S). The development of the adakitic signature in the late Miocene rocks is explained by arc magmas equilibrating with variable residual mineral assemblages in the lower crust related to increasing pressure conditions (Kay et al. 1991, 1999, 2005; Bissig et al. 2003; Litvak et al. 2007; Jones et al. 2016). The residual mineral assemblage develops from being pyroxene bearing in early–middle Miocene, to garnet bearing (higher pressure) by late Miocene times, reflecting an increase in crustal thickness (Kay et al. 1991, 1999, 2005; Bissig et al. 2003; Litvak et al. 2007; Goss et al. 2013; Jones et al. 2016). This is consistent with a more enriched isotopic signature; Nd and Sr isotope ratios are enriched and εNd decreases from the late Paleogene, suggesting a higher contribution of crustal-derived components with time (Fig. 4).

The Miocene arc magmas (Cerro de las Tórtolas Formation) equilibrated at the base of the crust and evolved toward intermediate depth magma chambers.

Equilibrium temperatures for phenocrysts assemblages were estimated using two-pyroxene, amphibole-plagioclase and amphibole geothermometers and show a consistent temperature range between 970 to 850 °C, while the equilibrium pressure estimated from amphibole compositions for the volcanic suites was close to 4 kbar (Litvak and Poma 2014). Thus, although the crustal thickness has increased from Early to Late Miocene, no change in pressure conditions for crystallization of phenocrysts assemblages has been recorded. The late Miocene volcanism ($\sim$8–6 Ma, Vacas Heladas Ignimbrites) is similar to the late Miocene intermediate magmas (Fig. 4), corresponding to high-K silicic rocks with arc-like geochemical signatures and high ratios of fluid-mobile/immobile incompatible trace elements (e.g., Ba/Nb, Pb/Ce, Ba/La), providing evidence for the influence of slab-derived fluids. Moreover, these arc magmas appear to have variably equilibrated with higher pressure assemblages, including amphibole, garnet, and lesser amounts of plagioclase (Fig. 4) (Litvak et al. 2007; Jones et al. 2016).

The highest $^{87}Sr/^{86}Sr$ values and lowest εNd within the entire late Oligocene–Late Miocene arc magmas are observed in the Vacas Heladas Ignimbrites rocks (Fig. 4), indicating they received the highest crustal contribution. It is suggested that these magmas represent small volumes of lower crustal melts, generated due the influence of heat and fluids derived from the subducting slab (Litvak et al. 2007; Kay et al. 1991; Kay and Abbruzzi 1996; Jones et al. 2016).

Middle-to-late Miocene arc-related volcanism is also developed in Argentina on the eastern side of the Cordillera del Colanguil, Valle de Iglesia, and western Precordillera (Fig. 2, Miocene Intrusives unit, Lomas del Campanario, and Las Trancas Formation). These magmatic rocks have mainly andesitic to dacitic compositions and ‘arc-like’ Ba/La ratios (Fig. 4). Ultimately before cessation of magmatism, volcanism migrated almost 700 km to the east of the Chilean trench, to the Pocho volcanic field in the Sierras Pampeanas. This has been associated with the arrival of the shallowly dipping subducting slab at these latitudes in the Chilean-Pampean flat-slab segment (Kay and Gordillo 1994).

5 Variable Crustal Contributions in Late Oligocene-to-Late Miocene Magmas Over the Pampean Flat-Slab Segment Based on the O and Hf Isotopic Composition of Zircon

Geochemical features of the arc-related Miocene magmatism over the Chilean-Pampean flat-slab region show a clear increase in crustal-derived contribution from the late Oligocene to late Miocene. A major question pertains to the origin of the crustal contaminants that have influenced the trace element and isotopic compositions of the arc magmas during petrogenesis and over time.

Kay et al. (1987, 1988, 1991) and Kay and Abbruzzi (1996) first correlated the chemical changes in the El Indio belt (Fig. 2) with an increase in crustal thickness, as reflected by the residual mineral assemblages equilibrating with the magmas at depth. Therefore, the main crustal contributions were related to crustal assimilation during magma ascent through the crust. On the other hand, geochemical variations in the Miocene-to-Pliocene volcanic suite of the El Teniente region, south of the Chilean-Pampean flat-slab ($\sim$34°), were explained by subducted sediment, along with continental crust entering the asthenospheric wedge from forearc subduction erosion (Stern and Skewes 2003; Skewes et al. 2002). Kay et al. (2005) concluded that both forearc erosion and crustal thickening are responsible for the geochemistry in the Oligocene–Miocene sequence of volcanic rocks at the latitude of the El Teniente deposit and in the northern SVZ. Moreover, trace element and isotopic ratios of Neogene andesitic magmatism in the Maricunga Belt, at the northern extreme of the Pampean flat-slab segment ($\sim$25.5°–28.2° S), were also attributed to magmas evolving under a thickened crust combined with a peak in forearc subduction erosion as the arc migrated east (Goss et al. 2013).

Similarly, crustal contributions in the Valle del Cura volcanic rocks have been related to both episodes of crustal thickening and increased crustal assimilation, and peaks in forearc subduction erosion (Kay and Mpodozis 2002; Kay et al. 2005; Litvak et al. 2007; Jones et al. 2016).

Studies based on in situ analysis of O and Hf isotopes in zircon combined with high-resolution U–Pb dating show across arc variations in age and composition of the crustal components that influenced late Oligocene-to-late Miocene arc volcanism (Jones et al. 2015). Late Oligocene-to-early Miocene volcanic rocks in the main arc, located in the Cordillera Frontal, show mantle-like $\delta^{18}O_{(zircon)}$ values (+4.8‰ ($\pm$0.2 (2σ)) to +5.8‰ ($\pm$0.5 (2σ))), but less radiogenic initial $\varepsilon Hf_{(zircon)}$ values than would be expected for recent mantle-derived melts (+1.0 ($\pm$1.1 (2σ)) to +4.0 ($\pm$0.6 (2σ))) compared to $\sim$+13 (Dhuime et al. 2011). Mixing models suggest that these arc magmas are derived from the mantle and have assimilated up to $\sim$20% Permo-Triassic basement (Jones et al. 2015). The same authors report that late Miocene rocks from the main arc in the Frontal Cordillera, together with the eastern sequences from late Oligocene to late Miocene times in the present Precordillera, require the assimilation of both the Permo-Triassic Andean crust and a Grenville-aged basement to produce the higher than mantle-like $\delta^{18}O_{(zircon)}$ values (+5.5‰ ($\pm$0.6 (2σ)) to +7.2‰ ($\pm$0.4 (2σ))) and unradiogenic, initial $Hf_{(zircon)}$ values (−3.9 ($\pm$1.0 (2σ)) to +1.6 ($\pm$4.4 (2σ))) (Fig. 5). It is important to notice that the isotopic variation correlates with the presence of Permo-Triassic and Grenville inherited zircon ages in the late Oligocene-to-late Miocene magmatic suites spanning the main Andes and Precordillera (Fig. 5). The observed isotopic changes appear to be related to their geographic position relative to the Chilean margin and the composition of the underlying basements, rather than the sample age (Fig. 5). Forearc subduction erosion might be an expected process during evolution of these arc magmas. However, any effect derived from subducted continental material affecting the melt source region appears to have been overprinted by the assimilation of the overlying crust en route to the surface, as O and Hf isotopic across arc variations reflect the distinct basement terranes (Jones et al. 2015). Thus, at least two types of basement, that differ in age and composition, are required to generate

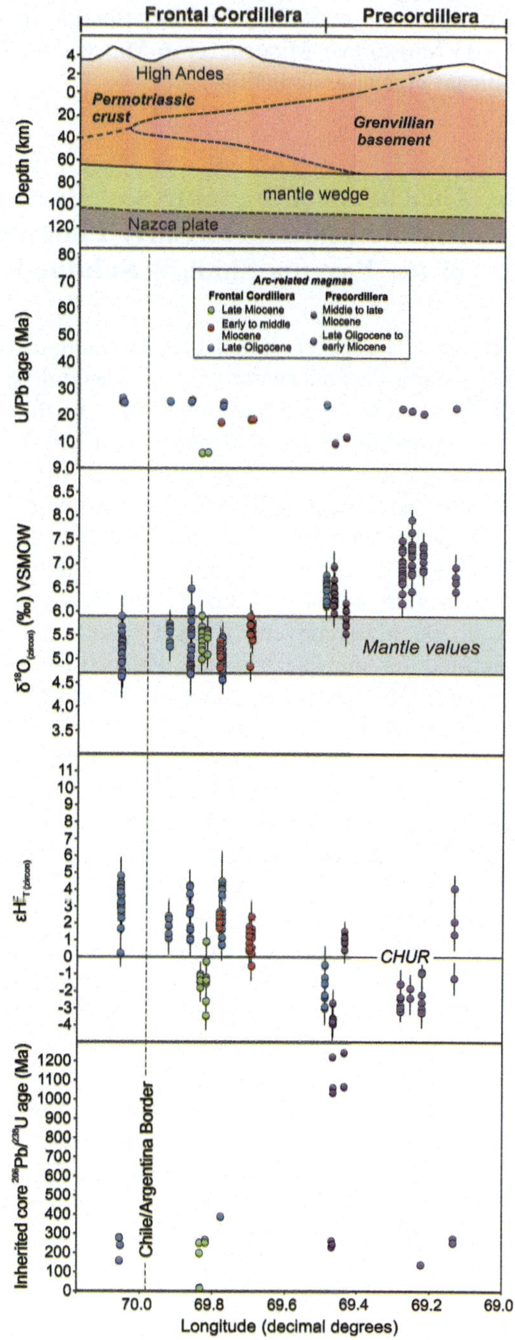

Fig. 5 Values of U–Pb crystallization ages, $\delta^{18}O_{(zircon)}$, $\varepsilon Hf_{(zircon)}$ from rims or zircons without inherited ages, and inherited $^{206}Pb/^{238}U$ core ages plotted against longitude. This demonstrates that the arc magmas have been affected by variable crustal contributions across the high Andes into the western Precordillera in the late Oligocene-to-late Miocene arc-related magmas over the Chilean-Pampean flat-slab segment (modified from Jones et al. 2015)

the observed variability in the isotopic composition of the late Oligocene (~23 Ma)-to-late Miocene (~6 Ma) arc magmas, highlighting a period of significant crustal assimilation.

6 Geochemical Signatures of Back-Arc Magmas During Middle Miocene to Early Pliocene as Evidence of the Payenia Shallow Subduction Development

The geochemical evolution of calc-alkaline back-arc magmas in the present-day retro-arc of Payenia can be evaluated according to the two time intervals that reflect the expansion of arc magmatism due to the shallowing of the subducting slab: (1) the middle to late Miocene (15–10 Ma), with volcanic rocks located in the near-to-middle retro-arc position and (2) the late Miocene-to-early Pliocene (9–3.5 Ma) middle-to-far retro-arc magmatic rocks.

Major element classifications show a range in composition from mid- to high-K basaltic to dacitic calc-alkaline rocks (Fig. 6). Some of the rocks are rhyolitic in composition, primarily from the late Miocene Chachahuén volcanic complex (Kay et al. 2006a, b); conversely, trachybasaltic to trachyandesitic composition has been reported for the late Miocene Palaoco volcanism (Dyhr et al. 2013b) (Fig. 6). Evidence for the influence of the subducting slab on the studied sequences is given by their Ba/Ta, La/Ta ratios, which are 'arc-like' regardless of silica content (Fig. 6). Normalized trace element patterns, also show depletions in HFSE (e.g., Nb, Ta, Zr) relative to LILE (e.g., Rb, Ba, Sr, Pb), suggesting that hydrous fluids coming from the subducted slab have enriched the melts in the asthenospheric mantle wedge. However, some geochemical differences are reported between the first volcanic stage (15–10 Ma) and the second one (9–3.5 Ma) that require changes in the geodynamic setting.

The middle-to-late Miocene arc-related sequences, representing the Payenia retro-arc, show lower SiO_2 versus FeO/MgO ratios when compared to the younger sequences within this stage (particularly Charilehue samples, ~14 Ma; Spagnuolo et al. 2012), which have a more tholeiitic differentiation trend and the lowest Ba/La and La/Ta ratios (Fig. 6). Younger samples, within this older stage (Cerro Negro, Huincán I and part of II lavas, and older San Rafael volcanic rocks, ~15–10 Ma), are more 'arc-like' based on the same ratios (Fig. 6) (Nullo et al. 2002; Kay et al. 2006b; Litvak et al. 2015). Consistently, the late Miocene Palaoco volcanic sequences (Dyhr et al. 2013a) also exhibit a stronger arc-related geochemical signature when compared to the older volcanic events.

The late Miocene-to-early Pliocene sequences continue the general trend toward an increase in slab-derived components in the retro-arc magmas with time. In particular, the youngest volcanic rocks within this stage (Early–Late Chachauhén and the younger San Rafael rocks), which represent the easternmost expression of the arc expansion, show the strongest arc-related signature when compared with

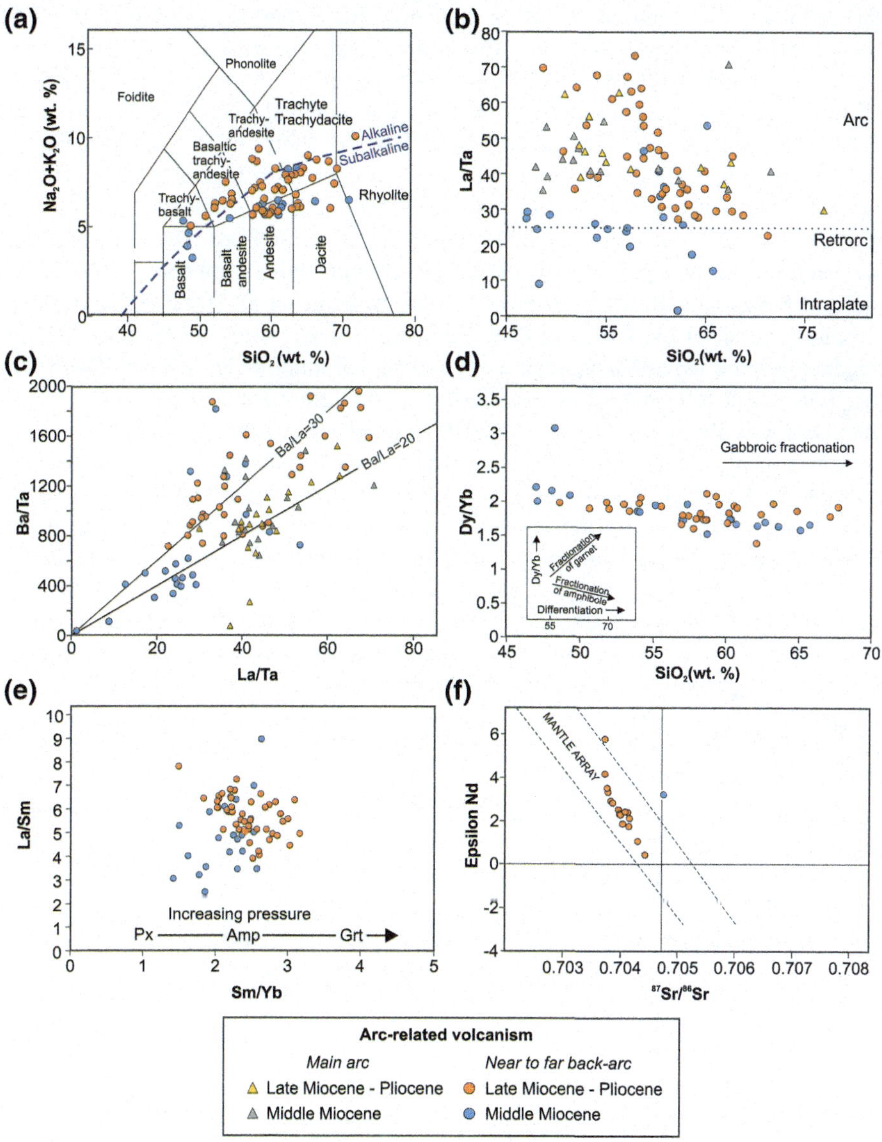

Fig. 6 Geochemical features of Late Cenozoic arc-related volcanism across the present-day Payenia retro-arc region. **a** Chemical classification based on total alkali versus silica content (Le Maitre et al. 1989); **b** 'arc-like' signatures given by La/Ta versus SiO_2; **c** Ba/Ta versus La/Ta for the middle Miocene-to-early Pliocene calc-alkaline sequences; **d** REE behavior reflected in Dy/Yb versus SiO_2; middle Miocene-to-early Pliocene main arc volcanism is included (**a**-**c**) for comparison; **e** variable HREE signatures demonstrated by La/Sm versus Sm/Yb ratios; **f** isotopic signatures that show a limited contribution from a crustal component. Compiled data came from Nullo et al. (2002), Kay et al. (2006a, b), Kay and Copeland (2006), Spagnuolo et al. (2012), Dyhr et al. (2013a, b) and Litvak et al. (2015)

earlier Miocene volcanism, based on their high Ba/Ta and La/Ta ratios (700–2000 and 27–50, respectively) (Fig. 6) (Kay et al. 2006a, b; Litvak et al. 2015).

Rare elements patterns from the whole series of middle Miocene-to-early Pliocene calc-alkaline back-arc magmas share a flattened overall REE pattern and similar La/Yb and Sm/Yb ratios, regardless of their silica content (Fig. 6). However, the late Miocene-to-early Pliocene samples show higher Sm/Yb, with the highest ratios reported for the youngest San Rafael volcanic sequences (Litvak et al. 2015; Fig. 6). This increase in Sm/Yb ratios for the arc-related retro-arc volcanism has been related to a period of back-arc crustal thickening between 12 and 10 Ma under the Neuquén Basin (Kay et al. 2006a, b; Nullo et al. 2002; Baldauf 1997). Despite this, all of the REE ratios are indicative of a low-to-intermediate pressure residual mineral assemblage, such as pyroxene and amphibole, in equilibrium with the Late Cenozoic magmas, demonstrating no significant increase in crustal thickness and a garnet-free lower crust (Litvak et al. 2015).

Lower crustal contamination was proposed as a plausible process that could explain the low REE, Cs, Rb, Th, and U contents and high values of Sr and Ba, such as those seen in the more silica-rich lavas of the Late Miocene San Rafael Block, which have Mg# similar to the more mafic lavas (Litvak et al. 2015). Kay et al. (2006b) proposed crustal contamination of the silicic samples from the early Miocene intraplate-like sequences within the Chachahuén volcanic complex (Vizcachas Group), based on whole rock chemistry and isotope composition. Isotopic data from the Miocene intermediate to silicic calc-alkaline retro-arc magmas show $^{87}Sr/^{86}Sr$ and εNd compositions (Fig. 6) relatively more radiogenic than the late Oligocene–early Miocene mafic back-arc alkaline-type volcanism (Kay et al. 2006a, b; Dyhr et al. 2013b). Overall, the Late Cenozoic magmas show an evolving trend with increasing $^{87}Sr/^{86}Sr$ ratios and decreasing εNd values as 'arc-like' signatures also increase. Contamination of the magmas may have occurred en route to the surface rather than by the addition of crust through forearc subduction erosion (Kay et al. 2006a, b). However, Dyhr et al. (2013a) advocated a role for subduction erosion on the petrogenesis of the late Miocene arc-related rocks (Early Palaoco andesites) in the present Payenia back-arc.

7 Tectonic and Geodynamic Implications for the Evolution of Arc-Related Magmas Over the Pampean Flat-Slab and Payenia Shallow Subduction Regimes

Geochemical features of the arc-related magmatism, developed between the late Oligocene to early Pliocene in both the Chilean-Pampean flat-slab and Payenia shallow subduction segments, provide evidence to refine the geodynamic evolution of the Andean margin in the Southern Central Andes (Figs. 7 and 8).

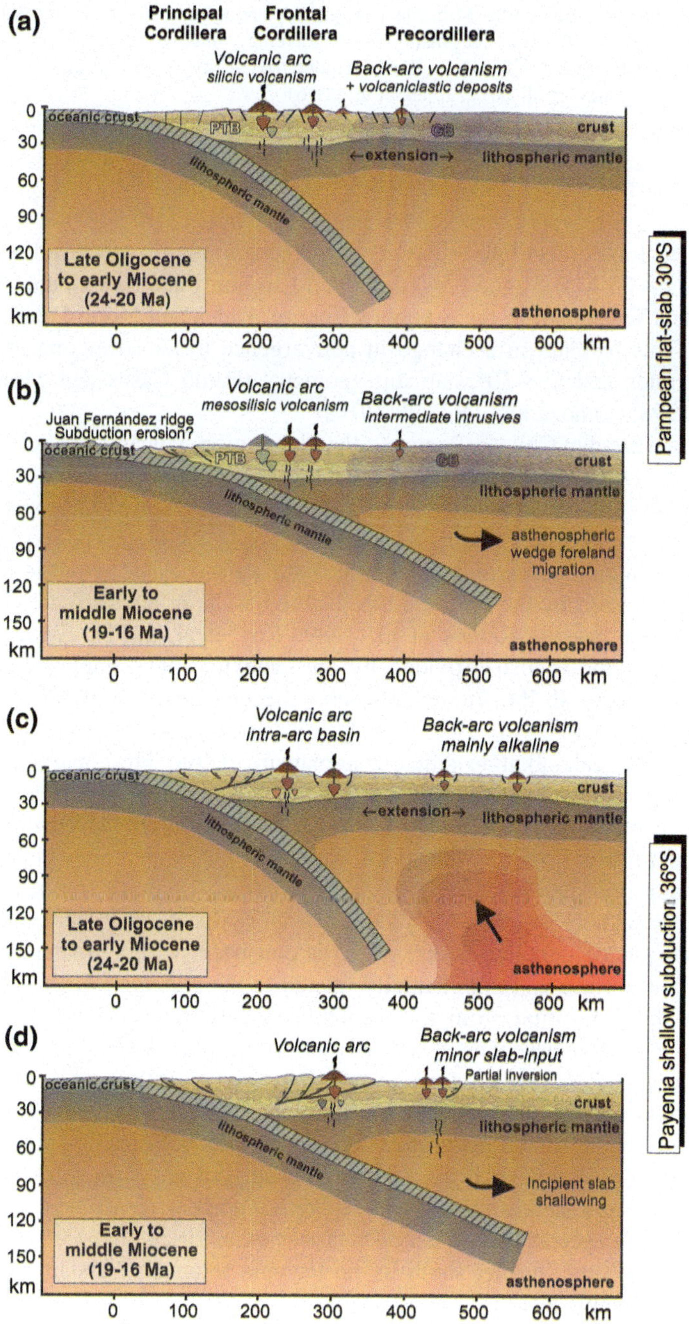
(a)
Principal Cordillera
Frontal Cordillera
Precordillera
Volcanic arc
silicic volcanism
Back-arc volcanism
+ volcaniclastic deposits
oceanic crust
PTB
CB
crust
←extension→
lithospheric mantle
lithospheric mantle
Late Oligocene to early Miocene (24-20 Ma)
asthenosphere
(b)
Volcanic arc
mesosilisic volcanism
Back-arc volcanism
intermediate intrusives
Juan Fernández ridge
Subduction erosion?
oceanic crust
PTB
CB
crust
lithospheric mantle
lithospheric mantle
asthenospheric wedge foreland migration
Early to middle Miocene (19-16 Ma)
asthenosphere
Pampean flat-slab 30°S
(c)
Volcanic arc
intra-arc basin
Back-arc volcanism
mainly alkaline
oceanic crust
crust
←extension→
lithospheric mantle
lithospheric mantle
Late Oligocene to early Miocene (24-20 Ma)
asthenosphere
(d)
Volcanic arc
Back-arc volcanism
minor slab-input
Partial inversion
oceanic crust
crust
lithospheric mantle
lithospheric mantle
Incipient slab shallowing
Early to middle Miocene (19-16 Ma)
asthenosphere
Payenia shallow subduction 36°S
km

◂**Fig. 7** Late Oligocene-to-middle Miocene geodynamic model and magmatic evolution for: **a–b** Chilean-Pampean flat-slab segment; **c–d** Payenia shallow subduction zone. PTB: Permian-Triassic basement, GB: Grenville basement (based on Kay et al. 2006a; Litvak et al. 2007, 2015; Spagnuolo et al. 2012; Jones et al. 2016)

7.1 Late Oligocene–Early Miocene (~24–20 Ma)

A major tectonic reconfiguration of the Andean margin occurred in the late Oligocene (~25 Ma) by the breakup of the Farallón Plate (see Chap. "Paleogene Arc-Related Volcanism in the Southern Central Andes and North Patagonia (39°–41° S)"). This resulted in a change in convergence to an orthogonal direction at increased rates and a ~30° slab dip geometry (Pardo Casas and Molnar 1987; Somoza 1998; Somoza and Ghidella 2012).

In the present-day Chilean-Pampean flat-slab region, volcanic arc activity during this period (~27–18 Ma) is represented by the calc-alkaline Doña Ana Group whose parental magmas formed in the asthenospheric mantle wedge and erupted through a crust of normal thickness (30–35 km) (Fig. 7a, b). These sequences were developed in a subduction-related setting, with an associated extensional tectonic regime, as reflected by the eruption of alkaline basalts in the retro-arc at ~21 Ma (Las Máquinas basalts; Ramos et al. 1989; Kay et al. 1991). Evidence for an extensional regime is also provided by the synextensional emplacement of volcaniclastic deposits in the former retro-arc, exposed in the Valle del Cura Basin (Winocur et al. 2015).

The high-K, calc-alkaline silicic magmatism (Tillito Fm.), particularly those closer to the Chilean trench, has trace element signatures and inherited zircon grains/cores that suggest assimilation of the Permo-Triassic basement; however, this assimilation is shown to be more limited in the contemporaneous and easternmost andesitic deposits (Jones et al. 2015) (Fig. 7). Furthermore, these latter andesites show more negative $\delta^{11}B$ values (from pyroxene melt inclusions) relative to the younger Miocene arc rocks, which is consistent with a greater depth to the slab-mantle interface and is potentially related to the widening of the volcanic arc and their distal position from the trench (Jones et al. 2014). Emplacement of subvolcanic bodies associated with pyroclastic and lava flows in the western Precordillera (~22–20 Ma, Miocene Intrusives and Las Trancas Formation) provides evidence for contemporaneous, although limited, volcanic activity in the easternmost sector of the retro-arc (Fig. 7a). While volcanic arc magmas show evidence for interaction with the Permo-Triassic Andean crust, these easternmost magmas also show contributions from a Grenville-age basement, as suggested by inherited zircon ages and Hf-O isotopes (Jones et al. 2015, 2016) (Figs. 5 and 7a).

At the same time, further south in the Payenia segment, the slab dip geometry was also relatively steeply dipping due to the prevailing Andean tectonic configuration and the main arc was characterized by volcanic and volcaniclastic activity as part of the infill of the intra-arc Cura Mallín Basin (Fig. 7c). Bimodal composition and trace element patterns obtained from these arc rocks reflect magmas evolving

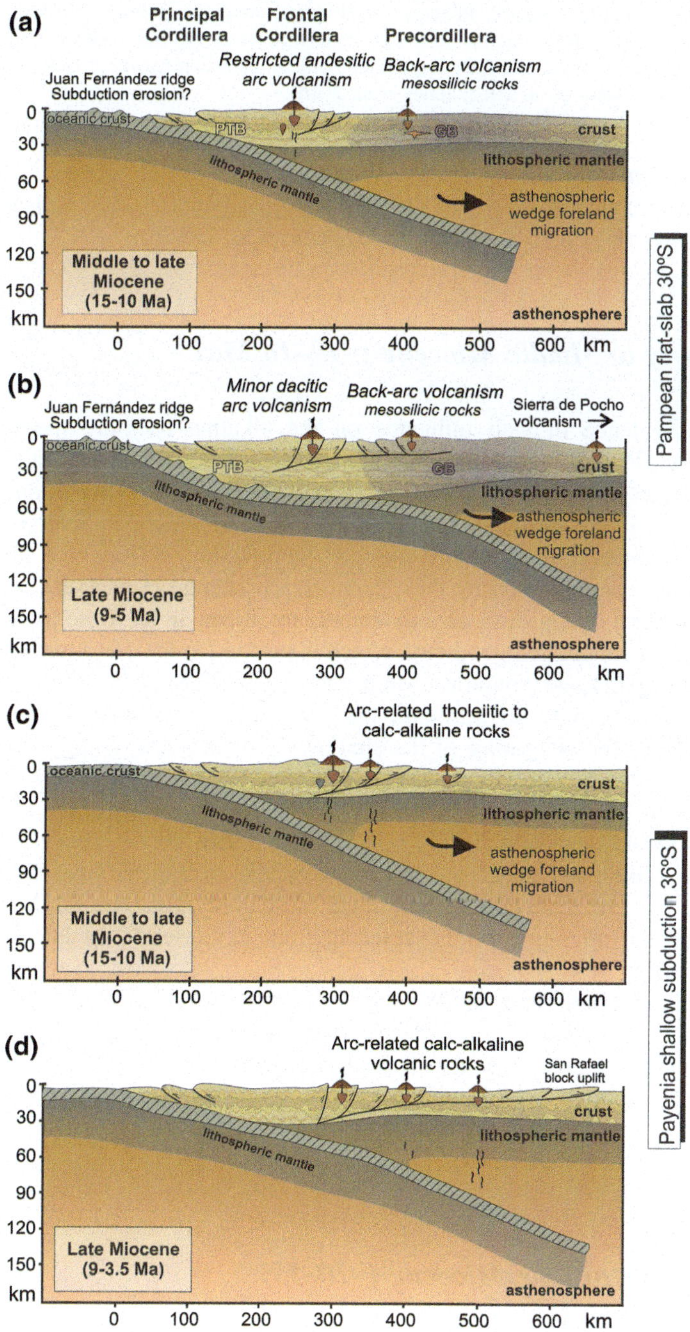
(a)
Principal Cordillera
Frontal Cordillera
Precordillera
Restricted andesitic arc volcanism
Back-arc volcanism mesosilicic rocks
Juan Fernández ridge Subduction erosion?
oceanic crust
PTB
GB
crust
lithospheric mantle
lithospheric mantle
asthenospheric wedge foreland migration
Middle to late Miocene (15-10 Ma)
asthenosphere
(b)
Minor dacitic arc volcanism
Back-arc volcanism mesosilicic rocks
Juan Fernández ridge Subduction erosion?
Sierra de Pocho volcanism
Late Miocene (9-5 Ma)
Pampean flat-slab 30°S
(c)
Arc-related tholeiitic to calc-alkaline rocks
Middle to late Miocene (15-10 Ma)
(d)
Arc-related calc-alkaline volcanic rocks
San Rafael block uplift
Late Miocene (9-3.5 Ma)
Payenia shallow subduction 36°S
km

◂**Fig. 8** Middle Miocene-to-early Pliocene geodynamic model and magmatic evolution over: **a–b** the present-day Chilean-Pampean flat-slab segment; **c–d** the Payenia shallow subduction zone. PTB: Permo-Triassic basement, GB: Grenville basement (based on Kay et al. 2006a; Litvak et al. 2007, 2015; Spagnuolo et al. 2012; Jones et al. 2016)

through a thin crust in an extensional setting (Kay et al. 2006a). Meanwhile, back-arc magmatism comprises a series of isotopically enriched intraplate, OIB-like basaltic flows (Kay and Copeland 2006; Dyhr et al. 2013b).

7.2 Early to Middle Miocene (~19–16 Ma)

The first evidence of slab influence on the alkaline back-arc magmas in the present-day Payenia region is registered in the early-to-middle Miocene Huantrainco region lavas (~20–19 Ma; Fig. 7d). These show depletions in HFSE, higher La/Ta ratios and an excess of fluid-mobile elements, indicative of a more hydrous component, relative to the previously erupted lavas (Kay et al. 2006a; Dyhr et al. 2013a). As explained by these authors, the change in back-arc lava composition took place over a short period, from 23 to 18 Ma, in response to the initiation of a shallower subduction geometry, which in turn led to the cessation of extension.

By the early to middle Miocene (~19–15 Ma) in the present-day Chilean-Pampean flat-slab segment, the prevailing extensional regime was also ceasing due to the beggining of the shallowing of the downgoing slab (Kay and Mpodozis 2002). Basaltic to andesitic lavas of early Miocene age (Escabroso Fm.) were erupted along the main arc. The main arc-volcanic activity also includes middle Miocene andesitic lavas (~16–14 Ma; Cerro de las Tórtolas Fm.) with clearer 'arc-like' signatures and REE ratios which indicate that they evolved through a crust of relatively normal thickness (30–35 km). The shallowing of the slab led to the migration of the volcanic front to the east and into present-day Argentina (Fig. 7b).

Overall, this period of andesitic arc volcanism shows evidence for increased influence of slab-derived fluids on the melt source region and a higher degree of partial melting, in comparison with the late Oligocene. This is indicated by higher $\delta^{11}B$ values reported from Miocene melt inclusions, suggesting an increase in the influence of serpentinite-derived fluids on the source of arc magmas after ~19.5 Ma (Jones et al. 2014).

7.3 Middle to Late Miocene (~10–15)

This period marks the main influence on evolution of arc magmatism (at ~29°-30° S) due to the arrival of the Juan Fernández Ridge. The gradual reduction of the subducted slab angle, in the present-day Chilean-Pampean flat-slab segment, led to the expansion of arc magmatism to the east and the development of a compressional regime, which resulted in the main phase of uplift of the Andean range (Kurtz et al. 1997; Gregory-Wodzicki 2000). Arc volcanism is represented by the andesitic to dacitic lavas and minor dacitic pyroclastic flows (upper Cerro de las Tórtolas and Tambo Fms.) which have adakitic signatures, resulting from the equilibration of the melts with a garnet-bearing lower crust (Kay et al. 1991; Litvak et al. 2007; Jones et al. 2016). This is consistent with tectonic shortening and an increase of crustal thickness (>50 km). In addition, the asthenospheric wedge was restricted in volume below the main arc (Kay and Abbruzzi 1996). Flattening of the slab is also evidenced by the eastwards migration of arc-related magmatism and the emplacement of shallow level, subvolcanic andesites and dacites in the eastern Frontal Cordillera and the western Precordillera (Tertiary Intrusives) (Fig. 8a).

Contemporaneously, arc magmatic rocks in the present-day Payenia back-arc region show an increase in slab-derived components at this time. While the main arc magmas show no significant geochemical differences to the early Miocene volcanic front (Kay et al. 2006a), retro-arc magmas from middle to late Miocene age (~ 15–10 Ma; Charilehue, Cerro Negro, Huincán I, part of Huincán II lavas, older San Rafael volcanic rocks, and Palaoco volcanism) show a progression from initial tholeiitic to more pronounced calc-alkaline signatures (Kay et al. 2006a, b; Dyhr et al. 2013a, b; Litvak et al. 2015). This eastern increase of slab-derived components present in back-arc magmas is consistent with the progressive shallowing of the slab at these latitudes and the expansion of the arc-related activity (Fig. 8c). The Pampean and Payenia segments show a spatial continuity (Fig. 1), while timing of slab shallowing in both segments is contemporaneous. Thus, the collision of the Juan Fernández Ridge is likely the cause of the progressive shallowing throughout both segments during Miocene times.

7.4 Late Miocene to Early Pliocene (~9–4 Ma)

As the subducting Nazca Plate continued to shallow, the volcanic front and arc-related products gradually migrated to the east during the late Miocene and volcanism ceased in the flat-slab segment (e.g., Ramos et al. 1989; Kay et al. 1991; Kay and Gordillo 1994; Kay and Mpodozis 2002; Litvak et al. 2007). Minor dacitic volcanism marked the end of the main arc activity (~ 8–6 Ma; Vacas Heladas Ignimbrites) as the remaining fluids and heat favored the melting of the crust.

The O-Hf zircon isotopic compositions, together with inherited zircon ages, in these late Miocene rocks overlap the values from the magmatic rocks present in the Precordillera (Kay et al. 1991; Jones et al. 2015). This provides evidence for the presence of the Grenville-aged basement under the Frontal Cordillera in this sector of the Southern Central Andes (Jones et al. 2015) as a result of increased compression and crustal shortening in the latest Miocene, as proposed by several structural models (e.g., Allmenndinger et al. 1990; Gans et al. 2011; Gilbert et al. 2006;

Ramos et al. 2004). Progressive asthenospheric wedge retraction led to the migration of the arc-related products further east into the western Precordillera, and then to the Sierra de Pocho (Sierras Pampeanas, Fig. 8b), 700 km away from the Chile trench (Kay and Gordillo 1994).

The distribution of late Miocene-to-early Pliocene arc-related products was also progressively expanded to the east in the present-day Payenia back-arc, as consequence of the shallowing subduction regime (Fig. 8d). Arc expansion continued between 10 and 4 Ma with the extrusion of the back-arc Chachahuén volcanic complex and the younger San Rafael volcanic rocks, representing the easternmost expression of the calc-alkaline into the back-arc (Litvak et al. 2015; Kay et al. 2006a, b). After ~4 Ma, no volcanism with clear arc-related signatures was developed in the back-arc at these latitudes. As the slab geometry changed drastically, late Pliocene to Recent back-arc volcanism comprised voluminous mafic eruptions with alkaline-type geochemical fingerprints in a reinitiated extensional setting due to new steepening of the subducted Nazca Plate (e.g., Ramos et al. 2014). The Juan Fernández Ridge collision with the Andean margin, and its variable influence from N to S along the Pampean and Payenia segments, could result in the subtle and progressive change in the angle of subduction around 33° S (Nacif et al. 2015). A re-steepening of the southern section of the slab might account for the different responses and evolutions of the two shallow subduction regimes in the late Miocene to early Pliocene times between both segments.

8 Conclusions

Arc-related magmatism in the Southern Central Andes is strongly controlled by changes in the geometry and composition of the subducted slab as reflected in the geochemical evolution of arc-related products over the late Miocene Payenia shallow subduction zone and the present-day Pampean flat-slab segment. Both segments show a progressive foreland expansion and migration of the volcanic arc, as shown by the geochemical signature of arc-related rocks, which is correlated with the more compressional style of deformation associated with the shallowing of the slab. Despite sharing this common general evolution, some differences can be appreciated between both the 'shallow' and 'flat' subduction regimes (Kay et al. 2006a, b; Litvak et al. 2015). Late Miocene shallow subduction is less pronounced in the Payenia segment, as the easternmost expression of arc volcanism is located 480–500 km away from the Chile trench, at ~36° S; while in the Chilean-Pampean zone, slab influence reaches almost 700 km away from the trench at ~30° S. A further and main difference between both segments is related to the increase in crustal thickness. Late Miocene magmas erupted in the main arc over the present-day Chilean-Pampean segment show geochemical features in their trace element compositions, which suggest magmas were equilibrating at the base of a thickened crust (>50 km); however, in the Payenia shallow subduction segment, the late Miocene arc rocks show a garnet-free residual mineral assemblages implying a crust of normal thickness (~40–45 km). Finally, while the re-steepening of the slab in the Payenia region promoted an increase of arc and retro-arc magmatism during late Neogene, the prevailing flat-slab geometry in the Chilean-Pampean segment defines the current volcanic arc gap in Southern Central Andes.

Acknowledgements This work has been funded by grants from CONICET (grant 11220150100426CO) and University of Buenos Aires (grant UBACYT 20020150100166BA and 20020130100628BA). R. Jones was funded by a NERC CASE studentship awarded by the University of Edinburgh (NE/G524128/1), EIMF grants (IMF425/1010 and IMF450/1011), and the Derek and Maureen Moss Scholarship. L. Kirstein acknowledges the support from the Carnegie Trust for the Universities of Scotland. We thank Drs. S. Kay and P. Leal for fruitful discussion on the subject.

References

Allmendinger R, Figueroa D, Snyder D, Beer J, Mpodozis C, Isacks B (1990) Foreland shortening and crustal balancing in the Andes at 30° S latitude. Tectonics 9:789–809

Alonso S, Limarino CO, Litvak V, Poma S, Suriano J, Remesal M (2011) Paleogeographic, magmatic and paleoenviromental scenarios at 30° SL during the Andean Orogeny: cross sections form the volcanic-arc to the orogenic front (San Juan province, Argentina). In: Salfity JA, Marquillas RA (eds) Cenozoic Geology of the Central Andes of Argentine. SGS Publisher, pp 23–45

Baldauf P (1997) Timing of the uplift of the Cordillera Principal, Mendoza Province, Argentina. MS thesis, George Washington University, p 356

Barckhausen U, Ranero CR, Cande SC, Engels M, Weinrebe W (2008) Birth of an intraoceanic spreading center. Geology 36(10):767–770

Bermúdez A (1991) Sierra del Nevado. El límite oriental del arco volcánico del Neógeno entre los 35°30′ y 36° L.S. Argentina. In: VI Congreso Geológico Chileno, Actas 1, Santiago de Chile, pp 318–322

Bermúdez A, Delpino D, Frey F, Saal A (1993) Los basaltos de retro-arcoextraandinos. In: Ramos VA (ed) Geología y Recursos Naturales de Mendoza. Asociación Geológica Argentina, Buenos Aires, pp 161–172

Bertotto GW, Cingolani CA, Bjerg EA (2009) Geochemical variations in Cenozoic back-arc basalts at the border of La Pampa and Mendoza provinces, Argentina. J South Am Earth Sci 28:360–373

Bissig B, Clark AH, Rainbow A, Montgomery A (2015) Physiographic and tectonic settings of high-sulfidation epithermalgold–silver deposits of the Andes and their controls onmineralizing processes. Ore Geol Rev 65:327–364

Bissig T, Clark AH, Lee JKW, Heather KB (2001) The Cenozoic history of volcanism and hydrothermal alteration in the Central Andean flat-slab region: new 40Ar–39Ar constrains from the El Indio-Pascua Au(–Ag, Cu) belt, 29°20–30°30″S. Int Geol Rev 43:312–340

Bissig T, Clark AH, Lee JKW, von Quadt A (2003) Petrogenetic and metallogenetic responses to Miocene slab flattenig: new constrains from the El Indio-Pascua Au–Ag–Cu Belt, Chile/Argentina. Miner. Deposita 38(7):844–862

Burns WM, Jordan T, Copeland P, Kelley S (2006) Extensional tectonics in the Oligo-Miocene Southern Andes as recorded in the Cura-Mallín basin. In: Kay SM, Ramos VA (eds) Evolution of an Andean margin: a tectonic and magmatic view from the Andes to the Neuqén basin (35–39° S). Geological Society of America, Special paper 407, pp 163–184

Cardó R, Díaz IN (1999) Hoja Geológica 3169-I, Rodeo, Provincias de San Juan. Instituto de Geología y Recursos Minerales, Servicio Geológico Minero Argentino, Buenos Aires

Cardó R, Díaz IN, Limarino CO, Litvak VD, Poma S, Santamaria G (2007) Hoja Geológica 2969-III, Malimán, provincias de San Juan y La Rioja, Boletín, 320th edn. Instituto de Geología y Recursos Minerales, Servicio Geológico Minero Argentino, Buenos Aires

Charrier R, Baeza O, Elgueta S, Flynn JJ, Gans P, Kay SM, Muñoz N, Wyss AR, Zurita E (2002) Evidence for Cenozoic extensional basin development and tectonic inversion south of the flat-slab segment, southern Central Andes, Chile (33°–36° S.L.). J South Am Earth Sci 15:117–139

Charrier R, Pinto L, Rodríguez MP (2007) Tectonostratigraphic evolution of the Andean Orogen in Chile. In: Moreno T, Gibbons W (eds) The Geology of Chile. Geological Society of London, Special Publications, pp 21–116

Chulick GS, Detweiler S, Mooney WD (2013) Seismic structure of the crust and uppermost mantle of South America and surrounding oceanic basins. J South Am Earth Sci 42:260–276

Cobbold PR, Rosello EA (2003) Aptian to recent compressional deformation, foothills of the Neuquén Basin Argentina. Mar Petrol Geol 20:429–443

Delpino DH, Bermúdez A (1985) Volcán Plateado. Vulcanismo andesítico de retro-arco en el sector extrandino de la Provincia de Mendoza, 35°42′ Lat. Sur. Argentina. In: IV Congreso Geológico Chileno, Actas 3, Antofagasta, 108–119

Dhuime B, Hawkesworth C, Cawood P (2011) When continents formed. Science 331:154–155

Drummond MS, Defant MJ (1990) A model for trondjemite-tonalite-dacite genesis and crustal growth via slab melting: Archean to modern comparisons. J Geophys Res 95:21503–21521

Dyhr CT, Holm PM, Llambías EJ (2013a) Geochemical constraints on the relationship between the Miocene-Pliocene volcanism and tectonics in the Palaoco and Fortunoso volcanic fields, Mendoza Region, Argentina: new insights from 40Ar/39Ar dating, Sr–Nd–Pb isotopes and trace elements. J Volcanol Geoth Res 266:50–68

Dyhr CT, Holm PM, Llambías EJ, Scherstén A (2013b) Subduction controls on Miocene back-arc lavas from Sierra de Huantraico and La Matancilla and new 40Ar/39Ar dating from the Mendoza Region, Argentina. Lithos 179:67–83

Folguera A, Naranjo JA, Orihashi Y, Sumino H, Nagzo K, Polanco E, Ramos VA (2009) Retro-arc volcanism in the northern San Rafael Block (34°–35°30′S), southern Central Andes: occurrence, age, and tectonic setting. J Volcanol Geoth Res 186:169–185

Folguera A, Orts D, Spagnuolo M, Rojas Vera E, Litvak VD, Sagripanti L, Ramos M, Ramos VA (2011) A review of Late Cretaceous to Quaternary paleogeography of the Southern Andes. Biol J Linn Soc 103:250–268

Fromm R, Zandt G, Beck SL (2004) Crustal thickness beneath the Andes and Sierras Pampeanas at 30° S inferred from Pn apparent phase velocities. Geophys Res Lett 31:L06625

Gans CR, Beck SL, Zandt G, Gilbert H, Alvarado P, Anderson M, Linkimer L (2011) Continental and oceanic crustal structure of the Pampean flat slab region, western Argentina, using receiver function analysis: new high-resolution results. Geophys J Int 186:45–58

Giambiagi LB, Bechis F, García V, Clark A (2005) Temporal and spatial relationship between thick- and thin-skinned deformation in the thrust front of the Malargüe fold and thrust belt, Southern Central Andes. In: VI International Symposium on Andean Geodynamics, Extended Abstracts, Niza, pp 315–318

Giambiagi L, Bechis F, García V, Clark A (2008) Temporal and spatial relationship between thick- and thin-skinned deformation in the Malargüe fold and thrust belt, Southern Central Andes. Tectonophysics 459:123–139

Gilbert H, Beck S, Zandt G (2006) Lithospheric and upper mantle structure of central Chile and Argentina. Geophys J Int 165:383–398

Goss AR, Kay SM, Mpodozis C (2013) Andean adakite-like high-Mg andesites on the northern margin of the Chilean-Pampean flat-slab (27–28.5° S) associated with frontal arc migration and fore-arc subduction erosion. J Petrol 54:2193–2234

Gregory-Wodzicki KM (2000) Uplift history of the Central and Northern Andes: a review. Geol Soc Am Bulletin 112:1091–1105

Groeber P (1946) Observaciones geológicas a lo largo del meridiano 70, I. Hoja Chos Malal. Revista Sociedad Geológica Argentina 1:177–208

Gudnason J, Holm PM, Søager N, Llambías EJ (2012) Geochronology of the late Pliocene to recent volcanic activity in the Payenia back-arc volcanic province, Mendoza Argentina. J South Am Earth Sci 37:191–201
Hernando IR, Aragón E, Frei R, González PD, Spakman W (2014) Constrains on the origin and evolution of the magmas in the PayúnMatrú Volcanic Field, Quaternary Andean Back-arc of Western Argentina. J Petrol 55(1):209–239
Jones RE, Hoog JC, Kirstein LA, Kasemann SA, Hinton R, Elliott T, Litvak VD, EIMF (2014) Temporal variations in the influence of the subducting slab on Central Andean arc magmas: evidence from boron isotope systematics. Earth Planet Sc Lett 408:390–401
Jones RE, Kirstein LA, Kasemann SA, Dhuime BR, Elliott T, Litvak VD, Alsonso R, EIMF (2015) Geodynamic controls on the contamination of Cenozoic arc magmas in the southern Central Andes: insigths from the O and Hf isotopic composition of zircon. Geochim Cosmochim Ac 164:386–402
Jones RE, Kirstein LA, Kasemann SA, Litvak VD, Poma S, Elliott T, Alonso R, Hinton R, EIMF (2016) The role of changing geodynamics in the progressive contamination of Late Cretaceous to Late Miocene arc magmas in the southern Central Andes. Lithos 262:169–191
Jordan TE, Isacks BL, Allmendinger RW, Brewer JA, Ramos VA, Ando CJ (1983) Andean tectonics related to geometry of subducted Nazca plate. Geol Soc Am Bull 94:341–361
Jordan TE, Burns WM, Veiga R, Pangaro F, Copeland P, Mpodozis C (2001) Mid-Cenozoic intra-arc basins in the southern Andes. Tectonics 20. https://doi.org/10.1029/1999TC001181
Kay SM, Abbruzzi JM (1996) Magmatic evidence for Neogene lithospheric evolution of the Central Andes "flat slab" between 30° S and 32° S. Tectonophysics 259:15–28
Kay SM, Copeland P (2006) Early to middle Miocene back-arc magmas of the Neuquén Basin: Geochemical consequences of slab shallowing and the westward drift of South America. In: Kay SM, Ramos VA (eds) Late Cretaceous to Recent magmatism and tectonism of the Southern Andean margin at the latitude of the Neuquen basin (36–39° S). Geological Society of America, Special Paper 407, pp 185–214
Kay SM, Gordillo CE (1994) Pocho volcanic rocks and the melting of depleted continental lithosphere above a shallowly dipping subduction zone in the central Andes. Contrib Mineral Petr 117:25–44
Kay SM, Mpodozis C (2002) Magmatism as a probe to Neogene shallowing of the Nazca plate beneath the modern Chilean flat-slab. J South Am Earth Sci 15(1):39–57
Kay SM, Maksaev VA, Moscoso R, Mpodozis C, Nasi C (1987) Probing the evolving Andean lithosphere: mid-late Tertiary Magmatism in Chile (29°-30°30′S) over the modern zone of subhorizontal subduction. J Geophys Res 92:6173–6189
Kay SM, Maksaev VA, Moscoso R, Mpodozis C, Nasi C, Gordillo CE (1988) Tertiary Andean Magmatism in Chile and Argentina between 28° S and 33° S: correlation of magmatic chemistry with changing Benioff zone. J South Am Earth Sci 1:21–38
Kay SM, Mpodozis C, Ramos VR, Munizaga F (1991) Magma source variations for mid-late Tertiary magmatic rocks associated with shallowing zone and thickening crust in the central Andes (28° to 33° S). In: Harmon RS, Rapela CW (eds). Andean magmatism and its tectonic setting. Geological Society of America, Special Paper 265, pp 113–137
Kay SM, Mpodozis C, Coira B (1999) Neogene magmatism, tectonism and mineral deposits of the Central Andes (22°–23° S Latitude). In: Skinner BJ (ed) Geology and Ore Deposits of the Central Andes. Soc of Economic Geol, Special Publication 7, pp 27–59
Kay SM, Godoy E, Kurtz A (2005) Episodic arc migration, crustal thickening, subduction erosion and magmatism in the South-Central Andes. Geol Soc Am Bulletin 117:67–88
Kay SM, Mancilla O, Copeland P (2006a) Evolution of the Back-arc Chachahuén volcanic complex at 37° S latitude over a transient Miocene shallow subduction zone under the Neuquén Basin. In: Kay SM, Ramos VA (eds) Late Cretaceous to Recent magmatism and tectonism of the Southern Andean margin at the latitude of the Neuquen basin (36–39° S). Geological Society of America, Special Paper 407, pp 215–246
Kay SM, Burns WM, Copeland P, Mancilla O (2006b) Upper Cretaceous to Holocene magmatism and evidence for transient Miocene shallowing of the Andean subduction zone under the

northern Neuquén Basin. In: Kay SM, Ramos VA (eds) Late Cretaceous to Recent magmatism and tectonism of the Southern Andean margin at the latitude of the Neuquen basin (36–39º S). Geological Society of America, Special Paper 407, pp. 19–60

Kurtz AC, Kay SM, Charrier R, Farrar E (1997) Geochronology of Miocene plutons and exhumation history of the El Teniente region, Central Chile (34–35° 8). Andean Geol 24:75–90

Le Maitre RW, Bateman P, Dudek A, Keller J, Lameyre MJ, Le Bas PA, Sabine R, Schmid H, Sorensen A, Streckeisen AR, Wooley Zanettin B (1989) A classification of igneous rocks and glossary of terms, pp 193, Blackwell, Oxford

Leveratto M (1976) Edad de intrusivos cenozoicos en la Precordillera de San Juan y su implicancia estratigráfica. Rev Asoc Geol Argent 31:53–58

Limarino CO, Gutiérrez PR, Malizia D, Barreda V, Page S, Ostera H, Linares E (1999) Edad de las secuencias paleógenas y neógenas de las cordilleras de la Brea y Zancarrón, Valle del Cura. San Juan. Rev Asoc Geol Argent 54(2):177–181

Litvak VD (2009) El volcanismo Oligoceno Superior—Mioceno Inferior del Grupo Doña Ana en la alta cordillera de San Juan. Rev Asoc Geol Argent 64(2):200–212

Litvak VD, Page S (2002) Nueva evidencia cronológica en el Valle del Cura, provincia de San Juan. Rev Asoc Geol Argent 57(4):483–486

Litvak VD, Poma S (2010) Geochemistry of mafic Paleocene volcanic rocks in Valle del Cura region: Implications for the petrogenesis of primary mantle-derived melts over the Pampean flat-slab. J South Am Earth Sci 29(3):705–716

Litvak VD, Poma S (2014) Petrogenesis of Miocene Volcanic arc rocks over the Chilean-Pampean flat-slab segment of the Central Andes constrained by mineral chemistry. Geologica Acta 12(2):151–171

Litvak VD, Poma S, K SM (2007) Paleogene and Neogene magmatism in the Valle del Cura region: a new perspective on the evolution of the Pampean flat slab, San Juan province, Argentina. J South Am Earth Sci 24(2–4):117–137

Litvak VD, Spagnuolo MG, Folguera A, Poma S, Jones R, Ramos VA (2015) Late Cenozoic calc-alkaline volcanism over the Payenia shallow subduction zone, South-Central Andean back-arc (34°30-37′S), Argentina. J South Am Earth Sci 64:365–380

Llambias EJ, Sato AM (1990) El Batolito de Colangüil (29–31° S) cordillera frontal de Argentina: estructura y marco tectonico. Andean Geol 17:89–108

Llambías EJ, Shaw S, Sato AM (1990). Lower Miocene plutons in the Eastern Cordillera Frontal of San Juan (29°75'S 69°30'W). In Congreso Geológico Argentino No. 11, Actas 1:83–86. San Juan

Llambías EJ, Bertotto GW, Risso C, Hernando I (2010) El volcanismo cuaternario en el retro-arco de Payenia: una revisión. Rev Asoc Geol Argent 67(2):278–300

Maksaev V, Moscoso R, Mpodozis C, Nasi C (1984) Las unidades volcánicas y plutónicas del Cenozoico superior entre la Alta Cordillera del Norte Chico (29°–31° S), Geología, alteración hidrotermal y mineralización. Rev Geol Chile 21(1):11–51

Martin MW, Clavero RJ, Mpodozis MC, Cuitiño L (1995) Estudio Geológico de la Franja El Indio, Cordillera de Coquimbo. Servicio Nacional de Geología y Minería, Santiago. Informe registrado IR-95–6, 1, 1–238

McGlashan N, Brown L, Kay S (2008) Crustal thickness in the central Andes from teleseismically recorded depth phase precursors. Geophys J Int 175:1013–1022

Muñoz J, Niemeyer RH (1984) Hoja Laguna del Maule, regiones del Maule y Biobio: Santiago, Chile. Servicio Nacional de Geología y Minería, Carta Geológica de Chile, no. 64, scale 1:250,000, 98 p

Muñoz Bravo J, Stern C, Bermúdez A, Delpino D, Dobbs MF, Frey FA (1989) El volcanismo plio-cuaternario a través de los 38° y 39° S de los Andes. Rev Asoc Geol Argent 44:270–286

Muñoz M, Fuentes F, Vergara M, Aguirr L, OlovNyström J, Féraud G, Demant A (2006) Abanico East Formation: petrology and geochemistry of volcanic rocks behind the Cenozoic arc front in the Andean Cordillera, central Chile (33°50′S). Rev Geol Chile 331:109–140

Nacif S, Triep E, Spagnotto S, Aragon E, Furlani R, Álvarez O (2015) The flat to normal subduction transition study to obtain the Nazca plate morphology using high resolution seismicity data fromthe Nazca plate in Central Chile. Tectonophysics 657:102–112

Nasi CP, Moscoso RD, Maksaev VJ (1990) Hoja Guanta, Región de Coquimbo. Carta Geológica de Chile N° 67. Santiago de Chile, Servicio de Geología y Minería, 140 p

Niemeyer RH, Muñoz J (1983) Hoja Laguna de la Laja, Región del Bio Bio: Santiago, Chile. Servicio Nacional de Geología y Minería, Carta Geológica de Chile, no. 57, scale 1:250,000, 52 p

Nullo FE, Stephens G, Otamendi J, Baldauf P (2002) El volcanismo del Terciario superior del sur de Mendoza. Rev Asoc Geol Argent 57(2):119–132

Nullo FE, Franchi M, González P, Herrero JC, Reinoso MS (1993) Mapa Geológico de la Provincia de Río Negro. Dirección Nacional del Servicio Geológico, Buenos Aires

Nullo FE, Panza JL, Blasco G (1999) El Jurásico y Cretácico de la Cuenca Austral. In: Caminos R (ed) Geología Argentina. Servicio Geológico Minero Argentino, pp 528–535

Ostera H, Linares E, Haller M (1999) Paramillos Altos intrusive belt, Southern Mendoza, Argentina. Ages, chemical and isotopic constraints. In: II South American Symposium on Isotope Geology, Actas 2, Córdoba, pp 256–260

Otamendi J, Nullo F, Godeas M, Pezzutti N (1994) Petrogenesis del volcanismo terciario del Valle del Cura, San Juan, Argentina. In: VII Congreso Geológico Chileno, Actas 2, Concepción, pp 1130–1135

Parada MA, Rivano S, Sepulveda P, Herve M, Herve F, Puig A, Munizaga F, Brook M, Pankhurst R, Snelling N (1988) Mesozoic and Cenozoic plutonic developmentin the Andes of central Chile (30°30′–32°30′S). J South Am Earth Sci 1:249–260

Pardo Casas F, Molnar P (1987) Relative motion of the Nazca (Farallón) and South America plates since Late Cretaceous time. Tectonics 6:233–248

Pilger RH (1981) Plate reconstructions, aseismic ridges, and low angle subduction beneath the Andes. Geol Soc Am Bulletin 92:448–456

Pilger RH (1984) Cenozoic plate kinematics, subduction and magmatism: South American Andes. J Geol Soc Lond 141:793–802

Pineda G, Emparan C (2006) Geología del área Vicuña-Pichasca, Región de Coquimbo. Carta Geológica de Chile, Serie Geología Básica. Servicio Nacional de Geología y Minería, Santiago

Poma S, Limarino C, Litvak V (2005) Formación Las Trancas: expresión del arcomagmático terciario en el faldeo occidental de la Precordilera de San Juan. In: XVI Congreso Geológico Argentino Actas 1, La Plata, pp 331–334

Poma S, Ramos A, Litvak VD, Quenardelle S, Maisonnave B (2014). Tertiary volcanism from Iglesia Valley Basin, San Juan Province, Argentina. In: XIX Congreso Geológico Argentino, Actas CD, Córdoba, 2 p

Quidelleur X, Carlut J, Tchilinguirian P, Germa A, Gillot PY (2009) Paleomagnetic directions from mid-latitude sites in the southern hemisphere (Argentina): contribution to time averaged field models. Phys Earth Planet Inter 172:199–209

Radic JP, Rojas L, Carpinelli A, Zurita E (2002) Evolución Tectónica de la cuenca Terciaria de Cura-Mallín Región Cordillerana Chileno Argentina (36°30′–39°00′S). In: XIV Congreso Geológico Argentino, Actas 3, Calafate, pp 233–237

Ramos VA, Barbieri M (1989) El volcanismo cenozoico de Huantraico: edad y relaciones isotópicas iniciales, provincia del Neuquén. Rev Asoc Geol Argent 43:210–223

Ramos VA, Folguera A (2005) Tectonic evolution of the Andes of Neuquén: Constraints derived from the magmatic arc and foreland deformation. In: Veiga G, Spalletti L, Howell J, Schwarz E (eds) The Neuquén Basin: a case study in sequence stratigraphy and basin dynamics. Geological Society of London, Special Publication 252, pp 15–35

Ramos VA, Folguera A (2011) Payenia volcanic province in the Southern Andes: an appraisal of an exceptional Quaternary tectonic setting. J Volcanol Geoth Res 201:53–64

Ramos VA, Page R, Kay S, Lapido O, Delpino D (1987) Geología de la región del Volcán Tórtolas, Valle del Cura, provincia de San Juan. In: X° Congreso Geológico Argentino, Actas IV, Tucumán, pp 260–263

Ramos VA, Kay SM, Page R, Munizaga F (1989) La Ignimbrita Vacas Heladas y el cese del volcanismo en el Valle del Cura, provincia de San Juan. Rev Asoc Geol Argent 44(1–2): 336–352

Ramos VA, Cristallini EO, Pérez DJ (2002) The Pampean flat-slab of the Central Andes. J South Am Earth Sci 15(1):59–78

Ramos VA, Zapata T, Cristallini E, Introcaso A (2004) The Andean thrust system—latitudinal variations in structural styles and orogenic shortening. Thrust Tectonics and Hydrocarbon Systems 82:30–50

Ramos VA, Folguera A, Litvak VD, Spagnuolo M (2014) Andean tectonic cycle: From crustal thickening to extension in a thin crust (34°–37° SL). Geosci Front 5:351–367

Skewes MA, Arévalo A, Floody R, Zuñiga H, Stern CR (2002) The giant El Teniente breccia deposit: hypogene copper distribution and emplacement. Society of Economic Geologists Special Publication 9:299–332

Søager N, Holm PM, Llambías EJ (2013) Payenia volcanic province, southern Mendoza, Argentina: OIB mantle upwelling in a back-arc environment. Chem Geol 349–350:36–53

Somoza R (1998) Updated azca (Farallon)—South America relative motions during the last 40 My: implications for mountain building in the central Andean region. J South Am Earth Sci 11:211–215

Somoza R, Ghidella ME (2012) Late Cretaceous to recent platemotions in western South America revisited. Earth Planet Sc Lett 331–332:152–163

Spagnuolo M, Litvak VD, Folguera A, Ramos VA (2012) Neogene magmatic expansion and mountain building at the southern Central Andes, 36°–37° S, Argentina. J Geodyn 53:81–94

Sruoga P, Rubinstein NA, Etcheverría MP, Cegarra M, Kay SM, Singer B, Lee J (2008) Estadio inicial del arco volcánico neógeno en la Cordillera Principal de Mendoza (35° S). Rev Asoc Geol Argent 63(3):454–469

Stern CR, Skewes MA (2003) Generation of giant Miocene and Pliocene copper deposits in Central Chile: Role of ridge subduction, decreased subduction angle, increased subduction erosion, crustal thickening, and mafic and adakite-like dacitic magma within long-lived, batholith size, open system magma chambers. In: X Congreso Geológico Chileno, Actas CD, Concepción

Suárez M, Emparan C (1995) The stratigraphy, geochronologyand paleophysiography of a Miocenefreshwater interarc basin, Southern Chile. J S Am Earth Sci 8:17–31

Wetten AF (2005) Andesita Cerro Bola: Nueva unidad vinculada al magmatismo mioceno de la Cordillera de Olivares, San Juan, Argentina (30°35′S; 69°30′O). Rev Asoc Geol Argent 60:003–008

Winocur DA, Litvak V, Ramos VA (2015) Magmatic and tectonic evolution of the Oligocene Valle del Cura basin, Main Andes of Argentina and Chile: evidence for generalized extension. Geological Society of London, Special Publications 399:109–130

Yañez GA, Ranero CR, von Huene R, Díaz J (2001) Magnetic anomaly interpretation across the southern central Andes (32°–34°): the role of the Juan Fernández Ridge in the late Tertiary evolution of the margin. J Geophys Res 106(B4):6325–6345

Yañez GA, Cembrano J, Pardo M, Ranero CR, Selles D (2002) The Challenger-Juan Fernández–Maipo major tectonic transition of the Nazca-Andean subduction system at 33–34° S: geodynamic evidence and implications. J South Am Earth Sci 15:28–38

Basin Thermal Structure in the Chilean-Pampean Flat Subduction Zone

Gilda Collo, Miguel Ezpeleta, Federico M. Dávila, Mario Giménez, Santiago Soler, Federico Martina, Pilar Ávila, Francisco Sánchez, Ricardo Calegari, Juan Lovecchio and Mario Schiuma

Abstract Flat-slab segments are considered refrigerated areas given that the asthenospheric wedge is forced to shift hundreds of kilometres away from the trench, and the flat and coupled subducting plate acts as a thermal insulator. Although lithospheric-scale thermal analysis based on numerical modelling and geophysical observations abound, studies on the thermal history of sedimentary basins are scarce. In this contribution, we present a temperature data compilation from more than 60 oil wells within the Chilean-Pampean flat-slab segment and the transitional zones to normal subduction to the north and south in the south-central Andes. The geothermal gradient data are correlated with basin-basal heat flow estimated from 1D modelling, Curie point depths derived from aeromagnetic surveys, and previous crustal and lithospheric thicknesses estimations. Their distribution evidences a quite good consistency and correlation from region-to-region. Our modelling demonstrates that sedimentation changes are not sufficient to explain the variations illustrated in the geothermal gradient map, and that basal heat flux variations are required to reproduce the reported values. According to our results, the coldest basins develop over the flat slab or cratonward regions, whereas the highest temperatures on areas where the slab plunges. This suggests that the flat-slab geometry as well as the lithospheric

G. Collo (✉) · M. Ezpeleta · F. M. Dávila · F. Martina · P. Ávila · F. Sánchez
CICTERRA, CONICET—FCEFyN—Universidad Nacional de Córdoba, Córdoba, Argentina
e-mail: gildacoll@gmail.com

F. M. Dávila
e-mail: fmd.geodinamica@gmail.com

M. Giménez
Instituto Geofísico Sismológico Ing. Volponi (IGSV), Universidad de Nacional San Juan, San Juan, Argentina

M. Giménez
Consejo Nacional de Investigaciones Científicas y Técnicas (CONICET), Buenos Aires, Argentina

S. Soler
Instituto Geofísico-Sismológico Volponi, San Juan, Argentina

R. Calegari · J. Lovecchio · M. Schiuma
YPF, Buenos Aires, Argentina

A. Folguera et al. (eds.), *The Evolution of the Chilean-Argentinean Andes*,
Springer Earth System Sciences, https://doi.org/10.1007/978-3-319-67774-3_21

structure affects the thermal state within the upper crust and particularly the sedimentary basins. Further studies will allow improving our database as well as the knowledge about the radiogenic contribution of the lithosphere and the asthenospheric heat input to the basins basal heat flow.

Keywords Chilean-Pampean flat subduction zone · Heat flow · Basins
Sub-lithospheric cooling · Thermal history · Bore hole data · Curie point depths
Magnetic data

1 Introduction

Flat subduction segments in South America have been deeply studied in the last decades since they were observed geophysically (Alvarado et al. 2007; Anderson et al. 2007; Gutscher et al. 2000; Marot et al. 2014; Tassara et al. 2006). The shallowing, and particularly the flattening of the oceanic subducting lithosphere, affects the overriding continental plate in different ways. These regions show a typical evolution from volcanic widening to shut-off (Kay and Mpodozis 2002) (see Chap. "The Late Paleogene to Neogene Volcanic Arc in the Southern Central Andes (28°-37° S)", increase of intraplate seismicity (Alvarado et al. 2009; Gans et al. 2011), basement foreland deformation (Ramos et al. 2002) (see Chap. "Cretaceous Orogeny and Marine Transgression in the Southern Central and Northern Patagonian Andes: Aftermath of a Large-Scale Flat-Subduction Event?") to long-wavelength dynamic uplift (Dávila and Lithgow-Bertelloni 2015) (see Chap. Mantle Influence on Andean and Pre-Andean Topography). Likewise, the regional thermal field is expected to change in relation to those from normal subduction regions (see Collo et al. 2011; Marot et al. 2014).

In South America, the highest geothermal gradients were described in normal subduction regions, i.e. where the slabs plunge ~30° E, or where lithosphere is abnormally thin. For example, the heat flows in Bolivia are in the order of 50–120° mW/m^2 at positions ~300–500 km away from the trench associated with a steeper subducting plate, the presence of an asthenosphere wedge and the local effect of volcanism driven by lithospheric thinning (Gutscher 2002; Hamza et al. 2005; Henry and Pollack 1988; Kay and Coira 2009). In the Neuquén Basin, where subduction is also interpreted as normal (Bohm et al. 2002), heat flow values are generally >46 mW m^{-2} and average ~160 mW m^{-2} (Hamza et al. 2005; Sigismondi 2012). In regions affected by flat subduction, in contrast, the thermal regimes show a substantial cooling. The overriding and subducting lithospheres conduct to regional refrigeration. This is because of a doble influence (Fig. 1), the asthenospheric wedge is forced to shift hundreds of kilometres away from the trench and the flat and coupled subducting plate acts as a thermal insulator (Álvarez et al. 2014; Gutscher et al. 2000; English et al. 2003; Manea and Manea 2011). In the flat slab of Peru at the same distance from the trench as in the Chilean-Pampean segment, for example, heat flows vary between 30 and 70 mW/m^2 (Gutscher 2002;

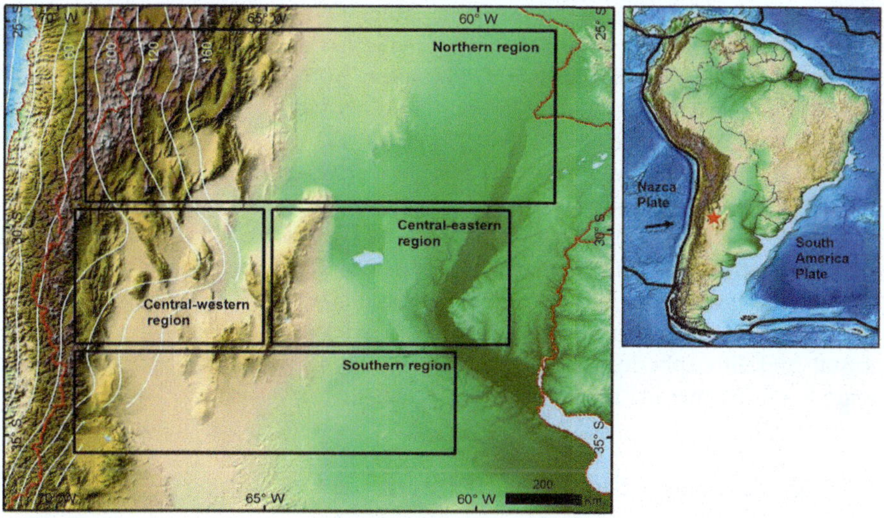

Fig. 1 Localization map of the study region with the Wadatti-Benioff isobaths (USGS), and the four discriminated regions analyzed in the present Chapter

Henry and Pollack 1988). Heat fluxes between 43 and 80 mW/m^2 were estimated by Ruiz and Introcaso (2004) for the Precordillera and the Western Pampean Ranges from Curie point depths values. Although lithospheric-scale thermal analysis based on numerical modelling and geophysical observations abound, studies on the thermal history of sedimentary basins are scarce (Collo et al. 2011, 2015; Dávila and Carter 2013). The basins can preserve in their sedimentary records the influence of heat derived from different sources, supplying valuable information about heat flow and its spatial and temporal variations. The Chilean-Pampean flat-slab segment in Argentina, and its northern and southern transitional zones to normal subduction, are in turn an indeed laboratory to analyze the main driving controls on basal heat flow at lithospheric scales. The knowledge of the current thermal regime in regions like these will contribute to the understanding of the shallow thermal structure within the different basins captured from borehole data or collected sensitive minerals (thermometers), and whether the thermal fluctuations from basin-to-basin respond either to the sedimentation history, post-depositional evolution and/or lithospheric structure and geodynamics.

Basin studies in the flat-slab segment have suggested extremely cold post-depositional thermal history, associated with refrigerated lithospheres. The Vinchina Basin in the northernmost part of the Chilean-Pampean flat-slab segment, for example; shows extremely low-heat flow values (likely the lowest in South American foreland region) below 30 mW/m^2 (Collo et al. 2011, 2015), related to a subduction dipping relatively low compared with other subduction-related foreland basins (Gutscher et al. 2000; Syracuse and Abers 2006). In foreland basins related to normal subduction, the thermal structure has been related to the influences of cold oceanic lithosphere underlying an upper continental lithosphere, but with an asthenospheric wedge as a heat source. Moreover, authors as Springer (1999) mention that low surface heat-flow density in the Andean foreland may be attributed to transient effects in the thermal field caused by recent sedimentation. Further evaluations, considering steady states and transient heat-flow regimes are necessary to evaluate these interpretations.

In this contribution, we present the first geothermal gradient map in the Pampean flat-slab segment of the south-central Andes. Considering the thermal history during slab shallowing and flattening, we correlate our new data with information in the flat-slab segment and the transitional zones to normal subduction (towards the north and south), basin-basal heat flow (BHF) estimated from petroleum exploration borehole data, curie point depths (CPD) derived from aeromagnetic surveys and crustal and lithospheric thicknesses. This will allow us evaluating the driving controls on the thermal records reported in sedimentary basins.

2 The Pampean Flat-Slab Segment

Approximately 10% of subduction margins are flat slabs (Gutscher 2002); in these settings, the oceanic plate, initially descending at a normal angle down to a depth of ~100 km, remains horizontally for several hundred of kilometres (Espurt et al. 2007; Gutscher et al. 2000; Kirby et al. 1996; Manea and Manea 2011). Gutscher et al. (2000) recognized ten flat subduction regions worldwide, being the best examples in South America (Peru and Chile-Argentina; see Barazangi and Isacks 1979; Cahill and Isacks 1992; Eakin et al. 2014; Engdahl et al. 1998; Gutscher et al. 2000; Hampel 2002; McGeary et al. 1985; Pardo et al. 2002; Pilger 1981; Sacks and y Okada 1974), associated with the subduction of the Nazca plate below the Sudamerican Plate. Although several shallow subduction segments were described in the Andean literature (modern and ancient, Ramos and Folguera 2009) (see Chap. "Cretaceous Orogeny and Marine Transgression in the Southern Central and Northern Patagonian Andes: Aftermath of a Large-Scale Flat-Subduction Event?"), the Chilean-Argentine analogous (Fig. 1) is probably the most studied worldwide. Based on eastward volcanic broadening (see Chap. "The Late Paleogene to Neogene Volcanic Arc in the Southern Central Andes (28°-37° S)"), elevation of the Main Cordillera, and formation of a basement broken foreland system, the change from shallowing to flattening subduction was proposed to begin at ~15–18 Ma (Gutscher et al. 2000; Jordan et al. 2001; Ramos and Folguera 2009; Ramos et al. 2002; Yañez et al. 2002), becoming completely flat (locally overturned, cf. Gans et al. 2011) at ~7–6 Ma (Gutscher et al. 2000; Kay et al. 1991; Kay and Mpodozis 2002).

The geometric subduction models (Fig. 1) of the Chilean-Pampean flat slab are based primarily on receiver functions and regional 3D seismic tomography (Gans et al. 2011; Gilbert et al. 2006; Marot et al. 2014), seismic and geophysical analysis (Barazangi and Isacks 1976; Alvarado et al. 2007; Anderson et al. 2007; Gimenez et al. 2009; Tassara et al. 2006) and show that in this area the Nazca plate lays sub-horizontally for ~400 km, from near the trench, before plunging into the mantle with "normal" angularity (~30–45° to the East).

Figure 1 shows the transition from flat to steep towards the north is more gradual than to the south, where it is sharper (about 32 S, Gans et al. 2011; Giambiagi and

Ramos 2002; Nacif et al. 2015; Pardo et al. 2002). The Pampean flat-slab geometry matches with the subduction and eastward projection to the continent of a thickened oceanic slab (15–20 km), known as the Juan Fernandez aseismic ridge (Anderson et al. 2007; Barazangi and Isacks 1976; Jordan et al. 1983; Pardo et al. 2002; among others). Recent work (e.g., van Hunen et al. 2008; Martinod et al. 2005; Marot et al. 2014; Tassara et al. 2006), however, suggested that the subduction of thickened ridges is a necessary but insufficient condition to drive a slab flattening.

In general, subduction zone thermal models show that the temperature distribution of the overriding plate is significantly affected by slab geometry (Currie et al. 2002; Manea et al. 2005). Most studies have shown substantial differences between normal and flat subduction segments along South America, with deeper geotherms where flat slabs occurs (e.g., Gutscher 2002; Marot et al. 2014). Across the flat-slab segments, the geotherms distribution is more spaced to the west than to the east where the astenospheric wedge occurs (Fig. 1). Faster seismic velocities and more recurrent hypocentres between the subducting plate and overriding lithosphere support this hypothesis.

Geothermal gradients and BHF are scarce. In Central Chile, between 26° and 33° S, Uyeda and Watanabe (1982) reported values between 10 and 79 mW/m^2 with the lowest values (10–36 mW/m^2) between 26.5° and 29° S; which are comparable with values estimated for the flat-slab segment of central Peru ($\sim$30 mW/m^2; Muñoz 2005). In the Mendoza region, Uyeda et al. (1978) reported data from more than 20 petroleum exploration wells in the 33° and 34° S segment, with geothermal gradient values between 28° and 44 °C/km. Data from water wells (with maximum depths of 500 m) were also reported for the region (Hamza and Muñoz 1996; Miranda and Pesce 1997; Pesce 1995, 2000, 2005; Uyeda et al. 1978), but inconsistences in temperatures trends likely associated with the shallowness were encountered. Grevemeyer et al. (2006), from vertical temperature gradient measured in the Chilean seafloor near the trench between 32.5° and 33.5° S, record heat flow values of 24–114 mW/m^2. In the Bermejo Basin, in the Chilean-Pampean flat-slab segment (between 30° and 31° S), studies in the Pozuelos and Mata gusanos wells calculated a similarly low BHF of $\sim$42 mW/m^2, associated with an extremely cold lithosphere (Collo et al. 2015). Further east, out of the flat-slab area (see Fig. 1), Calegari et al. (2014) reported a BHF $\sim$57 mW/m^2 in the General Levalle basin in the Hunt Cd.GL.x-1 well. The estimations from wells can be compared with upper crust heat flow data obtained from terrestrial and some aeromagnetic data to calculate the CPD. Ruiz and Introcaso (2004) estimated for the Precordillera segment the CPD between 24 and 40 km, where the slab is flat (which corresponds with heat flow values between 43 and 60 mW/m^2) and somewhat lower values between 20 and 36 km (heat fluxes between 50 and 80 mW/m^2) in the Western Pampean Ranges, where the slab is less flat (take into account that heat flow values were calculated using conductivity of $\sim$3 W/mK). These values show a maximum flow pattern in northeast—southwest direction, in agreement with the Juan Fernandez ridge projection under the south-central Andean lithosphere. These authors also mentioned an intriguing CPD anomaly of $\sim$20 km and $\sim$90 mW/m^2 near the Sierra de Valle Fertil. We revise, in this work, this anomaly. For the Chacoparaná basin, further east from the area where the flat slab becomes plunging (see Fig. 1), Sigismondi and Fantin (2014) calculated CPD between 23 and 54 km and, considering a reference value of 57 mW/m^2 obtained for the General Levalle Basin (cf. Calegari et al. 2014), proposing heat flow values between 42 and 54 mW/m^2. Their relative minimum values (42–44 mW/m^2) were placed in the central part of the Córdoba Province, within the leading edge of the flat subduction.

While from north to south in the Cordilleran region the lithospheric thicknesses show small changes (<10 km, Tassara and Echaurren 2012), crustal thicknesses in the flat subduction region are larger than in the normal subduction segments (Alvarado et al. 2007; Gans et al. 2011). Within the flat-slab segment, from west to east, crustal thicknesses vary from ~55–60 km (in the Argentine Precordillera region) to ~35–40 km (in the eastern Sierras Pampenas region, Alvarado et al. 2007, 2009; Álvarez et al. 2015; Chulick et al. 2013; Fromm et al. 2004; Gans et al. 2011; Gilbert et al. 2006; Gimenez et al. 2000; Tassara et al. 2006). The upper crustal thickness (estimated from gravity models, see Tassara and Echaurren 2012) is thin and relatively constant across the flat-slab segment compared to the normal subduction segments. According to the models, this might also contribute to the general cooling within the flat-slab segment, as radiogenic heat production would be lower than in the surrounding regions affected by normal subduction.

Although north-south lithospheric thickness variations are relatively small, west-east profiles (e.g., Tassara and Echaurren 2012) show an increase in thickness cratonward, from ~100 km in the region affected by the flat subduction to ~120 in the Río de la Plata region and ~160 km in the Brazilian craton to the East.

3 Factors Affecting the Thermal Structure of Sedimentary Basins

Changes in the thermal record reported for sedimentary basins can be attributed alternatively to variations in the depositional and post-depositional basin histories, in the thermal structure at crustal and lithospheric scales (associated with the geometry of subduction, variations in thicknesses, composition and ages, etc.) or by combination of them.

The distribution of the surface heat flow is determined by mechanical and thermal processes occurred within the crust and mantle (Allen and Allen 2005; Hantschel and Kaueraf 2009). Under steady state conditions, the surface heat flux is composed of the heat flow at the base of the lithosphere and the radiogenic heat production of the lithospheric mantle and crust (Fig. 2). Actually, there is a debate on the relative contribution of crustal and mantle heat flow into the observed variations in the surface heat flow of stable continents (Artemieva 2009). The average contribution of the asthenospheric flow is ~20 mW/m^2 (Waples 2001), although much lower values (11–18 mW/m^2) have been reported for cratonic regions (Jaupart et al. 2007; Mareschal and Jaupart 2013). In thick lithospheres, where the isotherms of 1,300–1,200 °C are more distant from surface, the heat flow recorded is expected to be low. The average heat productions for the lithospheric mantle and lower crust are 0.02 and 0.4 μW/m^3, respectively (Hasterak and

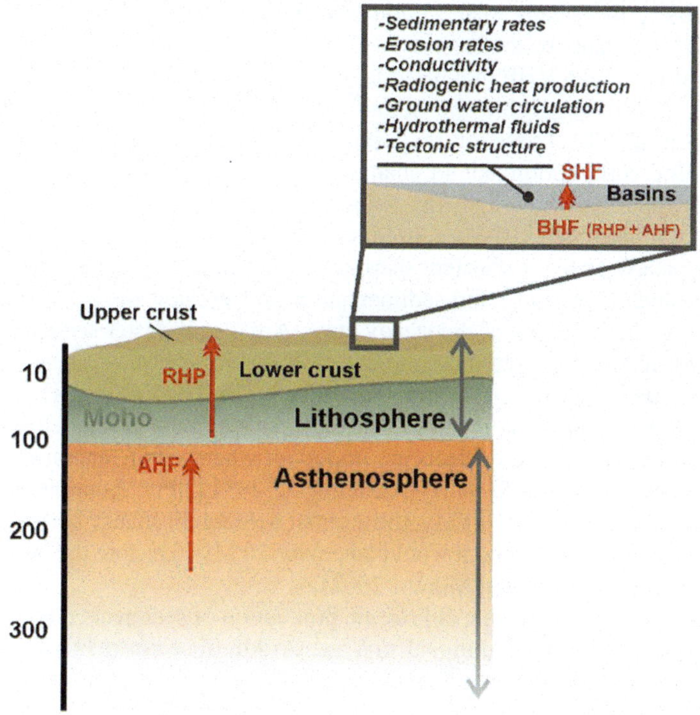

Fig. 2 Scheme of the factors affecting the thermal regime in sedimentary basins. AHF: asthenospheric heat flow; RHP: radiogenic heat production; BHF: basal heat flow; SHF: surface heat flow

Chapman 2011). Furthermore, the heat production within the upper crust might vary depending on composition and age, with values of 2 $\mu W/m^3$ for felsic rocks, 0.2 $\mu W/m^3$ for mafic rocks and 0.02 $\mu W/m^3$ for ultramafic rocks (Hasterak and Chapman 2011). As a result of natural decay of radiogenic element, old crusts should preserve lower relative heat production than younger crusts. While the production of lithospheric mantle, because of composition (mafic), is much lower than the production of the crust, it cannot be neglected in regions where the lithospheric thickness is important.

Variations in the surface heat flux can be also differentiated in wavelengths. While short wavelengths might be associated with variations in the composition and thicknesses within the crust, longer wavelengths have been interpreted as terranes with different ages or variations in heat flux to the base of the lithosphere. The influence of the basal lithosphere heat flow is associated with its thickness; the larger it is, the basal heat source is deeper and the time required to reach equilibrium is longer (Jaupart et al. 2007). For example, in a lithosphere

250 km thick, the steady state conditions could be reached after 200–500 My (Mareschal and Jaupart 2013).

To summarize, the BHF in a basin is mainly controlled by a combination of the radiogenic generation within the lithosphere (which depends in turn on the age, composition and thickness of the crust and the lithospheric mantle) and the astenospheric contribution. Rapid changes in the basin BHF generate transient or non-steady state basin thermal flux, which changes the geothermal gradients and generally stabilizes in ~10 My (Artemieva 2011).

A non-steady state or transient thermal flow could also be generated by sedimentation and/or erosion. Non-sedimentation and erosion hiatus, in turn, might contribute to the recovery of the steady state. Areas with faster sediment accumulation result in lower geothermal gradients (blanketing effect) and heat flow values decreasing towards the surface by the diffusion and advection (Hantschel and Kaueraf 2009). In constrast, increasing erosion rates might contribute to a rise in the geothermal gradients. These effects are important when sedimentation-erosion rates are larger than 100–1,000 m/My (Artemieva 2011). For example, with rates between ~200 and ~1000 m/My, the thermal field might change between ~1 and 3.5 mW/m^2, respectively, and it would take over 30 My to restore the normal steady state heat fluxes (Husson and Moretti 2002).

Within a sedimentary basin, radiogenic production of sediments might also affect on the basin heat flow record in ~1 mW/m^2 per km of sediments (Allen and Allen 2005). Other factors that might affect the basin thermal regime are the thermal conductivity and variations in surface temperature. The thermal conductivity depends mainly on porosity. A significant increment of porosity (~5%) can reduce the conductivity in ~25%. Considering the heat flow equation, the higher the thermal conductivity, the lower the geothermal gradient. Variations in surface temperature of about 10 °C (for example from icehouse to greenhouse, and viceversa) can be stabilized in ~10 My and influence much less (in magnitude) than the changes generated by variations in the basal heat flow (Hantschel and Kaueraf 2009). Ground water circulation, hydrothermal fluids as well as the tectonic structure can also affect the basin geothermal gradient, although their influences at the moment are difficult to estimate (Artemieva 2011).

4 Geothermal Gradients in the Segment Between 25 and 35° S

Temperature data were compiled from 46 wells between 25° and 35° S; from the Cordillera to the Chacoparaná basin. This database (Table 1) is the main source of the geothermal gradient map shown in Fig. 3.

Table 1 Integrated database of oil wells (this work; Uyeda et al. 1978) and geochemical data (Uyeda and Watanabe 1982) from the southern central Andes. *Wells removed in Fig. 3b

Well	Initials	Latitud W	Longitud S	Depth (mbs)	Measured T(°C)	Corrected T(°C)	Surface T(°C)	Geothermal gradient (°C/km)	BHF (mW/m^2)
This work									
Alto Pencoso es-2	AP	−33.481	−66.923	2791	78	94	21.5	26	
Árbol Blanco SE.AB. X-1	AB	−26.854	−61.957	3720	119	138	24.5	30	63
Beazley es-1	BZ1	−33.758	−66.775	3396	99	117	21.5	28	57
Beazley x − 2	BZ2	−33.759	−66.795	2483	70	84	21.5	25	
Bermejo	BE	−30.083	−68.194	5023	92	108	20.5	18	
Camilo Aldao es-1	CA	−33.122	−62.105	2212	67	79	21.5	26	50
Campo Gallo x-1	CG	−26.544	−62.772	2341	86	99	25.5	31	
Ceres es-1	Ce	−29.908	−61.775	1910	60	70	22.5	25	
Coronel rico x-1	CR	−26.462	−61.832	1703	65	73	25.5	28	
Corral de Totoras es-1	CT	−33.812	−66.615	2656	88	103	21.5	31	
El Cabure x-1	EC	−26.043	−62.395	2500	91	105	25.5	32	
El Rincón x-1*	ER	−26.783	−64.733	2059	110	121	24.5	47	
Estancia la Daisy x-1	ELD	−33.203	−66.630	2466	74	88	21.5	27	
Gancedo x-1	GA	−27.358	−61.549	2566	98	113	25.5	34	
General Levalle x-1	GL	−33.951	−63.970	5179	126	140	20.5	23	50
Guandacol es-1	GU	−29.510	−68.596	3693	74	93	20.5	20	
Ituzaingo x-1	IT	−34.758	−67.585	2912	91	108	19.5	30	
Josefina YPF.SF.J.es-	JO	−31.430	−61.996	4502	114	132	21.5	26	45
Las Breñas x-2	LB	−27.054	−61.128	1812	66	75	25.5	27	
Las Mochas x-1	LM	−28.429	−61.260	3204	99	117	24.5	29	

(continued)

Table 1 (continued)

Well	Initials	Latitud W	Longitud S	Depth (mbs)	Measured T(°C)	Corrected T(°C)	Surface T(°C)	Geothermal gradient (°C/km)	BHF (mW/m^2)
Las Salinas x-1	LS	−30.281	−68.460	2249	64	77	20.5	25	
Las Toscas	LTs	−31.800	−67.150	2300	78	90	21.5	30	68
Los Horcones x-1	LH1	−26.338	−64.024	2427	94	108	25.5	34	
Los Horcones x-2	LH2	−26.259	−64.136	2455	100	114	24.5	36	
Los Tigres x-1	LT	−25.760	−62.652	1298	67	70	25.5	34	
Los Tordillos es-1	LTO1	−34.024	−67.116	2460	91	105	19.5	35	
Los Tordillos x-3	LTO3	−33.977	−67.160	2210	85	97	19.5	35	
Mailin x-1	MA	−28.354	−63.149	2198	79	91	24.5	30	
Mariano Boedo 1	MB	−26.009	−58.392	1919	60	70	25.5	23	
Mata gusanos	MAT	−30.797	−68.593	5995	121	129	19.5	18	40
NEX-1	NEX	−30.442	−68.584	3451	72	91	19.5	26	
Nogoya X-1	NO	−32.398	−59.787	2078	71	82	20.5	29	
Nuestra Señora de Talavera	NST	−25.393	−63.752	2651	94	110	24.5	32	
Ordoñez YPF.Od.O. es-1	OR	−32.850	−62.827	3401	96	114	21.5	27	50
Pampa Bandera x-1001	PB	−26.408	−60.072	3971	107	125	25.5	25	
Pirane-1	PI	−25.683	−59.054	2403	77	90	25.5	27	
Pozuelos	PO	−29.972	−68.367	5132	107	122	17.5	20	40
Puesto Navarro x-1	PN	−35.422	−69.347	1871	68	77	14.5	33	
Rio Desaguadero X-1	RD	−33.286	−66.960	1755	72	80	21.5	34	
Rivadavia es-1	RI	−24.129	−62.863	3751	140	159	24.5	36	
Saira es-1	SA	−32.509	−62.069	2632	69	84	21.5	24	40
Salina de Mascasin es-1	SM	−31.382	−66.978	3562	90	109	21.5	24	58

(continued)

Table 1 (continued)

Well	Initials	Latitud W	Longitud S	Depth (mbs)	Measured T(°C)	Corrected T(°C)	Surface T(°C)	Geothermal gradient (°C/km)	BHF (mW/m^2)
Sierra de las Peñas	SP	−32.282	−68.856	4375	99	117	19.5	22	52
Telares x-1	TE	−28.864	−62.956	2234	74	87	24.5	28	
Varela x-1	VA	−34.056	−66.680	2000	77	87	21.5	33	
Uyeda et al. (1978)									
SJ-1	SJ1	−33.316	−69.183	4028	113	132	19.5	28	
TR	TR25	−33.244	−69.090	2279	83	96	19.5	34	
T-70	T70	−33.288	−69.065	3194	120	137	19.5	37	
MEIxp19	MEI	−33.210	−69.072	4484	144.4	162	19.5	32	
EI-12	EI	−33.220	−69.056	3057	110	127	19.5	35	
MPS.x1	MPS	−33.164	−68.992	3839	136	155	19.5	35	
MPC-xp104	MPC	−33.297	−69.047	4559	155.5	173	19.5	34	
CH-6	CH6	−33.389	−68.987	4005	160	179	19.5	40	
MCH-xp25	MCH	−33.375	−68.983	4650	159.4	177	19.5	34	
MBOx2	MBO	−33.183	−68.851	3898	132	151	19.5	34	
MBNa-1	MBN	−33.103	−68.783	2615	81.2	96	19.5	29	
MECPx-4	MECP	−33.074	−68.798	2992	90.5	107	19.5	29	
MEOPe-6	MEOP	−33.055	−68.807	3597	103.3	122	19.5	28	
B-153	B153	−33.200	−68.766	2634	104.4	120	19.5	38	
LC-79	LC79	−33.235	−68.727	2480	94.4	109	19.5	36	
LC-80	LC80	−33.222	−68.718	2483	83.3	98	19.5	31	
LV-80	LV80	−33.341	−68.704	2379	88.8	102	19.5	35	
VM-86	VM86	−33,501	−68,574	2447	112.2	126	19.5	44	
VM-146	VM146	−33.500	−68.573	2392	110	124	19.5	44	

(continued)

Table 1 (continued)

Well	Initials	Latitud W	Longitud S	Depth (mbs)	Measured T(°C)	Corrected T(°C)	Surface T(°C)	Geothermal gradient (°C/km)	BHF (mW/m^2)
PB-133	PB133	−33.529	−68.661	2315	94.4	107	19.5	38	
M-Via4	MVI	−33.568	−68.424	2519	81.1	96	19.5	30	
Vix-3	VI	−33.554	−68.415	2666	110	125	19.5	40	
LJx-1	LJ	−33.629	−68.384	2702	104.4	120	19.5	37	
MAVx-6	MAV	−33.466	−68.804	3073	108.8	126	19.5	35	
MNx-1	MN	−34.048	−67.908	3106	108.8	126	19.5	34	
MLTOes-1	MLTO	−34.008	−67.193	2942	95.5	112	19.5	32	
Uyeda and Watanabe (1982)									
EL SALVS*		−26.250	−69.567	104			25	75	
SANTA CL*		−26.533	−70.300	100			5	10	
CERRO NE*		−27.100	−70.350	325			7	36	
ELISA*		−27.267	−70.383	87			9	36	
VALLENAR*		−28.983	−70.883	202			10	21	
BOQUERON*		−28.083	−70.717	490			10	23	
LA AFRIC*		−33.333	−70.750	280			29	79	
DISPUTAD*		−33.467	−70.167	198			16	61	

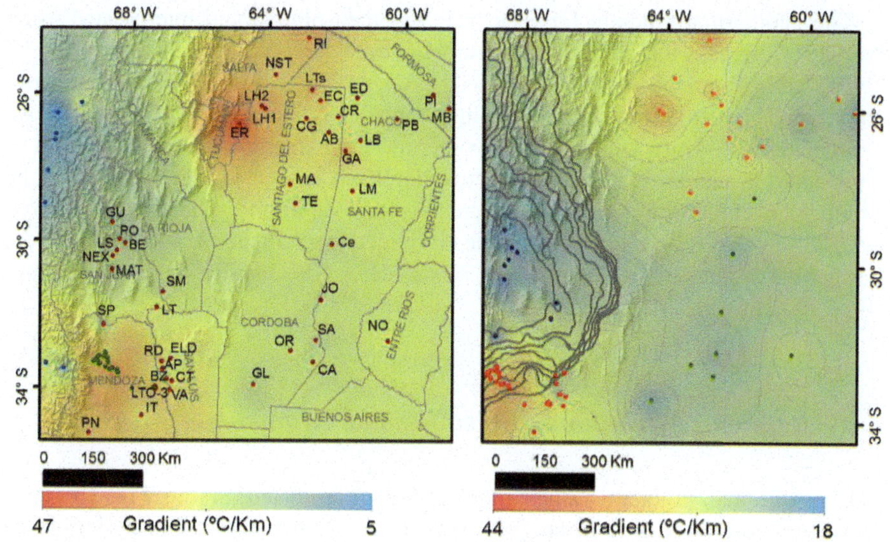

Fig. 3 **a** Geothermal gradient map of the Southern central Andes; red dots: data form this work; green dots: data from Uyeda et al. (1978) and blue dots: data from Uyeda and Watanabe (1982). Well initial following the Table 1; **b** Contour map of geothermal gradients without anomalous and shallower wells (see Table 1); asthenospheric wedge thicknesses (between 20 and 90 km from west to east; taken from Tassara and Echaurren 2012) are shown in black lines

4.1 The Database

A database composed of thermal information from 46 petroleum exploration wells, drilled since the 1960s in the area of study, was provided by YPF S.A. From the surface level, the wells are between 1,000 and 6,000 m deep (wells shallower than 1,000 m were disregarded). The geothermal gradient G (°C/km) was defined by the equation:

$$G = (\mathrm{BHT} - \mathrm{ST})/Z$$

Where BHT is the bottom-hole temperature (°C); ST is the surface temperature (°C) and Z is the distance (km) between the surface and the bottom of the well where the temperature was measured. Temperatures were extracted from the well log headers. Since the downhole measured temperature at the time of drilling is lower in 10–20 °C than the pre-drilling bottom-hole temperatures (cf. Artemieva 2011), pre-drilling temperatures were estimated using the Harrison et al. (1983) correction. Although other methods allow better corrections, like the Horner method (Horner 1951), they require data that are not available in the well log headers, given that most wells were drilled in the last century.

For those wells with various temperature-depth data, we used the deepest datum. Surface temperatures were compiled from the National Meteorological Service for each well and vary between 14.5 and 25.5 °C.

The geothermal gradient for each well was then plotted and interpolated using ArcGIS. Corrected temperature data from Central Chile and Mendoza provided by Uyeda and Watanabe (1982) and Uyeda et al. (1978) were also included in our database. The resulting map, comprising all available data is shown in Fig. 3a. An alternative map without anomalous geothermal gradients, as well as those data of Uyeda and Watanabe (1982; <500 metres depth), is shown in Fig. 3b, and will be useful to evaluate long wavelength and regional variations. The El Rincón well (–26.783 W; –64.783 S), for example, provided a gradient of 47 °C/km in the northernmost flat-slab segment, which sets aside ~10 °C above the nearest values (Fig. 3a).

4.2 *Regional Variations of Geothermal Gradients*

The geothermal gradient values plotted and interpolated in the Fig. 3b range between 7 and 44 °C/km. In order to better describe the results, four regions were delimited (Fig. 1). They represent four contrasting segments in the studied region:

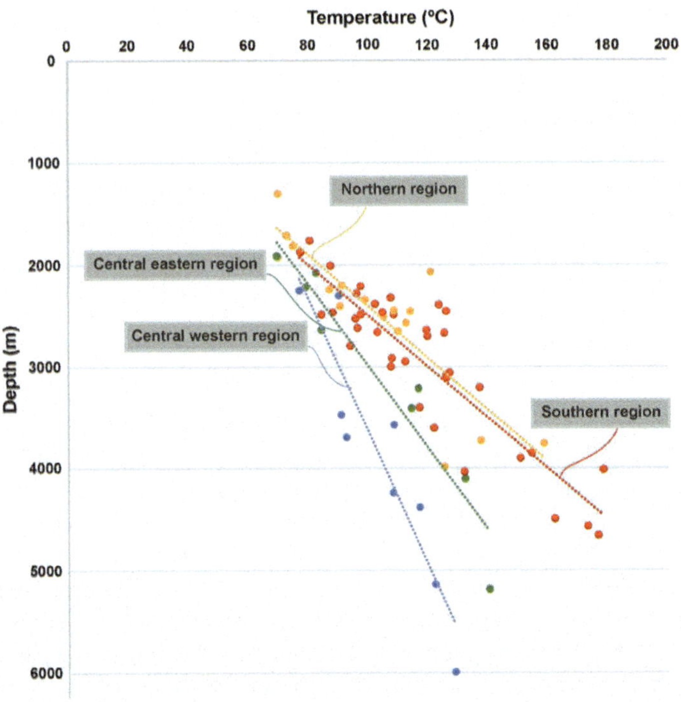

Fig. 4 Distribution of temperatures with depth. Uyeda et al. (1978) data are included. This plot allows separating four trend lines, corresponding to the four regions described in Fig. 1. Note the similar trends of the northern and southern regions. Reference colours as Fig. 3b

two normally dipping-slab segments to the north and south, one sub-horizontal-dipping segment in the central part and the leading edge of the flat slab to the east. The northern region is located to the north of the 29° S parallel, the central-western and the central-eastern regions between 29° and 33° S and the southern region to the south of 33° S. The highest values (above ~30 °C/km) come from wells drilled in the northern (western Santiago del Estero, southern Salta and south-western Chaco provinces) and southern regions (northern Mendoza and San Luis Provinces, Fig. 3a). The lower values (below 20 °C/km) were supplied by wells located in the central-western region (Mata gusanos, Bermejo, Pozuelos and Guandacol-1 wells drilled in the San Juan and La Rioja provinces; and data of the central Chilean region; Uyeda and Watanabe 1982). In this central segment the gradients rise from west to east, from <20 °C/km in the Bermejo basin, to values between ~24 and 29 °C/km in the eastern Chacoparaná basin (central-eastern region). Figure 4 illustrates the temperature-depth ratios for the four different regions.

5 1D Basin Models

5.1 The Database and Strategy

As stated above, sedimentary basins preserve a thermal record that is masked by surface and sub-surface factors. To weigh the influence of these factors, we performed 1D-basin models in 12 wells from the four areas delimitated in Fig. 1.

The main objective of basin modelling is to evaluate the thermal influence of the flat slab, which requires an estimation of the BHFs. It was performed using the PetroMod Schlumberger software, which calculates the BHF for each well as well as the heat flows at different depths (Fig. 5). The thermal upper boundary of the models corresponds to the average surface temperature through time, obtained from the National Meteorological Survey. The lower thermal boundary is the BHF in the basin, with its variations trough time based on a reliable geodynamic model, which is constrained by the current bottom-hole temperature data. Other inputs are: (1) the age of the sediments and rate of sedimentation, which are needed to estimate the compaction (and its consequent decrease in porosity) through time, the blanketing effect, etc.; (2) the identification and magnitude of erosional episodes; and (3) lithologies to estimate the radiogenic heat contribution. Other factors like fluid migration and volcanism are not considered in our models.

Given that the area of study comprises basins with diverse and complex sedimentation histories, we conducted a basin-by-basin analysis. Particularly, through the modelling we substrated the blanketing effect in the wells (and basins) to normalize values for comparison purposes. Our BHF results were plotted and interpolated in the map presented in Fig. 6. Data from central Chile of Uyeda and Watanabe (1982) were also included.

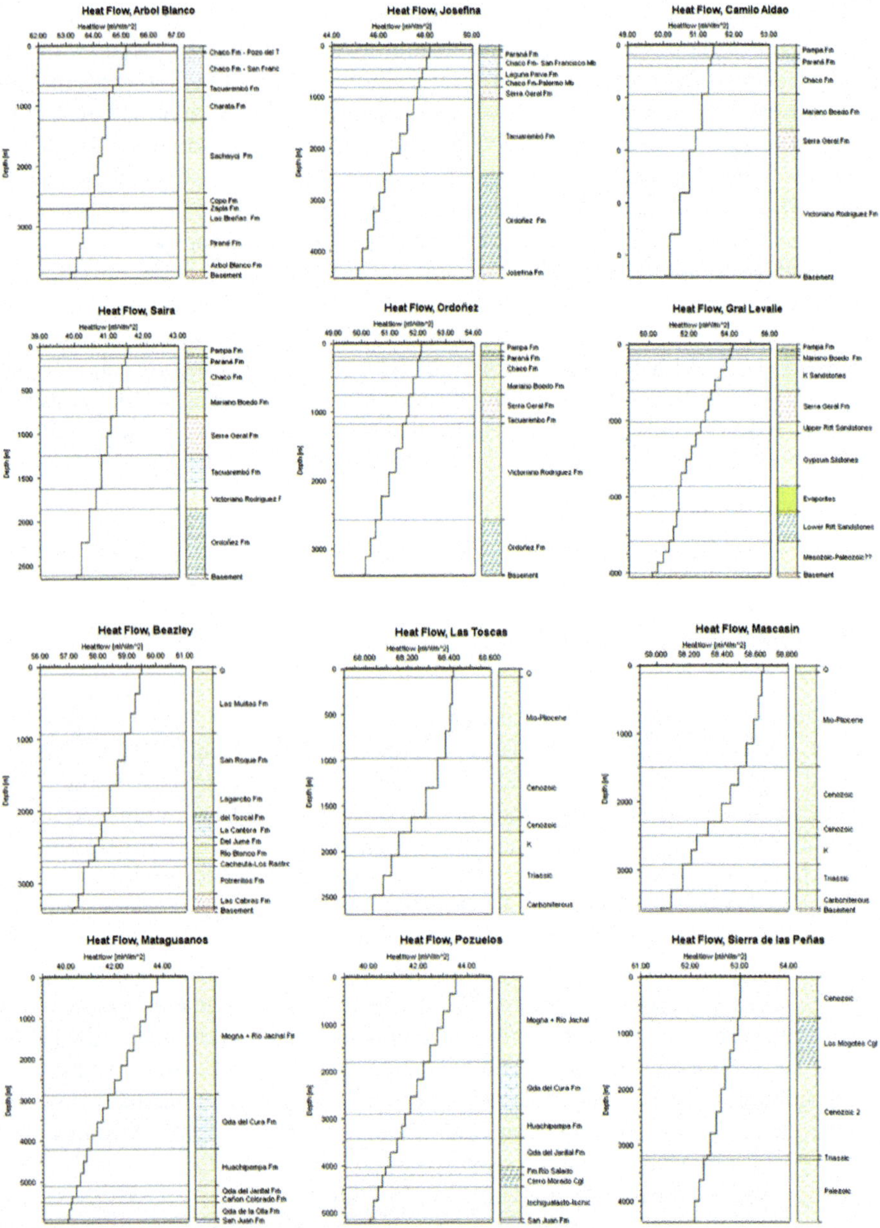

Fig. 5 Heat flow vs depth for the selected 12 wells. Stratigraphy and thicknesses from well files provided by YPF

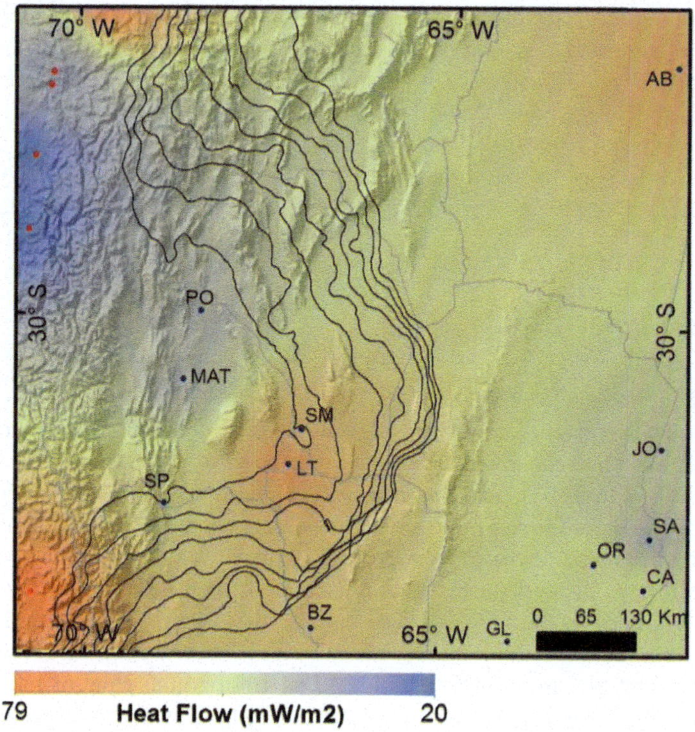

Fig. 6 Basal heat flow map. Data for Table 1. Notice that the four regions defined in the Fig. 1 can be also detected from the distribution of BHF values

5.2 *Regional Variations of Basal Heat Flow Values*

The 1D modelling results suggest that the thermal effects of sedimentation were not enough to reproduce the geothermal gradients observed in Fig. 3. This indicates that the BHF (Fig. 6) is likely the main driving control on the basins thermal state within the flat slab and nearby segments.

In the central-western region, along the Andean foredeep, the low geothermal gradients match with low BHF ($\sim$40 mW/m^2; Mata gusanos and Pozuelos wells). These heat flow values are comparable to those estimated in these same basins using magnetotelluric analysis (36–40 mW/m^2; Borzotta et al. 2009) and agree with the heat flows reported by Hamza et al. (2005). Within the flat-slab segment, relative high values of the BHF (between 58 and 68 mW/m^2) were obtained for Sierra de Mascasín, and Las Toscas wells, near Sierra de Valle Fértil.

In contrast, within the normal subduction segments to the north and south (northern and southern regions), both thermal gradient and BHF values are higher. Further south along the foredeep, the Sierra de las Peñas well shows higher geothermal gradient and BHF values (Figs. 3 and 6). Furthermore, a BHF value of ~60 mW/m^2 is proposed for the northern region (Árbol Blanco well). Within the central-eastern region, located eastward from the leading edge of the flat-slab segment, BHF values are between 40 and 50 mW/m^2 (wells in the Córdoba, Santa Fe and Entre Ríos provinces), with the lowest values to the northeast. The effect of the sedimentary history on the geothermal gradients (blanketing) is clearly observed, especially in those wells where sediment thicknesses are very different (e.g., General Levalle and Ordoñez wells) and the current BHF is similar.

6 Curie Point Depths

The base of magnetic sources in the Earth's crust is determined by the Curie isotherm. The magnetic anomalies caused by these deep sources are long wavelength and of relatively low amplitude with respect to those caused by superficial bodies (Ross et al. 2006). The conventional techniques to estimate the depths of the magnetic sources are based on examination of statistical properties of the anomaly configuration related to the power spectrum within the depths of the sources.

These factors depend on the geological terrain composition and its structural complexity since so little is known about the degree of variability of the magnetization.

The magnetized crust can be considered as a horizontal plate of infinite lateral extension, whose top and bottom are found at depths Zt and Zb, respectively. If its magnetization is uniformly random and not correlated, then the Power Spectrum of the magnetic anomalies that it generates can be related to its characteristic depths through the following equation (Blakely 1995),

$$\ln \bar{\Phi}_{\Delta T}(k) = \mathrm{LnA} - 2Z_t k + 2\ln(1 - e^{-k(z_b - z_t)})$$

where $_{\Delta \mathrm{T}}$ is the previously radially averaged Power Spectrum of the magnetic anomaly, k is the module of the wave vector and A is a constant with no physical significance. This magnetized slab can be thought as the layer that contains all the crustal ferromagnetic materials, and no magnetized bodies place below it. The CPD can be associated with the bottom of this magnetized slab or Zb. The results are depths expressed in kilometres. The Curie temperature depends consequently on the mineralogy of the magnetic bodies and, thus, it should not be considered as an isotherm. Nevertheless, Frost and Shive (1986) demonstrated that the magnetite is almost pure at lower crust levels and, in fact, it is the only magnetic source. Therefore, the base of the magnetic sources must be para-magnetic and evidence temperatures of ~580 °C. The Curie temperature of magnetite increases with pressure at a rate of 1.8 °C Kbar^{-1}. This suggests that rocks lose their magnetization at around 600 °C.

6.1 The Database

Aerial magnetic data supplied by the Argentine Geological Survey (SEGEMAR) were processed in order to estimate the CPD in the study area. We used the Soler (2015) method, a peak forward method (Ravat et al. 2007; Ross et al. 2006) and the software CuDePy. This method is based on the Levenberg–Marquardt least square-fitting algorithm for non-linear problems in order to adjust the equation stated above on the previously radially averaged power spectrum of the magnetic anomalies. It allows estimating the CPD in several points through an interactive

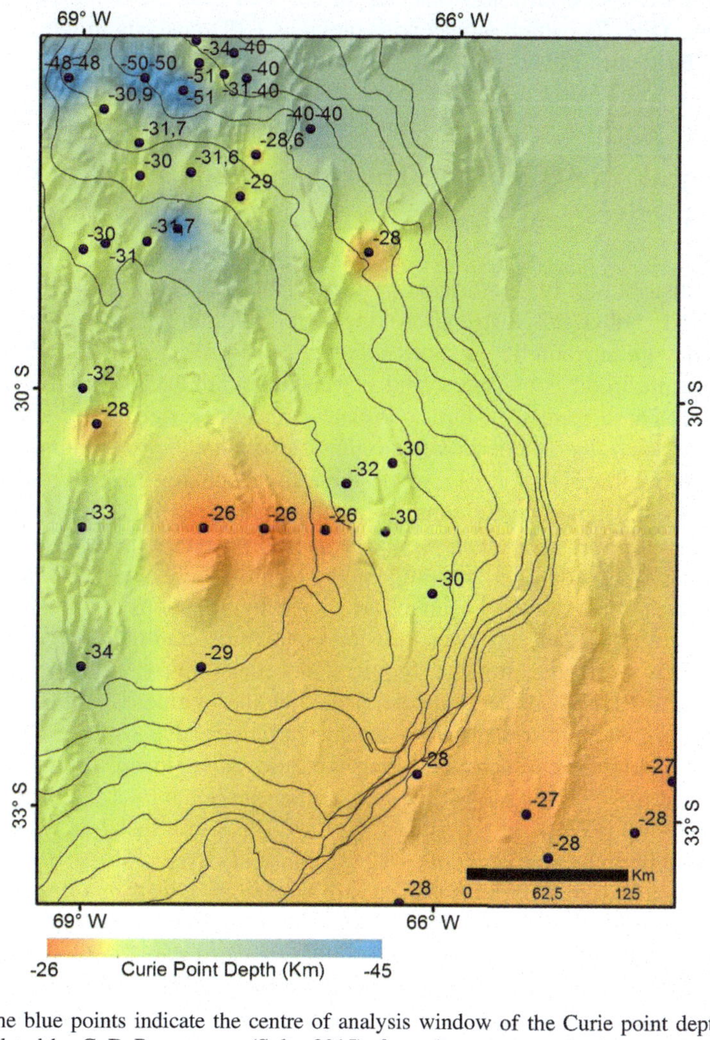

Fig. 7 The blue points indicate the centre of analysis window of the Curie point depths (CPD; km) calculated by CuDePy program (Soler 2015), from the aeromagnetic data

selection of square sub-regions of the magnetic anomaly grid. The database was plotted and interpolated using ArcGis (see Fig. 7). The low CPD coverage in some areas is due to the poor survey of short-wavelength magnetic data in the study area.

6.2 *Regional Variations of Curie Point Depths*

Figure 7 shows a long wavelength and regional decrease in the CPD from the northwest (~51 km depth) to the southeast (~27 km depth), with values between 30 and 34 km in the central-western region (between 28° and 32° S). An anomalous low local CPD (~26 km) was detected in the area near the Valle Fertil—Pie de Palo ranges (Fig. 7), which agrees with previous CPD analysis in the region (e.g., Ruiz and Introcaso 2004).

7 Integral Analysis of the Thermal Scheme and Future Perspectives

The map distribution of geothermal gradient data (Fig. 3) and thermal modelling results (BHF and CPD distribution) in the Argentine flat subduction segment (compared with surrounding regions affected by normal subduction, Fig. 1) evidences a quite good consistency and correlation among them as it is shown in Figs. 8a, b. These results, in turn, correlate remarkably well with the spatial distribution of the asthenospheric wedge along strike and the contours of the subducting flat oceanic slab (Fig. 1), where the asthenosphere strangles (i.e. where the oceanic slab is coupled to the continental lithosphere).

These new results show that variations in sedimentation histories are not sufficient to explain variations in temperature within the basins and support the hypothesis that major thermal changes are driven by variations at lithospheric scale. While the coldest basins lay over flat-slab or towards cratonic areas (pericratonic belts), basins with the highest temperatures develop on pluging-slab areas. Gradients and BHF inside and outside of the flat-slab segment, at similar distances to the modern oceanic trench, show strong variations along strike. For example, at ~390 km from the trench and along the Andean foredeep, the Mata gusanos well, located at ~31° S, shows low geothermal gradients (~18 °C/km) and very low BHF (~40 mW/m^2 at 6,000 m); south of 32° S the Sierra de las Peñas well records geothermal gradients of ~22 °C/km and BHF values of 52 mW/m^2 (at 4,000 m); and at ~35,5° S the Puesto Navarro x-1 well records gradients of ~33 °C/km (Figs. 3b and 6). Nevertheless, it is important to notice that some exceptions occur and local short-wavelength anomalies develop (see below).

These heat variations, with a north-south trend, could be associated with the flat-slab geometry: the installation of a cold sub-horizontal plate below the upper

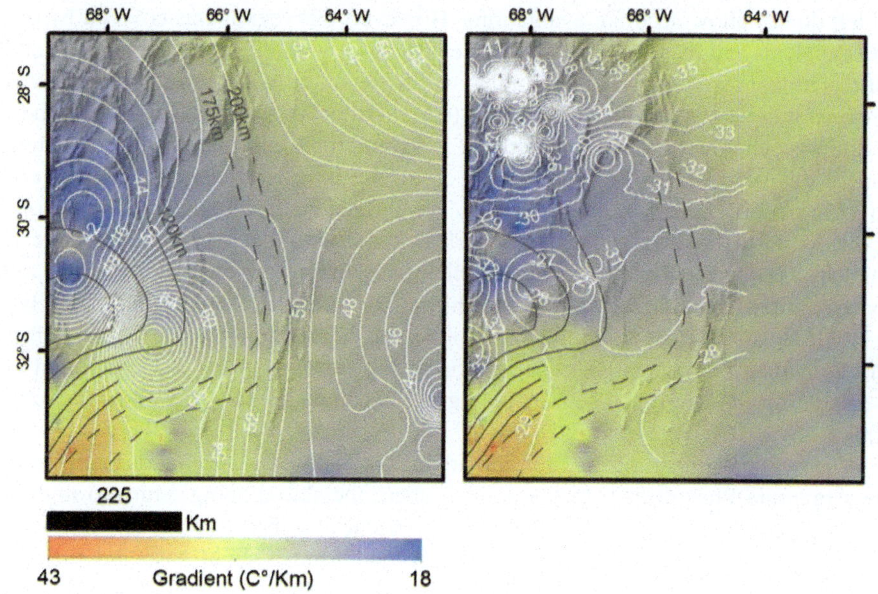

Fig. 8 Comparison of the geothermal gradient map with the heat flow isolines (mW/m^2; left) and the Curie point depths isolines (km; right). Wadatti-Benioff isobaths (taken from Alvarado et al. 2009) as a reference

lithosphere for more than 400 km and a thermal non-steady state generated by the modification of asthenospheric heat flow during the retreat of asthenosphere towards the cratonic region. Although the subduction geometry and lithospheric thicknesses seem to have had an important control on the thermal state in the study region, some regional variations in composition, thickness and structure of the upper lithosphere however cannot be disregarded. For example, some studies have mentioned that the complex structure of the continental crust, product of the continuous subduction from the west since the Cambrian, with the consequent amalgamation of different terranes and with a complex structure, might have influenced the thermal distribution in the region (e.g., Ruiz and Introcaso 2004; Tassara et al. 2006). Other studies (e.g., Muñoz 2005) have proposed that low-heat flows might be associated with low radiogenic heat production, which would be related to the location of allochthonous terranes and old tectonic structures. Heat production calculations indicate that the major contribution on the heat flow is from the upper crust (one order of magnitude more productive than the lower crust). According to Tassara and Echaurren (2012), the modelled upper crust in the flat-slab region is thinner than across normal subduction segments, which might explain partly as well the regional differences observed along strike.

Along the northern normal subduction segment (north of 30° S and <350 km from the trench, see Figs. 1 and 3a), the only reported wells in Chile are shallow (<500 m deep), with local anomalous gradient values (between 5–10 °C/km), which do not allow accurate evaluations. However, this region was covered by our CPD analysis and map (see Fig. 7). These data agree with the regional cooling trend estimated from well data. Future studies will focus on different correlation analyzes between these and other thermal proxies (e.g. thermochronology, clumped isotopes, organic geochemistry, clay mineralogy). It is important to notice that the southern part of this northern cold region located along the cordillera includes the Neogene Vinchina basin, recently interpreted as an anomalously cold depositional zone (Collo et al. 2011, 2015). This fossil refrigerated zone (detected by clay mineral analysis and thermochronology) was explained by heat flow variations in time and space, associated with the southward shifting of a shallow-to-flat slab (Gutscher et al. 2000; Husson and Moretti 2002; Collo et al. 2011; among others). This shift is based on Juan Fernandez Ridge reconstructions (Yañez et al. 2001), interpreted as the main driving control of slab flattening in the central Andes (Kay and Mpodozis 2002).

Across the segment affected by flat subduction, between ~29° and 33° S, from the Argentine Precordillera to the plains, where the slab dipping return to normal (leading edge of the flat slab), BHF values remain relatively constant (40–45 mW/m^2). The only variation is recorded from a slight eastward thermal gradient increase (comparing the Mata gusanos and Saira wells, Fig. 3; Table 1), likely associated with differences in the basin filling histories. While the Neogene basins along the foredeep, to the west, are >6 km deep, they record <<1 km to the east, along the pericratonic foreland (Fig. 9).

The low-heat fluxes reported in wells drilled in the Argentine plains, as well as the deep values of CPD (locally 54 km beneath the surface, Sigismondi and Fantin 2014) might be associated with a thicker lithosphere rather than the occurrence of a flat slab. This is because these data come from a segment with developed asthenospheric wedge (Fig. 1). Different models (e.g., Tassara and Echaurren 2012 or Lithos1.0) show that the lithospheric thickness increases from the trench towards the cratonic region, with values near 120 km in the Chacoparaná basin (between 29° and 33° S). The deeper location of the asthenosphere-lithosphere boundary (or LAB) might be the explanation of the lower thermal regime in this pericratonic region (central-eastern region). Southward in this region, the BHF values increase, as shown by the transition between the Saira (BHF 40 mW/m^2 at 2,600 m MD), Camilo Aldao (BHF 50 mW/m^2 at 2,100 m MD) and Ordoñez wells (BHF 50 mW/m^2 at 3,500 m MD). It is important to notice, however, that they have slightly different gradients despite their similar filling histories, particularly regarding to the Cenozoic, which is the main factor controlling the current geothermal gradient and the basal HF. 1D modelling shows that these differences in the gradients are, consequently, a result of along-strike BHF variations. Recent magnetometry studies (Sigismondi and Fantin 2014), combined with BHF calculations in General Levalle well, have proposed as well a regional north to south BHF increase.

To the south, as well as in the northern region within the Chacoparaná basin, the values of thermal gradients and basal heat fluxes also increase. We do not have CPD data in this area to compare the BHF results. Although the whole Chacoparaná basin rests over a zone where the asthenospheric wedge occurs and is in contact

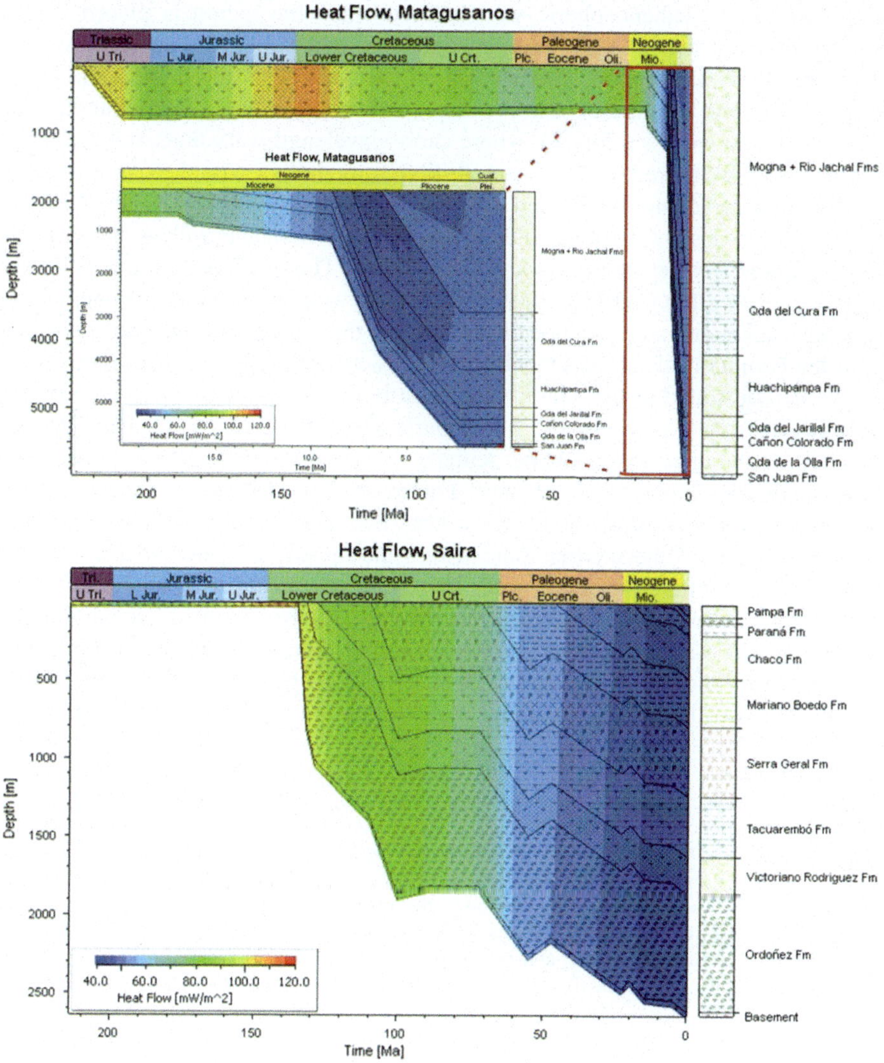

Fig. 9 1D models of the Mata gusanos and Saira wells showing the heat flow distribution with depth through time. Note the strong cooling effect associated with the high Cenozoic sedimentary rates in the Mata gusanos well. Stratigraphy and thicknesses were modified following from Milana et al. (2003) and Reinante et al. (2014)

with the continental lithosphere, a likely explanation for this southward and northward BHF increase might be a thinning of the lithosphere in these directions. While Tassara and Echaurren (2012) show a subtle thinning southward that might partly explain this hotter zone to the south within the plain; northwards the lithospheric thickness remains rather similar. An alternative might be to evaluate the geometry of the asthenospheric wedge, even when the region is placed several kilometres from the flat-slab leading edge, or variations in the composition and thicknesses within the crust. Moreover, we cannot disregard the influence of the El Rosario fault and circulation of hot springs, particularly in the western Santiago del Estero (Rincón well, see Fig. 3), where short-wavelength variations and local high anomalies of BHF are clearly observed. Future modelling will shed light on this.

Our data (gradients, BHF and CPD) and previous CPD analyzes (Ruiz and Introcaso 2004) show another anomaly in the intermontane valleys between the Mascasín and Beazley wells (57 and 68 mW/m^2). This thermal anomaly coincides with a shallow CPD (see Fig. 7). Although more studies are needed to arrive to a suitable explanation, we do not disregard that this anomaly might be associated with the evolution of the basement-involved Sierra de Valle Fértil megathrust.

We demonstrated in this contribution that regional variations of the subducting slab geometry as well as of the lithospheric structure strongly affect the thermal state within the upper crust and particularly the sedimentary basins. Evidently, this has strong exploratory implications. Future studies will allow improving our database as well as analyze the groundwater wells separately, in order to understand the stratification of the thermal state within the uppermost crust. BHF modelling will also be continued in those wells that were not included or analyzed in this work. The CPD map will be improved by increasing the data grid. New numerical modelling will assist us to discriminate other contributions to the basins basal heat flow, like the lithospheric radiogenic heat (associated mainly with the composition and thickness of the lithosphere) and the contribution from the asthenosphere.

Acknowledgements We are grateful to the Consejo Nacional de Investigaciones Científicas y Técnicas, the Agencia Nacional de Promoción Científica y Tecnológica (PICT 2015-1092), the Secretaría de Ciencia y Tecnología de la Universidad Nacional de Córdoba (Secyt-UNC 2016-2017 30720150100830CB), the Proyecto de Investigación UE 2016 - CONICET and the CAPES-MINCYT and the Conicet-Fapesp programs for financial support of our Research Projects in Argentina. We would like to acknowledge the help of Ignacio Brisson for it constructive discussion of the results during the research and the Sevicio Geologico Minero Nacional (SEGEMAR, Córdoba delegation) and the Gerencia de Exploración YPF SA. We acknowledge thorough reviews by Silvia Nassif and Francisco Ruiz, which helped to improve this work.

References

Allen PA, Allen JR (2005) Basin analysis: principles and applications, 2nd edn. Blackwell, Malden, MA

Alvarado P, Beck S, Zandt G (2007) Crustal structure of the south-central Andes Cordillera and backarc region from regional waveform modelling. Geophys J Int 170(2):858–875

Alvarado P, Pardo M, Gilbert H, Miranda S, Anderson M, Saez M, Beck S (2009) Flatslab subduction and crustal models for the seismically active Sierras Pampeanas region of Argentina. In: Kay SM, Ramos VA, Dickinson W (eds) Backbone of the Americas: shallow subduction, plateau uplift, and ridge and terrane collision. Geological Society of America, Memoir 204, pp 261–278

Álvarez O, Nacif S, Gimenez M, Folguera A, Braitenberg A (2014) Goce derived vertical gravity gradient delineates great earthquake rupture zones along the Chilean margin. Tectonophysics 622:198–215

Álvarez O, Lince Klinger F, Gimenez M, Ruiz F, Martinez P (2015) Density and thermal structure of the Southern Andes and adjacent foreland from 32° to 55° S using Earth gravity field models. In: Folguera A et al (eds) Growth of the Southern Andes, Springer Earth System Sciences. https://doi.org/10.1007/978-3-319-23060-3_2

Anderson M, Alvarado P, Zandt G, Beck S (2007) Geometry and brittle deformation of the subducting Nazca Plate, Central Chile and Argentina. Geophys J Int 171:419–434

Artemieva IM (2009) The continental lithosphere: reconciling thermal, seismic, and petrologic data. Lithos 109:23–46

Artemieva IM (2011) The lithosphere: an interdisciplinary approach. Cambridge University Press, Cambridge, 794 pp., ISBN 9780521843966

Barazangi M, Isacks BL (1976) Spatial distribution of earthquakes and subduction of the Nazca plate beneath South America. Geology 4:686–692

Barazangi M, Isacks BL (1979) Subduction of the Nazca plate beneath Peru: evidence from spatial distribution of earthquakes. Geophys J R astr Soc 57:537–555

Blakely RJ (1995) Potential theory in gravity and magnetic applications. Cambridge University Press, Cambridge

Bohm MS, Luth H, Echtler G, Asch K, Bataille C, Bruhn A, Rietbrock P, Wigger (2002) The Southern Andes between 36° and 40° S latitude: seismicity and average seismic velocities. Tectonophysics 356:275–289

Borzotta E, Mamaní MJ, Venencia JE (2009) Preliminary Magnetotelluric study of Ambato and Valle Fértil Lineaments in Bermejo Basin and Sierra de Valle Fértil, San Juan, Argentina. Acta Geodaetica et Geophysica Hungarica 44(2):157–166

Cahill T, Isacks BL (1992) Seismicity and shape of the subducted Nazca Plate. J Geophys Res 97:17503–17529

Calegari RJ, Chebli G, Manoni RS, Lázzari V (2014) Las cuencas cretácicas de la región central del país: General Levalle. In: Martino RD, Guereschi AB (eds) Geología y Recursos Naturales de la Provincia de Córdoba, Congreso Geológico Argentino. Asociación Geológica Argentina, Córdoba, pp 913–938

Chulick GS, Detweiler S, Mooney WD (2013) Seismic structure of the crust and uppermost mantle of South America and surrounding oceanic basins. J South Am Earth Sci 42:260–276. https://doi.org/10.1016/j.jsames.2012.06.002

Collo G, Dávila FM, Nóbile J, Astini RA, Gehrels G (2011) Clay mineralogy and thermal history of the Neogene Vinchina Basin, central Andes of Argentina: analysis of factors controlling the heating conditions. Tectonics 30:1–18

Collo G, Dávila FM, Teixeira W, Nóbile JC, Sant′ Anna LG, Carter A (2015) Isotopic and thermochronologic evidence of extremely cold lithosphere associated with a slab flattening in the Central Andes of Argentina. Basin Res. https://doi.org/10.1111/bre.12163

Currie CA, Hyndman RD, Wang K, Kostoglodov V (2002) Thermal models of the Mexico subduction zone: implications for the megathurst seismogenic zone. J Geophys Res 107:2370. https://doi.org/10.1029/2001JB000886

Dávila FM, Carter A (2013) Exhumation history of the Andean broken foreland revisited. Geology 41(4):443–446

Dávila FM, Lithgow-Bertelloni C (2015) Dynamic uplift during slab flattening. Earth and Planet Sci Lett 425:34–43

Eakin CM, Lithgow-Bertelloni C, Dávila FM (2014) Influence of Peruvian flat—subduction dynamics on the evolution of the Amazon basin. Earth and Planet Sci Lett 404:250–260. https://doi.org/10.1016/j.epsl.2014.07.027

Engdahl ER, van der Hilst RD, Buland R (1998) Global teleseismic earthquake relocation with improved travel times and procedures for depth determination. Bull Seismol Soc Am 88:722–743

English J, Johnston ST, Wang K (2003) Thermal modeling of the Laramide orogeny: testing the flat slab subduction hypothesis. Earth and Planet Sci Lett 214:619–632

Espurt N, Baby P, Brusset S, Roddaz M, Hermoza W, Regard V, Antoine P-O, Salas-Gismondi R, Bolaños R (2007) How does the Nazca Ridge subduction influence the modern Amazonian foreland basin? Geology 35:515–518

Fromm R, Zandt G, Beck SL (2004) Crustal thickness beneath the Andes and Sierras Pampeanas at 30° S inferred from Pn apparent phase velocities. Geophys Res Lett 31:L06625. https://doi.org/10.1029/2003GL019231

Frost BR, Shive PN (1986) Magnetic mineralogy of the lower continental crust. J Geophys Res 91: https://doi.org/10.1029/JB091iB06p06513

Gans CR, Beck SL, Zandt G, Gilbert H, Alvarado P, Anderson M, Linkimer L (2011) Continental and oceanic crustal structure of the Pampean flat slab region, western Argentina, using receiver function analysis: new high resolution results. Geophys J Int 186(1):45–58

Giambiagi LB, Ramos VA (2002) Structural evolution of the Andes between 33°30′ and 33°45′ S, above the transition zone between the flat and normal subduction segment, Argentina and Chile. J South Am Earth Sci 15(1):101–116

Gilbert H, Beck S, Zandt G (2006) Lithospheric and upper mantle structure of central Chile and Argentina. Geophys J Int 165:383–398

Gimenez ME, Martínez MP, Introcaso A (2000) A crustal model based mainly on gravity data in the area between the Bermejo Basin and the Sierras de Valle Fértil–Argentina. J South Am Earth Sci 13(3):275–286

Gimenez ME, Braithenberg C, Martinez MP, Introcaso A (2009) A comparative analysis of seismological and gravimetric crustal thicknesses below the Andean region with flat subduction of the Nazca Plate. Int J Geophys 2009:1–8

Grevemeyer I, Kaul N, Diaz-Naveas JL (2006) Geothermal evidence for fluid flow through the gas hydrate stability field off Central Chile—transient flow related to large subduction zone earthquakes? Geophys J Int 166:461–468

Gutscher MA (2002) Andean subduction styles and their effect on thermal structure and interplate coupling. J South Am Earth Sci 15:3–10

Gutscher MA, Maury R, Eissen JP, Bourdon E (2000) Can slab melting be caused by flat subduction? Geology 28:535–538

Hampel A (2002) The migration history of the Nazca Ridge along the Peruvian active margin: a re-evaluation. Earth Planet. Sci. Lett 203:665–679

Hamza VM, Muñoz M (1996) Heat flow map of South America. Geothermics 25:599–646

Hamza VM, Gomes AJL, Ferreira LET (2005) Status report on geothermal energy developments in Brazil. In: Proceedings of the World Geothermal Congress, Antalya

Hantschel T, Kaueraf AI (2009) Heat flow analysis. In: Kaueraf AI, Hantschel T (eds) Fundamentals of basin and petroleum systems modeling. Springer-Verlag Berlin Heidelberg, pp 103–150. https://doi.org/10.1007/978-3-540-72318-9

Harrison WE, Luza KV, Prater ML, Cheung PK (1983) Geothermal Resource Assessment in Oklahoma. Special Publications 83-1. Oklahoma Geological Survey

Hasterak D, Chapman DS (2011) Heat production and geotherms for the continental lithosphere. Earth Planet Sci Lett 307(1):59–70

Henry SG, Pollack H (1988) Terrestrial heat flow above the Andean subduction zone in Bolivia and Peru. J Geophys Res 93:15,153–15,162

Horner DR (1951) Pressure Build-up in Wells, 3 rd World Petroleum Congress, The Hague, NL, World Petroleum Congress, May 28–June 6, 1951

Husson L, Moretti S (2002) Thermal regime of fold and thrust belts–An application to the Bolivian sub Andean zone. Tectonophysics 345:253–280. https://doi.org/10.1016/S0040-1951(01)00216-5

Jaupart C, Labrosse S, Mareschal JC (2007) Temperatures, heat and energy in the mantle of the Earth. In: Shubert G, Bercovici D (eds) Treatise on geophysics: mantle dynamics 7(6):253–303

Jordan T, Isacks BL, Allmendinger RW, Brewer J, Ramos VA, Ando CJ (1983) Andean tectonics related to geometry of subducted Nazca Plate. Geol Soc Am Bull 94:341–361

Jordan TE, Schlunegger F, Cardozo N (2001) Unsteady and spatially variable evolution of the Neogene Andean Bermejo foreland basin, Argentina. J South Am Earth Sci 14:775–798
Kay SM, Coira B (2009) Shallowing and steepening subduction zones, continental lithosphere loss, magmatism and crustal flow under the Central Andean Altiplano-Puna Plateau. In: Kay SM, Ramos VA, Dickinson W (eds) Backbone of the Americas, Geological Society of América, GSA, Boulder: Memoir 404: 229–259
Kay SM, Mpodozis C (2002) Magmatism as a probe to the Neogene shallowing of the Nazca plate beneath the modern Chilean flat-slab. J South Am Earth Sci 15:39–57
Kay SM, Mpodozis C, Ramos VA, Munizaga F (1991) Magma source variations for mid-late Tertiary magmatic rocks associated with a shallowing subduction zone and a thickening crust in the central Andes (28° to 33°S) Argentina. In: Harmon RS, Rapela CW (eds) Andean magmatism and its tectonic setting: Boulder, Colorado, Geological Society of America Special Paper, vol 265, pp 113–137
Kirby SH, Engdahl ER, Denlinger R (1996) Intermediate-depth intraslab earthquakes and arc volcanism as physical expressions of crustal and uppermost mantle metamorphism in subducting slabs, in subduction: top to bottom. Geophys Monogr Ser 96:195–214
Manea VC, Manea M (2011) Flat-slab thermal structure and evolution beneath central Mexico. Pure App Geophys 168(8):1475–1487
Manea VC, Manea M, Kostoglodov V, Sewell G (2005) Thermo-mechanical model of the mantle wedge in Central Mexican subduction zone and a blob tracing approach for the magma transport. Phys Earth Planet Inter 149:165–186. https://doi.org/10.1016/JPEPI2004.08.024.
Mareschal JC, Jaupart C (2013) Radiogenic heat production, thermal regime and evolution of continental crust. Tectonophysics 609:524–534. https://doi.org/10.1016/j.tecto.2012.12.001
Marot M, Monfret T, Gerbault M, Nolet G, Ranalli G, Pardo M (2014) Flat vs. normal subduction zones: a comparison based on 3D regional traveltime tomography and petrological modeling of central Chile and western Argentina (29–35S). Geophys J Int 199:1633–1654
Martinod J, Funiciello F, Faccenna C, Regard V (2005) Dynamical effects of subducting ridges: insights from 3-D laboratory models. Geophys J Int 163:1137–1150
McGeary S, Nur A, Ben-Avraham Z (1985) Spatial gaps in arc volcanism: the effect of collision or subduction of oceanic plateaus. Tectonophysics 119:195–221
Milana JP, Bercowsky F, Jordan T (2003) Paleoambientes y magnetoestratigrafía de l Neógeno de la Sierra de Mogna, y su relación con la Cuenca de Antepaís Andina. Revista de la Asociación Geológica Argentina 58(3):447–473
Miranda FJ, Pesce AH (1997) Argentina geothermal resources: new trends in development. GRC Transactions 21:337–339
Muñoz M (2005) No flat Wadati–Benioff Zone in the central and Southern Central Andes. Tectonophysics 395:41–65. https://doi.org/10.1016/j.tecto.2004.09.002
Nacif S, Triep EG, Spagnotto SL, Aragon E, Furlani R, Álvarez O (2015) The flat to normal subduction transition study to obtain the Nazca plate morphology using high resolution seismicity data from the Nazca plate in Central Chile. Tectonophysics 657:102–112
Pardo M, Comte D, Monfret T (2002) Seismotectonic and stress distribution in the central Chile subduction zone. J South Am Earth Sci 15:11–22
Pesce AH (1995) Argentina Country update. In: Proceedings of the World Geothermal Congress, Florence, pp 35–43
Pesce AH (2000) Argentina Country update. In: Proceedings of the World Geothermal Congress, Kyushu-Tohoku, pp 35–43
Pesce AH (2005) Argentina Country update. In: Proceedings of the World Geothermal Congress, Antalya, pp 24–29
Pilger RH (1981) Plate reconstructions, aseismic ridges, and low angle subduction beneath the Andes. Geol Soc Am Bull 92:448–456
Ramos VA, Folguera A (2009) Andean flat slab subduction through time. In: Murphy B (ed) Ancient orogens and modern analogues. London, The Geological Society, Special Publication 327, pp31–54

Ramos VA, Cristallini EO, Perez DJ (2002) The Pampean flatslab of the central Andes. J S Am Earth Sci 15:59–78
Ravat D, Pignatelli A, Nicolosi I, Chiappini M (2007) A study of spectral methods of estimating the depth to the bottom of magnetic sources from near-surface magnetic anomaly data. Geophys J Int:421–434. https://doi.org/10.1111/j.1365-246x.2007.03305.x
Reinante SM, Olivieri G, Salinas A, Lovecchio JP, Basile Y (2014) La cuenca Chacoparaná: estratigrafía y recursos de hidrocarburos. In: Martino RD, Guereschi AB (eds) Geología de subsuelo, XIX Congreso Geológico Argentino. Asociación Geológica Argentina, Córdoba, pp 895–912
Ruiz F, Introcaso A (2004) Curie point depths beneath Precordillera Cuyana and Sierras Pampeanas obtained from spectral analysis of magnetic anomalies. J Gondwana Res 8:1133–1142
Ross HE, Blakely RJ, Zoback MD (2006) Testing the use of aeromagnetic data for the determination of Curie depth in California. Geophysics 71:L51. https://doi.org/10.1190/1.2335572
Sacks S, y Okada H (1974) A comparison of the anelasticity structure beneath western South America and Japan. Phys Earth Planet Inter 9:211–219
Sigismondi ME (2012) Estudio de la deformaciôn litosférica de la cuenca Neuquina: estructura termal, datos de gravedad y sísmica de reflexiôn. PhD thesis, Universidad de Buenos Aires, Buenos Aires, 1–367
Sigismondi ME, Fantin FA (2014) Estructura cortical y características geodinámicas. In: Martino RD, Guereschi AB (eds) Geología de subsuelo, XIX Congreso Geológico Argentino. Asociación Geológica Argentina, Córdoba, pp 01–23
Springer M (1999) Interpretation of heat-flow density in the central Andes. Tectonophysics 306:377–395
Soler SR (2015) Métodos Espectrales para la Determinación de la Profundidad del Punto de Curie y Espesor Elástico de la Corteza Terrestre. Facultad de Ciencias Exactas, Ingeniería y Agrimensura (FCEIA), Universidad Nacional de Rosario (UNR). Tesis de Grado para Título de Licenciado en Física
Syracuse EM, Abers GA (2006) Global compilation of variations in slab depth beneath arc volcanoes and implications. Geochem, Geophys and Geosyst 7(5). https://doi.org/10.1029/2005GC001045
Tassara A, Echaurren A (2012) Anatomy of the Andean subduction zone: three-dimensional density model upgraded and compared against global-scale models. Geophys J Int 189: 161–168
Tassara A, Gôtze HJ, Schmidt S, Hackney R (2006) Threedimensional density model of the Nazca plate and the Andean continental margin. J geophys Res 111. https://doi.org/10.1029/2005JB003976
Uyeda S, Watanabe T, Volponi F (1978) Report of heat flow measurements in San Juan and Mendoza, Argentina. Bull Earthq Res Inst 53:165–172
Uyeda S, Watanabe T (1982) Terrestrial heat flow in Western South America. Tectonophysics 83:63–70
van Hunen J, van Keken PE, Hynes A, Davies GF (2008) Tectonics of early Earth: some geodynamic considerations. In: Condie KC, Pease V (eds) When did plate tectonics begin on planet Earth?: Geological Society of America Special Paper, vol 440, pp 157–171. https://doi.org/10.1130/2008.2440(08)
Waples DW (2001) A new model for heat flow in extensional basins: radiogenic heat, asthenospheric heat and the McKenzie model. Natl Res Res 10
Yañez GA, Ranero CR, von Huene R, Diaz J (2001) Magnetic anomaly interpretation across the southern central Andes (32°– 34°S): the role of the Juan Fernández Ridge in the late Tertiary evolution of the margin. J Geophys Res 106:6325–6345
Yañez G, Cembrano J, Pardo M, Ranero C, Selle´s D (2002) The Challenger-Juan Fernández-Maipo major tectonic transition of the Nazca- Andean subduction system at 33–34° S: geodynamic evidence and implications. J South Am Earth Sci 15:23–38

The manufacturer's authorised representative in the EU is Springer Nature Customer Service Centre GmbH, Europaplatz 3, 69115 Heidelberg, Germany. If you have any concerns regarding our products, please contact ProductSafety@springernature.com

Printed and bound by CPI Group (UK) Ltd, Croydon, CR0 4YY

30/09/2026

02225438-0004